Consolidation Analyses of Soils

Consolidation Analyses of Soils

Jian-Hua Yin
Guofu Zhu

CRC Press
Taylor & Francis Group
Boca Raton London New York

CRC Press is an imprint of the
Taylor & Francis Group, an **informa** business

First edition published 2021
by CRC Press
2 Park Square, Milton Park, Abingdon, Oxon, OX14 4RN

and by CRC Press
6000 Broken Sound Parkway NW, Suite 300, Boca Raton, FL 33487-2742

CRC Press is an imprint of Informa UK Limited

British Library Cataloguing-in-Publication Data
A catalogue record for this book is available from the British Library

ISBN: 9780367555320 (hbk)
ISBN: 9781003093916 (ebk)

Typeset in Sabon
by KnowledgeWorks Global Ltd.

Dedication

Jian-Hua Yin

In memory of

my mother Ms LI Feng-Yi (李鳳儀),

my father WANG Ding-Zuo (汪鼎祚), and

my stepfather YIN Yang-Gu (殷陽谷)

To

my wife SUN Jun (孫浚) and son YIN Shao (殷劭)

To

my supervisors and co-workers Professor James Graham and Professor YUAN Jian-Xin (袁建新)

Guofu Zhu

In memory of

my father ZHU Yan-Bin (朱彥彬)

To

my mother Ms LIU Rong (劉 榮),

my wife REN Shan-Ling (任善玲), and
daughter ZHU Hui-Ying (朱慧瑩)

Contents

Preface xvii
Acknowledgements xix
Authors xxi

1 One-dimensional constitutive relations of soils 1

1.1 *Introduction* 1
1.2 *1-D Linear and Non-linear Elastic Models* 2
1.3 *1-D Elastic-Plastic Relationship* 5
1.4 *Two Simple Rheological Models* 6
1.4.1 *Maxwell model* 6
1.4.2 *Kelvin model* 7
1.5 *1-D EVP Model Using a Logarithmic Creep Function* 8
1.5.1 *Time-dependent stress-strain behavior of clayey soils* 8
1.5.2 *Equivalent time concept and deriving a 1-D EVP constitutive relationship* 11
1.5.2.1 *Instant time line (or κ-line)* 11
1.5.2.2 *Creep compression strain* 13
1.5.2.3 *Equivalent time line* 13
1.5.2.4 *General 1-D EVP constitutive relationship* 15
1.5.3 *Determination of all parameters in the 1-D EVP model* 16
1.5.3.1 *Determination of parameter κ (instant time line or κ-line)* 16
1.5.3.2 *Determination of two creep parameters Ψ/V and* t_o 16
1.5.3.3 *Determination of three parameters* ε^r_{zo}, σ'_{zo}, *and λ/V (reference time line or λ-line)* 19
1.5.4 *A simple approximate method for determination of all model parameters* 22
1.5.5 *Comparison of curves from measurements and 1-D EVP model predictions* 23

1.6 *Non-linear Creep and Compression of Soils* 26
1.7 *A 1-D EVP Model Incorporating the Non-linear Logarithmic Creep Function* 29
1.7.1 *Derivation of a 1-D EVPL model* 29
1.7.2 *Calibration and prediction of a 1-D EVPL model* 32
1.7.2.1 *Determination of parameter* κ *(instant time line or* κ*-line)* 34
1.7.2.2 *Determination of three parameters* ε^{r}_{zo}, σ'_{zo}, *and* λ/V *(reference time line or* λ*-line)* 34
1.7.2.3 *Determination of three creep parameters* $\Psi o/V$, *to and* ε^{vp}_{zl} 34
1.8 *Summary and Comparisons of 1-D Constitutive Models* 41
1.8.1 *Comparison of Maxwell model with 1-D EVP model and 1-D EVPL model* 41
1.8.2 *Relation of 1-D EVP model and 1-D EVPL model with a non-linear elastic model and an elastic-plastic model* 42
1.8.3 *Summary of all 1-D constitutive models, parameters, features, advantages, and limitations* 42

2 One-dimensional consolidation 47

2.1 *Introduction* 47
2.2 *Definitions of Excess Porewater Pressure* 47
2.3 *Governing Equations of Consolidation Problems* 49
2.3.1 *Basic assumptions for 1-D consolidation theory* 49
2.3.2 *Vertical total stress and effective stress* 49
2.3.3 *Constitutive equations* 49
2.3.4 *Darcy's law* 50
2.3.5 *Continuity equation* 50
2.4 *Consolidation of Soils With Constant Compressibility and Permeability* 51
2.4.1 *Solution for one-layered soil profile with depth-dependent ramp load* 52
2.4.1.1 *Convergence of the series* 56
2.4.1.2 *Special cases of the above solutions* 61
2.4.1.3 *Application* 70
2.4.2 *Solution for a double-layered soil profile under ramp load* 70
2.4.2.1 *Convergence of the series* 83
2.4.2.2 *Solution charts and tables for abruptly applied depth-independent loading* 84

2.4.2.3 *Example and discussion* 96

2.4.3 *Formal solution for n-layer soil profile* 113

2.4.3.1 *Construction of eigenfuctions* 115

2.4.3.2 *Determination of the eigenvalues by a sign-count method* 117

2.4.3.3 *Computation of eigenfunctions* 119

2.4.3.4 *Average degree of consolidation* 120

3 Analytical solutions to several one-dimensional consolidation problems **123**

3.1 *Introduction* 123

3.2 *Solution for a Soil layer increasing in thickness with time* 123

3.2.1 *Thickness of deposition proportional to* $\sqrt{t}$ 124

3.2.2 *Constant rate of deposition* 125

3.2.3 *Average degree of consolidation subsequent to cease of deposition* 125

3.2.4 *Consolidation of a saturated fill* 136

3.3 *Consolidation of Soils with depth-dependent compressibility and permeability* 136

3.3.1 *Basic equations and solutions* 136

3.3.2 *Special cases of depth-dependent permeability with constant compressibility* 140

3.3.3 *Special cases of depth-dependent compressibility with constant permeability* 149

3.3.4 *Special cases of depth-dependent permeability and compressibility resulting in constant* c_v 154

3.3.5 *Characteristics of the eigenvalue* λ_1 *for different* α 161

3.4 *Non-linear Consolidation of soils with constant coefficient of consolidation* 164

3.4.1 *Basic equations* 164

3.4.2 *Consolidation of a thin layer of soil under suddenly applied loading* 165

3.4.3 *Consolidation of a thin layer of soil under constant rate of loading* 167

3.5 *Consolidation of visco-elastic Soils* 171

3.5.1 *Three-parameter visco-elastic system* 172

3.5.2 *General visco-elastic soils* 176

3.6 *Observational procedure of Settlement prediction* 188

3.6.1 *Hyperbolic method for elastic soils* 188

3.6.2 *Asaoka's method for elastic soils* 190

3.7 *Consolidation of soil with non-linear Viscosity* 194

4 One-dimensional finite strain consolidation 199

4.1 *Introduction 199*
4.2 *Coordinate Systems 199*
4.3 *Governing Equations 201*
4.4 *Equations for Linear Finite Strain Consolidation 202*
4.5 *Solutions of Linear Finite Strain Consolidation 205*
4.5.1 *Self-weight consolidation with two-way drainage 205*
4.5.2 *Initially normally consolidated soil with two-way drainage 207*
4.5.3 *Self-weight consolidation with one-way drainage 211*
4.5.4 *Initially normally consolidated soil with one-way drainage 212*

5 Finite element method for consolidation analysis of one-dimensional problems 217

5.1 *Introduction 217*
5.2 *Governing Equations of Consolidation Problems 218*
5.3 *Finite Element Formulation 219*
5.4 *Numerical Characteristics of the Formulation 224*
5.5 *Analysis of a Clay Consolidation Test 227*
5.6 *Consolidation Modeling of a Case History 228*
5.6.1 *Soil profile and parameters for consolidation modeling 229*
5.6.2 *Result of case studies 233*

6 Consolidation of soil with vertical drain 239

6.1 *Introduction 239*
6.2 *Simplification of the Consolidation Problem 240*
6.2.1 *Basic assumptions 240*
6.2.2 *Unit cell 241*
6.2.3 *Equivalent drain radius of band-shaped vertical drain 241*
6.3 *Basic Equations and Solutions for the Case of Free Strain 243*
6.3.1 *Solutions for an ideal drain 244*
6.3.1.1 *Special case of a suddenly applied loading 248*
6.3.1.2 *Influence of N and L on average degree of consolidation 249*
6.3.1.3 *Design charts for vertical drains considering construction time 251*
6.3.1.4 *Design procedure 254*
6.3.1.5 *Example of application 256*

6.3.2 *Solutions for a vertical drain with smear zone 257*
6.3.2.1 *Special case of a suddenly applied loading 261*
6.3.2.2 *Influence of N, s, η, and L on average degree of consolidation 262*
6.4 *Basic Equations and Solutions for the Case of Equal Strain 267*
6.4.1 *Solutions for a vertical drain with smear zone 268*
6.4.2 *Cases of equal vertical strain with well resistance 270*
6.5 *Accuracy of Carrillo's Formula 272*
6.5.1 *Cases when U_r is calculated using free strain assumption 272*
6.5.2 *Cases when U_r is calculated using equal strain assumption 274*
6.6 *Comparison of Free Strain and Equal Strain Solutions 279*
6.7 *Calculation for Partially Penetrating Vertical Drain 282*

7 Finite difference consolidation analysis of one-dimensional problems **285**

7.1 *Introduction 285*
7.2 *General Finite Difference Method for 1-D Consolidation Analysis of Multiple Soil Layers 286*
7.3 *Finite Difference Scheme for 1-D Consolidation Analysis of One Clay Layer Using an Elastic Visco-Plastic Model 291*
7.4 *Simulation of Consolidation in a Clay Layer with Different Thicknesses and Comparison with Test Data 293*
7.5 *Influences of Thickness on Consolidation of a Soil Layer and Methods of Hypothesis A and Hypothesis B 302*
7.5.1 *Influences of thickness on consolidation of a soil layer by EVP consolidation modeling 302*
7.5.2 *Consolidation analyses of a soil layer using Hypothesis A method and Hypothesis B method 304*
7.5.3 *Proof of viscous compression in 'primary' consolidation 305*
7.6 *Excess Porewater Pressure Increases due to Creep, Cyclic Loading, and Mandel-Cryer Effects 307*
7.6.1 *Increase of excess porewater pressure due to creep compression 307*
7.6.2 *Increase of excess porewater pressure due to cyclic loading 311*
7.6.3 *Excess porewater pressure increases due to Mandel-Cryer effects 311*

8 Simplified methods for calculating consolidation settlements of soils exhibiting creep **315**

8.1 Introduction 315
8.2 Formulation of A General Simplified Hypothesis B Method for Calculating Consolidation Settlement of Multi-Layered Soils Exhibiting Creep 316
8.2.1 Formulation of a general simplified Hypothesis B method 316
8.2.2 Calculation of S_{fi} 319
8.2.3 Calculation of U_j and U 326
8.2.4 Calculation of S_{creep}, S_{creepj}, $S_{creep,fi}$, and $S_{creep,dj}$ 334
8.3 Consolidation Settlements of a clay layer with uniform stresses 338
8.4 Consolidation Settlements of a clay layer with OCR of 1 and 1.5 From simplified hypothesis b method and fully coupled consolidation analyses 345
8.5 Consolidation Settlements of double clay layers from simplified hypothesis b method and fully coupled consolidation analysis 353
8.6 Consolidation Settlements of multiple soil layers with vertical drains under staged loading from simplified hypothesis b method and fully coupled consolidation analysis 359
8.6.1 General equations for calculating consolidation settlements of multiple soil layers using simplified hypothesis b method 359
8.6.2 Calculation of primary consolidation settlements of soil layers under staged loadings 363
8.6.3 Calculation of creep settlements of soil layers under staged loadings 372
8.6.4 Calculation of total consolidation settlements of soil layers under staged loadings 381

9 Three-dimensional consolidation equations **389**

9.1 Introduction 389
9.2 Basic Assumptions for 3-D Consolidation Theory 389
9.3 Force Equilibrium Equations 390
9.3.1 Cartesian co-ordinates 390
9.3.2 Cylindrical co-ordinates 391
9.3.3 Spherical co-ordinates 392

9.4 *Deformation of Soil Skeleton* 393
9.4.1 *Cartesian co-ordinates* 393
9.4.2 *Cylindrical co-ordinates* 395
9.4.3 *Spherical co-ordinates* 396
9.5 *Effective Stress* 396
9.6 *Constitutive Equations* 397
9.7 *Darcy's Law* 398
9.7.1 *Cartesian co-ordinates* 398
9.7.2 *Cylindrical co-ordinates* 399
9.7.3 *Spherical co-ordinates* 399
9.8 *Continuity Equation* 400
9.8.1 *Cartesian co-ordinates* 400
9.8.2 *Cylindrical co-ordinates* 401
9.8.3 *Spherical co-ordinates* 401
9.9 *Boundary and Initial Conditions* 401
9.10 *Equations for an Elastic Skeleton and Constant Hydraulic Conductivity* 402
9.10.1 *Force equilibrium equations* 403
9.10.2 *Constitutive equations* 403
9.10.3 *Darcy's law* 405
9.10.4 *Boundary and initial conditions* 405
9.11 *Biot's consolidation equations and simple diffusion theory* 405
9.11.1 *Biot's theory* 405
9.11.2 *Simple diffusion theory* 407

10 Solutions of typical three-dimensional consolidation problems 409

10.1 *Introduction* 409
10.2 *Immediate Settlement and Long-Term Settlement* 409
10.3 *Mandel's Problem* 411
10.4 *Cryer's Problem* 420
10.5 *Consolidation of a Semi-infinite Layer Subject to Normal Circular Loading* 424
10.5.1 *Solution when the boundary $z = 0$ is permeable* 428
10.5.2 *Solution when the boundary $z = 0$ is impermeable* 432
10.6 *Consolidation of a Semi-infinite Layer Subject to Normal Rectangular Loading* 436
10.7 *Subsidence Due to a Point Sink in a Cross-anisotropic Porous Elastic Half Space* 445
10.7.1 *Problem definition and governing equations* 445
10.7.2 *Solutions* 447
10.7.3 *Results* 452

10.8 Consolidation of a Finite Layer Subject to Normal Strip Loading 457
10.8.1 Permeable rigid base 458
10.8.2 Impermeable rigid base 460
10.9 Consolidation of a Finite Layer Subject to Normal Loading on a Circular Area 461
10.9.1 Permeable rigid base 463
10.9.2 Impermeable rigid base 464

11 Simplified finite element consolidation analysis of soils with vertical drain 467

11.1 Introduction 467
11.2 Basic Equations 468
11.2.1 Darcy's law 469
11.2.2 The continuity equation 470
11.2.3 The constitutive equations 470
11.2.4 Vertical total stress 470
11.3 Finite Element Formulation 471
11.4 Comparison with Fully Coupled Analysis 475
11.4.1 Geometric models for the analysis 475
11.4.2 Comparison with Biot's porous elastic model 476
11.4.2.1 Normalized time 476
11.4.2.2 Average degree of consolidation 477
11.4.2.3 Excess porewater pressure 480
11.4.2.4 Vertical effective stress 484
11.4.3 Comparison with a non-linear porous elastic model 489
11.4.3.1 Normalized time 489
11.4.3.2 Average degree of consolidation 489
11.4.3.3 Excess porewater pressure 492
11.4.3.4 Vertical effective stress 496
11.5 Consolidation Analysis of a Case History 500
11.5.1 Background 500
11.5.2 Construction of the main test area 502
11.5.3 Soil profile and soil parameters 502
11.5.3.1 Upper marine clay 503
11.5.3.2 Upper alluvial crust 505
11.5.3.3 Lower marine clay 505
11.5.3.4 Lower alluvium 505
11.5.3.5 Location of instrumentation 505
11.5.3.6 Soil model and parameters 506
11.5.3.7 Initial and boundary conditions 506

11.5.4 Vertical drain characteristics and smear zone 506
11.5.5 Finite element analysis results and discussions 508
Appendix 11.A 513

12 Finite element method for three-dimensional consolidation problems 519

12.1 Introduction 519
12.2 Governing Equations 519
12.2.1 Force equilibrium equations 519
12.2.2 Strain-displacement relation 520
12.2.3 Effective stress 520
12.2.4 Constitutive equations 520
12.2.5 Darcy's law 520
12.2.6 Continuity equation 521
12.2.7 Boundary and initial conditions 521
12.3 Weak Formulation 521
12.4 Elements and Interpolation Functions 523
12.4.1 Linear element 523
12.4.2 Quadratic element 525
12.4.3 Evaluation of global derivatives 526
12.4.4 Evaluation of element integration 528
12.4.4.1 Newton-Cotes formulae 528
12.4.4.2 Gaussian quadrature 529
12.4.4.3 Calculation of element integration 531
12.5 Finite Element Formulation 531
12.6 Equivalent Modeling of Vertical Drains in Plane Strain Condition 536
12.7 Consolidation Analysis of an Elasto-plastic Soil 538

Index 547

Preface

Consolidation of a saturated soil is a coupled process of deformation of soil skeleton under the action of effective stresses and dissipation of excess porewater pressure. Deformation of soil skeleton is controlled by a constitutive law in a relationship of effective stresses, strains, and even strain rates if the stress-strain behavior is time-dependent. Dissipation of excess porewater pressure is controlled by Darcy's law. Mathematic equations for and solutions to one-dimensional (1-D) consolidation problem were first obtained by Terzaghi in 1925. Since then, mathematic equations for two- or three-dimensional (3-D) consolidation problems were presented and analytical or numerical solutions were obtained by many researchers.

This book focuses on the consolidation of fully saturated soils. It follows a classical approach by beginning with 1-D constitutive relations of soils and 1-D consolidation, then it moves on to analytical solutions to several 1-D consolidation problems, and 1-D finite strain consolidation. Afterwards, this book presents a finite element (FE) method for consolidation analysis of 1-D problems, analytical solutions to consolidation of soil with vertical drains, and a finite difference method for consolidation analysis of 1-D problems. Simplified methods for consolidation analysis of soils exhibiting creep are introduced and applied to different cases. 3-D consolidation equations, and solutions of typical 3-D consolidation problems are covered. Then a simplified FE consolidation analysis of soils with vertical drain and FE method for 3-D consolidation problems are presented. Uniquely, this book covers both classic solutions and state-of-the-art work in consolidation analyses of soils.

Preface

Acknowledgements

We would like to express sincere gratitude to Professor Jian-Xin YUAN, our supervisor and former Director of Institute of Rock and Soil Mechanics of Chinese Academy of Sciences and Professor James Graham, former supervisor of JH YIN and co-worker of a large number of papers with JH YIN and GF ZHU in constitutive modeling and consolidation analyses of soils. We would like to thank Dr Sam Wei-Qiang FENG, Dr Wen-Bo CHEN, and Mr. Ze-Jian CHEN for providing their assistance to us in preparation of chapters and example calculations in this book.

Authors

Jian-Hua Yin received a PhD from the University of Manitoba and joined as an Assistant Professor at Hong Kong Polytechnic University in 1995. He became Chair Professor of Soil Mechanics in 2014. Professor Yin also serves as Vice President of the International Association for Computer Methods and Advances in Geomechanics (IACMAG), co-editor of *International Journal of Geomechanics*, and a co-editor of *Geomechanics and Geoengineering* (UK). He has received the John Booker Medal, Chandra S. Desai Excellence Award, and Outstanding Contributions Medal from IACMAG.

Dr. Guofu Zhu received his PhD from Hong Kong Polytechnic University in 1999 and was appointed Professor in the Department of Engineering Structures and Mechanics at Wuhan University of Technology in 2005. His research interests include numerical methods and computer programming development in geotechnical engineering, consolidation analysis of soils, piled foundations, testing study of soil properties, and numerical modeling for various geotechnical structures. He has produced more than 70 publications in referred journals and conference proceedings.

Chapter 1

One-dimensional constitutive relations of soils

1.1 INTRODUCTION

It is commonly recognized that a constitutive relationship for the stress-strain-strength behaviour of a soil is a key equation, which is needed for a meaningful and accurate prediction of the performance of geotechnical soil structures (Chen and Mizuno, 1990; Yin, 1990; Chen, 1994). The research in constitutive modeling for soils has been a very active area with a long history (Chen and Mizuno, 1990). The stress-strain behavior of soils under static loading is non-linear, plastic, and time dependent.

Many geotechnical problems can be approximated as one-dimensional (1-D) compression/consolidation problems, in which, all deformation and water flow occur only in the vertical direction with zero lateral deformation and water flow. For example, marine reclamation by dumping sand fill uniformly over an extensive area of soils in horizontal strata can be considered a 1-D compression/consolidation problem. If the loading area of a shallow foundation or an embankment is three times or more of the thickness of the soil layer underneath, the compression/consolidation of the soil layer is approximate as 1-D.

The development of stress-strain relations for soils in 1-D compression has direct applications for 1-D straining condition. In addition, 1-D constitutive models involve fewer assumptions and fewer parameters than those in three-dimensional (3-D) models of the same type. Thus, 1-D constitutive models are not only simpler but also more reliable for applications to solving 1-D straining problems.

This chapter introduces a few 1-D constitutive relations for soils. These constitutive relations are divided into (a) linear and non-linear elastic models, (b) elastic-plastic models, (c) rheological models, and (d) elastic visco-plastic (EVP) models. Features, advantages, and limitations of these models are presented, summarized, and compared.

1.2 1-D LINEAR AND NON-LINEAR ELASTIC MODELS

A 1-D consolidation/compression test is normally carried out by using an oedometer with a soil specimen in a cylindrical confining ring. The sizes of the specimen (or inter space of the ring) are normally 50 to 100 mm in diameter and 20 mm in height with porous stones on the top and bottom. Since the lateral side is the specimen is confined by a metal ring which is much stiffer than soils, deformation and water flow will occur in the vertical direction only, where lateral strains are zero and there is only vertical stain. This kind of test condition is called 1-D straining. In comparison with the compression of an unconfined column under vertical stress loading, lateral stresses are zero and there is only vertical stress. This kind of test condition is called 1-D stressing.

The standard oedometer test is conducted at multi-stage vertical loading at a loading ratio of 2, that is, the new total loading is double of the previous total loading. The duration under each increment load is normally 24 hours. Figure 1.1 shows typical test results from an oedometer test on a marine clay with loading, unloading, and reloading. It is seen from Figure 1.1 that the relationship between vertical effective stress σ_z' (compression as positive) and vertical strain ε_z (compression as positive) is non-linear and plastic since there

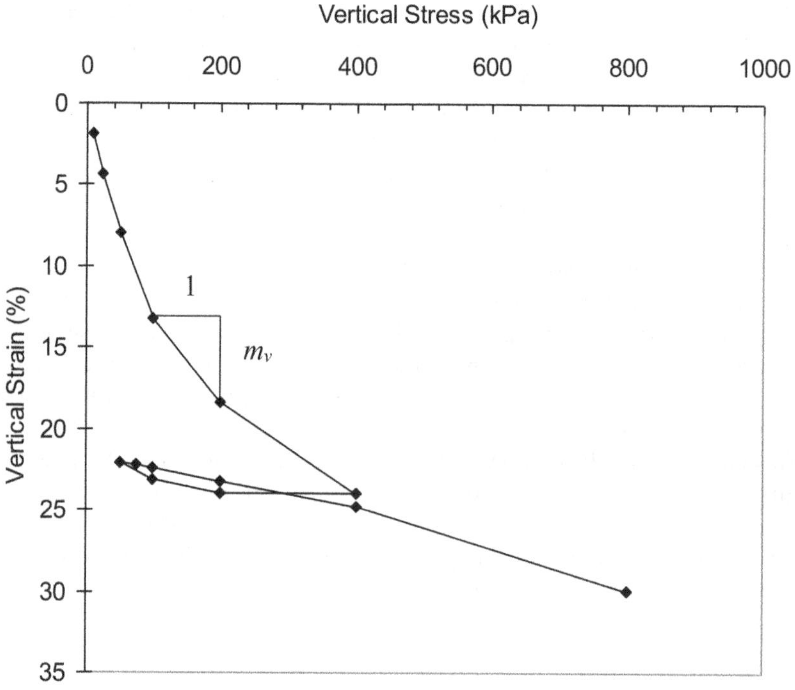

Figure 1.1 Results from an oedometer test: vertical effective stress versus vertical strain.

is an irreversible strain in unloading. The non-linear relationship is normally simplified as a linear one between vertical increment effective stress $\Delta\sigma_z'$ and vertical increment strain $\Delta\varepsilon_z$ is for the stress range of interests:

$$\Delta\varepsilon_z = m_v \Delta\sigma_z' \tag{1.1}$$

where m_v is the coefficient of volume compressibility as shown in Figure 1.1. (1.1) can be written as:

$$\Delta\sigma_z' = M\Delta\varepsilon_z \tag{1.2}$$

where $M = 1/m_v$ is the confined compression modulus. Equations (1.1) and (1.2) are considered a linear elastic relationship. Neither non-linearity nor plasticity is considered in (1.1) and (1.2). But the elastic relationship is simple and can be used with care for elastic deformation and consolidation analysis for the same stress increment range. It is noted that all constitutive equations in this chapter are formulated for relationships of effective stress-strain for time-independent behavior or effective stress-strain-strain rate for time-dependent behavior of saturated soil skeleton. The effective stress principle by Terzaghi (1925, 1943) is valid for saturated soils. Recently, Yin *et al.* (2020) invented a novel fiber Bragg grating (FBG)-based effective stress cell for direct measurement of effective stress in saturated soils.

The data in Figure 1.1 can be re-plotted with the vertical effective stress in a logarithmic scale in Figure 1.2. It is seen from Figure 1.2 that the relationship between vertical effective stress σ_z' and vertical strain ε_z in the main segment is linear with a constant slope $C_{c\varepsilon}$. This segment is called 'normally consolidated range'. This relationship can be expressed as:

$$\varepsilon_z = C_{c\varepsilon} \log \frac{\sigma_z'}{\sigma_{zo}'} \tag{1.3}$$

where σ_{zo}' is the vertical effective stress σ_z' resulting in zero vertical strain ε_z. The slope $C_{c\varepsilon}$ is called the compression index defined using the vertical strain, that is, $C_{c\varepsilon} = \Delta\varepsilon_z / \Delta\log\sigma_z'$. The compression index can also be defined using the void ratio e, that is, $C_{ce} = -\Delta e / \Delta\log\sigma_z'$. The relationship between $C_{c\varepsilon}$ and C_{ce} is: $C_{c\varepsilon} = C_{ce}/V$, where V is the specific volume, $V = (1 + e_o)$ and e_o is the initial void ratio corresponding zero vertical strain. Equation (1.3) has a problem at $\sigma_z' = 0$. If the loading is monotonically increasing, (1.3) can be considered to be a non-linear 'elastic' relationship. The confined compression modulus M can be obtained by differentiating (1.3):

$$\frac{d\varepsilon_z}{d\sigma_z'} = \frac{C_{c\varepsilon}}{\ln(10)\sigma_z'}$$

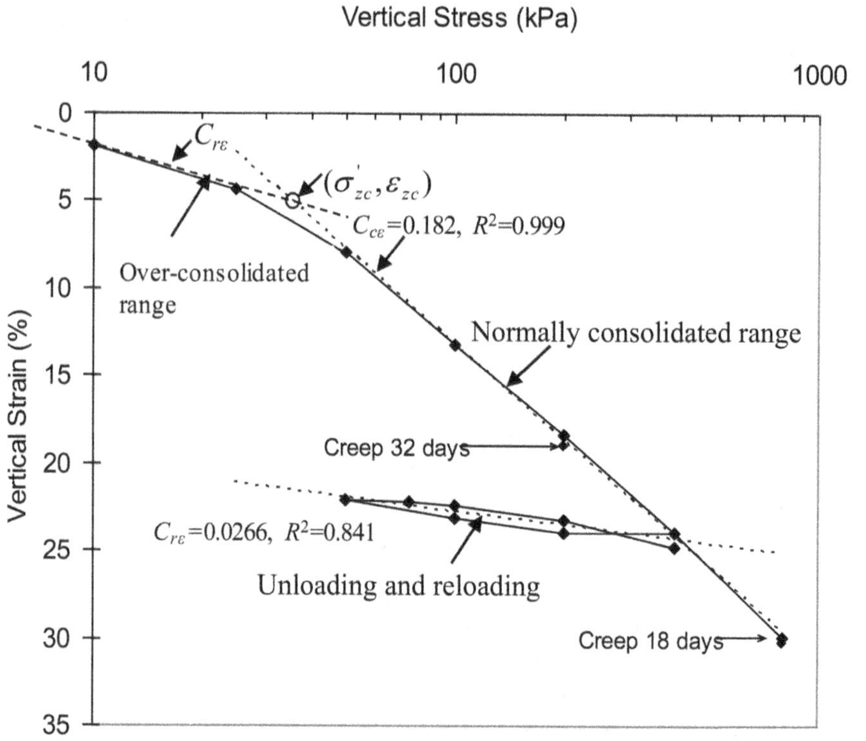

Figure 1.2 Results from an oedometer test: log (vertical effective stress) versus vertical strain.

From this expression, we have:

$$M = \frac{d\sigma'_z}{d\varepsilon_z} = \ln(10)\frac{\sigma'_z}{C_{c\varepsilon}} \tag{1.4}$$

It is seen that the confined compression modulus M is proportional to the vertical effective stress σ'_z. Care must be taken when using the non-linear 'elastic' model ensuring that the loading is monotonic in the normally consolidated range.

It is seen from Figure 1.2 that in the range of unloading and reloading or in the over-consolidated range, the relationship between vertical effective stress σ'_z and vertical strain ε_z is linear with slope $C_{r\varepsilon}$. This relationship can be expressed as:

$$\varepsilon_z = \varepsilon_{zi} + C_{r\varepsilon} \log\frac{\sigma'_z}{\sigma'_{zi}} \tag{1.5}$$

where σ'_{zi} is the vertical stress σ'_z resulting vertical strain ε_{zi}. The slope $C_{r\varepsilon}$ is called an unloading-reloading compression index defined using the vertical strain, that is, $C_{r\varepsilon} = \Delta\varepsilon_z / \Delta \log \sigma'_z$ for data points in unloading/reloading range. The compression index can also be defined using the void ratio e, that is, $C_{re} = -\Delta e / \Delta \log \sigma'_z$. The relationship between $C_{r\varepsilon}$ and C_{re} is $C_{e\varepsilon} = C_{re} / V$. The slope $C_{r\varepsilon}$ is normally considered to be the same as that in the over-consolidated range.

The stress-strain relationship in the over-consolidated range and the unloading/reloading range is normally considered to be non-linear elastic. The confined compression modulus M_r is:

$$M_r = \ln(10) \frac{\sigma'_z}{C_{r\varepsilon}} \tag{1.6}$$

The confined compression modulus M_r is also proportional to the vertical effective stress σ'_z.

1.3 1-D ELASTIC-PLASTIC RELATIONSHIP

Figures 1.1 and 1.2 indicate that the relationship between vertical effective stress σ'_z and vertical strain ε_z is non-linear and elastic-plastic. There is an irreversible strain in unloading. The whole stress-strain relationship can be described by a non-linear elastic-plastic model. A line can be drawn by fitting data points in the over-consolidated range and intercepting the pre-consolidation point(σ'_{zc}, ε_{zc}), where σ'_{zc} is the 'pre-consolidation pressure' and ε_{zc} is the corresponding strain. Using (1.3) and (1.5), the whole 1-D elastic-plastic relationship can be described as:

$$\begin{cases} \varepsilon_z = \varepsilon_{zc} + C_{r\varepsilon} \log \dfrac{\sigma'_z}{\sigma'_{zc}} & \textit{for } 0 < \sigma'_z < \sigma'_{zc} \textit{ in over-consolidation range} \\ \varepsilon_z = \varepsilon_{zc} + C_{c\varepsilon} \log \dfrac{\sigma'_z}{\sigma'_{zc}} & \textit{for } \sigma'_{zc} < \sigma'_z \textit{ for loading range only} \\ \varepsilon_z = \varepsilon_{z\max} + C_{r\varepsilon} \log \dfrac{\sigma'_z}{\sigma'_{z\max}} & \textit{for } 0 < \sigma'_z < \sigma'_{z\max} \textit{ for unloading-reloading range} \end{cases} \tag{1.7}$$

where the point ($\sigma'_{z\max}, \varepsilon_{z\max}$) is the maximum stress and strain reached in loading in the normally consolidated range, from which unloading starts. The vertical effective stress σ'_z in (1.7) shall be larger than zero, normally larger than 1 kPa, since if zero, the strain is $-\infty$. The above elastic-plastic relationship in (1.7) has been commonly used in practice in the calculation of the final compression of soil skeleton. One of the limitations is that the time-dependent effects are not considered in the above non-linear elastic-plastic relationship.

1.4 TWO SIMPLE RHEOLOGICAL MODELS

1.4.1 Maxwell model

Figure 1.3(a) shows a 1-D Maxwell model which consists of a linear elastic spring and a linear viscous dashpot connected in series. If Maxwell model is applied to a soil in 1-D straining, the compression of the soil skeleton is equal to time-independent elastic compression plus time-dependent viscous compression under the effective vertical stress, σ'_z. The constitutive equation of the Maxwell model is derived here. The elastic strain rate $\dot{\varepsilon}_z^e$ under the stress in the spring $\sigma_z'^e$ is:

$$\dot{\varepsilon}_z^e = \frac{\dot{\sigma}_z'^e}{M^e} \tag{1.8a}$$

or

$$\sigma_z'^e = M^e \varepsilon_z^e \tag{1.8b}$$

where M^e is the confined elastic modulus of the soil skeleton. The viscous strain rate $\dot{\varepsilon}_z^v$ under the stress $\sigma_z'^v$ in the dashpot is:

$$\dot{\varepsilon}_z^v = \frac{\sigma_z'^v}{\eta} \tag{1.9a}$$

or

$$\sigma_z'^v = \eta \dot{\varepsilon}_z^v \tag{1.9b}$$

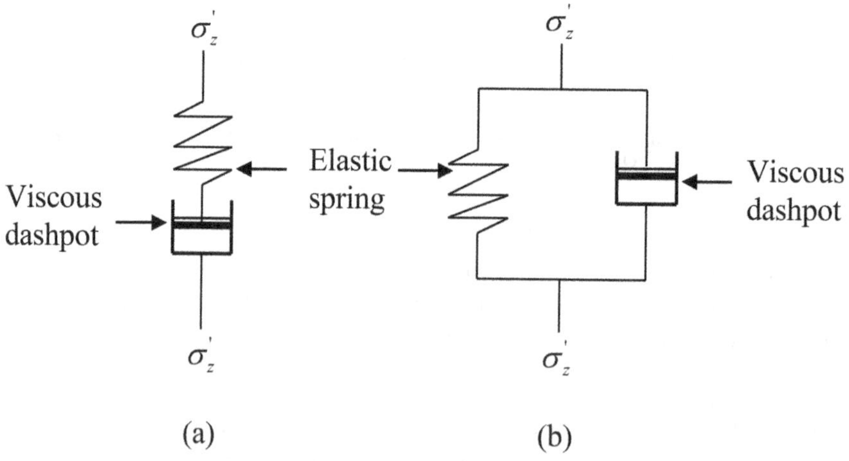

Figure 1.3 (a) Maxwell model and (b) Kelvin model.

where η is the viscosity of the soil skeleton in a confined condition. The total strain rate $\dot{\varepsilon}_z$ is:

$$\dot{\varepsilon}_z = \dot{\varepsilon}_z^e + \dot{\varepsilon}_z^v \tag{1.10}$$

The division of the total strain rates into an elastic strain rate and a visco-plastic strain rate was originally proposed by Perzyna (1963, 1966). Substituting (1.8) and (1.9) into (1.10), noting $\dot{\sigma}'_z = \dot{\sigma}'^e_z = \dot{\sigma}'^v_z$, we have

$$\dot{\varepsilon}_z = \frac{\dot{\sigma}'_z}{M^e} + \frac{\sigma'_z}{\eta} \tag{1.11}$$

Equation (1.11) is the differential constitutive equation of Maxwell model. It is interesting to note that after loading constant compressive effective stress σ'_z for a certain period of time, a certain strain ε_z is developed with time. When the effective stress σ'_z is unloaded to zero, the elastic strain ε_z^e is totally recovered, while the viscous strain ε_z^v is non-recoverable. In this regard, Maxwell model can be considered a linear elastic and linear visco-plastic model. Therefore, Maxwell model is not a 'visco-elastic model.' When the effective stress is unloaded to a certain negative value, the ε_z^v is recoverable and may become negative. However, negative values of σ'_z (extension stress) and ε_z^v(extension strain) have no physical meaning for the straining of the skeleton of a saturated soil. In addition, both the elastic compression and the visco compression of the soil skeleton are highly non-linear. Therefore, Maxwell model is not suitable for soils.

1.4.2 Kelvin model

Figure 1.3(b) shows a 1-D Kelvin model which consists of a linear elastic spring and a linear viscous dashpot connected in parallel. According to Kelvin model, the vertical effective stress σ'_z of soil skeleton in 1-D compression is equal to the effective stress from time-independent elastic compression plus the effective stress from time-dependent viscous compression, that is:

$$\sigma'_z = \sigma'^e_z + \sigma'^v_z \tag{1.12}$$

Substituting (1.8b) and (1.9b) into (1.12) and noting $\varepsilon_z = \varepsilon_z^e = \varepsilon_z^v$, we have:

$$\sigma'_z = M^e \varepsilon_z + \eta \dot{\varepsilon}_z \tag{1.13}$$

Equation (1.13) is the differential constitutive equation of Kelvin model.

It is interesting to note that, after loading a constant compression effective stress σ'_z for a certain period of time, strain ε_z will reach a limit. When the effective stress σ'_z is unloaded to zero, the strain ε_z is totally recovered. Therefore, Kelvin model is a 'visco-elastic model'. In relaxation where the strain ε_z is kept constant after loading, the effective stress is relaxed to zero. These features seem reflecting the time-dependent stress-strain behavior of soils. However, Kelvin model also has limitations for modeling soils. For example, in relaxation in 1-D compression, effective stress is not relaxed to zero normally since the viscosity of soil is non-linear. After unloading, the total strain is irreversible since a soil has plastic deformation. Therefore, Kelvin model is not suitable for soils.

Despite the limitations of Maxwell model and Kelvin model, rheological models consisting of a number of Maxwell model and Kelvin model connected in series and/or parallel have been proposed in the past. These rheological models suffer the same limitations as those of Maxwell model and Kelvin model, and their applications to modeling the time-dependent stress-strain behavior of soils are questionable and care must be taken when using all linear rheological models.

1.5 1-D EVP MODEL USING A LOGARITHMIC CREEP FUNCTION

1.5.1 Time-dependent stress-strain behavior of clayey soils

The time-dependency of the stress-strain behavior of soil skeleton is common for clayey soils. It is seen from Figure 1.4 that under the loading of a constant vertical stress σ_z in an oedometer test, the void ratio e continues to decrease with time (or vertical strain ε_z continues to increase with time). In Figure 1.4, the time is plotted in a logarithmic scale. Point A in Figure 1.4 is the End-of-Primary (EOP) consolidation with time t_{EOP} and void ratio e_{EOP}. The period from the beginning of loading to the point (t_{EOP}, e_{EOP}) is called 'primary' consolidation period with the excess porewater pressure larger than zero; while the period after the point (t_{EOP}, e_{EOP}) is called 'secondary' consolidation period with excess porewater pressure nearly zero. The continuing compression after point A is the creep of soil skeleton under a constant vertical effective stress σ_z'. The creep compression can be expressed:

$$e = e_{EOP} - C_{\alpha e}\log t \quad for\, t_{EOP} \le t \tag{1.14}$$

where $C_{\alpha e}$ is so-called 'secondary' consolidation coefficient and can be obtained by fitting (1.14) to data points after point A. If use data at point 1 and point 2, $C_{\alpha e}$ as the slope of the fitting line in Figure 1.4 can be calculated as:

$$C_{\alpha e} = \frac{-(e_2 - e_1)}{\log t_2 - \log t_1} = \frac{-(e_2 - e_1)}{\log(t_2/t_1)} \tag{1.15}$$

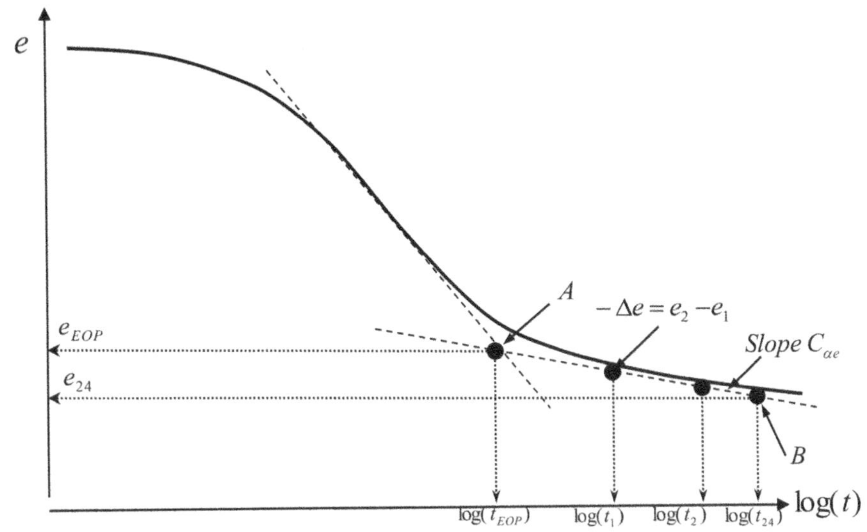

Figure 1.4 Relationship of void ratio e versus log(time) under a constant vertical effective stress σ'_z.

It is known that the relation between void ratio and the vertical strain ε_z is:

$$\varepsilon_z = -\frac{e - e_o}{1 + e_o} \tag{1.16}$$

where e_o is the initial void ratio corresponding to the initial zero strain. Using (1.14) and (1.16), the creep strain can be expressed:

$$\varepsilon_z = \varepsilon_{z,EOP} + C_{\alpha\varepsilon} \log t \tag{1.17}$$

where $\varepsilon_{z,EOP}$ is the strain at t_{EOP} and $C_{\alpha\varepsilon}$ is so-called 'secondary' consolidation coefficient defined using strain and is related to $C_{\alpha e}$:

$$C_{\alpha\varepsilon} = \frac{C_{\alpha e}}{1 + e_o} \tag{1.18}$$

Using (1.17), the creep strain rate can be calculated as:

$$\frac{d\varepsilon_z}{dt} = \frac{C_{\alpha\varepsilon}}{\ln(10)t} \tag{1.19}$$

It is seen that the creep strain rate decreases with time t. It shall be pointed out that the advocators of the 'secondary' consolidation consider creep occurs after the EOP consolidation, that is, t_{EOP}. Therefore, the time t in (1.17) and (1.19)

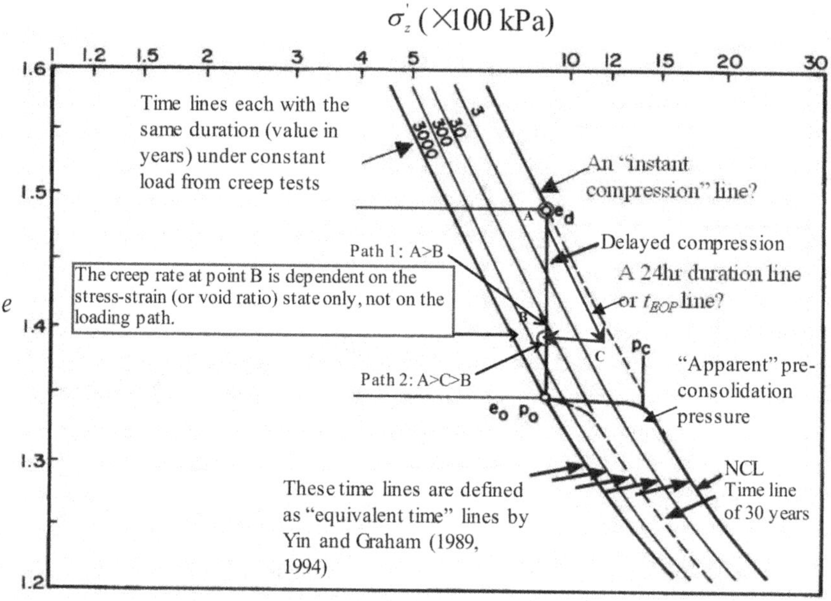

Figure 1.5 Relationship of void ratio e versus vertical effective stress σ_z' for different duration times, delayed compression, and 'apparent' pre-consolidation pressure pc.

shall be bigger than t_{EOP}. According to the mechanisms of viscous compression or creep of soil skeleton, viscous compression shall occur in both 'primary' and 'secondary' consolidation periods as long as the effective stresses exist. If the creep occurs from the beginning of the loading, that is, from $t = 0^+$, then (1.17) and (1.19) have no definition at $t = 0^+$.

Point B in Figure 1.4 is the point with a duration of 24 hours. It is noted that the value of t_{EOP} for the most clayey soils from conventional oedometer tests with specimen thickness normally 20 mm is approximately tens of minutes. The compression with a duration of 24 hours is more than that at t_{EOP}. The relations between e and the vertical effective stress σ_z' corresponding to different durations can be plotted in Figure 1.5. In Figure 1.5, p_c is 'apparent' pre-consolidation pressure at the compression turning point, first proposed by Bjerrum (1967) based on oedometer creep tests. The 'apparent' pre-consolidation pressure is caused by 'ageing' or 'delayed compression' (called creep) and is similar to the pre-consolidation pressure caused by previous loading or cementation.

The model in Figure 1.5 is a conceptual time line model proposed by Bjerrum (1967) since no mathematical constitutive equation was presented. Yin and Graham (1989, 1994) and Yin (1990) pointed out a few problems in Figure 1.5:

a. The 'instant compression' curve in Figure 1.5 is not a true instant compression. The true instant compression is an elastic compression, which dominates the compression in the over-consolidated range and in the unloading/reloading range. It is noted that creep still occurs in the over-consolidated and unloading/reloading ranges, though it is small. The 'instant compression' curve in Figure 1.5 is, in fact, a normal consolidation line (NCL), which has included both instant elastic compression and delayed visco-plastic (or creep) compression.

b. Curved time lines in Figure 1.5 with durations of 3, 30, 300, and 3000 years shall be obtained from oedometer tests on a soil specimen starting creep from the normally consolidated range, than is, NCL. If not, for example, if tests are done on a soil specimen in an over-consolidated range, then, the time lines will be different from those in Figure 1.5. The NCL here is similar to a 'reference time line' defined by Yin and Graham (1989, 1990) from which the time durations of all creep tests start to count. In this way, the 'time lines' in Figure 1.5 are similar to 'equivalent time lines' from which unique creep strain rates are obtained (Yin and Graham 1989, 1990).

1.5.2 Equivalent time concept and deriving a 1-D EVP constitutive relationship

The development of constitutive models for 1-D time-dependent stress-strain of soils has a long history. Based on the above understandings, Yin and Graham (1989, 1990) and Yin (1990) developed a 1-D EVP relationship for clayey soils. The derivation of the 1-D EVP relationship using logarithmic fitting functions is presented here.

1.5.2.1 Instant time line (or κ-line)

The instant time line is used to define instantaneous strains which are elastic and time independent. Instantaneous strains in this EVP modeling are not elastic-plastic strains as defined by others. Strains on the instant time line can be expressed:

$$\varepsilon_z^e = \varepsilon_{zi}^e + \frac{\kappa}{V}\ln\left(\frac{\sigma_z'}{\sigma_{zi}'}\right) \tag{1.20}$$

where σ_{zi}' is a reference vertical effective stress with corresponding vertical strain ε_{zi}^e. In (1.20), $V = 1 + e_o$ is the specific volume. The ratio κ/V together is a material parameter which has been used in the Cam-Clay models (Roscoe and Burland, 1968). (1.20) can be used to fit test data in an over-consolidated range or in unloading/reloading stages. This allows the parameter κ/V to be easily determined from fitting test data (see Figure 1.6).

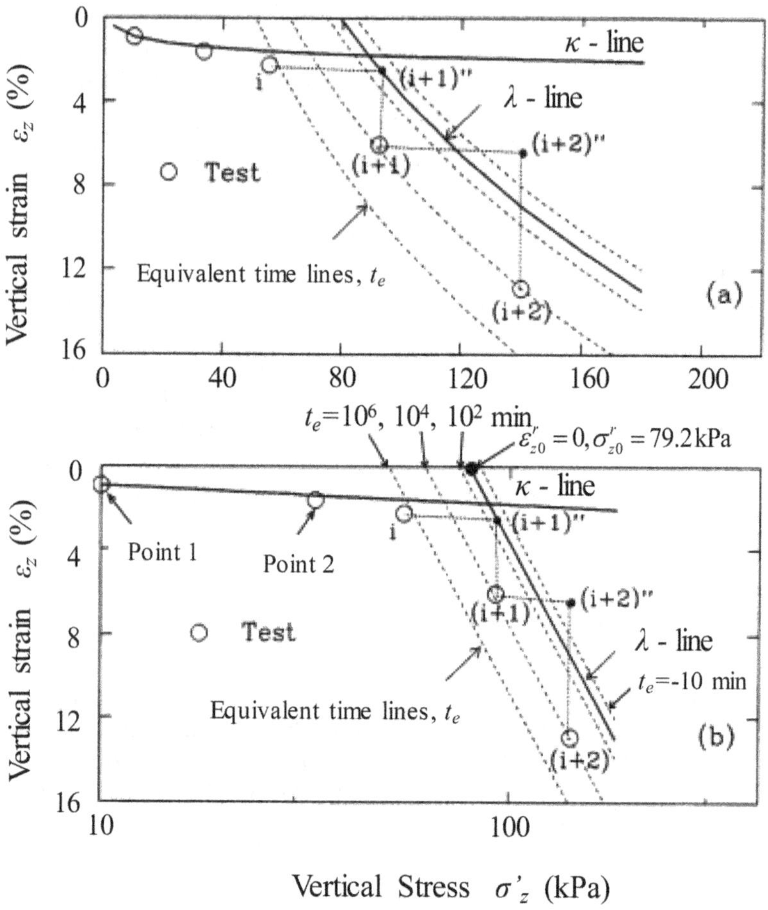

Figure 1.6 Relationships of vertical effective stress σ'_z versus vertical strain ε_z from an oedometer test, κ-line, λ-line, and 'equivalent time' t_e line: (a) σ'_z in arithmetic scale and (b) σ'_z in logarithmic scale.

Reference time line (or λ-line):
The reference time line is written:

$$\varepsilon^r_z = \varepsilon^r_{zo} + \frac{\lambda}{V} ln\left(\frac{\sigma'_z}{\sigma'_{zo}}\right) \tag{1.21}$$

where ε^r_{zo} is the vertical strain at effective stress $\sigma'_z = \sigma'_{zo}$. In (1.21), ε^r_{zo}, λ/V and σ'_{zo} are three material parameters. The superscript '*r*' is used here denoting the reference time line. The reference time line becomes an elastic-plastic compression line as that in the Cam-Clay models when there is no creep (Yin and Graham 1994). The term λ/V is similar to that used in the Cam-Clay models for defining the elastic-plastic line in isotropically consolidated specimens in a normally consolidated stress range. The parameter σ'_{zo} can be taken as a pre-consolidation effective pressure σ'_{zc} with the corresponding strain ε_{zc}, although another point may also be selected. In fact, the two values ε^r_{zo} and σ'_{zo} provide a point at which the λ-line passes through, for example, point (i+1)″ in Figure 1.6(b). In this way, the λ-line provides a reference for counting creep time and for calculating values of the 'equivalent time' which will be discussed later in this section. The later section will describe how the three parameters ε^r_{zo}, λ/V and σ'_{zo} can be determined.

1.5.2.2 Creep compression strain

Creep strains can be described by:

$$\varepsilon_z^{vp} = \frac{\psi}{V}\ln\left(\frac{t_o + t_e}{t_o}\right) \quad for - t_o < t_e \tag{1.22}$$

where ψ/V and t_o are two material parameters with positive values. Note that (1.22) uses equivalent time t_e counted from the reference time line in Figure 1.6(b) and not the time of the real loading duration. In this way, ψ/V becomes a material parameter and is a constant in (1.22). It is noted that (1.22) is different from (1.14), (1.17), and (1.19) which have no definition at $t = 0^+$. But (1.22) is valid for $t_e > -t_o$. Creep or viscous compression occurs from the beginning of loading so that the creep equation in (1.22) shall be valid for the time $t_e > -t_o$, including zero. This means that the equivalent time t_e may have a negative value as shown in Figure 1.6(b) and is meaningful. It is noted that (1.22) has definition at $t_e = 0$ and can be used to describe creep from beginning. Yin and Graham (1989, 1994) were among the first to use this logarithmic function in Soil Mechanics and this function may be called Yin and Graham's logarithmic function. This function can be extended to (1.5) and (1.7) for describing compression under stress.

1.5.2.3 Equivalent time line

'Equivalent time' was defined by Yin and Graham (1989, 1994) for creep behavior under 1-D straining in an oedometer test condition. Consider a soil specimen in a creep test is loaded elastically in Figure 1.6(b) from point i to point (i+1)″ and then allowed to creep to point (i+1) under constant vertical effective stress $\sigma'_{z(i+1)}$. Here we can select point (i+1)″ to be on the reference time line. The rationale is the same as that we chose the origin of a coordinate system. If point (i+1)″ is on the reference time line, creeping from point (i+1)″ to (i+1) takes place with the 'equivalent time' t_e, which is equal to the real creep duration t since loading from point i to point (i+1)″ takes zero time due to pure elastic compression.

Fitting (1.22) to creep test data allows the two parameters ψ/V and t_o be determined. Differentiating (1.22) gives the creep rate from point (i+1)″ to point (i+1) and beyond:

$$\frac{d\varepsilon_z^{vp}}{dt_e} = \frac{\psi}{V}\frac{1}{t_o + t} = \frac{\psi}{V}\frac{1}{t_o + t_e} = \dot{\varepsilon}_z^{vp} \tag{1.23}$$

Note that in this stage of creep, $t = t_e$ in value. Equation (1.23) indicates that the creep strain rate decreases with increasing time. This is true for creep pf a soil specimen in 1-D straining or oedometer test condition. It is also found from (1.23) that if the time value is the same, the creep strain rate is the same. Therefore, equivalent time lines with constant t_e values are also lines with constant creep strain rate values ($\dot{\varepsilon}_z^{vp}$ = constant). It is noted that all creep strains are visco-plastic strains since these creep strains are irreversible.

In (1.23), ψ/V and t_o have clear physical meaning. When $t_e = 0$, we have $\dot{\varepsilon}_{z,t_e=0}^{vp} = (\psi/V)/t_o$, which is the creep strain rate at the reference time line in (1.21). $(\psi/V)/t_o$ together gives the creep strain rate $\dot{\varepsilon}_{z,t_e=0}^{vp}$ on the reference time line. The smaller the t_o value, the larger the creep strain rate $\dot{\varepsilon}_{z,t_e=0}^{vp}$. Value of t_o is small, for example 1 day, compared to the creep time t_e, say 1 year to 100 years. From (1.22), considering t_o small, we have $\frac{\psi}{V} = \varepsilon_z^{vp}/\ln\left(\frac{t_o+t_e}{t_o}\right) \approx \varepsilon_z^{vp}/\ln(t_e)$. This means that $\frac{\psi}{V}$ is the slope of $\varepsilon_z^{vp}/\ln(t_e)$ when t_e is much larger than t_o.

Now consider what would happen to the creep rate $\dot{\varepsilon}_{z,(i+1)}^{vp}$ if point (i+1) had been reached by a different path or different loading history, say an unloading-and-reloading sequence. The 'equivalent time' concept says that the creep rate $\dot{\varepsilon}_{z,(i+1)}^{vp}$ at point (i+1) arrived from any loading path is the same as that obtained by creeping from (i+1)″ on the reference time line to point (i+1) for an 'equivalent time' t_e as that in (1.23). Referring to Figure 1.5, the creep strain rate $\dot{\varepsilon}_{z,B}^{vp}$ at point B can be obtained from a test creeping from point A to point B. But, point B can also be reached from loading from point A to point C and unloading from point C to point B. The creep strain rates from the two loading paths are the same. From this explanation, the following statement is true:

> ***The magnitude of creep strain rate at a state point*** $(\sigma_z', \varepsilon_z)$ ***is dependent on values of the effective stress and strain at this state point, not on the loading path how to reach this point.***

Since the creep rate $\dot{\varepsilon}_z^{vp}$ is uniquely related to the 'equivalent time' t_e, as in (1.23) so that the 'equivalent time' t_e is dependent on the stress-strain state $(\sigma_z', \varepsilon_z)$ only and is unique. If the creep strain rate can be calculated using (1.23) for any loading path, the total creep strain can also be calculated using (1.22) for any loading path. In this way, creep strains (or creep strain rates) in an over-consolidated or unloading/reloading stress range can be described by a single equation in (1.22) or integrating (1.23).

In other words, such 'equivalent time' lines result in a unique creep strain rate field in $(\sigma'_z, \varepsilon_z)$ space.

1.5.2.4 General 1-D EVP constitutive relationship

Using the 'equivalent time' t_e and assuming for the moment that the reference time line is known, the strain at any state point $(\sigma'_z, \varepsilon_z)$ in Figure 1.6 can be expressed, using (1.21) and (1.22):

$$\varepsilon_z = \varepsilon_z^r + \varepsilon_z^{vp} = \varepsilon_{zo}^r + \frac{\lambda}{V}\ln\left(\frac{\sigma'_z}{\sigma'_{zo}}\right) + \frac{\psi}{V}\ln\left(\frac{t_o + t_e}{t_o}\right) \tag{1.24}$$

Note that the equability of $\varepsilon_z = \varepsilon_z^r + \varepsilon_z^{vp}$ is valid based on the concept of 'equivalent time.' No matter what loading path, the total creep strain ε_z^{vp} can be calculated using (1.22) for any loading path starting from the reference time line with strain ε_z^r. The total strain ε_z is equal to the strain ε_z^r on the reference time line plus the creep strain ε_z^{vp} starting from the reference time line. Conversely, if stress and strain are known, the 'equivalent time' t_e is:

$$t_e = -t_o + t_o \exp\left[(\varepsilon_z - \varepsilon_{zo}^r)\frac{V}{\psi}\right]\left(\frac{\sigma'_z}{\sigma'_{zo}}\right)^{-\lambda/\psi} \tag{1.25}$$

Equation (1.25) shows that the equivalent time is uniquely related to the stress-strain state $(\sigma'_z, \varepsilon_z)$.

Continuous loading or unloading can be considered as a series of infinitesimal incremental loads $d\sigma'_z$ for real duration dt. The incremental strain $d\varepsilon_z$ is the sum of an elastic incremental strain $d\varepsilon_z^e$ and a creep (visco-plastic) incremental strain $d\varepsilon_z^{vp}$ or their rates:

$$d\varepsilon_z = d\varepsilon_z^e + d\varepsilon_z^{vp} \tag{1.26a}$$

$$\dot{\varepsilon}_z = \dot{\varepsilon}_z^e + \dot{\varepsilon}_z^{vp} \tag{1.26b}$$

where $d\varepsilon_z^e$ can be obtained by differentiating (1.20), that is, $d\varepsilon_z^e = \frac{\kappa/V}{\sigma'_z} d\sigma'_z$. The creep strain increment $d\varepsilon_z^{vp}$ is equal to the creep rate $\dot{\varepsilon}_z^{vp} = d\varepsilon_z^{vp}/dt$ multiplied by the incremental load duration dt, so that (1.26) becomes:

$$d\varepsilon_z = \frac{\kappa/V}{\sigma'_z} d\sigma'_z + \frac{d\varepsilon_z^{vp}}{dt} dt \tag{1.27}$$

Based on the concept of 'equivalent time' and the statement, the creep strain rate ε_z^{vp}/dt at a state point reached under any loading path shall be equal to the creep strain rate $d\varepsilon_z^{vp}/dt_e$ from a particular loading path of creep from the reference time line to the same state point, that is, $\varepsilon_z^{vp}/dt = d\varepsilon_z^{vp}/dt_e$. From (1.23) and (1.27), $d\varepsilon_z^{vp}/dt_e$ is:

$$\frac{d\varepsilon_z^{vp}}{dt_e} = \frac{\psi/V}{t_o + t_e} = \frac{\psi/V}{t_o}\exp\left[-(\varepsilon_z - \varepsilon_{zo}^r)\frac{V}{\psi}\right]\left(\frac{\sigma_z'}{\sigma_{zo}'}\right)^{\lambda/\psi} = \frac{d\varepsilon_z^{vp}}{dt} \tag{1.28}$$

This equation confirms the earlier statement that creep (or visco-plastic) strain rate is uniquely related to equivalent time and the stress-strain state point in Figures 1.5 and 1.6. Applying (1.28), (1.27) can be rewritten:

$$\frac{d\varepsilon_z}{dt} = \frac{\kappa/V}{\sigma_z'}\frac{d\sigma_z'}{dt} + \frac{\psi/V}{t_o}\exp\left[-(\varepsilon_z - \varepsilon_{zo}^r)\frac{V}{\psi}\right]\left(\frac{\sigma_z'}{\sigma_{zo}'}\right)^{\lambda/\psi} \tag{1.29a}$$

$$\dot{\varepsilon}_z = \frac{\kappa/V}{\sigma_z'}\dot{\sigma}_z' + \frac{\psi/V}{t_o}\exp\left[-(\varepsilon_z - \varepsilon_{zo}^r)\frac{V}{\psi}\right]\left(\frac{\sigma_z'}{\sigma_{zo}'}\right)^{\lambda/\psi} \tag{1.29b}$$

(1.29) is a general 1-D EVP relationship for any 1-D compression condition, for example, stepped-loading, constant-rate-of-strain loading, or even unloading/reloading. The structure of (1.23) is similar to the structure of the differential form of Maxwell model in (1.11). It is noted that the elastic strain rate $\dot{\varepsilon}_z^e$ and the visco-plastic strain rates $\dot{\varepsilon}_z^{vp}$ in (1.29) are all non-linear. The above Yin and Graham's 1-D EVP model (1989, 1994) is a non-linear rheological model and is an extension of the linear Maxwell rheological model.

1.5.3 Determination of all parameters in the 1-D EVP model

Data of a multi-stage oedometer test (test 7) on a clay from Drammen, Norway (Berre and Iversen, 1972) in Figures 1.6 through 1.8 is used to calibrate the 1-D EVP model.

1.5.3.1 Determination of parameter κ (instant time line or κ-line)

Equation (1.20) is used to fit test data in an over-consolidated range (or in unloading/reloading range if any). Figure 1.6 shows that the κ-line in (1.20) fits data at point 1 (9.936 kPa, 0.918%) and point 2 (33.534 kPa, 1.400%) in the over-consolidated range well. It is found from the fitting that the parameter κ/V is 0.004.

1.5.3.2 Determination of two creep parameters ψ/V and t_o

As shown in Figure 1.6, the vertical stress in increment 4 of test 7 was suddenly applied from σ_z = 55.3 to 92.5 kPa. Figure 1.7 shows that the

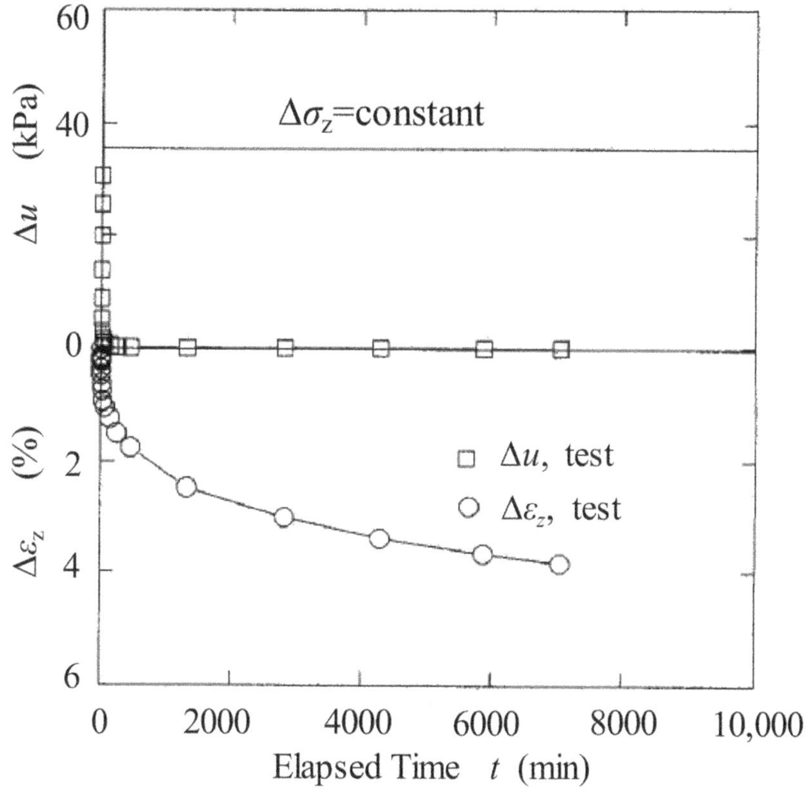

Figure 1.7 Relationships of $\Delta\sigma'_z, \Delta u, \Delta\varepsilon_z$ versus t for increment 4 in test 7.

excess pore water pressure Δu is negligible after an elapsed time of about 50 min, while creep compression continues due to the viscous nature of the clay. As mentioned and explained earlier, in this increment of load, the stress/strain state point moves elastically from point i to point (i+1)″ in Figure 1.6 and then allowed to creep to point (i+1) under constant vertical effective stress $\sigma'_{z(i+1)}$. The elastic strain increase is $\Delta\varepsilon^e_z = \kappa/V\ln(\sigma'_{z(i+1)}/\sigma'_{zi}) = 0.004\ln(92.5/55.3) = 0.206\%$.

Point (i+1)″ is selected to be on the reference time line with $t_e = 0$. Thus, the creep duration Δt is equal to the equivalent time t_e. The total strain increment $\Delta\varepsilon_z$ under the loading increment in Figure 1.7 has the relationship with the elastic strain increment $\Delta\varepsilon^e_z$ and creep strain increment ε^{vp}_z, that is, $\Delta\varepsilon_z = \Delta\varepsilon^e_z + \varepsilon^{vp}_z$. Here $\Delta\varepsilon^e_z$ of 0.206% has been calculated already. Therefore, the total strain increment $\Delta\varepsilon_z = \Delta\varepsilon^e_z + \varepsilon^{vp}_z = 0.206\% + \psi/V\ln[(t_o + \Delta t)/t_o]$ can be used to fit the compression test data in Figure 1.7. From the best fitting, the two parameters ψ/V and t_o are found to be $\psi/V = 0.007$ and $t_o = 40$ min. Figure 1.8 shows that the best-fitting curve is plotted in the coordinate of

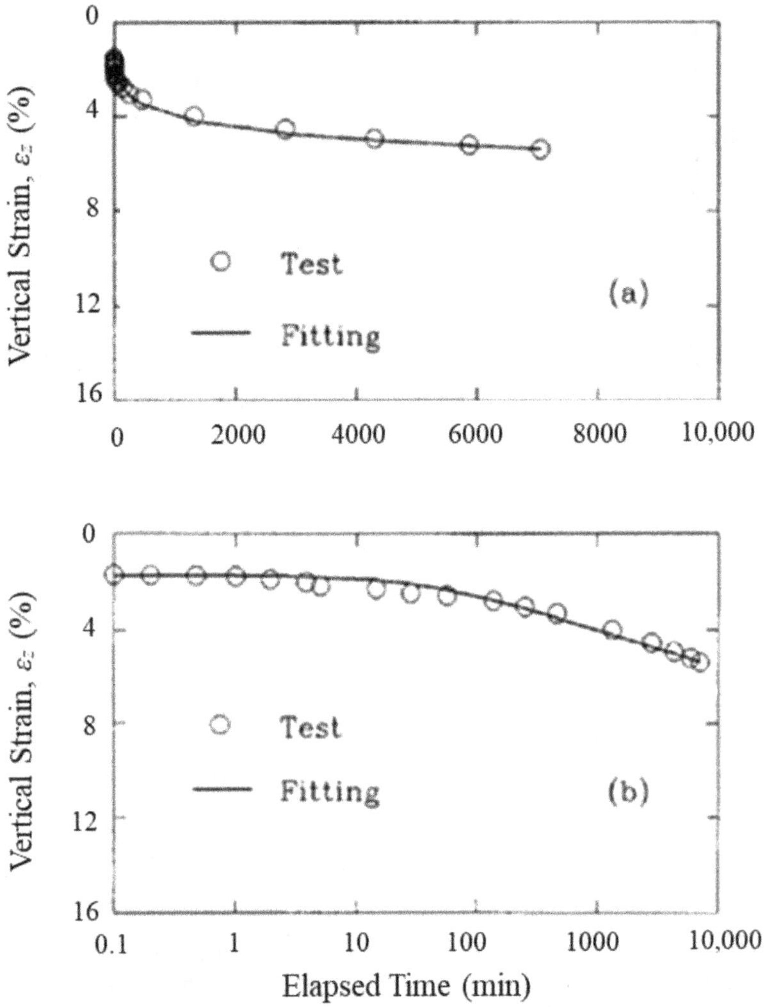

Figure 1.8 Test data and curve fitting data of ε_z versus t for increment 4 in test 7 (a) time in arithmetic scale and (b) time in logarithmic scale.

total vertical strain ε_z and time t in arithmetic scale and logarithmic scale. It is noted that $\varepsilon_z = \varepsilon_{zi} + \Delta\varepsilon_z$, where ε_{zi} is the total vertical strain under the previous loading in the over-consolidated range in Figure 1.6.

As discussed before, when $t_e = 0$, we have $\dot{\varepsilon}^{vp}_{z,t_e=0} = (\psi/V)/t_o$, which is the creep strain rate at the reference time line in (1.21). In this calibration, $\dot{\varepsilon}^{vp}_{z,t_e=0} = (\psi/V)/t_o = 0.007/40(\text{min}) = 1.75\times10^{-4}\,(1/\text{min})$, which is the creep strain rate at the reference time line. Therefore, ψ/V and t_o are soil creep parameters, which have clear physical meanings, are unique, and shall be determined from fitting creep test data.

1.5.3.3 Determination of three parameters ε_{zo}^r, σ'_{zo}, and λ/V (reference time line or λ-line)

Methods for determining the three parameters ε_{zo}^r, σ'_{zo}, and λ/V in (1.21) or (1.29) are now discussed. Consider two consecutive increments of creep testing – (a) the first increment from point i to point (i+1)″ with creep to point (i+1) under constant effective stress $\sigma'_{z(i+1)}$ and (b) the second increment from point (i+1) to point (i+2)″ with creep to point (i+2) under constant effective stress $\sigma'_{z(i+2)}$. As explained before, point (i+1)″ has been taken to lie on the reference time line and the data of this stage of creep test has been used to determine the two creep parameters in (1.22). Since point (i+1)″ lies on the reference time line, (1.21) can be expressed:

$$\varepsilon_{z,(i+1)''}^r = \varepsilon_{zo}^r + \frac{\lambda}{V}\ln\left(\frac{\sigma'_{z,i+1}}{\sigma'_{zo}}\right) \tag{1.30}$$

The physical meaning of $(\varepsilon_{zo}^r, \sigma'_{zo})$ is a point at which the reference time line (or λ-line) in Figure 1.6 passes through. Any point on the λ-line can be selected as this point $(\varepsilon_{zo}^r, \sigma'_{zo})$. If point (i+1)″ is selected, we then have $\varepsilon_{zo}^r = \varepsilon_{z(i+1)''}$ and $\sigma'_{zo} = \sigma'_{z(i+1)}$ so that the two parameters $(\varepsilon_{zo}^r, \sigma'_{zo})$ are determined. Alternately, we may select $\varepsilon_{zo}^r = 0$ and σ'_{zo} is the corresponding stress on the σ'_z-axis. In this second selection, we have:

$$\varepsilon_{z,(i+1)''}^r = \frac{\lambda}{V}\ln\left(\frac{\sigma'_{z,i+1}}{\sigma'_{zo}}\right) \tag{1.31}$$

Let us consider increment 5 of the creep test moving elastically from point (i+1) to point (i+2)″ followed by creep to point (i+2) for duration Δt under constant stress $\sigma'_{z(i+2)}$. Using (1.20), the strain $\varepsilon_{z(i+2)''}$ at Point (i+2)″ is:

$$\varepsilon_{z(i+2)''} = \varepsilon_{z,i+1} + \frac{\kappa}{V}\ln\left(\frac{\sigma'_{z,i+2}}{\sigma'_{z,i+1}}\right) \tag{1.32}$$

The strain $\varepsilon_{z(i+2)''}$ can also be calculated using the general Equation (1.24) with the corresponding equivalent time $t_{e,(i+2)''}$:

$$\varepsilon_{z,(i+2)''} = \varepsilon_{zo}^r + \frac{\lambda}{V}\ln\left(\frac{\sigma'_{z,i+2}}{\sigma'_{zo}}\right) + \frac{\psi}{V}\ln\left(\frac{t_o + t_{e,(i+2)''}}{t_o}\right) \tag{1.33}$$

Using (1.32) and (1.33) with equal $\varepsilon_{z(i+2)''}$, the equivalent time $t_{e,(i+2)''}$ at point (i+2)″ can be obtained:

$$t_{e(i+2)''} = t_o \exp\left[(\varepsilon_{z,i+1} - \varepsilon_{zo}^r)\frac{V}{\psi}\right]\left(\frac{\sigma'_{z,i+2}}{\sigma'_{z,i+1}}\right)^{\kappa/\psi}\left(\frac{\sigma'_{z,i+2}}{\sigma'_{zo}}\right)^{-\lambda/\psi} - t_o \tag{1.34}$$

The equivalent time $t_{e,(i+2)}$ at point (i+2) is:

$$t_{e,(i+2)} = t_{e,(i+2)''} + \Delta t \tag{1.35}$$

Substituting (1.34) into (1.35), we have:

$$t_{e,(i+2)} = t_o \left\{ \exp\left[(\varepsilon_{z,i+1} - \varepsilon^r_{zo}) \frac{V}{\psi} \right] \left(\frac{\sigma'_{z,i+2}}{\sigma'_{z,i+1}} \right)^{\kappa/\psi} \left(\frac{\sigma'_{z,i+2}}{\sigma'_{zo}} \right)^{-\lambda/\psi} - 1 \right\} + \Delta t \tag{1.36}$$

Using (1.24) again, the total vertical strain $\varepsilon_{z(i+2)}$ at point (i+2) is:

$$\varepsilon_{z(i+2)} = \varepsilon^r_{zo} + \frac{\lambda}{V} \ln\left(\frac{\sigma'_{z(i+2)}}{\sigma'_{zo}} \right) + \frac{\psi}{V} \ln\left(\frac{t_o + t_{e,(i+2)}}{t_o} \right) \tag{1.37}$$

Substituting (1.36) into (1.37), we have:

$$\varepsilon_{z(i+2)} = \varepsilon^r_{zo} + \frac{\lambda}{V} \ln\left(\frac{\sigma'_{z(i+2)}}{\sigma'_{zo}} \right) + \frac{\psi}{V} \ln\left\{ \frac{\Delta t}{t_o} + \exp\left[(\varepsilon_{z(i+1)} - \varepsilon^r_{zo}) \frac{V}{\psi} \right] \left(\frac{\sigma'_{z,i+2}}{\sigma'_{z,i+1}} \right)^{\kappa/\psi} \left(\frac{\sigma'_{z,i+2}}{\sigma'_{zo}} \right)^{-\lambda/\psi} \right\} \tag{1.38}$$

The ε^r_{zo} in (1.38) can be pre-selected, say, $\varepsilon^r_{zo} = 0$. Then, (1.38) is a non-linear equation for two unknowns λ/V and σ'_{z0}. Equation (1.31) has also two unknowns λ/V and σ'_{z0}. Both (1.32) and (1.38) can be used to determine values of λ/V and σ'_{z0} by a trial-error method easily.

Data of increment 4 of loading from point i to point (i+1)' and increment 5 of loading from point (i+1) to point (i+2)' is used to determine two unknowns λ/V and σ'_{z0} below:

a. At point i: $\sigma'_{zi} = 55.27\,kPa, \varepsilon_{zi} = 2.30\%$
   at point (i+1)': $\sigma'_{z(i+1)''} = 92.50\,kPa, \varepsilon_{z,(i+1)''} = 2.30\% + 0.004$
   $\ln(92.50/55.27) = 2.50\%$
   From (1.31), we have:

$$2.50\% = \frac{\lambda}{V} \ln\left(\frac{92.50}{\sigma'_{zo}} \right)$$

b. At point (i+1): $\varepsilon_{z(i+1)} = 5.8\%$ and $\sigma'_{z(i+1)} = 92.5$ kPa
   At point (i+2): $\sigma'_{z(i+2)} = 140.2$ kPa, final vertical strain $\varepsilon_{z(i+2)} = 12.9\%$, increment creep duration $\Delta t = 10{,}000$ min.

From (1.38), we have:

$$12.9\% = 0 + \frac{\lambda}{V}\ln\left(\frac{140.2}{\sigma'_{zo}}\right) +$$

$$+0.007\ln\left\{\frac{10000}{40} + \exp[(5.8\% - 0)/0.007]\left(\frac{140.2}{92.5}\right)^{0.004/0.007}\right.$$

$$\left.\left(\frac{140.2}{\sigma'_{zo}}\right)^{-\lambda/0.007}\right\}$$

Using the two equations in (a) and (b) above which are (1.32) and (1.38), we can find values of the two parameters $\lambda/V = 0.158$ and $\sigma'_{z0} = 79.2$ kPa by trial-error.

So far, values of all model parameters have been determined and summarized in Table 1.1. Using values of all parameters determined in Table 1.1 for the clay from Drammen, Norway (Berre and Iversen 1972), the κ-line, λ-line, and equivalent time lines with $t_e = -10$ min, 10^2 min, 10^4 min, and 10^6 min are plotted in Figure 1.6.

The above method is a rigorous method for determining the values of all parameters using data from a multi-staged loading oedometer test. Using the same method, model parameters for Bäckebol clay from Sweden (Sällfors, 1975), Batiscan clay from Quebec, Canada (Leroueil *et al.*, 1985), and a reconstituted illite (Yin and Graham, 1989) have been determined and listed in Table 1.1. Data from a few single-staged oedometer tests can also be used to determine all model parameters using the same approach.

The above method is an accurate method based on concept and definition of 'equivalent time' line, 'reference time lime', and 'instant time line' and natural logarithmic function. This method can be used without much difficulty. Nevertheless, an even simpler and approximate method is presented in Section 1.5.4 for a good and easy estimation of all model parameters.

Table 1.1 Parameters in the 1-D EVP model

Soil	κ/V	ψ/V	t_o *(min)*	ε'_{zo}	σ'_{zo} *(kPa)*	λ/V
Clay from Drammen, Norway (Berre and Iversen, 1972)	0.004	0.007	40	0	79.2	0.158
Bäckebol clay, Sweden (Sällfors, 1975)	0.010	0.0205	10	0	81.0	0.410
Batiscan clay Quebec, Canada (Leroueil et al., 1985)	0.020	0.022	1	0	128.0	0.550
Reconstituted illite (Yin and Graham, 1989)	0.025	0.004	10	0	47.0	0.100

1.5.4 A Simple approximate method for determination of all model parameters

In Civil Engineering practice, a common logarithmic function is used to fit data of void ratio and log(vertical effective stress) in ranges of over-consolidation or unloading-reloading to find the slope C_r (similar to κ), to fit data of void ratio and log(vertical effective stress) in a normal-consolidation range to find the slope C_c (similar to λ), and to fit void ratio and log(time) to find the slope $C_{\alpha e}$ (similar to Ψ) from a creep compression test. These common logarithmic function fittings are normally valid for stress ranges or time durations that are common in the design life of geotechnical structures.

The six parameters (κ, t_o, ψ, ε^r_{zo}, σ'_{zo}, λ) in (1.29) are similar to well-known conventional parameters. For example, from an oedometer test under a multi-staged loading with 24 hours of duration for each incremental load, a log(vertical effective stress)-void ratio (or strain) relationship with 24 hours ($t = 24\,hours = t_{24}$) of duration or at the EOP compression ($t = t_{EOP}$) for each increment load can be obtained. As explained before, for a normal soil specimen of 20 mm, $t_{EOP} < t_{24}$. From the relationship of $\log(\sigma'_z) vs e$ with unloading-reloading at $t = t_{24}$, we have obtain a pre-consolidation pressure $\sigma'_{zc,24}$ and the corresponding strain, $\varepsilon_{zc,24}$, the compression index $C_{c,24} = -\Delta e/\Delta \log(\sigma'_z)$, unloading-reloading compression index $C_{r,24} = -\Delta e/\Delta \log(\sigma'_z)$, and the secondary coefficient of consolidation $C_{\alpha e} = \frac{e_1 - e_2}{\log(t_2/t_1)}$ (or $C_{\alpha\varepsilon} = \frac{\varepsilon_{z2} - \varepsilon_{z1}}{\log(t_2/t_1)}$). From the relationship of $\log(\sigma'_z) vs e$ with unloading-reloading at $t = t_{EOP}$, we can obtain: $\sigma'_{zc,EOP}$, $\varepsilon_{zc,EOP}$, $C_{c,EOP}$, $C_{r,24}$, and $C_{\alpha e}$ (or $C_{\alpha\varepsilon}$). It is noted that $C_{\alpha e}$ (or $C_{\alpha\varepsilon}$) is not affected by the duration of 24 hours or the EOP if the logarithmic function is valid for the creep compression.

The approximate relationships between conventional consolidation parameters and the parameters in the 1-D EVP model are

$$\begin{cases} \kappa \approx \dfrac{C_{r,24}}{\ln(10)} \\ t_o = t_{24} = 1440\,\text{min}, \quad \psi \approx \dfrac{C_{\alpha e}}{\ln(10)} \\ \varepsilon^r_{zo} = \varepsilon_{zc,24}, \; \sigma'_{zo} = \sigma'_{zc,24}, \lambda \approx \dfrac{C_{c,24}}{\ln(10)} \qquad at\, t = 24\,hours = t_{24} \end{cases} \tag{1.39}$$

or

$$\begin{cases} \kappa \approx \dfrac{C_{r,EOP}}{\ln(10)} \\ t_o = t_{EOP}, \quad \psi \approx \dfrac{C_{\alpha e}}{\ln(10)} \\ \varepsilon^r_{zo} = \varepsilon_{zc,EOP}, \; \sigma'_{zo} = \sigma'_{zc,EOP}, \lambda \approx \dfrac{C_{c,EOP}}{\ln(10)} \qquad at\, t = t_{EOP} \end{cases} \tag{1.40}$$

In this way, all six model parameters in (1.29) can be estimated easily using the commonly used parameters from conventional consolidation tests.

1.5.5 Comparison of curves from measurements and 1-D EVP model predictions

Using parameters in Table 1.1, the calibrated 1-D EVP model in (1.29) has been used to predict the time-dependent stress-strain behavior of Bäckebol clay from Sweden (Sällfors, 1975), Batiscan clay from Quebec, Canada (Leroueil *et al.*, 1985), and reconstituted illite (Yin and Graham, 1989). It is noted that the 1-D EVP constitutive relationship in (1.29) is an ordinary differential equation. A predicted curve or simulation can be made by solving this ordinary differential equation in (1.29) using any numerical method, such Runge-Kutta methods under given an initial condition and a loading condition. It shall be pointed out that the initial condition is the initial state point (σ'_{zi} ,ε_{zi}) (or initial value of $\sigma'_z = \sigma'_{zi}$ and $\varepsilon_z = \varepsilon_{zi}$). This requirement is different from the prediction using an elastic-plastic model in which only initial stress is needed. The loading condition is that $\dot{\varepsilon}_z$ = constant for a constant rate of strain (CRSN) test; $\dot{\sigma}'_z$ = constant for a constant rate of stress (CRSS) test; and $\dot{\sigma}'_z$ = zero for a suddenly applied stress test (such as a multi-stage loading test).

Figure 1.9 shows measured and predicted relationships of vertical effective stress σ'_z *versus* vertical strain ε_z of Bäckebol clay compressed at a constant vertical strain rate $\dot{\varepsilon}_z = 7.5\times10^{-6}$/s, 2.0×10^{-6}/s, and 5.0×10^{-7}/s: (a) σ'_z in logarithmic scale and (b) σ'_z in arithmetic scale (Sällfors, 1975). It is seen that the 1-D EVP model prediction is in good agreement with the measured points. The model can simulate well the strain-rate effects on the stress-strain curves. In addition, it is seen that the model can also reproduce the dependency of the pre-consolidation pressure on strain rates.

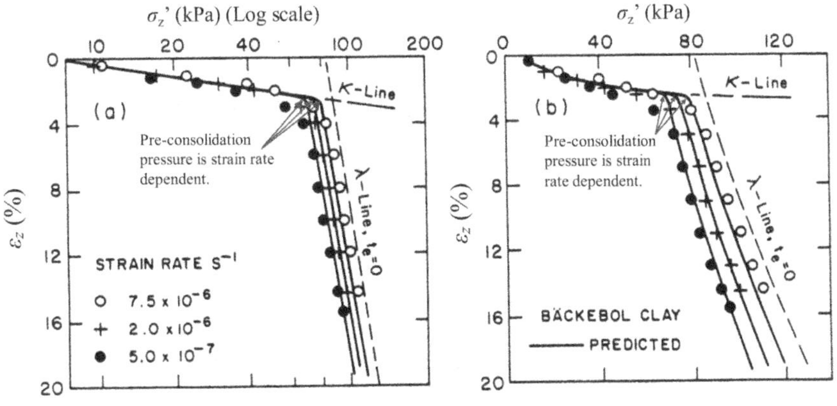

Figure 1.9 Measured and predicted relationships of vertical effective stress σ'_z versus vertical strain ε_z at strain rate $\dot{\varepsilon}_z = 7.5\times10^{-6}$/s, 2.0×10^{-6}/s, and 5.0×10^{-7}/s: (a) σ'_z in logarithmic scale and (b) σ'_z in arithmetic scale.

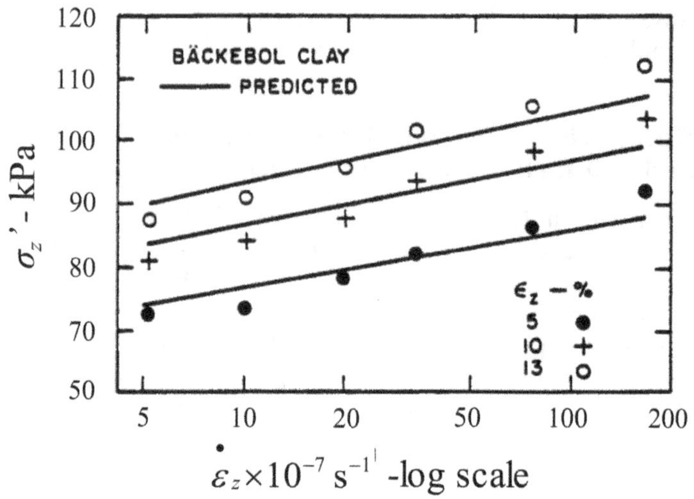

Figure 1.10 Measured and predicted relationships of vertical effective stress σ_z' versus vertical strain rate $\dot{\varepsilon}_z$ of Bäckebol clay for $\varepsilon_z = 5$, 10, and 13% from CRSN tests.

Using the data in Figure 1.9, the measured and predicted data of vertical effective stress σ_z' *versus* vertical strain rate $\dot{\varepsilon}_z$ of Bäckebol clay (Sällfors, 1975) can be plotted in Figure 1.10 for $\varepsilon_z = 5$, 10, and 13%. It is seen that for a given vertical strain value, the vertical effective stress increases almost linearly with the strain rate. The dependence of the pre-consolidation pressure (or effective stress) with the strain rate has the same trend as that in Figure 1.10.

The 1-D EVP model has been calibrated using single staged oedometer tests for a Batiscan clay reported by Leroueil *et al.* (1985). All model parameters are listed in Table 1.1. The calibrated model has been used to predict the strain rate dependent stress-strain curves from CRSN tests on the Batiscan clay. Figure 1.11 shows measured and predicted data of vertical effective stress σ_z' *versus* vertical strain rate $\dot{\varepsilon}_z$ of the Batiscan clay for $\varepsilon_z = 5$, 10, and 15%. It is seen again that for a given vertical strain value, the vertical effective stress is dependent on almost linearly with the strain rate.

Illite is one of three clay minerals. Montmorillonite is the most highly plastic and exhibits the largest creep under stress. Kaolinite has the lowest plasticity and the smallest creep among the three clay minerals. Illite is in the middle with moderate plasticity and creep. Yin (1990) did a multi-stage oedometer test, a stepped-changed CRSN test and a relaxation test. The oedometer test data was used to calibrate the 1-D EVP model with all parameters in Table 1.1. The calibrated model was used to predict the time-dependent behavior of the illite in the CRSN test and relaxation test. Figure 1.12 shows measured and predicted relationships of vertical effective stress σ_z' *versus* time t of illite from a relaxation test (Yin, 1990). It is seen that the vertical effective stress σ_z' decreases with time duration at a decreasing rate.

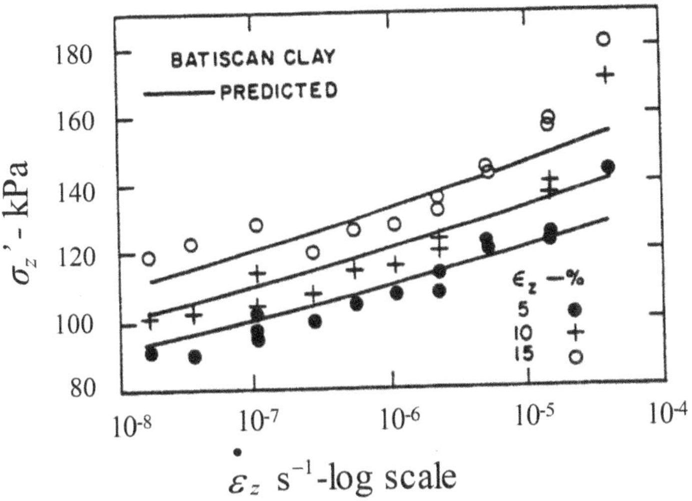

Figure 1.11 Measured and predicted relationships of vertical effective stress σ_z' versus vertical strain rate $\dot{\varepsilon}_z$ of Batiscan clay for $\varepsilon_z = 5$, 10, and 15% from CRSN tests.

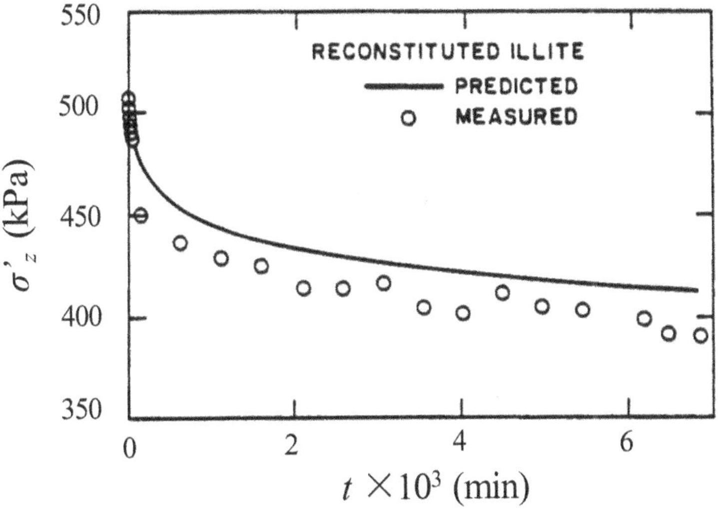

Figure 1.12 Measured and predicted relationships of vertical effective stress σ_z' versus time t of illite from a relaxation test.

1.6 NON-LINEAR CREEP AND COMPRESSION OF SOILS

A logarithmic function has been commonly used to fit data of $\log(\mathit{time})$ *vs void ratio*$(or\,\varepsilon_z)$ from a creep test in (1.14) and data of $\log(\sigma_z')$ *vs void ratio*$(or\,\varepsilon_z)$ from a compression test under loading in (1.13). The use of the logarithmic function may cause serious error for the estimation of the long-term settlement of soils. For example, from (1.14), $e = e_{EOP} - C_{\alpha e}\log t$. When time t is approaching to infinite, since $e_{EOP}, C_{\alpha e}$ are positive constants, the void ratio e can become negative. A negative value of void ratio has no physical meaning, even zero value of the void ratio is nearly impossible. From (1.13), $\varepsilon_z = C_{c\varepsilon}\log(\sigma_z'/\sigma_{zo}')$. Since e_o, C_{ce} are positive constants, when stress σ_z' is approaching infinite, strain ε_z is also getting infinite. This is impossible for confined compression or 1-D straining condition. Since $\varepsilon_z = -(e - e_o)/(1 + e_o)$, then we have $e = e_o - C_{c\varepsilon}(1 + e_o)\log(\sigma_z'/\sigma_{zo}') = e_o - C_{ce}\log(\sigma_z'/\sigma_{zo}')$. Again when stress σ_z' is approaching to infinite, void ratio e can become negative without any physical meaning. In fact, the void ratio or vertical strain will have a limit in infinite creep or stressing in 1-D straining condition as shown in Figure 1.5.

Based on the above observation and reasoning, a mathematical function with a limit shall be used to fit creep test data or stress-void ratio data. There are many functions with a limit such as a hyperbolic function, which was firstly used by Kondner (1963) to fit the relation of deviator stress $(\sigma_1 - \sigma_3)$ and axial strain ε_1 from a triaxial compressing test:

$$(\sigma_1 - \sigma_3) = \frac{\varepsilon_1}{a + b\varepsilon_1} \tag{1.41}$$

This hyperbolic function is called Kondner's hyperbolic function. The limit in (1.41) is $1/b$. This work was later extended by Duncan and Chang (1970) for developing a non-linear stress-strain model, called Duncan and Chang (1970). This hyperbolic function is not very suitable for fitting relations of time *vs* void ratio$e(or\,\varepsilon_z)$ from a creep test and *stress* σ_z' *vs* void ratio$e\ (or\,\varepsilon_z)$ from a compression test since the most initial segment of the curves is more close to logarithmic function fittings as in (1.14) and (1.13). Another function with a limit is an exponential function, for example, for fitting creep strain ε_z and creep time t:

$$\varepsilon_z = \varepsilon_{zo} + (\varepsilon_{z\infty} - \varepsilon_{zo})(1 - e^{-at}) \tag{1.42}$$

where ε_{zo} is the initial vertical strain when $t = 0$, $\varepsilon_{z\infty}$ is the limit vertical strain when $t \Rightarrow \infty$, and a is a constant. It is found that this exponential function is not suitable for fitting the data from a creep test nor from a compression test in 1-D straining condition since the decay by the e^{-at} is too fast and the most initial segment of the curves is more close to logarithmic function fittings.

Based on such observations, Yin (1999) proposed a new function with a limit which is a combination of the hyperbolic function and logarithmic function for creep test curves and is called Yin's function:

$$\varepsilon_z^{vp} = \frac{\frac{\psi_o}{V}\ln\frac{t_o+t_e}{t_o}}{1+\frac{\psi_o}{V\varepsilon_{zl}^{vp}}\ln\frac{t_o+t_e}{t_o}} = \left(\frac{\frac{\psi_o}{V}}{1+\frac{\psi_o}{V\varepsilon_{zl}^{vp}}\ln\frac{t_o+t_e}{t_o}}\right)\ln\frac{t_o+t_e}{t_o} = \frac{x}{a+bx} \tag{1.43}$$

where ε_z^{vp} is the time-dependent creep strain under current vertical effective stress, that is, creep strain increment, not accumulated creep strain under all previous loads, are ε_{zl}^{vp} is the creep limit when $t \Rightarrow \infty$. If (1.43) is compared to (1.22), we can find out the meaning of other parameters and ψ/V in (1.22) is:

$$\frac{\psi}{V} = \frac{\frac{\psi_o}{V}}{1+\frac{\psi_o}{V\varepsilon_{zl}^{vp}}\ln\frac{t_o+t_e}{t_o}} \tag{1.44}$$

It is seen that in (1.43) and (1.44), ψ/V is dependent on time t_e. ψ_o/V in (1.43) is the initial slope at time $t_e=0$. It is also seen that (1.43) is a hyperbolic function if we consider $a=V/\psi_o$ and $b=1/\varepsilon_{zl}^{vp}$. If the limit strain is $\varepsilon_{zl}^{vp} \Rightarrow \infty$, $\psi_0/V=\psi/V$, then (1.43) becomes (1.22). We found that (1.43) is very suitable for fitting non-linear creep test data (Yin, 1999).

Layer Yin (2011) extended (1.43) to fit compression test data of *stress* σ_z' *vs* void ratio e (*or* ε_z) in 1-D straining:

$$\varepsilon_z^r = \varepsilon_{zo}^r + \frac{\frac{\lambda_o}{V}\ln\frac{\sigma_z'}{\sigma_{zo}'}}{1+\frac{\lambda_o}{V\varepsilon_{zl}^r}\ln\frac{\sigma_z'}{\sigma_{zo}'}} = \left(\frac{\frac{\lambda_o}{V}}{1+\frac{\lambda_o}{V\varepsilon_{zl}^r}\ln\frac{\sigma_z'}{\sigma_{zo}'}}\right)\ln\frac{\sigma_z'}{\sigma_{zo}'} \tag{1.45a}$$

Comparing (1.45a) with (1.21), we can find out the meaning of all parameters in (1.45a). ε_{zl}^r is the strain limit when $\sigma_z' \Rightarrow \infty$. If (1.45a) is compared with (1.21), we can find out the meaning of other parameters and λ/V in (1.21) is:

$$\frac{\lambda}{V} = \frac{\lambda_o/V}{1+\frac{\lambda_o/V}{\varepsilon_{zl}^r}\ln\frac{\sigma_z'}{\sigma_{zo}'}} \tag{1.45b}$$

It is seen that in (1.45a) and (1.45b), λ/V is dependent on time σ_z'. λ_o/V in (1.45a) is the initial slope at $\sigma_z'=\sigma_{zo}'$. ε_{zl}^{vp} is the limit strain when $\sigma_z' \Rightarrow \infty$. If the limit strain is $\varepsilon_{zl}^{vp} \Rightarrow \infty$, $\lambda_0/V=\lambda/V$, then (1.45a) becomes (1.21).

It is noted that (1.43) and (1.45a) are expressed in terms of natural logarithmic function and strain. If common logarithmic function and void ratio are used, (1.43) can be written:

$$\varepsilon_z^{vp} = -\frac{e-e_i}{1+e_o} = \frac{\frac{\psi_o/V}{\log e}\log\frac{t_o+t_e}{t_o}}{1+\frac{\psi_o/V}{\frac{e_l-e_i}{1+e_o}\log e}\log\frac{t_o+t_e}{t_o}}$$

$$e = e_i - \frac{\frac{(\psi_o/V)(1+e_o)}{\log e}\log\frac{t_o+t_e}{t_o}}{1+\frac{\psi_o/V}{\log e\frac{(e_l-e_i)}{1+e_o}}\log\frac{t_o+t_e}{t_o}} = e_i - \frac{C_{\alpha eo}\log\frac{t_o+t_e}{t_o}}{1+\frac{C_{\alpha eo}}{(e_l-e_i)}\log\frac{t_o+t_e}{t_o}} \tag{1.46}$$

wheree_i is the initial void ratio before the new creep stress applied, e_o is the original void ratio in the field initial stress state with original zero strain, and e_l is the limit strain under current stress and beyond e_i when $t_e \Rightarrow \infty$. The e in $\log e$ is 2.718. From (1.46), we have $C_{\alpha eo} = (\psi_o/V)(1+e_o)/\log e = (\psi_o/V)(1+e_o)/\log 2.718 = 2.3026(\psi_o/V)(1+e_o)$. From (1.46), we have $C_{\alpha e} = C_{\alpha eo}/\left[1+\frac{C_{\alpha eo}}{(e_l-e_i)}\log\frac{t_o+t_e}{t_o}\right]$. It shall be pointed out that (1.46) is slightly different from (1.14) since $\log\frac{t_e+t_o}{t_o}$ is used in (1.47) and it has definition at $t_e = 0$.

If common logarithmic function and void ratio are used, (1.45a) can be written:

$$\varepsilon_z^r = -\frac{e^r-e_o}{1+e_o} = -\frac{e_o^r-e_o}{1+e_o} + \frac{\frac{\lambda_o/V}{\log e}\log\frac{\sigma_z'}{\sigma_{zo}'}}{1+\frac{\lambda_o/V}{\frac{e_l^r-e_o}{1+e_o}\log e}\log\frac{\sigma_z'}{\sigma_{zo}'}}$$

From the above relationship, we have:

$$e^r = e_o^r - \frac{\frac{(\lambda_o/V)(1+e_o)}{\log e}\log\frac{\sigma_z'}{\sigma_{zo}'}}{1+\frac{\lambda_o/V}{\frac{e_l^r-e_o}{1+e_o}\log e}\log\frac{\sigma_z'}{\sigma_{zo}'}} = e_o^r - \frac{C_{co}\log\frac{\sigma_z'}{\sigma_{zo}'}}{1+\frac{C_{co}}{(e_l^r-e_o)}\log\frac{\sigma_z'}{\sigma_{zo}'}} \tag{1.47}$$

where e_o^r is the initial reference at stress $\sigma_z' = \sigma_{zo}'$ at the reference time line which is similar to the NCL. e_l^r is the limit void ration of the reference time lime when $\sigma_z' \Rightarrow \infty$. The e in $\log e$ is 2.718. From (1.47), we have $C_{co} = (\lambda_o/V)(1+e_o)/\log e =$

$(\lambda_o/V)(1+e_o)/\log 2.718 = 2.3026(\lambda_o/V)(1+e_o)$ From (1.47), we have $C_c = C_{co}/\left[1+\frac{C_{co}}{(e_l^r-e_o)}\log\frac{\sigma_z'}{\sigma_{zo}'}\right]$. This means that the conventional compression index C_c is dependent on the vertical effective stress. Equations (1.21) and (1.45a) may have a problem for compression behavior when the vertical effective stress σ_z' is zero or near zero, for example, σ_z' at surface or near surface of seabed soils or soil ground. The authors propose an equation like (1.22):

To replace (1.21),

$$\varepsilon_z^r = \varepsilon_{zo}^r + \frac{\lambda}{V}\ln\left(\frac{\sigma_z' + \sigma_{unit}'}{\sigma_{zo}' + \sigma_{unit}'}\right) \tag{1.48a}$$

In (1.49), a new parameter σ_{unit}' is added in (1.21). σ_{unit}' can be taken as 0.01 to 1 kPa or checked by fitting test data at very small vertical effective stress. In this way, when σ_{zo}' or σ_z' is zero or near zero, (1.48a) can still give definite strain ε_z^r. Using the same approach, (1.45a) is revised:

$$\varepsilon_z^r = \varepsilon_{zo}^r + \frac{\frac{\lambda_o}{V}\ln\frac{\sigma_z' + \sigma_{unit}'}{\sigma_{zo}' + \sigma_{unit}'}}{1+\frac{\lambda_o}{V\varepsilon_{zl}^r}\ln\frac{\sigma_z' + \sigma_{unit}'}{\sigma_{zo}' + \sigma_{unit}'}} = \left(\frac{\frac{\lambda_o}{V}}{1+\frac{\lambda_o}{V\varepsilon_{zl}^r}\ln\frac{\sigma_z' + \sigma_{unit}'}{\sigma_{zo}' + \sigma_{unit}'}}\right)\ln\frac{\sigma_z' + \sigma_{unit}'}{\sigma_{zo}' + \sigma_{unit}'} \tag{1.48b}$$

When σ_{zo}' or σ_z' is zero or near zero, (1.48b) can still give definite strain ε_z^r.

1.7 A 1-D EVP MODEL INCORPORATING THE NON-LINEAR LOGARITHMIC CREEP FUNCTION

1.7.1 Derivation of a 1-D EVPL model

Adopting the non-linear logarithmic creep function in (1.43) and logarithmic function for reference time line in (1.21), a general 1-D EVP model is derived below. The total strain ε_z in any stress-strain state point $(\sigma_z', \varepsilon_z)$ can be calculated as using equivalent time t_e:

$$\varepsilon_z = \varepsilon_z^r + \varepsilon_z^{vp} = \varepsilon_{zo}^r + \frac{\lambda}{V}\ln\frac{\sigma_z'}{\sigma_{zo}'} + \frac{\frac{\psi_o}{V}}{1+\frac{\psi_o}{V\varepsilon_{zl}^{vp}}\ln\frac{t_o+t_e}{t_o}}\ln\left(\frac{t_o+t_e}{t_o}\right) \tag{1.49}$$

Noting that the creep parameter $\frac{\psi}{V} = \frac{\psi_o}{V}/\left(1+\frac{\psi_o}{V\varepsilon_{zl}^{vp}}\ln\frac{t_o+t_e}{t_o}\right)$ is no longer a constant, as mentioned above. Equation (1.49) means that the vertical strain under a certain vertical effective stress is equal to the volume strain at the reference time line plus the non-linear logarithmic creep strain. When the creep coefficient ψ/V approaches zero, (1.49) is reduced to the elastic-plastic strain in the modified Cam-Clay model. Based on (1.49), we have:

$$\left(\varepsilon_z - \varepsilon_{zo}^r - \frac{\lambda}{V}\ln\frac{\sigma_z'}{\sigma_{zo}'}\right)\left(1 + \frac{\psi_o}{V\varepsilon_{zl}^{vp}}\ln\frac{t_o + t_e}{t_o}\right) = \frac{\psi_o}{V}\ln\left(\frac{t_o + t_e}{t_o}\right)$$

$$\ln\left(\frac{t_o + t_e}{t_o}\right) = \frac{\frac{V}{\psi_o}\left(\varepsilon_z - \varepsilon_{zo}^r - \frac{\lambda}{V}\ln\frac{\sigma_z'}{\sigma_{zo}'}\right)}{\left[1 - \left(\varepsilon_z - \varepsilon_{zo}^r - \frac{\lambda}{V}\ln\frac{\sigma_z'}{\sigma_{zo}'}\right)/\varepsilon_{zl}^{vp}\right]} \tag{1.50a}$$

$$t_o + t_e = t_o \exp\frac{\left(\varepsilon_z - \varepsilon_{zo}^r - \frac{\lambda}{V}\ln\frac{\sigma_z'}{\sigma_{zo}'}\right)}{\frac{\psi_o}{V}\left[1 - \left(\varepsilon_z - \varepsilon_{zo}^r - \frac{\lambda}{V}\ln\frac{\sigma_z'}{\sigma_{zo}'}\right)/\varepsilon_{zl}^{vp}\right]} \tag{1.50b}$$

From (1.50b), the 'equivalent time' t_e is then given by:

$$t_e = t_o \exp\left[\frac{\left(\varepsilon_z - \varepsilon_{zo}^r - \frac{\lambda}{V}\ln\frac{\sigma_z'}{\sigma_{zo}'}\right)}{\frac{\psi_o}{V}\left(1 - \left(\varepsilon_z - \varepsilon_{zo}^r - \frac{\lambda}{V}\ln\frac{\sigma_z'}{\sigma_{zo}'}\right)/\varepsilon_{zl}^{vp}\right)}\right] - t_o \tag{1.50c}$$

Differentiating (1.43) with time t_e results in:

$$\frac{d\varepsilon_z^{vp}}{dt_e} = \frac{\left(1 + \frac{\psi_o}{V\varepsilon_{zl}^{vp}}\ln\frac{t_o + t_e}{t_o}\right)\frac{\psi_o}{V}\frac{1}{t_o + t_e} - \frac{\psi_o}{V}\ln\frac{t_o + t_e}{t_o}\frac{\psi_o}{V\varepsilon_{zl}^{vp}}\frac{1}{t_o + t_e}}{\left(1 + \frac{\psi_o}{V\varepsilon_{zl}^{vp}}\ln\frac{t_o + t_e}{t_o}\right)^2} = \tag{1.51a}$$

$$= \frac{\psi_o/V}{\left(1 + \frac{\psi_o}{V\varepsilon_{zl}^{vp}}\ln\frac{t_o + t_e}{t_o}\right)^2}\frac{1}{t_o + t_e}$$

Submitting (1.50a) and (1.50b) into (1.51b), we have:

$$\frac{d\varepsilon_z^{vp}}{dt_e} = \frac{\psi_o/V}{\left(1+\frac{\psi_o}{V\varepsilon_{zl}^{vp}}\frac{\frac{V}{\psi_o}\left(\varepsilon_z-\varepsilon_{zo}^r-\frac{\lambda}{V}\ln\frac{\sigma_z'}{\sigma_{zo}'}\right)}{\left[1-\left(\varepsilon_z-\varepsilon_{zo}^r-\frac{\lambda}{V}\ln\frac{\sigma_z'}{\sigma_{zo}'}\right)/\varepsilon_{zl}^{vp}\right]}\right)^2}$$

$$\frac{1}{t_o}\exp\frac{-\left(\varepsilon_z-\varepsilon_{zo}^r-\frac{\lambda}{V}\ln\frac{\sigma_z'}{\sigma_{zo}'}\right)}{\frac{\psi_o}{V}\left[1-\left(\varepsilon_z-\varepsilon_{zo}^r-\frac{\lambda}{V}\ln\frac{\sigma_z'}{\sigma_{zo}'}\right)/\varepsilon_{zl}^{vp}\right]}=$$

$$=\frac{\psi_o}{Vt_o}\frac{1}{\left(1+\frac{\left(\varepsilon_z-\varepsilon_{zo}^r-\frac{\lambda}{V}\ln\frac{\sigma_z'}{\sigma_{zo}'}\right)}{\left[\varepsilon_{zl}^{vp}-\left(\varepsilon_z-\varepsilon_{zo}^r-\frac{\lambda}{V}\ln\frac{\sigma_z'}{\sigma_{zo}'}\right)\right]}\right)^2}$$

$$\exp\frac{-\left(\varepsilon_z-\varepsilon_{zo}^r-\frac{\lambda}{V}\ln\frac{\sigma_z'}{\sigma_{zo}'}\right)}{\frac{\psi_o}{V}\left[1-\left(\varepsilon_z-\varepsilon_{zo}^r-\frac{\lambda}{V}\ln\frac{\sigma_z'}{\sigma_{zo}'}\right)/\varepsilon_{zl}^{vp}\right]}$$

The above expression can be further simplified as:

$$\frac{d\varepsilon_z^{vp}}{dt_e}=\frac{\psi_o}{Vt_o}\left(\frac{\varepsilon_{zl}^{vp}-\left(\varepsilon_z-\varepsilon_{zo}^r-\frac{\lambda}{V}\ln\frac{\sigma_z'}{\sigma_{zo}'}\right)}{\varepsilon_{zl}^{vp}}\right)^2\exp\frac{-\left(\varepsilon_z-\varepsilon_{zo}^r-\frac{\lambda}{V}\ln\frac{\sigma_z'}{\sigma_{zo}'}\right)}{\frac{\psi_o}{V}\left[1-\left(\varepsilon_z-\varepsilon_{zo}^r-\frac{\lambda}{V}\ln\frac{\sigma_z'}{\sigma_{zo}'}\right)/\varepsilon_{zl}^{vp}\right]} \tag{1.51b}$$

Equation (1.51b) indicates that the creep strain rate is dependent on the stress-strain state only $(\sigma_z',\varepsilon_z)$. Noting $\varepsilon_z^r=\varepsilon_{zo}^r+\frac{\lambda}{V}\ln\frac{\sigma_z'}{\sigma_{zo}'}$ in (1.21), the above equation can be simplified as:

$$\frac{d\varepsilon_z^{vp}}{dt_e}=\frac{d\varepsilon_z^{vp}}{dt}=\dot{\varepsilon}_z^{vp}=\frac{\psi_o}{Vt_o}\left(1+\frac{(\varepsilon_z^r-\varepsilon_z)}{\varepsilon_{zl}^{vp}}\right)^2\exp\left[\frac{(\varepsilon_z^r-\varepsilon_z)}{(1+(\varepsilon_z^r-\varepsilon_z)/\varepsilon_{zl}^{vp})}\frac{V}{\psi_o}\right] \tag{1.51c}$$

where ε_z^r is defined as in (1.21) and is related to stress σ_z'. Equations (1.50) and (1.51) show that the equivalent time t_e and creep strain rate $\dot{\varepsilon}_z^{vp}$ are uniquely related to the stress-strain state $(\sigma_z', \varepsilon_z)$.

Applying (1.51c), (1.27) can be rewritten:

$$\frac{d\varepsilon_z}{dt} = \frac{\kappa/V}{\sigma_z'}\frac{d\sigma_z'}{dt} + \frac{\psi_o}{Vt_o}\left(1+\frac{(\varepsilon_z^r - \varepsilon_z)}{\varepsilon_{zl}^{vp}}\right)^2 \exp\left[\frac{(\varepsilon_z^r - \varepsilon_z)}{(1+(\varepsilon_z^r - \varepsilon_z)/\varepsilon_{zl}^{vp})}\frac{V}{\psi_o}\right] \tag{1.52a}$$

$$\dot{\varepsilon}_z = \frac{\kappa/V}{\sigma_z'}\dot{\sigma}_z' + \frac{\psi_o}{Vt_o}\left(1+\frac{(\varepsilon_z^r - \varepsilon_z)}{\varepsilon_{zl}^{vp}}\right)^2 \exp\left[\frac{(\varepsilon_z^r - \varepsilon_z)}{(1+(\varepsilon_z^r - \varepsilon_z)/\varepsilon_{zl}^{vp})}\frac{V}{\psi_o}\right] \tag{1.52b}$$

Equation (1.52) is a general EVP model with a limit creep strain, called 1-D EVPL model for any 1-D compression loading condition, for example, stepped-loading, constant-rate-of-strain loading, or even unloading/reloading. The structure of (1.52) is similar to the structure of the differential form of 1-D EVP model in (1.29) but considering the creep coefficient decreases with time.

1.7.2 Calibration and prediction of a 1-D EVPL model

In order to apply the EVPL model, all values of parameters in this model must be determined. Data from a multi-staged creep test on Hong Kong Marine Clay can be used to determine all model parameters. The basic properties of the soil are listed in Table 1.2. Figure 1.13 shows the measured results of vertical strain *vs.* log(time) with unloading and reloading. The stress labels on Figure 1.13 refer to different constant vertical stresses in the multi-staged loading. Figure 1.13(a) shows curves of log(time) and vertical strain of compression/creep under staged loading process; Figure 1.13(b) shows the curves in staged unloading; and Figure 1.13(c) shows the curves of compression/creep under staged loading again. The time corresponding to this strain is commonly called the time at the EOP consolidation, that is, t_{EOP}. Corresponding to t_{EOP}, the total vertical strain is ε_{EOP}. As explained before, t_{EOP} for a soil specimen of 20 mm with double drainage is only tens of minutes. Therefore, strains after tens of minutes are all creep strains. It is estimated from Figure 1.13 that t_{EOP} for consolidation starting from NCL is 49 min. The relationship of vertical strain ε_{EOP} *vs.* vertical effective stress σ_z' (in log-scale) at the end of primary consolidation with loading and unloading-reloading is plotted in Figure 1.14 (all data points were taken from 'EOP').

Table 1.2 Basic properties of Hong Kong Marine Clay

Clay	*Silt*	*Fine sand*	*Gs*	w_L	w_P	*IP*	*w*	e_o
27.5%	58.4%	14.1%	2.664	60.0%	28.5%	31.5%	57.4%	1.529

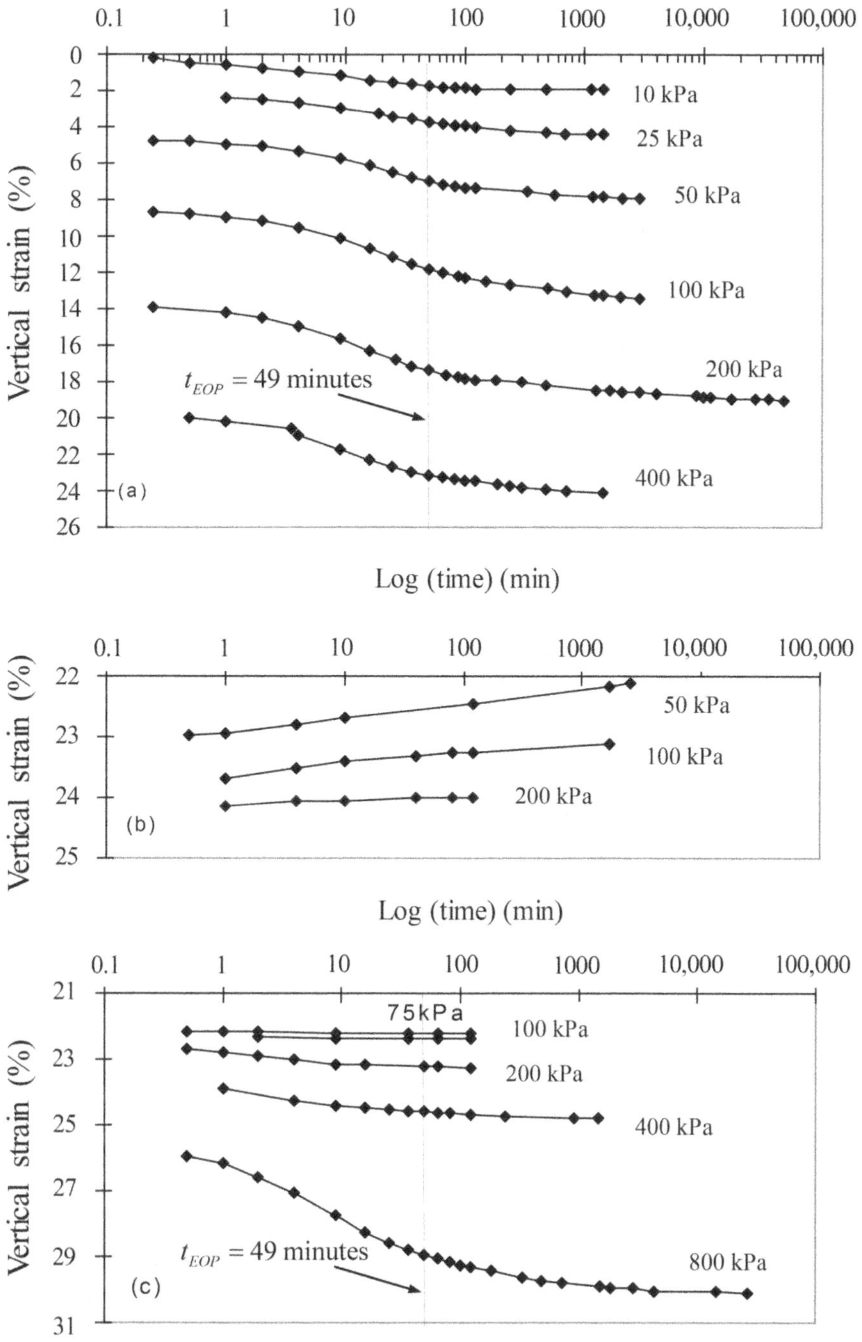

Figure 1.13 Vertical strain versus log(time) (min) – (a) first loading, (b) unloading, and (c) reloading except for stress 800 kPa being first loading.

Table 1.3 Parameters in the 1-D EVPL model

κ/V	t_o (min)	Average ψ_o/V	Average ε_{zl}^{vp}	ε_{zo}^{r}	σ'_{zo} (kPa)	λ/V
0.0116	49	0.00601	0.0252	0	24.0	0.0833

1.7.2.1 Determination of parameter κ (instant time line or κ-line)

Equation (1.20) is used to fit test data in the unloading/reloading range in Figure 1.14. The best fitting indicates that the parameter κ/V is 0.0116, which is listed in Table 1.3.

1.7.2.2 Determination of three parameters ε_{zo}^{r}, σ'_{zo}, and λ/V (reference time line or λ-line)

The simple approximate method in Section 1.5.4 is used to determine remaining parameters. The four data points for stresses 100, 200, 400, and 800 kPa are in the normally consolidated range and are almost on a straight line. Equation (1.21) is used to fit the four points. From the best fitting, $\lambda/V = 0.0833$, $\varepsilon_{zo}^{r} = 0$, and $\sigma'_{zo} = 24.0kPa$. Here we select $\varepsilon_{zo}^{r} = 0$. Values of the three parameters are listed in Table 1.3.

1.7.2.3 Determination of three creep parameters $\psi o/V$, to and ε_{zl}^{vp}

In order to find values of the free parameters from the best data fitting, (1.43) is written as:

$$\frac{\ln\frac{t_o+t_e}{t_o}}{\varepsilon_z^{vp}} = \frac{V}{\psi_o} + \frac{1}{\varepsilon_{zl}^{vp}}\ln\frac{t_o+t_e}{t_o} \qquad (1.53a)$$

where ε_z^{vp} is the strain increase due to current creep loading only, not including any instantaneous strain or creep strains under previous loads. The compression in oedometer tests is coupled with excess porewater pressure dissipation. However, (1.53a) shall fit the creep strain only, that is, the strain after porewater pressure is negligible or for $t > t_{EOP}$. Since we consider the line at time of t_{EOP} is the reference time line (also called NCL) as shown in Figure 1.14, the equivalent time in (1.53a) is $t_e = t - t_{EOP}$ and the creep strain is $\varepsilon_z^{vp} = \varepsilon_z - \varepsilon_{z,EOP}$. The time t is the time from the beginning of the current creep load, ε_z is the total strain under current creep load, and $\varepsilon_{z,EOP}$ is the accumulated strain up to

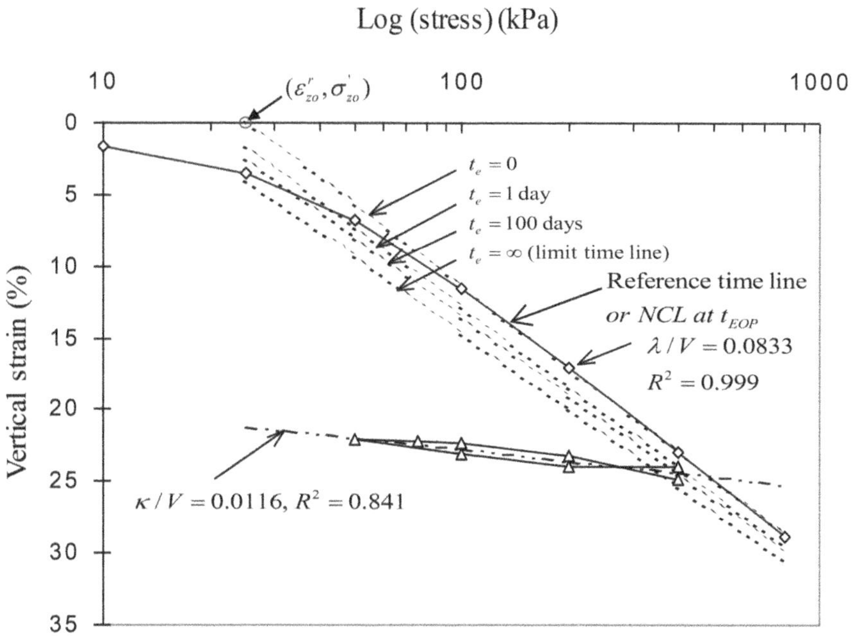

Figure 1.14 Vertical strain versus vertical effective stress in log scale: (i) test data of EOP and 24 hours, (ii) fitted lines of t_e=0, 1 day, 100 days and infinite, and (iii) unloading/reloading data and fitted line.

time t_{EOP}. According to the simple approximate method, we can take $t_o = t_{EOP}$. Considering these, (1.53a) can be written as:

$$\frac{\ln\dfrac{t}{t_{EOP}}}{\varepsilon_z - \varepsilon_{z,EOP}} = \frac{V}{\psi_o} + \frac{1}{\varepsilon^{vp}_{zl}} \ln\frac{t}{t_{EOP}} \tag{1.53b}$$

Equation (1.53a) or (1.53b) is a linear equation with slope $1/\varepsilon^{vp}_{zl}$ and intercept $1/\psi_o$. Equation (1.53a) or (1.53b) is used to fit data plotted in the coordinate of $\left(\ln\frac{t_o+t_e}{t_o}\right)/\varepsilon^{vp}_z$ $vs \ln\frac{t_o+t_e}{t_o}$ from a creep test. From the best fitting, ε^{vp}_{zl} and ψ_o can be determined.

For example, for the creep test under vertical stress of 100 kPa, the original data in Figure 1.13 is re-plotted in Figure 1.15(a). It is found that the data points of $\left(\ln\frac{t_o+t_e}{t_o}\right)/\varepsilon^{vp}_z$ $vs \ln\frac{t_o+t_e}{t_o}$ are almost on a straight line for a given $t_o = 49$ min. A linear line equation best fits these data points. The best fitting equation is

$$y = 29.925x + 129.86 \tag{1.54}$$

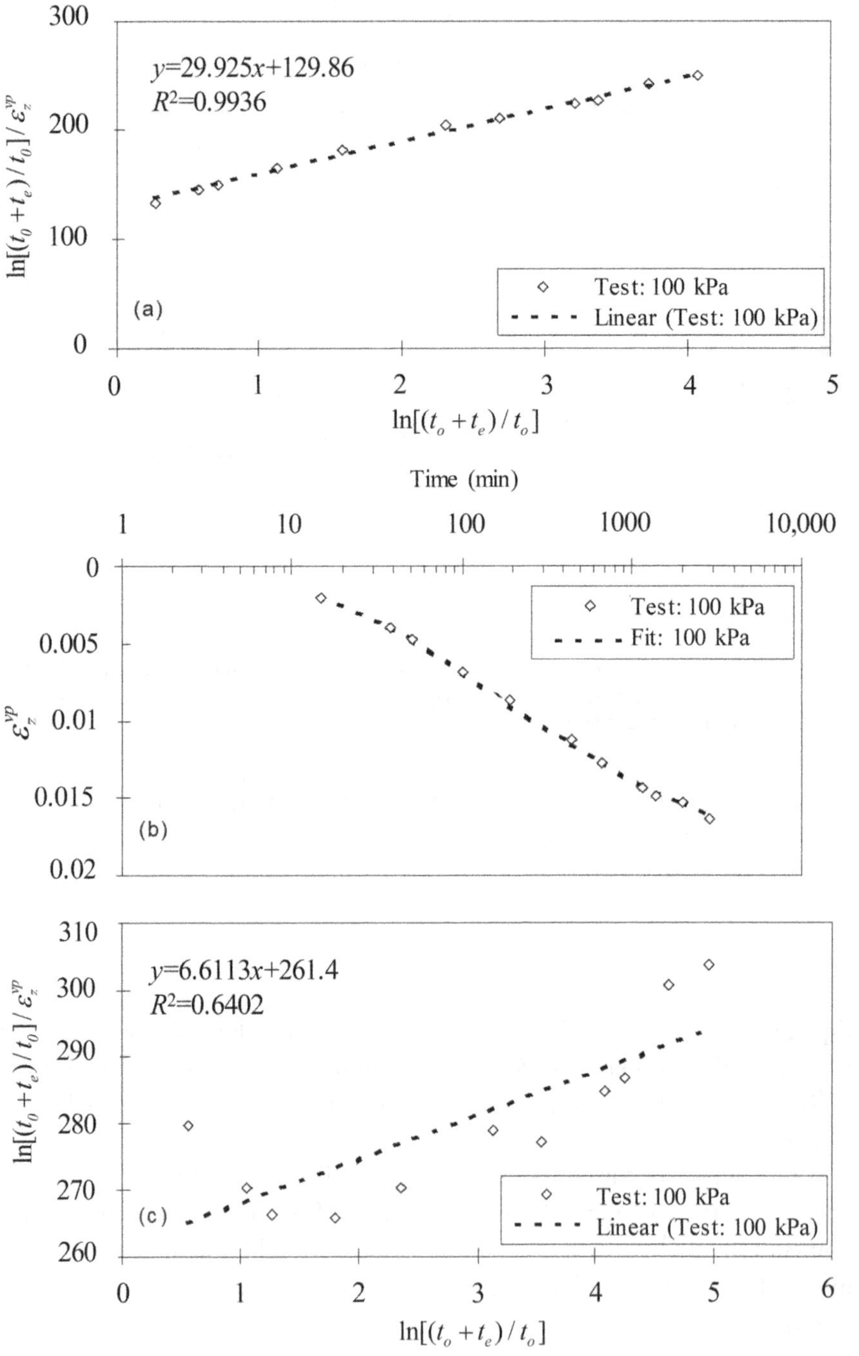

Figure 1.15 (a) Curve fitting by a straight line, (b) comparison of measured data and fitted vertical stress of 100 kPa and t_o = 49 min, and (c) t_o = 20 min.

with R^2 of 0.9936 indicating very good fitting. If t_o is chosen to be 20 min, the same data is plotted in Figure 1.15(c). It is seen that the data points are largely not on a straight line. The best-curve fitting to those points in Figure 1.15(c) produces R^2 to be only 0.6402. This demonstrates that the parameter t_o cannot be chosen arbitrarily. In general, t_o together with ψ_o and ε_{zl}^{vp} can be determined by best non-linear curve fitting. However, if t_o can be estimated in advance, the remaining two parameters with ψ_o and ε_{zl}^{vp} can be easily determined by the best line fitting. It is found in this study that choosing $t_o = t_{EOP}$ which obtained from oedometer tests can make the data points on a straight line and therefore produces the best line fitting based on (1.53). From (1.54), we have:

$$V/\psi_o = 129.86 \text{ and } \psi_o/V = 0.00770$$

$$1/\varepsilon_{zl}^{vp} = 29.925 \text{ and } \varepsilon_{zl}^{vp} = 0.03342 = 3.342\%.$$

Values of ψ_o, ε_{zl}^{vp}, and t_o for the creep test under stress 100 kPa in the first loading are presented in Table 1.4. Figure 1.15(b) shows the comparison of measured data and fitted curve with creep time t in log scale. It is seen that the curve calculated from (1.47) using $\psi_o/V = 0.00770$, $\varepsilon_{zl}^{vp} = 3.342\%$ and $t_o = 49$ min is in good agreement with test data.

The same procedure as described above is used to fit (1.53) to creep test data for vertical effective stresses 200, 400, and 800 kPa in first loading, respectively, using $t_o = 49$ min. The values of ψ_o/V and ε_{zl}^{vp} creep tests under the three stresses are presented in Table 1.4.

The fitted line (or curve) is shown in Figures 1.16, 1.17, and 1.18, respectively. In general, (1.47) or (1.53) fits test data very well. It shall be noted here that the unloading/reloading data are not recommended to be used for the determination of the creep parameters since creep in unloading or reloading is not very significant, especially when the over-consolidation ratio is large.

It is noted that both ψ_o/V and ε_{zl}^{vp} are dependent on the vertical effective stress, normally decreases as the stress increases. This is consistent with the description using (1.45). Yin (1999) found the following fitting relations:

$$\psi_o/V = 0.0139 - 0.0014\ln(\sigma_z'/\sigma_o') \tag{1.55a}$$

$$\varepsilon_{zl}^{vp} = 0.0617 - 0.0064\ln(\sigma_z'/\sigma_o') \tag{1.55b}$$

Table 1.4 Values of three parameters in (1.47) or (1.53) by best curve-fitting

	100 kPa	*200 kPa*	*400 kPa*	*800 kPa*
t_o (min)	49	49	49	49
ψ_o/V	0.00770	0.00637	0.00507	0.00489
ε_{zl}^{vp}	0.03342	0.02434	0.02347	0.01966

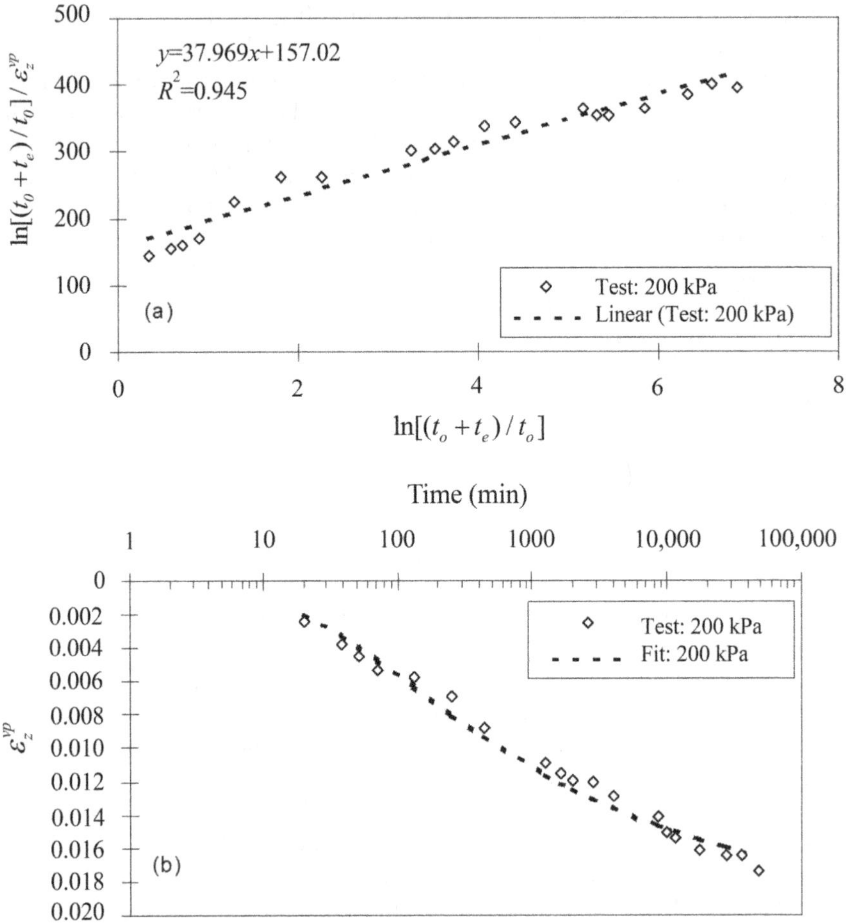

Figure 1.16 (a) Curve fitting by a straight line and (b) comparison of measured data and fitted curve – vertical stress of 200 kPa and t_o = 49 min.

where σ_o' is a unit stress and taken as 1 kPa. The averaged values of ψ_o/V and ε_{zl}^{vp} for four creep tests are listed in Table 1.3.

So far, all values of parameters in 1-D EVPL model have been determined. Regarding of ψ_o/V and ε_{zl}^{vp}, for a simple approach, the averaged values of ψ_o/V and ε_{zl}^{vp} can be used. However, for better consideration of the non-linear creep behavior, the two equations in (1.55a) and (1.55b) for ψ_o/V and ε_{zl}^{vp} are used.

Figure 1.19 shows simulated curves of $\log\sigma_z'$ $vs\,\varepsilon_z$ of a constant rate of strain (CRSN) tests at strain rate $\dot{\varepsilon}_z = 0.00053\%/\text{min}$ for $\psi_0/V = 0.02, 0.002$, and 0.0002 using the 1-D EVPL model with other parameters in Table 1.3. Instant time line (κ – line) and reference time lime (λ – line) are also shown in Figure 1.19.

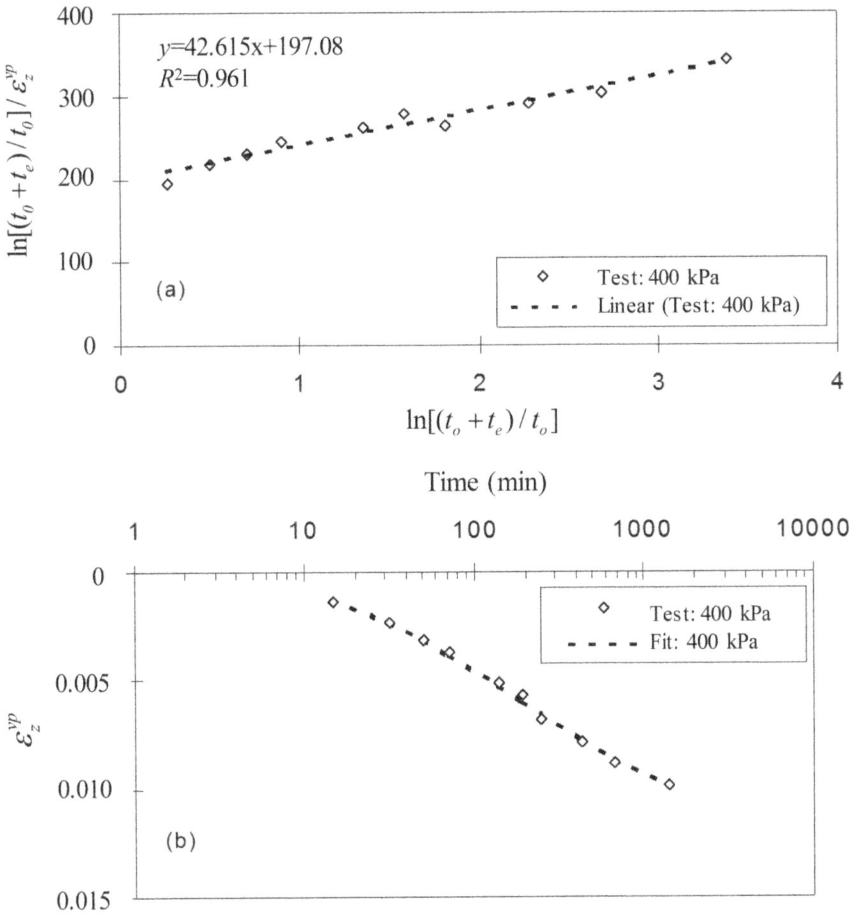

Figure 1.17 (a) Curve fitting by a straight line and (b) comparison of measured data and fitted curve – vertical stress of 400 kPa and t_o = 49 min.

Limit creep strain ε_{zl}^{vp} and limit compression strain ε_{zl}^{r} can be expressed:

$$\varepsilon_{zl}^{vp} = -\frac{e-e_i}{1+e_o} \tag{1.56a}$$

$$\varepsilon_{zl}^{r} = -\frac{e^r-e_o}{1+e_o} \tag{1.56b}$$

For the creep test under 100 kPa, the initial strain ε_{zi}^{r} before creep is 11.8%, found from Figure 1.13. The original initial void ratio e_o is 1.529 from Table 1.2. Thus, the initial void ratio e_i before creep

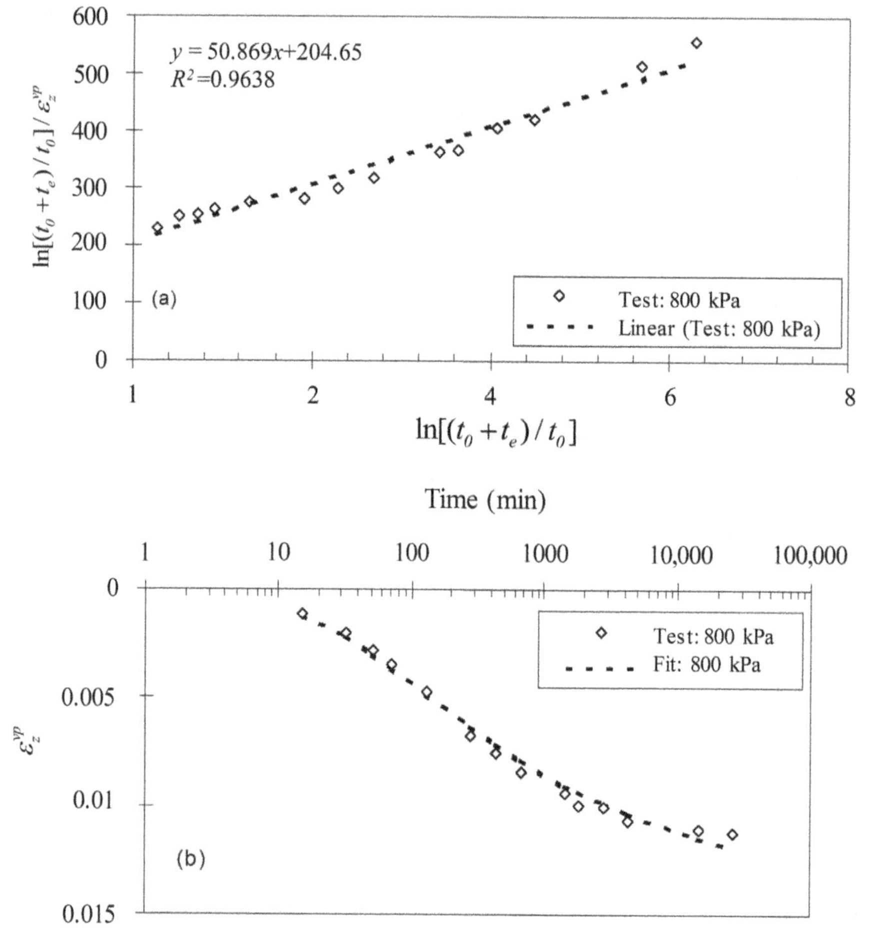

Figure 1.18 (a) Curve fitting by a straight line and (b) comparison of measured data and fitted curve – vertical stress of 800 kPa and $t_o = 49$ min.

is $e_i = e_o - \varepsilon^r_{zi}(1+e_o) = 1.529 - 0.118(1+1.529) = 1.231$. Thus, if the final void ratio is zero, that is, $e = 0$, the limit creep strain ε^{vp}_{zl} is:

$$\varepsilon^{vp}_{zl} = -\frac{e - e_i}{1+e_o} = -\frac{0-1.231}{1+1.529} = 0.4868 = 48.68\%$$

The limit compression strain ε^r_{zl}

$$\varepsilon^r_{zl} = -\frac{e^r - e_o}{1+e_o} = \frac{1.529}{1+1.529} = 0.6046 = 60.46\%$$

However, as seen in Table 1.4, the ε^{vp}_{zl} is 0.03342 = 3.342% for creep under 100 kPa, much smaller than the theoretical limit of 48.46%.

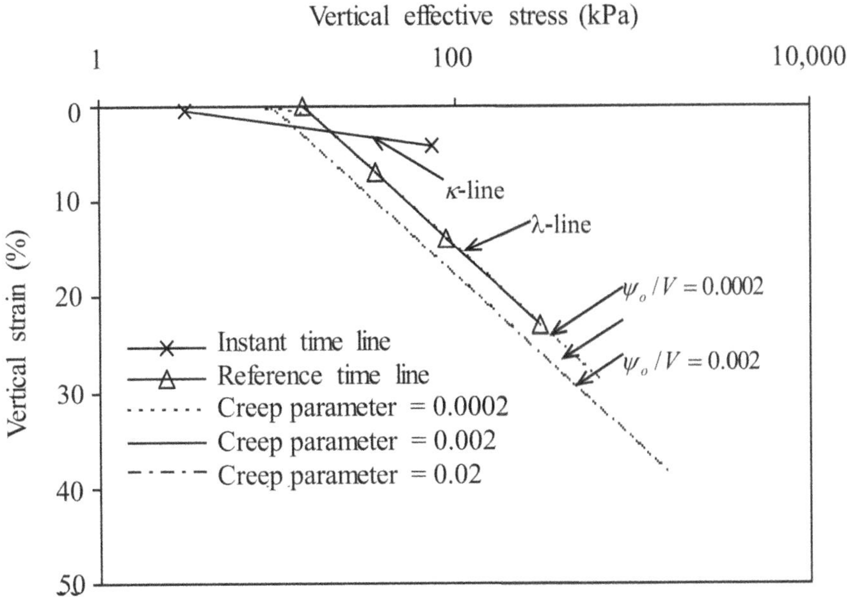

Figure 1.19 Simulated curves of $\log\sigma'_z$ vs ε_z of CRSN tests at a constant rate of strain $\dot{\varepsilon}_z$ = 0.00053%/min for ψ_0/V = 0.02, 0.002, and 0.0002 using the 1-D EVPL model.

1.8 SUMMARY AND COMPARISONS OF 1-D CONSTITUTIVE MODELS

1.8.1 Comparison of Maxwell model with 1-D EVP model and 1-D EVPL model

Maxwell's linear rheological model is (1.11) – the total strain rate $\dot{\varepsilon}_z$ is the sum of linear elastic strain rate $\dot{\varepsilon}_z^e$ and linear viscous strain rate $\dot{\varepsilon}_z^v$, that is, $\dot{\varepsilon}_z = \dot{\varepsilon}_z^e + \dot{\varepsilon}_z^v$. Constitutive equations of the 1-D EVP model and 1-D EVPL model are in (1.29) and (1.52). It is found that, in (1.29) and (1.52), the total strain rate $\dot{\varepsilon}_z$ is sum of non-linear elastic strain rate $\dot{\varepsilon}_z^e$ and non-linear visco-plastic strain rate $\dot{\varepsilon}_z^{vp}$, that is, $\dot{\varepsilon}_z = \dot{\varepsilon}_z^e + \dot{\varepsilon}_z^{vp}$. The structure of (1.11), (1.29), and (1.52) is the same consisting of two items on the right side. However, both the elastic strain rate $\dot{\varepsilon}_z^e$ and the visco-plastic strain rate $\dot{\varepsilon}_z^{vp}$ in the 1-D EVP model and the 1-D EVPL model are all non-linear, while $\dot{\varepsilon}_z^e$ and $\dot{\varepsilon}_z^v$ in Maxwell model are all linear. In addition, the strain rate $\dot{\varepsilon}_z^{vp}$ in the 1-D EVP and 1-D EVPL models are irreversible and therefore are called visco-plastic strain rate here. $\dot{\varepsilon}_z^v$ in Maxwell model is also irreversible when the stress is reduced to zero, but can be recovered and even becomes negative if the stress is unloaded to negative. This is contradictive to the real behavior of soils – non-linear and irreversible/plastic. Both Maxwell and Kelvin linear rheological models have serious limitations for applications to soils.

1.8.2 Relation of 1-D EVP model and 1-D EVPL model with a non-linear elastic model and an elastic-plastic model

It is interesting to note that the constitutive equations of both the 1-D EVP model and 1-D EVPL model in (1.29) and (1.52) can be reduced to a non-linear elastic model and an elastic-plastic model when the viscosity of the soil is zero.

Considering the creep parameter $\psi/V = 0$ in (1.29) or $\psi_o/V = 0$ in (1.52) for the state point $(\sigma'_z, \varepsilon_z)$ above the limit time line or considering the state point $(\sigma'_z, \varepsilon_z)$ below the limit time line, we have

$$\dot{\varepsilon}_z = \dot{\varepsilon}_z^e = \frac{\kappa}{V}\frac{\dot{\sigma}'_z}{\sigma'_z} \tag{1.57}$$

which is a non-linear elastic constitutive equation, which is the same as that used in the Camp-Clay model for calculation of the non-elastic strain. Integration of the above equation results in the non-linear elastic equation in (1.20).

Using the equivalent time, t_o, the 1-D EVP model and 1-D EVPL model are expressed in (1.24) and (1.49). Both (1.24) and (1.49) can be reduced to if $\psi/V = 0$ and $\psi_o/V = 0$:

$$\varepsilon_z = \varepsilon_{zo}^r + \frac{\lambda}{V}\ln\frac{\sigma'_z}{\sigma'_{zo}} \tag{1.58}$$

This is a non-linear elastic-plastic constitutive equation, which is the same as that used in the Camp-Clay model for calculation of the elastic-plastic strain in the normally consolidated range.

The above proof shows that the 1-D EVP model and 1-D EVPL model can become a time-independent 1-D non-linear elastic model or elastic-plastic model when the creep parameter ψ/V and ψ_o/V are zero, meaning zero viscosity.

1.8.3 Summary of all 1-D constitutive models, parameters, features, advantages, and limitations

As discussed before, the 1-D EVP and 1-D EVPL models can become an elastic model and an elastic-plastic model if the viscosity parameter is zero. This means that 1-D EVP and 1-D EVPL models are a more general constitutive relationship. The mathematical structure of the 1-D EVP and 1-D EVPL models can be written as one expression:

$$\frac{d\varepsilon_z}{dt} = \frac{df(\sigma'_z)}{dt} + g(\sigma'_z, \varepsilon_z) \tag{1.59}$$

Table 1.5 Summary of all 1-D models, parameters, features, advantages, and limitations

	1-D linear elastic model	***1-D non-linear elastic model***	***1-D elastic-plastic model***	***Maxwell model and Kelvin model***	***1-D EVP model (1-D EVP)***	***1-D EVP model with creep limit (1-D EVPL)***
Reference equation	(1.2)	(1.5) or (1.20)	(1.7)	(1.11) and (1.13)	(1.29)	(1.52)
Parameters	m_v	$C_{r\varepsilon}$ or κ/V	$C_{r\varepsilon}, C_{c\varepsilon}, \sigma'_{zc}$	M^e, η	κ/V $\psi_o/V, t_o \varepsilon^r_{zo}, \sigma'_{zo} \lambda/V$	κ/V $\psi_o/V, t_o\ \varepsilon^{vp}_{zl}, \varepsilon^r_{zo}, \sigma'_{zo}, \lambda/V$
Features	Linear elastic only	Non-linear elastic	Capturing non-linear elastic and non-linear plastic behavior	Linear elastic straining and linear viscous straining	Considering non-linear elastic and non-linear visco-plastic behavior	Considering non-linear elastic and non-linear visco-plastic behavior with limit strains
Advantages	This is the simplest model and obtaining analytical solutions for some boundary- value problems is easy.	This model is simple and good for non-elastic behavior in over-consolidated or unloading/ reloading cases.	This model is practical, simple, and capturing non-linear elastic and plastic straining.	This is the simplest rheologic model and obtaining analytical solutions is easy.	This is a non-linear EVP model suitable for clayey soils in 1-D straining.	This is a non-linear EVP model with creep limit suitable for clayey soils.
Limitations	Over-simplified	No consideration of plastic and time-simplification behavior	No consideration of creep and strain rate effects	No consideration of non-linear elastic and non-linear visco-plastic straining	Neither creep limit not compression limit	Not compression limit yet
Application Advice	Use with care	Use for right conditions	Good for most applications except for viscous effects	Use with care, no good for non-linear viscous plastic behavior of soils	Good for most applications	Good for most applications

where f is a function of the vertical effective stress and is an elastic part. The g is a function of the vertical effective stress and strain and is a visco-plastic part.

A summary of all 1-D models, parameters, features, advantages, and limitations is presented in Table 1.5. Advices on applications of all seven 1-D models are also given in Table 1.4. In summary, the elastic-plastic model in (1.7) is the most practical one and has been used commonly by engineers. The major limitation in (1.7) is that this model cannot consider viscous effects or time-dependent effects of the skeleton of clayey soils. Overall speaking, the 1-D EVP models without a limit (1-D EVP model) or with a creep limit (1-D EVPL model) are the most general constitutive relationships for the time-dependent stress-strain behavior of clayey soils in 1-D straining.

REFERENCES

Berre, T, and Iversen, K, 1972, Oedometer tests with different specimen heights on a clay exhibiting large secondary compression. *Géotechnique*, 22(1), pp. 81–118.

Bjerrum, L, 1967, Engineering geology of Norwegian normally consolidated marine clays as related to the settlements of buildings. *Géotechnique*, 17(2), pp. 83–118.

Chen, WF, 1994, *Constitutive Equations for Engineering Materials, Vol. 2: Plasticity and Modeling.* (Amsterdam: Elsevier).

Chen, WF, and Mizuno, E, 1990, *Nonlinear Analysis in Soil Mechanics – Theory and Implementation.* (Amsterdam: Elsevier).

Duncan, JM, and Chang, CY, 1970, Non-linear analysis of stress and strain in soils. *Journal of Soil Mechanics and Foundation Division, ASCE*, 96(5), pp. 1629–1654.

Kondner, RL, 1963. Hyperbolic stress-strain responses: Cohesive soils. *Journal of Soil Mechanics and Foundation Division, ASCE*, 89(1), pp. 115–143.

Leroueil, S, Kabbaj, M, Tavenas, F and Bouchard, R, 1985, Stress-strain-strain rate relation for the compressibility of sensitive natural clays. *Géotechnique*, 35(2), pp. 159–180.

Perzyna, P, 1963, The constitutive equations for working hardening and rate sensitive plastic materials. *Proceedings of Vibration Problems*, 4(3), pp. 281–290.

Perzyna, P, 1966, Fundamental problems in viscoplasticity. *Advances in Applied Mechanics*, 9, pp. 244–368.

Roscoe, KH, and Burland, JB, 1968, On the generalised stress-strain behaviour of wet clay, *Engineering Plasticity*, Heyman, J, Heyman, and Leckie, FP (Eds.) (Cambridge: Cambridge University Press), pp. 535–609.

Sällfors, G, 1975, Reconsolidation pressure of soft, high plastic clays. Ph.D. thesis submitted to Chalmers University of Technology, Göteborg, Sweden.

Terzaghi, K, 1925, *Erdbaumechanik auf Boden-physicalischen Grundlagen.* (Vienna: Deuticke).

Terzaghi, K, 1943, *Theoretical Soil Mechanics.* (New York: Wiley).

Yin, JH, 1990, Constitutive modelling of time-dependent stress-strain behaviour of soils. Ph.D. thesis submitted to University of Manitoba, Winnipeg, Canada.

Yin, JH, 1999, Non-linear creep of soils in oedometer tests. *Géotechnique*, 49(5), pp. 699–707.
Yin, JH, and Graham, J, 1989, Viscous-elastic-plastic modelling of one-dimensional time-dependent behaviour of clays. *Canadian Geotechnical Journal*, 26, pp. 199–209.
Yin, JH, and Graham, J, 1994, Equivalent times and one-dimensional elastic visco-plastic modelling of time-dependent stress-strain behaviour of clays. *Canadian Geotechnical Journal*, 31, pp. 42–52.
Yin, JH, 2011, From constitutive modeling to development of laboratory testing and optical fiber sensor monitoring technologies. *Chinese Journal of Geotechnical Engineering*, 33(1), pp. 1–15. (14th 'Huang Wen-Xi Lecture' in China).
Yin, JH, Qin, JQ, and Feng, WQ, 2020, Novel FBG-based effective stress cell for direct measurement of effective stress in saturated soil. *International Journal of Geomechanics, ASCE*, 20(8), 04020107-1~7.

Chapter 2

One-dimensional consolidation

2.1 INTRODUCTION

Consolidation of a fully saturated soil is the gradual reduction of its volume due to drainage of some of the porewater. The movement of water through and out of a soil mass requires time; the time rate of consolidation is controlled principally, though not entirely, by the permeability of soil. In the case of a clay soil in the field, the time required to reach equilibrium may be of the order of days, months, or years. The mathematical theory to describe this process occupies a unique place in the history of geotechnical engineering. Since Terzaghi's one-dimensional (1-D) consolidation theory was proposed based on the principle of effective stress (Terzaghi, 1925, 1943), extensive research has been carried out in this field. Until now, consolidation theory can be applied to simulate or predict the deformation and consolidation of soil for a wide range of geotechnical engineering practices.

In many situations, the consolidation of a soil layer can be simplified as 1-D straining such as the prediction of settlement of large area reclamation. Even if field problems are three-dimensional (3-D) in nature and the deformation and consolidation of the clay layer may not be 1-D in the vertical direction, the consolidation settlement estimation is still based largely on 1-D consolidation theory in practice, but is corrected for two- or 3-D effects using, for example, the Skempton-Bjerrum method (Skempton and Bjerrum, 1957).

The mathematical model of 1-D consolidation will be discussed in this chapter.

2.2 DEFINITIONS OF EXCESS POREWATER PRESSURE

In current soil mechanics textbooks, the excess porewater pressure in consolidation process is defined in at least two ways. First, the excess porewater pressure is defined as the excess over the hydrostatic pressure (the pressure distribution when the porewater is stationary). The second definition of excess porewater pressure states that it is the pore pressure in excess of a steady-state flow condition. These two definitions of excess pore pressure coincide when consolidation theory is used in cases where the water after consolidation is stationary. However, if water heads at boundaries are fluctuating with time or there is an initial hydraulic gradient, these two commonly used definitions of excess porewater pressure (Gibson *et al.*, 1989) are not applicable. In order to provide a broad concept, a more appropriate definition of the excess porewater pressure is given as follows.

As shown in Figure 2.1, the total water head h at an arbitrary point A can be expressed as (2.1) since seepage velocities in soils are normally so small that the velocity head can be neglected.

$$h = z + \frac{u_w}{\gamma_w} \tag{2.1}$$

where h = total head, u_w = total porewater pressure, γ_w = unit weight of water (9.81 kN/m³), and z = vertical coordinate.

The water head h_0 at a given initial time $t = 0^-$ before the change (or sudden change) of boundary conditions and at a given depth z_0 (point Q in Figure 2.1) is chosen as a reference water head. The *excess porewater pressure* u_e at any depth z including the positions on the boundaries and at any time after loading is defined as $u_e = \gamma_w(h - h_0)$. This definition agrees with the definition that the excess porewater pressure is the excess value over hydrostatic pressure for a stationary water pressure condition. According to this definition, the following (2.2) holds:

$$h = z + \frac{u_w}{\gamma_w} = h_0 + \frac{u_e}{\gamma_w} \tag{2.2}$$

In (2.2), the hydraulic stationary (or static) porewater pressure u_s at depth z is $u_s = \gamma_w(h_0 - z)$ from $h_0 = z + \frac{u_s}{\gamma_w}$ referring to h_0 at z_0. It is noted that from (2.2), $u_w = \gamma_w(h_0 - z) + u_e$. Using $\gamma_w(h_0 - z) = u_s$, we have $u_w = u_s + u_e$, which means the total porewater pressure u_w is the sum of static porewater pressure u_s and the excess porewater pressure u_e.

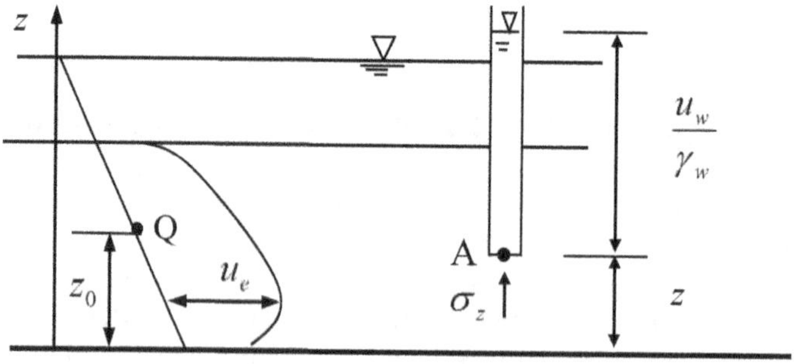

Figure 2.1 Definitions of terms by symbols.

2.3 GOVERNING EQUATIONS OF CONSOLIDATION PROBLEMS

2.3.1 Basic assumptions for 1-D consolidation theory

The 1-D consolidation theory is based the underlying assumptions:

a. The soil is completely saturated with water.
b. The porewater and soil particles are incompressible.
c. Darcy's law is valid.
d. Compression and flow are 1-D (vertical).
e. Strains are small.
f. Inertial force can be neglected.

2.3.2 Vertical total stress and effective stress

As shown in Figure 2.1, perpendicular to the horizontal plane across point A, there are acting a total stress σ_z(positive for compression) and a porewater pressure u_w. The vertical effective stress σ'_z is defined as equal to the total stress σ_z minus the porewater pressure u_w:

$$\sigma'_z = \sigma_z - u_w \tag{2.3}$$

Equation (2.3) is the effective stress equation.

The definition for effective stress of a soil is given by the *effective stress principle* (Terzaghi, 1925, 1943), which can be stated as follows:

a. The *effective stress* of a soil element is defined as the contact forces of particles over cross-section area of the element and equal to the total stress minus the pore pressure of the element.
b. The effective stress controls certain aspects of soil behavior, notably deformation and strength of the soil.

2.3.3 Constitutive equations

A central part in solving a physical problem is the description of the constitutive relations between physical quantities such as stress, stress-rate, strain, and strain-rate. As presented in Chapter 1, for soils, the relations between stress, stress-rate, strain, and strain-rate are complicated that different constitutive equations should be adopted to describe the linear, non-linear, irreversible, and/or time-dependent behavior of soils for different applications. In Chapter 1, an examination of the existing constitutive models can be written as a general constitutive relationship in (1.59) for 1-D stress-strain behavior of soils. For the consolidation analysis of an initial and boundary value problem under 1-D straining, (1.59) can be written as

$$\frac{\partial \varepsilon_z}{\partial t} = \frac{\partial f(\sigma'_z)}{\partial t} + g(\sigma'_z, \varepsilon_z) \tag{2.4}$$

where ε_z is vertical strain (positive for compression), $\frac{\partial \varepsilon_z}{\partial t}$ is the vertical strain rate as a partial differentiation of the strain with respect to time since z is another independent variable. In (2.4), σ'_z is the vertical effective stress; f and g are two functions of σ'_z and ε_z.

2.3.4 Darcy's law

In 1-D vertical consolidation, water flows through a fully saturated soil in accordance with Darcy's empirical law as assumed:

$$q_z = -k\frac{\partial h}{\partial z} \tag{2.5}$$

where k = coefficient of permeability and q_z = vertical flow rate. The negative sign means that the flow rate q_z decreases as the potential h increases with the vertical coordinate z in Figure 2.1.

From the definition of excess porewater pressure above, Darcy's law can be expressed as follows:

$$q_z = -\frac{k}{\gamma_w}\frac{\partial u_e}{\partial z} \tag{2.6}$$

2.3.5 Continuity equation

Consider an infinitesimal element having a unit cross-area and height dz as shown in Figure 2.2. In a differential time period dt, the net outflow of water is

$$\text{Net outflow of water} = (q_{z+dz} - q_z)dt = \frac{\partial q_z}{\partial z}dzdt \tag{2.7}$$

Since the flow and compression are 1-D (vertical), the volume change of the infinitesimal element after a differential time dt is

$$\text{Net volume compression} = d\varepsilon_z dz \tag{2.8}$$

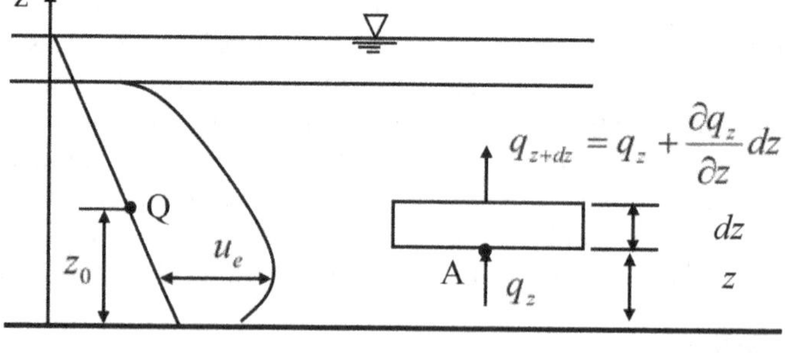

Figure 2.2 An infinitesimal element in a soil layer.

It is noted that for 1-D vertical compression or straining, lateral strains $d\varepsilon_x$ and $d\varepsilon_y$ are zero so that the volume strain increment $d\varepsilon_v$ is equal to vertical strain increment $d\varepsilon_z$. From the assumption that the water and soil particles are incompressible, this net volume change of the infinitesimal element is caused only by the net outflow of water and is equal to the net outflow:

$$\frac{\partial q_z}{\partial z} dzdt = d\varepsilon_z dz \tag{2.9}$$

Thus, the continuity equation becomes:

$$\frac{\partial q_z}{\partial z} = \frac{\partial \varepsilon_z}{\partial t} \tag{2.10}$$

Substituting (2.4) and (2.6) into (2.10) gives (2.11):

$$-\frac{\partial}{\partial z}\left(\frac{k}{\gamma_w}\frac{\partial u_e}{\partial z}\right) = \frac{\partial f(\sigma_z')}{\partial t} + g(\sigma_z', \varepsilon_z) \tag{2.11}$$

Equation (2.11) is the governing differential equation for 1-D consolidation problems while the porewater pressure u_w, excess pore pressure u_e, effective stress σ_z', and total stress σ_z are also related to (2.2) and (2.3).

2.4 CONSOLIDATION OF SOILS WITH CONSTANT COMPRESSIBILITY AND PERMEABILITY

In many situations, the applied load increment on a soil layer is small. The coefficient of permeability k remains essentially constant for the soil layer during the consolidation process. The constitutive relation of soils can be simplified as linear and time independent, that is, (2.4) can be expressed in (2.12):

$$\begin{cases} f = m_v \sigma_z' \\ g = 0 \end{cases} \tag{2.12}$$

where m_v = volume compressibility of soil.

Substituting (2.12) into (2.11) gives (2.13).

$$c_v \frac{\partial^2 u_e}{\partial z^2} = \frac{\partial u_w}{\partial t} - \frac{\partial \sigma_z}{\partial t} = \frac{\partial u_e}{\partial t} - \frac{\partial \sigma_z}{\partial t} \tag{2.13}$$

where

$$c_v = \frac{k}{m_v \gamma_w} \tag{2.14}$$

In (2.14), c_v is defined as the vertical *coefficient of consolidation*, suitable units being m²/year. Since k and m_v are assumed constant, c_v is constant during consolidation.

For a stationary water pressure reference condition, u_e is the excess value over hydrostatic pressure. The component in total stress σ_z that does not change with respect to this reference state will not result in any change in u_e, nor in the strain considering that the stress-strain relation is linear. Therefore, σ_z can be calculated only for the *total stress increment* $\Delta\sigma_z$ with respect to the reference condition. If the total stress increment $\Delta\sigma_z$ does not change with time, then (2.13) becomes the basic Terzaghi's consolidation equation (Terzaghi, 1925, 1943).

In deriving (2.13), the upward direction is selected as the positive direction. Sometimes it is more convenient to select the downward direction as the positive direction. However, (2.13) is the same for both coordinate forms.

2.4.1 Solution for one-layered soil profile with depth-dependent ramp load

Two typical soil profiles are shown in Figure 2.3. The vertical total stress increment $\Delta\sigma_z$ is assumed varying linearly with depth and time and remaining unchanged after time t_c (see Figure 2.4), that is,

$$\Delta\sigma_z(z,t) = \begin{cases} \left(\sigma_0 + \dfrac{\sigma_1 - \sigma_0}{H} z\right)\dfrac{t}{t_c} & t \le t_c \\ \left(\sigma_0 + \dfrac{\sigma_1 - \sigma_0}{H} z\right) & t > t_c \end{cases} \tag{2.15}$$

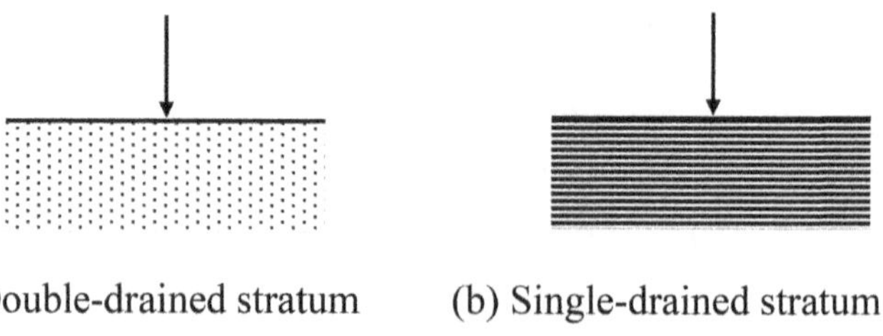

(a) Double-drained stratum (b) Single-drained stratum

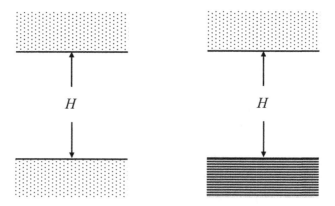

Figure 2.3 Typical soil layers (a) double-drained stratum and (b) single-drain stratum.

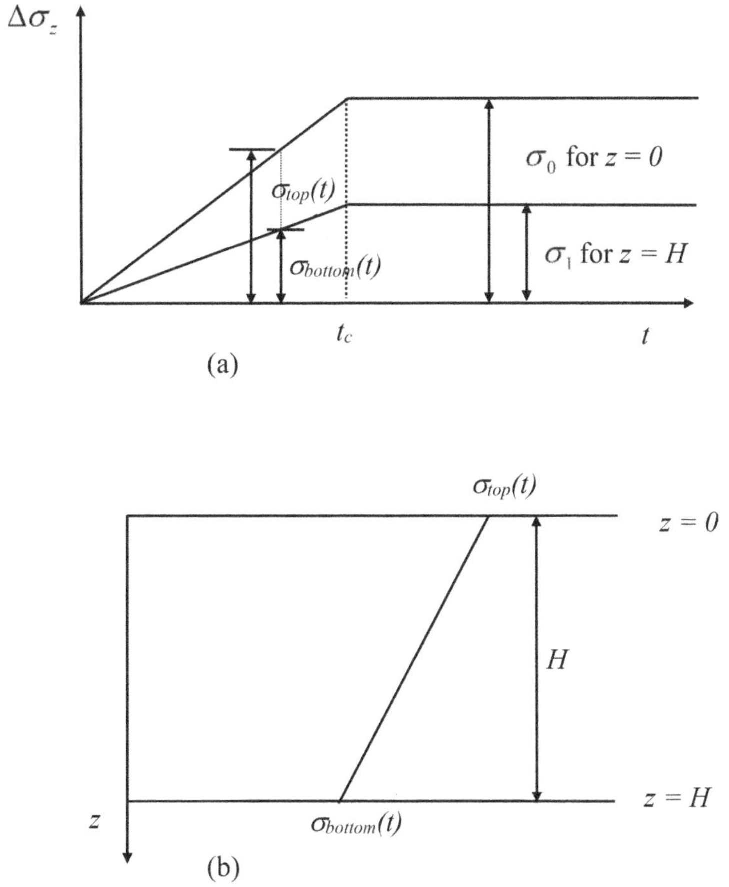

Figure 2.4 Variation of the vertical total stress (a) with time and (b) with depth.

where H is thickness of the soil layer; σ_0 is the vertical total stress increment at $z = 0$ and $t = t_c$; and σ_1 is vertical total stress increment at $z = H$ and $t = t_c$. Equation (2.15) provides a first-order approximation if the true 1-D simplification is not very good.

The boundary conditions for the soil layers in Figure 2.3(a) and (b) are expressed in (2.16) and (2.17), respectively:

$$\begin{cases} u_e(0,t) = 0 \\ u_e(H,t) = 0 \end{cases} \tag{2.16}$$

$$\begin{cases} u_e(0,t) = 0 \\ \dfrac{\partial u_e}{\partial z}\Big|_{z=H} = 0 \end{cases} \tag{2.17}$$

Equation (2.16) means that both sides of the soil layer are freely drained. Equation (2.17) means that the top side of the soil layer is freely drained and the bottom side is impermeable. The initial excess porewater pressure is assumed to be zero.

Using the method of separation of variables, the consolidation equation (2.13) under the given loading condition (2.15) and the boundary conditions (2.16) and (2.17) can be solved (Zhu and Yin, 1998). To express the solutions in a useful dimensionless form, the following dimensionless parameters are defined:

1. *Time factor*, T_v,

$$T_v = \frac{c_v t}{H^2} \tag{2.18}$$

2. *Construction time factor*, T_c,

$$T_c = \frac{c_v t_c}{H^2} \tag{2.19}$$

The solution of (2.13) becomes

$$u_e(z, T_v) = \sum_{n=1}^{+\infty} T_n(T_v) \sin\left(\lambda_n \frac{z}{H}\right) \tag{2.20}$$

where

$$T_n(T_v) = \begin{cases} \dfrac{b_n}{\lambda_n^3 T_c}[1 - \exp(-\lambda_n^2 T_v)] & T_v \le T_c \\ \dfrac{b_n}{\lambda_n^3 T_c}[1 - \exp(-\lambda_n^2 T_c)]\exp[-\lambda_n^2 (T_v - T_c)] & T_v > T_c \end{cases} \tag{2.21}$$

For boundary conditions (2.16), the constants λ_n and b_n in (2.21) are:

$$\begin{cases} \lambda_n = n\pi \\ b_n = 2[\sigma_0 - (-1)^n \sigma_1] \end{cases} \tag{2.22}$$

For boundary conditions (2.17),

$$\begin{cases} \lambda_n = n\pi - \dfrac{\pi}{2} \\ b_n = 2\left[\sigma_0 + \dfrac{2(\sigma_0 - \sigma_1)(-1)^n}{(2n-1)\pi}\right] \end{cases} \tag{2.23}$$

With the constants λ_n and b_n in (2.21) known, (2.20) can be used to calculate the excess porewater pressure distribution varying with depth and time.

In practical problems, the total compression of the stratum at each stage of the consolidation process is of particular interest. The compression can be found by summing the vertical compressions at the various depths, which is conveniently expressed by the *average degree of consolidation U*.

The *average degree of consolidation U* is defined as

$$U = \frac{compression\ at\ time\ t}{final\ compression} \tag{2.24}$$

Using this definition, the average degree of consolidation U for the one-layered soil profile can be expressed in (2.25)

$$U = \frac{S_t}{S_f} \tag{2.25}$$

where

$$S_t = \int_0^H \varepsilon_z(z,t)dz = \int_0^H [\Delta\sigma_z(z,t) - u_e(z,t)]\, m_v dz$$

$$S_f = \int_0^H \varepsilon_z(z,t=\infty)dz = \int_0^H \Delta\sigma_z(z,t=\infty)m_v dz$$

S_t is the compression of the soil layer at time t and S_f is the final compression when the excess porewater pressure is zero and the loading is constant at $t = \infty$. Substituting (2.15) and (2.20) into (2.25), the average degree of consolidation U can be obtained as follows.

For boundary conditions (2.16),

$$U(T_v) = \begin{cases} \dfrac{T_v}{T_c} - \displaystyle\sum_{n=1,3,5,\ldots} \frac{8}{n^4\pi^4 T_c}[1-\exp(-n^2\pi^2 T_v)] & T_v \le T_c \\ 1 - \displaystyle\sum_{n=1,3,5,\ldots} \frac{8}{n^4\pi^4 T_c}[1-\exp(-n^2\pi^2 T_c)]\exp[-n^2\pi^2(T_v - T_c)] & T_v > T_c \end{cases} \tag{2.26}$$

For boundary conditions (2.17),

$$U(T_v) = U_1(T_v) + \frac{\sigma_0 - \sigma_1}{\sigma_0 + \sigma_1} U_2(T_v) \tag{2.27}$$

where

$$U_1(T_v)=\begin{cases}\dfrac{T_v}{T_c}-\displaystyle\sum_{n=1,3,5,\ldots}\frac{32}{n^4\pi^4T_c}\left[1-\exp\left(-\frac{n^2\pi^2T_v}{4}\right)\right] & T_v\le T_c\\[2ex] 1-\displaystyle\sum_{n=1,3,5,\ldots}\frac{32}{n^4\pi^4T_c}\left[1-\exp\left(-\frac{n^2\pi^2T_c}{4}\right)\right]\exp\left[-\frac{n^2\pi^2}{4}(T_v-T_c)\right] & T_v>T_c\end{cases} \tag{2.28}$$

$$U_2(T_v)=\begin{cases}-\displaystyle\sum_{n=1,3,5,\ldots}\frac{32}{n^4\pi^4T_c}\left[1+(-1)^{\frac{n+1}{2}}\frac{4}{n\pi}\right]\left[1-\exp\left(-\frac{n^2\pi^2T_v}{4}\right)\right] & T_v\le T_c\\[2ex] -\displaystyle\sum_{n=1,3,5,\ldots}\frac{32}{n^4\pi^4T_c}\left[1+(-1)^{\frac{n+1}{2}}\frac{4}{n\pi}\right]\left[1-\exp\left(-\frac{n^2\pi^2T_c}{4}\right)\right] & \\ \qquad\exp\left[-\dfrac{n^2\pi^2}{4}(T_v-T_c)\right] & T_v>T_c\end{cases} \tag{2.29}$$

If H is reassigned as the length of the longest drainage path in the stratum, that is, if double drainage is assumed, a half of the thickness is used, (2.26) is the same as (2.28) where one side of the clay layer is assumed impermeable. (2.26) for the average degree of consolidation is the same as Olson's (1977) solution.

Equation (2.27) can be used to calculate the average degree of consolidation for any given values of σ_0 and σ_1 if U_1 and U_2 are known. Using (2.28) and (2.29), charts that show the relationships of T_v – U_1 and T_v – U_2 for an appropriate range in the construction time factor, T_c, are prepared in both arithmetic and semi-logarithmic scales, respectively (Figures 2.5 and 2.6, respectively). The relationship data are also presented in tabular form (Tables 2.1 and 2.2). These figures and tables are prepared for practical use of the calculation of the average degree of consolidation. If the vertical total stress σ is composed of several depth-dependent ramp loads, the superposition method may be used to calculate the excess porewater pressure and the average degree of consolidation.

2.4.1.1 Convergence of the series

In this section, it will be shown that the series for calculating the average degree of consolidation converges very rapidly. We first consider the following series for the truncation error

$$S_k=\sum_{n=k}^{+\infty}\frac{1}{(2n+1)^4}$$

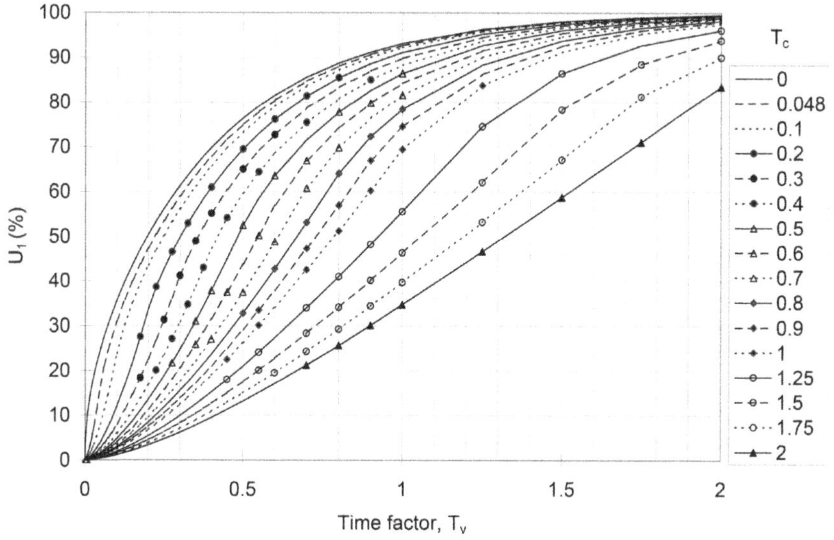

(a)

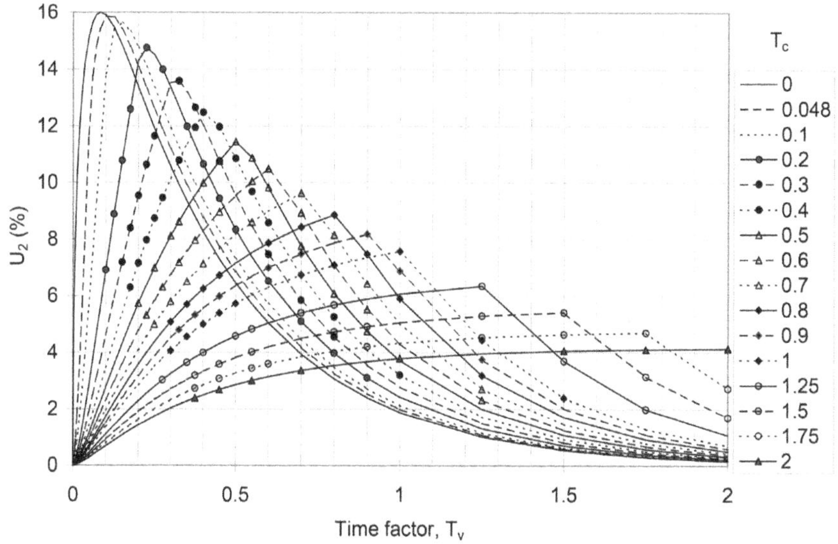

(b)

Figure 2.5 (a) $U_1 - T_v$ curves and (b) $U_2 - T_v$ curves for depth-dependent ramp loading in arithmetic scale.

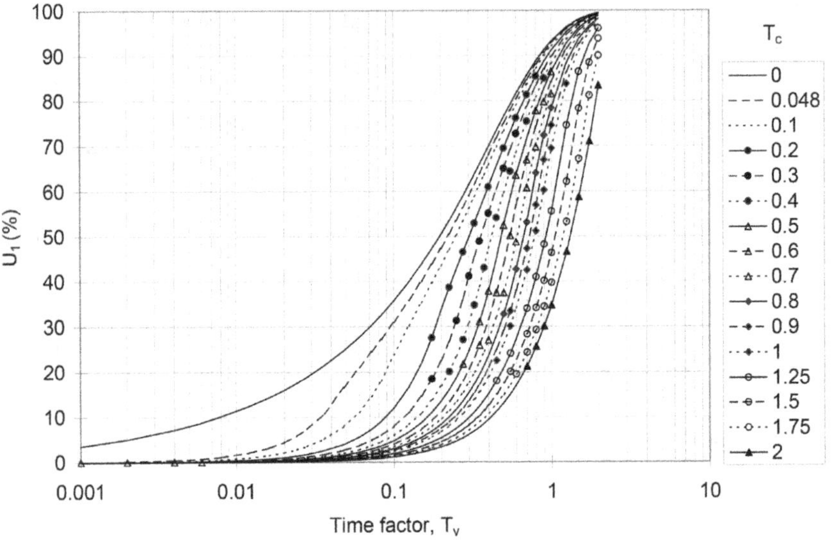

(a)

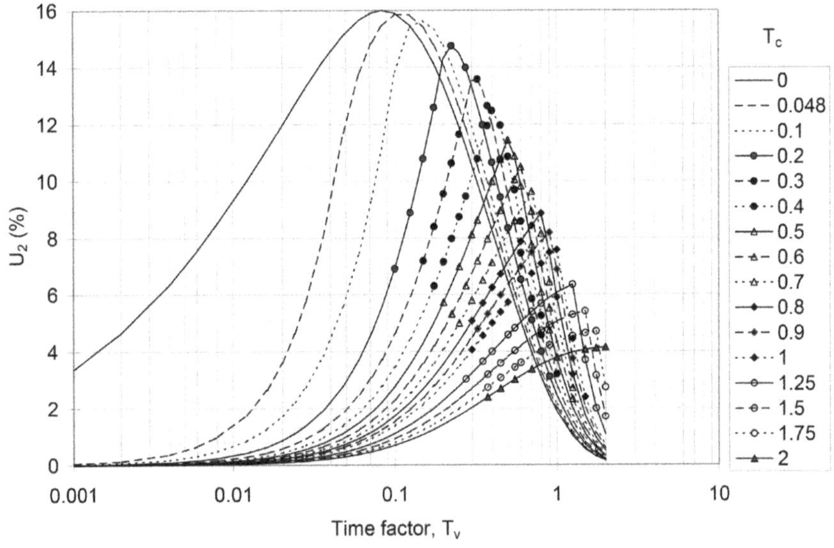

(b)

Figure 2.6 (a) $U_l - T_v$ curves and (b) $U_{2-} - T_v$ curves for depth-dependent ramp loading in semi-logarithmic scale.

Table 2.1 U_l values for various T_v (left column) and T_c (top row)

T_v	T_c=0.0	0.008	0.02	0.048	0.06	0.1	0.15	0.2	0.3	0.4	0.5	0.6	0.7	0.8	0.9	1	1.25	1.5	1.75	2
0.002	5.05	0.84	0.34	0.14	0.11	0.07	0.04	0.03	0.02	0.02	0.01	0.01	0.01	0.01	0.01	0.01	0.01	0	0	0
0.004	7.14	2.38	0.95	0.4	0.32	0.19	0.13	0.1	0.06	0.05	0.04	0.03	0.03	0.02	0.02	0.02	0.02	0.01	0.01	0.01
0.006	8.74	4.37	1.75	0.73	0.58	0.35	0.23	0.17	0.12	0.09	0.07	0.06	0.05	0.04	0.04	0.03	0.03	0.02	0.02	0.02
0.008	10.09	6.73	2.69	1.12	0.9	0.54	0.36	0.27	0.18	0.13	0.11	0.09	0.08	0.07	0.06	0.05	0.04	0.04	0.03	0.03
0.012	12.36	9.98	4.94	2.06	1.65	0.99	0.66	0.49	0.33	0.25	0.2	0.16	0.14	0.12	0.11	0.1	0.08	0.07	0.06	0.05
0.016	14.27	12.3	7.61	3.17	2.54	1.52	1.02	0.76	0.51	0.38	0.3	0.25	0.22	0.19	0.17	0.15	0.12	0.1	0.09	0.08
0.02	15.96	14.24	10.64	4.43	3.55	2.13	1.42	1.06	0.71	0.53	0.43	0.35	0.3	0.27	0.24	0.21	0.17	0.14	0.12	0.11
0.028	18.88	17.46	14.93	7.34	5.87	3.52	2.35	1.76	1.17	0.88	0.7	0.59	0.5	0.44	0.39	0.35	0.28	0.23	0.2	0.18
0.036	21.41	20.17	18.08	10.7	8.56	5.14	3.43	2.57	1.71	1.28	1.03	0.86	0.73	0.64	0.57	0.51	0.41	0.34	0.29	0.26
0.048	24.72	23.66	21.93	16.48	13.18	7.91	5.27	3.96	2.64	1.98	1.58	1.32	1.13	0.99	0.88	0.79	0.63	0.53	0.45	0.4
0.06	27.64	26.7	25.19	20.97	18.43	11.06	7.37	5.53	3.69	2.76	2.21	1.84	1.58	1.38	1.23	1.11	0.88	0.74	0.63	0.55
0.072	30.28	29.42	28.07	24.45	22.57	14.53	9.69	7.27	4.84	3.63	2.91	2.42	2.08	1.82	1.61	1.45	1.16	0.97	0.83	0.73
0.083	32.51	31.71	30.46	27.21	25.61	17.99	11.99	8.99	6	4.5	3.6	3	2.57	2.25	2	1.8	1.44	1.2	1.03	0.9
0.1	35.68	34.96	33.83	30.98	29.62	23.79	15.86	11.89	7.93	5.95	4.76	3.96	3.4	2.97	2.64	2.38	1.9	1.59	1.36	1.19
0.125	39.89	39.25	38.25	35.77	34.63	30.27	22.16	16.62	11.08	8.31	6.65	5.54	4.75	4.16	3.69	3.32	2.66	2.22	1.9	1.66
0.15	43.7	43.11	42.21	39.99	38.98	35.29	29.13	21.85	14.57	10.93	8.74	7.28	6.24	5.46	4.86	4.37	3.5	2.91	2.5	2.19
0.175	47.18	46.64	45.81	43.79	42.88	39.62	34.73	27.53	18.36	13.77	11.01	9.18	7.87	6.88	6.12	5.51	4.41	3.67	3.15	2.75
0.2	50.41	49.91	49.14	47.28	46.44	43.48	39.24	33.64	22.42	16.82	13.45	11.21	9.61	8.41	7.47	6.73	5.38	4.48	3.84	3.36
0.225	53.41	52.95	52.23	50.5	49.73	47.01	43.2	38.64	26.75	20.06	16.05	13.38	11.46	10.03	8.92	8.03	6.42	5.35	4.59	4.01
0.25	56.22	55.79	55.12	53.5	52.78	50.26	46.78	42.78	31.32	23.49	18.79	15.66	13.42	11.75	10.44	9.4	7.52	6.26	5.37	4.7
0.275	58.85	58.44	57.82	56.3	55.63	53.28	50.07	46.45	36.12	27.09	21.67	18.06	15.48	13.54	12.04	10.83	8.67	7.22	6.19	5.42
0.3	61.32	60.94	60.35	58.93	58.3	56.1	53.12	49.79	41.12	30.84	24.67	20.56	17.62	15.42	13.71	12.34	9.87	8.22	7.05	6.17
0.325	63.64	63.28	62.73	61.4	60.81	58.75	55.95	52.88	45.34	34.75	27.8	23.17	19.86	17.37	15.44	13.9	11.12	9.27	7.94	6.95
0.35	65.82	65.48	64.96	63.71	63.16	61.22	58.61	55.74	48.92	38.8	31.04	25.86	22.17	19.4	17.24	15.52	12.41	10.35	8.87	7.76
0.4	69.79	69.49	69.03	67.93	67.44	65.73	63.43	60.92	55.11	47.28	37.82	31.52	27.02	23.64	21.01	18.91	15.13	12.61	10.81	9.46
0.45	73.3	73.03	72.63	71.65	71.22	69.71	67.68	65.47	60.4	54.12	44.98	37.48	32.13	28.11	24.99	22.49	17.99	14.99	12.85	11.24
0.5	76.4	76.16	75.8	74.94	74.56	73.23	71.43	69.48	65.02	59.64	52.47	43.72	37.48	32.79	29.15	26.23	20.99	17.49	14.99	13.12
0.6	81.56	81.37	81.09	80.42	80.12	79.08	77.68	76.15	72.68	68.54	63.53	56.9	48.77	42.68	37.94	34.14	27.31	22.76	19.51	17.07
0.7	85.59	85.45	85.23	84.7	84.47	83.66	82.56	81.37	78.66	75.42	71.56	66.88	60.72	53.13	47.23	42.51	34.01	28.34	24.29	21.25
0.8	88.74	88.63	88.46	88.05	87.86	87.23	86.37	85.44	83.32	80.8	77.79	74.17	69.79	64.04	56.92	51.23	40.98	34.15	29.27	25.62
0.9	91.2	91.11	90.98	90.66	90.52	90.02	89.35	88.63	86.97	85	82.64	79.82	76.44	72.32	66.92	60.23	48.19	40.15	34.42	30.12
1	93.13	93.06	92.95	92.7	92.59	92.2	91.68	91.11	89.82	88.28	86.44	84.24	81.59	78.41	74.53	69.45	55.56	46.3	39.69	34.73
1.25	96.29	96.25	96.2	96.06	96	95.79	95.51	95.2	94.51	93.67	92.68	91.49	90.07	88.35	86.28	83.77	74.54	62.11	53.24	46.59
1.5	98	97.98	97.95	97.87	97.84	97.73	97.58	97.41	97.03	96.59	96.05	95.41	94.64	93.71	92.6	91.24	86.47	78.32	67.13	58.74
1.75	98.92	98.91	98.89	98.85	98.84	98.77	98.69	98.6	98.4	98.16	97.87	97.52	97.11	96.61	96	95.28	92.7	88.47	81.2	71.05
2	99.42	99.41	99.4	99.38	99.37	99.34	99.29	99.25	99.14	99.01	98.85	98.66	98.44	98.17	97.84	97.45	96.06	93.78	90	83.45

Table 2.2 U_2 values for various T_v (left column) and T_c (top row)

T_v	T_c=0.0	0.008	0.02	0.048	0.06	0.1	0.15	0.2	0.3	0.4	0.5	0.6	0.7	0.8	0.9	1	1.25	1.5	1.75	2
0.002	4.65	0.79	0.32	0.13	0.11	0.06	0.04	0.03	0.02	0.02	0.01	0.01	0.01	0.01	0.01	0.01	0.01	0	0	0
0.004	6.34	2.18	0.87	0.36	0.29	0.17	0.12	0.09	0.06	0.04	0.03	0.03	0.02	0.02	0.02	0.02	0.01	0.01	0.01	0.01
0.006	7.54	3.92	1.57	0.65	0.52	0.31	0.21	0.16	0.1	0.08	0.06	0.05	0.04	0.04	0.03	0.03	0.03	0.02	0.02	0.02
0.008	8.49	5.93	2.37	0.99	0.79	0.47	0.32	0.24	0.16	0.12	0.09	0.08	0.07	0.06	0.05	0.05	0.04	0.03	0.03	0.02
0.012	9.96	8.38	4.22	1.76	1.41	0.84	0.56	0.42	0.28	0.21	0.17	0.14	0.12	0.11	0.09	0.08	0.07	0.06	0.05	0.04
0.016	11.07	9.9	6.33	2.64	2.11	1.27	0.84	0.63	0.42	0.32	0.25	0.21	0.18	0.16	0.14	0.13	0.1	0.08	0.07	0.06
0.02	11.96	11.04	8.64	3.6	2.88	1.73	1.15	0.86	0.58	0.43	0.35	0.29	0.25	0.22	0.19	0.17	0.14	0.12	0.1	0.09
0.028	13.28	12.66	11.33	5.71	4.57	2.74	1.83	1.37	0.91	0.69	0.55	0.46	0.39	0.34	0.3	0.27	0.22	0.18	0.16	0.14
0.036	14.21	13.77	12.88	8	6.4	3.84	2.56	1.92	1.28	0.96	0.77	0.64	0.55	0.48	0.43	0.38	0.31	0.26	0.22	0.19
0.048	15.12	14.86	14.33	11.68	9.35	5.61	3.74	2.8	1.87	1.4	1.12	0.93	0.8	0.7	0.62	0.56	0.45	0.37	0.32	0.28
0.06	15.65	15.51	15.19	13.78	12.43	7.46	4.97	3.73	2.49	1.86	1.49	1.24	1.07	0.93	0.83	0.75	0.6	0.5	0.43	0.37
0.072	15.92	15.85	15.69	14.86	14.18	9.35	6.24	4.68	3.12	2.34	1.87	1.56	1.34	1.17	1.04	0.94	0.75	0.62	0.53	0.47
0.083	16	15.98	15.91	15.44	15.03	11.11	7.41	5.56	3.7	2.78	2.22	1.85	1.59	1.39	1.23	1.11	0.89	0.74	0.63	0.56
0.1	15.91	15.95	15.98	15.85	15.68	13.83	9.22	6.91	4.61	3.46	2.77	2.3	1.98	1.73	1.54	1.38	1.11	0.92	0.79	0.69
0.125	15.47	15.56	15.67	15.84	15.85	15.41	11.84	8.88	5.92	4.44	3.55	2.96	2.54	2.22	1.97	1.78	1.42	1.18	1.01	0.89
0.15	14.84	14.95	15.1	15.42	15.52	15.63	14.36	10.77	7.18	5.39	4.31	3.59	3.08	2.69	2.39	2.15	1.72	1.44	1.23	1.08
0.175	14.12	14.24	14.41	14.8	14.95	15.33	15.21	12.58	8.39	6.29	5.03	4.19	3.6	3.15	2.8	2.52	2.01	1.68	1.44	1.26
0.2	13.37	13.49	13.67	14.08	14.26	14.78	15.13	14.3	9.53	7.15	5.72	4.77	4.09	3.58	3.18	2.86	2.29	1.91	1.63	1.43
0.225	12.63	12.74	12.92	13.34	13.52	14.1	14.68	14.75	10.62	7.96	6.37	5.31	4.55	3.98	3.54	3.19	2.55	2.12	1.82	1.59
0.25	11.9	12.02	12.19	12.6	12.78	13.37	14.06	14.5	11.64	8.73	6.98	5.82	4.99	4.36	3.88	3.49	2.79	2.33	2	1.75
0.275	11.21	11.32	11.48	11.88	12.05	12.64	13.37	13.99	12.6	9.45	7.56	6.3	5.4	4.73	4.2	3.78	3.02	2.52	2.16	1.89
0.3	10.55	10.65	10.81	11.19	11.35	11.92	12.65	13.35	13.51	10.13	8.1	6.75	5.79	5.07	4.5	4.05	3.24	2.7	2.32	2.03
0.325	9.92	10.02	10.17	10.53	10.69	11.23	11.94	12.66	13.58	10.77	8.62	7.18	6.15	5.39	4.79	4.31	3.45	2.87	2.46	2.15
0.35	9.33	9.43	9.57	9.9	10.05	10.57	11.26	11.97	13.19	11.37	9.1	7.58	6.5	5.69	5.05	4.55	3.64	3.03	2.6	2.27
0.4	8.25	8.33	8.46	8.76	8.89	9.36	9.98	10.64	12.02	12.47	9.98	8.31	7.13	6.23	5.54	4.99	3.99	3.33	2.85	2.49
0.45	7.3	7.37	7.48	7.75	7.86	8.27	8.83	9.42	10.74	11.96	10.75	8.96	7.68	6.72	5.97	5.38	4.3	3.58	3.07	2.69
0.5	6.45	6.51	6.61	6.85	6.95	7.31	7.8	8.34	9.53	10.84	11.44	9.53	8.17	7.15	6.35	5.72	4.58	3.81	3.27	2.86
0.6	5.04	5.09	5.17	5.35	5.43	5.72	6.1	6.52	7.46	8.58	9.82	10.48	8.99	7.86	6.99	6.29	5.03	4.19	3.59	3.15
0.7	3.94	3.98	4.04	4.18	4.24	4.47	4.77	5.09	5.83	6.71	7.75	8.92	9.62	8.42	7.49	6.74	5.39	4.49	3.85	3.37
0.8	3.08	3.11	3.15	3.27	3.32	3.49	3.72	3.98	4.56	5.25	6.07	7.04	8.15	8.86	7.87	7.09	5.67	4.72	4.05	3.54
0.9	2.4	2.43	2.46	2.55	2.59	2.73	2.91	3.11	3.56	4.1	4.74	5.51	6.43	7.47	8.18	7.36	5.89	4.91	4.21	3.68
1	1.88	1.9	1.93	1.99	2.02	2.13	2.27	2.43	2.78	3.2	3.71	4.31	5.03	5.89	6.88	7.57	6.06	5.05	4.33	3.79
1.25	1.01	1.02	1.04	1.08	1.09	1.15	1.23	1.31	1.5	1.73	2	2.32	2.71	3.18	3.75	4.43	6.34	5.28	4.53	3.96
1.5	0.55	0.55	0.56	0.58	0.59	0.62	0.66	0.71	0.81	0.93	1.08	1.25	1.46	1.72	2.02	2.39	3.7	5.41	4.64	4.06
1.75	0.3	0.3	0.3	0.31	0.32	0.33	0.36	0.38	0.44	0.5	0.58	0.68	0.79	0.93	1.09	1.29	2	3.15	4.69	4.11
2	0.16	0.16	0.16	0.17	0.17	0.18	0.19	0.21	0.24	0.27	0.31	0.37	0.43	0.5	0.59	0.7	1.08	1.7	2.73	4.13

Obviously,

$$\int_{k}^{+\infty} \frac{dx}{(2x+1)^4} < S_k < \int_{k-1}^{+\infty} \frac{dx}{(2x+1)^4}$$

For $k = 2$ and 3, after integration, we have the truncation error S_2 and S_3 are

$$0.00133 < S_2 < 0.006173$$
$$0.000486 < S_3 < 0.00133$$

S_2 and S_3 are much smaller than 1. Taking into consideration of the exponent decaying parts in (2.28) and (2.29), we can say that for application purpose, at most 3 terms of the series will produce sufficient accuracy for the calculation of average degree of consolidation.

2.4.1.2 *Special cases of the above solutions*

The solution in (2.20), (2.26), and (2.27) above is a very general one. Many 1-D consolidation problems commonly used in soil mechanics (Das, 2018; Knappett, 2019) are only special cases of this solution. For easy utilization of the solution, several special cases are summarized as follows:

1. Suddenly applied uniform loading for a double-drained stratum
   The suddenly applied vertical total stress increment $\Delta\sigma_z$ is shown in Figure 2.7(a). The average degree of consolidation is as follows:

   $$U(T_v) = U_1(T_v)\ T_c = 0$$

   The normalized excess porewater pressure u_e/σ_0 is presented in Figure 2.8 and in Table 2.3. Since the excess porewater pressure is symmetric, the result is listed in Table 2.3 for one-half the thickness of the stratum.
2. Suddenly applied uniform loading for a single-drained stratum
   The suddenly applied vertical total stress increment $\Delta\sigma_z$ is shown in Figure 2.7(d). The average degree of consolidation is as follows:

   $$U(T_v) = U_1(T_v)\ T_c = 0$$

   The normalized excess porewater pressure u_e/σ_0 is the same as the result for a double-drained stratum for one-half the thickness.
3. Suddenly applied loading proportional to thickness in Figure 2.7(b)
   The average degree of consolidation is as follows:

   $$U(T_v) = U_1(T_v)\ T_c = 0$$

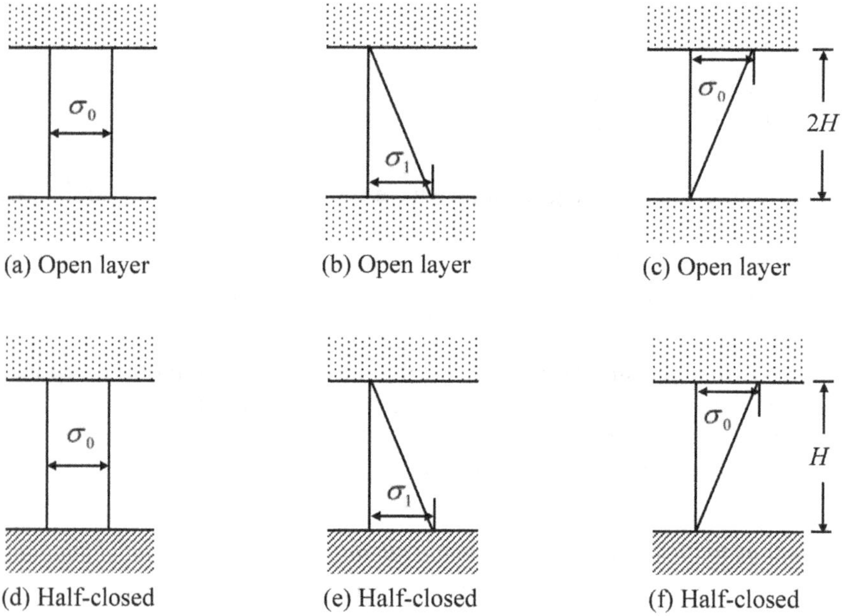

Figure 2.7 Suddenly applied vertical total stress increments $\Delta\sigma_z$ (a) uniform, (b) triangular, (c) trianglualr, (d) uinform, (e) triangualr, and (f) triangular.

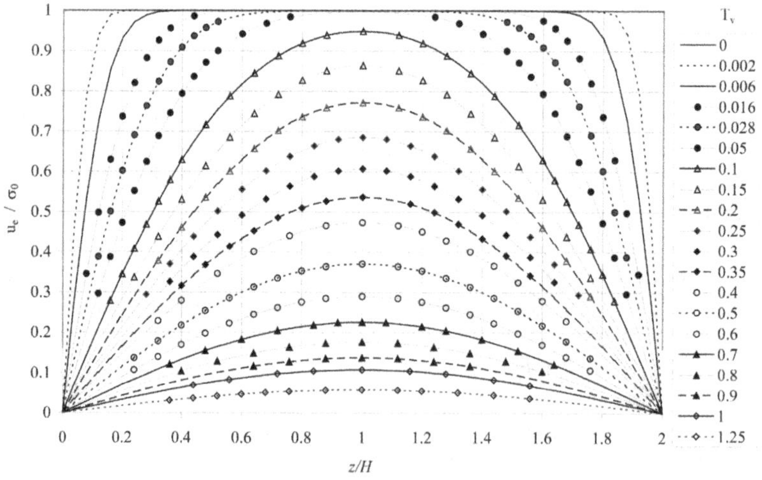

Figure 2.8 Isochrones of excess porewater pressure for suddenly applied uniform vertical total stress increments $\Delta\sigma_z = \sigma_0$ (see Figure 2.7(a), (d)).

Table 2.3 u_e/σ_0 for suddenly applied uniform vertical total stress increments (see Figure 2.7(a), (d))

T_v	$z/H=0$	0.04	0.08	0.12	0.16	0.2	0.24	0.28	0.32	0.36	0.4	0.44	0.48	0.52	0.56	0.6	0.64	0.68	0.72	0.76	0.8	0.84	0.88	0.92	0.96	1
0	1	1	1	1	1	1	1	1	1	1	1	1	1	1	1	1	1	1	1	1	1	1	1	1	1	1
0.002	0	0.473	0.794	0.942	0.989	0.998	1	1	1	1	1	1	1	1	1	1	1	1	1	1	1	1	1	1	1	1
0.004	0	0.345	0.629	0.82	0.926	0.975	0.993	0.998	1	1	1	1	1	1	1	1	1	1	1	1	1	1	1	1	1	1
0.006	0	0.285	0.535	0.727	0.856	0.932	0.972	0.989	0.997	0.999	1	1	1	1	1	1	1	1	1	1	1	1	1	1	1	1
0.008	0	0.248	0.473	0.657	0.794	0.886	0.942	0.973	0.989	0.996	0.998	0.999	1	1	1	1	1	1	1	1	1	1	1	1	1	1
0.012	0	0.204	0.394	0.561	0.698	0.803	0.879	0.929	0.961	0.98	0.99	0.995	0.998	0.999	1	1	1	1	1	1	1	1	1	1	1	1
0.016	0	0.177	0.345	0.498	0.629	0.736	0.82	0.882	0.926	0.956	0.975	0.986	0.993	0.996	0.998	0.999	1	1	1	1	1	1	1	1	1	1
0.02	0	0.159	0.311	0.451	0.576	0.683	0.77	0.838	0.89	0.928	0.954	0.972	0.984	0.991	0.995	0.997	0.999	0.999	1	1	1	1	1	1	1	1
0.028	0	0.134	0.265	0.388	0.501	0.602	0.69	0.763	0.824	0.872	0.909	0.937	0.957	0.972	0.982	0.989	0.993	0.996	0.998	0.999	0.999	1	1	1	1	1
0.036	0	0.119	0.234	0.345	0.449	0.544	0.629	0.703	0.767	0.82	0.864	0.899	0.926	0.947	0.963	0.975	0.983	0.989	0.993	0.995	0.997	0.998	0.999	0.999	1	1
0.048	0	0.103	0.204	0.301	0.394	0.481	0.561	0.634	0.698	0.755	0.803	0.844	0.879	0.907	0.929	0.947	0.961	0.972	0.98	0.986	0.99	0.993	0.995	0.997	0.997	0.998
0.06	0	0.092	0.183	0.271	0.356	0.436	0.512	0.581	0.644	0.701	0.752	0.796	0.834	0.867	0.894	0.917	0.935	0.95	0.962	0.971	0.979	0.984	0.988	0.99	0.992	0.992
0.072	0	0.084	0.167	0.248	0.327	0.402	0.473	0.539	0.601	0.657	0.708	0.754	0.794	0.829	0.86	0.886	0.908	0.926	0.941	0.954	0.963	0.971	0.976	0.98	0.982	0.983
0.083	0	0.078	0.156	0.232	0.305	0.376	0.444	0.508	0.568	0.623	0.674	0.72	0.761	0.798	0.83	0.859	0.883	0.904	0.921	0.936	0.947	0.956	0.963	0.968	0.971	0.972
0.1	0	0.071	0.142	0.212	0.279	0.345	0.408	0.469	0.526	0.579	0.629	0.674	0.716	0.754	0.788	0.819	0.845	0.868	0.888	0.905	0.919	0.93	0.939	0.945	0.948	0.949
0.125	0	0.064	0.127	0.19	0.251	0.311	0.368	0.424	0.477	0.527	0.575	0.619	0.661	0.699	0.733	0.765	0.793	0.818	0.84	0.858	0.874	0.887	0.897	0.903	0.908	0.909
0.15	0	0.058	0.116	0.173	0.229	0.284	0.337	0.389	0.439	0.486	0.531	0.574	0.614	0.651	0.685	0.716	0.744	0.77	0.792	0.811	0.827	0.841	0.851	0.858	0.863	0.864
0.175	0	0.054	0.107	0.16	0.212	0.263	0.312	0.36	0.407	0.452	0.494	0.535	0.573	0.608	0.641	0.672	0.699	0.724	0.746	0.765	0.781	0.794	0.805	0.812	0.817	0.818
0.2	0	0.05	0.099	0.148	0.197	0.244	0.291	0.336	0.379	0.421	0.462	0.5	0.536	0.57	0.601	0.63	0.657	0.681	0.702	0.721	0.736	0.749	0.759	0.767	0.771	0.772
0.225	0	0.046	0.093	0.138	0.184	0.228	0.272	0.314	0.355	0.394	0.432	0.468	0.502	0.535	0.564	0.592	0.617	0.64	0.661	0.678	0.693	0.706	0.715	0.722	0.727	0.728
0.25	0	0.043	0.087	0.13	0.172	0.214	0.254	0.294	0.333	0.37	0.405	0.439	0.472	0.502	0.53	0.556	0.58	0.602	0.621	0.638	0.652	0.664	0.674	0.68	0.684	0.685
0.275	0	0.041	0.081	0.122	0.161	0.2	0.239	0.276	0.312	0.347	0.381	0.413	0.443	0.472	0.498	0.523	0.546	0.566	0.584	0.6	0.614	0.625	0.634	0.64	0.644	0.645
0.3	0	0.038	0.076	0.114	0.151	0.188	0.224	0.259	0.293	0.326	0.358	0.388	0.416	0.443	0.468	0.492	0.513	0.532	0.549	0.564	0.577	0.588	0.596	0.602	0.606	0.607
0.325	0	0.036	0.072	0.107	0.142	0.177	0.21	0.243	0.275	0.306	0.336	0.364	0.391	0.416	0.44	0.462	0.482	0.5	0.517	0.531	0.543	0.553	0.561	0.566	0.57	0.571
0.35	0	0.034	0.067	0.101	0.134	0.166	0.198	0.229	0.259	0.288	0.316	0.342	0.368	0.391	0.414	0.434	0.453	0.47	0.486	0.499	0.51	0.52	0.527	0.532	0.536	0.537
0.4	0	0.03	0.059	0.089	0.118	0.147	0.175	0.202	0.229	0.254	0.279	0.303	0.325	0.346	0.366	0.384	0.401	0.416	0.429	0.441	0.451	0.46	0.466	0.471	0.474	0.474
0.45	0	0.026	0.053	0.079	0.104	0.13	0.154	0.179	0.202	0.225	0.247	0.267	0.287	0.306	0.323	0.339	0.354	0.368	0.38	0.39	0.399	0.406	0.412	0.416	0.419	0.419
0.5	0	0.023	0.046	0.069	0.092	0.115	0.137	0.158	0.179	0.199	0.218	0.236	0.254	0.27	0.286	0.3	0.313	0.325	0.335	0.345	0.353	0.359	0.364	0.368	0.37	0.371
0.6	0	0.018	0.036	0.054	0.072	0.09	0.107	0.123	0.14	0.155	0.17	0.185	0.198	0.211	0.223	0.234	0.245	0.254	0.262	0.269	0.276	0.281	0.285	0.287	0.289	0.29
0.7	0	0.014	0.028	0.042	0.056	0.07	0.083	0.096	0.109	0.121	0.133	0.144	0.155	0.165	0.174	0.183	0.191	0.198	0.205	0.21	0.215	0.219	0.222	0.225	0.226	0.226
0.8	0	0.011	0.022	0.033	0.044	0.055	0.065	0.075	0.085	0.095	0.104	0.113	0.121	0.129	0.136	0.143	0.149	0.155	0.16	0.164	0.168	0.171	0.174	0.175	0.177	0.177
0.9	0	0.009	0.017	0.026	0.034	0.043	0.051	0.059	0.067	0.074	0.081	0.088	0.095	0.101	0.106	0.112	0.117	0.121	0.125	0.128	0.131	0.134	0.136	0.137	0.138	0.138
1	0	0.007	0.014	0.02	0.027	0.033	0.04	0.046	0.052	0.058	0.063	0.069	0.074	0.079	0.083	0.087	0.091	0.095	0.098	0.1	0.103	0.105	0.106	0.107	0.108	0.108
1.25	0	0.004	0.007	0.011	0.014	0.018	0.021	0.025	0.028	0.031	0.034	0.037	0.04	0.042	0.045	0.047	0.049	0.051	0.053	0.054	0.055	0.056	0.057	0.058	0.058	0.058
1.5	0	0.002	0.004	0.006	0.008	0.01	0.012	0.013	0.015	0.017	0.018	0.02	0.022	0.023	0.024	0.025	0.027	0.028	0.028	0.029	0.03	0.03	0.031	0.031	0.031	0.031
1.75	0	0.001	0.002	0.003	0.004	0.005	0.006	0.007	0.008	0.009	0.01	0.011	0.012	0.012	0.013	0.014	0.014	0.015	0.015	0.016	0.016	0.016	0.017	0.017	0.017	0.017
2	0	0.001	0.001	0.002	0.002	0.003	0.003	0.004	0.004	0.005	0.005	0.006	0.006	0.007	0.007	0.007	0.008	0.008	0.008	0.009	0.009	0.009	0.009	0.009	0.009	0.009

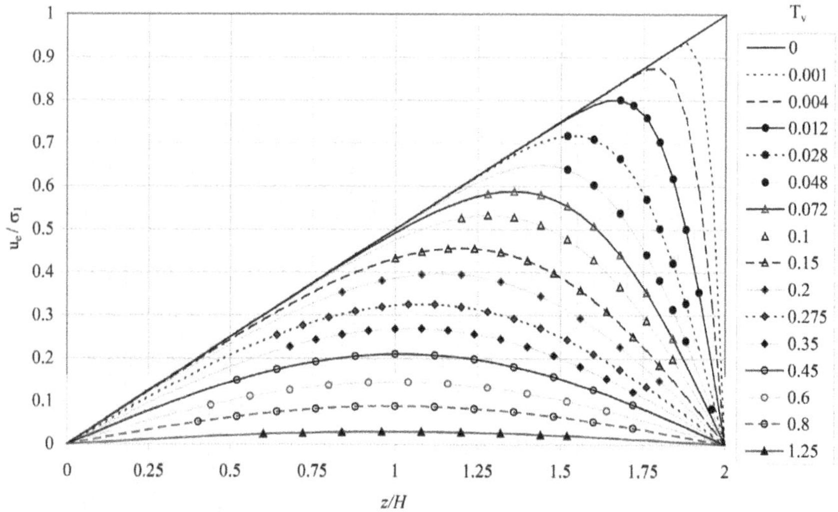

Figure 2.9 Isochrones of excess porewater pressure for suddenly applied vertical total stress increments $\Delta\sigma_z = \frac{z}{2H}\sigma_1$ (see Figure 2.7(b), (c)).

The normalized excess porewater pressure u_e/σ_1 is presented in Figure 2.9 and in Table 2.4.

4. Suddenly applied loading proportional to thickness in Figure 2.7(c)
   The average degree of consolidation is as follows:

$$U(T_v) = U_1(T_v) \; T_c = 0$$

The normalized excess porewater pressure u_e/σ_0 is the same as the result for the case in Figure 2.7(b) if the coordinate is reversed.

5. Suddenly applied loading proportional to thickness in Figure 2.7(e)
   The average degree of consolidation is as follows:

$$U(T_v) = U_1(T_v) - U_2(T_v) \; T_c = 0$$

The normalized excess porewater pressure u_e/σ_1 is presented in Figure 2.10 and in Table 2.5.

6. Suddenly applied loading proportional to thickness in Figure 2.7(f)
   The average degree of consolidation is as follows:

$$U(T_v) = U_1(T_v) + U_2(T_v) \; T_c = 0$$

The normalized excess porewater pressure u_e/σ_0 is presented in Figure 2.11 and in Table 2.6.

Table 2.4 u_e/σ_1 for suddenly applied loading proportional to thickness with double drainage (see Figure 2.7(b), (c))

T_v	z/H=0	0.04	0.08	0.12	0.16	0.2	0.24	0.28	0.32	0.36	0.4	0.44	0.48	0.52	0.56	0.6	0.64	0.68	0.72	0.76	0.8	0.84	0.88	0.92	0.96	1
0	0	0.02	0.04	0.06	0.08	0.1	0.12	0.14	0.16	0.18	0.2	0.22	0.24	0.26	0.28	0.3	0.32	0.34	0.36	0.38	0.4	0.42	0.44	0.46	0.48	0.5
0.001	0	0.02	0.04	0.06	0.08	0.1	0.12	0.14	0.16	0.18	0.2	0.22	0.24	0.26	0.28	0.3	0.32	0.34	0.36	0.38	0.4	0.42	0.44	0.46	0.48	0.5
0.002	0	0.02	0.04	0.06	0.08	0.1	0.12	0.14	0.16	0.18	0.2	0.22	0.24	0.26	0.28	0.3	0.32	0.34	0.36	0.38	0.4	0.42	0.44	0.46	0.48	0.5
0.004	0	0.02	0.04	0.06	0.08	0.1	0.12	0.14	0.16	0.18	0.2	0.22	0.24	0.26	0.28	0.3	0.32	0.34	0.36	0.38	0.4	0.42	0.44	0.46	0.48	0.5
0.006	0	0.02	0.04	0.06	0.08	0.1	0.12	0.14	0.16	0.18	0.2	0.22	0.24	0.26	0.28	0.3	0.32	0.34	0.36	0.38	0.4	0.42	0.44	0.46	0.48	0.5
0.008	0	0.02	0.04	0.06	0.08	0.1	0.12	0.14	0.16	0.18	0.2	0.22	0.24	0.26	0.28	0.3	0.32	0.34	0.36	0.38	0.4	0.42	0.44	0.46	0.48	0.5
0.012	0	0.02	0.04	0.06	0.08	0.1	0.12	0.14	0.16	0.18	0.2	0.22	0.24	0.26	0.28	0.3	0.32	0.34	0.36	0.38	0.4	0.42	0.44	0.46	0.48	0.5
0.016	0	0.02	0.04	0.06	0.08	0.1	0.12	0.14	0.16	0.18	0.2	0.22	0.24	0.26	0.28	0.3	0.32	0.34	0.36	0.38	0.4	0.42	0.44	0.46	0.48	0.5
0.02	0	0.02	0.04	0.06	0.08	0.1	0.12	0.14	0.16	0.18	0.2	0.22	0.24	0.26	0.28	0.3	0.32	0.34	0.36	0.38	0.4	0.42	0.44	0.46	0.48	0.5
0.028	0	0.02	0.04	0.06	0.08	0.1	0.12	0.14	0.16	0.18	0.2	0.22	0.24	0.26	0.28	0.3	0.32	0.34	0.36	0.38	0.4	0.42	0.44	0.46	0.48	0.5
0.036	0	0.02	0.04	0.06	0.08	0.1	0.12	0.14	0.16	0.18	0.2	0.22	0.24	0.26	0.28	0.3	0.32	0.34	0.36	0.38	0.4	0.42	0.44	0.46	0.48	0.5
0.048	0	0.02	0.04	0.06	0.08	0.1	0.12	0.14	0.16	0.18	0.2	0.22	0.24	0.26	0.28	0.3	0.32	0.34	0.36	0.38	0.4	0.42	0.44	0.46	0.479	0.499
0.06	0	0.02	0.04	0.06	0.08	0.1	0.12	0.14	0.16	0.18	0.2	0.22	0.24	0.26	0.28	0.3	0.32	0.34	0.36	0.38	0.399	0.419	0.439	0.458	0.477	0.496
0.072	0	0.02	0.04	0.06	0.08	0.1	0.12	0.14	0.16	0.18	0.2	0.22	0.24	0.26	0.28	0.3	0.32	0.339	0.359	0.379	0.398	0.418	0.437	0.456	0.474	0.492
0.083	0	0.02	0.04	0.06	0.08	0.1	0.12	0.14	0.16	0.18	0.2	0.22	0.24	0.26	0.28	0.299	0.319	0.339	0.358	0.378	0.397	0.416	0.434	0.452	0.469	0.486
0.1	0	0.02	0.04	0.06	0.08	0.1	0.12	0.14	0.16	0.18	0.2	0.22	0.239	0.259	0.279	0.298	0.318	0.337	0.356	0.374	0.393	0.411	0.428	0.444	0.46	0.475
0.125	0	0.02	0.04	0.06	0.08	0.1	0.12	0.139	0.159	0.179	0.199	0.218	0.238	0.257	0.276	0.295	0.313	0.332	0.35	0.367	0.384	0.4	0.415	0.429	0.442	0.454
0.15	0	0.02	0.04	0.06	0.079	0.099	0.119	0.138	0.158	0.177	0.197	0.216	0.234	0.253	0.271	0.289	0.307	0.324	0.341	0.356	0.372	0.386	0.399	0.411	0.422	0.432
0.175	0	0.02	0.039	0.059	0.078	0.098	0.117	0.136	0.156	0.174	0.193	0.212	0.23	0.248	0.265	0.282	0.298	0.314	0.33	0.344	0.357	0.37	0.382	0.392	0.401	0.409
0.2	0	0.019	0.039	0.058	0.077	0.096	0.115	0.134	0.152	0.171	0.189	0.206	0.224	0.241	0.257	0.273	0.289	0.303	0.317	0.33	0.342	0.353	0.363	0.372	0.38	0.386
0.225	0	0.019	0.038	0.057	0.075	0.094	0.112	0.13	0.148	0.166	0.183	0.2	0.217	0.233	0.248	0.263	0.277	0.291	0.304	0.316	0.326	0.336	0.345	0.353	0.359	0.364
0.25	0	0.018	0.037	0.055	0.073	0.091	0.109	0.126	0.144	0.16	0.177	0.193	0.209	0.224	0.239	0.253	0.266	0.278	0.29	0.301	0.31	0.319	0.327	0.333	0.339	0.343
0.275	0	0.018	0.035	0.053	0.07	0.088	0.105	0.122	0.138	0.154	0.17	0.186	0.2	0.215	0.228	0.241	0.254	0.265	0.276	0.286	0.295	0.302	0.309	0.315	0.319	0.323
0.3	0	0.017	0.034	0.051	0.068	0.084	0.101	0.117	0.133	0.148	0.163	0.178	0.192	0.205	0.218	0.23	0.242	0.252	0.262	0.271	0.279	0.286	0.292	0.297	0.301	0.303
0.325	0	0.016	0.033	0.049	0.065	0.081	0.096	0.112	0.127	0.141	0.156	0.169	0.183	0.195	0.207	0.219	0.229	0.239	0.248	0.257	0.264	0.27	0.276	0.28	0.283	0.285
0.35	0	0.016	0.031	0.047	0.062	0.077	0.092	0.107	0.121	0.135	0.148	0.161	0.174	0.186	0.197	0.208	0.218	0.227	0.235	0.243	0.249	0.255	0.26	0.264	0.267	0.268
0.4	0	0.014	0.028	0.042	0.056	0.07	0.083	0.096	0.109	0.122	0.134	0.145	0.156	0.167	0.177	0.186	0.195	0.203	0.21	0.216	0.222	0.227	0.231	0.234	0.236	0.237
0.45	0	0.013	0.025	0.038	0.05	0.063	0.075	0.086	0.098	0.109	0.12	0.13	0.14	0.149	0.158	0.166	0.174	0.181	0.187	0.192	0.197	0.201	0.205	0.207	0.209	0.21
0.5	0	0.011	0.023	0.034	0.045	0.056	0.067	0.077	0.087	0.097	0.107	0.116	0.125	0.133	0.141	0.148	0.154	0.161	0.166	0.171	0.175	0.178	0.181	0.183	0.185	0.185
0.6	0	0.009	0.018	0.027	0.036	0.044	0.053	0.061	0.069	0.077	0.084	0.091	0.098	0.105	0.111	0.116	0.122	0.126	0.13	0.134	0.137	0.14	0.142	0.144	0.144	0.145
0.7	0	0.007	0.014	0.021	0.028	0.035	0.041	0.048	0.054	0.06	0.066	0.072	0.077	0.082	0.087	0.091	0.095	0.099	0.102	0.105	0.107	0.109	0.111	0.112	0.113	0.113
0.8	0	0.006	0.011	0.017	0.022	0.027	0.032	0.038	0.043	0.047	0.052	0.056	0.06	0.064	0.068	0.071	0.075	0.077	0.08	0.082	0.084	0.086	0.087	0.088	0.088	0.088
0.9	0	0.004	0.009	0.013	0.017	0.021	0.025	0.029	0.033	0.037	0.041	0.044	0.047	0.05	0.053	0.056	0.058	0.061	0.062	0.064	0.066	0.067	0.068	0.069	0.069	0.069
1	0	0.003	0.007	0.01	0.013	0.017	0.02	0.023	0.026	0.029	0.032	0.034	0.037	0.039	0.042	0.044	0.046	0.047	0.049	0.05	0.051	0.052	0.053	0.054	0.054	0.054
1.25	0	0.002	0.004	0.005	0.007	0.009	0.011	0.012	0.014	0.016	0.017	0.019	0.02	0.021	0.022	0.024	0.025	0.026	0.026	0.027	0.028	0.028	0.029	0.029	0.029	0.029
1.5	0	0.001	0.002	0.003	0.004	0.005	0.006	0.007	0.008	0.008	0.009	0.01	0.011	0.011	0.012	0.013	0.013	0.014	0.014	0.015	0.015	0.015	0.015	0.016	0.016	0.016
1.75	0	0.001	0.001	0.002	0.002	0.003	0.003	0.004	0.004	0.005	0.005	0.005	0.006	0.006	0.007	0.007	0.007	0.007	0.008	0.008	0.008	0.008	0.008	0.008	0.008	0.008

(Continued)

Table 2.4 (Continued) u_e/σ_1 for suddenly applied loading proportional to thickness with double drainage (see Figure 2.7(b), (c))

T_v	z/H=1.04	1.08	1.12	1.16	1.2	1.24	1.28	1.32	1.36	1.4	1.44	1.48	1.52	1.56	1.6	1.64	1.68	1.72	1.76	1.8	1.84	1.88	1.92	1.96	2
0	0.52	0.54	0.56	0.58	0.6	0.62	0.64	0.66	0.68	0.7	0.72	0.74	0.76	0.78	0.8	0.82	0.84	0.86	0.88	0.9	0.92	0.94	0.96	0.98	1
0.001	0.52	0.54	0.56	0.58	0.6	0.62	0.64	0.66	0.68	0.7	0.72	0.74	0.76	0.78	0.8	0.82	0.84	0.86	0.88	0.9	0.92	0.933	0.886	0.609	0
0.002	0.52	0.54	0.56	0.58	0.6	0.62	0.64	0.66	0.68	0.7	0.72	0.74	0.76	0.78	0.8	0.82	0.84	0.86	0.88	0.898	0.909	0.882	0.754	0.453	0
0.004	0.52	0.54	0.56	0.58	0.6	0.62	0.64	0.66	0.68	0.7	0.72	0.74	0.76	0.78	0.8	0.82	0.84	0.858	0.873	0.875	0.846	0.76	0.589	0.325	0
0.006	0.52	0.54	0.56	0.58	0.6	0.62	0.64	0.66	0.68	0.7	0.72	0.74	0.76	0.78	0.8	0.819	0.837	0.849	0.852	0.832	0.776	0.667	0.495	0.265	0
0.008	0.52	0.54	0.56	0.58	0.6	0.62	0.64	0.66	0.68	0.7	0.72	0.74	0.76	0.779	0.798	0.816	0.829	0.833	0.822	0.786	0.714	0.597	0.433	0.228	0
0.012	0.52	0.54	0.56	0.58	0.6	0.62	0.64	0.66	0.68	0.7	0.72	0.739	0.758	0.775	0.79	0.8	0.801	0.789	0.759	0.703	0.618	0.501	0.354	0.184	0
0.016	0.52	0.54	0.56	0.58	0.6	0.62	0.64	0.66	0.68	0.699	0.718	0.736	0.753	0.766	0.775	0.776	0.766	0.742	0.7	0.636	0.549	0.438	0.305	0.157	0
0.02	0.52	0.54	0.56	0.58	0.6	0.62	0.64	0.659	0.679	0.697	0.715	0.731	0.744	0.752	0.754	0.748	0.73	0.698	0.65	0.583	0.496	0.391	0.271	0.139	0
0.028	0.52	0.54	0.56	0.58	0.599	0.619	0.638	0.656	0.673	0.689	0.702	0.712	0.717	0.717	0.709	0.692	0.664	0.623	0.57	0.502	0.421	0.328	0.225	0.114	0
0.036	0.52	0.539	0.559	0.578	0.597	0.615	0.633	0.649	0.663	0.675	0.683	0.687	0.686	0.679	0.664	0.64	0.607	0.563	0.509	0.444	0.369	0.285	0.194	0.099	0
0.048	0.518	0.537	0.555	0.573	0.59	0.606	0.62	0.632	0.641	0.647	0.649	0.647	0.639	0.624	0.603	0.575	0.538	0.494	0.441	0.381	0.314	0.241	0.164	0.083	0
0.06	0.514	0.532	0.549	0.565	0.579	0.592	0.602	0.61	0.615	0.617	0.614	0.607	0.594	0.576	0.552	0.521	0.484	0.441	0.392	0.336	0.276	0.211	0.143	0.072	0
0.072	0.509	0.525	0.54	0.553	0.565	0.575	0.582	0.587	0.588	0.586	0.58	0.569	0.554	0.534	0.508	0.477	0.441	0.399	0.353	0.302	0.247	0.188	0.127	0.064	0
0.083	0.502	0.516	0.529	0.541	0.55	0.558	0.563	0.565	0.564	0.559	0.551	0.538	0.521	0.5	0.474	0.443	0.408	0.368	0.324	0.276	0.225	0.172	0.116	0.058	0
0.1	0.488	0.5	0.511	0.52	0.526	0.531	0.533	0.532	0.528	0.52	0.51	0.495	0.477	0.455	0.429	0.399	0.366	0.329	0.288	0.245	0.199	0.152	0.102	0.051	0
0.125	0.465	0.474	0.482	0.487	0.49	0.491	0.49	0.486	0.479	0.47	0.457	0.442	0.423	0.401	0.376	0.348	0.318	0.285	0.249	0.211	0.171	0.13	0.087	0.044	0
0.15	0.44	0.447	0.452	0.455	0.456	0.455	0.451	0.446	0.437	0.427	0.413	0.398	0.379	0.358	0.335	0.309	0.281	0.251	0.219	0.185	0.15	0.113	0.076	0.038	0
0.175	0.415	0.42	0.423	0.424	0.424	0.421	0.416	0.41	0.401	0.39	0.376	0.361	0.343	0.323	0.301	0.277	0.251	0.224	0.195	0.165	0.133	0.101	0.068	0.034	0
0.2	0.391	0.394	0.396	0.396	0.394	0.391	0.385	0.378	0.368	0.357	0.344	0.329	0.312	0.293	0.273	0.251	0.227	0.202	0.176	0.148	0.12	0.09	0.061	0.03	0
0.225	0.368	0.37	0.37	0.37	0.367	0.363	0.357	0.349	0.34	0.329	0.316	0.302	0.286	0.268	0.249	0.228	0.207	0.184	0.159	0.134	0.109	0.082	0.055	0.028	0
0.25	0.345	0.347	0.347	0.345	0.342	0.338	0.331	0.324	0.315	0.304	0.292	0.278	0.263	0.246	0.228	0.209	0.189	0.168	0.146	0.123	0.099	0.075	0.05	0.025	0
0.275	0.325	0.325	0.325	0.323	0.319	0.315	0.308	0.301	0.292	0.282	0.27	0.257	0.243	0.227	0.21	0.193	0.174	0.154	0.134	0.113	0.091	0.069	0.046	0.023	0
0.3	0.305	0.305	0.304	0.302	0.298	0.294	0.287	0.28	0.271	0.261	0.25	0.238	0.225	0.21	0.194	0.178	0.16	0.142	0.123	0.104	0.084	0.063	0.042	0.021	0
0.325	0.286	0.286	0.285	0.283	0.279	0.274	0.268	0.261	0.253	0.243	0.233	0.221	0.208	0.195	0.18	0.165	0.149	0.132	0.114	0.096	0.077	0.058	0.039	0.02	0
0.35	0.269	0.269	0.267	0.265	0.261	0.256	0.251	0.244	0.236	0.227	0.217	0.206	0.194	0.181	0.167	0.153	0.138	0.122	0.106	0.089	0.072	0.054	0.036	0.018	0
0.4	0.238	0.237	0.235	0.233	0.229	0.225	0.219	0.213	0.206	0.198	0.189	0.179	0.169	0.157	0.145	0.133	0.12	0.106	0.092	0.077	0.062	0.047	0.031	0.016	0
0.45	0.21	0.209	0.207	0.205	0.202	0.198	0.193	0.187	0.18	0.173	0.165	0.157	0.147	0.137	0.127	0.116	0.104	0.092	0.08	0.067	0.054	0.041	0.027	0.014	0
0.5	0.185	0.184	0.183	0.181	0.178	0.174	0.17	0.164	0.159	0.152	0.145	0.137	0.129	0.12	0.111	0.101	0.091	0.081	0.07	0.059	0.047	0.036	0.024	0.012	0
0.6	0.145	0.144	0.143	0.141	0.138	0.135	0.132	0.128	0.123	0.118	0.112	0.106	0.1	0.093	0.086	0.078	0.071	0.062	0.054	0.045	0.036	0.027	0.018	0.009	0
0.7	0.113	0.112	0.111	0.11	0.108	0.105	0.103	0.099	0.096	0.092	0.088	0.083	0.078	0.072	0.067	0.061	0.055	0.048	0.042	0.035	0.028	0.021	0.014	0.007	0
0.8	0.088	0.088	0.087	0.086	0.084	0.082	0.08	0.078	0.075	0.072	0.068	0.065	0.061	0.056	0.052	0.047	0.043	0.038	0.033	0.027	0.022	0.017	0.011	0.006	0
0.9	0.069	0.069	0.068	0.067	0.066	0.064	0.063	0.061	0.058	0.056	0.053	0.05	0.047	0.044	0.041	0.037	0.033	0.029	0.025	0.021	0.017	0.013	0.009	0.004	0
1	0.054	0.054	0.053	0.052	0.051	0.05	0.049	0.047	0.046	0.044	0.042	0.039	0.037	0.034	0.032	0.029	0.026	0.023	0.02	0.017	0.013	0.01	0.007	0.003	0
1.25	0.029	0.029	0.029	0.028	0.028	0.027	0.026	0.026	0.025	0.024	0.022	0.021	0.02	0.019	0.017	0.016	0.014	0.012	0.011	0.009	0.007	0.005	0.004	0.002	0
1.5	0.016	0.016	0.015	0.015	0.015	0.015	0.014	0.014	0.013	0.013	0.012	0.011	0.011	0.01	0.009	0.008	0.008	0.007	0.006	0.005	0.004	0.003	0.002	0.001	0
1.75	0.008	0.008	0.008	0.008	0.008	0.008	0.008	0.007	0.007	0.007	0.007	0.006	0.006	0.005	0.005	0.005	0.004	0.004	0.003	0.003	0.002	0.002	0.001	0.001	0

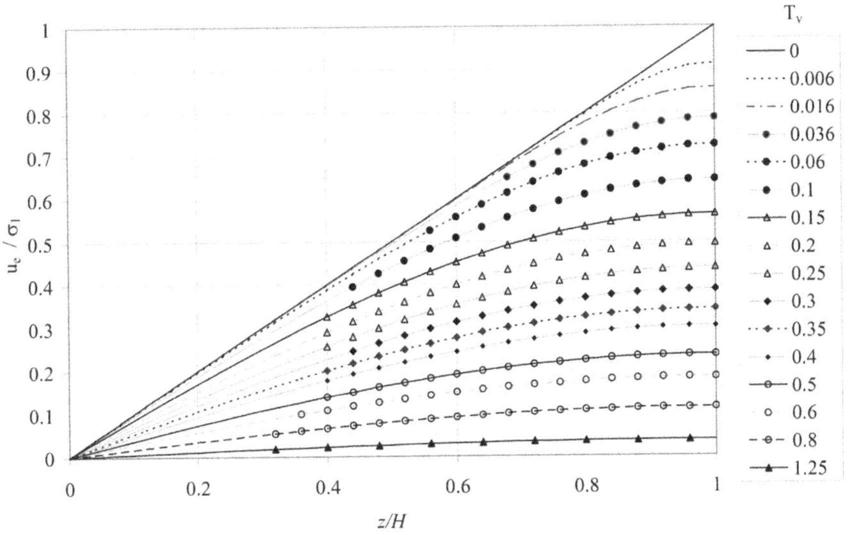

Figure 2.10 Isochrones of excess porewater pressure for suddenly applied vertical total stress increments $\Delta\sigma_z = \frac{z}{H}\sigma_1$ (see Figure 2.7(e)).

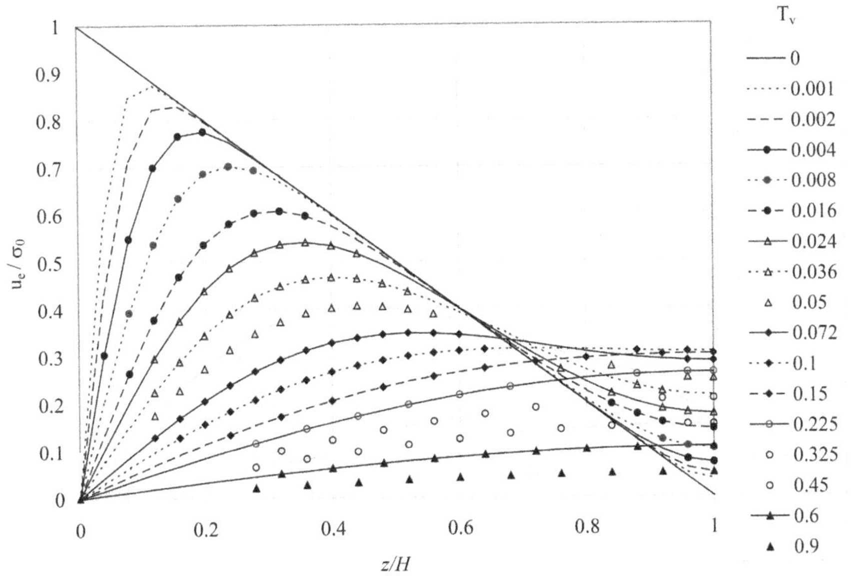

Figure 2.11 Isochrones of excess porewater pressure for suddenly applied vertical total stress increments $\Delta\sigma_z = \left(1 - \frac{z}{H}\right)\sigma_0$ (see Figure 2.7(f)).

Table 2.5 u_e/σ_l for suddenly applied loading proportional to thickness with single drainage (see Figure 2.7(e))

T_v	z/H=0	0.04	0.08	0.12	0.16	0.2	0.24	0.28	0.32	0.36	0.4	0.44	0.48	0.52	0.56	0.6	0.64	0.68	0.72	0.76	0.8	0.84	0.88	0.92	0.96	1
0	0	0.04	0.08	0.12	0.16	0.2	0.24	0.28	0.32	0.36	0.4	0.44	0.48	0.52	0.56	0.6	0.64	0.68	0.72	0.76	0.8	0.84	0.88	0.92	0.96	1
0.002	0	0.04	0.08	0.12	0.16	0.2	0.24	0.28	0.32	0.36	0.4	0.44	0.48	0.52	0.56	0.6	0.64	0.68	0.72	0.76	0.8	0.84	0.879	0.914	0.94	0.95
0.004	0	0.04	0.08	0.12	0.16	0.2	0.24	0.28	0.32	0.36	0.4	0.44	0.48	0.52	0.56	0.6	0.64	0.68	0.72	0.76	0.799	0.837	0.873	0.902	0.922	0.929
0.006	0	0.04	0.08	0.12	0.16	0.2	0.24	0.28	0.32	0.36	0.4	0.44	0.48	0.52	0.56	0.6	0.64	0.68	0.72	0.759	0.797	0.833	0.865	0.89	0.907	0.913
0.008	0	0.04	0.08	0.12	0.16	0.2	0.24	0.28	0.32	0.36	0.4	0.44	0.48	0.52	0.56	0.6	0.64	0.68	0.719	0.757	0.794	0.828	0.857	0.88	0.894	0.899
0.012	0	0.04	0.08	0.12	0.16	0.2	0.24	0.28	0.32	0.36	0.4	0.44	0.48	0.52	0.56	0.6	0.639	0.678	0.716	0.752	0.786	0.816	0.841	0.86	0.872	0.876
0.016	0	0.04	0.08	0.12	0.16	0.2	0.24	0.28	0.32	0.36	0.4	0.44	0.48	0.52	0.559	0.598	0.637	0.675	0.711	0.745	0.776	0.804	0.826	0.843	0.854	0.857
0.02	0	0.04	0.08	0.12	0.16	0.2	0.24	0.28	0.32	0.36	0.4	0.44	0.479	0.519	0.558	0.597	0.634	0.671	0.705	0.738	0.767	0.792	0.813	0.828	0.837	0.84
0.028	0	0.04	0.08	0.12	0.16	0.2	0.24	0.28	0.32	0.36	0.399	0.439	0.478	0.516	0.554	0.591	0.627	0.661	0.693	0.722	0.747	0.77	0.787	0.8	0.808	0.811
0.036	0	0.04	0.08	0.12	0.16	0.2	0.24	0.279	0.319	0.358	0.398	0.436	0.475	0.512	0.549	0.584	0.618	0.649	0.679	0.706	0.729	0.749	0.765	0.776	0.784	0.786
0.048	0	0.04	0.08	0.12	0.159	0.199	0.239	0.278	0.317	0.356	0.394	0.431	0.468	0.504	0.538	0.571	0.602	0.632	0.658	0.682	0.703	0.721	0.734	0.745	0.751	0.753
0.06	0	0.04	0.079	0.119	0.158	0.198	0.237	0.275	0.314	0.351	0.388	0.425	0.46	0.494	0.526	0.557	0.586	0.613	0.638	0.66	0.679	0.695	0.707	0.716	0.722	0.724
0.072	0	0.039	0.079	0.118	0.157	0.195	0.234	0.272	0.309	0.346	0.382	0.417	0.45	0.483	0.514	0.543	0.57	0.596	0.618	0.639	0.656	0.671	0.682	0.691	0.696	0.697
0.083	0	0.039	0.078	0.116	0.155	0.193	0.23	0.268	0.304	0.34	0.375	0.408	0.441	0.472	0.502	0.53	0.556	0.58	0.601	0.62	0.637	0.65	0.661	0.669	0.673	0.675
0.1	0	0.038	0.076	0.113	0.151	0.188	0.224	0.26	0.295	0.33	0.363	0.395	0.426	0.455	0.483	0.509	0.533	0.556	0.575	0.593	0.608	0.621	0.63	0.637	0.642	0.643
0.125	0	0.036	0.073	0.109	0.144	0.179	0.214	0.248	0.281	0.313	0.345	0.375	0.403	0.43	0.456	0.48	0.502	0.522	0.54	0.556	0.57	0.581	0.59	0.596	0.6	0.601
0.15	0	0.035	0.069	0.103	0.137	0.17	0.203	0.235	0.266	0.297	0.326	0.354	0.381	0.406	0.43	0.452	0.472	0.491	0.507	0.522	0.534	0.545	0.553	0.558	0.562	0.563
0.175	0	0.033	0.065	0.098	0.13	0.161	0.192	0.222	0.252	0.28	0.308	0.334	0.359	0.383	0.405	0.425	0.444	0.461	0.477	0.49	0.502	0.511	0.519	0.524	0.527	0.528
0.2	0	0.031	0.062	0.092	0.122	0.152	0.181	0.21	0.237	0.264	0.29	0.315	0.338	0.36	0.381	0.4	0.418	0.434	0.448	0.461	0.471	0.48	0.487	0.492	0.495	0.496
0.225	0	0.029	0.058	0.087	0.115	0.143	0.171	0.198	0.224	0.249	0.273	0.296	0.318	0.339	0.358	0.376	0.393	0.408	0.421	0.433	0.443	0.451	0.458	0.462	0.465	0.466
0.25	0	0.027	0.055	0.082	0.109	0.135	0.161	0.186	0.21	0.234	0.257	0.279	0.299	0.319	0.337	0.354	0.369	0.383	0.396	0.407	0.416	0.424	0.43	0.434	0.437	0.438
0.275	0	0.026	0.051	0.077	0.102	0.127	0.151	0.175	0.198	0.22	0.242	0.262	0.281	0.3	0.317	0.333	0.347	0.36	0.372	0.382	0.391	0.398	0.404	0.408	0.411	0.411
0.3	0	0.024	0.048	0.072	0.096	0.119	0.142	0.165	0.186	0.207	0.227	0.246	0.265	0.282	0.298	0.313	0.326	0.339	0.35	0.36	0.368	0.375	0.38	0.384	0.386	0.387
0.325	0	0.023	0.046	0.068	0.09	0.112	0.134	0.155	0.175	0.195	0.214	0.232	0.249	0.265	0.28	0.294	0.307	0.319	0.329	0.338	0.346	0.352	0.357	0.361	0.363	0.364
0.35	0	0.021	0.043	0.064	0.085	0.106	0.126	0.145	0.165	0.183	0.201	0.218	0.234	0.249	0.263	0.276	0.289	0.3	0.309	0.318	0.325	0.331	0.336	0.339	0.341	0.342
0.4	0	0.019	0.038	0.057	0.075	0.093	0.111	0.129	0.146	0.162	0.178	0.193	0.207	0.22	0.233	0.244	0.255	0.265	0.273	0.281	0.287	0.293	0.297	0.3	0.302	0.302
0.45	0	0.017	0.033	0.05	0.066	0.083	0.098	0.114	0.129	0.143	0.157	0.17	0.183	0.195	0.206	0.216	0.225	0.234	0.242	0.248	0.254	0.259	0.262	0.265	0.267	0.267
0.5	0	0.015	0.03	0.044	0.059	0.073	0.087	0.101	0.114	0.126	0.139	0.15	0.162	0.172	0.182	0.191	0.199	0.207	0.214	0.219	0.224	0.229	0.232	0.234	0.236	0.236
0.6	0	0.012	0.023	0.035	0.046	0.057	0.068	0.079	0.089	0.099	0.108	0.118	0.126	0.134	0.142	0.149	0.156	0.162	0.167	0.171	0.175	0.179	0.181	0.183	0.184	0.184
0.7	0	0.009	0.018	0.027	0.036	0.045	0.053	0.061	0.069	0.077	0.085	0.092	0.099	0.105	0.111	0.117	0.122	0.126	0.13	0.134	0.137	0.14	0.142	0.143	0.144	0.144
0.8	0	0.007	0.014	0.021	0.028	0.035	0.041	0.048	0.054	0.06	0.066	0.072	0.077	0.082	0.087	0.091	0.095	0.099	0.102	0.105	0.107	0.109	0.111	0.112	0.112	0.113
0.9	0	0.006	0.011	0.016	0.022	0.027	0.032	0.037	0.042	0.047	0.052	0.056	0.06	0.064	0.068	0.071	0.074	0.077	0.08	0.082	0.084	0.085	0.086	0.087	0.088	0.088
1	0	0.004	0.009	0.013	0.017	0.021	0.025	0.029	0.033	0.037	0.04	0.044	0.047	0.05	0.053	0.056	0.058	0.06	0.062	0.064	0.065	0.067	0.068	0.068	0.069	0.069
1.25	0	0.002	0.005	0.007	0.009	0.011	0.014	0.016	0.018	0.02	0.022	0.024	0.025	0.027	0.029	0.03	0.031	0.033	0.034	0.034	0.035	0.036	0.036	0.037	0.037	0.037
1.5	0	0.001	0.003	0.004	0.005	0.006	0.007	0.009	0.01	0.011	0.012	0.013	0.014	0.015	0.015	0.016	0.017	0.018	0.018	0.019	0.019	0.019	0.02	0.02	0.02	0.02
1.75	0	0.001	0.001	0.002	0.003	0.003	0.004	0.005	0.005	0.006	0.006	0.007	0.007	0.008	0.008	0.009	0.009	0.009	0.01	0.01	0.01	0.01	0.011	0.011	0.011	0.011
2	0	0	0.001	0.001	0.001	0.002	0.002	0.002	0.003	0.003	0.003	0.004	0.004	0.004	0.004	0.005	0.005	0.005	0.005	0.005	0.006	0.006	0.006	0.006	0.006	0.006

Table 2.6 u_e/σ_0 for suddenly applied loading proportional to thickness with single drainage (see Figure 2.7(f))

T_v	z/H=0	0.04	0.08	0.12	0.16	0.2	0.24	0.28	0.32	0.36	0.4	0.44	0.48	0.52	0.56	0.6	0.64	0.68	0.72	0.76	0.8	0.84	0.88	0.92	0.96	1
0	1	0.96	0.92	0.88	0.84	0.8	0.76	0.72	0.68	0.64	0.6	0.56	0.52	0.48	0.44	0.4	0.36	0.32	0.28	0.24	0.2	0.16	0.12	0.08	0.04	0
0.001	0	0.589	0.846	0.873	0.84	0.8	0.76	0.72	0.68	0.64	0.6	0.56	0.52	0.48	0.44	0.4	0.36	0.32	0.28	0.24	0.2	0.16	0.12	0.081	0.049	0.036
0.002	0	0.433	0.714	0.822	0.829	0.798	0.76	0.72	0.68	0.64	0.6	0.56	0.52	0.48	0.44	0.4	0.36	0.32	0.28	0.24	0.2	0.16	0.121	0.086	0.06	0.05
0.004	0	0.305	0.549	0.7	0.766	0.775	0.753	0.718	0.68	0.64	0.6	0.56	0.52	0.48	0.44	0.4	0.36	0.32	0.28	0.24	0.201	0.163	0.127	0.098	0.078	0.071
0.006	0	0.245	0.455	0.607	0.696	0.732	0.732	0.709	0.677	0.639	0.6	0.56	0.52	0.48	0.44	0.4	0.36	0.32	0.28	0.241	0.203	0.167	0.135	0.11	0.093	0.087
0.008	0	0.208	0.393	0.537	0.634	0.686	0.702	0.693	0.669	0.636	0.598	0.559	0.52	0.48	0.44	0.4	0.36	0.32	0.281	0.243	0.206	0.172	0.143	0.12	0.106	0.101
0.012	0	0.164	0.314	0.441	0.538	0.603	0.639	0.649	0.641	0.62	0.59	0.556	0.518	0.479	0.44	0.4	0.361	0.322	0.284	0.248	0.214	0.184	0.159	0.14	0.128	0.124
0.016	0	0.137	0.265	0.378	0.469	0.536	0.58	0.602	0.606	0.596	0.575	0.546	0.513	0.477	0.439	0.401	0.363	0.325	0.289	0.255	0.224	0.196	0.174	0.157	0.146	0.143
0.02	0	0.119	0.231	0.331	0.416	0.483	0.53	0.559	0.57	0.568	0.555	0.532	0.504	0.472	0.437	0.401	0.364	0.329	0.294	0.262	0.233	0.208	0.187	0.172	0.163	0.16
0.028	0	0.094	0.185	0.268	0.341	0.402	0.45	0.483	0.504	0.512	0.51	0.498	0.48	0.456	0.428	0.398	0.366	0.335	0.305	0.277	0.252	0.23	0.212	0.199	0.191	0.189
0.036	0	0.079	0.154	0.225	0.289	0.344	0.389	0.424	0.448	0.462	0.466	0.463	0.452	0.435	0.414	0.391	0.365	0.339	0.314	0.29	0.268	0.249	0.234	0.223	0.216	0.214
0.048	0	0.063	0.124	0.182	0.235	0.282	0.323	0.356	0.381	0.399	0.41	0.413	0.411	0.403	0.391	0.376	0.359	0.34	0.322	0.304	0.287	0.273	0.261	0.252	0.247	0.245
0.06	0	0.052	0.103	0.152	0.198	0.239	0.275	0.306	0.331	0.35	0.363	0.371	0.374	0.373	0.368	0.359	0.349	0.337	0.324	0.312	0.3	0.289	0.281	0.274	0.27	0.269
0.072	0	0.045	0.088	0.13	0.17	0.206	0.239	0.268	0.292	0.312	0.327	0.337	0.344	0.347	0.346	0.343	0.338	0.331	0.323	0.315	0.307	0.3	0.294	0.29	0.287	0.286
0.083	0	0.039	0.078	0.115	0.151	0.184	0.214	0.24	0.264	0.283	0.299	0.311	0.32	0.326	0.328	0.329	0.327	0.324	0.32	0.315	0.311	0.306	0.302	0.299	0.297	0.297
0.1	0	0.033	0.066	0.098	0.129	0.157	0.184	0.208	0.23	0.249	0.266	0.279	0.29	0.299	0.305	0.309	0.312	0.313	0.313	0.312	0.311	0.31	0.308	0.307	0.306	0.306
0.125	0	0.027	0.054	0.081	0.107	0.131	0.154	0.176	0.196	0.214	0.23	0.245	0.257	0.268	0.277	0.285	0.291	0.296	0.3	0.302	0.304	0.306	0.307	0.307	0.308	0.308
0.15	0	0.024	0.047	0.07	0.092	0.114	0.134	0.154	0.172	0.189	0.205	0.22	0.233	0.245	0.255	0.264	0.272	0.279	0.285	0.289	0.293	0.296	0.298	0.3	0.301	0.301
0.175	0	0.021	0.042	0.062	0.082	0.101	0.12	0.138	0.155	0.171	0.187	0.201	0.214	0.226	0.237	0.246	0.255	0.263	0.269	0.275	0.28	0.283	0.286	0.288	0.289	0.29
0.2	0	0.019	0.038	0.056	0.074	0.092	0.109	0.126	0.142	0.157	0.172	0.185	0.198	0.21	0.221	0.23	0.239	0.247	0.254	0.26	0.265	0.269	0.272	0.275	0.276	0.276
0.225	0	0.017	0.035	0.052	0.068	0.085	0.101	0.116	0.131	0.146	0.159	0.172	0.184	0.196	0.206	0.216	0.225	0.232	0.239	0.245	0.251	0.255	0.258	0.26	0.262	0.262
0.25	0	0.016	0.032	0.048	0.063	0.079	0.094	0.108	0.122	0.136	0.149	0.161	0.172	0.183	0.193	0.203	0.211	0.219	0.225	0.231	0.236	0.24	0.244	0.246	0.247	0.248
0.275	0	0.015	0.03	0.045	0.059	0.073	0.087	0.101	0.114	0.127	0.139	0.151	0.162	0.172	0.181	0.19	0.198	0.206	0.212	0.218	0.223	0.227	0.23	0.232	0.233	0.234
0.3	0	0.014	0.028	0.042	0.055	0.069	0.082	0.095	0.107	0.119	0.13	0.141	0.152	0.161	0.17	0.179	0.186	0.193	0.2	0.205	0.21	0.213	0.216	0.218	0.22	0.22
0.325	0	0.013	0.026	0.039	0.052	0.064	0.077	0.089	0.1	0.112	0.122	0.133	0.142	0.151	0.16	0.168	0.175	0.182	0.188	0.193	0.197	0.201	0.204	0.206	0.207	0.207
0.35	0	0.012	0.025	0.037	0.049	0.06	0.072	0.083	0.094	0.105	0.115	0.125	0.134	0.142	0.15	0.158	0.165	0.171	0.176	0.181	0.185	0.189	0.191	0.193	0.194	0.195
0.4	0	0.011	0.022	0.032	0.043	0.053	0.064	0.073	0.083	0.092	0.101	0.11	0.118	0.126	0.133	0.14	0.146	0.151	0.156	0.16	0.164	0.167	0.169	0.171	0.172	0.172
0.45	0	0.01	0.019	0.029	0.038	0.047	0.056	0.065	0.073	0.082	0.09	0.097	0.104	0.111	0.117	0.123	0.129	0.134	0.138	0.142	0.145	0.148	0.15	0.151	0.152	0.152
0.5	0	0.008	0.017	0.025	0.034	0.042	0.05	0.057	0.065	0.072	0.079	0.086	0.092	0.098	0.104	0.109	0.114	0.118	0.122	0.125	0.128	0.13	0.132	0.134	0.134	0.135
0.6	0	0.007	0.013	0.02	0.026	0.033	0.039	0.045	0.051	0.056	0.062	0.067	0.072	0.077	0.081	0.085	0.089	0.092	0.095	0.098	0.1	0.102	0.103	0.104	0.105	0.105
0.7	0	0.005	0.01	0.015	0.02	0.025	0.03	0.035	0.04	0.044	0.048	0.052	0.056	0.06	0.063	0.067	0.069	0.072	0.074	0.076	0.078	0.08	0.081	0.082	0.082	0.082
0.8	0	0.004	0.008	0.012	0.016	0.02	0.024	0.027	0.031	0.034	0.038	0.041	0.044	0.047	0.05	0.052	0.054	0.056	0.058	0.06	0.061	0.062	0.063	0.064	0.064	0.064
0.9	0	0.003	0.006	0.009	0.012	0.016	0.018	0.021	0.024	0.027	0.03	0.032	0.034	0.037	0.039	0.041	0.042	0.044	0.045	0.047	0.048	0.049	0.049	0.05	0.05	0.05
1	0	0.002	0.005	0.007	0.01	0.012	0.014	0.017	0.019	0.021	0.023	0.025	0.027	0.029	0.03	0.032	0.033	0.034	0.036	0.036	0.037	0.038	0.039	0.039	0.039	0.039
1.25	0	0.001	0.003	0.004	0.005	0.007	0.008	0.009	0.01	0.011	0.012	0.013	0.014	0.015	0.016	0.017	0.018	0.019	0.019	0.02	0.02	0.021	0.021	0.021	0.021	0.021
1.5	0	0.001	0.001	0.002	0.003	0.004	0.004	0.005	0.006	0.006	0.007	0.007	0.008	0.008	0.009	0.009	0.01	0.01	0.01	0.011	0.011	0.011	0.011	0.011	0.011	0.011
1.75	0	0	0.001	0.001	0.002	0.002	0.002	0.003	0.003	0.003	0.004	0.004	0.004	0.004	0.005	0.005	0.005	0.005	0.006	0.006	0.006	0.006	0.006	0.006	0.006	0.006

7. Uniformly distributed ramp loading
   Olson (1977) obtained an analytical solution taking into consideration the construction time effect ($T_c \neq 0$). This solution is a special case of $\sigma_0 = \sigma_1$ in Figure 2.4. The average degree of consolidation is as follows:

$$U(T_v) = U_1(T_v)$$

2.4.1.3 *Application*

Example 2.1: Determine the average degree of consolidation for the case shown in Figure 2.12 where $H = 3.6$ m, $t_c = 0.5$ year, $\sigma_0 = 120$ kPa, $\sigma_1 = 40$ kPa, and $c_v = 2.5$ m²/year.

Using given values of relevant parameters, the construction time factor T_c is $T_c = \frac{C_v t_c}{H^2} = 0.0964$. For $t = 0.25$, 0.5, 1, 2, and 4 years, U_1 are 8.44, 23.10, 42.63, 64.60, and 86.27, respectively, and U_2 are 5.98, 13.43, 14.89, 9.66, and 3.75, respectively, from Table 2.1 and Table 2.2. From (2.27), the average degree of consolidation U is 11.43, 29.82, 50.08, 69.43, and 88.15 for $t = 0.25$, 0.5, 1, 2, and 4 years. The settlement S_t can then be calculated using $S_t = S_f U$.

2.4.2 Solution for a double-layered soil profile under ramp load

The soil profile consisting of two contiguous layers is shown in Figure 2.13. An arbitrary layer is indexed with j ($j = 1$ or 2). The soil properties of the j-th layer are the coefficient of consolidation c_{vj}, the constrained compressibility m_{vj}, and the coefficient of permeability k_j. The compressible stratum has

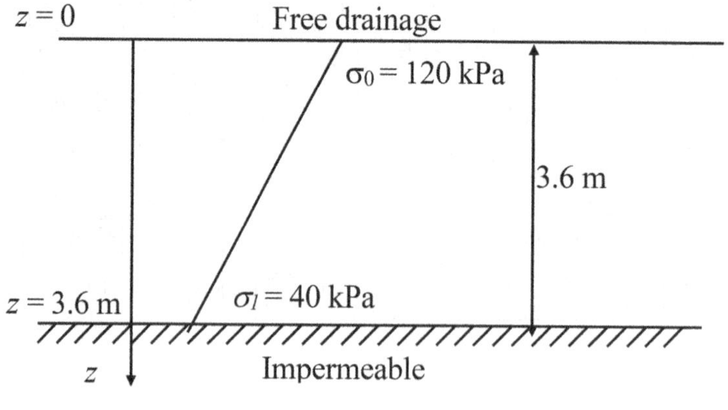

Figure 2.12 Variations of total stress when construction completed.

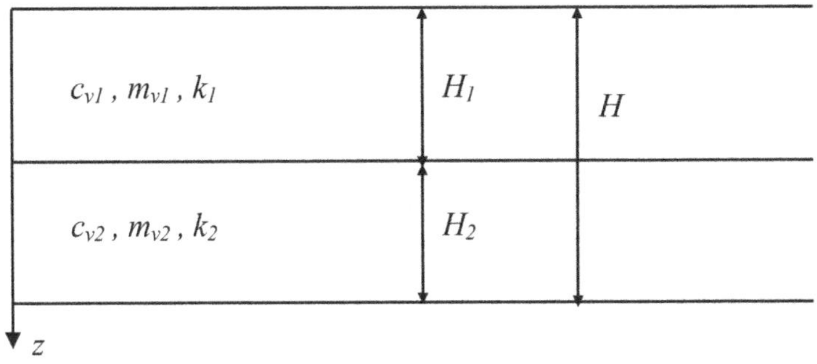

Figure 2.13 A double-layered soil profile.

a total thickness H. The consolidation equation for the two-layer system is rewritten from (2.13) as follows:

$$\frac{\partial u_e}{\partial t} = \begin{cases} c_{v1}\dfrac{\partial^2 u_e}{\partial z^2} + \dfrac{\partial \sigma_z}{\partial t} & 0 \le z \le H_1 \\ c_{v2}\dfrac{\partial^2 u_e}{\partial z^2} + \dfrac{\partial \sigma_z}{\partial t} & H_1 \le z \le H \end{cases} \tag{2.30}$$

The vertical total stress increment $\Delta\sigma_z$ is assumed to vary linearly with depth in each layer and with time and remain unchanged after time t_c (see Figure 2.14). That is,

$$\Delta\sigma_z(z,t) = \begin{cases} \left(\sigma_0 + \dfrac{\sigma_1 - \sigma_0}{H_1} z\right)\min\left(1, \dfrac{t}{t_c}\right) & 0 \le z \le H_1 \\ \left[\sigma_1 + \dfrac{\sigma_2 - \sigma_1}{H_2}(z - H_1)\right]\min\left(1, \dfrac{t}{t_c}\right) & H_1 \le z \le H \end{cases} \tag{2.31}$$

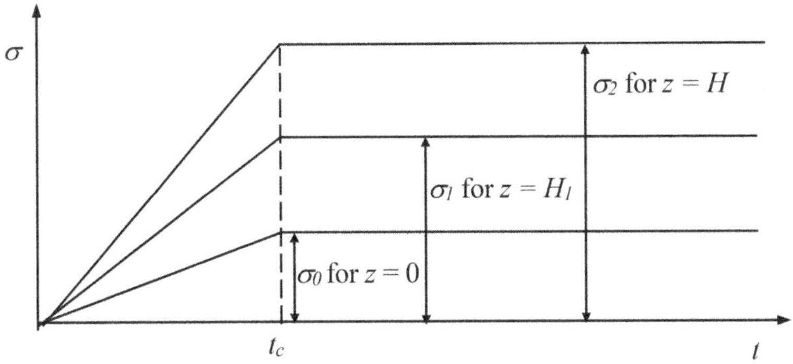

Figure 2.14 Variation of the vertical total stress.

where σ_0, σ_1, and σ_2 are the vertical total stress increments at $t = t_c$ and $z = 0$, $z = H_1$, and $z = H$, respectively. The 'min' means taking the minimum value between 1 and $\frac{t}{t_c}$.

As in Section 2.4.1, the boundary conditions in (2.16) and (2.17) are adopted. The initial excess porewater pressure is set to be zero. At the interface of the two contiguous layers, (2.32) should be satisfied.

$$u_e(z,t)\,|_{z=H_1^-} = u_e(z,t)\,|_{z=H_1^+} \tag{2.32a}$$

$$k_1 \frac{\partial u_e(z,t)}{\partial z}\,|_{z=H_1^-} = k_2 \frac{\partial u_e(z,t)}{\partial z}\,|_{z=H_1^+} \tag{2.32b}$$

Using the method of separation of variables, the consolidation equation (2.30) under the given loading condition (2.31), the boundary conditions (2.16) and (2.17) and the interface conditions (2.32) can be solved (Zhu and Yin, 1999, 2005). The following dimensionless parameters are introduced to express the solutions:

1. *Time factor,* T_v,

$$T_v = \frac{c_{v1}c_{v2}t}{\left(H_1\sqrt{c_{v2}} + H_2\sqrt{c_{v1}}\right)^2} \tag{2.33}$$

2. *Construction time factor,* T_c,

$$T_c = \frac{c_{v1}c_{v2}t_c}{\left(H_1\sqrt{c_{v2}} + H_2\sqrt{c_{v1}}\right)^2} \tag{2.34}$$

3. $$p = \frac{\sqrt{k_2 m_{v2}} - \sqrt{k_1 m_{v1}}}{\sqrt{k_2 m_{v2}} + \sqrt{k_1 m_{v1}}} \tag{2.35}$$

4. $$q = \frac{H_1\sqrt{c_{v2}} - H_2\sqrt{c_{v1}}}{H_1\sqrt{c_{v2}} + H_2\sqrt{c_{v1}}} \tag{2.36}$$

5. $$\alpha = \frac{H_1\sqrt{c_{v2}}}{H_1\sqrt{c_{v2}} + H_2\sqrt{c_{v1}}} = \frac{1}{2}(1+q) \tag{2.37}$$

6. $$\beta = \frac{H_2\sqrt{c_{v1}}}{H_1\sqrt{c_{v2}} + H_2\sqrt{c_{v1}}} = \frac{1}{2}(1-q) \tag{2.38}$$

The solution of (2.30) is

$$u(z,T_v) = \sum_{n=1}^{+\infty} T_n(T_v)Z_n(z) \tag{2.39}$$

where $T_n(T_v)$ is the same as in (2.21), but the parameters are re-defined in (2.33) and (2.34).

For boundary conditions (2.16), the constant b_n in (2.21) and the function Z_n in (2.39) are

$$b_n = \frac{\frac{m_{v1}H_1}{\alpha\sin(\lambda_n\alpha)}\sigma_0 + \frac{m_{v2}H_2}{\beta\sin(\lambda_n\beta)}\sigma_2 + \frac{m_{v1}H_1}{\alpha^2\lambda_n}(\sigma_1-\sigma_0) + \frac{m_{v2}H_2}{\beta^2\lambda_n}(\sigma_1-\sigma_2)}{\frac{m_{v1}H_1}{2\sin^2(\lambda_n\alpha)} + \frac{m_{v2}H_2}{2\sin^2(\lambda_n\beta)}},$$

$$Z_n = \begin{cases} \dfrac{\sin\left(\lambda_n\alpha\dfrac{z}{H_1}\right)}{\sin(\lambda_n\alpha)} & 0 \le z \le H_1 \\ \dfrac{\sin\left(\lambda_n\beta\dfrac{H-z}{H_2}\right)}{\sin(\lambda_n\beta)} & H_1 \le z \le H \end{cases},$$

and the constant λ_n in (2.21) is the n-th positive root of the following equation for variable θ

$$\sin\theta + p\sin(q\theta) = 0 \tag{2.40}$$

Letting $\lambda_n = \pi(n + \xi_n)$, the first 3 ξ_n values (n = 1, 2, 3) are shown in Figures 2.15–2.17 and tabulated in Tables 2.7–2.9.

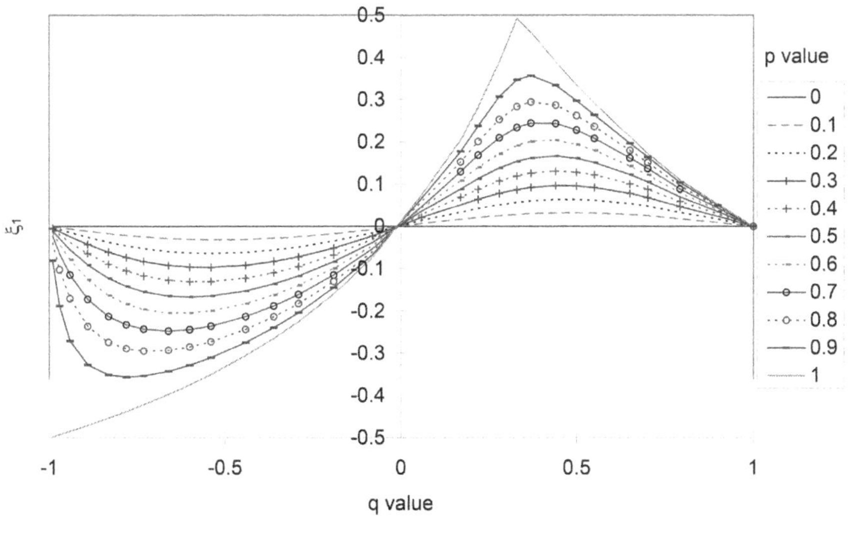

Figure 2.15 ξ_1 for different p and q.

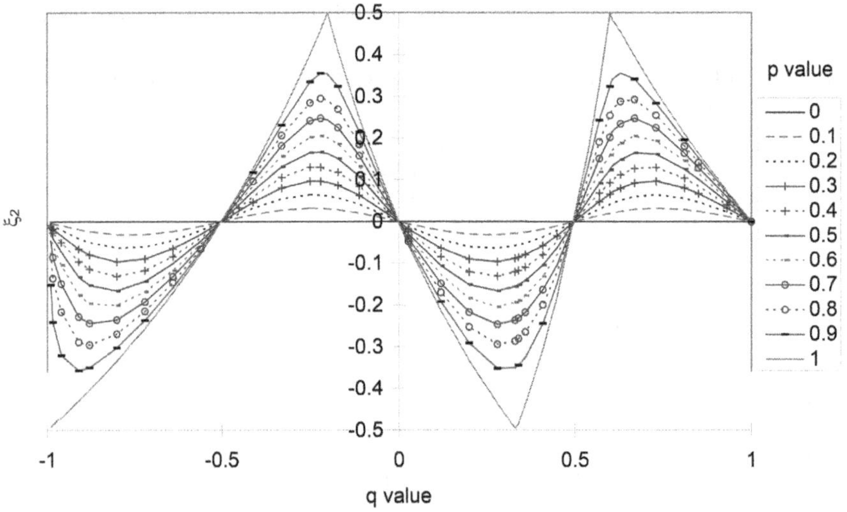

Figure 2.16 ξ_2 for different p and q.

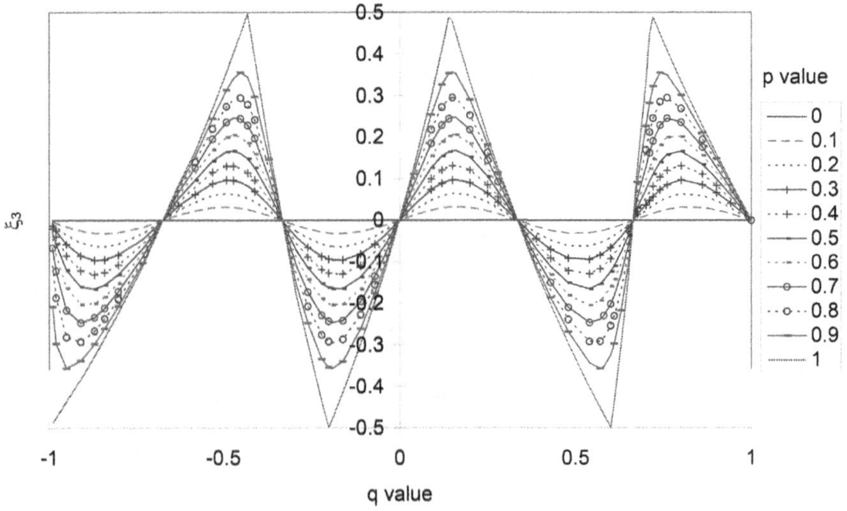

Figure 2.17 ξ_3 for different p and q.

Table 2.7 ξ_l values for various q (left column) and p (top row)

q	p=0.1	0.2	0.3	0.4	0.5	0.6	0.7	0.8	0.9	1
-0.990	-0.00111	-0.00249	-0.00427	-0.00662	-0.00989	-0.01476	-0.02274	-0.03821	-0.08065	-0.49749
-0.970	-0.00332	-0.00742	-0.01265	-0.01950	-0.02889	-0.04251	-0.06397	-0.10238	-0.18800	-0.49239
-0.940	-0.00657	-0.01462	-0.02471	-0.03769	-0.05497	-0.07899	-0.11435	-0.17058	-0.27041	-0.48454
-0.890	-0.01176	-0.02586	-0.04303	-0.06429	-0.09117	-0.12598	-0.17233	-0.23616	-0.32810	-0.47090
-0.830	-0.01744	-0.03772	-0.06149	-0.08960	-0.12317	-0.16366	-0.21309	-0.27433	-0.35183	-0.45355
-0.780	-0.02157	-0.04600	-0.07376	-0.10545	-0.14178	-0.18365	-0.23221	-0.28903	-0.35642	-0.43820
-0.730	-0.02507	-0.05272	-0.08326	-0.11704	-0.15447	-0.19609	-0.24254	-0.29470	-0.35386	-0.42197
-0.660	-0.02880	-0.05948	-0.09218	-0.12704	-0.16426	-0.20408	-0.24680	-0.29286	-0.34284	-0.39759
-0.600	-0.03084	-0.06285	-0.09610	-0.13067	-0.16667	-0.20423	-0.24356	-0.28489	-0.32855	-0.37500
-0.540	-0.03180	-0.06408	-0.09691	-0.13036	-0.16451	-0.19948	-0.23540	-0.27243	-0.31076	-0.35065
-0.440	-0.03103	-0.06168	-0.09205	-0.12226	-0.15240	-0.18257	-0.21287	-0.24339	-0.27425	-0.30556
-0.360	-0.02839	-0.05606	-0.08315	-0.10977	-0.13603	-0.16200	-0.18779	-0.21345	-0.23907	-0.26471
-0.290	-0.02473	-0.04871	-0.07207	-0.09489	-0.11727	-0.13928	-0.16099	-0.18244	-0.20370	-0.22481
-0.190	-0.01762	-0.03475	-0.05145	-0.06776	-0.08373	-0.09940	-0.11480	-0.12996	-0.14491	-0.15966
-0.080	-0.00786	-0.01560	-0.02323	-0.03076	-0.03819	-0.04553	-0.05279	-0.05996	-0.06705	-0.07407
0.059	0.00699	0.01408	0.02128	0.02859	0.03602	0.04358	0.04700	0.05911	0.06711	0.07527
0.170	0.01645	0.03344	0.05105	0.06938	0.08854	0.10869	0.13005	0.15288	0.17760	0.20482
0.220	0.02065	0.04211	0.06453	0.08810	0.11308	0.13982	0.16886	0.20103	0.23776	0.28205
0.280	0.02499	0.05104	0.07839	0.10734	0.13836	0.17210	0.20965	0.25301	0.30675	0.38889
0.330	0.02789	0.05692	0.08735	0.11956	0.15408	0.19174	0.23390	0.28326	0.34675	0.49254
0.370	0.02967	0.06042	0.09250	0.12628	0.16226	0.20122	0.24442	0.29423	0.35642	0.45985
0.440	0.03155	0.06377	0.09684	0.13099	0.16651	0.20380	0.24342	0.28625	0.33376	0.38889
0.500	0.03184	0.06377	0.09585	0.12819	0.16086	0.19397	0.22764	0.26198	0.29715	0.33333
0.550	0.03118	0.06185	0.09204	0.12177	0.15105	0.17988	0.20825	0.23613	0.26350	0.29032
0.650	0.02754	0.05349	0.07792	0.10090	0.12251	0.14280	0.16184	0.17970	0.19644	0.21212
0.700	0.02472	0.04750	0.06849	0.08784	0.10568	0.12215	0.13736	0.15142	0.16443	0.17300
0.793	0.01756	0.03307	0.04684	0.05911	0.07010	0.07999	0.08892	0.09701	0.10438	0.11111
0.900	0.00906	0.01679	0.02345	0.02925	0.03434	0.03885	0.04286	0.04646	0.04970	0.05263

Table 2.8 ξ_2 values for various q (left column) and p (top row)

q	p=0.1	0.2	0.3	0.4	0.5	0.6	0.7	0.8	0.9	1
-0.990	-0.00222	-0.00498	-0.00852	-0.01321	-0.01973	-0.02937	-0.04508	-0.07498	-0.15128	-0.49246
-0.960	-0.00873	-0.01943	-0.03283	-0.05006	-0.07287	-0.10421	-0.14912	-0.21636	-0.31966	-0.46939
-0.910	-0.01846	-0.04015	-0.06581	-0.09636	-0.13284	-0.17641	-0.22808	-0.28833	-0.35637	-0.42932
-0.880	-0.02326	-0.04973	-0.07986	-0.11408	-0.15271	-0.19595	-0.24364	-0.29517	-0.34932	-0.40426
-0.800	-0.03100	-0.06332	-0.09673	-0.13097	-0.16571	-0.20056	-0.23515	-0.26905	-0.30190	-0.33333
-0.720	-0.03082	-0.06068	-0.08946	-0.11709	-0.14347	-0.16858	-0.19237	-0.21484	-0.23598	-0.25581
-0.640	-0.02356	-0.04531	-0.06542	-0.08403	-0.10128	-0.11728	-0.13213	-0.14594	-0.15878	-0.17073
-0.560	-0.01114	-0.02123	-0.03040	-0.03879	-0.04647	-0.05354	-0.06007	-0.06611	-0.07171	-0.07692
-0.450	0.00943	0.01812	0.02616	0.03361	0.04054	0.04700	0.05303	0.05868	0.06398	0.06897
-0.410	0.01649	0.03193	0.04642	0.06007	0.07293	0.08509	0.09660	0.10750	0.11784	0.12766
-0.330	0.02748	0.05424	0.08041	0.10612	0.13146	0.15653	0.18141	0.20617	0.23089	0.25564
-0.250	0.03187	0.06401	0.09670	0.13026	0.16513	0.20185	0.24128	0.28482	0.33523	0.40000
-0.220	0.03144	0.06342	0.09624	0.13032	0.16617	0.20457	0.24677	0.29512	0.35537	0.45902
-0.200	0.03050	0.06165	0.09375	0.12300	0.16257	0.20064	0.24278	0.29156	0.35368	0.50000
-0.170	0.02816	0.05700	0.08679	0.11789	0.15076	0.18609	0.22499	0.26950	0.32437	0.40964
-0.110	0.02048	0.04139	0.06286	0.08501	0.10802	0.13209	0.15751	0.18469	0.21425	0.24719
-0.030	0.00598	0.01200	0.01806	0.02417	0.03032	0.03651	0.04276	0.04907	0.05543	0.06500
0.027	-0.00595	-0.01186	-0.01775	-0.02360	-0.02943	-0.03524	-0.04102	-0.04678	-0.05253	-0.05825
0.120	-0.02162	-0.04295	-0.06410	-0.08515	-0.10618	-0.12728	-0.14853	-0.17004	-0.19192	-0.21429
0.200	-0.03013	-0.06012	-0.09020	-0.12063	-0.15167	-0.18366	-0.21701	-0.25233	-0.29056	-0.33333
0.280	-0.03147	-0.06352	-0.09645	-0.13061	-0.16650	-0.20482	-0.24669	-0.29412	-0.35171	-0.43750
0.330	-0.02837	-0.05784	-0.08870	-0.12130	-0.15620	-0.19421	-0.23670	-0.28638	-0.35016	-0.49624
0.340	-0.02740	-0.05598	-0.08603	-0.11790	-0.15213	-0.18954	-0.23151	-0.28071	-0.34409	-0.48485
0.360	-0.02512	-0.05156	-0.07958	-0.10953	-0.14192	-0.17755	-0.21770	-0.26492	-0.32550	-0.43750
0.410	-0.01767	-0.03669	-0.05728	-0.07971	-0.10438	-0.13186	-0.16301	-0.19936	-0.24392	-0.30508
0.450	-0.01028	-0.02152	-0.03389	-0.04757	-0.06282	-0.07997	-0.10500	-0.12208	-0.14885	-0.18182
0.540	0.00835	0.01768	0.02818	0.04007	0.05369	0.06947	0.08802	0.11030	0.13790	0.17391
0.570	0.01429	0.03023	0.04813	0.06841	0.09164	0.11866	0.15079	0.19036	0.24250	0.32558
0.600	0.01966	0.04144	0.06571	0.09297	0.12392	0.15961	0.20177	0.25373	0.32394	0.50000
0.606	0.02129	0.04478	0.07085	0.10000	0.13290	0.17060	0.21479	0.26868	0.34034	0.48447
0.630	0.02425	0.05076	0.07987	0.11200	0.14775	0.18795	0.23396	0.28813	0.35561	0.45399
0.670	0.02881	0.05953	0.09227	0.12718	0.16442	0.20420	0.24677	0.29248	0.34176	0.39521
0.730	0.03184	0.06408	0.09653	0.12899	0.16124	0.19309	0.22430	0.25467	0.28401	0.31214
0.810	0.02870	0.05558	0.08062	0.10386	0.12535	0.14519	0.16345	0.18026	0.19572	0.20994
0.850	0.02450	0.04662	0.06659	0.08462	0.10089	0.11562	0.12896	0.14108	0.15210	0.16216
0.930	0.01250	0.02317	0.03237	0.04038	0.04741	0.05362	0.05914	0.06408	0.06852	0.07254

Table 2.9 ξ_3 values for various q (left column) and p (top row)

q	p=0.1	0.2	0.3	0.4	0.5	0.6	0.7	0.8	0.9	1
-0.990	-0.00332	-0.00746	-0.01275	-0.01976	-0.02945	-0.04372	-0.06668	-0.10912	-0.20731	-0.48744
-0.870	-0.03080	-0.06312	-0.09666	-0.13098	-0.16562	-0.20004	-0.23375	-0.26626	-0.29718	-0.32620
-0.800	-0.02950	-0.05740	-0.08362	-0.10815	-0.13100	-0.15221	-0.17186	-0.19001	-0.20677	-0.22222
-0.730	-0.01687	-0.03189	-0.04534	-0.05743	-0.06834	-0.07823	-0.08722	-0.09542	-0.10293	-0.10983
-0.628	0.01283	0.02437	0.03479	0.04424	0.05285	0.06073	0.06795	0.07460	0.08074	0.08642
-0.580	0.02232	0.04299	0.06217	0.08001	0.09662	0.11211	0.12657	0.14008	0.15272	0.16456
-0.490	0.03185	0.06409	0.09684	0.13021	0.16436	0.19948	0.23580	0.27363	0.31340	0.35570
-0.430	0.02585	0.05322	0.08241	0.11377	0.14788	0.18556	0.22827	0.27883	0.34452	0.49650
-0.410	0.02173	0.04494	0.06990	0.09693	0.12653	0.15945	0.19689	0.24113	0.29754	0.38983
-0.370	0.01117	0.02319	0.03617	0.05026	0.06563	0.08253	0.10128	0.12233	0.14639	0.17460
-0.293	-0.01299	-0.02673	-0.04131	-0.05685	-0.07349	-0.09142	-0.11089	-0.13227	-0.15606	-0.18310
-0.260	-0.02072	-0.04238	-0.06515	-0.08925	-0.11499	-0.14277	-0.17325	-0.20746	-0.24733	-0.29730
-0.220	-0.02822	-0.05727	-0.08740	-0.11899	-0.15253	-0.18879	-0.22898	-0.27544	-0.33392	-0.43590
-0.200	-0.03050	-0.06165	-0.09375	-0.12721	-0.16257	-0.20064	-0.24278	-0.29156	-0.35368	-0.50000
-0.160	-0.03178	-0.06380	-0.09637	-0.12985	-0.16472	-0.20163	-0.24158	-0.28629	-0.33952	-0.41379
-0.110	-0.02728	-0.05444	-0.08168	-0.10920	-0.13721	-0.16596	-0.19580	-0.22714	-0.26063	-0.29730
-0.040	-0.01168	-0.02328	-0.03483	-0.04634	-0.05783	-0.06930	-0.08078	-0.09228	-0.10381	-0.11538
0.040	0.01176	0.02363	0.03562	0.04776	0.06006	0.07255	0.08525	0.09821	0.11144	0.12500
0.090	0.02404	0.04851	0.07358	0.09945	0.12636	0.15466	0.18482	0.21755	0.25406	0.29670
0.140	0.03098	0.06247	0.09480	0.12837	0.16373	0.20169	0.24359	0.29199	0.35344	0.48837
0.200	0.03013	0.06012	0.09020	0.12063	0.15167	0.18366	0.21701	0.25233	0.29056	0.33333
0.280	0.01497	0.02926	0.04294	0.05605	0.06865	0.08076	0.09242	0.10367	0.11452	0.12500
0.390	-0.01568	-0.03039	-0.04423	-0.05728	-0.06961	-0.08128	-0.09234	-0.10283	-0.11281	-0.12230
0.480	-0.03100	-0.06154	-0.09170	-0.12156	-0.15117	-0.18060	-0.20989	-0.23908	-0.26821	-0.29730
0.600	-0.01966	-0.04144	-0.06571	-0.09297	-0.12392	-0.15961	-0.20177	-0.25373	-0.32394	-0.50000
0.610	-0.01710	-0.03621	-0.05770	-0.08209	-0.11009	-0.14276	-0.18181	-0.23052	-0.29696	-0.43590
0.630	-0.01146	-0.02446	-0.03934	-0.05653	-0.07667	-0.10500	-0.12986	-0.16680	-0.21670	-0.29730
0.650	-0.00532	-0.01143	-0.01850	-0.02680	-0.03667	-0.04861	-0.06339	-0.08221	-0.10722	-0.14286
0.676	0.00428	0.00922	0.01500	0.02185	0.03010	0.04021	0.05294	0.06947	0.09198	0.12500
0.700	0.01054	0.02269	0.03685	0.05355	0.07358	0.09806	0.12884	0.16923	0.22673	0.33333
0.720	0.01636	0.03506	0.05661	0.08169	0.11123	0.14662	0.19004	0.24553	0.32287	0.48837
0.760	0.02575	0.05413	0.08540	0.11983	0.15772	0.19936	0.24507	0.29513	0.34978	0.40909
0.800	0.03100	0.06332	0.09673	0.13097	0.16571	0.20056	0.23515	0.26905	0.30190	0.33333
0.860	0.03013	0.05872	0.08561	0.11073	0.13406	0.15561	0.17547	0.19372	0.21046	0.22581
0.940	0.01579	0.02937	0.04113	0.05141	0.06043	0.06842	0.07553	0.08000	0.08761	0.09278

For boundary conditions (2.17),

$$b_n = \frac{\frac{m_{v1}H_1}{\alpha \sin(\lambda_n \alpha)}\sigma_0 + \frac{m_{v1}H_1}{\alpha^2 \lambda_n}(\sigma_1 - \sigma_0) + \frac{m_{v2}H_2[\cos(\lambda_n \beta) - 1]}{\beta^2 \lambda_n \cos(\lambda_n \beta)}(\sigma_1 - \sigma_2)}{\frac{m_{v1}H_1}{2\sin^2(\lambda_n \alpha)} + \frac{m_{v2}H_2}{2\cos^2(\lambda_n \beta)}},$$

$$Z_n = \begin{cases} \dfrac{\sin\left(\lambda_n \alpha \frac{z}{H_1}\right)}{\sin(\lambda_n \alpha)} & 0 \le z \le H_1 \\ \dfrac{\cos\left(\lambda_n \beta \frac{H-z}{H_2}\right)}{\cos(\lambda_n \beta)} & H_1 \le z \le H \end{cases},$$

and the constant λ_n is the n-th positive root of the following equation

$$\cos\theta - p\cos(q\theta) = 0 \qquad (2.41)$$

Letting $\lambda_n = \pi\left(n - \frac{1}{2} + \varsigma_n\right)$, the first 3 ς_n values (n = 1, 2, 3) are shown in Figures 2.18–2.20 and tabulated in Tables 2.10–2.12.

With the constants λ_n and b_n known, (2.39) can be used to calculate the excess porewater pressure distribution varying with depth and time for the double-layered soil profile.

The average degree of consolidation U is calculated using (2.25) and (2.39). In the integration in (2.25), the compressibility m_v assumes m_{v1} for layer 1 and m_{v2} for layer 2, respectively, as shown in Figure 2.13.

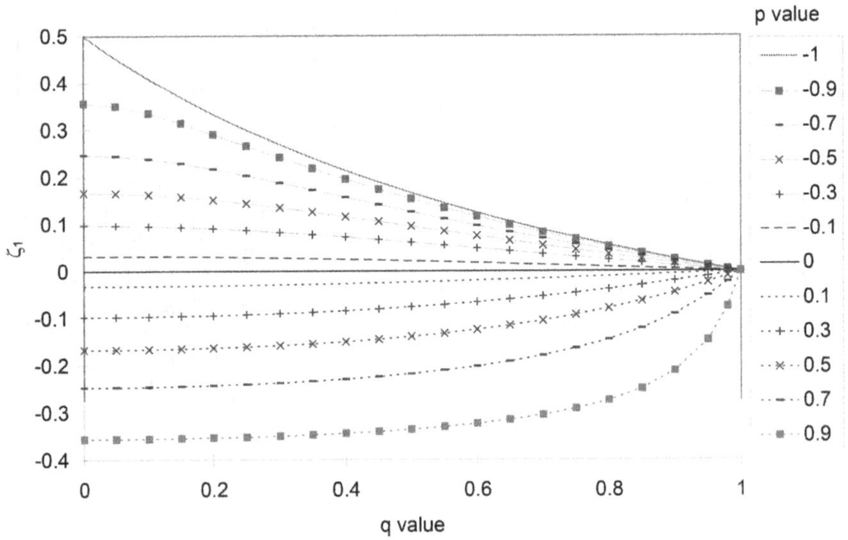

Figure 2.18 ς_1 for different p and q.

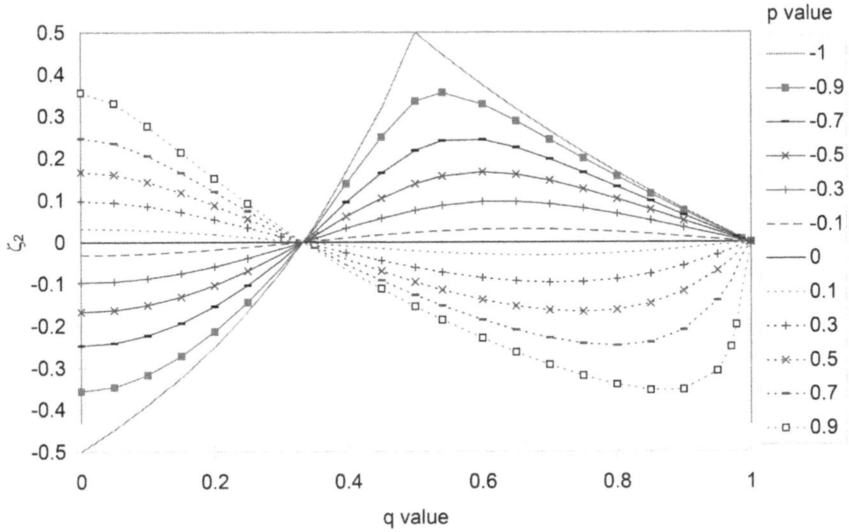

Figure 2.19 ς_2 for different p and q.

For boundary conditions (2.16),

$$U(T_v, T_c) = \min\left(1,\ \frac{T_v}{T_c}\right) - \sum_{n=1}^{\infty} \frac{2\left[\frac{m_{v1}H_1}{\alpha \sin(\lambda_n \alpha)} + \frac{m_{v2}H_2}{\beta \sin(\lambda_n \beta)}\right] T_n(T_v)}{\lambda_n \left[m_{v1}H_1(\sigma_0 + \sigma_1) + m_{v2}H_2(\sigma_1 + \sigma_2)\right]} \tag{2.42}$$

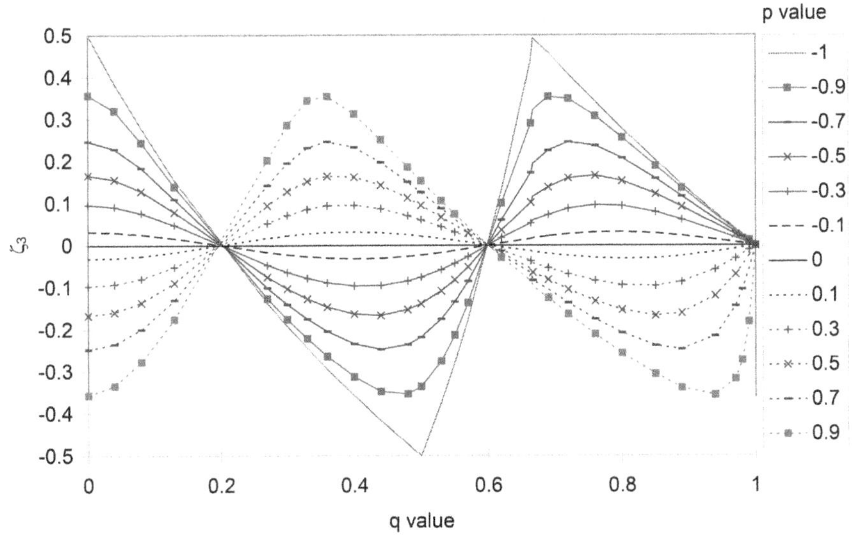

Figure 2.20 ς_3 for different p and q.

Table 2.10 ς_l values for various q (left column) and p (top row)

q	p= -1	-0.9	-0.8	-0.7	-0.6	-0.5	-0.4	-0.3	-0.2	-0.1	0.1	0.2	0.3	0.4	0.5	0.6	0.7	0.8	0.9	1
0.00	0.5000	0.3564	0.2952	0.2468	0.2048	0.1667	0.1310	0.0970	0.0641	0.0319	-0.0319	-0.0641	-0.0970	-0.1310	-0.1667	-0.2048	-0.2468	-0.2952	-0.3564	-0.5000
0.05	0.4524	0.3507	0.2919	0.2447	0.2034	0.1657	0.1303	0.0965	0.0638	0.0318	-0.0318	-0.0639	-0.0968	-0.1308	-0.1664	-0.2046	-0.2466	-0.2949	-0.3563	-0.5000
0.10	0.4091	0.3354	0.2827	0.2386	0.1991	0.1627	0.1283	0.0952	0.0631	0.0314	-0.0315	-0.0635	-0.0962	-0.1301	-0.1657	-0.2038	-0.2458	-0.2943	-0.3558	-0.5000
0.15	0.3696	0.3143	0.2690	0.2291	0.1924	0.1580	0.1250	0.0931	0.0618	0.0309	-0.0311	-0.0627	-0.0952	-0.1289	-0.1644	-0.2025	-0.2446	-0.2932	-0.3549	-0.5000
0.20	0.3333	0.2906	0.2523	0.2170	0.1837	0.1517	0.1206	0.0902	0.0601	0.0301	-0.0305	-0.0616	-0.0937	-0.1272	-0.1626	-0.2006	-0.2428	-0.2916	-0.3537	-0.5000
0.25	0.3000	0.2661	0.2340	0.2032	0.1733	0.1441	0.1152	0.0866	0.0580	0.0292	-0.0297	-0.0603	-0.0919	-0.1250	-0.1602	-0.1982	-0.2404	-0.2894	-0.3521	-0.5000
0.30	0.2692	0.2419	0.2150	0.1883	0.1619	0.1355	0.1090	0.0824	0.0554	0.0280	-0.0288	-0.0586	-0.0896	-0.1223	-0.1572	-0.1952	-0.2374	-0.2868	-0.3500	-0.5000
0.35	0.2407	0.2184	0.1958	0.1729	0.1497	0.1262	0.1022	0.0776	0.0525	0.0267	-0.0277	-0.0565	-0.0869	-0.1191	-0.1537	-0.1914	-0.2338	-0.2835	-0.3475	-0.5000
0.40	0.2143	0.1959	0.1769	0.1574	0.1372	0.1163	0.0948	0.0725	0.0493	0.0252	-0.0264	-0.0542	-0.0837	-0.1152	-0.1494	-0.1870	-0.2294	-0.2795	-0.3444	-0.5000
0.45	0.1897	0.1744	0.1585	0.1419	0.1244	0.1062	0.0870	0.0669	0.0458	0.0235	-0.0249	-0.0515	-0.0800	-0.1107	-0.1444	-0.1817	-0.2242	-0.2747	-0.3406	-0.5000
0.50	0.1667	0.1541	0.1407	0.1266	0.1117	0.0958	0.0790	0.0611	0.0421	0.0217	-0.0233	-0.0485	-0.0758	-0.1056	-0.1385	-0.1755	-0.2179	-0.2689	-0.3361	-0.5000
0.55	0.1452	0.1348	0.1237	0.1118	0.0991	0.0855	0.0708	0.0551	0.0382	0.0198	-0.0216	-0.0451	-0.0710	-0.0997	-0.1317	-0.1682	-0.2105	-0.2620	-0.3306	-0.5000
0.60	0.1250	0.1165	0.1073	0.0974	0.0867	0.0751	0.0626	0.0490	0.0341	0.0178	-0.0197	-0.0414	-0.0657	-0.0930	-0.1239	-0.1596	-0.2018	-0.2537	-0.3239	-0.5000
0.65	0.1061	0.0991	0.0916	0.0835	0.0746	0.0649	0.0544	0.0427	0.0299	0.0158	-0.0176	-0.0374	-0.0598	-0.0854	-0.1149	-0.1496	-0.1913	-0.2437	-0.3157	-0.5000
0.70	0.0882	0.0827	0.0766	0.0701	0.0628	0.0549	0.0462	0.0365	0.0257	0.0136	-0.0154	-0.0330	-0.0532	-0.0768	-0.1045	-0.1378	-0.1787	-0.2313	-0.3054	-0.5000
0.75	0.0714	0.0671	0.0624	0.0572	0.0514	0.0451	0.0381	0.0302	0.0214	0.0114	-0.0131	-0.0283	-0.0460	-0.0671	-0.0926	-0.1239	-0.1634	-0.2158	-0.2921	-0.5000
0.80	0.0556	0.0523	0.0487	0.0448	0.0404	0.0356	0.0301	0.0240	0.0171	0.0091	-0.0106	-0.0232	-0.0382	-0.0563	-0.0788	-0.1073	-0.1446	-0.1961	-0.2744	-0.5000
0.85	0.0405	0.0382	0.0357	0.0329	0.0297	0.0263	0.0223	0.0179	0.0127	0.0069	-0.0081	-0.0178	-0.0296	-0.0443	-0.0629	-0.0874	-0.1210	-0.1699	-0.2496	-0.5000
0.90	0.0263	0.0249	0.0232	0.0215	0.0195	0.0172	0.0147	0.0118	0.0085	0.0046	-0.0055	-0.0121	-0.0204	-0.0308	-0.0446	-0.0635	-0.0909	-0.1339	-0.2118	-0.5000
0.95	0.0128	0.0121	0.0114	0.0105	0.0096	0.0085	0.0072	0.0058	0.0042	0.0023	-0.0028	-0.0062	-0.0105	-0.0161	-0.0237	-0.0346	-0.0515	-0.0812	-0.1459	-0.5000
0.98	0.0051	0.0048	0.0045	0.0042	0.0038	0.0034	0.0029	0.0023	0.0017	0.0009	-0.0011	-0.0025	-0.0042	-0.0066	-0.0098	-0.0145	-0.0222	-0.0368	-0.0749	-0.5000

Table 2.11 ς_2 values for various q (left column) and p (top row)

q	p= -1	-0.9	-0.8	-0.7	-0.6	-0.5	-0.4	-0.3	-0.2	-0.1	0.1	0.2	0.3	0.4	0.5	0.6	0.7	0.8	0.9	1
0.000	-0.5000	-0.3564	-0.2952	-0.2468	-0.2048	-0.1667	-0.1310	-0.0970	-0.0641	-0.0319	0.0319	0.0641	0.0970	0.1310	0.1667	0.2048	0.2468	0.2952	0.3564	0.5000
0.050	-0.4474	-0.3460	-0.2876	-0.2408	-0.1999	-0.1626	-0.1278	-0.0946	-0.0624	-0.0310	0.0310	0.0621	0.0939	0.1265	0.1605	0.1965	0.2355	0.2792	0.3314	0.4048
0.100	-0.3889	-0.3166	-0.2654	-0.2229	-0.1852	-0.1506	-0.1182	-0.0873	-0.0575	-0.0285	0.0283	0.0565	0.0849	0.1137	0.1432	0.1737	0.2056	0.2395	0.2765	0.3182
0.150	-0.3235	-0.2716	-0.2298	-0.1936	-0.1610	-0.1308	-0.1025	-0.0755	-0.0496	-0.0245	0.0240	0.0476	0.0711	0.0944	0.1178	0.1412	0.1650	0.1891	0.2137	0.2391
0.200	-0.2500	-0.2135	-0.1819	-0.1536	-0.1277	-0.1035	-0.0808	-0.0593	-0.0388	-0.0190	0.0184	0.0363	0.0537	0.0708	0.0874	0.1038	0.1198	0.1357	0.1513	0.1667
0.250	-0.1667	-0.1435	-0.1226	-0.1035	-0.0858	-0.0694	-0.0539	-0.0394	-0.0256	-0.0125	0.0119	0.0233	0.0342	0.0447	0.0548	0.0645	0.0739	0.0829	0.0916	0.1000
0.300	-0.0714	-0.0615	-0.0525	-0.0441	-0.0364	-0.0293	-0.0226	-0.0164	-0.0106	-0.0051	0.0048	0.0094	0.0137	0.0178	0.0217	0.0254	0.0289	0.0322	0.0354	0.0385
0.350	0.0385	0.0328	0.0278	0.0232	0.0190	0.0151	0.0116	0.0084	0.0054	0.0026	-0.0024	-0.0047	-0.0068	-0.0088	-0.0106	-0.0124	-0.0141	-0.0156	-0.0171	-0.0185
0.397	0.1667	0.1388	0.1154	0.0950	0.0771	0.0610	0.0465	0.0333	0.0213	0.0102	-0.0095	-0.0183	-0.0265	-0.0342	-0.0414	-0.0481	-0.0545	-0.0604	-0.0661	-0.0714
0.450	0.3182	0.2498	0.2022	0.1642	0.1320	0.1040	0.0790	0.0565	0.0361	0.0173	-0.0160	-0.0309	-0.0448	-0.0578	-0.0699	-0.0813	-0.0921	-0.1022	-0.1117	-0.1207
0.500	0.5000	0.3361	0.2689	0.2179	0.1755	0.1385	0.1056	0.0758	0.0485	0.0233	-0.0217	-0.0421	-0.0611	-0.0790	-0.0958	-0.1117	-0.1266	-0.1407	-0.1541	-0.1667
0.540	0.4481	0.3564	0.2931	0.2414	0.1967	0.1568	0.1205	0.0871	0.0561	0.0272	-0.0256	-0.0497	-0.0724	-0.0940	-0.1144	-0.1337	-0.1520	-0.1693	-0.1857	-0.2013
0.600	0.3750	0.3286	0.2849	0.2436	0.2042	0.1667	0.1307	0.0961	0.0629	0.0308	-0.0297	-0.0583	-0.0859	-0.1124	-0.1379	-0.1624	-0.1858	-0.2082	-0.2296	-0.2500
0.650	0.3182	0.2878	0.2567	0.2251	0.1932	0.1610	0.1286	0.0963	0.0640	0.0319	-0.0316	-0.0627	-0.0934	-0.1235	-0.1530	-0.1817	-0.2097	-0.2367	-0.2628	-0.2879
0.700	0.2647	0.2435	0.2211	0.1974	0.1725	0.1464	0.1191	0.0907	0.0613	0.0311	-0.0318	-0.0641	-0.0968	-0.1298	-0.1630	-0.1960	-0.2288	-0.2611	-0.2928	-0.3235
0.750	0.2143	0.1992	0.1829	0.1654	0.1464	0.1260	0.1041	0.0806	0.0554	0.0286	-0.0302	-0.0621	-0.0956	-0.1305	-0.1667	-0.2039	-0.2420	-0.2805	-0.3190	-0.3571
0.800	0.1667	0.1560	0.1444	0.1317	0.1177	0.1025	0.0857	0.0672	0.0469	0.0246	-0.0270	-0.0566	-0.0889	-0.1241	-0.1622	-0.2032	-0.2468	-0.2927	-0.3404	-0.3889
0.850	0.1216	0.1144	0.1064	0.0977	0.0880	0.0772	0.0651	0.0517	0.0365	0.0194	-0.0221	-0.0473	-0.0761	-0.1092	-0.1470	-0.1901	-0.2390	-0.2938	-0.3542	-0.4189
0.900	0.0789	0.0745	0.0696	0.0641	0.0581	0.0512	0.0436	0.0348	0.0249	0.0134	-0.0157	-0.0343	-0.0567	-0.0838	-0.1170	-0.1582	-0.2096	-0.2738	-0.3531	-0.4474
0.950	0.0385	0.0364	0.0341	0.0315	0.0286	0.0254	0.0217	0.0174	0.0125	0.0068	-0.0082	-0.0182	-0.0308	-0.0469	-0.0682	-0.0975	-0.1398	-0.2042	-0.3076	-0.4744
0.970	0.0228	0.0216	0.0203	0.0188	0.0171	0.0151	0.0130	0.0104	0.0075	0.0041	-0.0050	-0.0111	-0.0189	-0.0291	-0.0429	-0.0628	-0.0934	-0.1458	-0.2494	-0.4848
0.978	0.0152	0.0143	0.0135	0.0125	0.0113	0.0101	0.0086	0.0070	0.0050	0.0027	-0.0033	-0.0074	-0.0127	-0.0196	-0.0292	-0.0431	-0.0653	-0.1057	-0.1974	-0.4899
0.986	0.0075	0.0071	0.0067	0.0062	0.0056	0.0050	0.0043	0.0035	0.0025	0.0014	-0.0017	-0.0037	-0.0064	-0.0099	-0.0148	-0.0221	-0.0340	-0.0569	-0.1177	-0.4950

Table 2.12 ς_3 values for various q (left column) and p (top row)

q	p= -1	-0.9	-0.8	-0.7	-0.6	-0.5	-0.4	-0.3	-0.2	-0.1	0.1	0.2	0.3	0.4	0.5	0.6	0.7	0.8	0.9	1
0.000	0.5000	0.3564	0.2952	0.2468	0.2048	0.1667	0.1310	0.0970	0.0641	0.0319	-0.0319	-0.0641	-0.0970	-0.1310	-0.1667	-0.2048	-0.2468	-0.2952	-0.3564	-0.5000
0.040	0.3846	0.3199	0.2709	0.2291	0.1915	0.1566	0.1235	0.0917	0.0608	0.0303	-0.0304	-0.0611	-0.0925	-0.1249	-0.1588	-0.1950	-0.2345	-0.2793	-0.3339	-0.4167
0.080	0.2778	0.2443	0.2133	0.1841	0.1562	0.1292	0.1028	0.0769	0.0512	0.0257	-0.0259	-0.0522	-0.0792	-0.1070	-0.1360	-0.1666	-0.1994	-0.2354	-0.2763	-0.3261
0.130	0.1549	0.1400	0.1250	0.1099	0.0948	0.0795	0.0641	0.0485	0.0326	0.0165	-0.0168	-0.0341	-0.0518	-0.0702	-0.0892	-0.1091	-0.1299	-0.1520	-0.1756	-0.2011
0.270	-0.1378	-0.1260	-0.1139	-0.1014	-0.0884	-0.0751	-0.0612	-0.0468	-0.0318	-0.0163	0.0170	0.0349	0.0538	0.0739	0.0953	0.1183	0.1434	0.1712	0.2027	0.2397
0.300	-0.1923	-0.1752	-0.1578	-0.1399	-0.1217	-0.1029	-0.0836	-0.0638	-0.0433	-0.0221	0.0230	0.0471	0.0726	0.0997	0.1288	0.1606	0.1960	0.2366	0.2863	0.3571
0.330	-0.2444	-0.2212	-0.1980	-0.1745	-0.1509	-0.1270	-0.1027	-0.0779	-0.0526	-0.0267	0.0276	0.0564	0.0866	0.1186	0.1529	0.1904	0.2324	0.2815	0.3448	0.4851
0.360	-0.2941	-0.2635	-0.2336	-0.2042	-0.1752	-0.1463	-0.1175	-0.0886	-0.0594	-0.0300	0.0306	0.0622	0.0949	0.1290	0.1652	0.2040	0.2466	0.2951	0.3540	0.4412
0.400	-0.3571	-0.3125	-0.2718	-0.2337	-0.1976	-0.1628	-0.1292	-0.0963	-0.0639	-0.0319	0.0319	0.0639	0.0963	0.1292	0.1628	0.1976	0.2337	0.2718	0.3125	0.3571
0.440	-0.4167	-0.3476	-0.2934	-0.2467	-0.2047	-0.1660	-0.1297	-0.0953	-0.0624	-0.0307	0.0299	0.0591	0.0877	0.1158	0.1435	0.1709	0.1979	0.2247	0.2514	0.2778
0.480	-0.4730	-0.3535	-0.2872	-0.2354	-0.1914	-0.1525	-0.1172	-0.0848	-0.0547	-0.0265	0.0251	0.0488	0.0714	0.0929	0.1134	0.1329	0.1516	0.1694	0.1864	0.2027
0.500	-0.5000	-0.3361	-0.2689	-0.2179	-0.1755	-0.1385	-0.1056	-0.0758	-0.0485	-0.0233	0.0217	0.0421	0.0611	0.0790	0.0958	0.1117	0.1266	0.1407	0.1541	0.1667
0.530	-0.3723	-0.2755	-0.2182	-0.1746	-0.1388	-0.1082	-0.0815	-0.0578	-0.0366	-0.0174	0.0159	0.0305	0.0439	0.0564	0.0678	0.0785	0.0884	0.0977	0.1063	0.1144
0.550	-0.2778	-0.2131	-0.1688	-0.1343	-0.1059	-0.0819	-0.0613	-0.0431	-0.0271	-0.0128	0.0116	0.0221	0.0317	0.0405	0.0486	0.0560	0.0629	0.0692	0.0751	0.0806
0.570	-0.1744	-0.1364	-0.1080	-0.0855	-0.0670	-0.0515	-0.0382	-0.0267	-0.0167	-0.0079	0.0070	0.0134	0.0191	0.0243	0.0291	0.0334	0.0375	0.0412	0.0446	0.0478
0.620	0.1316	0.1009	0.0784	0.0611	0.0473	0.0359	0.0264	0.0183	0.0113	0.0053	-0.0047	-0.0089	-0.0126	-0.0160	-0.0191	-0.0218	-0.0244	-0.0267	-0.0289	-0.0309
0.664	0.4412	0.2904	0.2218	0.1726	0.1340	0.1022	0.0755	0.0526	0.0327	0.0154	-0.0136	-0.0259	-0.0368	-0.0467	-0.0557	-0.0639	-0.0650	-0.0782	-0.0845	-0.0904
0.667	0.4940	0.3218	0.2479	0.1941	0.1514	0.1161	0.0860	0.0601	0.0375	0.0176	-0.0157	-0.0298	-0.0425	-0.0540	-0.0645	-0.0740	-0.0827	-0.0907	-0.0980	-0.1048
0.690	0.4586	0.3533	0.2821	0.2259	0.1791	0.1390	0.1041	0.0734	0.0461	0.0218	-0.0196	-0.0374	-0.0535	-0.0681	-0.0814	-0.0936	-0.1047	-0.1150	-0.1244	-0.1331
0.720	0.4070	0.3483	0.2949	0.2460	0.2013	0.1602	0.1226	0.0879	0.0561	0.0269	-0.0247	-0.0474	-0.0682	-0.0874	-0.1050	-0.1212	-0.1362	-0.1500	-0.1627	-0.1744
0.760	0.3409	0.3072	0.2724	0.2371	0.2015	0.1660	0.1310	0.0966	0.0632	0.0310	-0.0295	-0.0576	-0.0841	-0.1090	-0.1324	-0.1542	-0.1745	-0.1934	-0.2110	-0.2273
0.800	0.2778	0.2559	0.2325	0.2076	0.1812	0.1533	0.1243	0.0942	0.0633	0.0318	-0.0318	-0.0633	-0.0942	-0.1243	-0.1533	-0.1812	-0.2076	-0.2325	-0.2559	-0.2778
0.850	0.2027	0.1894	0.1749	0.1590	0.1416	0.1226	0.1019	0.0794	0.0549	0.0284	-0.0303	-0.0624	-0.0961	-0.1309	-0.1664	-0.2023	-0.2378	-0.2726	-0.3061	-0.3378
0.890	0.1455	0.1368	0.1273	0.1167	0.1050	0.0920	0.0775	0.0613	0.0432	0.0229	-0.0257	-0.0544	-0.0864	-0.1218	-0.1606	-0.2025	-0.2468	-0.2926	-0.3386	-0.3836
0.940	0.0773	0.0730	0.0683	0.0631	0.0572	0.0506	0.0431	0.0345	0.0247	0.0133	-0.0158	-0.0346	-0.0574	-0.0853	-0.1197	-0.1624	-0.2154	-0.2799	-0.3556	-0.4381
0.970	0.0381	0.0360	0.0338	0.0312	0.0284	0.0252	0.0216	0.0174	0.0125	0.0068	-0.0082	-0.0183	-0.0310	-0.0474	-0.0693	-0.0998	-0.1440	-0.2115	-0.3172	-0.4695
0.980	0.0253	0.0239	0.0224	0.0208	0.0189	0.0168	0.0143	0.0116	0.0083	0.0045	-0.0055	-0.0123	-0.0210	-0.0324	-0.0479	-0.0702	-0.1046	-0.1628	-0.2727	-0.4798
0.990	0.0126	0.0119	0.0112	0.0103	0.0094	0.0084	0.0072	0.0058	0.0042	0.0023	-0.0028	-0.0062	-0.0106	-0.0165	-0.0246	-0.0366	-0.0560	-0.0924	-0.1812	-0.4899

For boundary conditions (2.17),

$$U(T_v,T_c)=\min\left(1,\frac{T_v}{T_c}\right)-\sum_{n=1}^{\infty}\frac{2m_{v1}H_1T_n(T_v)}{\lambda_n\alpha\sin(\lambda_n\alpha)\left[m_{v1}H_1(\sigma_0+\sigma_1)+m_{v2}H_2(\sigma_1+\sigma_2)\right]} \quad (2.43)$$

If the vertical total stress increment is uniform with depth, the average degree of consolidation U can be simplified as follows.

$$U(T_v,T_c)=\begin{cases}\dfrac{T_v}{T_c}-\displaystyle\sum_{n=1}^{\infty}\frac{c_n}{\lambda_n^4T_c}\left[1-\exp(-\lambda_n^2T_v)\right] & T_v\le T_c\\ 1-\displaystyle\sum_{n=1}^{\infty}\frac{c_n}{\lambda_n^4T_c}\left[1-\exp(-\lambda_n^2T_c)\right]\exp\left[-\lambda_n^2(T_v-T_c)\right] & T_v>T_c\end{cases} \quad (2.44)$$

where

$$c_n=\frac{2[m_{v1}H_1\beta\sin(\lambda_n\beta)+m_{v2}H_2\alpha\sin(\lambda_n\alpha)]^2}{\alpha^2\beta^2(m_{v1}H_1+m_{v2}H_2)[m_{v1}H_1\sin^2(\lambda_n\beta)+m_{v2}H_2\sin^2(\lambda_n\alpha)]} \quad (2.44a)$$

for boundary conditions (2.16), and

$$c_n=\frac{2[m_{v1}H_1\cos(\lambda_n\beta)]^2}{\alpha^2(m_{v1}H_1+m_{v2}H_2)[m_{v1}H_1\cos^2(\lambda_n\beta)+m_{v2}H_2\sin^2(\lambda_n\alpha)]} \quad (2.44b)$$

for boundary conditions (2.17).

Considering the definitions of p and q given in (2.35–2.36), the following relationship exists:

$$\frac{m_{v1}H_1}{m_{v2}H_2}=\frac{(1+q)(1-p)}{(1-q)(1+p)}, \quad (2.45)$$

The parameters α and β are related to q only. Thus, the average degree of consolidation for a double-layered soil profile depends only on the parameter p, q, T_v, and the loading condition.

2.4.2.1 Convergence of the series

As in the one-layer situation, the series for calculating the average degree of consolidation converges very rapidly. It can be easily shown that the coefficient in the general term in (2.42) and (2.43) is less than $\frac{c}{\lambda_n^4}$, where c is a constant and λ_n is of the order of n. Taking into consideration the exponent decaying parts in (2.42) and (2.43), we can say that for application purposes, 3 terms of the series at most will produce sufficient accuracy for the calculation of the average degree of consolidation.

2.4.2.2 Solution charts and tables for abruptly applied depth-independent loading

The average degree of consolidation of double soil layers is determined by the compressibility, permeability, and thickness of the material comprising each layer and the loading condition. For abruptly applied depth-independent loading ($T_c = 0$), the influences of compressibility, permeability, and thickness are included in the parameters p and q and the normalized time T_v as expressed in (2.44). In this section, (2.44) is used to prepare the solution charts and tables for the case of suddenly applied depth-independent loading ($T_c = 0$) in order to reduce the number of charts and tables prepared. The effect of construction time can be estimated from the approximation relation in (2.46).

$$U(T_v, T_c) = \begin{cases} \dfrac{1}{T_c}\displaystyle\int_0^{T_v} U(T_v, 0)dT_v \cong \dfrac{T_v}{T_c} U\left(\dfrac{T_v}{2}, 0\right) & T_v \le T_c \\ \dfrac{1}{T_c}\displaystyle\int_{T_v - T_c}^{T_v} U(T_v, 0)dT_v \cong U\left(T_v - \dfrac{T_c}{2}, 0\right) & T_v > T_c \end{cases} \tag{2.46}$$

To facilitate preparing the solution charts and tables, the influence of p and q on the average degree of consolidation is first examined. From the definitions of p and q in (2.35–2.36), it is clear that $|p| < 1$ and $|q| < 1$. For example, if $m_{v2} = m_{v1}$ and $k_2 = 100k_1$, then $p = 9/11 \cong 0.818$.

For the two-way drainage condition (2.16), it is clear that $p \ge 0$ can be satisfied by choosing the coordinate direction. When $p = 0$ or $q = 0$, the relationship between the average degree of consolidation and normalized time is the same as that for the single-layer situation even though the two layers may be dissimilar. For other cases, the relationship is different. Figure 2.21 shows typical results of the effect of p on the average degree of consolidation for constant q. As shown in Figure 2.21, the curves are somewhat different from those for the single layer while clear two-step consolidation is observed in some cases. For example, the elapsed times required to reach an average degree of consolidation of 10, 50, 90, and 95% for the curve of $q = 0.8$ and $p = 0.90$ are $Tv_{10} = 0.000154$, $Tv_{50} = 0.004$, $Tv_{90} = 0.0853$, and $Tv_{95} = 0.143$, respectively. For a single soil layer, the corresponding times are 0.00196, 0.0492, 0.212, and 0.282, respectively (a factor of 4 should be applied in comparison with the results in soil mechanics textbook since the thickness is not divided by 2). The time ratios of T_{v90}/T_{v50} for the single and double soil layers are 4.31 and 21.33, respectively. Several time ratios are listed in Table 2.13 for the single layer and two double layer cases for comparison. It can also be seen in Figure 2.21 that the difference in average degree of consolidation between the single layer ($q = 0$ line) and double layer can be as high as 60% even for the same normalized time. The average degree of consolidation monotonically decreases with the increase of p for $q < 0$ and constant

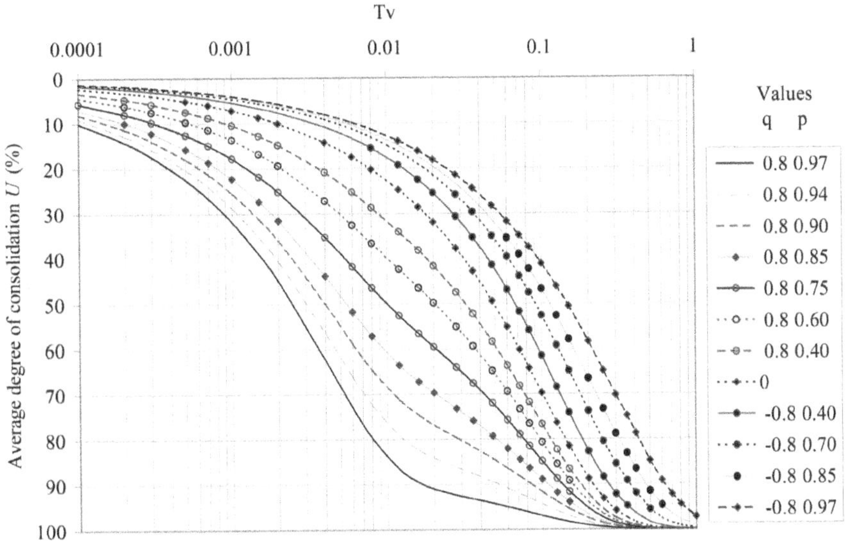

Figure 2.21 U vs. T_v for different p for two-way drainage.

time and monotonically increases for $q > 0$. However, the difference from the single layer situation ($p = 0$) is normally within 10% if $p \leq 0.4$.

The influence of q on the average degree of consolidation is illustrated in Figures 2.22 and 2.23. Obviously, the curve of average degree of consolidation vs. normalized time oscillates with the change of q. The oscillation is also intensified by larger p. Clear trends are not observed in Figure 2.22 although the curve is above the one for a single layer ($q = 0$) for $q < 0$ and below it for $q > 0$. It is favorable to express the relation between the average degree of consolidation and normalized time in the form of Figure 2.23 since the oscillation due to q can be clearly determined and the interpolation between charts is ensured because of the monotonicity with p. The isochrones in Figure 2.23 are more or less like *sine* functions.

For the one-way drainage condition (2.17), the effect of p on the average degree of consolidation is illustrated in Figure 2.24. Similar to the two-way drainage situation, the curves are somewhat different from those for the single layer ($p = 0$ line) while clear two-step consolidation is observed in some cases

Table 2.13 Consolidation time for single and double soil layers with two-way drainage

Soil layer	T_{v10}	T_{v50}	T_{v90}	T_{v95}	$\frac{T_{v50}}{T_{v10}}$	$\frac{T_{v90}}{T_{v50}}$	$\frac{T_{v95}}{T_{v50}}$
Single	1.96×10^{-3}	0.0492	0.212	0.282	25.10	4.31	5.73
Double $q = 0.8, p = 0.9$	1.54×10^{-4}	0.0040	0.0853	0.143	25.97	21.33	35.75
Double $q = -0.8, p = 0.85$	5.5×10^{-3}	0.113	0.463	0.614	20.55	4.10	5.43

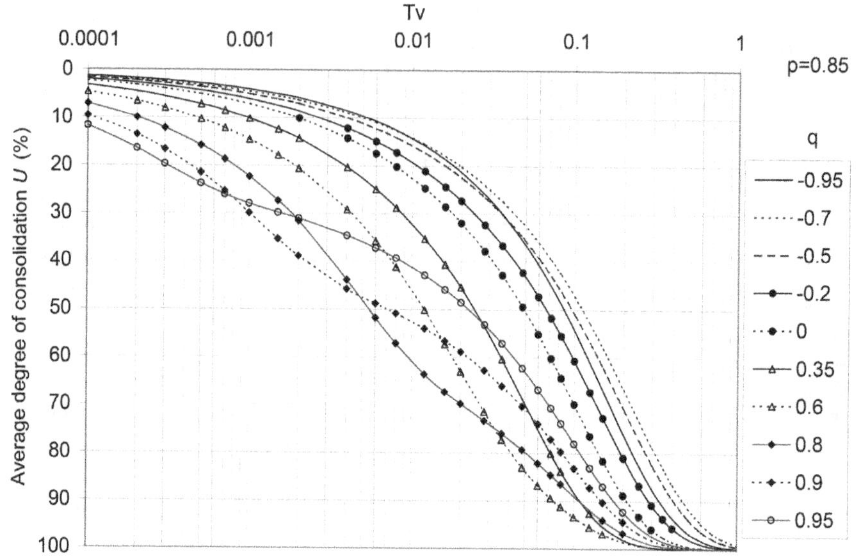

Figure 2.22 U vs. T_v for different *q* for two-way drainage.

as shown in Figure 2.24. The elapsed times required to reach an average degree of consolidation of 10, 50, 90, and 95% for the curve of $q = -0.5$ and $p = -0.82$ are $Tv_{10} = 8.25 \times 10^{-4}$, $Tv_{50} = 0.021$, $Tv_{90} = 0.224$, and $Tv_{95} = 0.392$, respectively. For single soil layer, the corresponding times are 7.84×10^{-3}, 0.197, 0.848, and 1.13, respectively. The time ratios of T_{v90}/T_{v50} for the single and

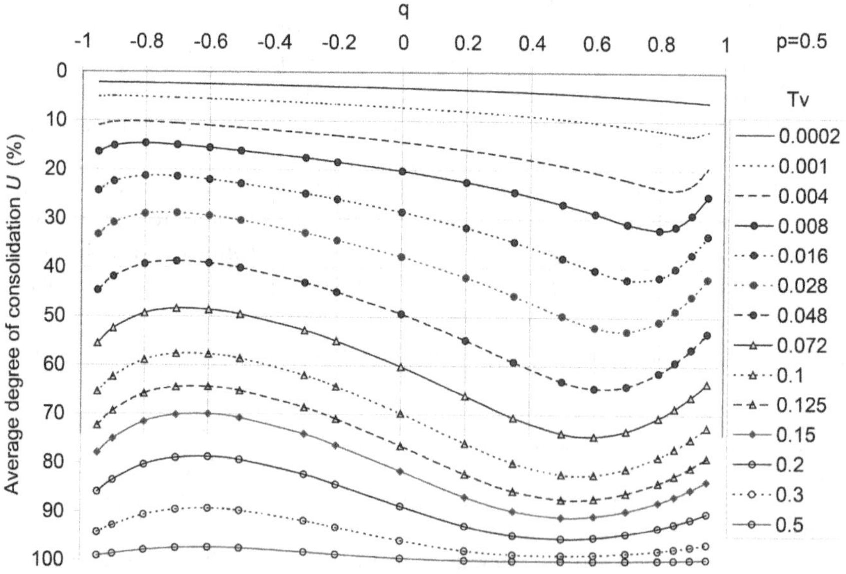

Figure 2.23 Isochrones of U vs. *q* with $p = 0.5$ for two-way drainage.

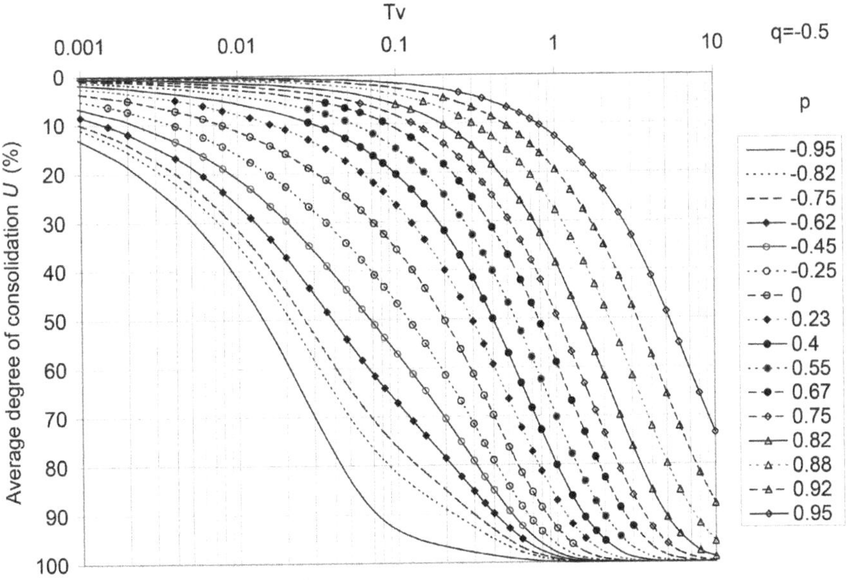

Figure 2.24 U vs. T_v for different p for one-way drainage.

double soil layers are 4.31 and 10.67, respectively. For the case of $q = -0.5$ and $p = 0.82$, the time ratio of T_{v50}/T_{v10} is 7.37, which is much smaller than 25.1 for the single layer. Several other time ratios are listed in Table 2.14 for comparison. Figure 2.24 also shows that the difference in average degree of consolidation between the single layer ($p = 0$ line) and double layer can be as high as 80% even for same normalized time. In addition, the average degree of consolidation monotonically decreases with the increase of p for constant q and time as shown in Figure 2.24. Thus, the curves of average degree of consolidation vs. time are above the one for single layer ($p = 0$) for $p > 0$ and below it for $p < 0$.

The influence of q on the average degree of consolidation for one-way drainage is like the situation in two-way drainage. As in the two-way drainage, the curve of average degree of consolidation vs. normalized time oscillates with the change of q. The oscillation is also intensified by larger p. However, the average degree of consolidation monotonically decreases with q for $q > 0$, $p < 0$, and constant time, and monotonically increases with q for $q > 0$, $p > 0$, and constant time.

Table 2.14 Consolidation time for single and double soil layers with one-way drainage

Soil layer	T_{v10}	T_{v50}	T_{v90}	T_{v95}	$\frac{T_{v50}}{T_{v10}}$	$\frac{T_{v90}}{T_{v50}}$	$\frac{T_{v95}}{T_{v50}}$
Single	7.84×10^{-3}	0.197	0.848	1.13	25.10	4.31	5.73
Double $q = -0.5, p = -0.82$	8.25×10^{-4}	0.021	0.224	0.392	25.45	10.67	18.67
Double $q = -0.5, p = 0.82$	0.194	1.43	4.82	6.28	7.37	3.37	4.39

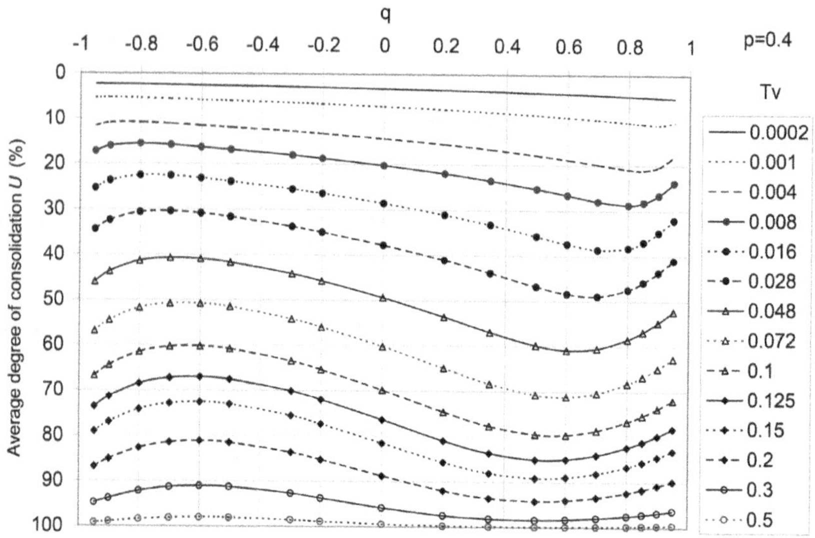

Figure 2.25 (a) Isochrones of U vs. q for two-way drainage for $p = 0.4$.

Considering the properties discussed above, charts that express the relationship of U-p-q-T_v are prepared using (2.44) for abruptly applied loading ($T_c = 0$). For a two-way drainage condition (2.16), the average degree of consolidation is shown in Figure 2.25 for $p = 0.4$, 0.6, 0.75, 0.85, 0.9, and 0.95. For a one-way drainage condition (2.17), the average degree

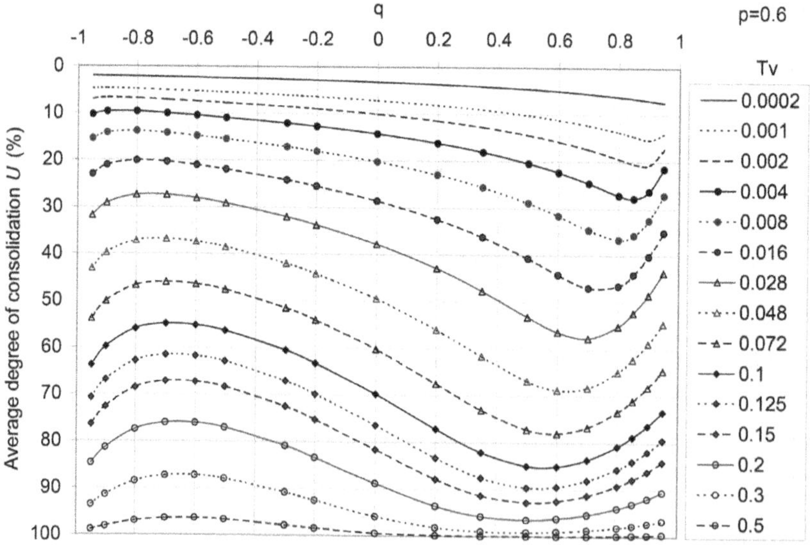

Figure 2.25 (b) Isochrones of U vs. q for two-way drainage for $p = 0.6$.

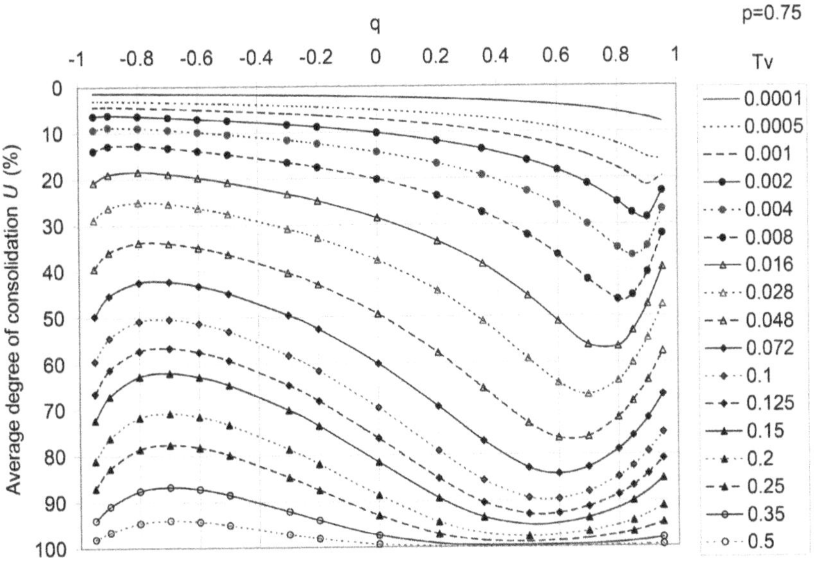

Figure 2.25 (c) Isochrones of U vs. q for two-way drainage for $p = 0.75$.

of consolidation is shown in Figure 2.26 for $p = \pm0.3$, ±0.6, ±0.8, ±0.9, and ±0.95. For $p = 0$, the solution is the same as that for the single layer and shown in Figure 2.27. The solution for single layer is also implied in Figure 2.25 since the intersection of isochrones and $q = 0$ corresponds the average degree of consolidation.

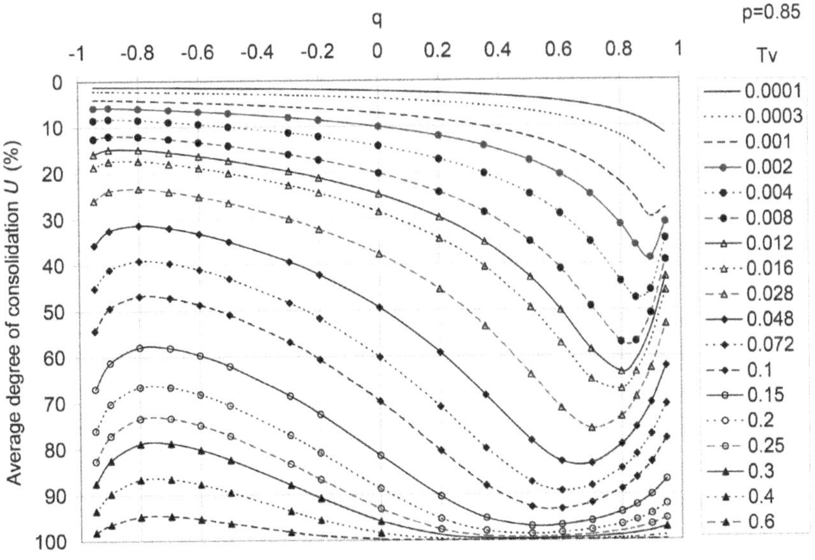

Figure 2.25 (d) Isochrones of U vs. q for two-way drainage for $p = 0.85$.

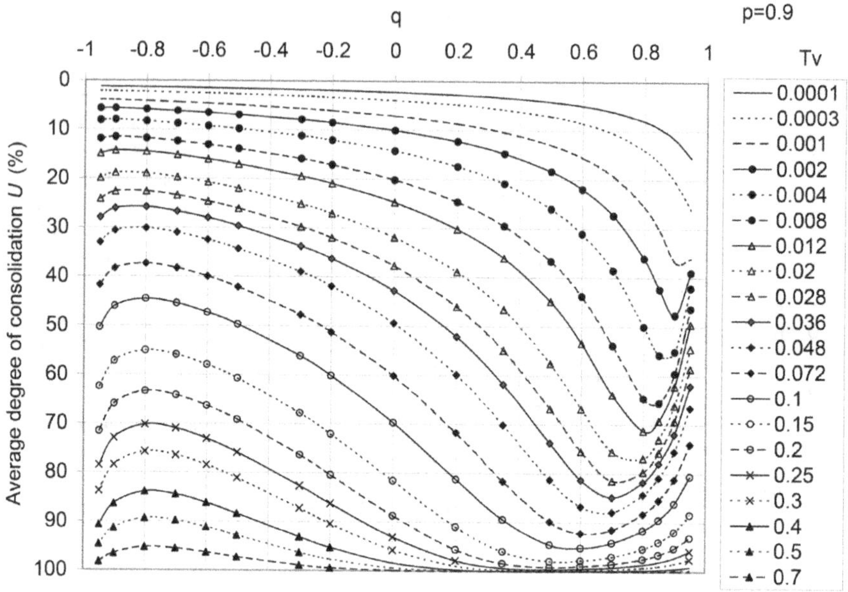

Figure 2.25 (e) Isochrones of U vs. q for two-way drainage for $p = 0.9$.

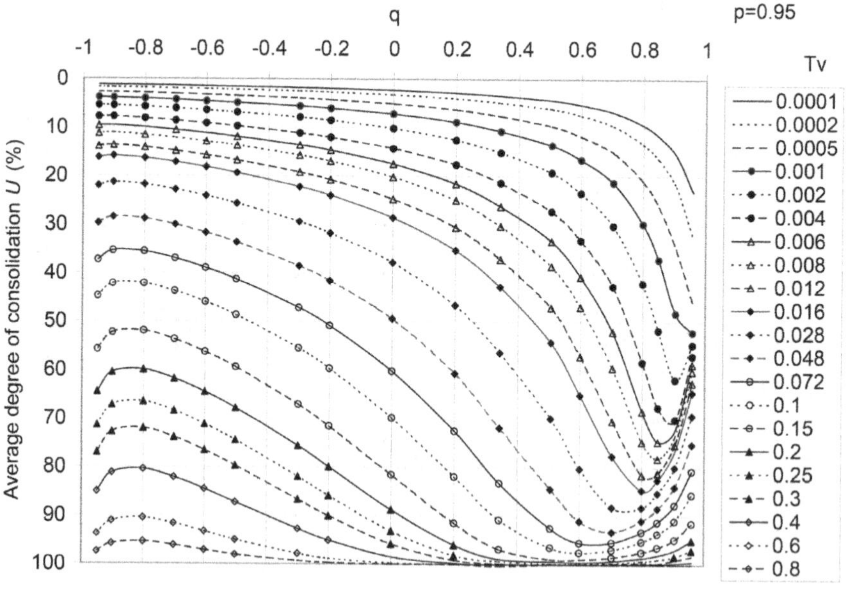

Figure 2.25 (f) Isochrones of U vs. q for two-way drainage for $p = 0.95$.

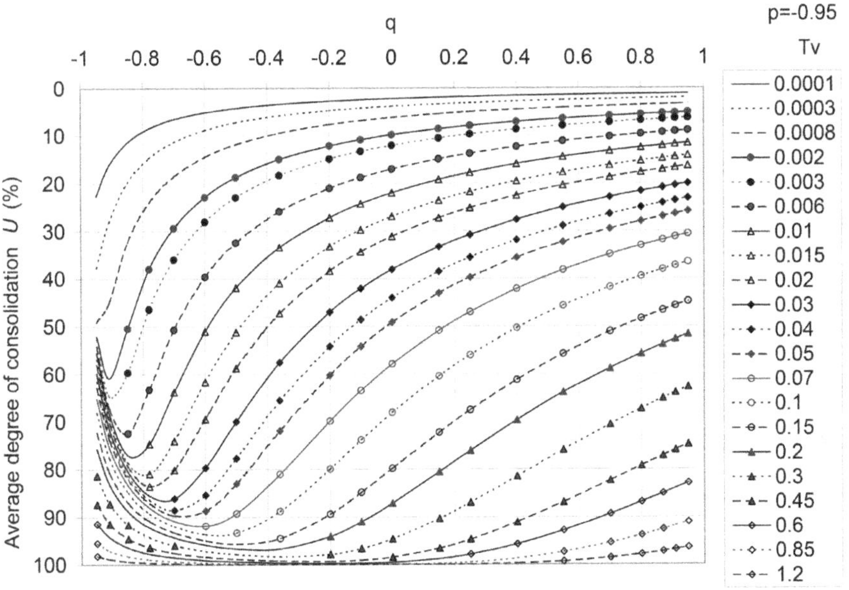

Figure 2.26 (a) Isochrones of U vs. q for one-way drainage for $p = -0.95$.

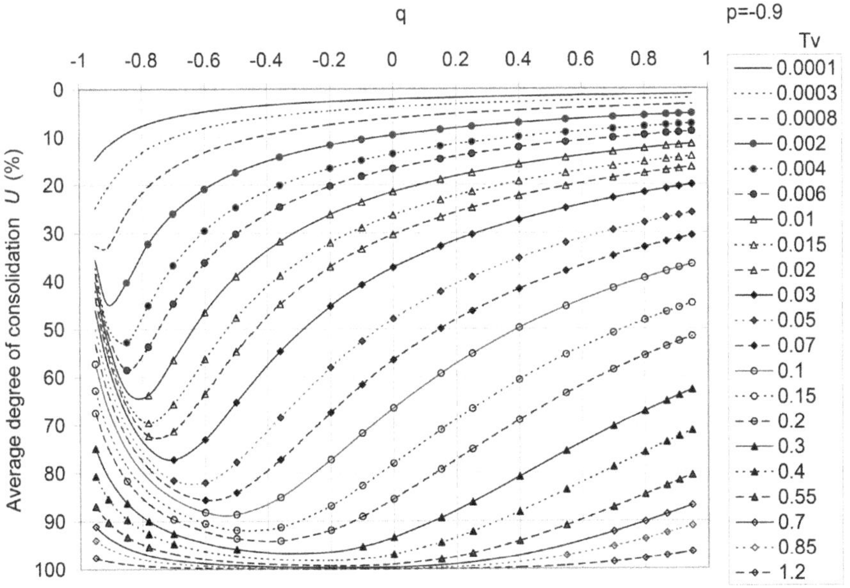

Figure 2.26 (b) Isochrones of U vs. q for one-way drainage for $p = -0.9$.

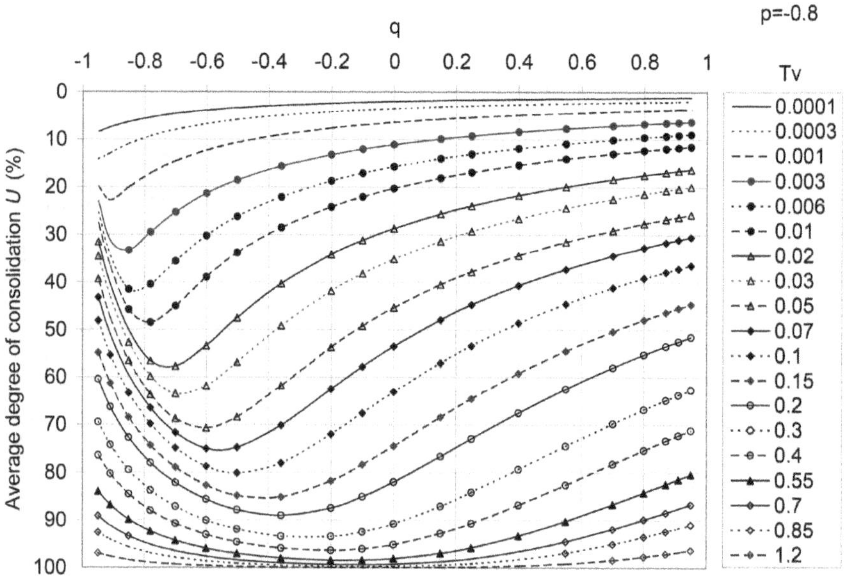

Figure 2.26 (c) Isochrones of U vs. q for one-way drainage for $p = -0.8$.

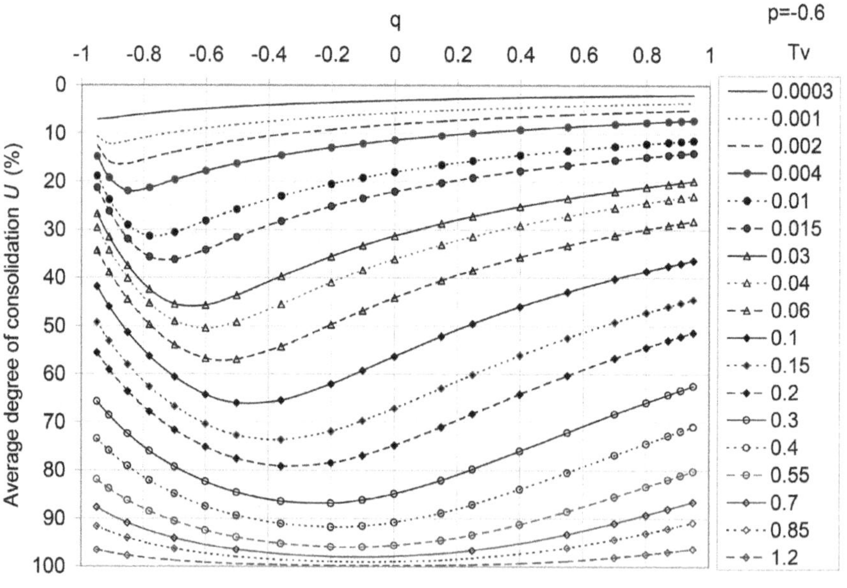

Figure 2.26 (d) Isochrones of U vs. q for one-way drainage for $p = -0.6$.

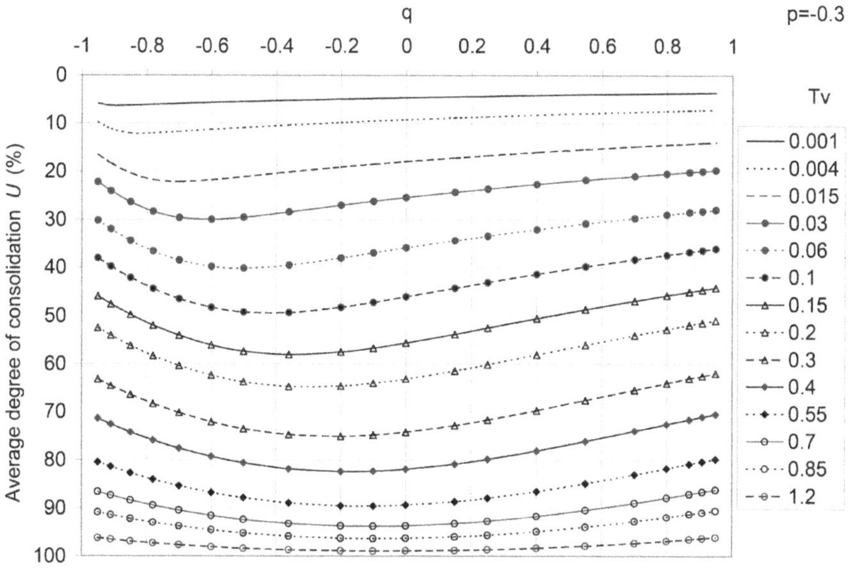

Figure 2.26 (e) Isochrones of U vs. q for one-way drainage for $p = -0.3$.

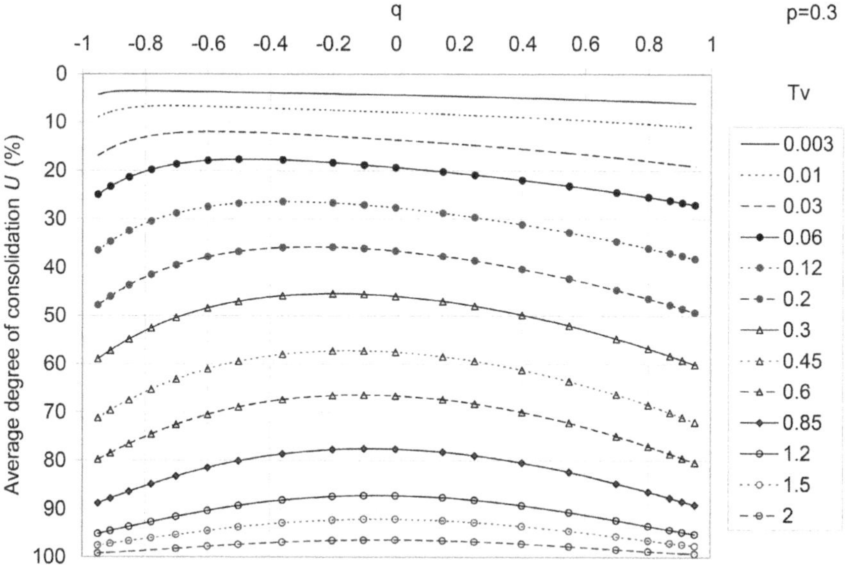

Figure 2.26 (f) Isochrones of U vs. q for one-way drainage for $p = 0.3$.

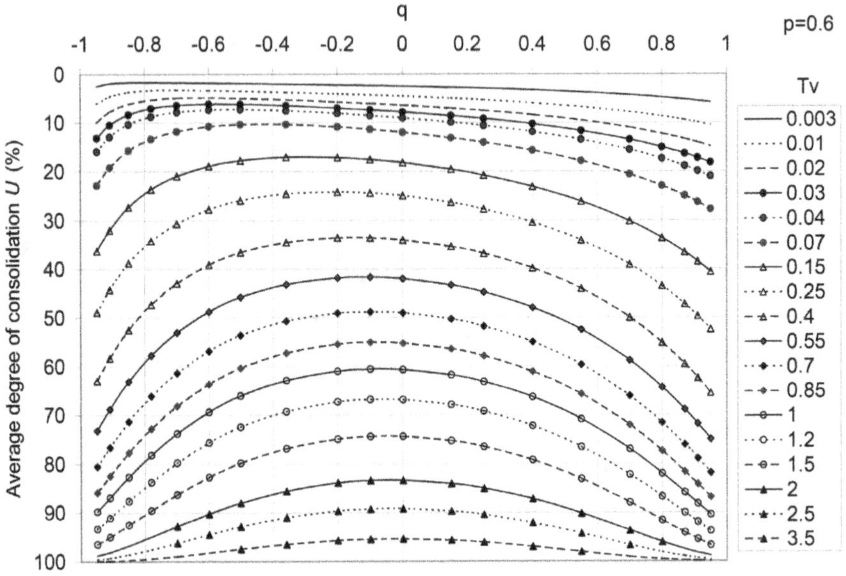

Figure 2.26 (g) Isochrones of U vs. q for one-way drainage for $p = 0.6$.

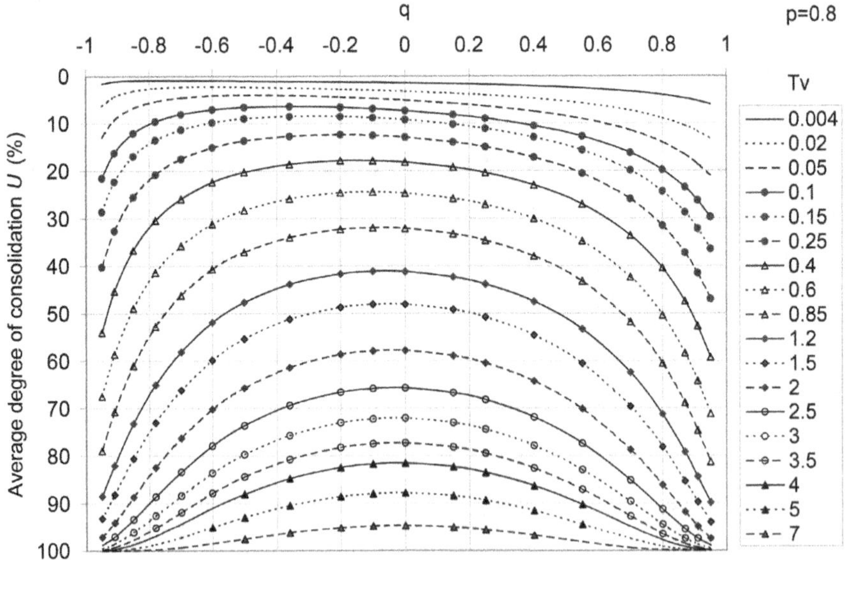

Figure 2.26 (h) Isochrones of U vs. q for one-way drainage for $p = 0.8$.

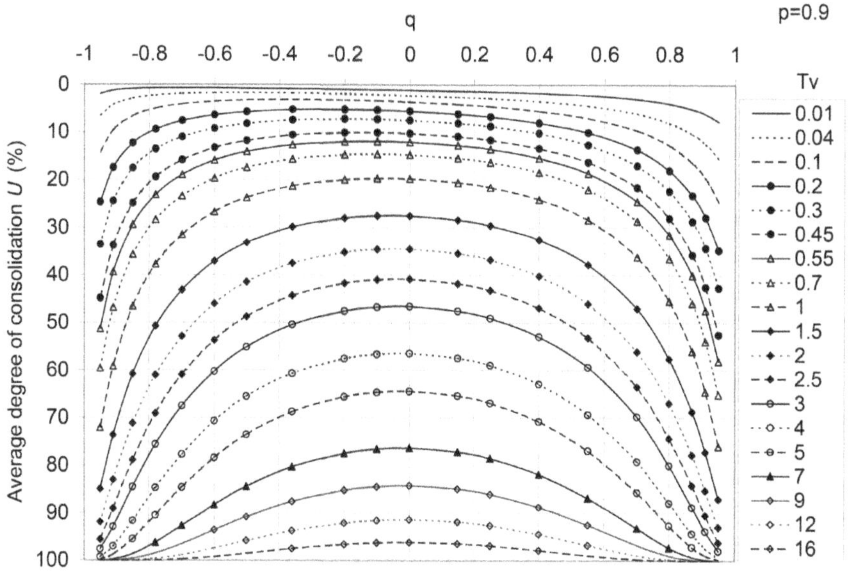

Figure 2.26 (i) Isochrones of U vs. q for one-way drainage for $p = 0.9$.

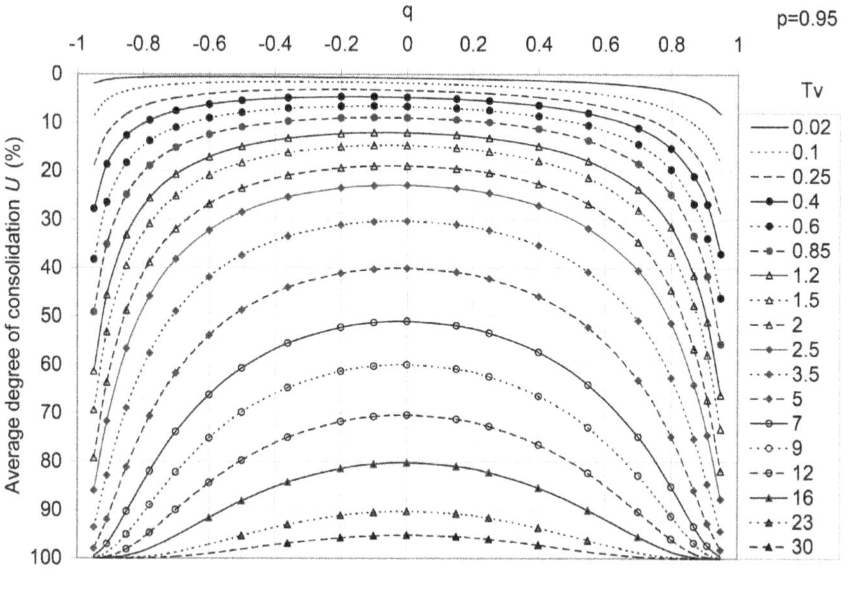

Figure 2.26 (j) Isochrones of U vs. q for one-way drainage for $p = 0.95$.

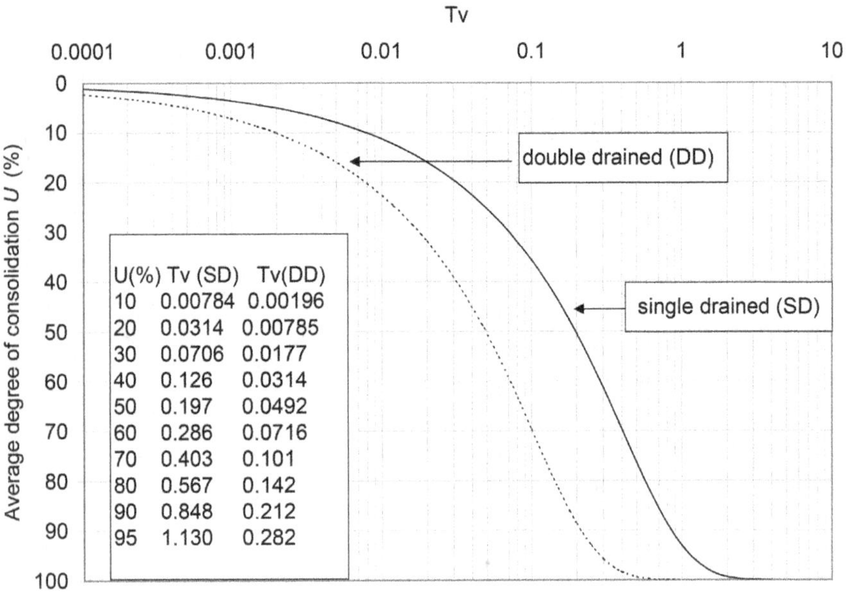

Figure 2.27 U vs. T_v for $p = 0$ for single and double drainage.

The elapsed times for achieving an average degree of consolidation of 10, 20, 30, 40, 50, 60, 70, 80, 90, and 95 for two-way drainage and one-way drainage are listed in Table 2.15 and Table 2.16, respectively.

2.4.2.3 Example and discussion

Example 2.2: The soil strata are shown in Figure 2.13. Calculate the average degree of consolidation for the following three cases:

a. $c_{v1} = 2.22 \times 10^{-7}$ m^2/s, $m_{v1} = 2.87 \times 10^{-4}$ m^2/kN, $H_1 = 2$ m, $c_{v2} = 3.4 \times 10^{-7}$ m^2/s, $m_{v2} = 6.6 \times 10^{-4}$ m^2/kN, and $H_2 = 1.5$ m. Determine the average degree of consolidation after a loading time of 60 days with two-way drainage.
b. $c_{v1} = 1.0 \times 10^{-7}$ m^2/s, $m_{v1} = 1.0 \times 10^{-4}$ m^2/kN, $H_1 = 2$ m, $c_{v2} = 1.0 \times 10^{-7}$ m^2/s, $m_{v2} = 10.0 \times 10^{-4}$ m^2/kN, and $H_2 = 1$ m. Determine the average degree of consolidation after a loading time of 365 days with one-way drainage.
c. $c_{v1} = 1.0 \times 10^{-7}$ m^2/s, $m_{v1} = 1.0 \times 10^{-4}$ m^2/kN, $H_1 = 2$ m, $c_{v2} = 10.0 \times 10^{-7}$ m^2/s, $m_{v2} = 1.0 \times 10^{-4}$ m^2/kN, and $H_2 = 1$ m. Determine the average degree of consolidation after a loading time of 365 days with one-way drainage.

Table 2.15 Elapsed times for certain average degree of consolidation for two-way drainage

	T_{v10}	T_{v20}	T_{v30}	T_{v40}	T_{v50}	T_{v60}	T_{v70}	T_{v80}	T_{v90}	T_{v95}
$pq = 0$	1.96E-3	7.85E-3	1.77E-2	3.14E-2	4.92E-2	7.16E-2	1.01E-1	1.42E-1	2.12E-1	2.82E-1
$p = 0.4$										
q	T_{v10}	T_{v20}	T_{v30}	T_{v40}	T_{v50}	T_{v60}	T_{v70}	T_{v80}	T_{v90}	T_{v95}
−0.95	3.04E-3	1.05E-2	2.18E-2	3.70E-2	5.62E-2	8.02E-2	1.11E-1	1.55E-1	2.30E-1	3.05E-1
−0.90	3.40E-3	1.19E-2	2.44E-2	4.09E-2	6.13E-2	8.68E-2	1.20E-1	1.66E-1	2.46E-1	3.25E-1
−0.80	3.41E-3	1.29E-2	2.69E-2	4.51E-2	6.75E-2	9.53E-2	1.31E-1	1.82E-1	2.68E-1	3.55E-1
−0.70	3.21E-3	1.27E-2	2.74E-2	4.64E-2	6.99E-2	9.90E-2	1.37E-1	1.90E-1	2.81E-1	3.72E-1
−0.60	3.02E-3	1.21E-2	2.66E-2	4.59E-2	6.98E-2	9.94E-2	1.38E-1	1.92E-1	2.85E-1	3.78E-1
−0.50	2.82E-3	1.13E-2	2.53E-2	4.42E-2	6.79E-2	9.74E-2	1.36E-1	1.90E-1	2.82E-1	3.74E-1
−0.30	2.46E-3	9.85E-3	2.22E-2	3.93E-2	6.13E-2	8.88E-2	1.24E-1	1.75E-1	2.61E-1	3.47E-1
−0.20	2.29E-3	9.16E-3	2.06E-2	3.66E-2	5.73E-2	8.33E-2	1.17E-1	1.65E-1	2.46E-1	3.27E-1
0.20	1.66E-3	6.65E-3	1.50E-2	2.66E-2	4.17E-2	6.08E-2	8.56E-2	1.21E-1	1.81E-1	2.41E-1
0.35	1.45E-3	5.81E-3	1.31E-2	2.33E-2	3.66E-2	5.37E-2	7.64E-2	1.09E-1	1.64E-1	2.20E-1
0.50	1.25E-3	5.03E-3	1.13E-2	2.02E-2	3.22E-2	4.83E-2	7.02E-2	1.02E-1	1.57E-1	2.12E-1
0.60	1.13E-3	4.53E-3	1.02E-2	1.85E-2	3.02E-2	4.64E-2	6.90E-2	1.02E-1	1.58E-1	2.15E-1
0.70	1.02E-3	4.07E-3	9.29E-3	1.74E-2	2.96E-2	4.70E-2	7.11E-2	1.06E-1	1.65E-1	2.24E-1
0.80	9.08E-4	3.66E-3	8.88E-3	1.80E-2	3.21E-2	5.14E-2	7.72E-2	1.14E-1	1.76E-1	2.39E-1
0.85	8.54E-4	3.57E-3	9.38E-3	1.97E-2	3.49E-2	5.51E-2	8.17E-2	1.19E-1	1.84E-1	2.48E-1
0.90	8.10E-4	3.86E-3	1.10E-2	2.26E-2	3.87E-2	5.97E-2	8.71E-2	1.26E-1	1.92E-1	2.59E-1
0.95	8.90E-4	5.23E-3	1.38E-2	2.65E-2	4.35E-2	6.52E-2	9.35E-2	1.33E-1	2.02E-1	2.70E-1
$p = 0.6$										
q	T_{v10}	T_{v20}	T_{v30}	T_{v40}	T_{v50}	T_{v60}	T_{v70}	T_{v80}	T_{v90}	T_{v95}
−0.95	3.77E-3	1.25E-2	2.53E-2	4.20E-2	6.29E-2	8.88E-2	1.22E-1	1.69E-1	2.50E-1	3.31E-1
−0.90	4.26E-3	1.46E-2	2.95E-2	4.85E-2	7.18E-2	1.01E-1	1.38E-1	1.91E-1	2.81E-1	3.71E-1

(*Continued*)

Table 2.15 (Continued) Elapsed times for certain average degree of consolidation for two-way drainage

	T_{v10}	T_{v20}	T_{v30}	T_{v40}	T_{v50}	T_{v60}	T_{v70}	T_{v80}	T_{v90}	T_{v95}
−0.80	4.26E-3	1.59E-2	3.29E-2	5.46E-2	8.13E-2	1.14E-1	1.57E-1	2.18E-1	3.21E-1	4.25E-1
−0.70	3.96E-3	1.55E-2	3.31E-2	5.59E-2	8.39E-2	1.19E-1	1.64E-1	2.29E-1	3.38E-1	4.49E-1
−0.60	3.63E-3	1.45E-2	3.17E-2	5.45E-2	8.27E-2	1.18E-1	1.64E-1	2.29E-1	3.40E-1	4.50E-1
−0.50	3.32E-3	1.33E-2	2.96E-2	5.16E-2	7.91E-2	1.13E-1	1.58E-1	2.21E-1	3.29E-1	4.38E-1
−0.30	2.73E-3	1.09E-2	2.46E-2	4.36E-2	6.79E-2	9.84E-2	1.38E-1	1.94E-1	2.89E-1	3.85E-1
−0.20	2.46E-3	9.85E-3	2.22E-2	3.94E-2	6.16E-2	8.95E-2	1.26E-1	1.77E-1	2.64E-1	3.52E-1
0.20	1.52E-3	6.08E-3	1.37E-2	2.43E-2	3.81E-2	5.56E-2	7.84E-2	1.11E-1	1.66E-1	2.21E-1
0.35	1.22E-3	4.90E-3	1.10E-2	1.96E-2	3.08E-2	4.53E-2	6.46E-2	9.24E-2	1.41E-1	1.89E-1
0.50	9.61E-4	3.85E-3	8.66E-3	1.54E-2	2.45E-2	3.69E-2	5.43E-2	8.07E-2	1.28E-1	1.77E-1
0.60	8.04E-4	3.22E-3	7.24E-3	1.30E-2	2.11E-2	3.28E-2	5.07E-2	7.89E-2	1.30E-1	1.82E-1
0.70	6.60E-4	2.64E-3	5.97E-3	1.10E-2	1.88E-2	3.16E-2	5.23E-2	8.42E-2	1.40E-1	1.96E-1
0.80	5.30E-4	2.13E-3	4.97E-3	1.02E-2	2.03E-2	3.72E-2	6.16E-2	9.67E-2	1.57E-1	2.17E-1
0.85	4.72E-4	1.91E-3	4.88E-3	1.16E-2	2.43E-2	4.32E-2	6.89E-2	1.05E-1	1.68E-1	2.31E-1
0.90	4.15E-4	1.84E-3	6.18E-3	1.58E-2	3.07E-2	5.09E-2	7.78E-2	1.16E-1	1.81E-1	2.46E-1
0.95	3.90E-4	3.19E-3	1.06E-2	2.25E-2	3.90E-2	6.04E-2	8.83E-2	1.28E-1	1.95E-1	2.63E-1
$p = 0.75$										
q	T_{v10}	T_{v20}	T_{v30}	T_{v40}	T_{v50}	T_{v60}	T_{v70}	T_{v80}	T_{v90}	T_{v95}
−0.95	4.55E-3	1.50E-2	3.00E-2	4.92E-2	7.27E-2	1.02E-1	1.40E-1	1.93E-1	2.84E-1	3.75E-1
−0.90	5.06E-3	1.75E-2	3.54E-2	5.80E-2	8.57E-2	1.20E-1	1.65E-1	2.27E-1	3.35E-1	4.42E-1
−0.80	4.99E-3	1.86E-2	3.89E-2	6.49E-2	9.70E-2	1.37E-1	1.89E-1	2.63E-1	3.89E-1	5.15E-1
−0.70	4.57E-3	1.79E-2	3.83E-2	6.51E-2	9.85E-2	1.40E-1	1.95E-1	2.72E-1	4.03E-1	5.35E-1
−0.60	4.13E-3	1.64E-2	3.61E-2	6.22E-2	9.51E-2	1.36E-1	1.90E-1	2.66E-1	3.95E-1	5.25E-1
−0.50	3.71E-3	1.48E-2	3.31E-2	5.78E-2	8.90E-2	1.28E-1	1.79E-1	2.51E-1	3.74E-1	4.97E-1
−0.30	2.94E-3	1.18E-2	2.65E-2	4.70E-2	7.32E-2	1.06E-1	1.49E-1	2.10E-1	3.13E-1	4.16E-1

–0.20	2.59E-3	1.04E-2	2.34E-2	4.15E-2	6.49E-2	9.44E-2	1.33E-1	1.87E-1	2.79E-1	3.71E-1
0.20	1.42E-3	5.67E-3	1.28E-2	2.27E-2	3.56E-2	5.18E-2	7.31E-2	1.03E-1	1.55E-1	2.06E-1
0.35	1.07E-3	4.27E-3	9.61E-3	1.71E-2	2.68E-2	3.94E-2	5.60E-2	8.01E-2	1.22E-1	1.65E-1
0.50	7.68E-4	3.07E-3	6.90E-3	1.23E-2	1.94E-2	2.89E-2	4.23E-2	6.32E-2	1.03E-1	1.46E-1
0.60	5.93E-4	2.38E-3	5.34E-3	9.54E-3	1.52E-2	2.32E-2	3.56E-2	5.72E-2	1.02E-1	1.50E-1
0.70	4.43E-4	1.77E-3	3.99E-3	7.18E-3	1.18E-2	1.92E-2	3.30E-2	6.05E-2	1.13E-1	1.67E-1
0.80	3.15E-4	1.26E-3	2.85E-3	5.39E-3	1.01E-2	2.07E-2	4.23E-2	7.63E-2	1.35E-1	1.94E-1
0.85	2.58E-4	1.03E-3	2.41E-3	5.07E-3	1.20E-2	2.77E-2	5.23E-2	8.81E-2	1.50E-1	2.11E-1
0.90	2.07E-4	8.42E-4	2.32E-3	7.51E-3	1.98E-2	3.88E-2	6.51E-2	1.03E-1	1.67E-1	2.31E-1
0.95	1.63E-4	1.07E-3	6.23E-3	1.67E-2	3.23E-2	5.33E-2	8.11E-2	1.20E-1	1.88E-1	2.55E-1
$p = 0.85$										
q	T_{v10}	T_{v20}	T_{v30}	T_{v40}	T_{v50}	T_{v60}	T_{v70}	T_{v80}	T_{v90}	T_{v95}
–0.95	5.32E-3	1.79E-2	3.58E-2	5.85E-2	8.62E-2	1.21E-1	1.65E-1	2.28E-1	3.35E-1	4.42E-1
–0.90	5.74E-3	2.04E-2	4.17E-2	6.90E-2	1.03E-1	1.44E-1	1.98E-1	2.75E-1	4.06E-1	5.37E-1
–0.80	5.50E-3	2.09E-2	4.43E-2	7.49E-2	1.13E-1	1.61E-1	2.24E-1	3.12E-1	4.63E-1	6.14E-1
–0.70	4.99E-3	1.96E-2	4.26E-2	7.32E-2	1.12E-1	1.60E-1	2.23E-1	3.13E-1	4.65E-1	6.18E-1
–0.60	4.48E-3	1.78E-2	3.94E-2	6.85E-2	1.05E-1	1.52E-1	2.12E-1	2.98E-1	4.43E-1	5.90E-1
–0.50	3.99E-3	1.60E-2	3.56E-2	6.25E-2	9.66E-2	1.40E-1	1.96E-1	2.75E-1	4.10E-1	5.45E-1
–0.30	3.09E-3	1.24E-2	2.78E-2	4.94E-2	7.70E-2	1.12E-1	1.57E-1	2.21E-1	3.30E-1	4.40E-1
–0.20	2.68E-3	1.08E-2	2.42E-2	4.30E-2	6.72E-2	9.78E-2	1.37E-1	1.93E-1	2.89E-1	3.85E-1
0.20	1.35E-3	5.41E-3	1.22E-2	2.16E-2	3.39E-2	4.94E-2	6.96E-2	9.81E-2	1.47E-1	1.96E-1
0.35	9.70E-4	3.87E-3	8.72E-3	1.55E-2	2.43E-2	3.56E-2	5.04E-2	7.19E-2	1.09E-1	1.48E-1
0.50	6.49E-4	2.60E-3	5.85E-3	1.04E-2	1.63E-2	2.41E-2	3.48E-2	5.13E-2	8.33E-2	1.20E-1
0.60	4.72E-4	1.89E-3	4.24E-3	7.55E-3	1.19E-2	1.78E-2	2.64E-2	4.12E-2	7.57E-2	1.19E-1

(Continued)

Table 2.15 (Continued) Elapsed times for certain average degree of consolidation for two-way drainage

	T_{v10}	T_{v20}	T_{v30}	T_{v40}	T_{v50}	T_{v60}	T_{v70}	T_{v80}	T_{v90}	T_{v95}
0.70	3.22E-4	1.29E-3	2.90E-3	5.17E-3	8.27E-3	1.27E-2	2.02E-2	3.69E-2	8.42E-2	1.36E-1
0.80	2.01E-4	8.04E-4	1.81E-3	3.27E-3	5.48E-3	9.51E-3	2.06E-2	5.09E-2	1.09E-1	1.66E-1
0.85	1.51E-4	6.05E-4	1.37E-3	2.55E-3	4.69E-3	1.09E-2	3.07E-2	6.55E-2	1.26E-1	1.87E-1
0.90	1.09E-4	4.34E-4	1.01E-3	2.20E-3	7.15E-3	2.23E-2	4.74E-2	8.46E-2	1.49E-1	2.12E-1
0.95	7.27E-5	3.11E-4	1.58E-3	8.73E-3	2.24E-2	4.26E-2	7.01E-2	1.09E-1	1.76E-1	2.43E-1
$p = 0.9$										
q	T_{v10}	T_{v20}	T_{v30}	T_{v40}	T_{v50}	T_{v60}	T_{v70}	T_{v80}	T_{v90}	T_{v95}
–0.95	5.83E-3	2.02E-2	4.07E-2	6.68E-2	9.88E-2	1.39E-1	1.90E-1	2.63E-1	3.88E-1	5.12E-1
–0.90	6.13E-3	2.23E-2	4.64E-2	7.75E-2	1.16E-1	1.64E-1	2.27E-1	3.16E-1	4.68E-1	6.21E-1
–0.80	5.78E-3	2.22E-2	4.77E-2	8.16E-2	1.24E-1	1.78E-1	2.48E-1	3.47E-1	5.16E-1	6.86E-1
–0.70	5.23E-3	2.06E-2	4.51E-2	7.81E-2	1.20E-1	1.73E-1	2.42E-1	3.39E-1	5.05E-1	6.72E-1
–0.60	4.66E-3	1.86E-2	4.12E-2	7.20E-2	1.11E-1	1.61E-1	2.26E-1	3.17E-1	4.73E-1	6.29E-1
–0.50	4.13E-3	1.65E-2	3.69E-2	6.50E-2	1.01E-1	1.46E-1	2.05E-1	2.88E-1	4.30E-1	5.73E-1
–0.30	3.16E-3	1.27E-2	2.85E-2	5.06E-2	7.90E-2	1.15E-1	1.61E-1	2.27E-1	3.39E-1	4.52E-1
–0.20	2.73E-3	1.09E-2	2.46E-2	4.37E-2	6.84E-2	9.95E-2	1.40E-1	1.97E-1	2.94E-1	3.92E-1
0.20	1.32E-3	5.28E-3	1.19E-2	2.11E-2	3.31E-2	4.82E-2	6.79E-2	9.56E-2	1.43E-1	1.91E-1
0.35	9.20E-4	3.69E-3	8.29E-3	1.47E-2	2.31E-2	3.37E-2	4.77E-2	6.78E-2	1.03E-1	1.39E-1
0.50	5.93E-4	2.38E-3	5.34E-3	9.50E-3	1.49E-2	2.19E-2	3.14E-2	4.56E-2	7.27E-2	1.04E-1
0.60	4.15E-4	1.66E-3	3.74E-3	6.65E-3	1.05E-2	1.55E-2	2.25E-2	3.39E-2	5.98E-2	9.67E-2
0.70	2.69E-4	1.07E-3	2.42E-3	4.31E-3	6.81E-3	1.02E-2	1.55E-2	2.55E-2	6.06E-2	1.11E-1
0.80	1.54E-4	6.16E-4	1.39E-3	2.48E-3	4.00E-3	6.33E-3	1.10E-2	2.91E-2	8.53E-2	1.43E-1
0.85	1.09E-4	4.34E-4	9.77E-4	1.76E-3	2.95E-3	5.17E-3	1.34E-2	4.43E-2	1.05E-1	1.65E-1
0.90	7.09E-5	2.84E-4	6.42E-4	1.22E-3	2.40E-3	8.67E-3	3.01E-2	6.65E-2	1.30E-1	1.94E-1
0.95	4.12E-5	1.67E-4	4.40E-4	2.56E-3	1.27E-2	3.12E-2	5.82E-2	9.73E-2	1.64E-1	2.31E-1

$p = 0.95$

q	T_{v10}	T_{v20}	T_{v30}	T_{v40}	T_{v50}	T_{v60}	T_{v70}	T_{v80}	T_{v90}	T_{v95}
−0.95	6.51E-3	2.36E-2	4.90E-2	8.17E-2	1.22E-1	1.73E-1	2.39E-1	3.33E-1	4.93E-1	6.53E-1
−0.90	6.55E-3	2.48E-2	5.31E-2	9.05E-2	1.38E-1	1.97E-1	2.74E-1	3.83E-1	5.70E-1	7.57E-1
−0.80	6.06E-3	2.37E-2	5.19E-2	9.02E-2	1.39E-1	2.01E-1	2.81E-1	3.94E-1	5.87E-1	7.81E-1
−0.70	5.44E-3	2.16E-2	4.79E-2	8.38E-2	1.30E-1	1.88E-1	2.64E-1	3.71E-1	5.53E-1	7.36E-1
−0.60	4.84E-3	1.93E-2	4.31E-2	7.60E-2	1.18E-1	1.71E-1	2.41E-1	3.38E-1	5.05E-1	6.72E-1
−0.50	4.26E-3	1.71E-2	3.83E-2	6.78E-2	1.06E-1	1.53E-1	2.15E-1	3.03E-1	4.53E-1	6.02E-1
−0.30	3.24E-3	1.30E-2	2.92E-2	5.18E-2	8.11E-2	1.18E-1	1.66E-1	2.33E-1	3.49E-1	4.64E-1
−0.20	2.78E-3	1.11E-2	2.50E-2	4.45E-2	6.96E-2	1.01E-1	1.42E-1	2.01E-1	3.00E-1	3.99E-1
0.20	1.29E-3	5.15E-3	1.16E-2	2.06E-2	3.23E-2	4.70E-2	6.61E-2	9.32E-2	1.39E-1	1.86E-1
0.35	8.75E-4	3.50E-3	7.87E-3	1.40E-2	2.19E-2	3.20E-2	4.51E-2	6.38E-2	9.62E-2	1.29E-1
0.50	5.42E-4	2.16E-3	4.87E-3	8.66E-3	1.36E-2	1.98E-2	2.81E-2	4.03E-2	6.25E-2	8.69E-2
0.60	3.63E-4	1.45E-3	3.27E-3	5.81E-3	9.12E-3	1.34E-2	1.91E-2	2.78E-2	4.51E-2	6.82E-2
0.70	2.20E-4	8.82E-4	1.98E-3	3.53E-3	5.54E-3	8.17E-3	1.18E-2	1.78E-2	3.28E-2	6.72E-2
0.80	1.13E-4	4.53E-4	1.02E-3	1.81E-3	2.86E-3	4.29E-3	6.45E-3	1.08E-2	4.05E-2	9.72E-2
0.85	7.27E-5	2.91E-4	6.55E-4	1.17E-3	1.86E-3	2.85E-3	4.54E-3	1.01E-2	6.12E-2	1.22E-1
0.90	4.12E-5	1.65E-4	3.71E-4	6.66E-4	1.09E-3	1.80E-3	3.98E-3	2.83E-2	9.09E-2	1.54E-1
0.95	1.87E-5	7.46E-5	1.70E-4	3.27E-4	7.36E-4	7.84E-3	3.03E-2	6.85E-2	1.35E-1	2.02E-1

Table 2.16 Elapsed times for certain average degree of consolidation for one-way drainage

	T_{v10}	T_{v20}	T_{v30}	T_{v40}	T_{v50}	T_{v60}	T_{v70}	T_{v80}	T_{v90}	T_{v95}
$p = 0$	7.84E-3	3.14E-2	7.06E-2	1.26E-1	1.97E-1	2.86E-1	4.03E-1	5.67E-1	8.48E-1	1.13
$p = -0.95$										
q	T_{v10}	T_{v20}	T_{v30}	T_{v40}	T_{v50}	T_{v60}	T_{v70}	T_{v80}	T_{v90}	T_{v95}
−0.95	1.96E-5	7.84E-5	1.78E-4	3.50E-4	9.39E-4	3.05E-2	1.20E-1	2.73E-1	5.40E-1	8.08E-1
−0.91	3.79E-5	1.51E-4	3.41E-4	6.15E-4	1.03E-3	1.88E-3	1.79E-2	1.36E-1	3.91E-1	6.48E-1
−0.85	7.66E-5	3.06E-4	6.88E-4	1.23E-3	1.96E-3	3.05E-3	5.09E-3	2.48E-2	2.38E-1	4.80E-1
−0.78	1.38E-4	5.54E-4	1.25E-3	2.22E-3	3.51E-3	5.27E-3	7.99E-3	1.40E-2	1.22E-1	3.44E-1
−0.70	2.31E-4	9.26E-4	2.08E-3	3.71E-3	5.83E-3	8.65E-3	1.27E-2	1.99E-2	5.42E-2	2.37E-1
−0.60	3.82E-4	1.53E-3	3.43E-3	6.11E-3	9.60E-3	1.41E-2	2.04E-2	3.05E-2	5.63E-2	1.52E-1
−0.50	5.70E-4	2.28E-3	5.12E-3	9.11E-3	1.43E-2	2.10E-2	3.00E-2	4.39E-2	7.38E-2	1.29E-1
−0.36	8.96E-4	3.58E-3	8.05E-3	1.43E-2	2.24E-2	3.28E-2	4.67E-2	6.73E-2	1.07E-1	1.58E-1
−0.20	1.35E-3	5.42E-3	1.22E-2	2.17E-2	3.40E-2	4.96E-2	7.02E-2	1.00E-1	1.55E-1	2.15E-1
0.00	2.07E-3	8.26E-3	1.86E-2	3.30E-2	5.18E-2	7.55E-2	1.07E-1	1.51E-1	2.29E-1	3.08E-1
0.25	3.16E-3	1.27E-2	2.85E-2	5.06E-2	7.93E-2	1.15E-1	1.63E-1	2.29E-1	3.44E-1	4.59E-1
0.50	4.49E-3	1.80E-2	4.05E-2	7.19E-2	1.13E-1	1.64E-1	2.31E-1	3.25E-1	4.86E-1	6.46E-1
0.70	5.74E-3	2.29E-2	5.15E-2	9.16E-2	1.43E-1	2.09E-1	2.94E-1	4.13E-1	6.19E-1	8.23E-1
0.80	6.40E-3	2.56E-2	5.76E-2	1.02E-1	1.60E-1	2.33E-1	3.28E-1	4.62E-1	6.91E-1	9.20E-1
0.87	6.89E-3	2.76E-2	6.20E-2	1.10E-1	1.73E-1	2.51E-1	3.53E-1	4.98E-1	7.44E-1	9.91E-1
0.95	7.48E-3	2.99E-2	6.73E-2	1.20E-1	1.87E-1	2.73E-1	3.83E-1	5.40E-1	8.07E-1	1.08
$p = -0.9$										
q	T_{v10}	T_{v20}	T_{v30}	T_{v40}	T_{v50}	T_{v60}	T_{v70}	T_{v80}	T_{v90}	T_{v95}
−0.95	4.57E-5	1.85E-4	5.33E-4	9.47E-3	5.02E-2	1.24E-1	2.32E-1	3.88E-1	6.56E-1	9.24E-1
−0.91	7.11E-5	2.85E-4	6.52E-4	1.33E-3	4.51E-3	4.47E-2	1.37E-1	2.85E-1	5.43E-1	8.00E-1
−0.85	1.20E-4	4.81E-4	1.08E-3	1.97E-3	3.41E-3	6.91E-3	4.48E-2	1.71E-1	4.13E-1	6.56E-1

–0.78	1.93E-4	7.72E-4	1.74E-3	3.11E-3	5.06E-3	8.18E-3	1.57E-2	8.27E-2	3.02E-1	5.27E-1
–0.70	2.97E-4	1.19E-3	2.68E-3	4.77E-3	7.61E-3	1.17E-2	1.87E-2	4.04E-2	2.11E-1	4.18E-1
–0.60	4.60E-4	1.84E-3	4.14E-3	7.37E-3	1.16E-2	1.75E-2	2.63E-2	4.37E-2	1.38E-1	3.20E-1
–0.50	6.57E-4	2.63E-3	5.92E-3	1.05E-2	1.66E-2	2.46E-2	3.61E-2	5.58E-2	1.16E-1	2.57E-1
–0.36	9.94E-4	3.98E-3	8.94E-3	1.59E-2	2.50E-2	3.68E-2	5.31E-2	7.87E-2	1.35E-1	2.25E-1
–0.20	1.46E-3	5.85E-3	1.32E-2	2.34E-2	3.67E-2	5.38E-2	7.68E-2	1.11E-1	1.78E-1	2.56E-1
0.00	2.18E-3	8.71E-3	1.96E-2	3.48E-2	5.46E-2	7.98E-2	1.13E-1	1.61E-1	2.47E-1	3.36E-1
0.25	3.26E-3	1.31E-2	2.94E-2	5.22E-2	8.18E-2	1.19E-1	1.68E-1	2.38E-1	3.57E-1	4.77E-1
0.50	4.57E-3	1.83E-2	4.12E-2	7.32E-2	1.15E-1	1.67E-1	2.35E-1	3.31E-1	4.95E-1	6.59E-1
0.70	5.79E-3	2.31E-2	5.20E-2	9.25E-2	1.45E-1	2.11E-1	2.96E-1	4.18E-1	6.25E-1	8.31E-1
0.80	6.44E-3	2.57E-2	5.80E-2	1.03E-1	1.61E-1	2.35E-1	3.30E-1	4.65E-1	6.95E-1	9.25E-1
0.87	6.92E-3	2.77E-2	6.22E-2	1.11E-1	1.73E-1	2.52E-1	3.55E-1	5.00E-1	7.47E-1	9.95E-1
0.95	7.48E-3	3.00E-2	6.73E-2	1.20E-1	1.88E-1	2.73E-1	3.84E-1	5.41E-1	8.08E-1	1.08
$p = -0.8$										
q	T_{v10}	T_{v20}	T_{v30}	T_{v40}	T_{v50}	T_{v60}	T_{v70}	T_{v80}	T_{v90}	T_{v95}
–0.95	1.40E-4	9.90E-4	1.55E-2	5.27E-2	1.13E-1	1.96E-1	3.06E-1	4.63E-1	7.32E-1	1.00
–0.91	1.79E-4	7.35E-4	2.38E-3	1.83E-2	6.36E-2	1.38E-1	2.43E-1	3.95E-1	6.54E-1	9.13E-1
–0.85	2.48E-4	9.93E-4	2.32E-3	5.15E-3	1.97E-2	7.31E-2	1.67E-1	3.09E-1	5.54E-1	7.99E-1
–0.78	3.43E-4	1.37E-3	3.10E-3	5.83E-3	1.11E-2	3.07E-2	1.02E-1	2.32E-1	4.60E-1	6.88E-1
–0.70	4.69E-4	1.87E-3	4.23E-3	7.66E-3	1.30E-2	2.32E-2	5.77E-2	1.66E-1	3.76E-1	5.87E-1
–0.60	6.55E-4	2.62E-3	5.89E-3	1.06E-2	1.71E-2	2.73E-2	4.77E-2	1.14E-1	2.99E-1	4.89E-1
–0.50	8.72E-4	3.49E-3	7.85E-3	1.40E-2	2.24E-2	3.43E-2	5.41E-2	9.91E-2	2.47E-1	4.18E-1
–0.36	1.23E-3	4.91E-3	1.11E-2	1.97E-2	3.11E-2	4.66E-2	6.97E-2	1.11E-1	2.17E-1	3.57E-1
–0.20	1.71E-3	6.84E-3	1.54E-2	2.74E-2	4.31E-2	6.38E-2	9.27E-2	1.39E-1	2.33E-1	3.45E-1

(Continued)

Table 2.16 (Continued) Elapsed times for certain average degree of consolidation for one-way drainage

	T_{v10}	T_{v20}	T_{v30}	T_{v40}	T_{v50}	T_{v60}	T_{v70}	T_{v80}	T_{v90}	T_{v95}
0.00	2.42E-3	9.69E-3	2.18E-2	3.88E-2	6.09E-2	8.94E-2	1.28E-1	1.85E-1	2.87E-1	3.95E-1
0.25	3.48E-3	1.40E-2	3.14E-2	5.58E-2	8.75E-2	1.28E-1	1.81E-1	2.57E-1	3.87E-1	5.17E-1
0.50	4.75E-3	1.90E-2	4.27E-2	7.60E-2	1.19E-1	1.73E-1	2.44E-1	3.44E-1	5.15E-1	6.87E-1
0.70	5.90E-3	2.36E-2	5.31E-2	9.44E-2	1.48E-1	2.15E-1	3.03E-1	4.26E-1	6.38E-1	8.49E-1
0.80	6.53E-3	2.61E-2	5.87E-2	1.04E-1	1.63E-1	2.38E-1	3.34E-1	4.71E-1	7.04E-1	9.38E-1
0.87	6.98E-3	2.79E-2	6.27E-2	1.12E-1	1.75E-1	2.54E-1	3.58E-1	5.04E-1	7.53E-1	1.00
0.95	7.52E-3	3.01E-2	6.76E-2	1.20E-1	1.88E-1	2.74E-1	3.85E-1	5.42E-1	8.11E-1	1.08
$p = -0.6$										
q	T_{v10}	T_{v20}	T_{v30}	T_{v40}	T_{v50}	T_{v60}	T_{v70}	T_{v80}	T_{v90}	T_{v95}
−0.95	7.71E-4	1.21E-2	4.17E-2	8.94E-2	1.55E-1	2.41E-1	3.53E-1	5.11E-1	7.82E-1	1.05
−0.91	6.39E-4	4.67E-3	2.50E-2	6.58E-2	1.27E-1	2.09E-1	3.17E-1	4.70E-1	7.33E-1	9.95E-1
−0.85	7.36E-4	3.16E-3	1.15E-2	3.95E-2	9.14E-2	1.67E-1	2.70E-1	4.16E-1	6.66E-1	9.16E-1
−0.78	8.69E-4	3.50E-3	8.88E-3	2.33E-2	6.16E-2	1.28E-1	2.24E-1	3.62E-1	5.99E-1	8.35E-1
−0.70	1.03E-3	4.13E-3	9.59E-3	1.97E-2	4.33E-2	9.59E-2	1.83E-1	3.12E-1	5.33E-1	7.55E-1
−0.60	1.25E-3	5.03E-3	1.14E-2	2.14E-2	3.89E-2	7.48E-2	1.46E-1	2.63E-1	4.67E-1	6.72E-1
−0.50	1.50E-3	6.01E-3	1.35E-2	2.47E-2	4.17E-2	7.10E-2	1.27E-1	2.30E-1	4.18E-1	6.05E-1
−0.36	1.88E-3	7.54E-3	1.70E-2	3.04E-2	4.94E-2	7.78E-2	1.25E-1	2.10E-1	3.74E-1	5.42E-1
−0.20	2.38E-3	9.49E-3	2.14E-2	3.81E-2	6.07E-2	9.21E-2	1.39E-1	2.15E-1	3.59E-1	5.09E-1
0.00	3.06E-3	1.23E-2	2.76E-2	4.91E-2	7.75E-2	1.15E-1	1.67E-1	2.45E-1	3.83E-1	5.24E-1
0.25	4.05E-3	1.62E-2	3.65E-2	6.49E-2	1.02E-1	1.49E-1	2.12E-1	3.02E-1	4.57E-1	6.12E-1
0.50	5.18E-3	2.07E-2	4.67E-2	8.30E-2	1.30E-1	1.90E-1	2.67E-1	3.77E-1	5.65E-1	7.52E-1
0.70	6.19E-3	2.47E-2	5.56E-2	9.89E-2	1.55E-1	2.26E-1	3.18E-1	4.47E-1	6.69E-1	8.91E-1
0.80	6.71E-3	2.69E-2	6.05E-2	1.08E-1	1.68E-1	2.45E-1	3.45E-1	4.85E-1	7.26E-1	9.66E-1
0.87	7.12E-3	2.84E-2	6.39E-2	1.14E-1	1.78E-1	2.59E-1	3.65E-1	5.13E-1	7.68E-1	1.02

0.91	7.34E-3	2.93E-2	6.60E-2	1.17E-1	1.84E-1	2.67E-1	3.76E-1	5.30E-1	7.92E-1	1.05
0.95	7.57E-3	3.02E-2	6.81E-2	1.21E-1	1.89E-1	2.76E-1	3.88E-1	5.46E-1	8.17E-1	1.09
$p = -0.3$										
q	T_{v10}	T_{v20}	T_{v30}	T_{v40}	T_{v50}	T_{v60}	T_{v70}	T_{v80}	T_{v90}	T_{v95}
–0.95	4.20E-3	2.35E-2	5.91E-2	1.11E-1	1.80E-1	2.67E-1	3.81E-1	5.42E-1	8.16E-1	1.09
–0.91	2.95E-3	1.87E-2	5.14E-2	1.01E-1	1.68E-1	2.53E-1	3.65E-1	5.22E-1	7.92E-1	1.06
–0.85	2.63E-3	1.42E-2	4.21E-2	8.79E-2	1.51E-1	2.34E-1	3.43E-1	4.96E-1	7.58E-1	1.02
–0.78	2.73E-3	1.22E-2	3.49E-2	7.59E-2	1.36E-1	2.15E-1	3.20E-1	4.68E-1	7.21E-1	9.75E-1
–0.70	2.90E-3	1.20E-2	3.10E-2	6.65E-2	1.21E-1	1.97E-1	2.98E-1	4.40E-1	6.84E-1	9.29E-1
–0.60	3.12E-3	1.26E-2	3.00E-2	6.06E-2	1.10E-1	1.80E-1	2.76E-1	4.12E-1	6.45E-1	8.78E-1
–0.50	3.36E-3	1.34E-2	3.10E-2	5.94E-2	1.04E-1	1.70E-1	2.60E-1	3.91E-1	6.13E-1	8.36E-1
–0.36	3.70E-3	1.48E-2	3.35E-2	6.16E-2	1.03E-1	1.63E-1	2.48E-1	3.71E-1	5.82E-1	7.94E-1
–0.20	4.10E-3	1.64E-2	3.70E-2	6.67E-2	1.08E-1	1.66E-1	2.47E-1	3.64E-1	5.65E-1	7.67E-1
0.00	4.66E-3	1.86E-2	4.18E-2	7.47E-2	1.19E-1	1.78E-1	2.57E-1	3.72E-1	5.69E-1	7.65E-1
0.25	5.38E-3	2.15E-2	4.84E-2	8.60E-2	1.35E-1	1.99E-1	2.83E-1	4.02E-1	6.06E-1	8.10E-1
0.50	6.15E-3	2.46E-2	5.53E-2	9.84E-2	1.54E-1	2.25E-1	3.17E-1	4.48E-1	6.71E-1	8.93E-1
0.70	6.80E-3	2.72E-2	6.12E-2	1.09E-1	1.70E-1	2.48E-1	3.49E-1	4.92E-1	7.36E-1	9.81E-1
0.80	7.14E-3	2.86E-2	6.43E-2	1.14E-1	1.79E-1	2.61E-1	3.67E-1	5.16E-1	7.72E-1	1.03
0.87	7.39E-3	2.96E-2	6.65E-2	1.18E-1	1.85E-1	2.70E-1	3.79E-1	5.34E-1	7.98E-1	1.06
0.91	7.52E-3	3.02E-2	6.78E-2	1.20E-1	1.89E-1	2.75E-1	3.86E-1	5.44E-1	8.13E-1	1.08
0.95	7.66E-3	3.07E-2	6.91E-2	1.23E-1	1.92E-1	2.80E-1	3.94E-1	5.54E-1	8.29E-1	1.10
$p = 0.3$										
q	T_{v10}	T_{v20}	T_{v30}	T_{v40}	T_{v50}	T_{v60}	T_{v70}	T_{v80}	T_{v90}	T_{v95}
–0.95	1.22E-2	4.03E-2	8.37E-2	1.43E-1	2.17E-1	3.11E-1	4.32E-1	6.04E-1	8.97E-1	1.19
–0.91	1.48E-2	4.63E-2	9.31E-2	1.55E-1	2.33E-1	3.29E-1	4.55E-1	6.32E-1	9.35E-1	1.24

(Continued)

Table 2.16 (Continued) Elapsed times for certain average degree of consolidation for one-way drainage

	T_{v10}	T_{v20}	T_{v30}	T_{v40}	T_{v50}	T_{v60}	T_{v70}	T_{v80}	T_{v90}	T_{v95}
–0.85	1.77E-2	5.39E-2	1.05E-1	1.71E-1	2.53E-1	3.55E-1	4.87E-1	6.73E-1	9.90E-1	1.31
–0.78	2.00E-2	6.10E-2	1.17E-1	1.88E-1	2.75E-1	3.82E-1	5.20E-1	7.16E-1	1.05	1.38
–0.70	2.13E-2	6.69E-2	1.28E-1	2.04E-1	2.95E-1	4.08E-1	5.54E-1	7.60E-1	1.11	1.46
–0.60	2.17E-2	7.18E-2	1.38E-1	2.19E-1	3.17E-1	4.36E-1	5.91E-1	8.09E-1	1.18	1.55
–0.50	2.11E-2	7.41E-2	1.45E-1	2.30E-1	3.32E-1	4.58E-1	6.20E-1	8.48E-1	1.24	1.63
–0.36	1.97E-2	7.41E-2	1.49E-1	2.39E-1	3.47E-1	4.79E-1	6.49E-1	8.90E-1	1.30	1.71
–0.20	1.80E-2	7.08E-2	1.48E-1	2.42E-1	3.53E-1	4.90E-1	6.67E-1	9.16E-1	1.34	1.77
0.00	1.60E-2	6.40E-2	1.39E-1	2.34E-1	3.48E-1	4.87E-1	6.67E-1	9.20E-1	1.35	1.78
0.25	1.37E-2	5.49E-2	1.23E-1	2.13E-1	3.24E-1	4.59E-1	6.34E-1	8.81E-1	1.30	1.72
0.50	1.16E-2	4.63E-2	1.04E-1	1.84E-1	2.85E-1	4.11E-1	5.73E-1	8.01E-1	1.19	1.58
0.70	1.00E-2	4.00E-2	9.00E-2	1.60E-1	2.50E-1	3.62E-1	5.08E-1	7.14E-1	1.07	1.42
0.80	9.26E-3	3.70E-2	8.33E-2	1.48E-1	2.32E-1	3.37E-1	4.74E-1	6.66E-1	9.95E-1	1.33
0.87	8.75E-3	3.50E-2	7.87E-2	1.40E-1	2.19E-1	3.19E-1	4.49E-1	6.31E-1	9.44E-1	1.26
0.91	8.47E-3	3.39E-2	7.62E-2	1.36E-1	2.12E-1	3.09E-1	4.34E-1	6.11E-1	9.15E-1	1.22
0.95	8.20E-3	3.28E-2	7.38E-2	1.31E-1	2.05E-1	2.99E-1	4.20E-1	5.92E-1	8.85E-1	1.18
$p = 0.6$										
q	T_{v10}	T_{v20}	T_{v30}	T_{v40}	T_{v50}	T_{v60}	T_{v70}	T_{v80}	T_{v90}	T_{v95}
–0.95	2.02E-2	5.70E-2	1.09E-1	1.77E-1	2.60E-1	3.64E-1	4.99E-1	6.88E-1	1.01	1.34
–0.91	2.82E-2	7.45E-2	1.36E-1	2.13E-1	3.07E-1	4.22E-1	5.71E-1	7.82E-1	1.14	1.50
–0.85	3.82E-2	9.72E-2	1.72E-1	2.62E-1	3.70E-1	5.03E-1	6.74E-1	9.15E-1	1.33	1.74
–0.78	4.77E-2	1.20E-1	2.08E-1	3.12E-1	4.36E-1	5.88E-1	7.84E-1	1.06	1.53	2.00
–0.70	5.59E-2	1.41E-1	2.42E-1	3.61E-1	5.02E-1	6.74E-1	8.96E-1	1.21	1.74	2.28
–0.60	6.28E-2	1.62E-1	2.78E-1	4.12E-1	5.72E-1	7.67E-1	1.02	1.37	1.98	2.59

−0.50	6.63E-2	1.76E-1	3.04E-1	4.53E-1	6.28E-1	8.43E-1	1.12	1.51	2.18	2.84
−0.36	6.62E-2	1.88E-1	3.29E-1	4.93E-1	6.85E-1	9.22E-1	1.23	1.66	2.39	3.12
−0.20	6.02E-2	1.90E-1	3.41E-1	5.15E-1	7.21E-1	9.73E-1	1.30	1.76	2.54	3.32
0.00	4.91E-2	1.76E-1	3.31E-1	5.10E-1	7.23E-1	9.82E-1	1.32	1.79	2.60	3.40
0.25	3.55E-2	1.39E-1	2.86E-1	4.57E-1	6.60E-1	9.08E-1	1.23	1.68	2.45	3.22
0.50	2.40E-2	9.63E-2	2.12E-1	3.59E-1	5.34E-1	7.49E-1	1.03	1.42	2.08	2.75
0.70	1.65E-2	6.61E-2	1.49E-1	2.61E-1	4.01E-1	5.73E-1	7.95E-1	1.11	1.64	2.18
0.80	1.33E-2	5.31E-2	1.19E-1	2.12E-1	3.29E-1	4.75E-1	6.64E-1	9.31E-1	1.39	1.84
0.87	1.12E-2	4.49E-2	1.01E-1	1.79E-1	2.80E-1	4.07E-1	5.71E-1	8.02E-1	1.20	1.59
0.91	1.01E-2	4.05E-2	9.10E-2	1.62E-1	2.53E-1	3.68E-1	5.17E-1	7.28E-1	1.09	1.45
0.95	9.09E-3	3.63E-2	8.16E-2	1.45E-1	2.27E-1	3.31E-1	4.65E-1	6.55E-1	9.79E-1	1.30
$p = 0.8$										
q	T_{v10}	T_{v20}	T_{v30}	T_{v40}	T_{v50}	T_{v60}	T_{v70}	T_{v80}	T_{v90}	T_{v95}
−0.95	3.61E-2	9.07E-2	1.61E-1	2.48E-1	3.52E-1	4.81E-1	6.47E-1	8.81E-1	1.28	1.68
−0.91	5.49E-2	1.31E-1	2.24E-1	3.34E-1	4.65E-1	6.25E-1	8.32E-1	1.12	1.62	2.12
−0.85	7.99E-2	1.84E-1	3.07E-1	4.50E-1	6.20E-1	8.27E-1	1.09	1.47	2.12	2.76
−0.78	1.05E-1	2.39E-1	3.94E-1	5.73E-1	7.85E-1	1.04	1.38	1.85	2.65	3.46
−0.70	1.29E-1	2.93E-1	4.80E-1	6.97E-1	9.53E-1	1.27	1.67	2.24	3.21	4.18
−0.60	1.54E-1	3.50E-1	5.73E-1	8.30E-1	1.13	1.51	1.99	2.67	3.82	4.98
−0.50	1.71E-1	3.95E-1	6.48E-1	9.40E-1	1.29	1.71	2.26	3.02	4.34	5.66
−0.36	1.86E-1	4.39E-1	7.25E-1	1.06	1.45	1.93	2.54	3.41	4.90	6.39
−0.20	1.88E-1	4.63E-1	7.74E-1	1.13	1.56	2.08	2.75	3.70	5.31	6.93
0.00	1.71E-1	4.55E-1	7.78E-1	1.15	1.59	2.13	2.82	3.80	5.48	7.15
0.25	1.24E-1	3.89E-1	6.95E-1	1.05	1.46	1.97	2.63	3.56	5.14	6.73

(Continued)

Table 2.16 (Continued) Elapsed times for certain average degree of consolidation for one-way drainage

	T_{v10}	T_{v20}	T_{v30}	T_{v40}	T_{v50}	T_{v60}	T_{v70}	T_{v80}	T_{v90}	T_{v95}
0.50	7.08E-2	2.67E-1	5.20E-1	8.13E-1	1.16	1.58	2.13	2.90	4.21	5.53
0.70	3.81E-2	1.52E-1	3.28E-1	5.44E-1	8.00E-1	1.11	1.52	2.09	3.06	4.03
0.80	2.55E-2	1.02E-1	2.27E-1	3.92E-1	5.92E-1	8.37E-1	1.15	1.60	2.36	3.12
0.87	1.81E-2	7.26E-2	1.63E-1	2.88E-1	4.43E-1	6.35E-1	8.82E-1	1.23	1.83	2.42
0.91	1.45E-2	5.81E-2	1.31E-1	2.32E-1	3.60E-1	5.21E-1	7.28E-1	1.02	1.52	2.02
0.95	1.13E-2	4.52E-2	1.02E-1	1.81E-1	2.83E-1	4.11E-1	5.77E-1	8.11E-1	1.21	1.61
$p = 0.9$										
q	T_{v10}	T_{v20}	T_{v30}	T_{v40}	T_{v50}	T_{v60}	T_{v70}	T_{v80}	T_{v90}	T_{v95}
−0.95	6.54E-2	1.53E-1	2.58E-1	3.81E-1	5.29E-1	7.09E-1	9.41E-1	1.27	1.83	2.39
−0.91	1.05E-1	2.35E-1	3.86E-1	5.61E-1	7.69E-1	1.02	1.35	1.81	2.60	3.39
−0.85	1.58E-1	3.47E-1	5.62E-1	8.11E-1	1.11	1.47	1.93	2.59	3.71	4.83
−0.78	2.13E-1	4.64E-1	7.49E-1	1.08	1.47	1.94	2.56	3.42	4.90	6.38
−0.70	2.68E-1	5.83E-1	9.41E-1	1.35	1.84	2.44	3.21	4.29	6.14	8.00
−0.60	3.26E-1	7.12E-1	1.15	1.65	2.25	2.98	3.92	5.24	7.51	9.77
−0.50	3.73E-1	8.17E-1	1.32	1.90	2.59	3.43	4.52	6.04	8.66	11.3
−0.36	4.18E-1	9.28E-1	1.50	2.17	2.96	3.92	5.17	6.92	9.92	12.9
−0.20	4.43E-1	1.00	1.63	2.36	3.22	4.28	5.64	7.56	10.8	14.1
0.00	4.34E-1	1.01	1.67	2.43	3.32	4.42	5.84	7.83	11.2	14.6
0.25	3.62E-1	9.08E-1	1.53	2.24	3.08	4.12	5.45	7.33	10.5	13.7
0.50	2.28E-1	6.71E-1	1.17	1.76	2.44	3.29	4.37	5.90	8.51	11.1
0.70	1.07E-1	3.96E-1	7.52E-1	1.16	1.65	2.25	3.02	4.10	5.96	7.81
0.80	6.16E-2	2.43E-1	5.04E-1	8.10E-1	1.17	1.62	2.19	3.00	4.38	5.75
0.87	3.69E-2	1.48E-1	3.25E-1	5.48E-1	8.14E-1	1.14	1.56	2.15	3.16	4.18

0.91	2.57E-2	1.03E-1	2.31E-1	4.00E-1	6.07E-1	8.61E-1	1.19	1.65	2.44	3.23
0.95	1.65E-2	6.60E-2	1.49E-1	2.63E-1	4.08E-1	5.88E-1	8.20E-1	1.15	1.71	2.27
$p = 0.95$										
q	T_{v10}	T_{v20}	T_{v30}	T_{v40}	T_{v50}	T_{v60}	T_{v70}	T_{v80}	T_{v90}	T_{v95}
–0.95	1.21E-1	2.69E-1	4.40E-1	6.38E-1	8.72E-1	1.16	1.53	2.05	2.94	3.83
–0.91	2.00E-1	4.33E-1	6.99E-1	1.01	1.37	1.81	2.39	3.19	4.57	5.95
–0.85	3.08E-1	6.61E-1	1.06	1.52	2.07	2.74	3.60	4.82	6.90	8.97
–0.78	4.23E-1	9.04E-1	1.45	2.08	2.83	3.74	4.91	6.57	9.41	12.2
–0.70	5.40E-1	1.15	1.85	2.66	3.61	4.77	6.28	8.39	12.0	15.6
–0.60	6.66E-1	1.43	2.29	3.29	4.47	5.91	7.77	10.4	14.9	19.3
–0.50	7.70E-1	1.66	2.66	3.82	5.19	6.87	9.03	12.1	17.3	22.5
–0.36	8.78E-1	1.90	3.06	4.39	5.98	7.91	10.4	13.9	19.9	26.0
–0.20	9.50E-1	2.07	3.35	4.81	6.55	8.68	11.4	15.3	21.9	28.5
0.00	9.62E-1	2.13	3.45	4.98	6.79	9.00	11.9	15.9	22.8	29.6
0.25	8.53E-1	1.95	3.20	4.63	6.33	8.41	11.1	14.9	21.3	27.8
0.50	6.14E-1	1.50	2.50	3.66	5.03	6.71	8.88	11.9	17.1	22.4
0.70	3.32E-1	9.46E-1	1.64	2.45	3.40	4.57	6.07	8.19	11.8	15.4
0.80	1.80E-1	6.11E-1	1.12	1.70	2.39	3.24	4.33	5.86	8.49	11.1
0.87	9.44E-2	3.63E-1	7.18E-1	1.13	1.62	2.21	2.98	4.06	5.91	7.76
0.91	5.76E-2	2.30E-1	4.86E-1	7.92E-1	1.15	1.60	2.17	2.98	4.36	5.74
0.95	2.99E-2	1.19E-1	2.67E-1	4.60E-1	6.94E-1	9.80E-1	1.35	1.87	2.76	3.65

CASE (A):

$$p = \frac{m_{v2}\sqrt{c_{v2}} - m_{v1}\sqrt{c_{v1}}}{m_{v2}\sqrt{c_{v2}} + m_{v1}\sqrt{c_{v1}}} = \frac{6.6\sqrt{3.4} - 2.87\sqrt{2.22}}{6.6\sqrt{3.4} + 2.87\sqrt{2.22}} = 0.48,$$

$$q = \frac{H_1\sqrt{c_{v2}} - H_2\sqrt{c_{v1}}}{H_1\sqrt{c_{v2}} + H_2\sqrt{c_{v1}}} = \frac{2\sqrt{3.4} - 1.5\sqrt{2.22}}{2\sqrt{3.4} + 1.5\sqrt{2.22}} = 0.245,$$

$$T_v = \frac{c_{v1}c_{v2}t}{\left(H_1\sqrt{c_{v2}} + H_2\sqrt{c_{v1}}\right)^2} = \frac{2.22 \times 3.4 \times 6 \times 864}{\left(200\sqrt{3.4} + 150\sqrt{2.22}\right)^2} = 0.1115$$

From Figure 2.25(a), the average degree of consolidation for $q = 0.245$ and $T_v = 0.1115$ is $U = 78\%$ for $p = 0.4$. From Figure 2.25(b), the average degree of consolidation for $q = 0.245$ and $T_v = 0.1115$ is $U = 82\%$ for $p = 0.6$. So, for $p = 0.48$, by interpolation, the average degree of consolidation U of the double soil layer after a loading time of 60 days is

$$\frac{0.6 - 0.48}{0.6 - 0.4} \times 78\% + \frac{0.48 - 0.4}{0.6 - 0.4} \times 82\% = 79.6\%$$

CASE (B):

$$p = 0.818, \quad q = 0.333, \quad T_v = 0.3504$$

From Figure 2.26(h), the average degree of consolidation for $q = 0.333$ and $T_v = 0.3504$ is $U = 20\%$ for $p = 0.8$. From Figure 2.26(i), the average degree of consolidation for $q = 0.333$ and $T_v = 0.3504$ is $U = 11\%$ for $p = 0.9$. For $p = 0.818$, by interpolation, the average degree of consolidation U of the double soil layer after a loading time of 365 days is

$$\frac{0.9 - 0.818}{0.9 - 0.8} \times 20\% + \frac{0.818 - 0.8}{0.9 - 0.8} \times 11\% = 18.4\%$$

CASE (C):

$$p = 0.52, \quad q = 0.727, \quad T_v = 0.5878$$

From Figure 2.26(f), the average degree of consolidation for $q = 0.727$ and $T_v = 0.5878$ is $U = 75\%$ for $p = 0.3$. From Figure 2.26(g), the average degree of consolidation for $q = 0.727$ and $T_v = 0.5878$ is $U = 62\%$ for $p = 0.6$. By interpolation, the average degree of consolidation U of the double soil layer after a loading time of 365 days is

$$\frac{0.6 - 0.52}{0.6 - 0.3} \times 75\% + \frac{0.52 - 0.3}{0.6 - 0.3} \times 62\% = 65.5\%$$

If the coefficients of consolidation are averaged for case (b) and case (c), that is $c_v = \frac{H_1 c_{v1} + H_2 c_{v2}}{H}$, as a simplification, the time factors T_v for Cases (b) and (c) corresponding the one-layer simplification are 0.3504 and 1.402, respectively. Thus, the average degrees of consolidation for case (b) and case (c) are $U = 65.85\%$ and $U = 97.45\%$, respectively. These figures are far different from $U = 18.4\%$ for case (b) and $U = 65.5\%$ for case (c) from the double-layer consolidation solutions. This means that the above simplification is not accurate.

If both coefficients of hydraulic conductivity and compressibility are averaged for case (b) and case (c), that is $c_v = \frac{H}{\frac{H_1}{c_{v1} m_{v1}} + \frac{H_2}{c_{v2} m_{v2}}} \times \frac{H}{H_1 m_{v1} + H_2 m_{v2}}$, as another simplification, the time factors T_v for Cases (b) and (c) corresponding the one-layer simplification are 0.1251 and 0.5006, respectively. The average degrees of consolidation for case (b) and case (c) are $U = 39.91\%$ and $U = 76.43\%$, respectively. Significant errors are still generated in this simplification.

Example 2.3: Examine the average degrees of consolidation for the four cases with same coefficient of consolidation as shown in Figure 2.28 under abruptly applied loading and ramp loading.

For case (1) in Figure 2.28, substitution of the values of k_1, m_{v1}, c_{v1}, k_2, m_{v2}, c_{v2}, H_1, and H_2 into (2.35–2.36) for p, q, and (2.44b) for c_n lead to $p = \frac{9}{11}$, $q = 0$, $c_n = \frac{4}{11}$(for any n). The eigenvalues can be expressed analytically for this particular case:

$$\lambda_n = \left[n - \frac{1-(-1)^n}{2} \right] \pi + (-1)^{n-1} \arccos\left(\frac{9}{11} \right).$$

For case (2) in Figure 2.28, similarly, we have $p = -\frac{9}{11}$, $q = 0$, $c_n = \frac{40}{11}$, and the eigenvalues are

$$\lambda_n = \left[n - \frac{1-(-1)^n}{2} \right] \pi + (-1)^{n-1} \arccos\left(-\frac{9}{11} \right).$$

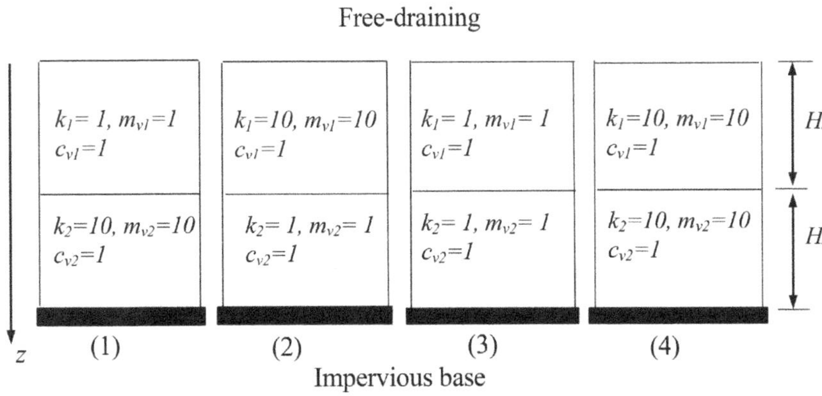

Figure 2.28 Assumed four soil profiles. (After Pyrah 1996.)

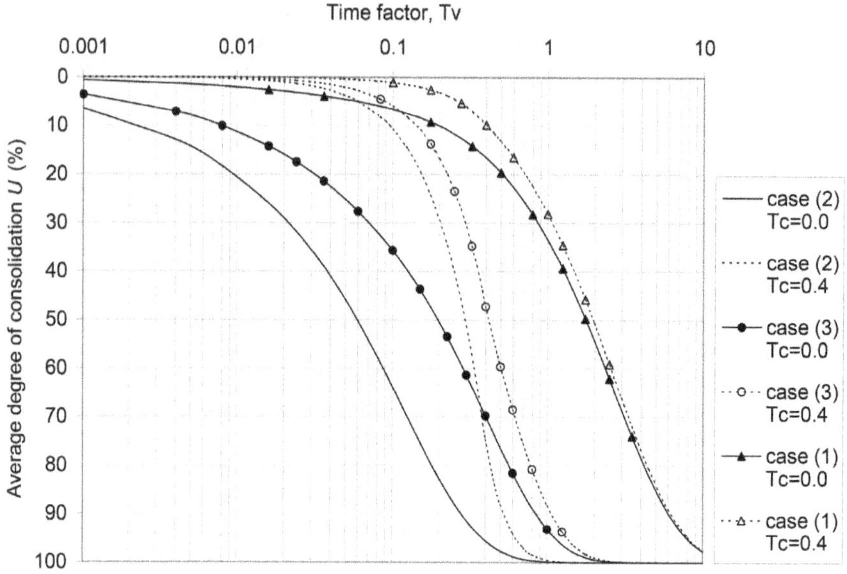

Figure 2.29 Average degree of consolidation for different soil profiles.

For case (3) and case (4) in Figure 2.28, we have $p = q = 0$, $c_n = 2$, and the eigenvalues are

$$\lambda_n = n\pi - \frac{1}{2}\pi.$$

The λ_n values above can also be obtained using the relationship $\lambda_n = \pi\left(n - \frac{1}{2} + \varsigma_n\right)$ where the first 3 ς_n values (n = 1, 2, 3) are shown in Figures 2.18–2.20.

Case (3) and case (4) are not truly layered soils and can be represented by case (3) only. In this situation, Olson's (1977) solution can be applied directly.

Using the λ_n values obtained above for different n-index and (2.44), the relationship between the average degree of consolidation U and T_v can be calculated, and the excess porewater pressure can be calculated using the analytical solution in (2.39). Two loading conditions are considered: (1) a suddenly applied loading with $T_c = 0$ and (2) a ramp loading with $T_c = 0.4$. The relationships of U *vs.* T_v for the three soil profile cases in Figure 2.28 under two loading conditions ($T_c = 0$ and $T_c = 0.4$) are shown in Figure 2.29.

The time factor T_{v50} for $U = 50\%$ is 1.76 in case (1) and 0.060 in case (2) for the suddenly applied loading condition. To achieve the same average degree of consolidation $U = 50\%$, the T_{v50} in case (2) is much smaller than that in case (1). The ratio of $T_{v50,case(2)}$ over $T_{v50,case(1)}$ is only 3.4%. This indicates that the dissipation of excess porewater pressure in case (2) is much faster than the dissipation in case (1). This is because the permeability k in case (2) is 10 times that of the upper layer in case (1).

For the ramp-loading condition, the time factor T_{v50} for $U = 50\%$ is 1.97 in case (1) and 0.29 in case (2). To achieve the same average degree of consolidation $U = 50\%$, the T_{v50} in case (2) is smaller than that in case (1). However, the ratio of $T_{v50,case(2)}$ over $T_{v50,case(1)}$ increased to 14.7%. This indicates that the dissipation of excess porewater pressure in case (2) is faster than the dissipation in case (1) due to the larger permeability of the upper layer in case (2). But, the difference of pressure dissipation between case (2) and case (1) for the ramp-loading condition is smaller than the difference in the case of the suddenly applied loading condition. The difference is small in ramp-loading cases because (a) excess porewater pressure dissipates during ramp loading and (b) the dissipation rate is higher at the beginning becoming smaller and smaller with time. Since the hydraulic gradient in ramp-loading cases is smaller than that in suddenly loading cases, according to Darcy's law, the pressure dissipation in ramp-loading cases is smaller than that in suddenly loading cases.

2.4.3 Formal solution for *n*-layer soil profile

The n-layer soil profile is shown in Figure 2.30. An arbitrary layer is indexed with j ($j = 1, 2, \ldots, n$). The soil properties of the j-th layer are the coefficient of consolidation c_{vj}, the constrained compressibility m_{vj}, and the coefficient of permeability k_j. The compressible stratum has a total

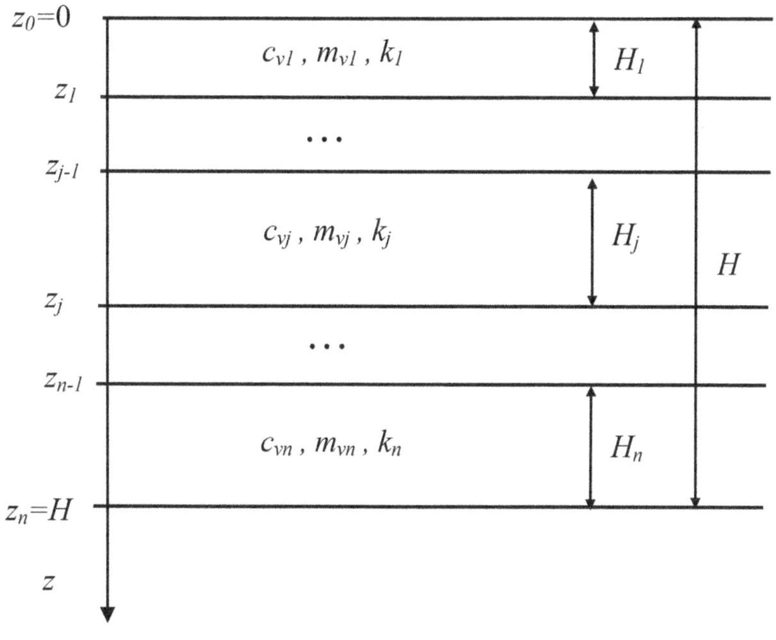

Figure 2.30 An *n*-layer soil profile.

thickness H. The consolidation equation for the n-layer system is rewritten from (2.13) as

$$\frac{\partial u_e}{\partial t} = c_{vj}\frac{\partial^2 u_e}{\partial z^2} + \frac{\partial \sigma_z}{\partial t} \qquad z_{j-1} \le z \le z_j, \quad j = 1, 2, ..., n \tag{2.47}$$

subject to the initial condition

$$u_e(z,0) = u_0(z) \tag{2.48}$$

and the boundary and interface conditions

$$\left[\alpha_0 u_e(z,t) - \beta_0 \frac{\partial u_e(z,t)}{\partial z}\right]\Big|_{z=0} = c_0(t) \tag{2.49a}$$

$$u_e(z,t)\,|_{z=z_j^-} = u_e(z,t)\,|_{z=z_j^+}, \quad j = 1, 2, ..., n-1 \tag{2.49b}$$

$$k_j \frac{\partial u_e(z,t)}{\partial z}\bigg|_{z=z_j^-} = k_{j+1}\frac{\partial u_e(z,t)}{\partial z}\bigg|_{z=z_j^+}, \quad j = 1, 2, ..., n-1 \tag{2.49c}$$

$$\left[\alpha_n u_e(z,t) + \beta_n \frac{\partial u_e(z,t)}{\partial z}\right]\Big|_{z=H} = c_n(t) \tag{2.49d}$$

where α_0, β_0, α_n, and β_n are constants; c_0 and c_n are functions of t.

We require that $\alpha_j \ge 0$, $\beta_j \ge 0$, and $\alpha_j + \beta_j > 0$ for $j = 0$ and n. If (1) $\alpha_j \ne 0$, $\beta_j = 0$, (2) $\alpha_j = 0$, $\beta_j \ne 0$, (3) $\alpha_j \ne 0$, $\beta_j \ne 0$, the corresponding boundary conditions are known as boundary conditions of the *first*, *second*, and *third kind*, respectively. Boundary conditions of the first and second kind are often referred to as *Dirichlet* and *Neumann conditions*.

The formal solution to (2.47) is

$$u_e(z,t) = \sum_{i=1}^{+\infty} T_i(t) Z_i(z) \tag{2.50}$$

where

$$T_i(t) = \frac{\exp(-\mu_i^2 t)\left[\int_0^H m_v u_0(z) Z_i(z)dz + \int_0^t \varphi_i(\tau)\exp(\mu_i^2\tau)d\tau\right]}{\int_0^H m_v Z_i^2 dz} \tag{2.51}$$

$$\phi_i(t) = \frac{\int_0^H m_v \frac{\partial \sigma_z}{\partial t} Z_i(z)dz + \frac{k_n c_n(t)}{\gamma_w(\alpha_n + \beta_n)}[Z_i(H) - Z_i'(H)] + \frac{k_1 c_0(t)}{\gamma_w(\alpha_0 + \beta_0)}[Z_i(0) + Z_i'(0)]}{} \tag{2.52}$$

The function $Z_i(z)$ in (2.50) and the constant μ_i in (2.51) are the i-th solution pair of the following eigenvalue problem

$$\frac{d^2 Z(z)}{dz^2} + \frac{\mu^2}{c_{vj}} Z(z) = 0 \quad z_{j-1} \leq z \leq z_j, \quad j = 1, 2, ..., n \tag{2.53}$$

subject to the boundary and interface conditions

$$\left[\alpha_0 Z(z) - \beta_0 \frac{dZ(z)}{dz}\right]\Big|_{z=0} = 0 \tag{2.54a}$$

$$Z(z)\big|_{z=z_j^-} = Z(z)\big|_{z=z_j^+}, \quad j = 1, 2, ..., n-1 \tag{2.54b}$$

$$k_j \frac{dZ(z)}{dz}\Big|_{z=z_j^-} = k_{j+1} \frac{dZ(z)}{dz}\Big|_{z=z_j^+}, \quad j = 1, 2, ..., n-1 \tag{2.54c}$$

$$\left[\alpha_n Z(z) + \beta_n \frac{dZ(z)}{dz}\right]\Big|_{z=H} = 0 \tag{2.54d}$$

In the integration in (2.51–2.52), the compressibility m_v assumes m_{vj} for layer j as shown in Figure 2.30.

2.4.3.1 Construction of eigenfuctions

The solution of the eigenvalue problem defined in (2.53) is

$$Z(z) = \frac{\sin\left(\frac{\mu(z_j - z)}{\sqrt{c_{vj}}}\right)}{\sin\left(\frac{\mu H_j}{\sqrt{c_{vj}}}\right)} Z^{j-1} + \frac{\sin\left(\frac{\mu(z - z_{j-1})}{\sqrt{c_{vj}}}\right)}{\sin\left(\frac{\mu H_j}{\sqrt{c_{vj}}}\right)} Z^j \quad z_{j-1} \leq z \leq z_j \quad j = 1, 2, ..., n \tag{2.55}$$

where $\sin\left(\frac{\mu H_j}{\sqrt{c_{vj}}}\right) \neq 0$; $Z^j = Z(z_j)$.

If the solution, (2.55), should satisfy the boundary and interface conditions in (2.54), we have, respectively

$$\left(\alpha_0+\frac{\beta_0\mu}{\sqrt{c_{v1}}}\frac{\cos\left(\frac{\mu H_1}{\sqrt{c_{v1}}}\right)}{\sin\left(\frac{\mu H_1}{\sqrt{c_{v1}}}\right)}\right)Z^0-\frac{\beta_0\mu}{\sqrt{c_{v1}}\sin\left(\frac{\mu H_1}{\sqrt{c_{v1}}}\right)}Z^1=0 \tag{2.56a}$$

$$-\frac{k_jZ^{j-1}}{\sqrt{c_{vj}}\sin\left(\frac{\mu H_j}{\sqrt{c_{vj}}}\right)}+\left[\frac{k_j\cos\left(\frac{\mu H_j}{\sqrt{c_{vj}}}\right)}{\sqrt{c_{vj}}\sin\left(\frac{\mu H_j}{\sqrt{c_{vj}}}\right)}+\frac{k_{j+1}\cos\left(\frac{\mu H_{j+1}}{\sqrt{c_{vj+1}}}\right)}{\sqrt{c_{vj+1}}\sin\left(\frac{\mu H_{j+1}}{\sqrt{c_{vj+1}}}\right)}\right]Z^j-\frac{k_{j+1}Z^{j+1}}{\sqrt{c_{vj+1}}\sin\left(\frac{\mu H_{j+1}}{\sqrt{c_{vj+1}}}\right)}=0$$

$$j=1,2,\ldots,n \tag{2.56b}$$

$$-\frac{\beta_n\mu}{\sqrt{c_{vn}}\sin\left(\frac{\mu H_n}{\sqrt{c_{vn}}}\right)}Z^{n-1}+\left(\alpha_n+\frac{\beta_n\mu}{\sqrt{c_{vn}}}\frac{\cos\left(\frac{\mu H_n}{\sqrt{c_{vn}}}\right)}{\sin\left(\frac{\mu H_n}{\sqrt{c_{vn}}}\right)}\right)Z^n=0 \tag{2.56c}$$

The eigenfunctions defined by (2.53) are introduced to (2.56) to yield $(n + 1)$ homogeneous equations for the determination of the $(n + 1)$ constants Z^j ($j = 0, 1, \ldots, n$). This system of equations can be expressed in the matrix form as

$$\mathbf{A}(\mu)\mathbf{Z}^*=0 \tag{2.57a}$$

where

$$Z^{*T}=\{Z^0, Z^1, \ldots, Z^n\} \tag{2.57b}$$

is the transpose of $\mathbf{Z}^*$. The matrix $A(\mu)$ is given by

$$A(\mu)=\begin{bmatrix} a_0 & b_1 & & \cdots & 0 & 0 \\ b_1 & a_1 & b_2 & & & \vdots \\ 0 & b_2 & a_2 & & & \vdots \\ \vdots & & & & & 0 \\ \vdots & & & b_{n-1} & a_{n-1} & b_n \\ 0 & 0 & \cdots & 0 & b_n & a_n \end{bmatrix} \tag{2.57c}$$

where

$$a_0=\begin{cases} \alpha_0 & \beta_0=0 \\ \frac{k_1\alpha_0}{\mu\beta_0}+\frac{k_1}{\sqrt{c_{v1}}}ctg\left(\frac{\mu H_1}{\sqrt{c_{v1}}}\right) & \beta_0\neq 0 \end{cases} \tag{2.57d}$$

$$b_1 = \begin{cases} 0 & \beta_0 = 0 \\ -\dfrac{k_1}{\sqrt{c_{v1}}\sin\left(\frac{\mu H_1}{\sqrt{c_{v1}}}\right)} & \beta_0 \neq 0 \end{cases} \tag{2.57e}$$

$$a_j = \frac{k_j}{\sqrt{c_{vj}}} ctg\left(\frac{\mu H_j}{\sqrt{c_{vj}}}\right) + \frac{k_{j+1}}{\sqrt{c_{vj+1}}} ctg\left(\frac{\mu H_{j+1}}{\sqrt{c_{vj+1}}}\right), \; j = 1, 2, \ldots, n-1 \tag{2.57f}$$

$$b_j = -\frac{k_j}{\sqrt{c_{vj}}\sin\left(\frac{\mu H_j}{\sqrt{c_{vj}}}\right)}, \; j = 2, \ldots, n-1 \tag{2.57g}$$

$$a_n = \begin{cases} \alpha_n & \beta_n = 0 \\ \dfrac{k_n \alpha_n}{\mu \beta_n} + \dfrac{k_n}{\sqrt{c_{vn}}} ctg\left(\dfrac{\mu H_n}{\sqrt{c_{vn}}}\right) & \beta_n \neq 0 \end{cases} \tag{2.57h}$$

$$b_n = \begin{cases} 0 & \beta_n = 0 \\ -\dfrac{k_n}{\sqrt{c_{vn}}\sin\left(\frac{\mu H_n}{\sqrt{c_{vn}}}\right)} & \beta_n \neq 0 \end{cases} \tag{2.57i}$$

If the system of (2.57) has a nontrivial solution, the determinant of the matrix $A(\mu)$ should vanish, that is,

$$det(A(\mu)) = 0 \tag{2.58}$$

The infinite number of real roots of this transcendental equation give the eigenvalues μ_i of the eigenvalue problem, (2.53).

In the above formulations, it is required that $\sin\left(\frac{\mu H_j}{\sqrt{c_{vj}}}\right) \neq 0$. If $\sin\left(\frac{\mu H_j}{\sqrt{c_{vj}}}\right) = 0$, similar equations can be derived. However, the symmetry of the matrix $A(\mu)$ is not preserved. In real computations, this situation is handled by choosing a slightly different μ so that the above formulations hold. The continuity of the eigenvalues and eigenfunctions depending on μ will entail the correct solutions.

2.4.3.2 *Determination of the eigenvalues by a sign-count method*

Central to the formulations above is the determination of the eigenvalues. Wittrick and Williams (1971) developed an efficient procedure, *sign-count method*, to compute the eigenfrequencies of linear elastic structures. Mikhailov and Vulchanov (1983) adapted the sign-count method to Sturm-Liouville problems. Since (2.53) is a Sturm-Liouville problem, the sign-count method is equally applied (Mikhailov and Özişik, 1994).

Wittrick and Williams (1971) show that the number of positive eigenvalues $J(\mu^*)$ lying between zero and some prescribed value of $\mu = \mu^*$ is equal to:

$$J(\mu^*) = J_0(\mu^*) + \textit{Sign-count}\ (\mathbf{A}(\mu^*)) \tag{2.59}$$

where

$J_0(\mu^*)$ = number of positive eigenvalues not exceeding μ^* when all components of the vector $\mathbf{Z}^*$ corresponding to $A(\mu^*)$ are zero (i.e., decoupled system).

Sign-count $(A(\mu^*))$ = number of changes of sign between consecutive members of the Sturm sequence

$$\{1,\ det(A_1(\mu^*)),\ det(A_2(\mu^*)),\ \ldots,\ det(A_n(\mu^*)),\ det(A_{n+1}(\mu^*))\}$$

Here, $\det(A_j(\mu^*))$ $(j = 1, 2, \ldots, n+1)$ is the leading principal minor determinant of order j of $A(\mu^*)$, i.e. the determinant of the $j \times j$ matrix formed from the elements in the first j rows and columns of $A(\mu^*)$. If $\det(A_j(\mu^*)) = 0$, its sign assumes the sign of the term just before it.

The leading principal minor determinants can be computed using the following relation

$$det(\mathbf{A}_1(\mu^*)) = a_0 \tag{2.60a}$$

$$det(\mathbf{A}_2(\mu^*)) = a_0 a_1 - b_1^2 \tag{2.60b}$$

$$det(\mathbf{A}_j(\mu^*)) = det(\mathbf{A}_{j-1}(\mu^*))\, a_{j-1} - b_{j-1}^2\, det(\mathbf{A}_{j-2}(\mu^*)) \tag{2.60c}$$

In order to avoid excessive large values of the Sturm sequence, the sign-count is normally computed by counting the number of negative elements in the following sequence

$$\left\{ \det(A_1(\mu^*)),\ \frac{\det(A_2(\mu^*))}{\det(A_1(\mu^*))},\ \frac{\det(A_3(\mu^*))}{\det(A_2(\mu^*))},\ \ldots,\ \frac{\det(A_{n+1}(\mu^*))}{\det(A_n(\mu^*))} \right\}$$

It must be pointed out that difficulties arise if one or more of the minors are zero. In practice, however, it is exceeding unlikely that such a case will ever be encountered. If it is, the difficulties can be avoided by choosing a slightly different μ^*.

To find $J_0(\mu^*)$, one takes into account the fact that components of the vector $\mathbf{Z}^*$ are zero. Then (2.53) degenerates into a decoupled set of equations:

$$\frac{d^2Z(z)}{dz^2} + \frac{\mu^2}{c_{vj}} Z(z) = 0 \qquad z_{j-1} \le z \le z_j, \qquad j = 1, 2, \ldots, n \tag{2.61}$$

subject to the boundary conditions

$$Z^{j-1} = 0; \; Z^j = 0 \tag{2.62}$$

The eigencondition of (2.61) is:

$$\sin\left(\frac{\mu H_j}{\sqrt{c_{vj}}}\right) = 0 \tag{2.63}$$

Namely,

$$\frac{\mu H_j}{\sqrt{c_{vj}}} = i\pi \;\; (i = 1, 2, 3, \ldots) \tag{2.64}$$

The number of eigenvalues not exceeding μ^*, for the considered interval $z_{j-1} < z < z_j$, is given by '$\mathrm{int}\left(\frac{\mu^* H_j}{\pi\sqrt{c_{vj}}}\right)$.' Then the total number of eigenvalues for all the intervals is given by

$$J_0(\mu^*) = \sum_{j=1}^{n} \mathrm{int}\left(\frac{\mu^* H_j}{\pi\sqrt{c_{vj}}}\right) \tag{2.65}$$

By computing $J(\mu^*)$, the number of eigenvalues in any given interval can be determined. Thus, the above fast computation of eigenvalues will not miss any of the eigenvalues.

2.4.3.3 Computation of eigenfunctions

Once the eigenvalue μ_i is available, the corresponding eigenfunction Z_i can be determined if the Z^j ($j = 0, 1, \ldots, n$) are known. They can be computed by utilizing (2.57) as described in the following.

From (2.57), we have

$$a_0 Z^0 + b_1 Z^1 = 0 \tag{2.66a}$$

$$b_{j+1} Z^{j+1} = -a_j Z^j + b_j Z^{j-1} \;\; j = 1, 2, \ldots, n-1 \tag{2.66b}$$

$$b_n Z^{n-1} + a_n Z^n = 0 \tag{2.66c}$$

Since the eigenfunctions are arbitrary within a multiplication constant, one can set

$$Z^0 = -b_1,\ Z^1 = a_0 \tag{2.66d}$$

The rest of Z^j ($j = 2, 3, \ldots, n$) is obtained from the recurrence (2.66b). Finally, (2.66c) is used to estimate the magnitude of the global error involved in the computation of Z^j. For the special case of $b_n = 0$ (*Dirichlet* boundary condition at $z = H$), we have $Z^n = 0$. The accuracy estimate can be made from (2.66b), written for $j = n - 1$.

2.4.3.4 Average degree of consolidation

The average degree of consolidation U for the n-layer soil profile can be expressed in (2.67) according to the definition in (2.24)

$$U = \frac{\int_0^H m_v\left[\sigma_z(z,t) - \sigma_z(z,0) + u_0(z)\right]dz - \sum_{i=1}^{+\infty} T_i(t)\int_0^H m_v Z_i dz}{\int_0^H m_v\left[\sigma_z(z,t=\infty) - \sigma_z(z,0) + u_0(z)\right]dz} \tag{2.67}$$

REFERENCES

Das, BM, 2018, *Advanced Soil Mechanics*, 5th ed., (Oxford: Taylor & Francis).

Gibson, RL, Schiffman, RL, and Whitman, RV, 1989, On two definitions of excess pore water pressure. *Géotechnique*, 39(1), pp. 169–171.

Knappett, J, 2019, *Craig's Soil Mechanics*, 9th ed., (Oxford: Taylor & Francis).

Mikhailov, MD, and Özişik, MN, 1994, *Unified Analysis and Solutions of Heat and Mass Diffusion*, (New York: Dover Publications, Inc.)

Mikhailov, MD, and Vulchanov, NL, 1983, Computational procedure for Sturm-Liouville problems. *Journal of Computational Physics*, 50, pp. 323–336.

Olson, RE, 1977, Consolidation under time-dependent loading. *Journal of Geotechnical Engineering Division*, ASCE, 103(GT1), pp. 55–60.

Pyrah, IC, 1996, One-dimensional consolidation of layered soils. *Géotechnique*, 46(3), pp. 555–560.

Skempton, AW, and Bjerrum, L, 1957, A contribution to the settlement analysis of foundations on clay. *Géotechnique*, 7, pp. 168–178.

Terzaghi, K, 1925, *Erdbaumechanik auf Boden-physicalischen Grundlagen*, (Vienna: Deuticke).

Terzaghi, K, 1943, *Theoretical Soil Mechanics*, (New York: Wiley).

Wittrick, WH, and Williams, FW, 1971, A general algorithm for computing natural frequencies of elastic structures. *Quarterly Journal of Mechanics and Applied Mathematics*, 24(3), pp. 263–284.

Zhu, GF, and Yin, J-H, 1998, Consolidation of soil under depth-dependent ramp load. *Canadian Geotechnical Journal,* 35(3), pp. 344–350.

Zhu, GF, and Yin, J-H, 1999, Consolidation of double soil layers under depth-dependent ramp load. *Géotechnique,* 49(3), pp. 415–421.

Zhu, GF, and Yin, J-H, 2005, Solution charts for the consolidation of double soil layers. *Canadian Geotechnical Journal*, 42(3), pp. 949–956.

Chapter 3

Analytical solutions to several one-dimensional consolidation problems

3.1 INTRODUCTION

The general theory of one-dimensional (1-D) consolidation was presented in Chapter 2. This chapter presents several typical solutions for 1-D consolidation problems. The finite element method for 1-D consolidation analysis will be discussed in Chapter 5.

3.2 SOLUTION FOR A SOIL LAYER INCREASING IN THICKNESS WITH TIME

It is occasionally required to estimate the progress of consolidation in a soil layer which is increasing in thickness with time. Some typical examples are the dissipation of porewater pressure in newly dumped soils on the seabed, mine tailings deposited in a disposal pond. In these cases, the rate of deposition is known. When the deposition is fairly uniform over an area whose dimensions are large compared to the thickness of the layer, the consolidation will be approximately 1-D.

In the mathematical treatment of this kind of problem, it is necessary to satisfy a condition on the porewater pressure at a *moving* boundary; and for this reason, analytical solutions are difficult to obtain. The solutions derived by Gibson (1958) for two particular cases are recalculated in this section.

As shown in Figure 3.1, sedimentation takes place through still water of depth H_0, the current thickness of the deposit being $H(t)$. The initial thickness $H(0)$ of the layer is taken as zero.

The equation governing the consolidation of the soil layer can be rewritten from (2.13) as

$$c_v \frac{\partial^2 u_e}{\partial z^2} = \frac{\partial u_e}{\partial t} - \gamma' \frac{dH}{dt} \tag{3.1}$$

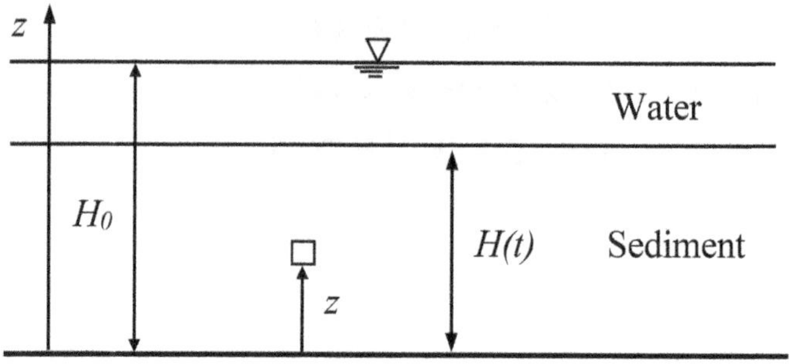

Figure 3.1 A soil layer increasing in thickness with time.

where γ' is the buoyant unit weight and H_0 is selected as the reference water head.

The boundary conditions for a permeable base and an impermeable base are expressed in (3.2) and (3.3), respectively.

$$\begin{cases} u_e(0,t) = 0 \\ u_e(H,t) = 0 \end{cases} \tag{3.2}$$

$$\begin{cases} \dfrac{\partial u_e}{\partial z}(0,t) = 0 \\ u_e(H,t) = 0 \end{cases} \tag{3.3}$$

The term $-\gamma'\frac{dH}{dt}$ in (3.1) arises because the total vertical stress varies with time, this variation being caused by the increasing weight of superincumbent sediment. Analytical solutions were found for the time-thickness relations of (a) $H(t) = R\sqrt{t}$ and (b)$H(t) = Qt$, where R and Q are constants.

3.2.1 Thickness of deposition proportional to $\sqrt{t}$

For the time-thickness relation of $H(t) = R\sqrt{t}$, the solutions to (3.1) with boundary conditions (3.2) and (3.3) are expressed in (3.4) and (3.5), respectively.

$$u_e = \gamma' R\sqrt{t}\left\{1 - \exp\left(-\frac{z^2}{4c_v t}\right) + \frac{z}{R\sqrt{t}}\left[\exp\left(-\frac{R^2}{4c_v t}\right) - 1\right] + \frac{z}{2}\sqrt{\frac{\pi}{c_v t}}\left[erf\left(\frac{R}{2\sqrt{c_v}}\right) - erf\left(\frac{z}{2\sqrt{c_v t}}\right)\right]\right\} \tag{3.4}$$

$$u_e = \gamma' R\sqrt{t}\left[1 - \frac{\exp\left(-\dfrac{z^2}{4c_v t}\right) + \dfrac{z}{2}\sqrt{\dfrac{\pi}{c_v t}}\,erf\left(\dfrac{z}{2\sqrt{c_v t}}\right)}{\exp\left(-\dfrac{R^2}{4c_v}\right) + \dfrac{R}{2}\sqrt{\dfrac{\pi}{c_v}}\,erf\left(\dfrac{R}{2\sqrt{c_v}}\right)}\right] \tag{3.5}$$

Numerical values of excess porewater pressure have been evaluated from these expressions and presented in Tables 3.1 and 3.2. Curves of $u_e/\gamma' H$ against z/H are given in Figures 3.2–3.3 for the various values of $R/\sqrt{c_v}$. It is of interest to note that for this particular rate of deposition the distribution of excess porewater pressure through the thickness of the deposit is controlled only by the time independent parameter $R/\sqrt{c_v}$.

3.2.2 Constant rate of deposition

For the time-thickness relation of $H(t) = Qt$, the solutions to (3.1) with boundary conditions (3.2) and (3.3) are expressed in (3.6) and (3.7), respectively.

$$u_e = -\gamma' z\left(1 + \frac{Qz}{2c_v}\right) + \frac{\gamma' Q}{2c_v\sqrt{\pi c_v t}}\exp\left(-\frac{z^2}{4c_v t}\right)\int_0^{+\infty} \xi^2 \coth\left(\frac{Q\xi}{2c_v}\right)$$

$$\sinh\left(\frac{z\xi}{2c_v t}\right)\exp\left(-\frac{\xi^2}{4c_v t}\right)d\xi \tag{3.6}$$

$$u_e = \gamma' Qt - \frac{\gamma'}{\sqrt{\pi c_v t}}\exp\left(-\frac{z^2}{4c_v t}\right)\int_0^{+\infty} \xi \tanh\left(\frac{Q\xi}{2c_v}\right)\cosh\left(\frac{z\xi}{2c_v t}\right)\exp\left(-\frac{\xi^2}{4c_v t}\right)d\xi \tag{3.7}$$

The integrals in (3.6–3.7) must be calculated numerically. Results of excess porewater pressure from (3.6–3.7) are presented in Tables 3.3 and 3.4. Curves of $u_e/\gamma' H$ against z/H are given in Figures 3.4–3.5 for the various values of $c_v t/H^2$.

3.2.3 Average degree of consolidation subsequent to cease of deposition

If the layer increases in thickness up to a time t_0 and then deposition ceases, the subsequent average degree of consolidation can be calculated using (2.67) (Zhu and Yin, 2005). At the moment when deposition ceases, the distribution of excess porewater pressure expressed in (3.4–3.7) will form the 'initial' condition for its subsequent dissipation.

Table 3.1 $u_e/\gamma'H$ values for different $R/\sqrt{c_v}$ (left column) and z/H (top row) for $H(t) = R\sqrt{t}$ (permeable base)

$\frac{R}{\sqrt{c_v}}$	*z/H* 0	0.05	0.1	0.15	0.2	0.25	0.3	0.35	0.4	0.45	0.5	0.55	0.6	0.65	0.7	0.75	0.8	0.85	0.9	0.95	1
0.5	0	0.003	0.006	0.008	0.01	0.012	0.013	0.014	0.015	0.015	0.015	0.015	0.015	0.014	0.013	0.011	0.01	0.008	0.005	0.003	0
1	0	0.011	0.022	0.03	0.038	0.044	0.05	0.054	0.056	0.058	0.058	0.057	0.055	0.052	0.048	0.043	0.036	0.029	0.02	0.011	0
1.5	0	0.024	0.046	0.065	0.081	0.094	0.104	0.112	0.117	0.12	0.12	0.118	0.113	0.106	0.097	0.086	0.072	0.057	0.04	0.021	0
1.8	0	0.034	0.064	0.089	0.111	0.129	0.143	0.153	0.16	0.163	0.162	0.158	0.151	0.141	0.129	0.113	0.095	0.074	0.052	0.027	0
2	0	0.041	0.076	0.107	0.133	0.154	0.17	0.181	0.189	0.192	0.191	0.186	0.177	0.165	0.149	0.131	0.11	0.086	0.059	0.031	0
2.5	0	0.059	0.11	0.153	0.189	0.217	0.239	0.253	0.262	0.264	0.26	0.251	0.237	0.219	0.197	0.171	0.142	0.11	0.075	0.039	0
3	0	0.078	0.145	0.201	0.246	0.281	0.306	0.322	0.33	0.329	0.322	0.308	0.288	0.264	0.235	0.202	0.166	0.127	0.087	0.044	0
3.5	0	0.098	0.18	0.248	0.301	0.341	0.369	0.384	0.389	0.385	0.373	0.353	0.327	0.296	0.261	0.223	0.182	0.139	0.094	0.047	0
4	0	0.117	0.215	0.293	0.353	0.397	0.424	0.438	0.439	0.43	0.412	0.386	0.355	0.319	0.279	0.236	0.191	0.145	0.097	0.049	0
4.5	0	0.137	0.249	0.336	0.402	0.446	0.473	0.483	0.479	0.465	0.441	0.41	0.373	0.333	0.289	0.243	0.196	0.148	0.099	0.05	0
5	0	0.156	0.281	0.377	0.446	0.491	0.514	0.52	0.511	0.491	0.461	0.425	0.385	0.341	0.295	0.247	0.198	0.149	0.1	0.05	0
6	0	0.193	0.343	0.452	0.524	0.564	0.579	0.574	0.554	0.523	0.485	0.442	0.396	0.348	0.299	0.25	0.2	0.15	0.1	0.05	0
7	0	0.23	0.4	0.517	0.587	0.62	0.624	0.608	0.578	0.539	0.495	0.448	0.399	0.35	0.3	0.25	0.2	0.15	0.1	0.05	0
8	0	0.265	0.453	0.574	0.638	0.661	0.654	0.628	0.59	0.546	0.498	0.449	0.4	0.35	0.3	0.25	0.2	0.15	0.1	0.05	0
10	0	0.331	0.546	0.664	0.711	0.711	0.685	0.645	0.598	0.55	0.5	0.45	0.4	0.35	0.3	0.25	0.2	0.15	0.1	0.05	0
14	0	0.45	0.687	0.774	0.778	0.745	0.699	0.65	0.6	0.55	0.5	0.45	0.4	0.35	0.3	0.25	0.2	0.15	0.1	0.05	0
20	0	0.596	0.811	0.835	0.798	0.75	0.7	0.65	0.6	0.55	0.5	0.45	0.4	0.35	0.3	0.25	0.2	0.15	0.1	0.05	0
30	0	0.764	0.885	0.85	0.8	0.75	0.7	0.65	0.6	0.55	0.5	0.45	0.4	0.35	0.3	0.25	0.2	0.15	0.1	0.05	0

Table 3.2 $u_e/\gamma' H$ values for different $R/\sqrt{c_v}$ (left column) and z/H (top row) for $H(t) = R\sqrt{t}$ (impermeable base)

$\frac{R}{\sqrt{c_v}}$	z/H: *0*	*0.05*	*0.1*	*0.15*	*0.2*	*0.25*	*0.3*	*0.35*	*0.4*	*0.45*	*0.5*	*0.55*	*0.6*	*0.65*	*0.7*	*0.75*	*0.8*	*0.85*	*0.9*	*0.95*	*1*
0.4	0.038	0.038	0.038	0.037	0.037	0.036	0.035	0.034	0.032	0.03	0.029	0.027	0.024	0.022	0.019	0.017	0.014	0.011	0.007	0.004	0
0.6	0.081	0.081	0.081	0.08	0.078	0.076	0.074	0.071	0.068	0.065	0.061	0.057	0.052	0.047	0.041	0.035	0.029	0.022	0.015	0.008	0
0.8	0.135	0.135	0.133	0.132	0.129	0.126	0.122	0.118	0.113	0.107	0.1	0.093	0.085	0.077	0.068	0.058	0.048	0.037	0.025	0.013	0
1	0.194	0.193	0.192	0.189	0.186	0.181	0.176	0.169	0.162	0.153	0.144	0.133	0.122	0.11	0.097	0.083	0.068	0.052	0.036	0.018	0
1.2	0.254	0.253	0.251	0.248	0.243	0.237	0.23	0.221	0.211	0.2	0.187	0.174	0.159	0.143	0.126	0.107	0.088	0.068	0.046	0.023	0
1.4	0.312	0.311	0.309	0.304	0.299	0.291	0.282	0.271	0.259	0.245	0.229	0.213	0.194	0.174	0.153	0.131	0.107	0.082	0.056	0.028	0
1.6	0.367	0.366	0.363	0.358	0.351	0.342	0.331	0.318	0.303	0.287	0.268	0.248	0.226	0.203	0.178	0.152	0.124	0.095	0.064	0.033	0
1.8	0.417	0.416	0.413	0.407	0.399	0.388	0.375	0.36	0.343	0.324	0.303	0.28	0.255	0.229	0.2	0.17	0.139	0.106	0.072	0.036	0
2	0.463	0.461	0.457	0.451	0.441	0.43	0.415	0.398	0.379	0.358	0.334	0.308	0.28	0.251	0.219	0.186	0.151	0.115	0.078	0.04	0
2.4	0.54	0.538	0.533	0.525	0.514	0.499	0.481	0.461	0.438	0.412	0.383	0.353	0.32	0.285	0.249	0.21	0.171	0.13	0.087	0.044	0
2.7	0.587	0.585	0.579	0.57	0.557	0.541	0.521	0.498	0.472	0.443	0.412	0.378	0.342	0.304	0.264	0.223	0.18	0.137	0.092	0.046	0
3	0.626	0.624	0.618	0.607	0.593	0.575	0.553	0.527	0.499	0.467	0.433	0.397	0.358	0.318	0.276	0.232	0.187	0.142	0.095	0.048	0
4	0.718	0.715	0.707	0.693	0.674	0.65	0.622	0.59	0.555	0.516	0.475	0.432	0.387	0.341	0.294	0.246	0.198	0.149	0.099	0.05	0
5	0.774	0.771	0.76	0.743	0.72	0.691	0.658	0.621	0.58	0.537	0.491	0.444	0.397	0.348	0.299	0.249	0.2	0.15	0.1	0.05	0
7	0.839	0.834	0.819	0.796	0.766	0.729	0.688	0.643	0.596	0.548	0.499	0.45	0.4	0.35	0.3	0.25	0.2	0.15	0.1	0.05	0
10	0.887	0.88	0.86	0.829	0.79	0.746	0.698	0.649	0.6	0.55	0.5	0.45	0.4	0.35	0.3	0.25	0.2	0.15	0.1	0.05	0
20	0.944	0.93	0.895	0.849	0.8	0.75	0.7	0.65	0.6	0.55	0.5	0.45	0.4	0.35	0.3	0.25	0.2	0.15	0.1	0.05	0

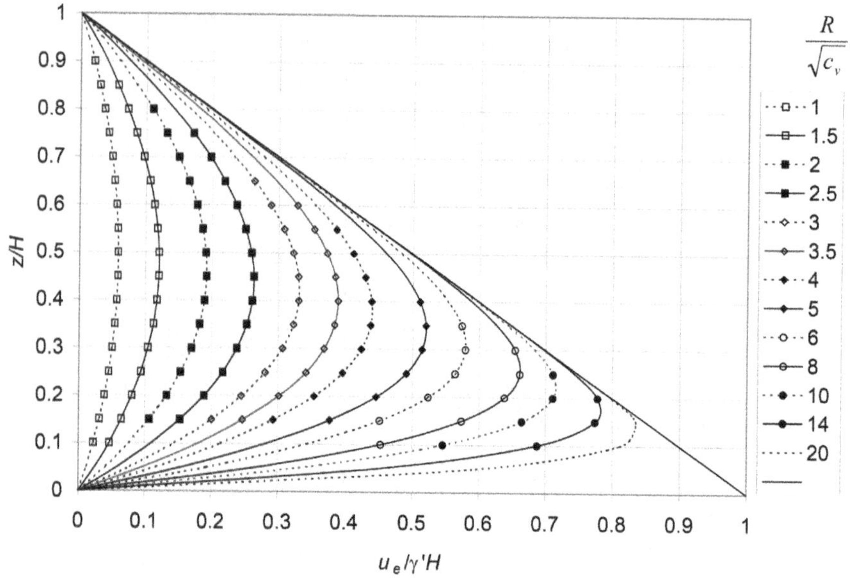

Figure 3.2 Relation between $u_e/\gamma'H$ and z/H for $H(t) = R\sqrt{t}$ (permeable base).

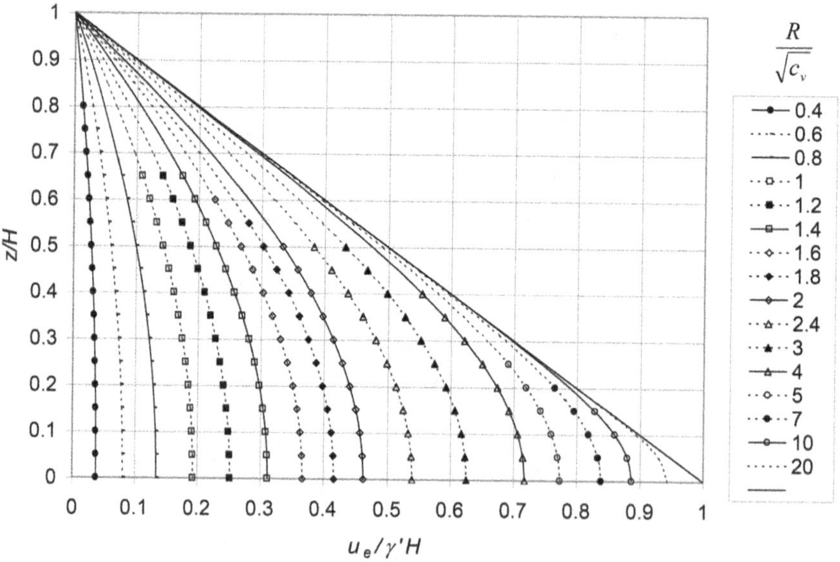

Figure 3.3 Relation between $u_e/\gamma'H$ and z/H for $H(t) = R\sqrt{t}$ (impermeable base).

Table 3.3 $u_e/\gamma'H$ values for different c_vt/H^2 (left column) and z/H (top row) for $H(t) = Qt$ (permeable base)

$\frac{c_vt}{H^2}$	*z/H*																				
	0	*0.05*	*0.1*	*0.15*	*0.2*	*0.25*	*0.3*	*0.35*	*0.4*	*0.45*	*0.5*	*0.55*	*0.6*	*0.65*	*0.7*	*0.75*	*0.8*	*0.85*	*0.9*	*0.95*	*1*
0.002	0	0.714	0.863	0.846	0.8	0.75	0.7	0.65	0.6	0.55	0.5	0.45	0.4	0.35	0.3	0.25	0.2	0.15	0.1	0.05	0
0.005	0	0.531	0.749	0.804	0.788	0.748	0.7	0.65	0.6	0.55	0.5	0.45	0.4	0.35	0.3	0.25	0.2	0.15	0.1	0.05	0
0.01	0	0.401	0.62	0.719	0.743	0.728	0.692	0.647	0.599	0.55	0.5	0.45	0.4	0.35	0.3	0.25	0.2	0.15	0.1	0.05	0
0.02	0	0.291	0.481	0.594	0.65	0.666	0.655	0.627	0.589	0.545	0.498	0.449	0.4	0.35	0.3	0.25	0.2	0.15	0.1	0.05	0
0.03	0	0.237	0.404	0.513	0.578	0.607	0.611	0.596	0.569	0.532	0.49	0.445	0.397	0.349	0.299	0.25	0.2	0.15	0.1	0.05	0
0.05	0	0.179	0.315	0.412	0.478	0.517	0.535	0.535	0.521	0.498	0.466	0.428	0.387	0.342	0.295	0.247	0.199	0.149	0.1	0.05	0
0.07	0	0.147	0.263	0.35	0.412	0.453	0.475	0.483	0.477	0.461	0.437	0.406	0.37	0.33	0.287	0.242	0.195	0.148	0.099	0.05	0
0.1	0	0.118	0.214	0.288	0.345	0.384	0.409	0.421	0.422	0.413	0.397	0.373	0.344	0.31	0.272	0.231	0.188	0.143	0.096	0.049	0
0.15	0	0.091	0.166	0.227	0.274	0.31	0.334	0.348	0.354	0.351	0.341	0.325	0.303	0.276	0.245	0.211	0.173	0.133	0.09	0.046	0
0.2	0	0.074	0.137	0.188	0.23	0.261	0.284	0.298	0.305	0.305	0.299	0.287	0.269	0.247	0.221	0.191	0.158	0.122	0.083	0.043	0
0.3	0	0.055	0.102	0.142	0.175	0.2	0.22	0.233	0.24	0.242	0.239	0.232	0.219	0.203	0.183	0.16	0.133	0.104	0.072	0.037	0
0.4	0	0.044	0.082	0.115	0.142	0.163	0.18	0.192	0.199	0.201	0.2	0.194	0.185	0.172	0.156	0.137	0.115	0.09	0.062	0.032	0
0.6	0	0.032	0.06	0.084	0.104	0.12	0.133	0.142	0.148	0.151	0.151	0.147	0.141	0.132	0.12	0.106	0.089	0.07	0.049	0.026	0
1	0	0.021	0.039	0.054	0.068	0.079	0.088	0.094	0.099	0.101	0.101	0.099	0.096	0.09	0.082	0.073	0.062	0.049	0.034	0.018	0
2.5	0	0.009	0.017	0.024	0.03	0.035	0.039	0.042	0.044	0.045	0.046	0.045	0.043	0.041	0.038	0.034	0.029	0.023	0.016	0.008	0

Table 3.4 $u_e/\gamma' H$ values for different $c_v t/H^2$ (left column) and z/H (top row) for $H(t) = Qt$ (impermeable base)

$\frac{c_v t}{H^2}$	z/H																				
	0	*0.05*	*0.1*	*0.15*	*0.2*	*0.25*	*0.3*	*0.35*	*0.4*	*0.45*	*0.5*	*0.55*	*0.6*	*0.65*	*0.7*	*0.75*	*0.8*	*0.85*	*0.9*	*0.95*	*1*
0.01	0.888	0.881	0.861	0.83	0.79	0.746	0.698	0.649	0.6	0.55	0.5	0.45	0.4	0.35	0.3	0.25	0.2	0.15	0.1	0.05	0
0.02	0.843	0.838	0.823	0.799	0.768	0.731	0.689	0.644	0.597	0.549	0.499	0.45	0.4	0.35	0.3	0.25	0.2	0.15	0.1	0.05	0
0.04	0.781	0.778	0.767	0.749	0.726	0.696	0.662	0.624	0.583	0.539	0.493	0.446	0.398	0.349	0.299	0.25	0.2	0.15	0.1	0.05	0
0.06	0.736	0.733	0.724	0.71	0.69	0.665	0.635	0.602	0.565	0.525	0.482	0.438	0.392	0.345	0.297	0.248	0.199	0.149	0.1	0.05	0
0.1	0.668	0.666	0.659	0.648	0.632	0.611	0.587	0.559	0.528	0.494	0.457	0.417	0.376	0.332	0.288	0.242	0.195	0.147	0.098	0.049	0
0.2	0.559	0.557	0.552	0.544	0.532	0.517	0.499	0.478	0.454	0.428	0.399	0.367	0.333	0.297	0.259	0.219	0.178	0.135	0.091	0.046	0
0.25	0.52	0.518	0.514	0.506	0.496	0.482	0.466	0.447	0.426	0.401	0.375	0.346	0.314	0.281	0.246	0.208	0.17	0.129	0.087	0.044	0
0.3	0.487	0.486	0.482	0.475	0.465	0.453	0.438	0.421	0.401	0.378	0.354	0.327	0.297	0.266	0.233	0.198	0.162	0.123	0.084	0.042	0
0.35	0.459	0.458	0.454	0.448	0.439	0.428	0.414	0.398	0.379	0.358	0.335	0.31	0.282	0.253	0.222	0.189	0.154	0.118	0.08	0.041	0
0.4	0.435	0.434	0.43	0.424	0.416	0.405	0.392	0.377	0.36	0.34	0.318	0.295	0.269	0.241	0.212	0.18	0.147	0.113	0.077	0.039	0
0.5	0.394	0.393	0.39	0.385	0.377	0.368	0.356	0.343	0.327	0.31	0.29	0.269	0.246	0.221	0.194	0.166	0.135	0.104	0.071	0.036	0
0.6	0.361	0.36	0.358	0.353	0.346	0.338	0.327	0.315	0.301	0.285	0.267	0.248	0.226	0.204	0.179	0.153	0.125	0.096	0.065	0.033	0
0.8	0.311	0.31	0.308	0.304	0.298	0.291	0.282	0.271	0.259	0.246	0.231	0.214	0.196	0.176	0.155	0.133	0.109	0.084	0.057	0.029	0
1	0.274	0.273	0.271	0.267	0.262	0.256	0.248	0.239	0.229	0.217	0.204	0.189	0.173	0.156	0.138	0.118	0.097	0.074	0.051	0.026	0
1.5	0.212	0.211	0.21	0.207	0.203	0.198	0.193	0.186	0.177	0.168	0.158	0.147	0.135	0.121	0.107	0.092	0.075	0.058	0.04	0.02	0
2.5	0.147	0.147	0.146	0.144	0.141	0.138	0.134	0.129	0.123	0.117	0.11	0.102	0.094	0.085	0.075	0.064	0.053	0.041	0.028	0.014	0
4	0.102	0.101	0.1	0.099	0.097	0.095	0.092	0.089	0.085	0.081	0.076	0.071	0.065	0.059	0.052	0.044	0.036	0.028	0.019	0.01	0
8	0.056	0.056	0.055	0.055	0.054	0.052	0.051	0.049	0.047	0.045	0.042	0.039	0.036	0.032	0.028	0.024	0.02	0.015	0.011	0.005	0

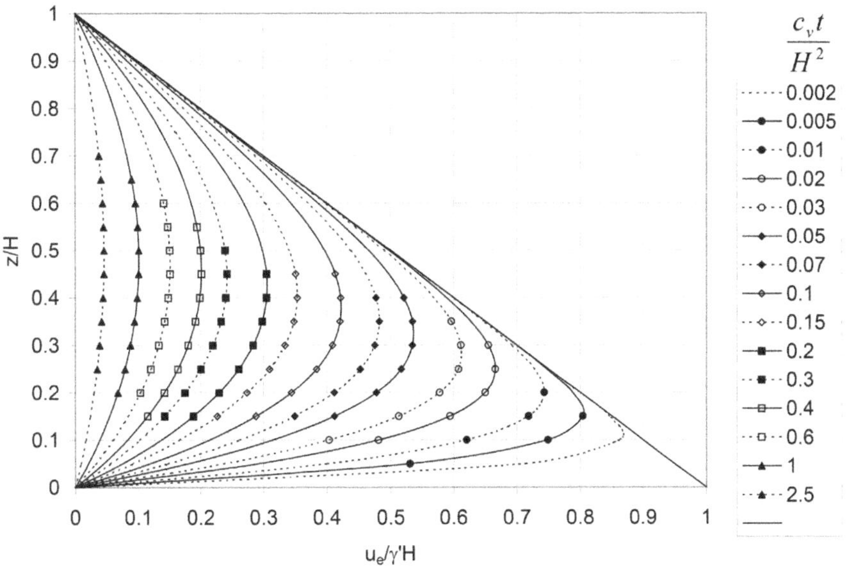

Figure 3.4 Relation between $u_e/\gamma'H$ and z/H for $H(t) = Qt$ (permeable base).

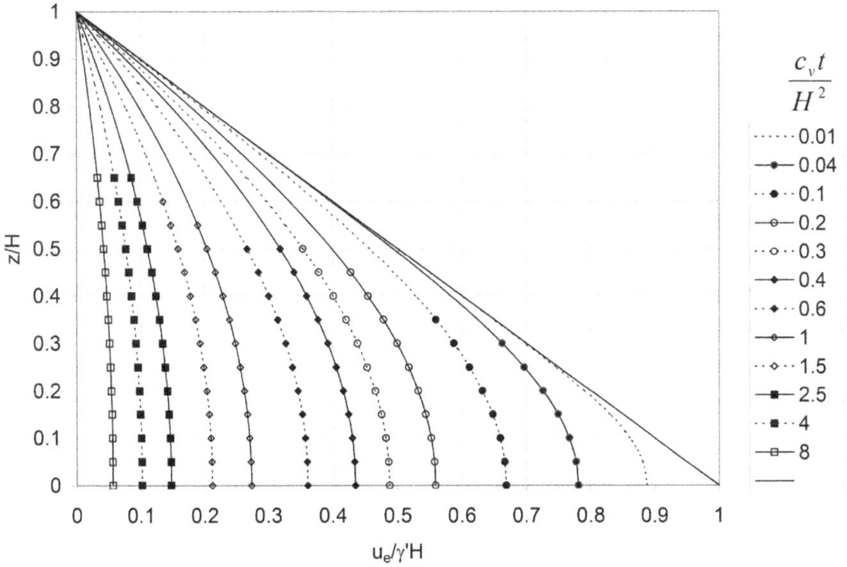

Figure 3.5 Relation between $u_e/\gamma'H$ and z/H for $H(t) = Qt$ (impermeable base).

From (2.67), the average degree of consolidation for boundary conditions (3.2) and (3.3) is expressed in (3.8) and (3.9), respectively.

$$U = 1 - \sum_{i=1}^{+\infty} \frac{4}{(2i-1)\pi} \frac{\int_0^H u_e(z,t_0)\sin\left((2i-1)\pi\frac{z}{H}\right)dz}{\int_0^H u_e(z,t_0)dz} \exp\left(-(2i-1)^2\pi^2\frac{c_v(t-t_0)}{H^2}\right) \tag{3.8}$$

$$U = 1 - \sum_{i=1}^{+\infty} \frac{4(-1)^{i-1}}{(2i-1)\pi} \frac{\int_0^H u_e(z,t_0)\cos\left(\left(\frac{2i-1}{2}\right)\pi\frac{z}{H}\right)dz}{\int_0^H u_e(z,t_0)dz} \exp\left(-\left(\frac{2i-1}{2}\right)^2\pi^2\frac{c_v(t-t_0)}{H^2}\right) \tag{3.9}$$

Substituting (3.4) and (3.6) into (3.8), and (3.5) and (3.7) into (3.9), the average degree of consolidation can be obtained. The integrals in (3.6–3.9) are calculated numerically. For the purposes of comparison, the average degrees of consolidation for the initial excess pore pressures (A) and (B) in Figure 3.6 are also calculated and denoted as U_A and U_B as follows (see (2.27))

$$U_A = 1 - \sum_{n=1,3,5,\ldots} \frac{8}{n^2\pi^2}\exp\left(-\frac{n^2\pi^2}{4}T_v\right) \tag{3.10}$$

$$U_B = 1 - \sum_{n=1,3,5,\ldots} (-1)^{\frac{n-1}{2}} \frac{32}{n^3\pi^3}\exp\left(-\frac{n^2\pi^2}{4}T_v\right) \tag{3.11}$$

where

$$T_v = \frac{c_v(t-t_0)}{H^2(t_0)} \tag{3.12}$$

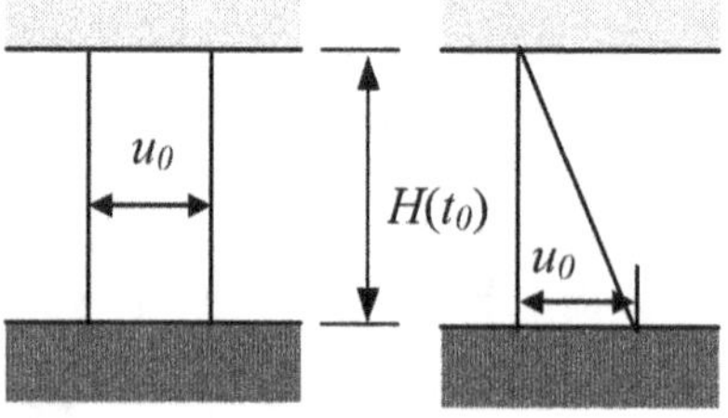

Figure 3.6 Two initial excess pore pressure distributions: (a) uniform, and (b) triangular.

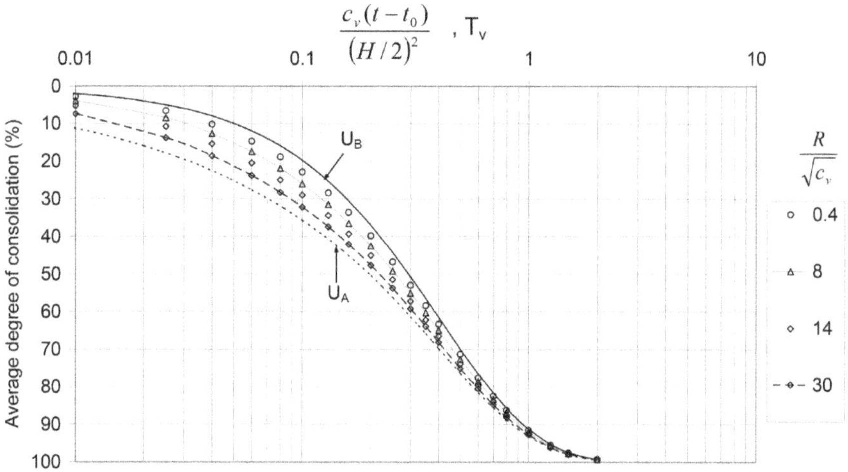

Figure 3.7 U for different $R/\sqrt{c_v}$ and initial deposition $H(t) = R\sqrt{t}$ (permeable base).

Numerical values of the average degrees of consolidation are presented in Table 3.5 and in Figures 3.7–3.10. Clearly, the average degrees of consolidation fall in a narrow band for a wide range of $R/\sqrt{c_v}$ or $c_v t_0/H^2$.

For the case of two-way drainage, the average degrees of consolidation are bounded by U_A and U_B as shown in Figures 3.7 and 3.9. If the soil layer is quickly deposited, the curve of average degree of consolidation is close to U_A. If the soil layer is slowly deposited, the curve of average degree of consolidation is close to U_B. Considering the wide range of the two time-thickness relations can express, it could be expected that the average degree

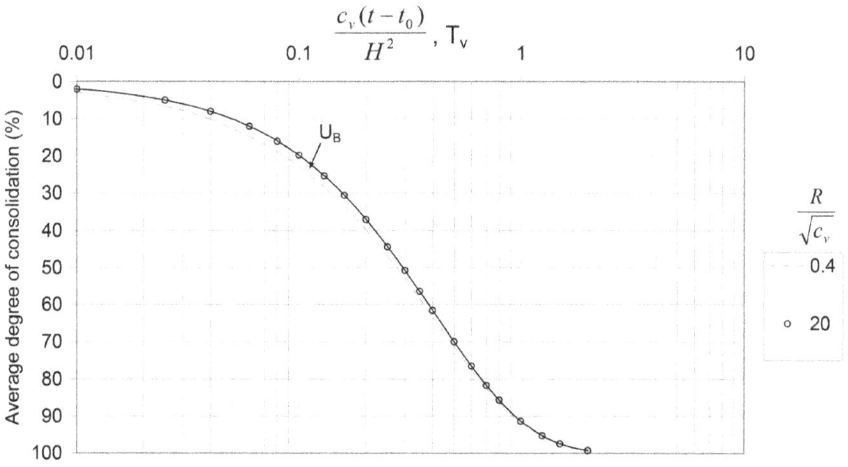

Figure 3.8 U for different $R/\sqrt{c_v}$ and initial deposition $H(t) = R\sqrt{t}$ (impermeable base).

Table 3.5 Average degree of consolidation U subsequent to cease of deposition for thickness increases of $H(t)=R\sqrt{t}$ and $H(t)=Qt$

$\frac{c_v(t-t_0)}{4H^2}$	$R/\sqrt{c_v}$ ($H=R\sqrt{t}$, permeable base)							$c_v t_0/H^2$ ($H=Qt$, permeable base)				$\frac{c_v(t-t_0)}{H^2}$	$R/\sqrt{c_v}$ ($H=R\sqrt{t}$, impermeable base)					$c_v t_0/H^2$ ($H=Qt$, impermeable base)			
	0.4	*4*	*8*	*10*	*14*	*20*	*30*	*0.002*	*0.01*	*0.03*	*2.5*		*0.4*	*1*	*2*	*4*	*20*	*0.01*	*0.1*	*1*	*8*
0.01	2.8	2.9	3.8	4.3	5.2	6.2	7.4	7.1	4.8	3.6	2.8	0.01	2.8	2.7	2.6	2.3	2	2	2.3	2.7	2.8
0.025	6.6	6.8	8.5	9.4	10.8	12.3	13.8	13.4	10.2	8.2	6.6	0.025	6.6	6.5	6.2	5.6	5	5.1	5.8	6.5	6.6
0.04	10.2	10.5	12.7	13.7	15.4	17	18.6	18.2	14.7	12.2	10.2	0.04	10.2	10.1	9.7	8.9	8	8.2	9.1	10.1	10.2
0.06	14.7	15	17.5	18.7	20.5	22.2	23.8	23.4	19.8	17	14.7	0.06	14.7	14.5	14.1	13.1	12	12.2	13.4	14.6	14.7
0.08	18.9	19.2	21.9	23.2	25	26.7	28.3	27.8	24.2	21.4	18.9	0.08	18.9	18.7	18.3	17.2	16	16.2	17.6	18.8	18.9
0.1	22.9	23.2	26	27.2	29	30.7	32.2	31.8	28.2	25.4	22.9	0.1	22.8	22.7	22.2	21.2	19.8	20.1	21.5	22.7	22.9
0.13	28.4	28.8	31.5	32.7	34.4	36	37.4	37.1	33.7	30.9	28.4	0.13	28.4	28.3	27.8	26.7	25.4	25.6	27.1	28.3	28.4
0.16	33.6	33.9	36.5	37.6	39.3	40.8	42.1	41.8	38.6	35.9	33.6	0.16	33.5	33.4	33	31.9	30.6	30.9	32.3	33.4	33.6
0.2	39.8	40.1	42.5	43.6	45.1	46.5	47.7	47.4	44.5	42	39.8	0.2	39.8	39.7	39.3	38.3	37.1	37.3	38.6	39.7	39.8
0.25	46.8	47.1	49.2	50.2	51.5	52.8	53.8	53.5	50.9	48.8	46.8	0.25	46.8	46.7	46.3	45.5	44.4	44.6	45.7	46.7	46.8
0.3	53	53.2	55.1	56	57.2	58.3	59.2	58.9	56.7	54.7	53	0.3	53	52.9	52.6	51.8	50.8	51	52	52.9	53
0.35	58.4	58.7	60.3	61.1	62.2	63.1	63.9	63.7	61.7	60	58.4	0.35	58.4	58.3	58.1	57.4	56.5	56.7	57.6	58.4	58.4
0.4	63.3	63.5	64.9	65.6	66.5	67.4	68.1	67.9	66.1	64.6	63.3	0.4	63.3	63.2	62.9	62.3	61.6	61.7	62.5	63.2	63.3
0.5	71.3	71.4	72.6	73.1	73.9	74.5	75.1	74.9	73.5	72.4	71.3	0.5	71.3	71.2	71	70.6	70	70.1	70.7	71.2	71.3
0.6	77.6	77.7	78.6	79	79.6	80.1	80.5	80.4	79.3	78.4	77.6	0.6	77.6	77.5	77.4	77	76.5	76.6	77.1	77.5	77.6
0.7	82.5	82.6	83.3	83.6	84	84.4	84.8	84.7	83.8	83.1	82.5	0.7	82.5	82.4	82.3	82	81.7	81.7	82.1	82.4	82.5
0.8	86.3	86.4	86.9	87.2	87.5	87.8	88.1	88	87.4	86.8	86.3	0.8	86.3	86.3	86.2	86	85.7	85.7	86	86.3	86.3
1	91.6	91.7	92	92.2	92.4	92.6	92.7	92.7	92.3	92	91.6	1	91.6	91.6	91.6	91.4	91.3	91.3	91.5	91.6	91.6
1.25	95.5	95.5	95.7	95.8	95.9	96	96.1	96.1	95.8	95.7	95.5	1.25	95.5	95.5	95.4	95.4	95.3	95.3	95.4	95.5	95.5
1.5	97.6	97.6	97.7	97.7	97.8	97.8	97.9	97.9	97.8	97.7	97.6	1.5	97.6	97.6	97.5	97.5	97.5	97.5	97.5	97.6	97.6
2	99.3	99.3	99.3	99.3	99.4	99.4	99.4	99.4	99.3	99.3	99.3	2	99.3	99.3	99.3	99.3	99.3	99.3	99.3	99.3	99.3

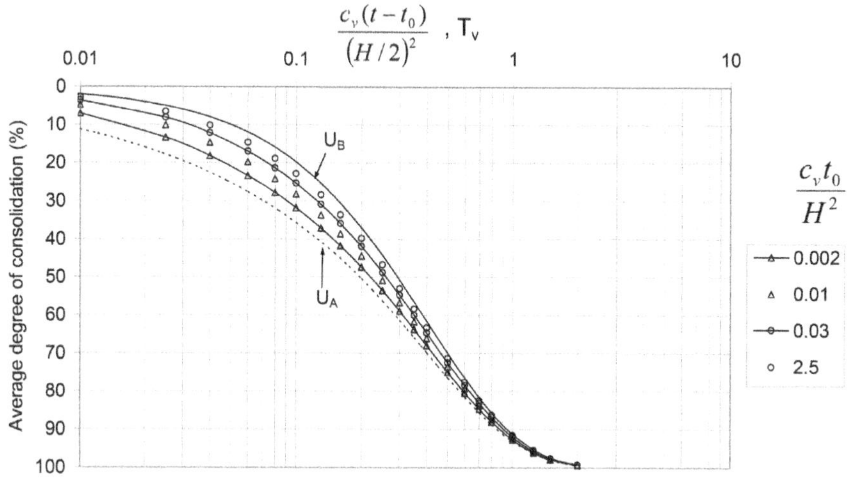

Figure 3.9 U for different $c_v t_0/H^2$ and initial deposition $H(t) = Qt$ (permeable base).

of consolidation subsequent to cease of deposition still fall in the range bounded by U_A and U_B if the thickness monotonically increases with time in the deposition stage.

For the case of one-way drainage, the average degrees of consolidation are very close to U_B as shown in Figures 3.8 and 3.10 for both time-thickness relations. Thus, U_B can be used to estimate the average degree of consolidation subsequent to cease of deposition for one-way drainage boundaries if the thickness monotonically increases with time in the deposition stage.

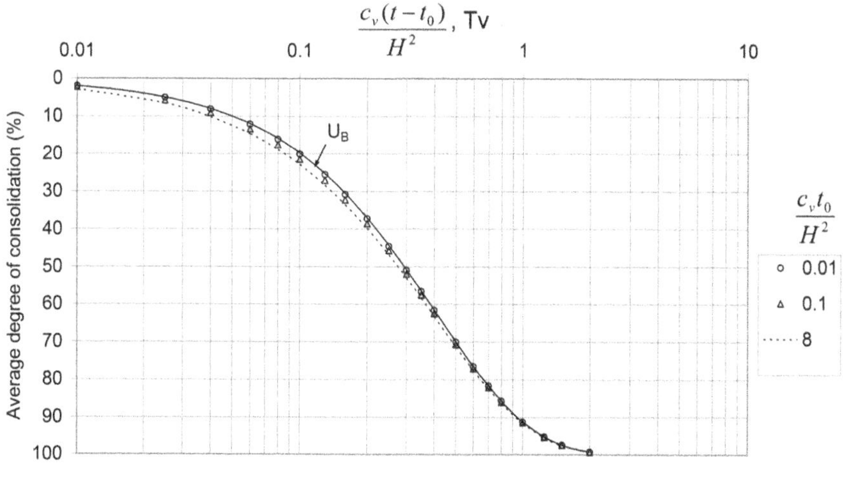

Figure 3.10 U for different $c_v t_0/H^2$ and initial deposition $H(t) = Qt$ (impermeable base).

3.2.4 Consolidation of a saturated fill

The prediction of the progress of consolidation in a fill of soft saturated clay is closely related to the problem in this section. Suppose the fill of thickness *H(t)* is to be placed. The water level will be at the top of the layer. The equation governing the water pressure in this case will be:

$$c_v \frac{\partial^2 u_w}{\partial z^2} = \frac{\partial u_w}{\partial t} - \gamma \frac{dH}{dt} \tag{3.13}$$

where γ is the unit weight of the fill.

Comparing (3.13) with (3.1), it is clear that the two equations have the same form. Therefore, the solutions to (3.1) apply to (3.13) with same boundary conditions.

3.3 CONSOLIDATION OF SOILS WITH DEPTH-DEPENDENT COMPRESSIBILITY AND PERMEABILITY

Field and laboratory tests on clay deposits have shown that the permeability and compressibility of loaded clay layers are generally not constants. These consolidation parameters may vary spatially due to variations of soil type, initial state of stress, and stress history, among other reasons. Two types of spatial variation of the permeability and compressibility exist. In the first case, the soil stratum consists of discrete layers of different soils, in each of which the soil properties are sensibly uniform as discussed in Section 2.4.3. In the second case, there is a known spatially continuous variation of the consolidation parameters in the soil stratum. It is the late problem with fairly general laws for the depth-dependent variation of soil parameters that is analyzed herein.

3.3.1 Basic equations and solutions

The soil stratum is shown in Figure 3.11. The constitutive relation is assumed to be linear and time independent. However, there is a spatially continuous variation of consolidation parameters through the soil layer, that is, (2.4) can be expressed as (3.14):

$$\begin{cases} f = m_v(z)\sigma'_z \\ g = 0 \end{cases} \tag{3.14}$$

A constant vertical total stress increment $\Delta\sigma_z = \sigma_0$ is suddenly applied. A special form of (2.11) is thus expressed as (3.15):

$$\frac{\partial}{\partial z}\left[\frac{k(z)}{\gamma_w}\frac{\partial u_e}{\partial z}\right] = m_v(z)\frac{\partial u_e}{\partial t} \tag{3.15a}$$

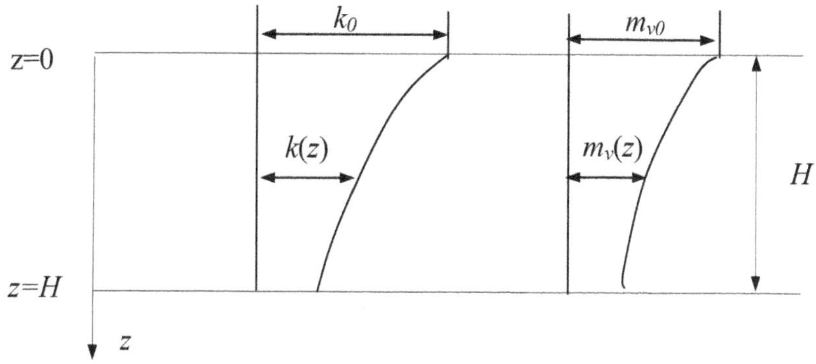

Figure 3.11 Soil profile.

$$k(z) = k_0\left(1+\alpha\frac{z}{H}\right)^p \tag{3.15b}$$

$$m_v(z) = m_{v0}\left(1+\alpha\frac{z}{H}\right)^q \tag{3.15c}$$

where p, q, α, k_0, and m_{v0} are constants.

(3.15a) can also be rewritten as

$$c_v(z)\frac{\partial^2 u_e}{\partial z^2}+\frac{1}{\gamma_w m_v(z)}\frac{\partial k(z)}{\partial z}\frac{\partial u_e}{\partial z}=\frac{\partial u_e}{\partial t} \tag{3.16a}$$

$$c_v(z) = c_{v0}\left(1+\alpha\frac{z}{H}\right)^n \tag{3.16b}$$

where $c_{v0} = \frac{k_0}{\gamma_w m_{v0}}$ and $n = p\text{-}q$.

The boundary conditions for (3.15) are the same as (2.16) and (2.17), respectively.

$$\begin{cases} u_e(0,t)=0 \\ u_e(H,t)=0 \end{cases} \tag{2.16}$$

$$\begin{cases} u_e(0,t)=0 \\ \dfrac{\partial u_e}{\partial z}\Big|_{z=H}=0 \end{cases} \tag{2.17}$$

For the boundary condition (2.16), the solution has some symmetric features. Actually, substituting $\alpha = -\alpha'/(1+\alpha')$ and $z = H - z'$ into (3.15), the following (3.17), which is similar to (3.15), is obtained:

$$\frac{\partial}{\partial z'}\left\{\frac{k_0}{\gamma_w}\left(1+\alpha'\frac{z'}{H}\right)^p\frac{\partial u_e}{\partial z'}\right\} = m_0\left(1+\alpha'\frac{z'}{H}\right)^q\frac{\partial u_e}{\partial[(1+\alpha')^{-n}t]} \tag{3.17}$$

When $n \neq 2$ and $\alpha \neq 0$, the following changes of variable,

$$y = \left(1+\alpha\frac{z}{H}\right)^{1-\frac{n}{2}} \tag{3.18a}$$

and

$$u_e = y^{\frac{1-p}{2-n}}w(y,t) \tag{3.18b}$$

lead to

$$\frac{\partial^2 w}{\partial y^2} + \frac{1}{y}\frac{\partial w}{\partial y} - \frac{v^2}{y^2}w = \frac{4H^2}{\alpha^2(2-n)^2 c_{v0}}\frac{\partial w}{\partial t} \tag{3.19}$$

where $v = \left|\frac{1-p}{2-n}\right|$.

Equation (3.19) can be solved by using the method of separation of variables (Zhu and Yin, 2012). The excess pore pressure is

$$\frac{u_e}{\sigma_0} = y^{\frac{1-p}{2-n}}\sum_{m=1}^{+\infty} c_m Z_v^m(\lambda_m y)\exp\left(-\frac{\pi^2\lambda_m^2}{4\lambda_1^2}T\right) \tag{3.20}$$

where

$$T = \frac{c_{v0}(2-n)^2\alpha^2\lambda_1^2 t}{\pi^2 H^2} \tag{3.21}$$

$$c_m = \frac{\int_1^b y^{1+\frac{p-1}{2-n}} Z_v^m(\lambda_m y)\,dy}{\int_1^b y[Z_v^m(\lambda_m y)]^2\,dy} \tag{3.22}$$

$$b = (1+\alpha)^{1-\frac{n}{2}} \tag{3.23}$$

and

$$Z_\mu^m(x) = Y_v(\lambda_m)J_\mu(x) - J_v(\lambda_m)Y_\mu(x) \tag{3.24}$$

where J_v and Y_v denote, respectively, Bessel functions of the first and second kind of order v ; J_μ and Y_μ denote, respectively, Bessel functions of the first and second kind of order μ ; λ_m are eigenvalues arranged in increasing order ($m = 1, 2, 3, \ldots\infty$).

For boundary conditions (2.16), the constant c_m in (3.22) is

$$c_m = \begin{cases} \dfrac{2\pi[2+\pi\lambda_m b^{1-v} Z_{v-1}^m(\lambda_m b)]}{4-[b\pi\lambda_m Z_{1+v}^m(\lambda_m b)]^2} & \dfrac{p-1}{2-n} = -v \\ \dfrac{2\pi[2-\pi\lambda_m b^{1+v} Z_{1+v}^m(\lambda_m b)]}{4-[b\pi\lambda_m Z_{1+v}^m(\lambda_m b)]^2} & \dfrac{p-1}{2-n} = v \end{cases} \tag{3.25}$$

and the constant λ_m is the m-th positive root of the following (3.26) for variable λ

$$Z_v(\lambda b) = 0 \tag{3.26}$$

where

$$Z_\mu(x) = Y_v(\lambda)J_\mu(x) - J_v(\lambda)Y_\mu(x) \tag{3.27}$$

For boundary conditions (2.17), the constant c_m in (3.22) is

$$c_m = \frac{4\pi}{4-[b\pi\lambda_m Z_v^m(\lambda_m b)]^2} \tag{3.28}$$

and the constant λ_m is the m-th positive root of the following (3.29) for variable λ

$$\frac{1-p}{2-n} Z_v(\lambda b) + \lambda b Z_v'(\lambda b) = 0 \tag{3.29}$$

The average degree of consolidation $U(T,\alpha)$ defined as (2.24) is presented as follows.

For boundary conditions (2.16),

$$U = \begin{cases} 1 - \dfrac{4(1+q)}{(2-n)[(1+\alpha)^{1+q}-1]} \displaystyle\sum_{m=1}^{\infty} \dfrac{[2+\pi\lambda_m b^{1-v} Z_{v-1}^m(\lambda_m b)]^2}{\lambda_m^2\{[b\pi\lambda_m Z_{1+v}^m(\lambda_m b)]^2-4\}} \exp\left(-\dfrac{\pi^2\lambda_m^2}{4\lambda_1^2}T\right) & \dfrac{p-1}{2-n} = -v \\ 1 - \dfrac{4(1+q)}{(2-n)[(1+\alpha)^{1+q}-1]} \displaystyle\sum_{m=1}^{\infty} \dfrac{[2-\pi\lambda_m b^{1+v} Z_{1+v}^m(\lambda_m b)]^2}{\lambda_m^2\{[b\pi\lambda_m Z_{1+v}^m(\lambda_m b)]^2-4\}} \exp\left(-\dfrac{\pi^2\lambda_m^2}{4\lambda_1^2}T\right) & \dfrac{p-1}{2-n} = v \end{cases} \tag{3.30}$$

For boundary conditions (2.17),

$$U(T,\alpha)=1-\frac{16(1+q)}{(2-n)[(1+\alpha)^{1+q}-1]}\sum_{n=1}^{\infty}\frac{1}{\lambda_m^2\{[b\pi\lambda_m Z_v^m(\lambda_m b)]^2-4\}}\exp\left(-\frac{\pi^2\lambda_m^2}{4\lambda_1^2}T\right) \tag{3.31}$$

Equations (3.30) and (3.31) can be used to calculate the average degree of consolidation for a double-drained soil stratum and a single-drained stratum, respectively. Several specific variations of the consolidation parameters will be considered in the following.

3.3.2 Special cases of depth-dependent permeability with constant compressibility

In order to assess the effect of variation of permeability, a series of computations were performed at $q = 0$. Figures 3.12–3.13 show the average degree of consolidation for $p = 0.5$ and $p = 1$, respectively. The result for $\alpha = 0$ (conventional Terzaghi's solution) is also plotted in Figures 3.12–3.13 for comparison. As shown in Figures 3.12–3.13, the average degree of consolidation is almost the same as the conventional Terzaghi's solution for double-drained condition for a wide range of α, if the time is normalized using (3.21). For the single-drained condition, the average degree of consolidation falls in a very narrow band, and thus Figures 3.12–3.13 can be used to estimate the average degree of consolidation accurately.

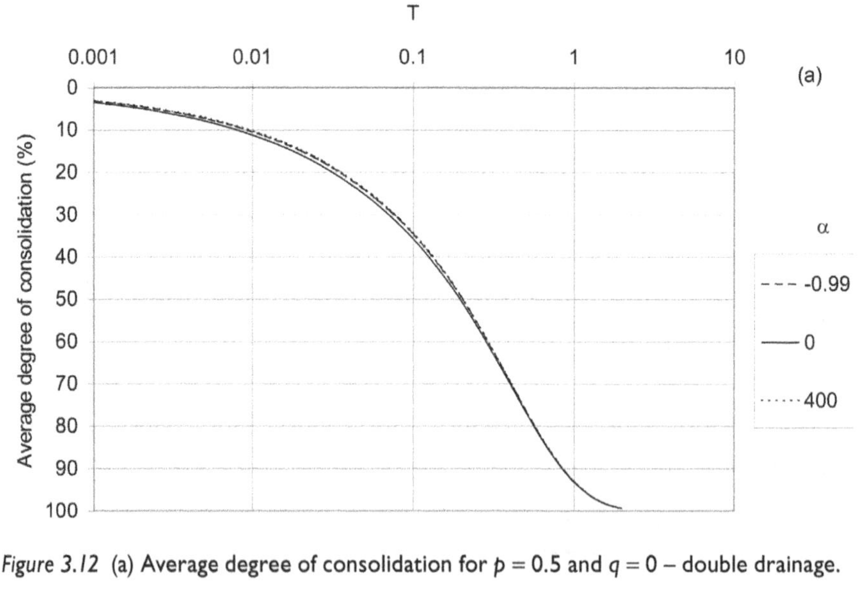

Figure 3.12 (a) Average degree of consolidation for $p = 0.5$ and $q = 0$ – double drainage.

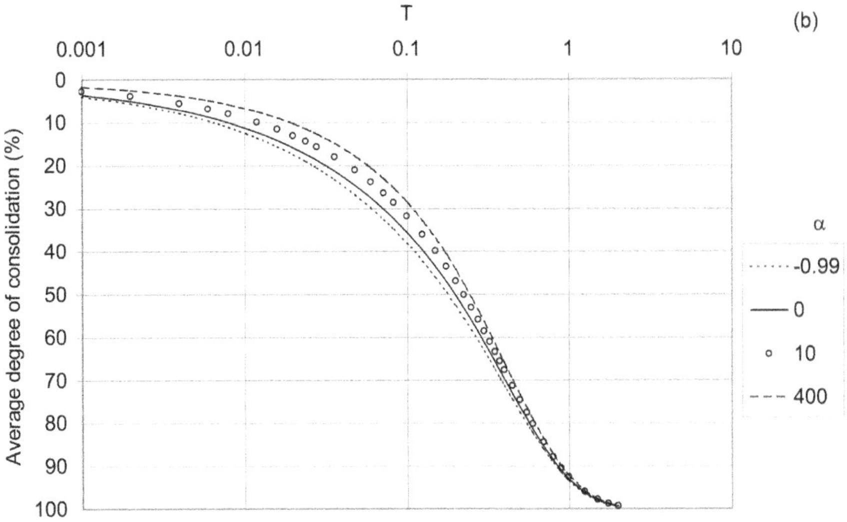

Figure 3.12 (b) Average degree of consolidation for $p = 0.5$ and $q = 0$ – single drainage.

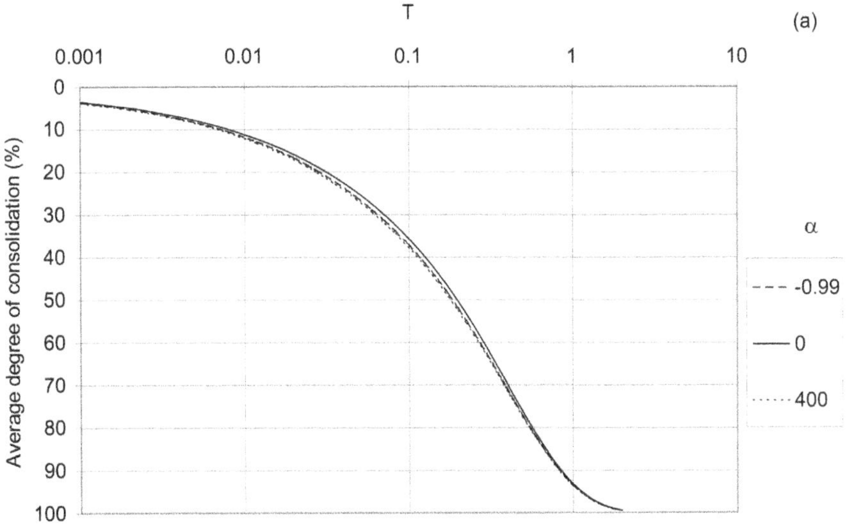

Figure 3.13 (a) Average degree of consolidation for $p = 1$ and $q = 0$ – double drainage.

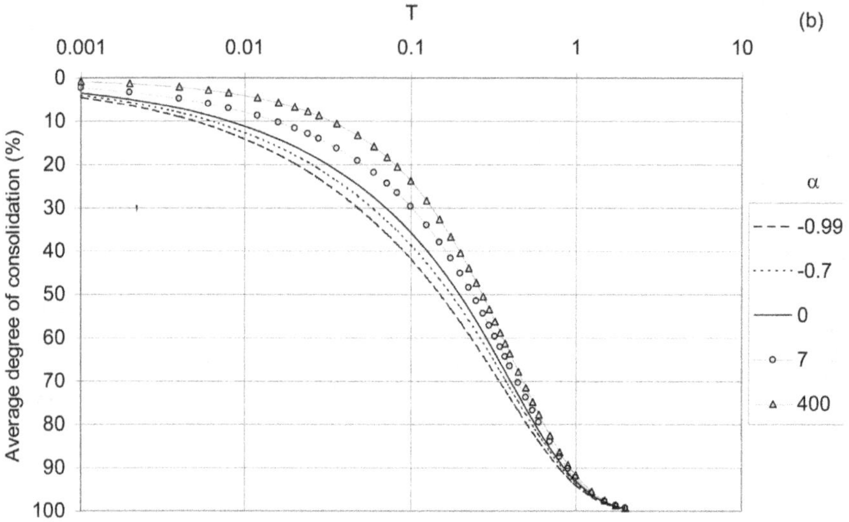

Figure 3.13 (b) Average degree of consolidation for $p = 1$ and $q = 0$ – single drainage.

For the parabolic variation $p = 2$ and $\alpha \neq 0$, we must return to the differential equation (3.15). The solution in this case is expressible in terms of elementary functions. By modifying (3.18a) and (3.21) as (3.32) and (3.33), respectively,

$$y = \begin{cases} \left(1+\alpha\dfrac{z}{H}\right)^{1-\frac{n}{2}} & n \neq 2 \\ \left(1+\alpha\dfrac{z}{H}\right) & n = 2 \end{cases} \tag{3.32}$$

$$T = \begin{cases} \dfrac{c_{v0}(2-n)^2\alpha^2\lambda_1^2 t}{\pi^2 H^2} & n \neq 2 \\ \dfrac{4c_{v0}\alpha^2\lambda_1^2 t}{\pi^2 H^2} & n = 2 \end{cases} \tag{3.33}$$

the results for $p = 2$ and $\alpha \neq 0$ can be expressed as follows:

For boundary condition (2.16),

$$\frac{u}{\sigma_0} = \sum_{m=1}^{+\infty} \frac{2[1-(-1)^m\sqrt{1+\alpha}\,]}{m\pi\left[1+\left(\dfrac{\ln\sqrt{1+\alpha}}{m\pi}\right)^2\right]} \frac{\sin\left[m\pi\dfrac{\ln y}{\ln(1+\alpha)}\right]}{\sqrt{y}} \exp\left(-\frac{\pi^2\lambda_m^2}{4\lambda_1^2}T\right) \tag{3.34}$$

$$U(T,\alpha)=1-\frac{2\ln(1+\alpha)}{\alpha}\sum_{m=1}^{+\infty}\frac{[1-(-1)^m\sqrt{1+\alpha}]^2}{m^2\pi^2\left[1+\left(\frac{\ln\sqrt{1+\alpha}}{m\pi}\right)^2\right]^2}\exp\left(-\frac{\pi^2\lambda_m^2}{4\lambda_1^2}T\right) \tag{3.35}$$

$$\lambda_m=\sqrt{\frac{1}{4}+\left[\frac{m\pi}{\ln(1+\alpha)}\right]^2} \tag{3.36}$$

For boundary condition (2.17),

$$\frac{u}{\sigma_0}=\begin{cases}\displaystyle\sum_{m=1}^{+\infty}\frac{\sqrt{4\lambda_m^2-1}\sin[\sqrt{4\lambda_m^2-1}\ln(\sqrt{y})]}{\lambda_m^2\{\ln(1+\alpha)-2[\cos(\sqrt{4\lambda_m^2-1}\ln\sqrt{1+\alpha})]^2\}\sqrt{y}}\exp\left(-\frac{\pi^2\lambda_m^2}{4\lambda_1^2}T\right) & \lambda_1>\frac{1}{2}\\ \displaystyle\frac{\sqrt{1-4\lambda_1^2}(y^{\sqrt{1-4\lambda_1^2}}-y^{-\sqrt{1-4\lambda_1^2}})}{[1-2\lambda_1^2\ln(1+\alpha)]\sqrt{y}}\exp\left(-\frac{\pi^2}{4}T\right)+ & \\ \displaystyle\sum_{m=2}^{+\infty}\frac{\sqrt{4\lambda_m^2-1}\sin[\sqrt{4\lambda_m^2-1}\ln(\sqrt{y})]}{\lambda_m^2\{\ln(1+\alpha)-2[\cos(\sqrt{4\lambda_m^2-1}\ln\sqrt{1+\alpha})]^2\}\sqrt{y}}\exp\left(-\frac{\pi^2\lambda_m^2}{4\lambda_1^2}T\right) & \lambda_1<\frac{1}{2}\end{cases} \tag{3.37}$$

$$U(T,\alpha)=\begin{cases}\displaystyle 1-\frac{1}{2\alpha}\sum_{m=1}^{+\infty}\frac{4\lambda_m^2-1}{\lambda_m^4\{\ln(1+\alpha)-2[\cos(\sqrt{4\lambda_m^2-1}\ln\sqrt{1+\alpha})]^2\}}\exp\left(-\frac{\pi^2\lambda_m^2}{4\lambda_1^2}T\right) & \lambda_1>\frac{1}{2}\\ \displaystyle 1-\frac{1-4\lambda_1^2}{\alpha\lambda_1^2[1-2\lambda_1^2\ln(1+\alpha)]}\exp\left(-\frac{\pi^2}{4}T\right)-\frac{1}{2\alpha} & \lambda_1<\frac{1}{2}\\ \displaystyle\sum_{m=2}^{+\infty}\frac{4\lambda_m^2-1}{\lambda_m^4\{\ln(1+\alpha)-2[\cos(\sqrt{4\lambda_m^2-1}\ln\sqrt{1+\alpha})]^2\}}\exp\left(-\frac{\pi^2\lambda_m^2}{4\lambda_1^2}T\right) & \end{cases} \tag{3.38}$$

where λ_m is the m-th positive root of the following (3.39) for variable λ

$$\begin{cases}\sqrt{4\lambda^2-1}\cos(\sqrt{4\lambda^2-1}\ln\sqrt{1+\alpha})-\sin(\sqrt{4\lambda^2-1}\ln\sqrt{1+\alpha})=0 & \lambda>\frac{1}{2}\\ (1-\sqrt{1-4\lambda^2})(1+\alpha)^{\sqrt{1-4\lambda^2}}-(1+\sqrt{1-4\lambda^2})=0 & \lambda<\frac{1}{2}\end{cases} \tag{3.39}$$

The average degree of consolidation for $p=2$ and $\alpha\neq 0$ is shown in Figure 3.14. As shown in Figure 3.14, the average degree of consolidation still falls in a very narrow band considering that $k(H)$ changes from $k_0/400$ to $10201k_0$, which is much larger than that for the cases of $p=0.5$ and $p=1$.

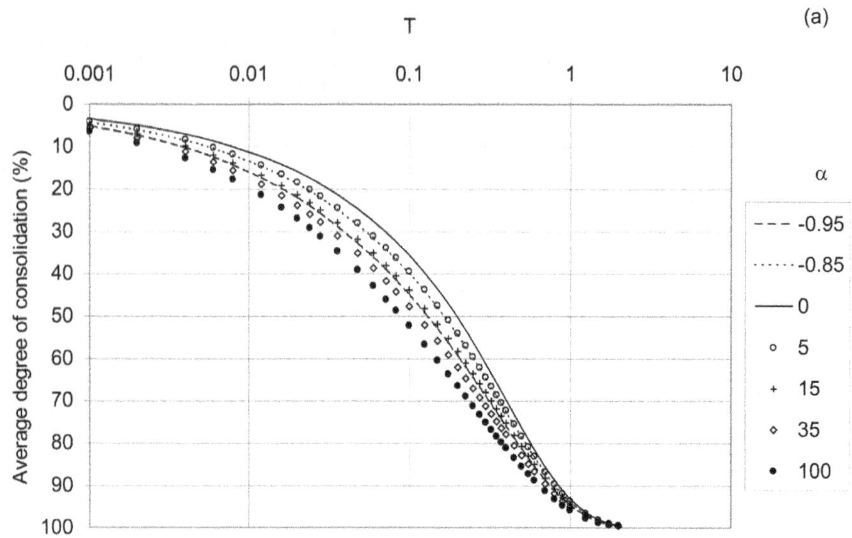

Figure 3.14 (a) Average degree of consolidation for $p = 2$ and $q = 0$ – double drainage.

The average degree of consolidation shows very little sensitivity to α if the time is normalized using (3.33). This is because the influence of α is included in the normalized time T in (3.33). The eigenvalue λ_1 for a wide range of α is listed in Table 3.6. The average degrees of consolidation for different α are listed in Tables 3.7–3.8.

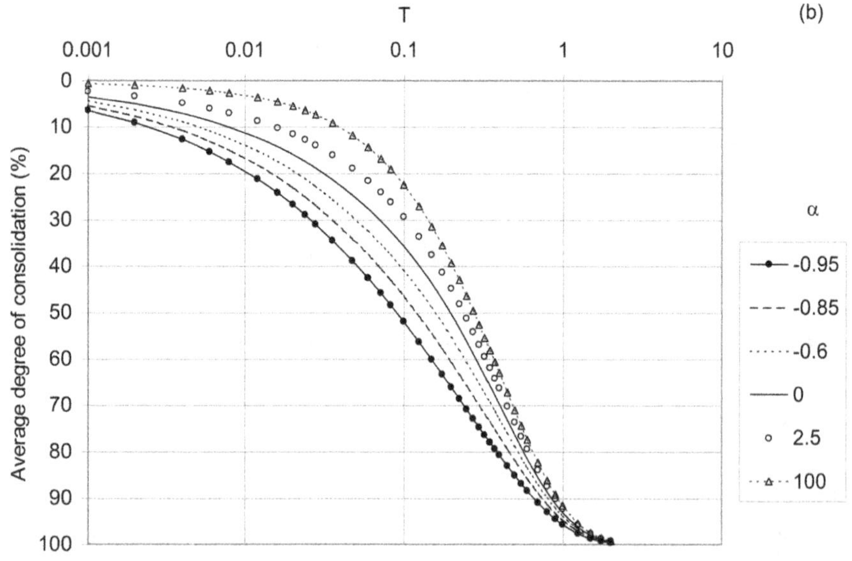

Figure 3.14 (b) Average degree of consolidation for $p = 2$ and $q = 0$ – single drainage.

Table 3.6 Eigenvalue λ_l for different α for cases of depth-dependent permeability with m_v = constant

	Single-drained condition			*Double-drained condition*		
α	$p = 0.5$ $q = 0$	$p = 1$ $q = 0$	$p = 2$ $q = 0$	$p = 0.5$ $q = 0$	$p = 1$ $q = 0$	$p = 2$ $q = 0$
–0.99	1.88954	2.4481		3.09474	3.31394	
–0.95	1.98556	2.61537	0.881293	3.41972	3.9454	1.16179
–0.9	2.11512	2.82935	1.03132	3.75411	4.5232	1.45311
–0.8	2.41721	3.30921	1.31665	4.44009	5.6384	2.015
–0.7	2.80005	3.90218	1.64007	5.25441	6.915	2.65683
–0.6	3.30608	4.67562	2.0458	6.3004	8.5251	3.46486
–0.5	4.01052	5.74399	2.59458	7.73495	10.7099	4.55986
–0.4	5.06324	7.33292	3.40083	9.86079	13.9262	6.17032
–0.35	5.81377	8.46309	3.97104	11.3706	16.2031	7.30987
–0.3	6.81349	9.96676	4.72759	13.378	19.2257	8.82218
–0.25	8.21202	12.0684	5.78277	16.1824	23.443	10.9318
–0.2	10.3086	15.2168	7.36104	20.3823	29.7529	14.0877
–0.17	12.1578	17.9927	8.7513	24.0846	35.3121	16.8678
–0.14	14.799	21.9565	10.7354	29.3707	43.2468	20.8357
–0.12	17.2931	25.699	12.6081	34.3613	50.7363	24.5808
–0.1	20.7846	30.9374	15.2287	41.3464	61.2175	29.8217
0.1	21.0963	31.8726	16.164	42.3954	64.3634	32.9656
0.12	17.6049	26.6346	13.5439	35.4111	53.884	27.7256
0.14	15.111	22.8926	11.6719	30.4215	46.3967	23.9817
0.17	12.47	18.9298	9.68891	25.1371	38.4661	20.0159
0.2	10.6211	16.155	8.29999	21.4369	32.9117	17.2383
0.25	8.52525	13.0088	6.72449	17.2413	26.6118	14.0877
0.3	7.12754	10.9101	5.67278	14.4423	22.407	11.9846
0.35	6.12882	9.40988	4.92044	12.4414	19.3996	10.4803
0.4	5.37947	8.28382	4.35527	10.9395	17.1408	9.35023
0.5	4.32971	6.7053	3.56198	8.8338	13.9712	7.76424
0.6	3.62915	5.65086	3.03103	7.4271	11.8508	6.70286
0.7	3.1282	4.89605	2.65015	6.42005	10.3304	5.94159
0.8	2.75204	4.32863	2.36319	5.66296	9.1856	5.36812
1	2.22443	3.5314	1.95867	4.59923	7.573	4.55986
1.2	1.87173	2.99712	1.68627	3.88639	6.4884	4.01573
1.4	1.61907	2.6134	1.48967	3.37452	5.7067	3.62314
1.7	1.35057	2.20432	1.27884	2.829	4.8695	3.20221
2	1.16175	1.9155	1.12892	2.44408	4.2754	2.90298
2.5	0.946469	1.58454	0.955559	2.00345	3.5901	2.55709
3	0.801836	1.36078	0.837041	1.70599	3.123	2.32068
3.5	0.697741	1.19874	0.75032	1.49098	2.7823	2.14773
4	0.619092	1.0756	0.683764	1.3279	2.5216	2.015
5	0.507849	0.900035	0.587618	1.09608	2.1464	1.82325
6	0.432675	0.780164	0.520876	0.938484	1.8873	1.69011
7	0.378295	0.692624	0.471405	0.823901	1.6961	1.59138
8	0.337021	0.625598	0.433013	0.736552	1.5485	1.5147
10	0.27831	0.529145	0.376796	0.611634	1.3334	1.40231
12	0.238369	0.462572	0.337165	0.526127	1.1829	1.32294
15	0.19754	0.393456	0.295104	0.438194	1.0244	1.2385
20	0.155453	0.320697	0.249526	0.346895	0.8544	1.14664
25	0.129333	0.274486	0.219682	0.289833	0.7443	1.08617

(Continued)

Table 3.6 (Continued) Eigenvalue λ_i for different α for cases of depth-dependent permeability with m_v = constant

	Single-drained condition			Double-drained condition		
α	$p = 0.5$ $q = 0$	$p = 1$ $q = 0$	$p = 2$ $q = 0$	$p = 0.5$ $q = 0$	$p = 1$ $q = 0$	$p = 2$ $q = 0$
30	0.111411	0.242156	0.19828	0.250475	0.6661	1.04257
35	0.0982838	0.218072	0.182001	0.221526	0.607	1.00924
40	0.0882156	0.199323	0.169097	0.199248	0.5605	0.982688
50	0.073714	0.171802	0.149732	0.167036	0.4913	0.942564
65	0.059772	0.14462	0.130013	0.135912	0.4217	0.901259
80	0.0506865	0.126412	0.116404	0.115538	0.3742	0.8724
100	0.0424958	0.109565	0.103468	0.0971024	0.3295	0.844616
200	0.0247095	0.0709148		0.0568083	0.223578	
300	0.0180484	0.0552894		0.0416095	0.178937	
400	0.0144584	0.0464384		0.0333888	0.153013	

Table 3.7 U for cases of depth-dependent permeability and double drainage with m_v = constant

		$p = 0.5$		$p = 1$		$p = 2$				
T	$\alpha = 0$	−0.99	400	−0.99	400	−0.95	−0.85	8.00	30	100
0.001	3.57	3.26	3.15	3.76	3.85	5.24	4.34	4.57	5.62	6.60
0.004	7.14	6.58	6.41	7.51	7.69	10.31	8.61	9.04	11.02	12.86
0.012	12.37	11.55	11.32	13.00	13.30	17.41	14.70	15.40	18.55	21.45
0.02	15.96	15.02	14.77	16.76	17.13	22.09	18.81	19.65	23.47	26.99
0.028	18.89	17.87	17.60	19.81	20.24	25.78	22.09	23.03	27.33	31.26
0.036	21.42	20.35	20.07	22.45	22.91	28.88	24.88	25.91	30.55	34.81
0.048	24.73	23.62	23.34	25.88	26.40	32.83	28.49	29.60	34.64	39.24
0.06	27.65	26.51	26.24	28.90	29.46	36.20	31.62	32.79	38.11	42.96
0.072	30.28	29.15	28.87	31.62	32.21	39.16	34.41	35.63	41.15	46.17
0.083	32.52	31.38	31.11	33.91	34.52	41.61	36.74	37.99	43.64	48.78
0.1	35.69	34.58	34.31	37.16	37.79	44.99	40.02	41.29	47.07	52.33
0.125	39.90	38.83	38.58	41.45	42.10	49.32	44.28	45.57	51.43	56.76
0.15	43.71	42.69	42.46	45.29	45.96	53.08	48.06	49.34	55.18	60.50
0.2	50.42	49.53	49.33	52.01	52.67	59.40	54.57	55.81	61.43	66.56
0.25	56.23	55.46	55.29	57.77	58.39	64.60	60.09	61.24	66.50	71.34
0.3	61.33	60.66	60.51	62.77	63.34	68.99	64.85	65.91	70.75	75.23
0.35	65.83	65.24	65.11	67.15	67.67	72.76	69.01	69.96	74.36	78.47
0.4	69.80	69.28	69.17	70.99	71.46	76.03	72.65	73.51	77.48	81.21
0.45	73.30	72.85	72.75	74.38	74.80	78.88	75.85	76.62	80.19	83.55
0.5	76.40	76.01	75.92	77.36	77.73	81.38	78.67	79.35	82.55	85.57
0.6	81.56	81.26	81.19	82.32	82.61	85.50	83.35	83.89	86.42	88.85
0.7	85.59	85.35	85.30	86.19	86.42	88.69	87.00	87.42	89.42	91.34
0.9	91.21	91.06	91.03	91.57	91.71	93.10	92.06	92.32	93.55	94.75
1	93.13	93.01	92.99	93.41	93.52	94.61	93.80	94.00	94.97	95.90
1.25	96.29	96.23	96.22	96.45	96.50	97.09	96.65	96.76	97.28	97.79
1.75	98.92	98.90	98.90	98.97	98.98	99.15	99.03	99.06	99.21	99.36

Table 3.8 U for cases of depth-dependent permeability and single drainage with m_v = constant

		$p = 0.5$			$p = 1$				$p = 2$				
T	$\alpha = 0$	-0.99	10	400	-0.99	-0.7	7	400	-0.95	-0.85	-0.6	2.5	100
0.001	3.57	3.98	2.75	1.69	4.58	4.08	2.38	0.85	6.53	5.52	4.52	2.4	0.67
0.004	7.14	7.93	5.63	3.84	9.08	8.11	4.91	2.13	12.73	10.83	8.94	4.91	1.63
0.012	12.36	13.65	10.03	7.52	15.52	13.94	8.86	4.66	21.26	18.25	15.24	8.8	3.66
0.02	15.96	17.54	13.18	10.33	19.84	17.89	11.74	6.82	26.75	23.12	19.46	11.62	5.56
0.028	18.88	20.68	15.81	12.76	23.29	21.06	14.17	8.8	31.01	26.93	22.82	14.01	7.41
0.036	21.41	23.38	18.13	14.94	26.22	23.78	16.35	10.68	34.53	30.13	25.68	16.14	9.22
0.048	24.72	26.89	21.22	17.93	29.99	27.3	19.29	13.37	38.94	34.18	29.35	19.02	11.87
0.06	27.64	29.95	24	20.65	33.25	30.37	21.97	15.93	42.64	37.63	32.53	21.66	14.44
0.072	30.28	32.69	26.56	23.19	36.14	33.12	24.46	18.4	45.84	40.64	35.35	24.12	16.94
0.083	32.51	34.99	28.75	25.4	38.55	35.42	26.62	20.6	48.45	43.12	37.69	26.26	19.16
0.1	35.68	38.24	31.91	28.61	41.91	38.66	29.77	23.86	51.99	46.53	40.97	29.38	22.48
0.125	39.89	42.5	36.18	32.99	46.25	42.9	34.06	28.42	56.4	50.86	45.22	33.67	27.12
0.15	43.69	46.29	40.09	37.05	50.05	46.68	38.05	32.7	60.13	54.59	48.97	37.66	31.48
0.2	50.41	52.88	47.12	44.39	56.52	53.22	45.27	40.51	66.17	60.79	55.38	44.91	39.43
0.25	56.22	58.49	53.27	50.85	61.9	58.79	51.62	47.42	70.92	65.82	60.77	51.31	46.46
0.3	61.32	63.37	58.7	56.56	66.5	63.63	57.24	53.52	74.79	70.05	65.42	56.96	52.67
0.35	65.82	67.65	63.5	61.6	70.47	67.87	62.2	58.91	78.02	73.66	69.47	61.96	58.17
0.4	69.79	71.41	67.73	66.06	73.94	71.61	66.59	63.68	80.75	76.79	73.03	66.37	63.02
0.45	73.29	74.73	71.48	70	76.99	74.91	70.47	67.9	83.1	79.53	76.17	70.27	67.31
0.5	76.39	77.67	74.79	73.48	79.67	77.82	73.89	71.62	85.13	81.93	78.94	73.72	71.11
0.6	81.56	82.55	80.3	79.28	84.12	82.67	79.6	77.83	88.45	85.89	83.55	79.47	77.43
0.7	85.59	86.37	84.61	83.81	87.6	86.46	84.06	82.68	91	88.98	87.15	83.96	82.36
0.9	91.2	91.68	90.6	90.12	92.43	91.73	90.27	89.42	94.52	93.28	92.15	90.21	89.23
1	93.13	93.5	92.66	92.28	94.08	93.54	92.4	91.74	95.72	94.75	93.87	92.35	91.59
1.25	96.29	96.49	96.04	95.83	96.81	96.52	95.9	95.54	97.69	97.16	96.69	95.87	95.46
1.75	98.92	98.98	98.85	98.79	99.07	98.99	98.81	98.7	99.33	99.17	99.04	98.8	98.68

The excess pore pressure isochrones at 50% average degree of consolidation are shown in Figure 3.15. As shown in Figure 3.15, the excess pore pressure distribution is very different from the conventional Terzaghi's solution even though the average degree of consolidation is the same. The distribution of excess pore pressure for $\alpha \neq 0$ and the double-drainage condition is considerably skewed from the conventional Terzaghi's solution. For the single-drainage situation, however, the excess pore pressure increases with the increase of z. Also shown in Figure 3.15, the smaller the permeability, the larger the excess pore pressure.

Another typical feature shown in Figure 3.15(a) is the symmetric behavior for the double-drainage condition. Actually, it can be easily verified that if $(1+\alpha_1)(1+\alpha_2)=1$, the average degree of consolidation $U(T,\alpha_1)$ is equal to $U(T,\alpha_2)$, and the excess pore pressure is symmetric about $z/H = 0.5$.

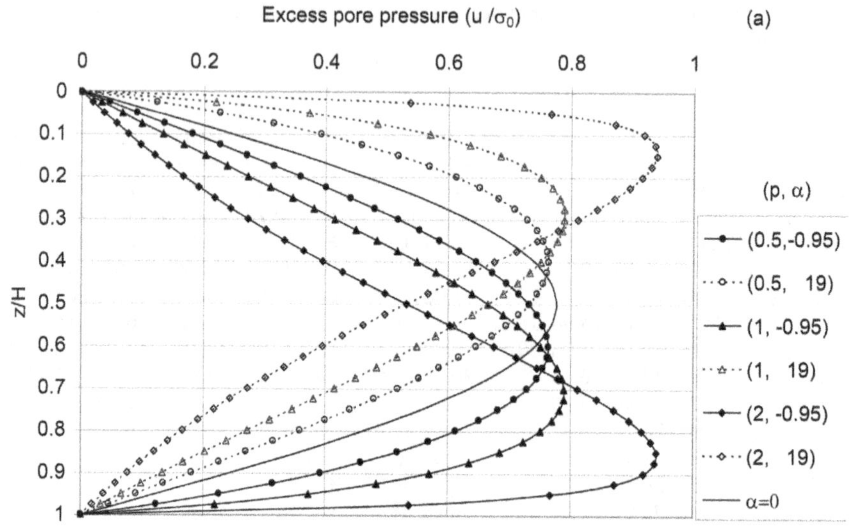

Figure 3.15 (a) Excess pore pressure isochrones for 50% average degree of consolidation for $p = 0.5, 1, 2$ and $q = 0$ – double drainage.

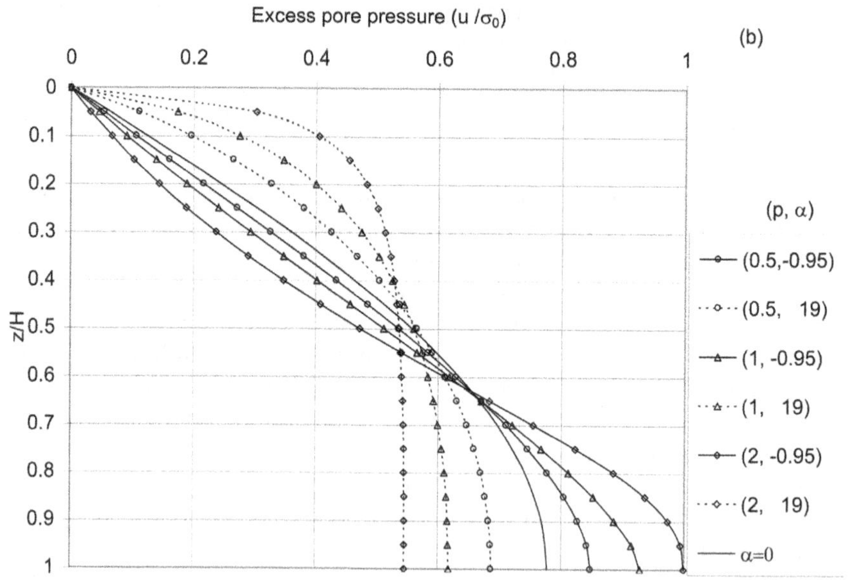

Figure 3.15 (b) Excess pore pressure isochrones for 50% average degree of consolidation for $p = 0.5, 1, 2$ and $q = 0$ – single drainage.

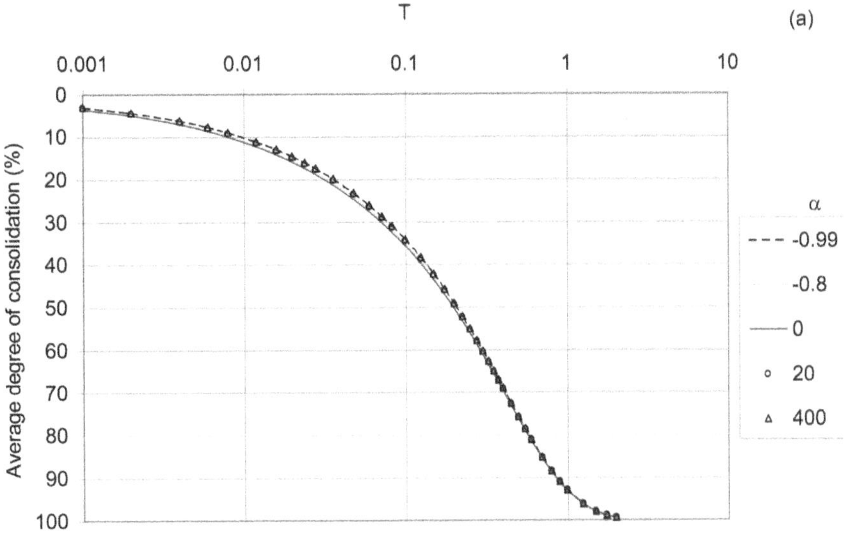

Figure 3.16 (a) Average degree of consolidation for $p = 0$ and $q = 0.5$ – double drainage.

3.3.3 Special cases of depth-dependent compressibility with constant permeability

Figures 3.16–3.18 show the average degree of consolidation for $q = 0.5$, 1, and 2, respectively, with k being held constant ($p = 0$). Although it was shown in Schiffman and Gibson (1964) that the variations in the average degree of consolidation are considerably greater than in the constant m_v situation for

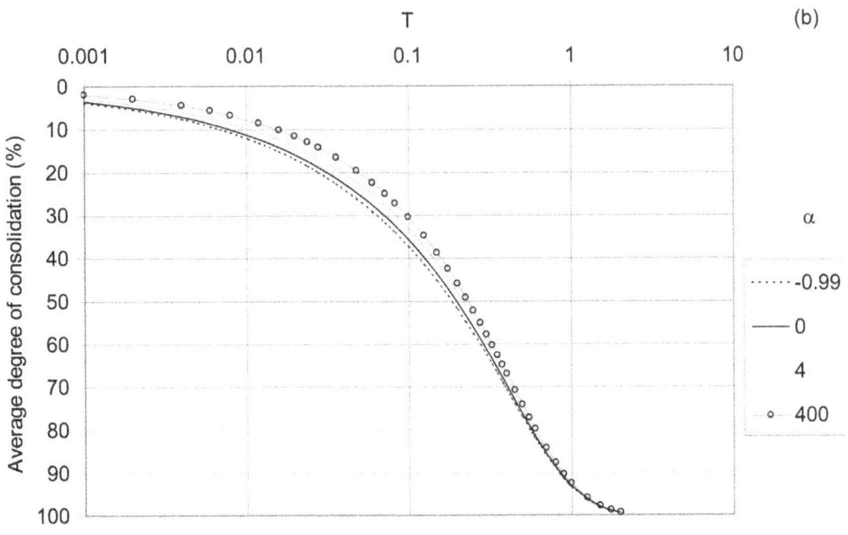

Figure 3.16 (b) Average degree of consolidation for $p = 0$ and $q = 0.5$ – single drainage.

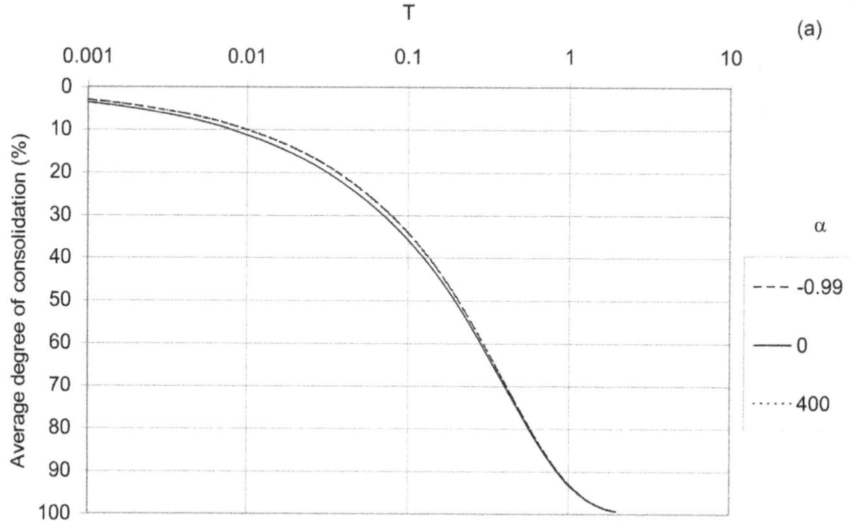

Figure 3.17 (a) Average degree of consolidation for $p = 0$ and $q = 1$ – double drainage.

the double-drainage condition, almost the same result as the conventional Terzaghi's solution is still obtained for a wide range of α if time is normalized using (3.33). For the single-drained condition, the average degree of consolidation is close to the conventional Terzaghi's solution. The eigenvalue λ_1 for different α is presented in Table 3.9 for easy calculation. The average degrees of consolidation for different α are presented in Tables 3.10–3.11.

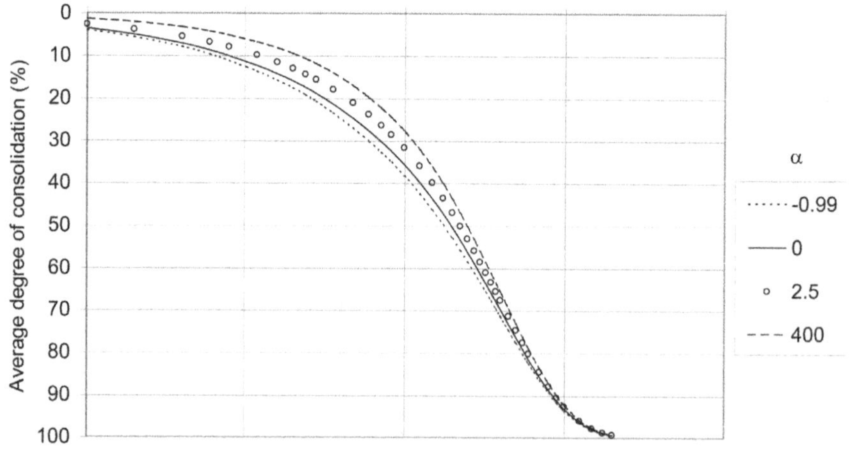

Figure 3.17 (b) Average degree of consolidation for $p = 0$ and $q = 1$ – single drainage.

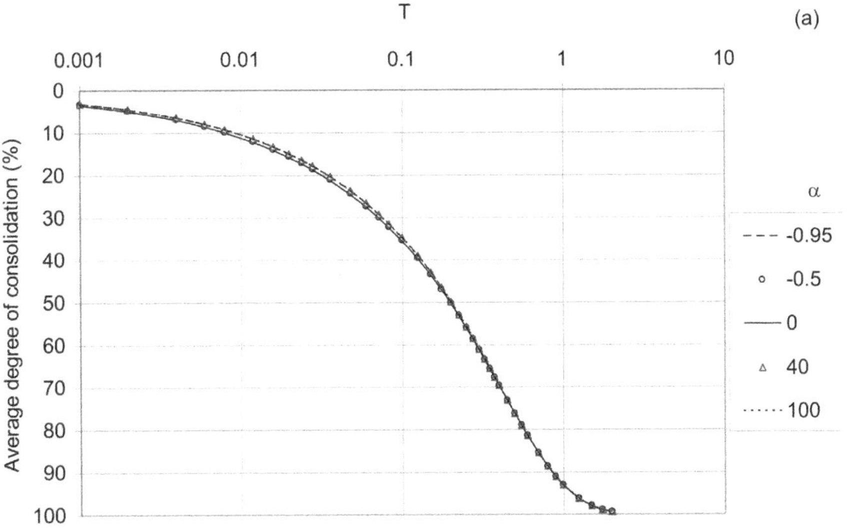

Figure 3.18 (a) Average degree of consolidation for $p = 0$ and $q = 2$ – double drainage.

The excess pore pressure isochrones at 50% average degree of consolidation are shown in Figure 3.19. For the double-drained condition, the excess pore pressure is clearly symmetric about $z/H = 0.5$ if $(1+\alpha_1)(1+\alpha_2) = 1$. The deviation from the conventional Terzaghi's solution is less than as shown in Figure 3.15(a) for the same α, but skewed in opposite direction. For the

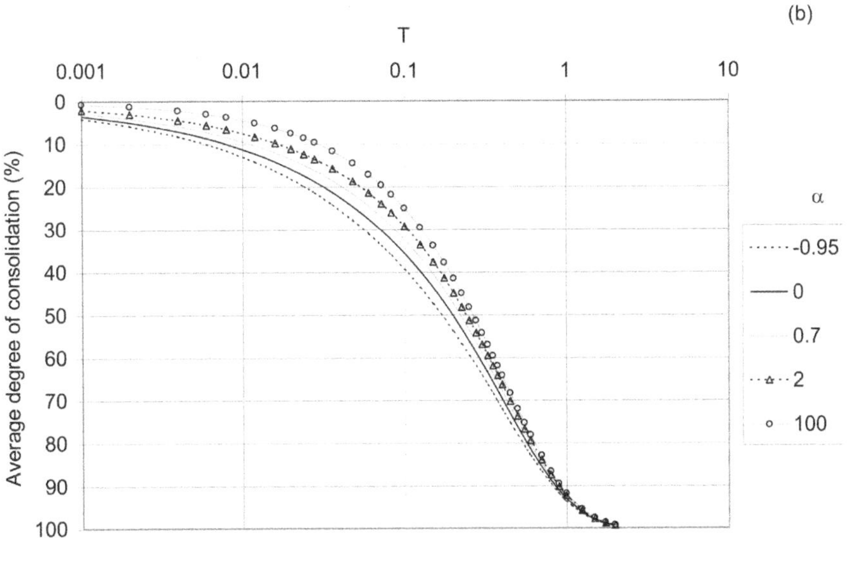

Figure 3.18 (b) Average degree of consolidation for $p = 0$ and $q = 2$ – single drainage.

Table 3.9 Eigenvalue λ_l for different α for cases of depth-dependent compressibility with k = constant

	Single–drained condition			*Double–drained condition*		
α	$p = 0$ $q = 0.5$	$p = 0$ $q = 1$	$p = 0$ $q = 2$	$p = 0$ $q = 0.5$	$p = 0$ $q = 1$	$p = 0$ $q = 2$
–0.99	1.75297	1.86658		3.02111	2.91953	
–0.95	1.77342	1.87213	2.00664	3.11613	2.9921	2.8406
–0.9	1.81557	1.88954	2.00905	3.24944	3.09474	2.9092
–0.8	1.94377	1.96099	2.0283	3.57425	3.34843	3.08038
–0.7	2.13411	2.08823	2.08095	4.002	3.68835	3.31397
–0.6	2.4066	2.28842	2.18688	4.58184	4.15579	3.64186
–0.5	2.80371	2.59639	2.37584	5.40306	4.82527	4.12071
–0.4	3.4138	3.08511	2.70409	6.64481	5.84613	4.86312
–0.35	3.85453	3.44358	2.95535	7.53551	6.58176	5.40337
–0.3	4.44543	3.92779	3.30194	8.72572	7.56721	6.13115
–0.25	5.27621	4.61245	3.79987	10.3949	8.95206	7.15867
–0.2	6.52632	5.64705	4.5612	12.9022	11.0355	8.71034
–0.17	7.6314	6.56388	5.24049	15.1163	12.8771	10.0851
–0.14	9.21182	7.87697	6.21729	18.2809	15.5109	12.0541
–0.12	10.7055	9.11917	7.14376	21.2707	18.0001	13.9169
–0.1	12.7976	10.8602	8.44445	25.4573	21.4868	16.5277
0.1	12.3551	10.1222	7.33635	24.8282	20.4382	14.9548
0.12	10.2624	8.38022	6.03373	20.6412	16.951	12.343
0.14	8.76817	7.13687	5.10497	17.6511	14.461	10.4791
0.17	7.18671	5.82173	4.12406	14.4857	11.826	8.50818
0.2	6.08037	4.90243	3.43984	12.2708	9.98285	7.13104
0.25	4.82769	3.86275	2.66845	9.76172	7.89635	5.57462
0.3	3.9937	3.17179	2.15812	8.0903	6.50772	4.54133
0.35	3.3989	2.67995	1.79669	6.89746	5.51777	3.80679
0.4	2.95355	2.31247	1.52813	6.00369	4.77691	3.25881
0.5	2.3317	1.80106	1.15756	4.75434	3.74333	2.49821
0.6	1.91878	1.46316	0.915822	3.92344	3.05799	1.99784
0.7	1.62509	1.22412	0.747085	3.3315	2.57137	1.64563
0.8	1.40582	1.04663	0.623521	2.8888	2.20872	1.38558
1	1.10089	0.801836	0.456544	2.27171	1.70599	1.03018
1.2	0.899552	0.642091	0.350648	1.86295	1.37565	0.801457
1.4	0.757135	0.530413	0.278674	1.57292	1.14322	0.64394
1.7	0.608117	0.415196	0.206856	1.26841	0.901712	0.484528
2	0.505287	0.337021	0.160012	1.05746	0.736552	0.378965
2.5	0.390803	0.251768	0.111294	0.821566	0.554865	0.267409
3	0.316129	0.19754	0.0820187	0.666946	0.438194	0.199236
3.5	0.263867	0.160455	0.0630068	0.558286	0.357785	0.154371
4	0.22541	0.133744	0.0499445	0.478049	0.299493	0.123215
5	0.172912	0.0982838	0.0336244	0.36806	0.221526	0.0838204
6	0.139008	0.0761691	0.0241855	0.296698	0.172501	0.0607423
7	0.115466	0.0612786	0.0182339	0.246969	0.139283	0.0460525
8	0.0982541	0.0506865	0.0142393	0.210506	0.115538	0.0361212
10	0.0749288	0.0368205	0.00937509	0.160932	0.0842865	0.0239338
12	0.059983	0.0283012	0.00663743	0.129059	0.064977	0.0170197
15	0.0456353	0.0204653	0.00433231	0.0983711	0.0471322	0.0111596
20	0.0320329	0.0134365	0.00248622	0.0691853	0.0310447	0.0064344
25	0.0243203	0.0096778	0.0016107	0.0525921	0.0224051	0.00418061

(Continued)

Table 3.9 (Continued) Eigenvalue λ_l for different α for cases of depth-dependent compressibility with k = constant

	Single–drained condition			*Double–drained condition*		
α	$p=0\ q=0.5$	$p=0\ q=1$	$p=0\ q=2$	$p=0\ q=0.5$	$p=0\ q=1$	$p=0\ q=2$
30	0.0194094	0.00739483	0.00112776	0.0420078	0.0171432	0.00293286
35	0.0160348	0.00588698	0.000833463	0.0347254	0.0136612	0.00217057
40	0.0135871	0.00483001	0.000640964	0.0294384	0.0112168	0.00167103
50	0.0102984	0.00346779	0.000412791	0.0223278	0.0080619	0.0010778
65	0.00743136	0.00234695	0.000245677	0.0161219	0.0054616	0.000642364
80	0.00573858	0.00172225	0.000162777	0.0124544	0.00401038	0.000425985
100	0.00434575	0.00123446	0.000104508	0.00943481	0.00287611	0.000273706
200	0.00183053	0.000437958		0.00397695	0.00102151	
300	0.0011034	0.00023867		0.00239777	0.00055689	
400	0.000770357	0.00015511		0.00167425	0.000361988	

Table 3.10 U for cases of depth-dependent compressibility and double drainage with k = constant

		$q=0.5$		$q=1$		$q=2$	
T	$\alpha=0$	−0.99	400	−0.99	400	−0.95	100
0.001	3.57	3.09	3.03	3.02	2.99	3.22	3.2
0.004	7.14	6.34	6.25	6.22	6.18	6.55	6.53
0.012	12.36	11.24	11.13	11.08	11.03	11.55	11.53
0.02	15.96	14.68	14.57	14.51	14.46	15.06	15.04
0.028	18.88	17.52	17.4	17.34	17.28	17.94	17.92
0.036	21.41	19.99	19.87	19.8	19.75	20.45	20.43
0.048	24.72	23.25	23.13	23.07	23.01	23.75	23.74
0.06	27.64	26.14	26.03	25.97	25.91	26.69	26.68
0.072	30.28	28.78	28.66	28.61	28.55	29.35	29.35
0.083	32.51	31.01	30.9	30.85	30.8	31.61	31.61
0.1	35.68	34.21	34.1	34.06	34.01	34.84	34.85
0.125	39.89	38.47	38.36	38.35	38.3	39.14	39.16
0.15	43.69	42.33	42.24	42.24	42.19	43.03	43.06
0.2	50.41	49.19	49.11	49.14	49.1	49.91	49.94
0.25	56.22	55.15	55.07	55.12	55.09	55.85	55.88
0.3	61.32	60.38	60.31	60.37	60.34	61.04	61.07
0.35	65.82	64.99	64.93	64.99	64.97	65.6	65.63
0.4	69.79	69.06	69	69.06	69.04	69.61	69.64
0.45	73.29	72.65	72.6	72.66	72.64	73.15	73.18
0.5	76.39	75.83	75.79	75.84	75.82	76.27	76.3
0.6	81.56	81.11	81.08	81.12	81.11	81.47	81.49
0.7	85.59	85.24	85.22	85.25	85.24	85.52	85.54
0.9	91.2	90.99	90.98	91	90.99	91.16	91.17
1	93.13	92.96	92.95	92.97	92.96	93.09	93.1
1.25	96.29	96.2	96.2	96.2	96.2	96.27	96.28
1.75	98.92	98.89	98.89	98.89	98.89	98.91	98.92

Table 3.11 U for cases of depth-dependent compressibility and single drainage with k = constant

		$q = 0.5$			$q = 1$			$q = 2$		
T	$\alpha = 0$	−0.99	4	400	−0.99	2.5	400	−0.95	1.2	100
0.001	3.57	3.83	2.98	1.97	3.99	2.71	1.31	4.17	2.56	0.82
0.004	7.14	7.64	6.06	4.47	7.94	5.55	3.27	8.29	5.26	2.29
0.012	12.36	13.18	10.72	8.59	13.68	9.92	6.78	14.23	9.44	5.19
0.02	15.96	16.97	14.01	11.65	17.58	13.06	9.51	18.26	12.47	7.58
0.028	18.88	20.04	16.73	14.24	20.73	15.67	11.9	21.5	15.01	9.74
0.036	21.41	22.67	19.12	16.55	23.43	17.98	14.06	24.28	17.27	11.75
0.048	24.72	26.11	22.28	19.65	26.94	21.07	17.02	27.87	20.3	14.56
0.06	27.64	29.12	25.11	22.46	30	23.85	19.75	30.99	23.03	17.2
0.072	30.28	31.83	27.69	25.04	32.75	26.4	22.29	33.79	25.56	19.7
0.083	32.51	34.1	29.9	27.27	35.05	28.59	24.5	36.12	27.74	21.9
0.1	35.68	37.33	33.06	30.48	38.3	31.75	27.73	39.41	30.9	25.15
0.125	39.89	41.57	37.31	34.83	42.56	36.02	32.15	43.7	35.18	29.66
0.15	43.69	45.36	41.19	38.82	46.36	39.95	36.25	47.49	39.13	33.88
0.2	50.41	51.99	48.12	45.99	52.95	46.99	43.68	54.04	46.24	41.56
0.25	56.22	57.67	54.17	52.28	58.56	53.16	50.22	59.57	52.49	48.34
0.3	61.32	62.62	59.5	57.82	63.43	58.6	56	64.36	58.01	54.34
0.35	65.82	66.98	64.2	62.72	67.7	63.4	61.11	68.53	62.88	59.64
0.4	69.79	70.82	68.36	67.05	71.46	67.65	65.62	72.2	67.19	64.32
0.45	73.29	74.21	72.03	70.87	74.77	71.41	69.61	75.44	71	68.46
0.5	76.39	77.2	75.28	74.25	77.7	74.73	73.14	78.29	74.36	72.12
0.6	81.56	82.19	80.68	79.88	82.58	80.25	79.01	83.04	79.97	78.22
0.7	85.59	86.08	84.91	84.28	86.39	84.57	83.6	86.75	84.35	82.98
0.9	91.2	91.5	90.79	90.4	91.69	90.58	89.99	91.91	90.45	89.61
1	93.13	93.36	92.8	92.5	93.51	92.64	92.18	93.68	92.53	91.88
1.25	96.29	96.42	96.11	95.95	96.5	96.03	95.78	96.59	95.97	95.62
1.75	98.92	98.96	98.87	98.82	98.98	98.84	98.77	99.01	98.83	98.72

single-drained condition, smaller compressibility will generate larger excess pore pressure as shown in Figure 3.19(b) while smaller permeability will produce smaller excess pore pressure near the drained boundary, but generate larger excess pore pressure near the undrained boundary as compared to results of the constant compressibility shown in Figure 3.15(b).

3.3.4 Special cases of depth-dependent permeability and compressibility resulting in constant c_v

When there is little change of soil type through the thickness of a soil layer, the changes of permeability with depth are often accompanied with similar changes in compressibility. The coefficient of consolidation can be approximated as a constant. In this section, the analytical solutions are presented for several cases of constant coefficient of consolidation.

Figures 3.20–3.22 show the average degree of consolidation for the cases of $p = q = 0.5$, 1, and 2, respectively. As shown in Figures 3.20–3.22, good

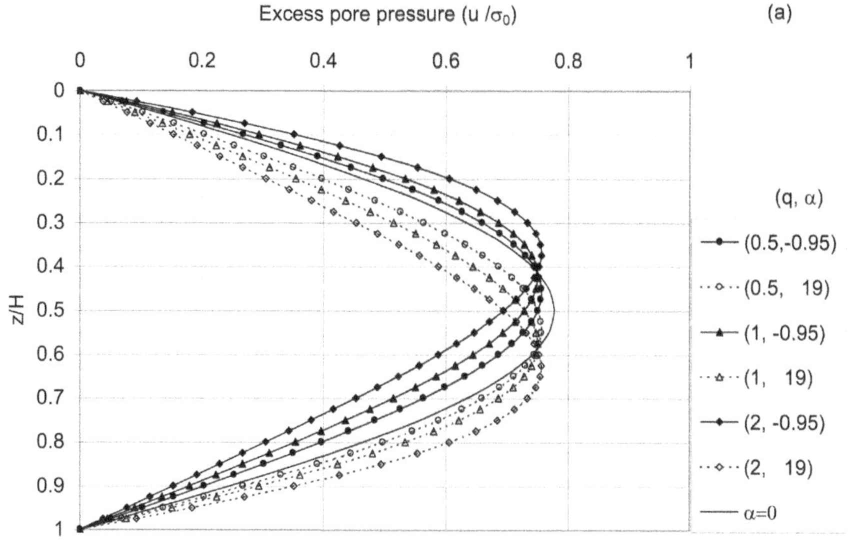

Figure 3.19 (a) Excess pore pressure isochrones for 50% average degree of consolidation for $p = 0$ and $q = 0.5, 1, 2$ – double drainage.

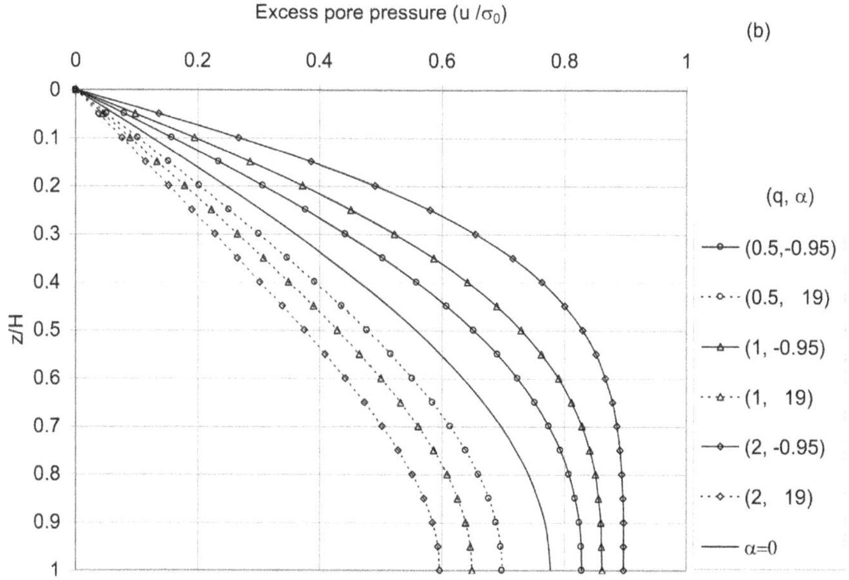

Figure 3.19 (b) Excess pore pressure isochrones for 50% average degree of consolidation for $p = 0$ and $q = 0.5, 1, 2$ – single drainage.

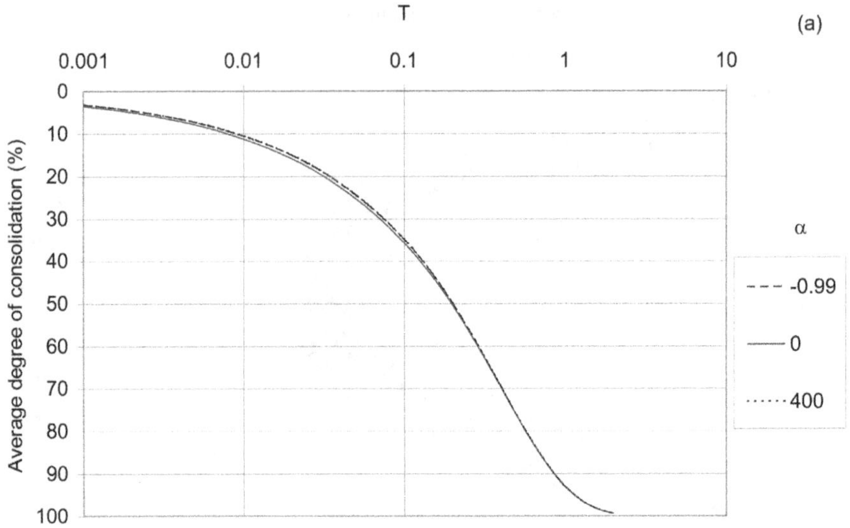

Figure 3.20 (a) Average degree of consolidation for $p = q = 0.5$ – double drainage.

results are obtained for a wide range of α if time is normalized using (3.33), especially for the double-drained condition. The eigenvalue λ_1 for different α is presented in Table 3.12. The average degrees of consolidation are also listed in Tables 3.13–3.14. Tables 3.13–3.14 and Figures 3.20–3.22 can be used to calculate the average degree of consolidation for a variety of α.

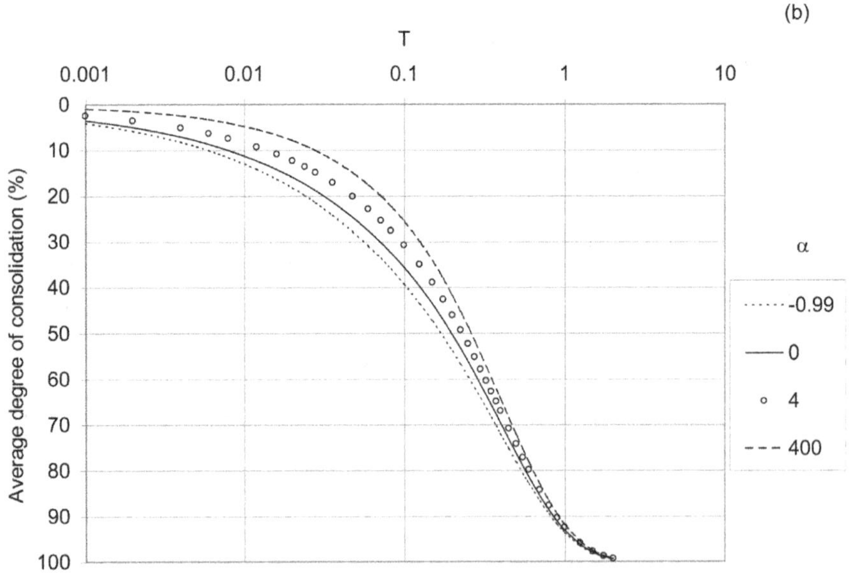

Figure 3.20 (b) Average degree of consolidation for $p = q = 0.5$ – single drainage.

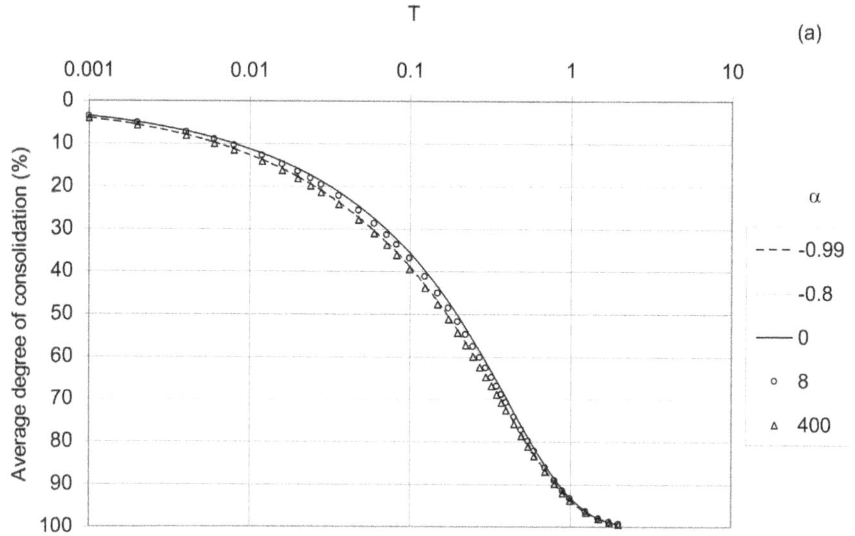

Figure 3.21 (a) Average degree of consolidation for $p = q = 1$ – double drainage.

Figure 3.23 illustrates the excess pore pressure isochrones at 50% average degree of consolidation. For the double-drained condition, the excess pore pressure is symmetric about $z/H = 0.5$ if $(1+\alpha_1)(1+\alpha_2) = 1$. The maximum excess pore pressure is a little larger than as shown in Figure 3.15(a) for the same α, but skewed less in the same direction. For the single-drained

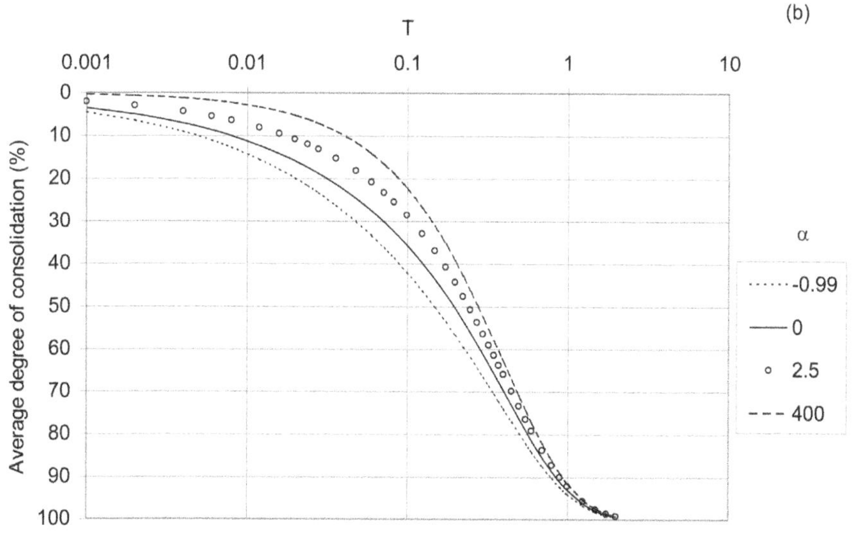

Figure 3.21 (b) Average degree of consolidation for $p = q = 1$ – single drainage.

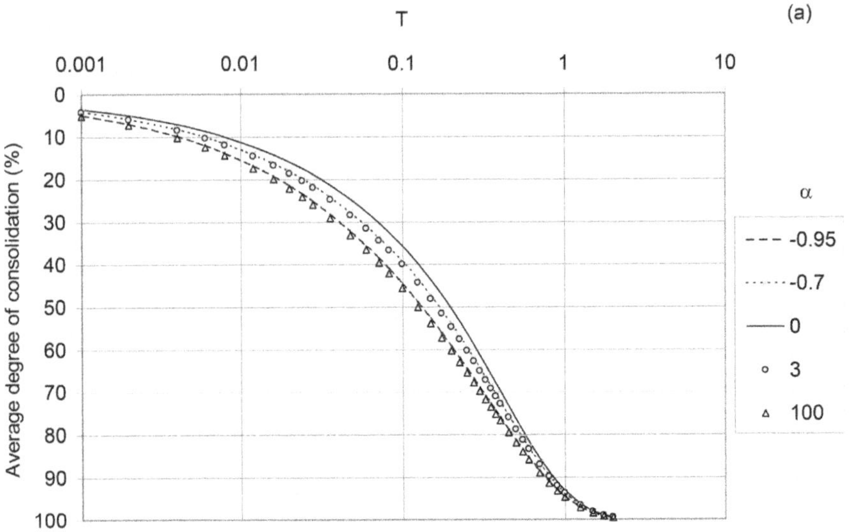

Figure 3.22 (a) Average degree of consolidation for $p = q = 2$ – double drainage.

condition, the excess pore pressure distribution combines the features of Figures 3.15(b) and 3.19(b). For $\alpha > 0$, the excess pore pressure distribution is similar to but smaller than the results of the constant compressibility shown in Figure 3.15(b). For $\alpha < 0$, the excess pore pressure distribution is similar to but greater than the results of the constant permeability shown in Figure 3.19(b).

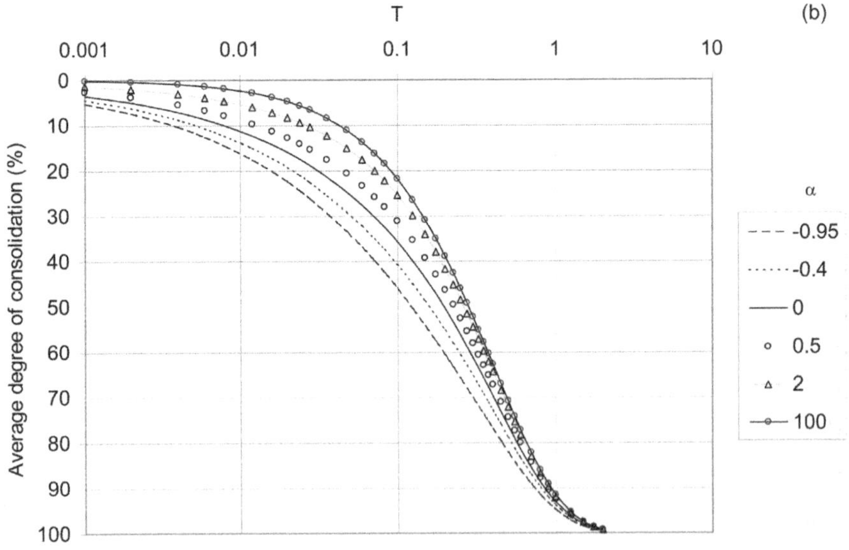

Figure 3.22 (b) Average degree of consolidation for $p = q = 2$ – single drainage.

Table 3.12 Eigenvalue λ_l for different α for cases of depth-dependent permeability and compressibility resulting in constant c_v

	Single-drained condition			Double-drained condition		
α	$p = q = 0.5$	$p = q = 1$	$p = q = 2$	$p = q = 0.5$	$p = q = 1$	$p = q = 2$
–0.99	2.00905	2.40527		2.9092	2.80092	
–0.95	2.03706	2.41584	3.14287	3.12897	3.06441	3.30694
–0.9	2.09391	2.4481	3.15144	3.35968	3.31394	3.49066
–0.8	2.26323	2.57363	3.21304	3.84414	3.81596	3.92699
–0.7	2.50981	2.78559	3.36025	4.43144	4.41239	4.48799
–0.6	2.85828	3.10667	3.62438	5.19636	5.18307	5.23599
–0.5	3.36172	3.58802	4.05752	6.25537	6.24606	6.28319
–0.4	4.13066	4.3388	4.76773	7.83483	7.82844	7.85398
–0.35	4.68449	4.88471	5.29578	8.96034	8.95512	8.97598
–0.3	5.42589	5.61883	6.01345	10.4594	10.4552	10.472
–0.25	6.467	6.65323	7.03266	12.5565	12.5533	12.5664
–0.2	8.03216	8.21217	8.57754	15.7006	15.6981	15.708
–0.17	9.41496	9.59146	9.94888	18.4739	18.4718	18.48
–0.14	11.3919	11.565	11.9149	22.4351	22.4335	22.4399
–0.12	13.2599	13.4309	13.7758	26.1759	26.1745	26.1799
–0.1	15.876	16.0449	16.3851	31.4126	31.4115	31.4159
0.1	15.5567	15.4061	15.1069	31.4132	31.4123	31.4159
0.12	12.9401	12.7911	12.4955	26.1767	26.1757	26.1799
0.14	11.0716	10.9241	10.632	22.4363	22.4351	22.4399
0.17	9.09367	8.94846	8.66138	18.4756	18.4742	18.48
0.2	7.70969	7.56667	7.28446	15.703	15.7014	15.708
0.25	6.14213	6.00258	5.72813	12.5604	12.5585	12.5664
0.3	5.098	4.96176	4.69462	10.4652	10.4629	10.472
0.35	4.35292	4.21981	3.9596	8.96834	8.9658	8.97598
0.4	3.79471	3.66457	3.41092	7.8456	7.8428	7.85398
0.5	3.0145	2.88989	2.64839	6.27348	6.27024	6.28319
0.6	2.49565	2.37607	2.14557	5.22515	5.22154	5.23599
0.7	2.12601	2.01103	1.79054	4.4762	4.47226	4.48799
0.8	1.84954	1.73878	1.52744	3.91438	3.91017	3.92699
1	1.46403	1.36078	1.16556	3.12768	3.12303	3.14159
1.2	1.20848	1.11169	0.930235	2.60313	2.59815	2.61799
1.4	1.02697	0.935803	0.766245	2.22842	2.2232	2.24399
1.7	0.836056	0.752112	0.597654	1.83171	1.82624	1.848
2	0.703472	0.625598	0.483701	1.55408	1.54846	1.5708
2.5	0.55464	0.485007	0.359947	1.23961	1.23387	1.25664
3	0.456544	0.393456	0.281577	1.03018	1.02442	1.0472
3.5	0.387194	0.329443	0.228176	0.880753	0.87504	0.897598
4	0.335667	0.282358	0.189827	0.768828	0.763191	0.785398
5	0.26439	0.218072	0.13909	0.612434	0.607	0.628319
6	0.217559	0.17651	0.107565	0.508455	0.503245	0.523599
7	0.184523	0.147598	0.0863971	0.43438	0.429394	0.448799
8	0.160012	0.126412	0.0713727	0.378965	0.374192	0.392699
10	0.126149	0.097592	0.0517513	0.301648	0.297259	0.314159
12	0.103918	0.0790199	0.0397256	0.250324	0.246265	0.261799
15	0.0820187	0.0610626	0.0286878	0.199236	0.195583	0.20944
20	0.060517	0.0438462	0.0188085	0.148455	0.145317	0.15708
25	0.0478463	0.0339475	0.0135352	0.118182	0.11542	0.125664

(Continued)

Table 3.12 (Continued) Eigenvalue λ_l for different α for cases of depth-dependent permeability and compressibility resulting in constant c_v

	Single-drained condition			*Double-drained condition*		
α	$p=q=0.5$	$p=q=1$	$p=q=2$	$p=q=0.5$	$p=q=1$	$p=q=2$
30	0.0395131	0.0275632	0.0103361	0.0981023	0.0956295	0.10472
35	0.0336244	0.0231241	0.00822499	0.0838204	0.0815767	0.0897598
40	0.0292466	0.019869	0.00674605	0.0731471	0.07109	0.0785398
50	0.0231797	0.0154326	0.00484121	0.0582681	0.0564976	0.0628319
65	0.0176527	0.0114815	0.00327505	0.0446113	0.0431377	0.0483322
80	0.0142393	0.00909486	0.00240266	0.0361212	0.034853	0.0392699
100	0.0113105	0.00708699	0.00172175	0.0287967	0.0277221	0.0314159
200	0.00555259	0.00328554		0.0142667	0.0136394	
300	0.00367075	0.00210391		0.00947	0.00901846	
400	0.00273907	0.00153594		0.0070836	0.00672782	

Table 3.13 U for cases of depth-dependent permeability and compressibility resulting in constant c_v with double drainage

		$p=q=0.5$		$p=q=1$			$p=q=2$			
T	$\alpha=0$	−0.99	400	−0.99	8	400	−0.95	−0.7	3	100
0.001	3.57	3.25	3.22	4.04	3.74	4.15	5.03	4.17	4.3	5.23
0.004	7.14	6.6	6.55	8.05	7.49	8.27	9.94	8.29	8.54	10.31
0.012	12.36	11.61	11.55	13.88	12.95	14.24	16.89	14.22	14.62	17.49
0.02	15.96	15.12	15.06	17.86	16.7	18.29	21.51	18.24	18.73	22.25
0.028	18.88	18	17.94	21.06	19.75	21.54	25.18	21.47	22.03	26.01
0.036	21.41	20.5	20.45	23.81	22.37	24.33	28.27	24.23	24.84	29.18
0.048	24.72	23.81	23.75	27.38	25.8	27.95	32.23	27.81	28.48	33.22
0.06	27.64	26.73	26.69	30.5	28.81	31.11	35.63	30.92	31.63	36.69
0.072	30.28	29.39	29.35	33.3	31.53	33.93	38.63	33.71	34.45	39.73
0.083	32.51	31.65	31.61	35.65	33.82	36.29	41.11	36.04	36.81	42.24
0.1	35.68	34.87	34.84	38.96	37.06	39.62	44.55	39.33	40.11	45.72
0.125	39.89	39.16	39.14	43.3	41.34	43.97	48.96	43.62	44.42	50.16
0.15	43.69	43.04	43.03	47.15	45.19	47.82	52.79	47.43	48.24	53.99
0.2	50.41	49.9	49.91	53.83	51.91	54.46	59.23	54.03	54.82	60.39
0.25	56.22	55.84	55.85	59.47	57.67	60.06	64.5	59.63	60.36	65.59
0.3	61.32	61.02	61.04	64.33	62.68	64.86	68.94	64.45	65.13	69.94
0.35	65.82	65.58	65.6	68.56	67.07	69.03	72.73	68.66	69.28	73.65
0.4	69.79	69.59	69.61	72.25	70.92	72.68	76.01	72.35	72.9	76.83
0.45	73.29	73.13	73.15	75.5	74.31	75.88	78.87	75.59	76.08	79.6
0.5	76.39	76.26	76.27	78.36	77.3	78.69	81.36	78.44	78.88	82.02
0.6	81.56	81.45	81.47	83.11	82.28	83.36	85.48	83.17	83.52	85.99
0.7	85.59	85.51	85.52	86.8	86.15	87.01	88.67	86.85	87.13	89.07
0.9	91.2	91.15	91.16	91.95	91.55	92.07	93.09	91.98	92.14	93.33
1	93.13	93.09	93.09	93.71	93.4	93.8	94.6	93.73	93.86	94.79
1.25	96.29	96.27	96.27	96.6	96.44	96.66	97.09	96.62	96.69	97.19
1.75	98.92	98.91	98.91	99.01	98.96	99.03	99.15	99.01	99.04	99.18

Table 3.14 U for cases of depth-dependent permeability and compressibility resulting in constant c_v with single drainage

		$p=q=0.5$			$p=q=1$			$p=q=2$				
T	$\alpha=0$	-0.99	4	400	-0.99	2.5	400	-0.95	-0.4	0.5	2	100
0.001	3.57	4.17	2.54	0.96	4.62	2.15	0.38	5.28	4.46	2.72	1.46	0.25
0.004	7.14	8.28	5.22	2.52	9.15	4.47	1.26	10.4	8.83	5.52	3.16	0.99
0.012	12.36	14.23	9.38	5.51	15.63	8.17	3.35	17.64	15.08	9.79	6.09	2.92
0.02	15.96	18.26	12.4	7.95	19.98	10.93	5.29	22.43	19.29	12.84	8.41	4.82
0.028	18.88	21.5	14.93	10.14	23.44	13.28	7.17	26.22	22.64	15.39	10.48	6.68
0.036	21.41	24.27	17.18	12.17	26.39	15.4	8.99	29.41	25.5	17.64	12.4	8.51
0.048	24.72	27.86	20.21	14.99	30.18	18.29	11.65	33.47	29.17	20.65	15.1	11.17
0.06	27.64	30.98	22.94	17.63	33.46	20.94	14.23	36.95	32.35	23.37	17.65	13.77
0.072	30.28	33.78	25.47	20.14	36.36	23.41	16.73	40.01	35.18	25.88	20.1	16.28
0.083	32.51	36.11	27.65	22.33	38.78	25.57	18.96	42.53	37.54	28.04	22.25	18.52
0.1	35.68	39.4	30.8	25.58	42.15	28.72	22.29	46.01	40.83	31.17	25.46	21.87
0.125	39.89	43.69	35.09	30.06	46.5	33.05	26.94	50.46	45.1	35.43	29.93	26.54
0.15	43.69	47.48	39.04	34.26	50.31	37.09	31.31	54.29	48.86	39.35	34.12	30.94
0.2	50.41	54.03	46.17	41.9	56.79	44.4	39.28	60.68	55.31	46.43	41.77	38.95
0.25	56.22	59.56	52.42	48.64	62.16	50.86	46.33	65.87	60.72	52.66	48.53	46.04
0.3	61.32	64.35	57.95	54.6	66.74	56.56	52.56	70.2	65.38	58.16	54.5	52.3
0.35	65.82	68.52	62.83	59.87	70.7	61.6	58.06	73.87	69.45	63.01	59.78	57.84
0.4	69.79	72.2	67.14	64.53	74.15	66.06	62.93	77.04	73.01	67.31	64.45	62.73
0.45	73.29	75.43	70.96	68.64	77.17	70	67.23	79.78	76.15	71.1	68.57	67.06
0.5	76.39	78.29	74.33	72.28	79.84	73.48	71.04	82.18	78.93	74.46	72.22	70.88
0.6	81.56	83.04	79.94	78.34	84.26	79.28	77.37	86.12	83.54	80.04	78.3	77.25
0.7	85.59	86.75	84.33	83.08	87.7	83.81	82.32	89.17	87.14	84.41	83.04	82.22
0.9	91.2	91.91	90.43	89.67	92.49	90.12	89.21	93.4	92.15	90.48	89.65	89.15
1	93.13	93.68	92.52	91.93	94.13	92.28	91.57	94.84	93.86	92.56	91.91	91.52
1.25	96.29	96.59	95.97	95.64	96.83	95.83	95.45	97.22	96.69	95.99	95.63	95.42
1.75	98.92	99.01	98.83	98.73	99.08	98.79	98.67	99.19	99.04	98.83	98.73	98.67

It should be noted that for the case of $p=q=2$, the solution is also expressible in terms of elementary functions since $v=1/2$ and (3.40) holds.

$$Z_v^m(\lambda_m y)=\frac{\sin[\lambda_m(y-1)]}{\sqrt{\lambda_m y}} \tag{3.40}$$

where $\lambda_m=\frac{m\pi}{|\alpha|}$ for boundary condition (2.16); and λ_m is the m-th positive root of the following (3.41) for variable λ for boundary condition (2.17)

$$(1+\alpha)\lambda\cos(\alpha\lambda)-\sin(\alpha\lambda)=0 \tag{3.41}$$

3.3.5 Characteristics of the eigenvalue λ_1 for different α

As illustrated in the above sections, the eigenvalue λ_1 plays an important role in normalizing the time. To show the characteristics of λ_1 vs. α the following parameter β is defined.

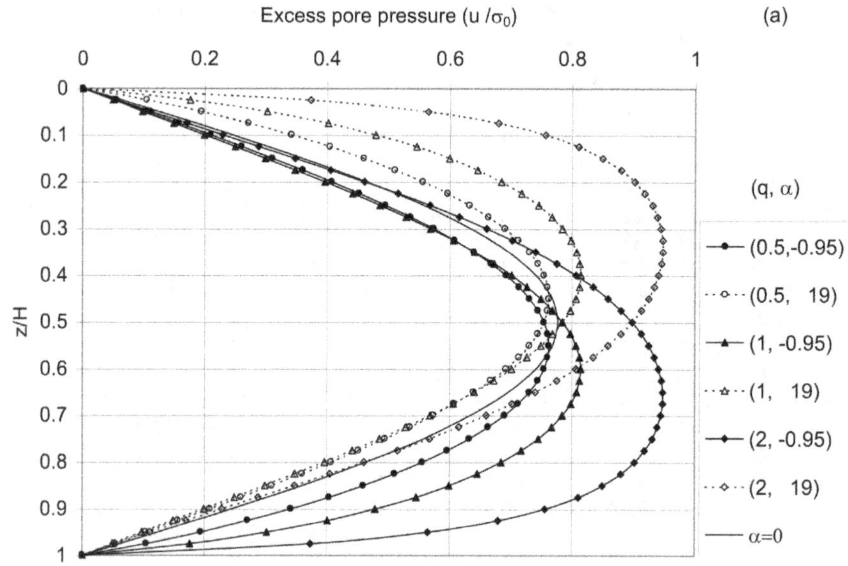

Figure 3.23 (a) Excess pore pressure isochrones for 50% average degree of consolidation for $p = q = 0.5$, 1, 2 – double drainage.

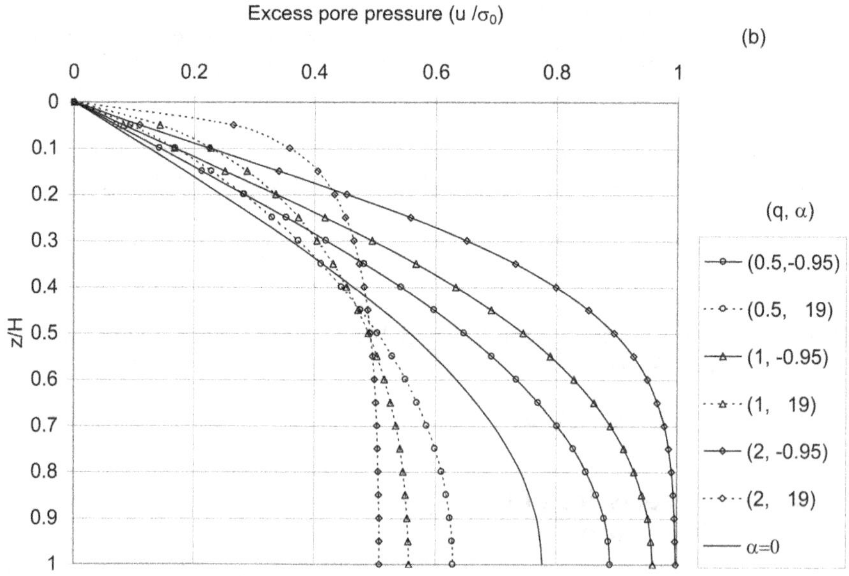

Figure 3.23 (b) Excess pore pressure isochrones for 50% average degree of consolidation for $p = q = 0.5$, 1, 2 – single drainage.

$$\beta = \begin{cases} |(2-n)\alpha|\lambda_1 & n \neq 2 \\ 2|\alpha|\lambda_1 & n = 2 \end{cases} \tag{3.42}$$

Clearly, this parameter is a measure of the time range that is of importance to the consolidation behavior. The relationship of β vs. α is shown in Figure 3.24. For the double-drained condition, the curves intersect at $\beta = 2\pi$ and $\alpha = 0$. For the single-drained condition, the curves intersect at $\beta = \pi$ and $\alpha = 0$. β is almost a constant for the cases of constant coefficient of consolidation and the double-drained condition, and thus the average degree of consolidation in this case will fall in a narrow band whether or not the time is normalized. As shown in Figure 3.24, β increases with the increase of α for $q = 0$, but decreases with the increase of α for $p = 0$ and constant coefficient of consolidation. β changes normally from its minimum to about 4 to 20 times larger for the selected value of α. However, a larger range is also calculated. For example, β changes from 10.8 for $\alpha = -0.95$ to 0.109 (about 100 times smaller) for $\alpha = 100$ for $p = 0$ and $q = 2$. In these situations, normalized time will greatly increase the accuracy and reduce the number of solution charts.

The average degree of consolidation is very close to the conventional Terzaghi's solution if the time is normalized. This shows that the Asaoka (1978) and hyperbolic (Tan and Chew, 1996) observational methods for monitoring and predicting the primary consolidation settlement are still applicable to the nonhomogeneous soils discussed above. However, the excess pore pressure distribution may be quite different depending on the soil conditions and the back-calculated coefficient of consolidation is $c_{v0}\beta^2/\pi^2$.

(a)

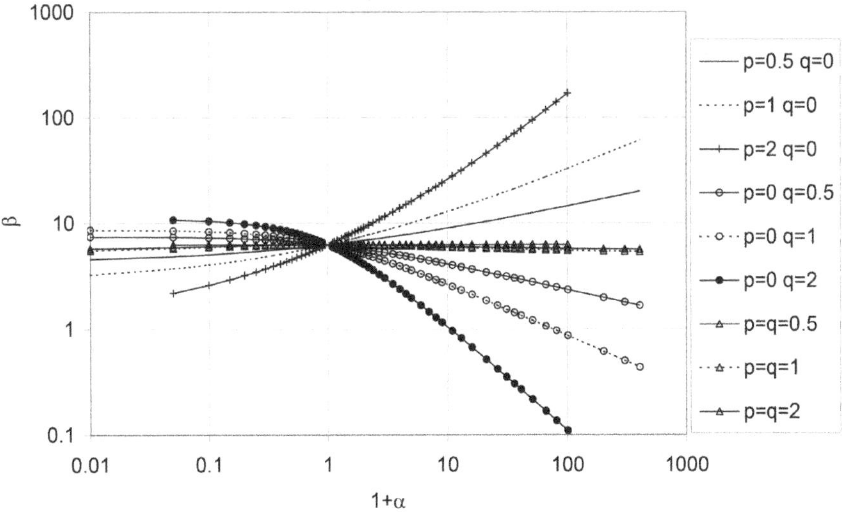

Figure 3.24 (a) Relationship of β vs. α – double drainage.

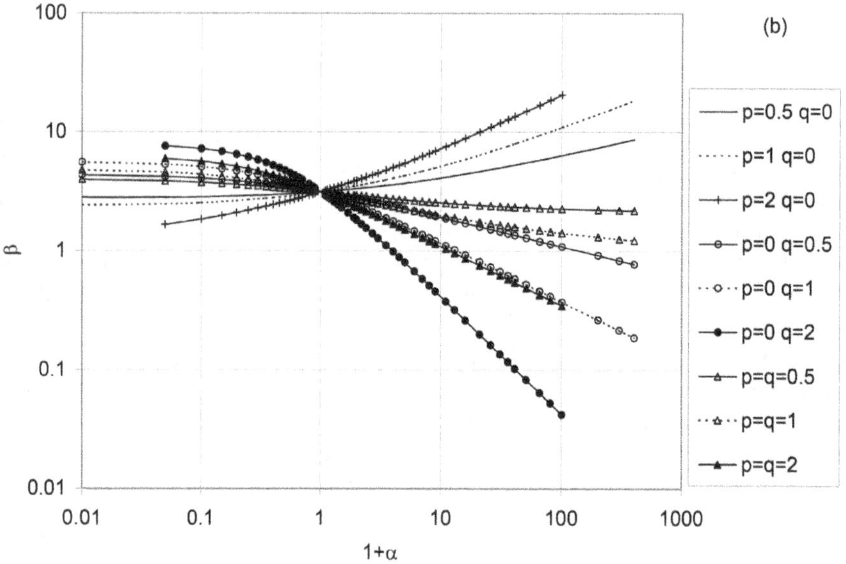

Figure 3.24 (b) Relationship of β vs. α – single drainage.

3.4 NON-LINEAR CONSOLIDATION OF SOILS WITH CONSTANT COEFFICIENT OF CONSOLIDATION

In a mass of real soil, the compressibility, permeability, and coefficient of consolidation vary during the consolidation process. The least variable factor for normally consolidated clays is the coefficient of consolidation. If the applied load increment on a soil layer is significant, the relationship between ε_z and $\ln(\sigma_z')$ is essentially linear over a wide range. Therefore, the non-linear consolidation behavior of clays can be partially accounted by assuming the coefficient of consolidation to be constant while the compressibility and permeability are both allowed to decrease with increasing pressure.

3.4.1 Basic equations

The constitutive equation (2.4) can be expressed in (3.43):

$$\begin{cases} f = \lambda \ln(\sigma_z') \\ g = 0 \end{cases} \tag{3.43}$$

where λ = compression index.

The coefficient of permeability k is expressed as (3.44)

$$k = \gamma_w c_v \frac{\partial f}{\partial \sigma_z'} = \gamma_w c_v \frac{\lambda}{\sigma_z'} \tag{3.44}$$

Substituting (3.43) and (3.44) into (2.11) gives (3.45).

$$-c_v \frac{\partial}{\partial z}\left(\frac{1}{\sigma_z'}\frac{\partial u_e}{\partial z}\right) = \frac{\partial \ln(\sigma_z')}{\partial t} \tag{3.45}$$

3.4.2 Consolidation of a thin layer of soil under suddenly applied loading

For a thin layer of soil, the weight of the soil can be ignored so that (3.46) is valid.

$$\frac{\partial \sigma_z}{\partial z} = \frac{\partial \sigma_z'}{\partial z} + \frac{\partial u_e}{\partial z} = 0 \tag{3.46}$$

The suddenly applied vertical total stress increment $\Delta\sigma_z = \sigma_f' - \sigma_i'$
where σ_f' = final effective stress; σ_i' = initial effective stress at time $t = 0$.
Using the substitution (Davis and Raymond, 1965):

$$w = \ln\left(\frac{\sigma_z'}{\sigma_f'}\right) / \ln\left(\frac{\sigma_i'}{\sigma_f'}\right) \tag{3.47}$$

and (3.46), (3.45) becomes

$$c_v \frac{\partial^2 w}{\partial z^2} = \frac{\partial w}{\partial t} \tag{3.48}$$

Only the boundary condition (2.17) is considered here. The solution for boundary condition (2.16) can be obtained by reassigning H as the length of the longest drainage path in the stratum (no flow at the middle of the stratum for the double-drainage condition) since the coefficient of (3.48) and initial condition are depth independent. The boundary condition (2.17) and initial condition are easily expressed as

$$\begin{cases} w(0,t) = 0 \\ \dfrac{\partial w}{\partial z}\Big|_{z=H} = 0 \\ w(z,0) = 1 \end{cases} \tag{3.49}$$

Equations (3.48) and (3.49) are special forms of the equations discussed in Section 2.4.1 (see Figure 2.7(d)). The solution is:

$$w(z,t) = \sum_{n=1}^{\infty} \frac{2}{\left(n-\frac{1}{2}\right)\pi} \sin\left[\left(n-\frac{1}{2}\right)\pi\frac{z}{H}\right] \exp\left[-\left(n-\frac{1}{2}\right)^2 \pi^2 T_v\right] \tag{3.50}$$

$$\text{Therefore } \frac{\sigma'_z}{\sigma'_f} = \frac{\sigma'_f - u_e}{\sigma'_f} = \left(\frac{\sigma'_i}{\sigma'_f}\right)^w \tag{3.51}$$

Arranging (3.51):

$$\frac{u_e}{\sigma'_f - \sigma'_i} = \frac{\sigma'_f}{\sigma'_f - \sigma'_i}\left[1 - \left(\frac{\sigma'_i}{\sigma'_f}\right)^w\right] \tag{3.52}$$

Values of excess pore pressures were computed for $z/H = 1$ and for different ratios of σ'_f/σ'_i. The results are tabulated in Table 3.15 and plotted graphically in Figure 3.25. Clearly, the excess pore pressure is close to the linear solution $(\sigma'_f/\sigma'_i = 1)$ if $\sigma'_f/\sigma'_i < 2$.

The average degree of consolidation for the assumed ideal soil is

$$U = \frac{\int_0^H \varepsilon_z(z,t)dz}{\int_0^H \varepsilon_z(z,t=\infty)dz} = \frac{\int_0^H \ln(\sigma'_z/\sigma'_i)dz}{\int_0^H \ln(\sigma'_f/\sigma'_i)dz} \tag{3.53}$$

Table 3.15 $u_e/(\sigma'_f - \sigma'_i)$ for $z/H = 1$ and for different ratios of σ'_f/σ'_i

T_v	$\sigma'_f/\sigma'_i = 1$	1.5	2	4	8	16
0.04	0.999	0.999	0.999	1	1	1
0.08	0.975	0.98	0.983	0.988	0.992	0.995
0.1	0.949	0.958	0.964	0.976	0.984	0.99
0.15	0.864	0.887	0.901	0.931	0.953	0.97
0.2	0.772	0.807	0.829	0.876	0.913	0.941
0.25	0.685	0.728	0.756	0.818	0.868	0.907
0.3	0.607	0.654	0.687	0.758	0.819	0.868
0.35	0.537	0.587	0.621	0.7	0.768	0.826
0.4	0.474	0.525	0.561	0.643	0.717	0.78
0.45	0.419	0.469	0.505	0.588	0.665	0.733
0.5	0.371	0.419	0.453	0.536	0.614	0.685
0.55	0.328	0.373	0.406	0.487	0.565	0.637
0.6	0.29	0.332	0.364	0.441	0.517	0.589
0.7	0.226	0.263	0.29	0.359	0.429	0.497
0.8	0.177	0.208	0.231	0.29	0.352	0.413
0.9	0.138	0.163	0.183	0.232	0.285	0.34
1	0.108	0.129	0.144	0.185	0.23	0.276
1.25	0.058	0.07	0.079	0.103	0.13	0.159
1.5	0.031	0.038	0.043	0.057	0.072	0.089
2	0.009	0.011	0.013	0.017	0.022	0.027

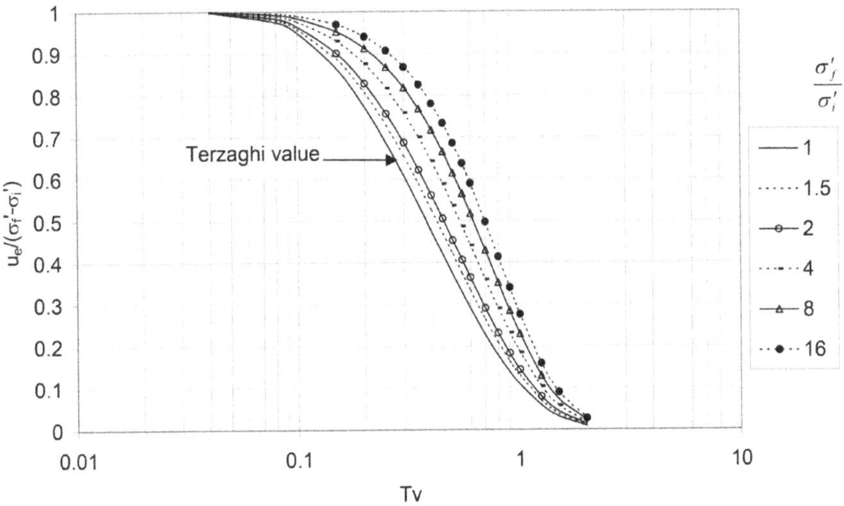

Figure 3.25 $u_e/(\sigma_f'-\sigma_i')$ for $z/H = 1$ and for different ratios of σ_f'/σ_i'.

Substituting (3.51) gives:

$$U = \frac{\int_0^H \ln\left[\frac{\sigma_f'}{\sigma_i'} \cdot \left(\frac{\sigma_i'}{\sigma_f'}\right)^w\right] dz}{\int_0^H \ln\left(\frac{\sigma_f'}{\sigma_i'}\right) dz} = \frac{1}{H}\int_0^H (1-w)dz \tag{3.54}$$

Equation (3.54) is the special case of U_l (2.28) for $T_c = 0$. Thus, the average degree of consolidation for the non-linear soil is independent of the ratio of σ_f'/σ_i' and equal to the linear solution.

3.4.3 Consolidation of a thin layer of soil under constant rate of loading

In addition to the standard incremental loading technique of consolidation test, the consolidation behavior of soil can also be determined using a constant rate of loading (Aboshi *et al.*, 1970). In this section, the response of the assumed ideal soil (3.45) under constant rate of loading is analyzed. Using the substitution (Vaid, 1985):

$$w^* = \ln\left(\frac{\sigma_z'}{\sigma_i'}\right) \tag{3.55}$$

and (3.46), (3.45) becomes

$$c_v \frac{\partial^2 w^*}{\partial z^2} = \frac{\partial w^*}{\partial t} \tag{3.56}$$

As in the above section, only the boundary condition (2.17) is considered here. For consolidation under the constant rate of loading, the boundary and initial conditions are:

$$\begin{cases} \sigma_z'(0,t) = \sigma_i' + Rt \\ \dfrac{\partial \sigma_z'}{\partial z}\Big|_{z=H} = 0 \\ \sigma_z'(z,0) = \sigma_i' \end{cases} \tag{3.57}$$

in which R = rate of total stress increase. Expressing (3.57) in terms of w^* gives:

$$\begin{cases} w^*(0,t) = \ln(1 + Rt/\sigma_i') \\ \dfrac{\partial w^*}{\partial z}\Big|_{z=H} = 0 \\ w^*(z,0) = \sigma_i' \end{cases} \tag{3.58}$$

The solution of (3.56) with boundary and initial conditions (3.58) (Vaid, 1985) is

$$w^* = \ln(1+\alpha T_v) - \frac{4}{\pi} \sum_{n=1,3,5...}^{\infty} \frac{\alpha}{n} \sin\left(\frac{n\pi z}{2H}\right) \int_0^{T_v} \exp\left[-\frac{n^2\pi^2(T_v - \tau)}{4}\right] \frac{1}{1+\alpha\tau} d\tau \tag{3.59}$$

where $\alpha = RH^2/(c_v\sigma_i')$ and is a dimensionless number.

Thus,

$$\sigma_z' = \sigma_i' + Rt - u_e = \sigma_i' \exp(w^*) \tag{3.60}$$

and the ratio u_e/Rt becomes

$$\frac{u_e}{Rt} = \frac{1+\alpha T_v}{\alpha T_v} \left\{ 1 - \exp\left[-\frac{4}{\pi} \sum_{n=1,3,5...}^{\infty} \frac{\alpha}{n} \sin\left(\frac{n\pi z}{2H}\right) \int_0^{T_v} \exp\left[-\frac{n^2\pi^2(T_v - \tau)}{4}\right] \frac{1}{1+\alpha\tau} d\tau \right] \right\} \tag{3.61}$$

The corresponding pore pressure ratio for linear situation (a special form of (2.20)) is governed by the following equation (Aboshi *et al.*, 1970)

$$\frac{u_e}{Rt} = \frac{16}{\pi^3 T_v} \sum_{n=1,3,5...}^{\infty} \frac{1}{n^3} \sin\left(\frac{n\pi z}{2H}\right)\left[1 - \exp\left(-\frac{n^2\pi^2 T_v}{4}\right)\right] \tag{3.62}$$

Isochrones of excess pore pressure are illustrated in Figure 3.26 and tabulated in Table 3.16 for both non-linear (3.61) and linear (3.62) solutions. Unlike the linear case, the excess pore pressure isochrones in the non-linear solution depend on the value of parameter α in addition to the time factor T_v. The

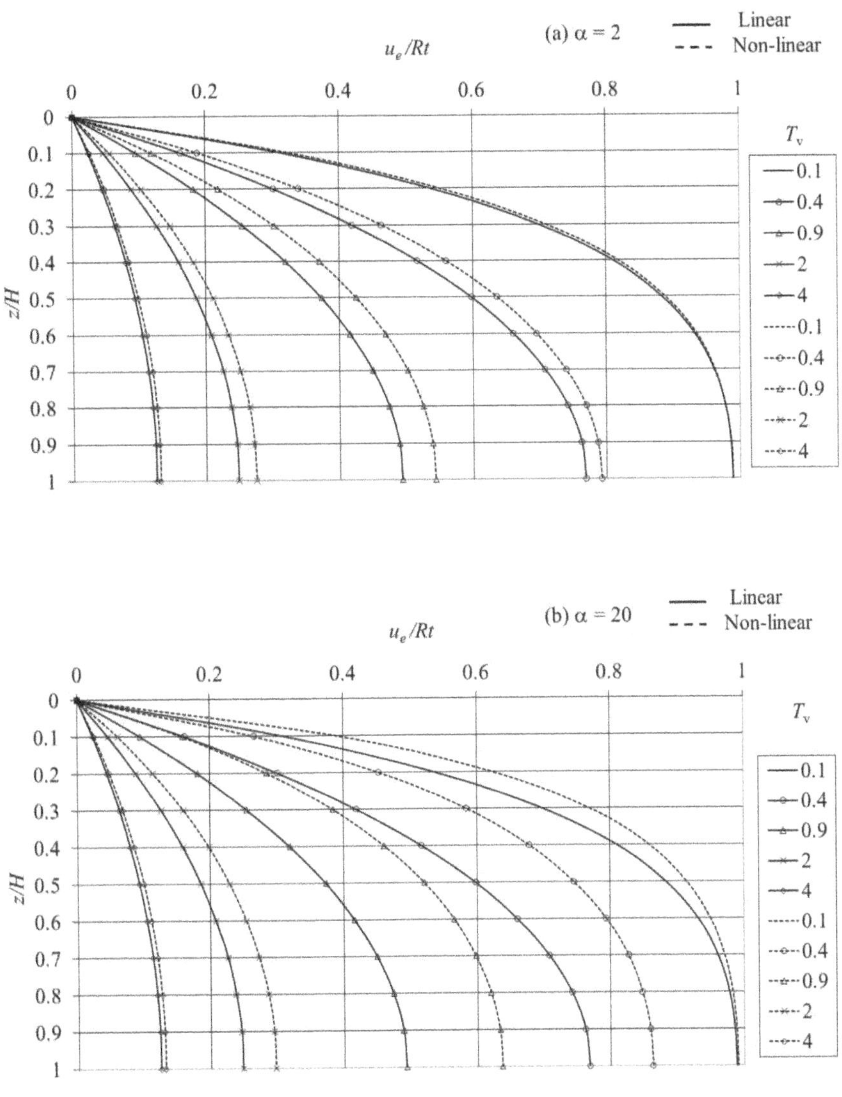

Figure 3.26 Excess pore pressure isochrones for linear and non-linear solutions: (a) $\alpha = 2$; (b) $\alpha = 20$.

Table 3.16 u_e/Rt for linear and non-linear ($\alpha = 2, 20$) solutions

Linear solution

	z/H										
T_v	0	0.1	0.2	0.3	0.4	0.5	0.6	0.7	0.8	0.9	1
0.1	0	0.31	0.537	0.699	0.81	0.884	0.932	0.961	0.978	0.986	0.989
0.4	0	0.162	0.301	0.419	0.517	0.598	0.661	0.709	0.743	0.763	0.769
0.9	0	0.096	0.181	0.255	0.319	0.373	0.416	0.45	0.474	0.489	0.493
2	0	0.047	0.089	0.127	0.159	0.186	0.208	0.226	0.238	0.246	0.248
4	0	0.024	0.045	0.064	0.08	0.094	0.105	0.114	0.12	0.124	0.125

Non-linear solution ($\alpha = 2$)

	z/H										
T_v	0	0.1	0.2	0.3	0.4	0.5	0.6	0.7	0.8	0.9	1
0.1	0	0.322	0.551	0.71	0.818	0.889	0.935	0.962	0.978	0.987	0.989
0.4	0	0.188	0.34	0.463	0.56	0.637	0.696	0.74	0.77	0.787	0.793
0.9	0	0.119	0.219	0.302	0.37	0.425	0.469	0.502	0.525	0.539	0.543
2	0	0.056	0.104	0.146	0.181	0.211	0.234	0.252	0.265	0.272	0.275
4	0	0.026	0.049	0.068	0.085	0.099	0.111	0.12	0.126	0.13	0.131

Non-linear solution ($\alpha = 20$)

	z/H										
T_v	0	0.1	0.2	0.3	0.4	0.5	0.6	0.7	0.8	0.9	1
0.1	0	0.39	0.624	0.769	0.859	0.915	0.95	0.971	0.983	0.989	0.991
0.4	0	0.266	0.453	0.585	0.679	0.746	0.794	0.827	0.848	0.86	0.864
0.9	0	0.159	0.285	0.384	0.461	0.521	0.566	0.598	0.621	0.634	0.638
2	0	0.062	0.115	0.16	0.198	0.229	0.254	0.273	0.287	0.295	0.297
4	0	0.026	0.049	0.069	0.086	0.101	0.112	0.121	0.127	0.131	0.132

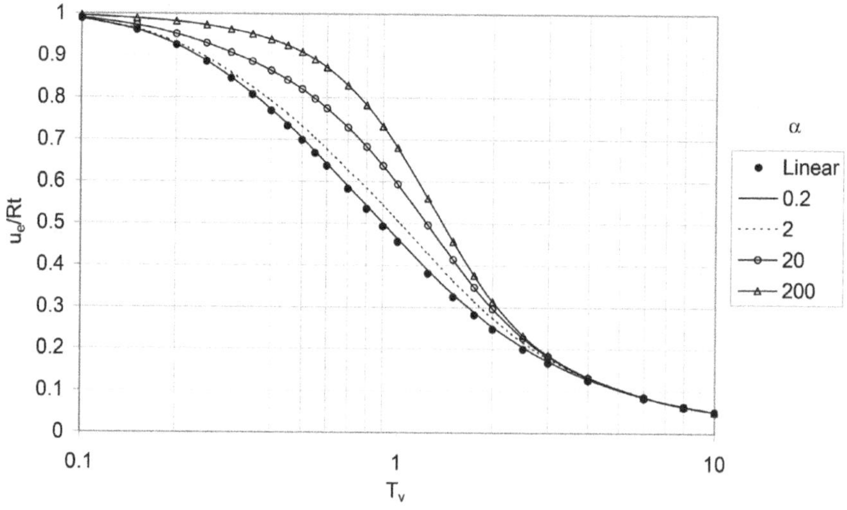

Figure 3.27 Excess pore pressure at $z/H = 1$ for linear and non-linear solutions.

Table 3.17 u_e/Rt at $z/H = 1$ for linear and non-linear solutions

		α (Non-linear)						α (Non-linear)			
T_v	Linear	0.2	2	20	200	T_v	Linear	0.2	2	20	200
0.1	0.989	0.989	0.989	0.991	0.996	0.9	0.493	0.501	0.543	0.638	0.732
0.15	0.962	0.962	0.964	0.973	0.989	1	0.456	0.464	0.506	0.594	0.681
0.2	0.926	0.927	0.931	0.952	0.982	1.25	0.381	0.389	0.427	0.496	0.56
0.25	0.886	0.887	0.896	0.93	0.973	1.5	0.325	0.332	0.364	0.414	0.457
0.3	0.846	0.848	0.86	0.908	0.963	1.75	0.282	0.289	0.315	0.348	0.376
0.35	0.807	0.809	0.826	0.886	0.952	2	0.248	0.254	0.275	0.297	0.314
0.4	0.769	0.772	0.793	0.864	0.939	2.5	0.2	0.205	0.217	0.227	0.233
0.45	0.733	0.737	0.762	0.842	0.925	3	0.167	0.171	0.179	0.183	0.185
0.5	0.699	0.704	0.733	0.82	0.909	4	0.125	0.128	0.131	0.132	0.133
0.55	0.668	0.672	0.704	0.797	0.891	6	0.083	0.085	0.086	0.086	0.086
0.6	0.638	0.643	0.678	0.774	0.872	8	0.062	0.063	0.064	0.064	0.064
0.7	0.583	0.59	0.628	0.729	0.829	10	0.05	0.051	0.051	0.051	0.051
0.8	0.535	0.543	0.584	0.683	0.782						

excess pore pressure at $z/H = 1$ is plotted as a function of T_v for different α in Figure 3.27 and tabulated in Table 3.17. For small α values, linear and non-linear solutions coincide at all T_v values. This would be expected since the effective stress increment ratio $\alpha T_v\ (Rt/\sigma'_i)$ will be small. Thus, little error will arise whether or not non-linear constitutive relation is adopted. With increasing α values, the range of effective stress for similar range of T_v increases, and therefore linear and non-linear solutions depart. The excess pore pressure ratios at equal T_v are higher in the non-linear than the linear case. However, for all T_v values in excess of about 4, the excess pore pressure ratio is almost the same as the linear solution for all α values.

3.5 CONSOLIDATION OF VISCO-ELASTIC SOILS

The compression of a saturated clay layer is time dependent. This is primarily due to the dissipation of the excess pore pressures caused by external loading. However, many soils, the organic soils in particular, exhibit a 1-D (odometer) compression behavior which, in the later stages of compression, becomes approximately proportional to the logarithm of time even after the excess pore pressures have substantially dissipated. This compression is due to the creep nature of the soil skeleton. The compression that occurs while the excess pore pressures dissipate is generally called *primary compression*. The slow compression that continues after the excess pore pressures have substantially dissipated is called *secondary compression*.

The relative importance of primary and secondary compression is affected by the thickness of a soil layer as shown in Figure 3.28. As the soil layer becomes thinner, the time required to dissipate excess pore pressures becomes

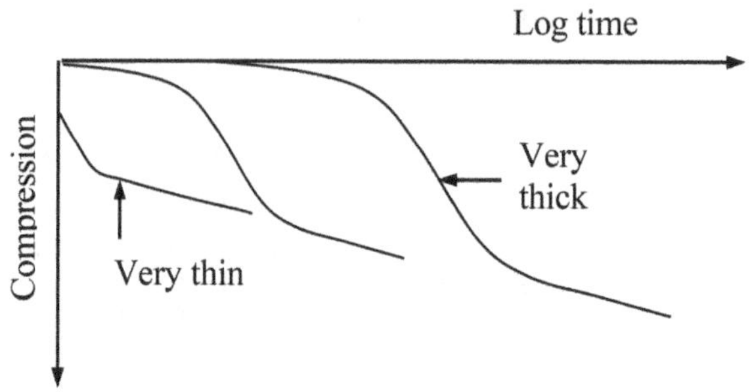

Figure 3.28 Effect of soil layer thickness on primary and secondary compressions.

shorter. If it were possible to load a very thin soil layer, compression could indeed occur in two distinct phases: (a) an *instantaneous compression* which occurred simultaneously with the increase in effective pressure and caused a reduction in void ratio until an equilibrium value was reached at which the structure effectively supported the overburden pressure; and (b) a *delayed compression* representing the reduction in volume at unchanged effective stress. For soil layers of finite thickness, the instantaneous and delayed effects are both present during the so-called primary compression. For the very thick layers of clay, much of the compression that occurs as excess pore pressures dissipate may actually be delayed compression.

First rational theories taking into consideration of the structural viscosity of soil assumed visco-elastic effective stress-strain relationships for the soil (Tan, 1957; Gibson and Lo, 1961; Schiffman *et al.*, 1964). In this section, the consolidation behavior of visco-elastic soils is analyzed.

3.5.1 Three-parameter visco-elastic system

The constitutive equation (2.4) for a three-parameter visco-elastic system shown in Figure 3.29 can be expressed as (3.63):

$$\begin{cases} f = m_{vp}\sigma'_z \\ g = \varsigma(m_v\sigma'_z - \varepsilon_z) \end{cases} \tag{3.63}$$

where $m_v = m_{vp} + m_{vs}$ = soil compressibility; m_{vp} = primary soil compressibility; m_{vs} = secondary soil compressibility; ς = ratio of fluidity.

Therefore, (2.11) becomes

$$-\frac{\partial}{\partial z}\left(\frac{k}{\gamma_w}\frac{\partial u_e}{\partial z}\right) = m_{vp}\frac{\partial \sigma'_z}{\partial t} + \varsigma(m_v\sigma'_z - \varepsilon_z) \tag{3.64a}$$

The parameters k, m_v, m_{vp}, and ς are assumed constant. The excess pore pressure and strain just before loading are assumed zero, and the

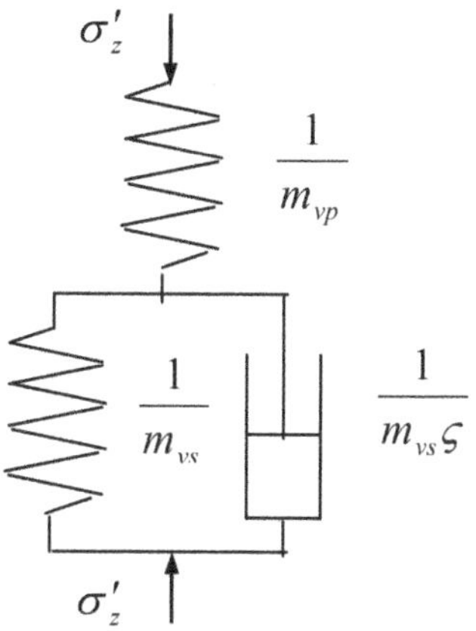

Figure 3.29 Three-parameter effective stress-strain system.

effective stress equation for depth-independent vertical total stress increment $\sigma_z(t)$ becomes

$$u_e = \sigma_z(t) - \sigma_z' \tag{3.64b}$$

Equation (3.64) has the following solution for suddenly applied constant vertical total stress increment σ_0 for boundary conditions (2.16) and (2.17),

$$\frac{u_e}{\sigma_0} = \sum_{n=1}^{\infty} \frac{2(1-\cos\lambda_n)}{\lambda_n \xi}\left[\frac{\varsigma + \xi\eta_{1n}}{\eta_{1n} - \eta_{2n}}\exp(\eta_{1n}t) - \frac{\varsigma + \xi\eta_{2n}}{\eta_{1n} - \eta_{2n}}\exp(\eta_{2n}t)\right]\sin\left(\lambda_n \frac{z}{H}\right) \tag{3.65a}$$

$$\frac{\varepsilon_z}{m_v\sigma_0} = 1 + \sum_{n=1}^{\infty} \frac{2(1-\cos\lambda_n)}{\lambda_n \xi}\sin\left(\lambda_n \frac{z}{H}\right)$$
$$\left[\frac{(\varsigma + \xi\eta_{1n})\eta_{2n}}{\eta_{1n} - \eta_{2n}}\exp(\eta_{1n}t) - \frac{(\varsigma + \xi\eta_{2n})\eta_{1n}}{\eta_{1n} - \eta_{2n}}\exp(\eta_{2n}t)\right] \tag{3.65b}$$

$$U = 1 + \sum_{n=1}^{\infty} \frac{2(1-\cos\lambda_n)^2}{\lambda_n^2 \xi}\left[\frac{(\varsigma + \xi\eta_{1n})\eta_{2n}}{\eta_{1n} - \eta_{2n}}\exp(\eta_{1n}t) - \frac{(\varsigma + \xi\eta_{2n})\eta_{1n}}{\eta_{1n} - \eta_{2n}}\exp(\eta_{2n}t)\right] \tag{3.65c}$$

$$\eta_{1n} = \frac{1}{2\xi}\left[-\left(\varsigma + \frac{c_v\lambda_n^2}{H^2}\right) + \sqrt{\left(\varsigma + \frac{c_v\lambda_n^2}{H^2}\right)^2 - 4\varsigma\xi\frac{c_v\lambda_n^2}{H^2}}\right] \tag{3.65d}$$

$$\eta_{2n} = \frac{1}{2\xi}\left[-\left(\varsigma + \frac{c_v\lambda_n^2}{H^2}\right) - \sqrt{\left(\varsigma + \frac{c_v\lambda_n^2}{H^2}\right)^2 - 4\varsigma\xi\frac{c_v\lambda_n^2}{H^2}}\right] \tag{3.65e}$$

$$\xi = \frac{m_{vp}}{m_v} \tag{3.65f}$$

where $\lambda_n = n\pi$ for boundary conditions (2.16), and $\lambda_n = (n-1/2)\pi$ for boundary conditions (2.17).

The solution (3.65) for boundary conditions (2.16) can also be obtained by reassigning the soil layer thickness H as the length of the longest drainage path in the stratum. Thus, the solution behavior of the three-parameter viscoelastic system is only analyzed for boundary conditions (2.17) by examining the effects of ξ and the following parameter ω,

$$\omega = \frac{H^2\varsigma}{c_v\lambda_1^2}. \tag{3.66}$$

Specification of these two parameters will completely specify the consolidation of the compressible layer.

Figure 3.30 illustrates the effects of ξ and ω on the average degree of consolidation U. As shown in Figure 3.30, fluidity will prolong the compression process, and this effect increases with the decrease of ξ. A clear two-step consolidation process (primary compression and secondary compression) is

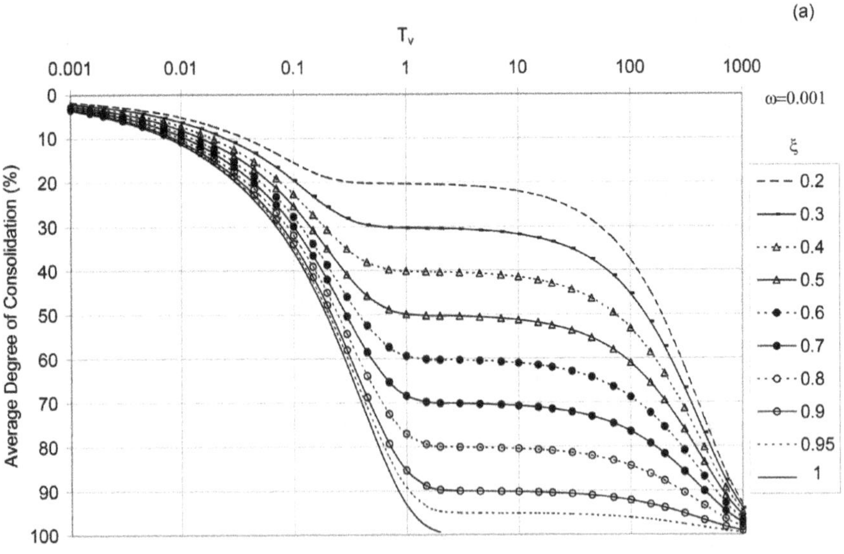

Figure 3.30 (a) Effect of soil skeleton compressibility ratios on time-consolidation relations – $\omega = 0.001$.

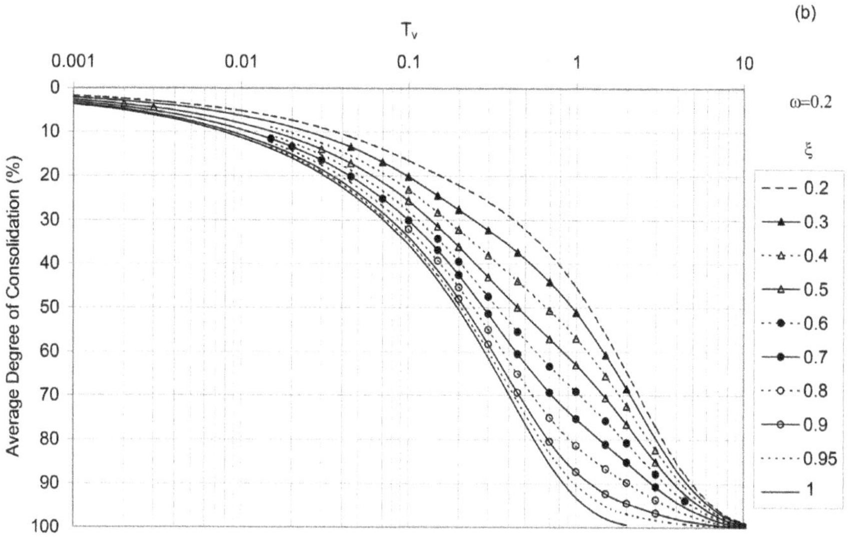

Figure 3.30 (b) Effect of soil skeleton compressibility ratios on time-consolidation relations – $\omega = 0.2$.

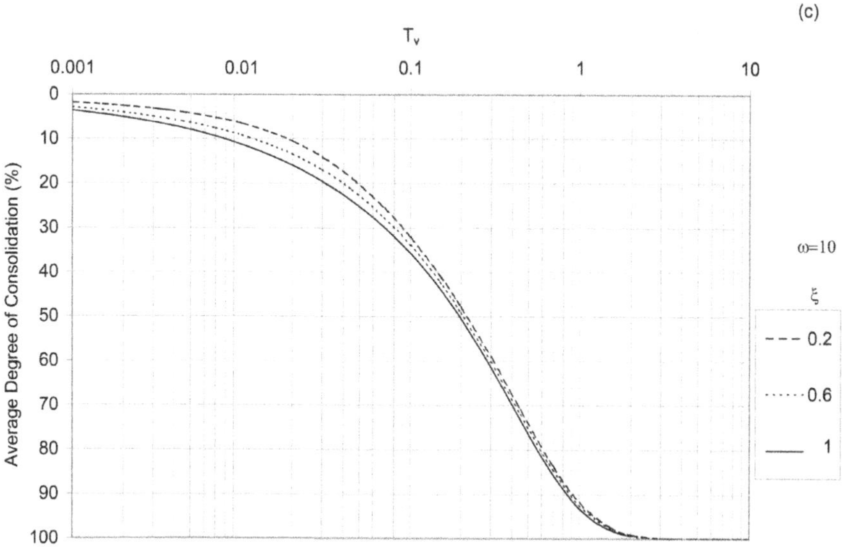

Figure 3.30 (c) Effect of soil skeleton compressibility ratios on time-consolidation relations – $\omega = 10$.

observed for very small fluidity ($\omega = 0.001$). For large fluidity ($\omega = 10$), the average degree of consolidation is close to the value of Terzaghi's solution ($\xi = 1$). For intermediate fluidity ($\omega = 0.2$), the primary compression and secondary compression are not easily distinguished from Figure 3.30.

Figure 3.31 shows the influence of ξ and ω on the excess pore pressure at the impermeable boundary. In the initial stage of loading, fluidity will speed up the dissipation of excess pore pressure, and this effect increases with the decrease of ξ. For large fluidity ($\omega = 10$), the excess pore pressure is close to the value of Terzaghi's solution ($\omega = \infty$). However, in the late stage of consolidation, intermediate fluidity will delay the dissipation of excess pore pressure, and a residual excess pore pressure exists comparing with the Terzaghi's solution ($\omega = \infty$).

Values of average degree of consolidation were computed for different values of ξ and ω. The results are plotted graphically in Figure 3.32. The elapsed times for achieving an average degree of consolidation of 10, 20, 30, 40, 50, 60, 70, 80, 90, and 95 for one-way drainage are tabulated in Table 3.18.

3.5.2 General visco-elastic soils

Owing to the complicated behaviour of soils, the visco-elastic relationship between the vertical effective stress and the vertical strain is generally expressed as a combination of spring and dashpot elements (Schiffman *et al.*, 1964;

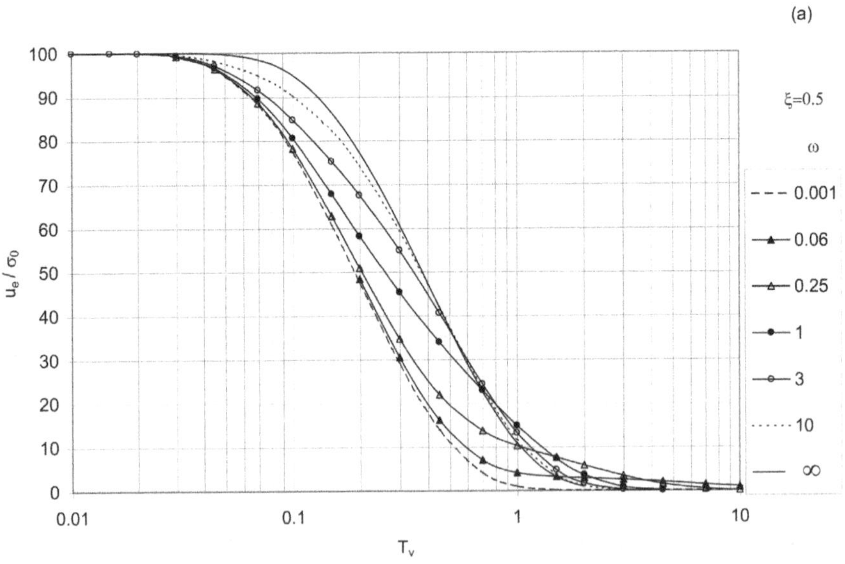

Figure 3.31 (a) Effect of soil skeleton compressibility ratios on excess pore pressure at the impermeable boundary ($z/H = 1$) – $\xi = 0.5$.

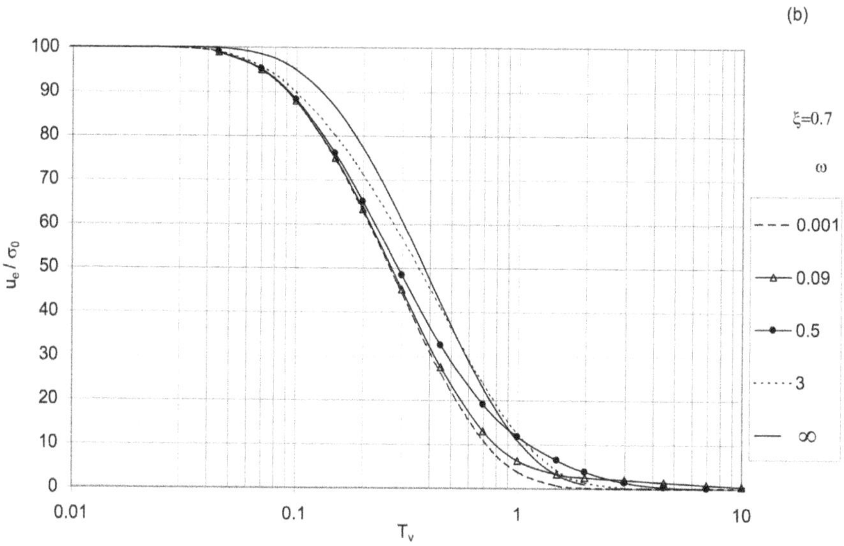

Figure 3.31 (b) Effect of soil skeleton compressibility ratios on excess pore pressure at the impermeable boundary ($z/H = 1$) – $\xi = 0.7$.

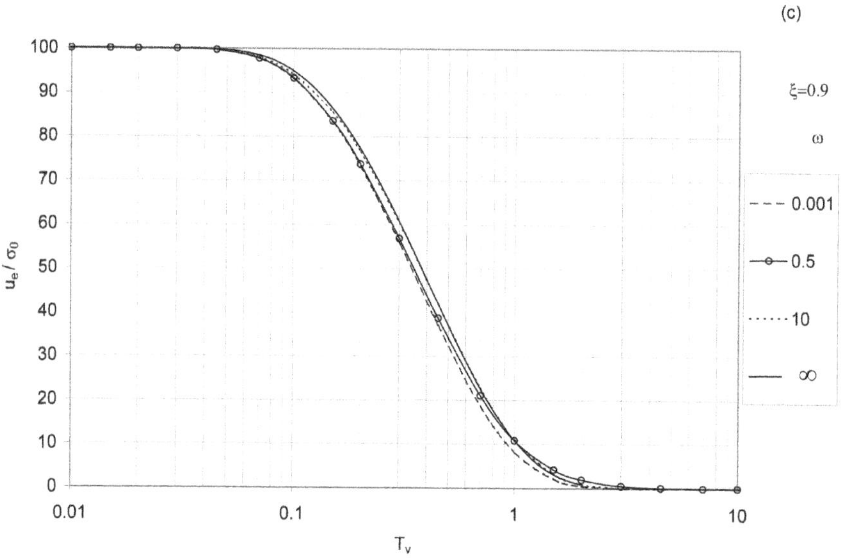

Figure 3.31 (c) Effect of soil skeleton compressibility ratios on excess pore pressure at the impermeable boundary ($z/H = 1$) – $\xi = 0.9$.

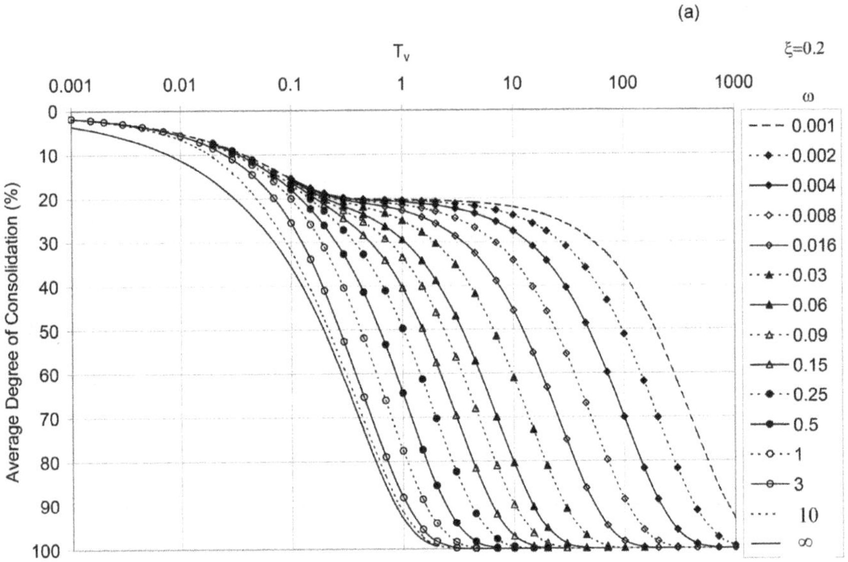

Figure 3.32 (a) Average degree of consolidation for different ξ and $\omega - \xi = 0.2$.

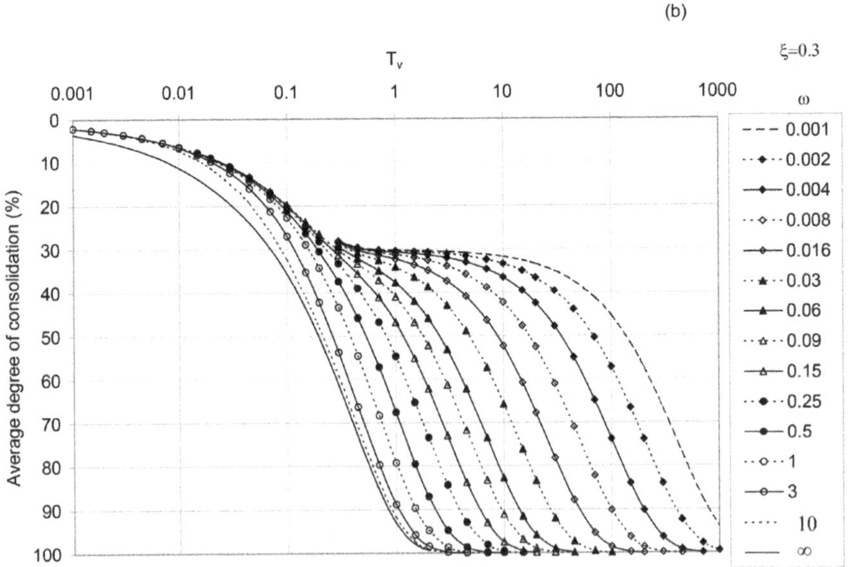

Figure 3.32 (b) Average degree of consolidation for different ξ and $\omega - \xi = 0.3$.

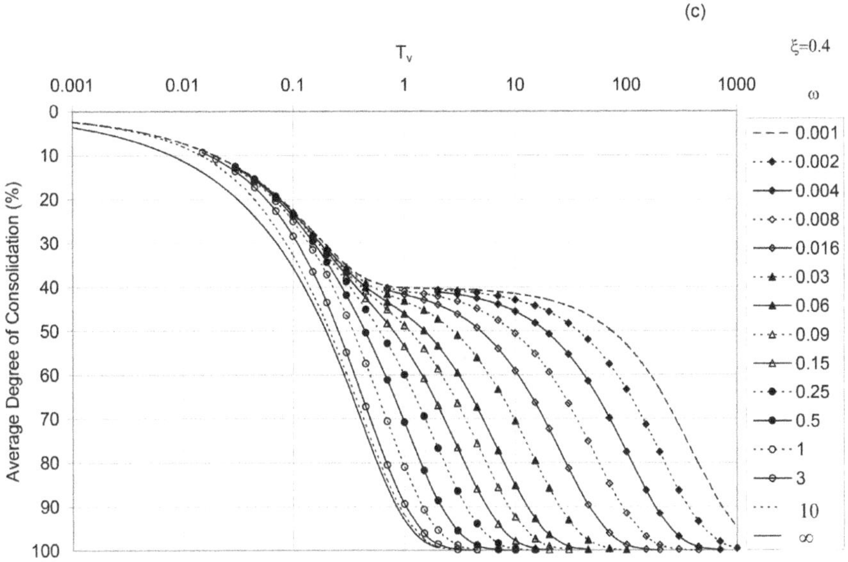

Figure 3.32 (c) Average degree of consolidation for different ξ and $\omega - \xi = 0.4$.

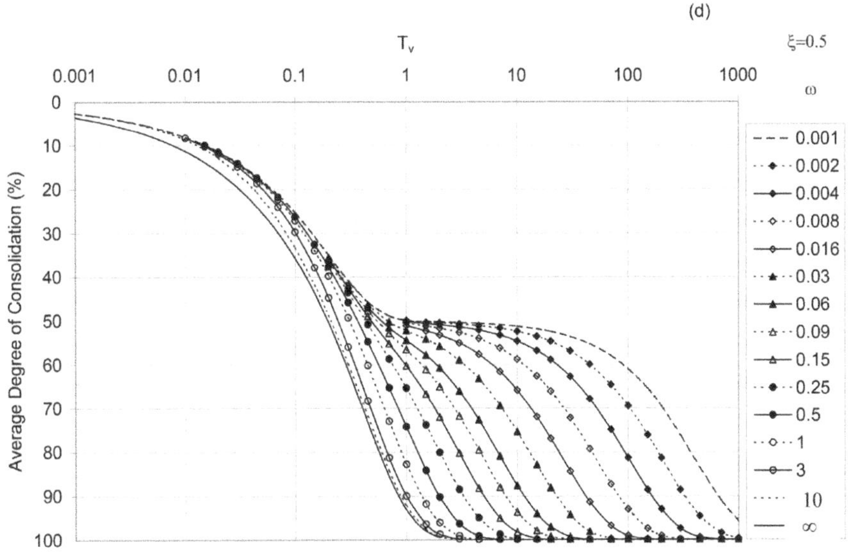

Figure 3.32 (d) Average degree of consolidation for different ξ and $\omega - \xi = 0.5$.

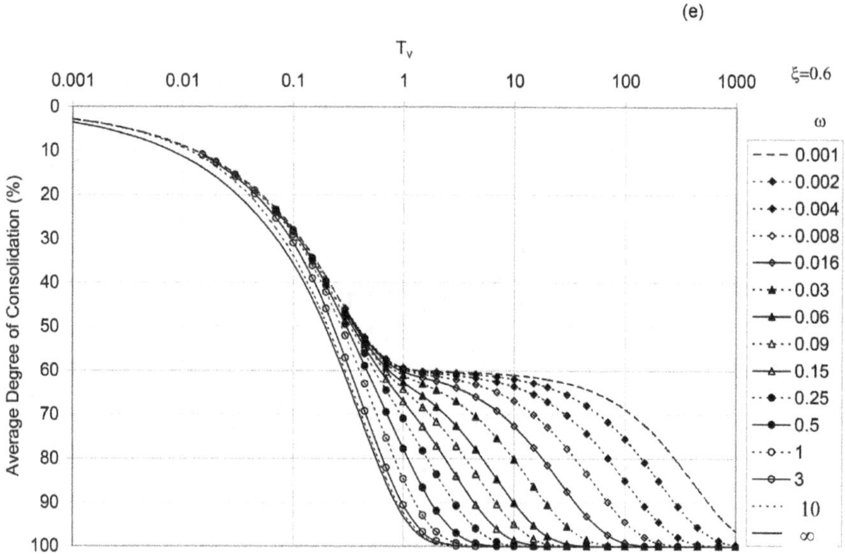

Figure 3.32 (e) Average degree of consolidation for different ξ and $\omega - \xi = 0.6$.

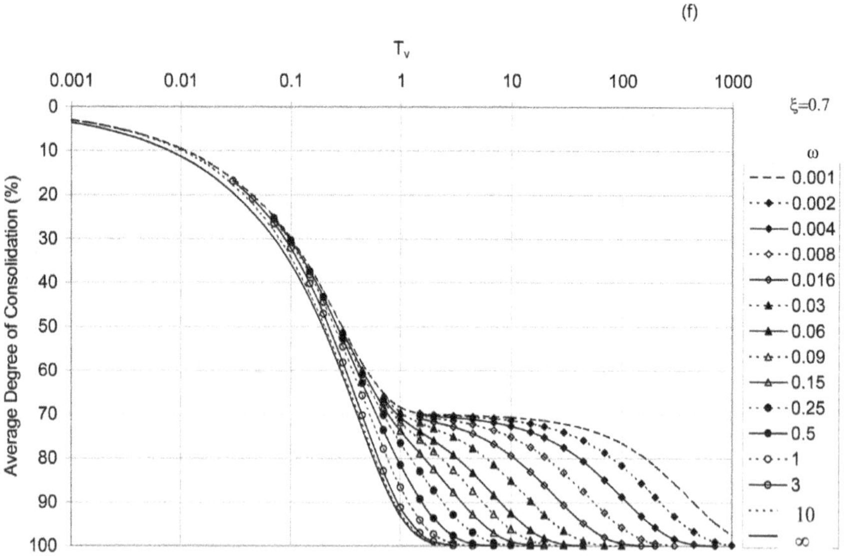

Figure 3.32 (f) Average degree of consolidation for different ξ and $\omega - \xi = 0.7$.

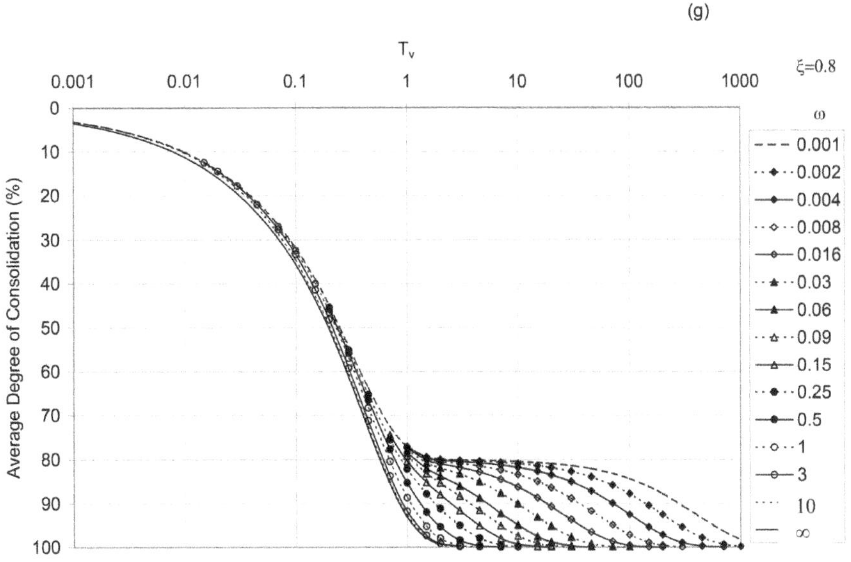

Figure 3.32 (g) Average degree of consolidation for different ξ and $\omega - \xi = 0.8$.

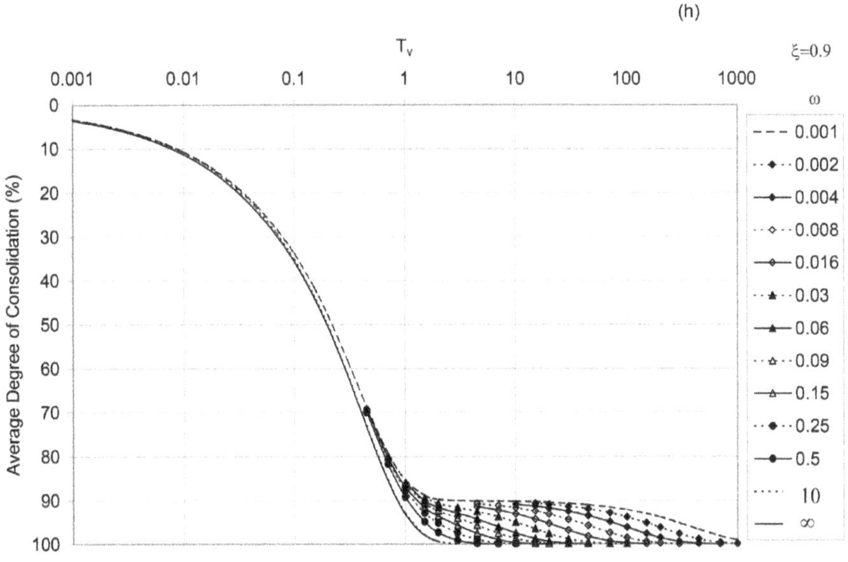

Figure 3.32 (h) Average degree of consolidation for different ξ and $\omega - \xi = 0.9$.

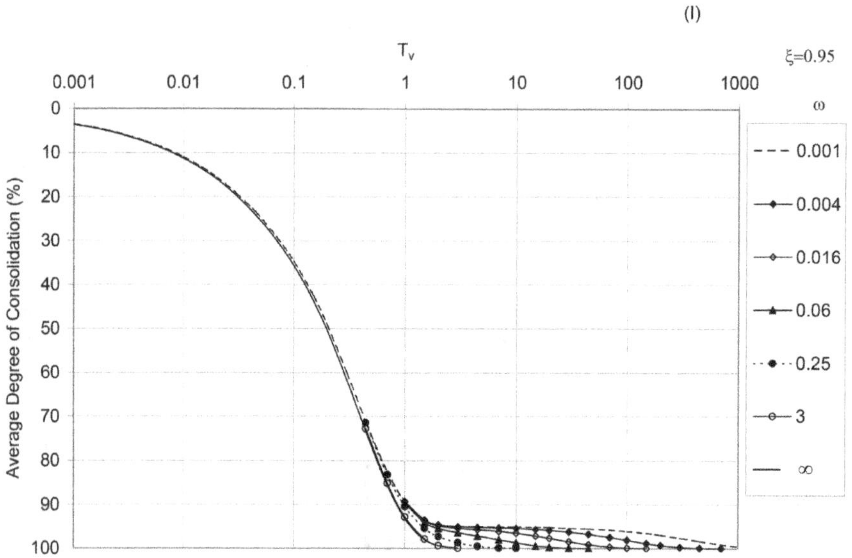

Figure 3.32 (l) Average degree of consolidation for different ξ and $\omega - \xi = 0.95$.

Table 3.18 Elapsed times for certain average degree of consolidation for three-parameter visco-elastic soils with one-way drainage

ω	T_{v10}	T_{v20}	T_{v30}	T_{v40}	T_{v50}	T_{v60}	T_{v70}	T_{v80}	T_{v90}	T_{v95}
					$\xi = 0.2$					
0.001	0.0381	0.354	53.5	116	190	280	397	562	843	1120
0.002	0.0381	0.34	26.8	58.1	95.1	140	199	281	422	562
0.004	0.038	0.32	13.5	29.2	47.7	70.3	99.6	141	211	282
0.008	0.038	0.294	6.83	14.7	24	35.3	50	70.6	106	141
0.016	0.0379	0.265	3.5	7.44	12.1	17.8	25.2	35.6	53.3	71.1
0.03	0.0378	0.237	1.94	4.07	6.58	9.65	13.6	19.2	28.8	38.3
0.06	0.0375	0.206	1.05	2.14	3.42	4.99	7.01	9.86	14.7	19.6
0.09	0.0373	0.189	0.757	1.49	2.36	3.43	4.8	6.74	10.1	13.4
0.15	0.0367	0.167	0.518	0.976	1.52	2.18	3.04	4.25	6.31	8.38
0.25	0.036	0.147	0.372	0.663	1.01	1.43	1.98	2.75	4.08	5.41
0.5	0.0343	0.122	0.256	0.423	0.623	0.868	1.19	1.64	2.41	3.19
1	0.0316	0.0996	0.188	0.295	0.422	0.58	0.786	1.08	1.59	2.11
3	0.0257	0.0708	0.126	0.192	0.274	0.379	0.52	0.725	1.08	1.44
10	0.0185	0.0489	0.0899	0.145	0.217	0.311	0.436	0.613	0.917	1.22
					$\xi = 0.3$					
ω	T_{v10}	T_{v20}	T_{v30}	T_{v40}	T_{v50}	T_{v60}	T_{v70}	T_{v80}	T_{v90}	T_{v95}
0.001	0.0254	0.106	0.579	61.9	136	226	343	507	788	1070
0.002	0.0254	0.106	0.55	31.1	68.1	113	172	254	395	535
0.004	0.0254	0.106	0.512	15.6	34.2	56.8	86	127	198	268
0.008	0.0254	0.106	0.467	7.94	17.2	28.6	43.2	63.9	99.1	134
0.016	0.0254	0.106	0.418	4.09	8.75	14.5	21.8	32.2	49.9	67.6

(Continued)

Table 3.18 (Continued) Elapsed times for certain average degree of consolidation for three-parameter visco-elastic soils with one-way drainage

0.03	0.0254	0.105	0.373	2.29	4.8	7.86	11.8	17.4	26.9	36.4
0.06	0.0253	0.104	0.324	1.26	2.54	4.1	6.11	8.94	13.8	18.6
0.09	0.0252	0.102	0.296	0.918	1.78	2.84	4.21	6.13	9.42	12.7
0.15	0.0251	0.0999	0.263	0.643	1.18	1.84	2.68	3.88	5.92	7.97
0.25	0.0249	0.0964	0.232	0.477	0.813	1.23	1.77	2.53	3.83	5.14
0.5	0.0243	0.0897	0.193	0.342	0.534	0.773	1.08	1.52	2.28	3.04
1	0.0234	0.0806	0.16	0.26	0.383	0.537	0.738	1.02	1.52	2.03
3	0.0208	0.0636	0.117	0.183	0.264	0.367	0.506	0.706	1.06	1.41
10	0.0165	0.0467	0.0877	0.143	0.215	0.308	0.432	0.607	0.908	1.21

$\xi = 0.4$

ω	T_{v10}	T_{v20}	T_{v30}	T_{v40}	T_{v50}	T_{v60}	T_{v70}	T_{v80}	T_{v90}	T_{v95}
0.001	0.0192	0.0777	0.189	0.822	73.4	164	281	445	726	1010
0.002	0.0192	0.0777	0.188	0.774	36.8	82.1	140	223	363	504
0.004	0.0192	0.0777	0.188	0.716	18.6	41.2	70.4	112	182	252
0.008	0.0192	0.0776	0.188	0.65	9.44	20.8	35.4	56	91.3	127
0.016	0.0192	0.0775	0.187	0.581	4.87	10.6	17.9	28.3	46	63.7
0.03	0.0191	0.0773	0.185	0.518	2.74	5.8	9.74	15.3	24.8	34.3
0.06	0.0191	0.0769	0.182	0.45	1.52	3.07	5.08	7.9	12.7	17.5
0.09	0.0191	0.0764	0.179	0.412	1.12	2.16	3.52	5.43	8.7	12
0.15	0.019	0.0756	0.174	0.367	0.794	1.44	2.28	3.46	5.48	7.51
0.25	0.019	0.0744	0.167	0.325	0.6	1	1.53	2.28	3.56	4.84
0.5	0.0187	0.0716	0.154	0.273	0.442	0.669	0.971	1.4	2.14	2.88
1	0.0184	0.0672	0.137	0.228	0.344	0.492	0.687	0.965	1.45	1.94
3	0.0171	0.057	0.109	0.174	0.254	0.355	0.491	0.687	1.03	1.37
10	0.0146	0.0444	0.0854	0.14	0.212	0.305	0.427	0.602	0.9	1.2

$\xi = 0.5$

ω	T_{v10}	T_{v20}	T_{v30}	T_{v40}	T_{v50}	T_{v60}	T_{v70}	T_{v80}	T_{v90}	T_{v95}
0.001	0.0154	0.0622	0.142	0.281	1.08	90	207	371	652	933
0.002	0.0154	0.0622	0.142	0.281	1.02	45.2	104	186	326	467
0.004	0.0154	0.0622	0.142	0.281	0.935	22.8	52	93.1	163	234
0.008	0.0154	0.0622	0.142	0.28	0.848	11.6	26.2	46.8	82	117
0.016	0.0154	0.0621	0.142	0.278	0.757	5.97	13.3	23.6	41.3	59
0.03	0.0154	0.062	0.141	0.276	0.676	3.36	7.29	12.8	22.3	31.8
0.06	0.0154	0.0618	0.14	0.27	0.589	1.87	3.86	6.67	11.5	16.3
0.09	0.0154	0.0617	0.139	0.265	0.54	1.37	2.71	4.61	7.86	11.1
0.15	0.0153	0.0613	0.137	0.257	0.483	0.983	1.8	2.97	4.97	6.97
0.25	0.0153	0.0607	0.135	0.245	0.429	0.751	1.25	1.99	3.25	4.51
0.5	0.0152	0.0594	0.129	0.225	0.363	0.562	0.845	1.26	1.97	2.69
1	0.015	0.0572	0.12	0.202	0.308	0.446	0.632	0.9	1.37	1.84
3	0.0144	0.0511	0.102	0.165	0.243	0.343	0.476	0.668	1	1.34
10	0.013	0.0421	0.0831	0.138	0.209	0.302	0.423	0.596	0.892	1.19

$\xi = 0.6$

ω	T_{v10}	T_{v20}	T_{v30}	T_{v40}	T_{v50}	T_{v60}	T_{v70}	T_{v80}	T_{v90}	T_{v95}
0.001	0.0129	0.0519	0.117	0.215	0.382	1.37	116	281	562	843
0.002	0.0129	0.0519	0.117	0.215	0.382	1.28	58.3	141	281	422
0.004	0.0129	0.0519	0.117	0.215	0.381	1.17	29.4	70.5	141	211

(Continued)

Table 3.18 (Continued) Elapsed times for certain average degree of consolidation for three-parameter visco-elastic soils with one-way drainage

0.008	0.0129	0.0519	0.117	0.215	0.38	1.06	14.9	35.5	70.7	106
0.016	0.0129	0.0519	0.117	0.214	0.378	0.952	7.68	18	35.7	53.3
0.03	0.0129	0.0519	0.117	0.213	0.374	0.851	4.3	9.83	19.3	28.8
0.06	0.0129	0.0518	0.117	0.212	0.367	0.745	2.37	5.17	9.95	14.7
0.09	0.0129	0.0517	0.116	0.21	0.36	0.685	1.74	3.62	6.84	10.1
0.15	0.0129	0.0515	0.115	0.207	0.348	0.614	1.24	2.38	4.35	6.33
0.25	0.0128	0.0512	0.114	0.203	0.333	0.549	0.947	1.64	2.87	4.11
0.5	0.0128	0.0506	0.111	0.194	0.307	0.47	0.715	1.1	1.79	2.48
1	0.0127	0.0494	0.106	0.181	0.277	0.404	0.577	0.831	1.28	1.73
3	0.0124	0.0459	0.0946	0.156	0.233	0.331	0.461	0.648	0.972	1.3
10	0.0116	0.0399	0.0807	0.135	0.207	0.299	0.419	0.59	0.883	1.18

$\xi = 0.7$

ω	T_{v10}	T_{v20}	T_{v30}	T_{v40}	T_{v50}	T_{v60}	T_{v70}	T_{v80}	T_{v90}	T_{v95}
0.001	0.0111	0.0446	0.101	0.18	0.295	0.49	1.68	164	445	726
0.002	0.0111	0.0446	0.101	0.18	0.295	0.49	1.57	82.3	223	363
0.004	0.0111	0.0446	0.101	0.18	0.295	0.489	1.44	41.4	112	182
0.008	0.0111	0.0446	0.101	0.18	0.294	0.488	1.31	20.9	56.1	91.3
0.016	0.0111	0.0446	0.1	0.18	0.294	0.485	1.17	10.7	28.4	46
0.03	0.0111	0.0446	0.1	0.18	0.293	0.481	1.05	5.96	15.4	24.8
0.06	0.0111	0.0445	0.1	0.179	0.291	0.472	0.926	3.24	8	12.8
0.09	0.0111	0.0445	0.1	0.178	0.288	0.464	0.855	2.34	5.54	8.73
0.15	0.0111	0.0444	0.0996	0.177	0.285	0.45	0.771	1.64	3.57	5.53
0.25	0.0111	0.0443	0.0989	0.175	0.279	0.432	0.693	1.23	2.41	3.62
0.5	0.011	0.044	0.0974	0.171	0.267	0.401	0.6	0.93	1.57	2.23
1	0.011	0.0434	0.0949	0.164	0.252	0.366	0.524	0.758	1.18	1.6
3	0.0109	0.0414	0.0879	0.148	0.224	0.319	0.446	0.628	0.942	1.26
10	0.0104	0.0376	0.0782	0.133	0.204	0.295	0.415	0.584	0.874	1.16

$\xi = 0.8$

ω	T_{v10}	T_{v20}	T_{v30}	T_{v40}	T_{v50}	T_{v60}	T_{v70}	T_{v80}	T_{v90}	T_{v95}
0.001	0.00975	0.0391	0.0881	0.157	0.25	0.381	0.604	2.05	281	562
0.002	0.00975	0.0391	0.0881	0.157	0.25	0.381	0.604	1.92	141	281
0.004	0.00975	0.0391	0.0881	0.157	0.249	0.38	0.603	1.76	70.6	141
0.008	0.00975	0.0391	0.0881	0.157	0.249	0.38	0.602	1.6	35.6	70.8
0.016	0.00975	0.0391	0.0881	0.157	0.249	0.379	0.599	1.44	18.1	35.7
0.03	0.00975	0.0391	0.088	0.157	0.249	0.378	0.595	1.3	9.93	19.3
0.06	0.00974	0.0391	0.0879	0.156	0.248	0.376	0.586	1.15	5.26	10
0.09	0.00974	0.0391	0.0878	0.156	0.247	0.374	0.577	1.07	3.72	6.89
0.15	0.00974	0.039	0.0877	0.156	0.246	0.37	0.563	0.97	2.49	4.41
0.25	0.00974	0.039	0.0874	0.155	0.243	0.364	0.544	0.879	1.79	2.95
0.5	0.00973	0.0388	0.0867	0.153	0.238	0.351	0.512	0.772	1.31	1.92
1	0.00971	0.0385	0.0854	0.149	0.23	0.335	0.477	0.687	1.07	1.46
3	0.00964	0.0376	0.0816	0.14	0.214	0.308	0.431	0.608	0.912	1.22
10	0.00943	0.0354	0.0757	0.131	0.202	0.292	0.411	0.579	0.866	1.15

$\xi = 0.9$

ω	T_{v10}	T_{v20}	T_{v30}	T_{v40}	T_{v50}	T_{v60}	T_{v70}	T_{v80}	T_{v90}	T_{v95}
0.001	0.0087	0.0348	0.0784	0.14	0.219	0.324	0.472	0.724	2.55	281

(Continued)

Table 3.18 (Continued) Elapsed times for certain average degree of consolidation for three-parameter visco-elastic soils with one-way drainage

0.002	0.0087	0.0348	0.0784	0.14	0.219	0.324	0.472	0.724	2.38	141
0.004	0.0087	0.0348	0.0784	0.14	0.219	0.324	0.471	0.723	2.2	70.7
0.008	0.0087	0.0348	0.0784	0.14	0.219	0.324	0.471	0.722	2.01	35.6
0.016	0.0087	0.0348	0.0784	0.139	0.219	0.324	0.471	0.72	1.82	18.1
0.03	0.0087	0.0348	0.0784	0.139	0.219	0.323	0.47	0.717	1.66	9.97
0.06	0.0087	0.0348	0.0784	0.139	0.219	0.323	0.468	0.71	1.48	5.31
0.09	0.0087	0.0348	0.0783	0.139	0.219	0.322	0.466	0.704	1.38	3.77
0.15	0.0087	0.0348	0.0783	0.139	0.218	0.321	0.463	0.693	1.27	2.58
0.25	0.0087	0.0348	0.0782	0.139	0.217	0.319	0.459	0.678	1.17	1.94
0.5	0.00869	0.0347	0.0779	0.138	0.215	0.315	0.449	0.652	1.04	1.51
1	0.00869	0.0346	0.0774	0.137	0.212	0.309	0.436	0.622	0.955	1.3
3	0.00866	0.0343	0.0759	0.133	0.205	0.297	0.417	0.587	0.88	1.17
10	0.00858	0.0334	0.0732	0.128	0.199	0.289	0.407	0.573	0.857	1.14

$\xi = 0.95$

ω	T_{v10}	T_{v20}	T_{v30}	T_{v40}	T_{v50}	T_{v60}	T_{v70}	T_{v80}	T_{v90}	T_{v95}
0.001	0.00826	0.033	0.0744	0.132	0.207	0.304	0.433	0.629	1.05	2.93
0.002	0.00826	0.033	0.0744	0.132	0.207	0.304	0.433	0.629	1.05	2.74
0.004	0.00826	0.033	0.0744	0.132	0.207	0.304	0.433	0.629	1.05	2.55
0.008	0.00826	0.033	0.0744	0.132	0.207	0.303	0.433	0.629	1.05	2.34
0.016	0.00826	0.033	0.0743	0.132	0.207	0.303	0.433	0.628	1.04	2.14
0.03	0.00826	0.033	0.0743	0.132	0.207	0.303	0.432	0.628	1.04	1.96
0.06	0.00826	0.033	0.0743	0.132	0.207	0.303	0.432	0.626	1.03	1.77
0.09	0.00826	0.033	0.0743	0.132	0.207	0.303	0.431	0.624	1.02	1.66
0.15	0.00826	0.033	0.0743	0.132	0.207	0.302	0.43	0.62	0.998	1.54
0.25	0.00826	0.033	0.0742	0.132	0.206	0.302	0.428	0.616	0.974	1.43
0.5	0.00825	0.033	0.0741	0.132	0.206	0.3	0.424	0.606	0.937	1.3
1	0.00825	0.0329	0.0739	0.131	0.204	0.297	0.419	0.594	0.9	1.21
3	0.00824	0.0328	0.0732	0.129	0.201	0.292	0.41	0.577	0.864	1.15
10	0.00821	0.0324	0.072	0.127	0.198	0.288	0.405	0.57	0.853	1.14

Flugge, 1975). The effective stress-strain relationship can be written in terms of operator equations as follows (Zhu and Yuan, 1994).

$$P(D)\sigma_z' = Q(D)\varepsilon_z \tag{3.67}$$

where D is time derivative operator; $P(x)$ and $Q(x)$ are polynomials with constant coefficients, respectively. The roots α_i of $P(x)$ and the roots β_i of $Q(x)$ are all real, single non-positive values and satisfy:

$$0 \geq \beta_1 > \alpha_1 > \beta_2 > \alpha_2 > ..., \quad P(+\infty) > 0, \ Q(+\infty) > 0 \tag{3.68}$$

The order of polynomial $P(x)$ is equal to either the order of $Q(x)$, or the order of $Q(x)$ minus 1.

Since $P(x)$ and $Q(x)$ have the above properties, the roots of the following $\psi(x)$ are all real, single non-positive values for any real λ:

$$\psi(x) = xP(x) + \lambda^2 Q(x) \tag{3.69}$$

In fact,

$$\psi(0) = \lambda^2 Q(0) \geq 0 \tag{3.70a}$$

$$Sign\ [\psi(\alpha_i)] = Sign\ [\lambda^2 Q(\alpha_i)] = (-1)^i \quad i = 1,\ 2, \ldots,\ \partial P \tag{3.70b}$$

$$Sign\ [\psi(-\infty)] = (-)^{1+\partial P} \tag{3.70c}$$

where ∂P denotes the order of $P(x)$. Thus, the roots of $\psi(x)$ are all real, single non-positive values and isolated by the roots of $P(x)$.

Equation (3.67), when substituting into (2.10), will result in the following equation governing the consolidation of a visco-elastic soil layer

$$-\frac{k}{\gamma_w}\frac{\partial^2}{\partial z^2}[Q(D)u_e] = \frac{\partial}{\partial t}[P(D)\sigma_z'] \tag{3.71a}$$

The parameters in (3.71a) are assumed constant. The excess pore pressure, effective stress and strain (increments) just before loading are assumed zero, and the effective stress equation for vertical total stress increment $\Delta\sigma_z(z,t)$ becomes

$$u_e = \Delta\sigma_z(z,t) - \sigma_z' \tag{3.71b}$$

The solution of (3.71) is expressed as:

$$u_e = \sum_{n=1}^{\infty} T_n(t)\sin\left(\lambda_n \frac{z}{H}\right) \tag{3.72}$$

where $\lambda_n = n\pi$ for boundary conditions (2.16), and $\lambda_n = (n-1/2)\pi$ for boundary conditions (2.17). Substituting (3.72) into (3.71) gives

$$P(D)T_n' + \frac{k\lambda_n^2}{\gamma_w H^2}Q(D)T_n = P(D)\frac{\partial}{\partial t}\left[\frac{2}{H}\int_0^H \Delta\sigma_z(z,t)\sin\left(\lambda_n \frac{z}{H}\right)dz\right] \tag{3.73}$$

Applying the Laplace transform to (3.73) gives

$$\overline{T_n} = \frac{sP(s)}{\psi_n(s)}\frac{2}{H}\int_0^H \Delta\overline{\sigma_z}(z,s)\sin\left(\lambda_n \frac{z}{H}\right)dz \tag{3.74}$$

where

$$\psi_n(s) = sP(s) + \frac{k\lambda_n^2}{\gamma_w H^2}Q(s) \tag{3.75}$$

and the overbar denotes the Laplace transform of the function, that is,

$$\overline{T_n}(s) = \int_0^{\infty} e^{-st} T_n(t) dt \tag{3.76}$$

The roots η_{jn} of $\psi_n(s)$ are all real, single non-positive values as discussed above and thus (3.74) can be written as

$$\overline{T_n} = \left(a_{0n} + \sum_{j=1}^{1+\partial P} \frac{a_{jn}}{s - \eta_{jn}} \right) \frac{2}{H} \int_0^H \Delta\overline{\sigma_z}(z,s) \sin\left(\lambda_n \frac{z}{H} \right) dz \tag{3.77}$$

where

$$a_{0n} = \lim_{s \to +\infty} \frac{sP(s)}{\psi_n(s)}, \quad a_{jn} = \frac{\eta_{jn} P(\eta_{jn})}{\psi'_n(\eta_{jn})}, \quad j = 1, \ldots, \quad 1 + \partial P \tag{3.78}$$

Therefore,

$$T_n(t) = \frac{2}{H} \int_0^H \left[a_{0n} \Delta\sigma_z(z,t) + \sum_{j=1}^{1+\partial P} a_{jn} \int_0^t e^{\eta_{jn}(t-\tau)} \Delta\sigma_z(z,\tau) d\tau \right] \sin\left(\lambda_n \frac{z}{H} \right) dz \tag{3.79}$$

Taking the Laplace transform of ε_z gives

$$\begin{aligned} \overline{\varepsilon_z} &= \sum_{n=1}^{\infty} \frac{k\lambda_n^2}{\gamma_w H^2} \sin\left(\lambda_n \frac{z}{H} \right) \left[\frac{P(s)}{\psi_n(s)} \frac{2}{H} \int_0^H \Delta\overline{\sigma_z}(z,s) \sin\left(\lambda_n \frac{z}{H} \right) dz \right] \\ &= \sum_{n=1}^{\infty} \frac{k\lambda_n^2}{\gamma_w H^2} \sin\left(\lambda_n \frac{z}{H} \right) \frac{2}{H} \int_0^H \left[\sum_{j=1}^{1+\partial P} \frac{P(\eta_{jn})}{\psi'_n(\eta_{jn})(s - \eta_{jn})} \right] \Delta\overline{\sigma_z}(z,s) \sin\left(\lambda_n \frac{z}{H} \right) dz \end{aligned} \tag{3.80}$$

Thus,

$$\varepsilon_z = \sum_{n=1}^{\infty} \frac{k\lambda_n^2}{\gamma_w H^2} \sin\left(\lambda_n \frac{z}{H} \right) \frac{2}{H} \int_0^H \left[\sum_{j=1}^{1+\partial P} \frac{P(\eta_{jn})}{\psi'_n(\eta_{jn})} \int_0^t e^{\eta_{jn}(t-\tau)} \Delta\sigma_z(z,\tau) d\tau \right] \sin\left(\lambda_n \frac{z}{H} \right) dz \tag{3.81}$$

The compression S_t of the soil layer at time t then becomes

$$S_t = \sum_{n=1}^{\infty} \frac{2k\lambda_n(1 - \cos\lambda_n)}{\gamma_w H^2} \int_0^H \left[\sum_{j=1}^{1+\partial P} \frac{P(\eta_{jn})}{\psi'_n(\eta_{jn})} \int_0^t e^{\eta_{jn}(t-\tau)} \Delta\sigma_z(z,\tau) d\tau \right] \sin\left(\lambda_n \frac{z}{H} \right) dz \tag{3.82}$$

If the vertical stress increment $\Delta\sigma_z(z,t)$ is only a non-zero function of z after a designated time, the final compression S_f is

$$S_f = \begin{cases} \int_0^H \varepsilon_z(z,t=\infty)dz = \dfrac{P(0)}{Q(0)}\int_0^H \Delta\sigma_z(z,t=\infty)dz & Q(0) \neq 0 \\ \infty & Q(0) = 0 \end{cases} \tag{3.83}$$

3.6 OBSERVATIONAL PROCEDURE OF SETTLEMENT PREDICTION

Deformation is one of the design criteria for structures founded on cohesive soils. The usual practice for settlement prediction is to multiply the calculated values of average degree of consolidation and the ultimate settlement from 1-D compression properties of the sublayers. Nevertheless, the use of analytical solution of consolidation equation for settlement prediction is not always effective since such conditions as an initial distribution of excess pore pressure, drainage length, loading conditions, the compressibility, and the permeability of soils are sometimes quite uncertain in practical engineering problems. Therefore, some means of settlement prediction based on measured results are useful. In this section, two observational procedures for settlement prediction, that is, the hyperbolic method (Sridharan *et al.*, 1987; Tan, 1995, 1996) and Asaoka's (1978) method, are discussed.

3.6.1 Hyperbolic method for elastic soils

The basis of hyperbolic method is the property that Terzaghi's U-T_v relationship is a rectangular hyperbola over a fairly wide range of T_v. As shown in Figure 3.33(a), when the average degree of consolidation U for suddenly applied uniform loading for a single-drained stratum (2.27) is plotted as T_v/U vs T_v, there is an initial concave segment up to $U = 0.6$, and followed essentially by a linear segment between $U = 0.6$ and $U = 0.9$. A second linear portion exists for T_v greater than 1.0, which approaches a slope value of 1.0. The least squares line for the first linear segment can be expressed as

$$\frac{T_v}{U} = 0.8272T_v + 0.2407 \tag{3.84a}$$

with a regression coefficient r^2 greater than 0.9999. When lines joining the origin to the points on the curve corresponding to $U = 0.6$ and $U = 0.9$ are drawn, the slopes of these lines are $\alpha_{60} = 1/0.6$ and $\alpha_{90} = 1/0.9$, respectively.

For predicting settlement from measured values, (3.84a) is expressed in terms of settlement versus time as

$$\frac{t}{S_t} = 0.8272\frac{t}{S_f} + 0.2407\frac{H^2}{c_v S_f} \tag{3.84b}$$

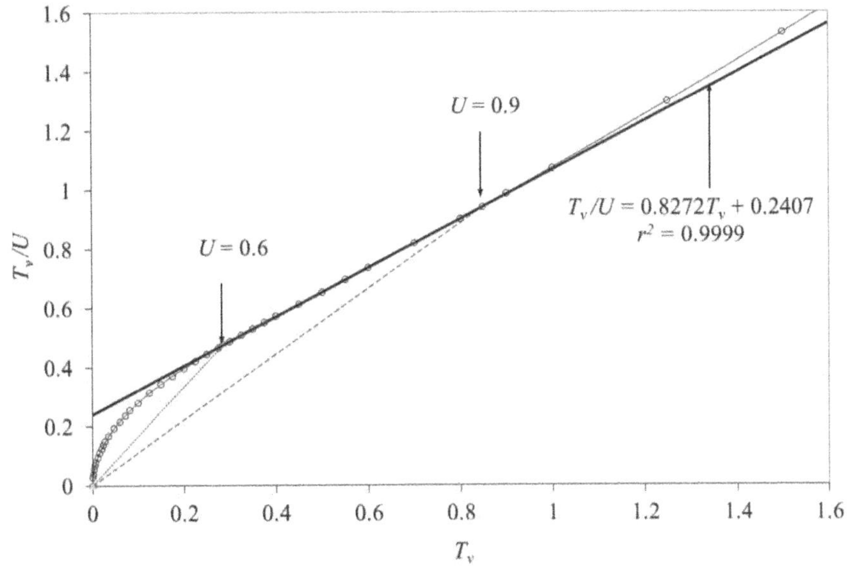

Figure 3.33 Hyperbolic plots – Terzaghi's solution.

In field monitoring of the consolidations of soils, plots of settlement versus time are recorded. When a field settlement record is plotted in the form of t/S_t vs t, as in Figure 3.33, the same features as in Figure 3.33, namely an initial concave segment followed by a linear segment between 60 and 90% consolidation state are observed. The straight-line portion can be represented by the equation

$$\frac{t}{S_t} = \alpha t + \beta \tag{3.85}$$

where α is the slope of the line and β is the intercept on the axis t/S_t. Comparing (3.84a) with (3.85), the following equations are derived

$$S_f = \frac{0.8272}{\alpha} \tag{3.86a}$$

$$c_v = 0.291\frac{\alpha H^2}{\beta} \tag{3.86b}$$

Thus, the final settlement S_f and the coefficient of consolidation can be obtained from (3.86).

The above method is based on the solution with instantaneous loading. Since the method attempts to fit measured settlement data between 60 and 90% consolidation state, it is not sensitive to the initial loading condition as long as the loading period is small compared to the time required for an average degree of consolidation of 60% to occur. When this method is used to calculate the coefficient of consolidation from odometer test results, the initial compression due to imperfect contact should be deducted. Experiences from various odometer tests show that the coefficient of consolidation obtained from this method is normally larger than the value obtained from Casagrande's method, and smaller than the value obtained from Taylor's method (Sridharan *et al.*, 1987).

3.6.2 Asaoka's method for elastic soils

Asaoka (1978) noticed that the 1-D consolidation settlement-time relationship under constant external loading can be approximated by a difference equation with constant coefficients. The coefficients of this difference equation are determined from early settlement data by using least squares method. The ultimate settlement is then predicted based on the difference equation. This method is becoming increasingly popular because of its simplicity and lack of requirements for detailed sampling and laboratory testing to determine soil properties or for the monitoring of field pore pressure behavior. The method is derived and examined in this section.

The average degree of consolidation for one-way drainage under depth-dependent ramp load is expressed in (2.27). From (2.27), the following (3.87) is obtained for $T_v \geq T_c$.

$$1-U_j = \sum_{n=1,3,5,\ldots} \frac{32}{n^4\pi^4 T_c}\left[1-\exp\left(-\frac{n^2\pi^2 T_c}{4}\right)\right]\left\{1+\frac{\sigma_0-\sigma_1}{\sigma_0+\sigma_1}\left[1+(-1)^{\frac{n+1}{2}}\frac{4}{n\pi}\right]\right\}\mu_n^j \tag{3.87a}$$

$$U_j = U(T_c + j\Delta T_v) \quad j = 0,\ 1,\ 2,\ \ldots,\infty \tag{3.87b}$$

$$\mu_n = \exp\left(-\frac{n^2\pi^2}{4}\Delta T_v\right) \tag{3.87c}$$

where ΔT_v is a constant increment.

The series in (3.87a) converges rapidly. Then the following (3.88) is adopted as an m-th order approximation to (3.87a).

$$1-U_j = \sum_{n=1,3,5,\ldots}^{2m-1} \frac{32}{n^4\pi^4 T_c}\left[1-\exp\left(-\frac{n^2\pi^2 T_c}{4}\right)\right]\left\{1+\frac{\sigma_0-\sigma_1}{\sigma_0+\sigma_1}\left[1+(-1)^{\frac{n+1}{2}}\frac{4}{n\pi}\right]\right\}\mu_n^j \tag{3.88}$$

Equation (3.88) is the solution of the following m-th order finite difference equation (3.89)

$$1-U_j + \beta_1(1-U_{j-1}) + \cdots + \beta_m(1-U_{j-m}) = 0 \quad j = m,\ m+1,\ldots\ \infty \tag{3.89}$$

where the coefficients β_i are the coefficients of the m-th polynomial (3.90)

$$P(x) = x^m + \beta_1 x^{m-1} + \cdots + \beta_m = (x - \mu_1)(x - \mu_3)\cdots(x - \mu_{2m-1}) \tag{3.90}$$

Substituting $S_j = U_j S_f$ into (3.89) gives

$$S_j + \beta_1 S_{j-1} + \cdots + \beta_m S_{j-m} = \beta_0 \tag{3.91a}$$

$$\beta_0 = (1 + \beta_1 + \cdots + \beta_m) S_f \tag{3.91b}$$

$$S_f = \frac{\beta_0}{1 + \beta_1 + \cdots + \beta_m} \tag{3.91c}$$

Equation (3.91) is the basic equation for consolidation settlement prediction. The coefficients in (3.91a) are determined by the least square method using measured settlement data S_j at equal time intervals. Once the coefficients are determined, (3.91a) can be used to predict settlement of a future time under situations in which a coefficient of consolidation and both initial and boundary conditions are quite uncertain. The final settlement S_f can be estimated using (3.91c). In practical applications, the first-order and second-order difference equations are normally utilized. The accuracy of the method is analyzed in the following.

For the first-order finite difference approximation, (3.91) becomes

$$S_j + \beta_1 S_{j-1} = \beta_0 \tag{3.92a}$$

$$\beta_1 = -\mu_1 \tag{3.92b}$$

$$S_f = \frac{\beta_0}{1 + \beta_1} \tag{3.92c}$$

$$c_v = -\frac{4H^2 \ln(-\beta_1)}{\pi^2 \Delta t} \tag{3.92d}$$

For the second-order finite difference approximation, (3.91) becomes

$$S_j + \beta_1 S_{j-1} + \beta_2 S_{j-2} = \beta_0 \tag{3.93a}$$

$$\beta_1 = -(\mu_1 + \mu_3) = -(\mu_1 + \mu_1^9) \tag{3.93b}$$

$$\beta_2 = \mu_1 \mu_3 = \mu_1^{10} \tag{3.93c}$$

$$S_f = \frac{\beta_0}{1 + \beta_1 + \beta_2} \tag{3.93d}$$

$$c_v = -\frac{4H^2}{\pi^2 \Delta t} \ln\left(\frac{-\beta_1 + \sqrt{\beta_1^2 - 4\beta_2}}{2}\right) \tag{3.93e}$$

Table 3.19 Accuracy of Asaoka's method for different data ranges (theoretical values: $c_v = S_f = 1$)

$T_c = 0$		First-order difference				Second-order difference				
ΔT_v	*U range*	β_0	$-\beta_1$	c_v	S_f	β_0	$-\beta_1$	β_2	c_v	S_f
0.01	0.3–0.6	0.0291	0.967	1.368	0.877	0.0052	1.762	0.767	1.016	0.994
0.01	0.6–0.9	0.0244	0.976	1.001	1.000	0.0049	1.777	0.781	1.000	1.000
0.02	0.3–0.6	0.0573	0.935	1.366	0.879	0.0183	1.574	0.592	1.013	0.995
0.02	0.6–0.9	0.0482	0.952	1.001	1.000	0.0173	1.593	0.611	1.000	1.000
0.04	0.3–0.6	0.1105	0.875	1.359	0.880	0.0571	1.300	0.358	1.007	0.997
0.04	0.6–0.9	0.0941	0.906	1.001	1.000	0.0553	1.317	0.373	1.000	1.000
$T_c = 0.2$		First-order difference				Second-order difference				
ΔT_v	*U range*	β_0	$-\beta_1$	c_v	S_f	β_0	$-\beta_1$	β_2	c_v	S_f
0.01	0.3–0.6	0.0332	0.959	1.680	0.818	0.0113	1.560	0.572	1.219	0.926
0.01	0.6–0.9	0.0244	0.976	1.005	0.996	0.0049	1.776	0.781	1.000	1.000
0.02	0.3–0.6	0.0657	0.919	1.714	0.810	0.0306	1.364	0.397	1.169	0.941
0.02	0.6–0.9	0.0483	0.952	1.003	1.000	0.0173	1.593	0.610	1.000	1.000
0.04	0.3–0.6	0.128	0.839	1.774	0.797	0.0757	1.137	0.217	1.116	0.958
0.04	0.6–0.9	0.0942	0.906	1.004	0.999	0.0553	1.317	0.373	1.000	1.000
$T_c = 0.4$		First-order difference				Second-order difference				
ΔT_v	*U range*	β_0	$-\beta_1$	c_v	S_f	β_0	$-\beta_1$	β_2	c_v	S_f
0.005	0.3–0.6	0.0182	0.978	1.836	0.813	0.0054	1.630	0.636	1.395	0.890
0.005	0.6–0.9	0.0124	0.988	1.011	1.000	0.0013	1.882	0.883	1.000	1.000
0.01	0.3–0.6	0.036	0.956	1.836	0.813	0.0147	1.464	0.480	1.330	0.905
0.01	0.6–0.9	0.0246	0.975	1.009	1.000	0.0049	1.776	0.780	1.000	1.000
0.02	0.3–0.6	0.07	0.914	1.822	0.814	0.0368	1.276	0.315	1.248	0.926
0.02	0.6–0.9	0.0485	0.951	1.010	0.998	0.0173	1.592	0.610	1.000	1.000

Equation (3.92a) represents a straight line in a S_j vs S_{j-1} plot. Equation (3.93a) can take some non-linear effects of the S_j vs S_{j-1} plot into account although the graphical explanation is not clear. To evaluate the accuracy and the appropriate range of settlement data for least squares linear regression to determine the coefficients of the difference (3.92a) and (3.93a), the analytical solution (3.87a) for $\sigma_0 = \sigma_1$ is used as standard data for different time increments ΔT_v and construction time T_c. Typical plots of U_j vs U_{j-1} are illustrated in Figure 3.34. Least squares linear regression results for the data range of $U = 0.3$ to 0.6 and $U = 0.6$ to 0.9 are shown in Table 3.19. The exact values for both c_v and S_f are 1. Using the data range from $U = 0.3$ to 0.6 consolidation, S_f is underestimated by more than 10%, with c_v overestimated by more than 35% if the first-order finite difference is adopted. However, much better estimates are obtained if the second-order finite difference is used. If the data range from $U = 0.6$ to 0.9 consolidation is used in the regression,

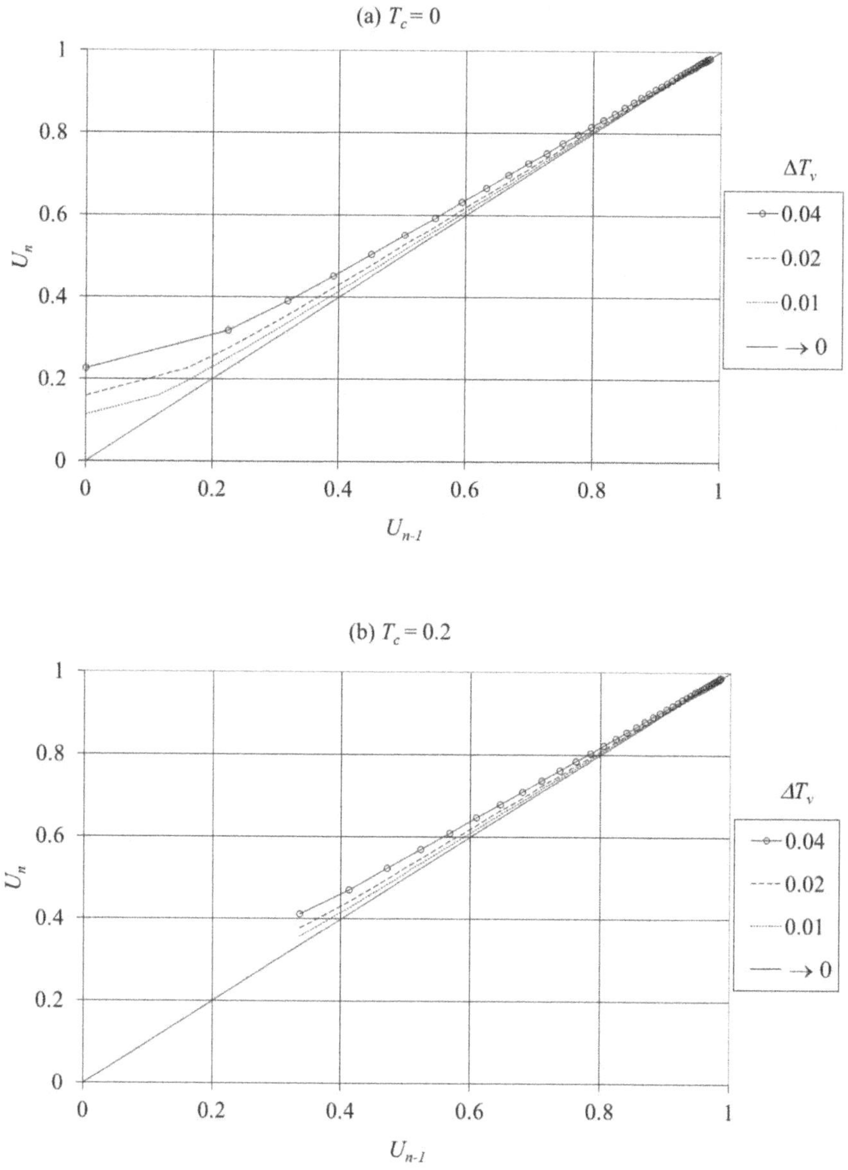

Figure 3.34 U_j vs U_{j-1} plots of analytical solutions for different time increment – (a) $T_c = 0$; (b) $T_c = 0.2$.

accurate values are obtained for both first-order and second-order finite difference approximations. In practical applications of the Asaoka's method, enough measured data should be used in order to give a reliable prediction, and both the first- and second-order finite difference approximations should be adopted to crosscheck the results.

3.7 CONSOLIDATION OF SOIL WITH NON-LINEAR VISCOSITY

In Section 3.5, the consolidation of visco-elastic soil is analyzed. The consolidation of soil with non-linear viscosity is briefly discussed in this section.

The time-dependent behavior of soil skeleton is extremely complex. Different forms of $\varepsilon_z - \sigma'_z - t$ relationship as shown in Figure 3.35 have been reported in the conflicting literature. The magnitude of the creep behavior of soil skeleton is often expressed by the *coefficient of secondary consolidation*, $C_{\alpha\varepsilon}$, defined as $C_{\alpha\varepsilon} = \frac{\partial \varepsilon_z}{\partial \log t}$. Figure 3.35(a) gives $C_{\alpha\varepsilon}$ decreasing with t and independent of $\Delta\sigma'_z$; Figure 3.35(b), $C_{\alpha\varepsilon}$ decreasing with t and increasing with $\Delta\sigma'_z$; Figure 3.35(c), $C_{\alpha\varepsilon}$ independent of t and increasing with $\Delta\sigma'_z$; and Figure 3.35(d), $C_{\alpha\varepsilon}$ independent of both t and $\Delta\sigma'_z$. However, for normally consolidated soils, the form in Figure 3.35(d) will give a fairly good approximation over realistic ranges of time and effective stress increment. One of the typical constitutive relations (Wu *et al.*, 1966; Barden, 1969) giving general agreement with Figure 3.35(d) is expressed in (3.94) and shown in Figure 3.36.

$$\frac{\partial \varepsilon_z}{\partial t} = b \sinh\left[a\left(\sigma'_z - \frac{\varepsilon_z}{m_v}\right)\right] \tag{3.94}$$

where a and b are two soil parameters. The model is essentially Terzaghi's basic model, but with a non-linear dashpot in parallel with the linear spring.

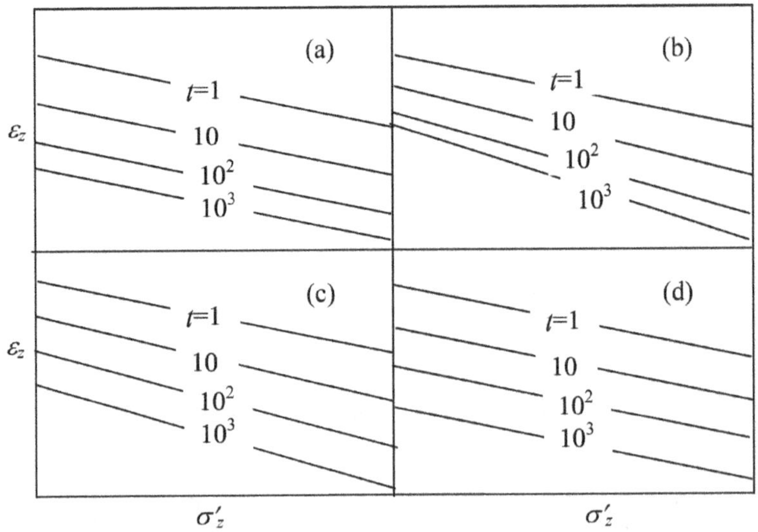

Figure 3.35 Different forms of $\varepsilon_z - \sigma'_z - t$ relationship (a) $C_{\alpha\varepsilon}$ decreasing with t and independent of σ'_z; (b) $C_{\alpha\varepsilon}$ decreasing with t and increasing with σ'_z; (c) $C_{\alpha\varepsilon}$ independent of t and increasing with σ'_z; and (d) $C_{\alpha\varepsilon}$ independent of both t and σ'_z.

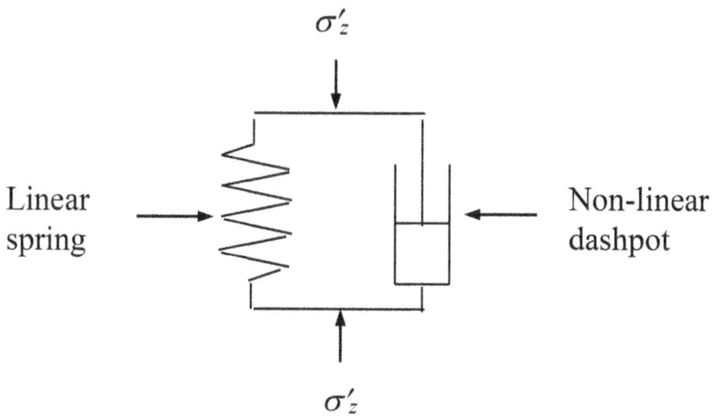

Figure 3.36 Non-linear rheological model.

The behavior of the skeleton when loaded by an increment of effective stress equal to σ_0 is given by

$$\frac{\varepsilon_z}{m_v \sigma_0} = 1 - \frac{1}{a\sigma_0} \ln \left[\frac{1 + \tanh\left(\frac{a\sigma_0}{2}\right) \exp\left(-\frac{ab}{m_v} t\right)}{1 - \tanh\left(\frac{a\sigma_0}{2}\right) \exp\left(-\frac{ab}{m_v} t\right)} \right] \tag{3.95}$$

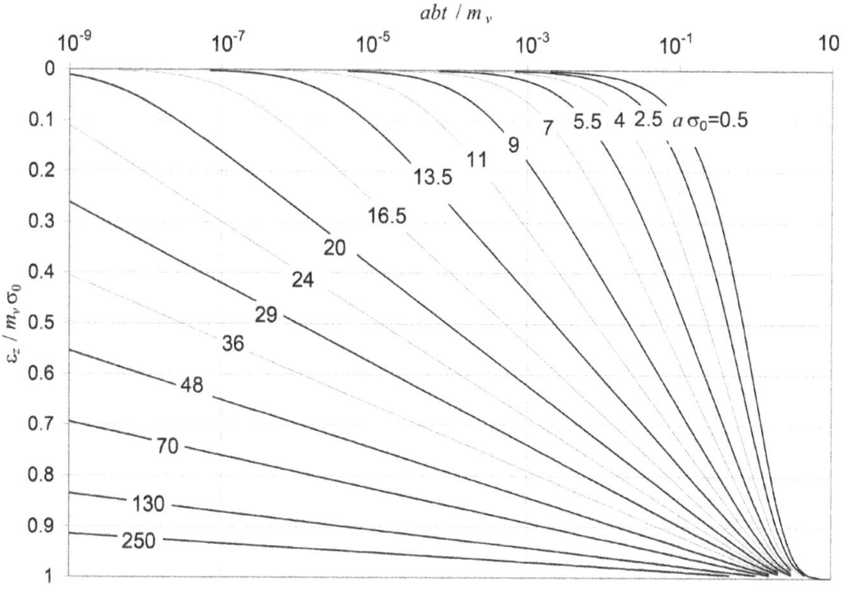

Figure 3.37 Compression time behavior for skeleton.

and plotted in Figure 3.37. It is important to include a realistic number of cycles of time in the creep life of clay. Figure 3.37 covers 10 cycles on the assumption that $t = 0.5$ second is essentially 'immediate' compression and that for all engineering purposes $t = 100$ years defines the 'final' compression. As shown in Figure 3.37, the soil will creep linearly with $\log t$ after the initial transition period.

To analyze the consolidation process of this soil, the integration of (3.94) and (2.11) is required. The solution is not presented here.

$$-\frac{\partial}{\partial z}\left(\frac{k}{\gamma_w}\frac{\partial u_e}{\partial z}\right)=\frac{\partial \varepsilon_z}{\partial t} \tag{2.11}$$

REFERENCES

Aboshi, H, Yoshikuni, H, and Maruyama, S, 1970, Constant loading rate consolidation test. *Soils and Foundations*, **10**(1), pp. 43–56.

Asaoka, A, 1978, Observational procedure of settlement prediction. *Soils and Foundations*, **18**(4), pp. 87–101.

Barden, L, 1996, Time dependent deformation of normally consolidated clays and peats. *Journal of Soil Mechanics & Foundations Division*, **95**, pp. 1–31.

Davis, EH, and Raymond, GP, 1965, A non-linear theory of consolidation. *Géotechnique*, **15**, pp. 161–173.

Flugge, W, 1975, *Viscoelasticity*, 2nd ed., (New York: Springer).

Gibson, RE, 1958, The progress of consolidation in a clay layer increasing in thickness with time. *Géotechnique*, **8**(4), pp. 171–182.

Gibson, RE, and Lo, KY, 1961, A theory of consolidation of soils exhibiting secondary consolidation. *Norges Geotekniske Institutt Publication*, 41.

Schiffman, RL, and Gibson, RE, 1964, Consolidation of nonhomogeneous clay layers. *Journal of Soil Mechanics & Foundations Division*, **90** (SM5), pp. 1–30.

Schiffman, RL, Ladd, CC, and Chen, ATF, 1964, The secondary consolidation of clay. *Rheology and Soil Mechanics, In Proceedings of the International Union of Theoretical and Applied Mechanics Symposium*, pp. 273–303. (Berlin: Grenoble).

Sridharan, A, Murthy, NS, and Prakash, K, 1987, Rectangular hyperbola method of consolidation analysis. *Géotechnique*, **37**(3), pp. 355–368.

Tan, TK, 1957, Three-dimensional theory on the consolidation and flow of the clay-layers. *Scientia Sinica*, **6**(1), pp. 203–215.

Tan, S, 1995, Validation of hyperbolic method for settlement in clays with vertical drains. *Soils and Foundations* ,**35**(1), pp. 101–113.

Tan, S, and Chew, S, 1996, Comparison of the hyperbolic and Asaoka observational method of monitoring consolidation with vertical drains. *Soils and Foundations*, **36**(3), pp. 31–42.

Vaid, YP, 1985, Constant rate of loading nonlinear consolidation. *Soils and Foundations*, **25**(1), pp. 105–108.

Wu, TH, Resendiz, D, and Neukirchner, RJ, 1966, Analysis of consolidation by rate process theory. *Journal of Soil Mechanics & Foundations Division*, ASCE, **90**(SM6), pp. 125–157.

Zhu, GF, and Yin, J-H, 2005, Consolidation of a soil layer subsequent to cessation of deposition. *Canadian Geotechnical Journal*, **42**(2), pp. 678–682.

Zhu, GF, and Yin, J-H, 2012, Analysis and mathematical solutions for consolidation analysis of a soil layer with depth-dependent parameters under confined compression. *International Journal of Geomechanics*, **12**(4), pp. 451–461.

Zhu, GF, and Yuan, JX, 1994, The eigenvalue properties of differential relations for viscoelastic rheological models. *In Proceedings of 8th International Conference on International Association for Numerical Methods and Advances in Geomechanics*, May 22–28, **1**, pp. 411–414.

Chapter 4

One-dimensional finite strain consolidation

4.1 INTRODUCTION

Terzaghi's one-dimensional (1-D) solution has least applicability to soft soils that exhibit large strains or compressions. The reason is clear since Terzaghi's solution assumes infinitesimal strain. Another reason except for the constitutive relations for the breakdown of Terzaghi's solution is that in soft soil, consolidation due to its own weight is very significant and may be the only load to cause consolidation. An example is the consolidation of land reclaimed using marine clay. However, this is neglected in Terzaghi's formulation. In this chapter, the finite strain theory of consolidation is discussed (Gibson *et al.*, 1981). Analytical solutions are derived for a linear form of the governing equations (Morris, 2002).

4.2 COORDINATE SYSTEMS

Infinitesimal strain theories of consolidation assume that the thickness of the compressible layer is constant; the deformation of the layer, during consolidation, is assumed to be small compared to its thickness. There is no need to distinguish an Eulerian system and a Lagrangian system. If the deformations are large compared to the thickness of the compressible layer, the situation is different. For example, a piezometer fixed in space but within the zone of settlement, i.e. near the boundary of the clay layer, may be outside of the layer after some time. A real measuring system is one which convects with the material particles. The piezometer would always be surrounded by the same material points and would measure the pore pressure of this part of the skeleton as a function of its momentary position and time. This type of system is termed as a convective coordinate system.

Consider the 1-D consolidation situation shown in Figure 4.1. A clay layer has an initial configuration as shown in Figure 4.1(a). The top boundary (datum plane) is at $a = 0$. The lower boundary is at $a = a_0$. An infinitesimal

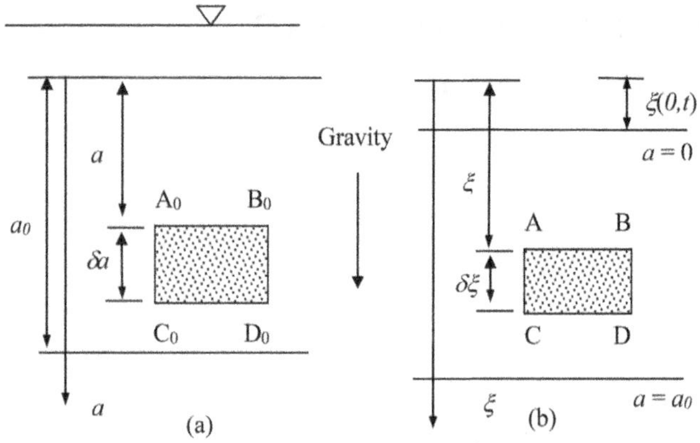

Figure 4.1 Lagrangian and convective coordinates: (a) initial configuration at $t = 0$; (b) configuration at time t.

element $A_0B_0C_0D_0$ has a coordinate position a and has thickness δa. This is the configuration that would exist before consolidation begins. With consolidation, the clay layer in the configuration shown by Figure 4.1(a) will have a new configuration shown in Figure 4.1(b). The top boundary has moved down a distance of $\xi(0, t)$ with respect to the datum plane and the infinitesimal element has deformed to a new position ABCD. A new coordinate ξ locates a material point as a function of time. The coordinate ξ is the convective coordinate.

Another set of coordinates is based on the volume of soil particles in a prism of unit cross-sectional area lying between the datum plane and the point analyzed (McNabb, 1960). A coordinate z is defined as

$$z(a) = \int_0^a \frac{da}{1+e(a,0)} \tag{4.1}$$

where e is the void ratio.

The following relationships exist between different coordinates

$$\frac{\partial z}{\partial a} = \frac{1}{1+e_0} \tag{4.2}$$

$$\frac{\partial z}{\partial \xi} = \frac{1}{1+e} \tag{4.3}$$

$$\frac{\partial \xi}{\partial a} = \frac{1+e}{1+e_0} \tag{4.4}$$

where e_0 is the void ratio at time $t = 0$.

4.3 GOVERNING EQUATIONS

All the basic assumptions in Section 2.3 are retained except the small strain assumption. The vertical equilibrium of the infinitesimal element ABCD in Figure 4.1 requires that

$$\frac{\partial \sigma}{\partial \xi} = \gamma_m = \frac{G_s + e}{1+e}\gamma_w \tag{4.5}$$

where σ is the vertical total stress, γ_m is the unit weight of saturated soil, and G_s is the specific gravity of the soil particles.

The pore fluid moves through the soil skeleton in accordance with Darcy's law and is expressed as

$$q = -\frac{k}{\gamma_w}\left(\frac{\partial u_w}{\partial \xi} - \gamma_w\right) \tag{4.6}$$

where q is vertical flow rate.

The rate of outflow of water from ABCD is

$$\text{Net outflow rate of water} = \frac{\partial q}{\partial \xi} d\xi \tag{4.7}$$

which must equal the volume compression rate of the element, so that

$$\frac{\partial q}{\partial \xi} = -\frac{1}{1+e}\frac{\partial e}{\partial t} \tag{4.8}$$

If the soil skeleton is homogeneous and possesses no creep effects and the consolidation is monotonic, then the permeability k may be expected to depend on the void ratio alone, thus

$$k = k(e) \tag{4.9}$$

while the vertical effective stress

$$\sigma' = \sigma - u_w \tag{4.10}$$

controls the void ratio, so that

$$\sigma' = \sigma'(e) \tag{4.11}$$

Equations (4.5), (4.6), and (4.8) can now be united to yield the following equation for the void ratio:

$$\frac{1}{1+e}\frac{\partial e}{\partial t} = \frac{\partial}{\partial \xi}\left[\frac{k(e)}{\gamma_w}\left(\frac{G_s - 1}{1+e}\gamma_w - \frac{d\sigma'}{de}\frac{\partial e}{\partial \xi}\right)\right] \tag{4.12}$$

Expressed in variable z, (4.12) becomes:

$$\frac{\partial e}{\partial t} = (G_s - 1)\frac{d}{de}\left[\frac{k(e)}{1+e}\right]\frac{\partial e}{\partial z} - \frac{\partial}{\partial z}\left[\frac{k(e)}{\gamma_w(1+e)}\frac{d\sigma'}{de}\frac{\partial e}{\partial z}\right] \tag{4.13}$$

4.4 EQUATIONS FOR LINEAR FINITE STRAIN CONSOLIDATION

Equation (4.13) is highly non-linear. However, it can be reduced to a linear form if the following c_f is a constant,

$$c_f = -\frac{k(e)}{\gamma_w(1+e)}\frac{d\sigma'}{de} \tag{4.14}$$

and the relation between void ratio and effective stress satisfies

$$e = (e_{00} - e_\infty)\exp(-\lambda\sigma') + e_\infty \tag{4.15}$$

where e_{00}, e_∞, and λ are constants.

Substituting (4.14) and (4.15) into (4.13) gives

$$\frac{\partial e}{\partial t} = c_f\frac{\partial^2 e}{\partial z^2} + (G_s - 1)c_f\lambda\gamma_w\frac{\partial e}{\partial z} \tag{4.16}$$

The parameter c_f is related to c_v by (Gibson *et al.*, 1981)

$$c_f = \frac{c_v}{(1+e)^2} \tag{4.17}$$

Consider a layer of initial thickness H fully consolidated and in equilibrium under both its own weight and a vertical effective stress q_0' acting on its upper surface as shown in Figure 4.2. The layer is fully saturated, with a water level at a distance H_0 above the top of the clay in this equilibrium state. The upper surface of the clay layer is in contact with a pervious soil. The lower surface of the clay can be in contact with (a) a pervious soil or (b) may be impervious.

The initial vertical effective stress can be expressed in reduced coordinate z as

$$\sigma'(z,0) = q_0' + (G_s - 1)\gamma_w z \tag{4.18}$$

Thus, by virtue of (4.15), the void ratio varies initially through the layer according to (4.19)

$$e(z,0) = (e_{00} - e_\infty)\exp\{-\lambda[q_0' + (G_s - 1)\gamma_w z]\} + e_\infty \tag{4.19}$$

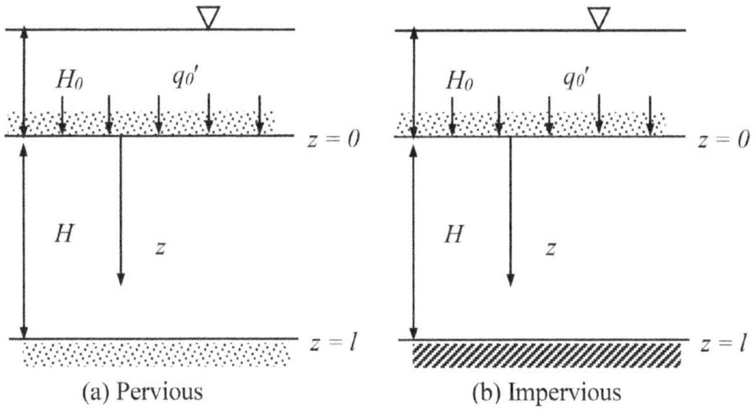

Figure 4.2 Soil layer: (a) two-way drainage; (b) one-way drainage.

If the total vertical stress is suddenly augmented by an amount Δq and maintained thereafter, this increased total stress immediately becomes fully effective on a pervious boundary. Therefore, the persistent boundary conditions for the soil layer with two-way drainage in Figure 4.2 are

$$\begin{cases} e(0,t) = (e_{00} - e_\infty)\exp[-\lambda(q_0' + \Delta q)] + e_\infty \\ e(l,t) = (e_{00} - e_\infty)\exp\{-\lambda[q_0' + \Delta q + (G_s - 1)\gamma_w l]\} + e_\infty \end{cases} \tag{4.20}$$

where l is the layer thickness in z coordinate.

For an impervious boundary, (4.6) requires

$$\frac{\partial u_w}{\partial \xi} - \gamma_w = 0 \tag{4.21}$$

which means

$$\frac{\partial \sigma'}{\partial z} - (G_s - 1)\gamma_w = 0 \tag{4.22}$$

Substituting (4.15) into (4.22) gives

$$\frac{\partial e}{\partial z} + (G_s - 1)\lambda\gamma_w(e - e_\infty) = 0 \tag{4.23}$$

Therefore, the persistent boundary conditions for the soil layer with one-way drainage in Figure 4.2 become

$$\begin{cases} e(0,t) = (e_{00} - e_\infty)\exp[-\lambda(q_0' + \Delta q)] + e_\infty \\ \left[\dfrac{\partial e}{\partial z} + (G_s - 1)\lambda\gamma_w(e - e_\infty)\right]_{z=l} = 0 \end{cases} \tag{4.24}$$

All relationships governing the void ratio e above are in terms of reduced coordinates. For practical use to be made of these relationships, it is necessary to relate the Lagrangian coordinate a of a plane of grains in the clay layer to the reduced coordinate z of this plane. From (4.2) and (4.19), the following relation between z and a can be established

$$a = (1+e_\infty)z + \frac{(e_{00} - e_\infty)}{\lambda(G_s - 1)\gamma_w} \exp(-\lambda q_0')\{1 - \exp[-\lambda(G_s - 1)\gamma_w z]\} \tag{4.25}$$

The settlement of the whole stratum is

$$S(t) = \int_0^H \left(1 - \frac{\partial \xi}{\partial a}\right) da = \int_0^l [e(z,0) - e(z,t)] dz \tag{4.26}$$

Furthermore, the average degree of consolidation can be expressed as

$$U(t) = \frac{S(t)}{S(\infty)} = \frac{\int_0^l [e(z,0) - e(z,t)] dz}{\int_0^l [e(z,0) - e(z,\infty)] dz} \tag{4.27}$$

The solution of the appropriate boundary value problem for e as a function of z and t can be obtained by a variety of methods. Using these void ratio results, the effective stress may be evaluated from

$$\sigma'(z,t) = \frac{1}{\lambda} \ln\left[\frac{e_{00} - e_\infty}{e(z,t) - e_\infty}\right] \tag{4.28}$$

The total stress is determined from (4.5)

$$\sigma(z,t) = \gamma_w \left[H_0 + S(t) + G_s z + \int_0^z e(z,t) dz\right] + q_0' + \Delta q \tag{4.29}$$

where it is implied that the settlement of the lower surface of the clay layer is zero in the calculation of the total stress.

The excess pore pressure can be calculated from (4.30)

$$u_e(z,t) = \sigma - \sigma' - \gamma_w(\xi + H_0) \tag{4.30}$$

Using (4.3), the excess pore pressure becomes

$$u_e(z,t) = \sigma - \sigma' - \gamma_w \left[z + \int_0^z e(z,t) dz + S(t) + H_0\right] \tag{4.31}$$

4.5 SOLUTIONS OF LINEAR FINITE STRAIN CONSOLIDATION

In this section, analytical solutions for several typical problems of linear finite strain consolidation are derived. To this end, the following parameters are defined

$$Z = \frac{z}{l} \tag{4.32}$$

$$T_f = \frac{c_f t}{l^2} \tag{4.33}$$

$$N = (G_s - 1)\lambda l \gamma_w \tag{4.34}$$

$$R = (e_{00} - e_\infty)\exp(-\lambda q_0') \tag{4.35}$$

$$B = (e_{00} - e_\infty)\exp[-\lambda(q_0' + \Delta q)] \tag{4.36}$$

in terms of which (4.16) becomes

$$\frac{\partial e}{\partial T_f} = \frac{\partial^2 e}{\partial Z^2} + N\frac{\partial e}{\partial Z} \tag{4.37}$$

with the initial condition

$$e(Z,0) = e_\infty + R\exp(-NZ) \tag{4.38}$$

and the two-way drainage boundary conditions

$$\begin{cases} e(0,T_f) = e_\infty + B \\ e(1,T_f) = e_\infty + B\exp(-N) \end{cases} \tag{4.39}$$

or the one-way drainage boundary conditions

$$\begin{cases} e(0,T_f) = e_\infty + B \\ \left[\dfrac{\partial e}{\partial Z} + N(e - e_\infty)\right]_{Z=1} = 0 \end{cases} \tag{4.40}$$

4.5.1 Self-weight consolidation with two-way drainage

A layer of soil deposited at a uniform initial e corresponding to $\sigma' = 0$ is subjected to self-weight consolidation with drainage at both the top and bottom.

The governing equation is (4.37). The initial and boundary conditions to be applied to (4.37) are, respectively,

$$e(Z,0) = e_{00} \tag{4.41}$$

and

$$\begin{cases} e(0,T_f) = e_{00} \\ e(1,T_f) = e_\infty + (e_{00} - e_\infty)\exp(-N) \end{cases} \tag{4.42}$$

Letting

$$e = e_\infty + (e_{00} - e_\infty)\exp(-NZ) + \exp\left(-\frac{NZ}{2} - \frac{N^2T_f}{4}\right)e^* \tag{4.43}$$

The governing equation and initial and boundary conditions are transformed to the following

$$\frac{\partial e^*}{\partial T_f} = \frac{\partial^2 e^*}{\partial Z^2} \tag{4.44}$$

$$e^*(Z,0) = (e_{00} - e_\infty)\left[\exp\left(\frac{NZ}{2}\right) - \exp\left(-\frac{NZ}{2}\right)\right] \tag{4.45}$$

$$\begin{cases} e^*(0,T_f) = 0 \\ e^*(1,T_f) = 0 \end{cases} \tag{4.46}$$

Separation of variables leads to

$$e^* = (e_{00} - e_\infty)\sum_{n=1}^{\infty} \frac{8n\pi(-1)^{n+1}\left[\exp\left(\frac{N}{2}\right) - \exp\left(-\frac{N}{2}\right)\right]}{N^2 + 4n^2\pi^2}\sin(n\pi Z)\exp(-n^2\pi^2T_f) \tag{4.47}$$

Substituting (4.47) into (4.43) gives the solution for e

$$\begin{aligned} e = {} & e_\infty + (e_{00} - e_\infty)\exp(-NZ) + (e_{00} - e_\infty)\exp\left(-\frac{NZ}{2} - \frac{N^2T_f}{4}\right) \\ & \times \sum_{n=1}^{\infty} \frac{8n\pi(-1)^{n+1}\left[\exp\left(\frac{N}{2}\right) - \exp\left(-\frac{N}{2}\right)\right]}{N^2 + 4n^2\pi^2}\sin(n\pi Z)\exp(-n^2\pi^2T_f) \end{aligned} \tag{4.48}$$

Therefore, the settlement of the whole layer is

$$S(T_f) = (e_{00} - e_\infty) l \left\{ 1 + \frac{\exp(-N) - 1}{N} - \left[\exp\left(\frac{N}{2}\right) - \exp\left(-\frac{N}{2}\right) \right] \right.$$
$$\left. \times \sum_{n=1}^{\infty} \frac{32 n^2 \pi^2 \left[(-1)^{n+1} + \exp\left(-\frac{N}{2}\right) \right]}{(N^2 + 4n^2\pi^2)^2} \exp\left(-\frac{N^2 T_f}{4} - n^2 \pi^2 T_f \right) \right\} \tag{4.49}$$

and the average degree of consolidation is

$$U(T_f) = 1 - \frac{32\pi^2 N \left[\exp\left(\frac{N}{2}\right) - \exp\left(-\frac{N}{2}\right) \right]}{N + \exp(-N) - 1} \exp\left(-\frac{N^2 T_f}{4}\right)$$
$$\times \sum_{n=1}^{\infty} \frac{n^2 \left[(-1)^{n+1} + \exp\left(-\frac{N}{2}\right) \right]}{(N^2 + 4n^2\pi^2)^2} \exp(-n^2\pi^2 T_f) \tag{4.50}$$

For the limiting case of $N = 0$, (4.50) reduces to

$$U(T_f) = 1 - \frac{8}{\pi^2} \sum_{n=1,3,5,\cdots} \frac{1}{n^2} \exp(-n^2 \pi^2 T_f) \tag{4.51}$$

which is a special case of (2.26).

The solution chart for different N values corresponding to (4.50) and (4.51) is shown in Figure 4.3. The elapsed times for achieving an average degree of consolidation of 10, 20, 30, 40, 50, 60, 70, 80, 90, and 95 are listed in Table 4.1.

4.5.2 Initially normally consolidated soil with two-way drainage

A layer of soil with drainage at the top and bottom is initially fully consolidated and in equilibrium under both its own weight and a vertical effective stress q_0' acting on its upper surface, as shown in Figure 4.2. The total vertical stress is suddenly augmented at $t = 0$ by an amount Δq and maintained thereafter. The governing equation is (4.37). The initial and boundary conditions to be applied to (4.37) are (4.38) and (4.39).

Letting

$$e = e_\infty + B\exp(-NZ) + \exp\left(-\frac{NZ}{2} - \frac{N^2 T_f}{4}\right) e^* \tag{4.52}$$

Table 4.1 Elapsed times for certain average degree of consolidation for initially unconsolidated soil layers with two-way drainage by linear finite strain theory

N	T_{f10}	T_{f20}	T_{f30}	T_{f40}	T_{f50}	T_{f60}	T_{f70}	T_{f80}	T_{f90}	T_{f95}
0	1.96E-3	7.85E-3	1.77E-2	3.14E-2	4.92E-2	7.16E-2	1.01E-1	1.42E-1	2.12E-1	2.82E-1
0.5	2.26E-3	8.84E-3	1.95E-2	3.40E-2	5.22E-2	7.49E-2	1.04E-1	1.45E-1	2.15E-1	2.85E-1
1	2.54E-3	9.75E-3	2.11E-2	3.60E-2	5.45E-2	7.70E-2	1.06E-1	1.46E-1	2.15E-1	2.83E-1
1.5	2.82E-3	1.06E-2	2.24E-2	3.75E-2	5.59E-2	7.80E-2	1.06E-1	1.45E-1	2.12E-1	2.78E-1
2	3.07E-3	1.12E-2	2.33E-2	3.85E-2	5.65E-2	7.80E-2	1.05E-1	1.43E-1	2.07E-1	2.71E-1
3	3.48E-3	1.22E-2	2.43E-2	3.90E-2	5.59E-2	7.57E-2	1.00E-1	1.34E-1	1.92E-1	2.49E-1
4	3.74E-3	1.26E-2	2.44E-2	3.81E-2	5.36E-2	7.15E-2	9.36E-2	1.24E-1	1.74E-1	2.24E-1
5	3.89E-3	1.26E-2	2.38E-2	3.65E-2	5.06E-2	6.66E-2	8.60E-2	1.12E-1	1.56E-1	1.99E-1
7	3.94E-3	1.20E-2	2.18E-2	3.25E-2	4.41E-2	5.68E-2	7.20E-2	9.21E-2	1.25E-1	1.56E-1
9	3.83E-3	1.11E-2	1.96E-2	2.87E-2	3.83E-2	4.87E-2	6.08E-2	7.65E-2	1.01E-1	1.25E-1
12	3.57E-3	9.81E-3	1.68E-2	2.41E-2	3.17E-2	3.97E-2	4.87E-2	6.02E-2	7.78E-2	9.42E-2
15	3.29E-3	8.69E-3	1.46E-2	2.06E-2	2.68E-2	3.33E-2	4.05E-2	4.93E-2	6.25E-2	7.47E-2

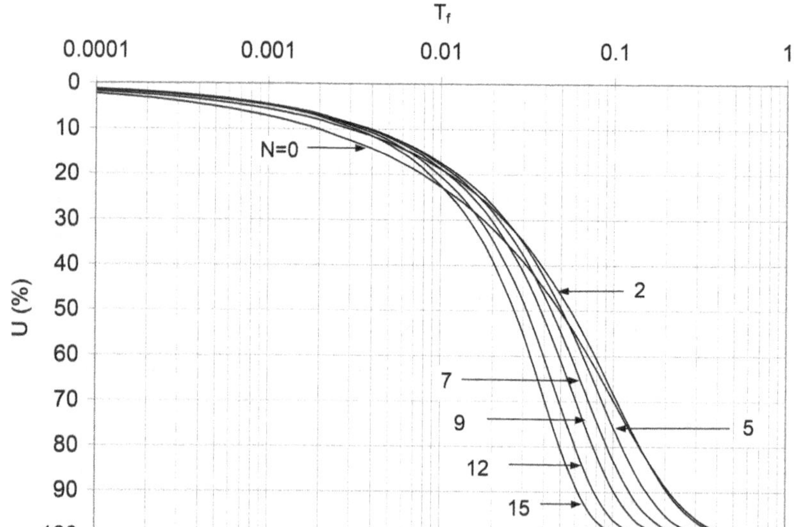

Figure 4.3 Degree of consolidation versus dimensionless time factor for initially unconsolidated soil layers with two-way drainage by linear finite strain theory.

The governing equation and boundary conditions are transformed to (4.44) and (4.46), respectively. The initial condition becomes

$$e^*(Z,0) = (R-B)\exp\left(-\frac{NZ}{2}\right) \tag{4.53}$$

As for the previous case, the solution for e is

$$e = e_\infty + B\exp(-NZ) + (R-B)\exp\left(-\frac{NZ}{2} - \frac{N^2T_f}{4}\right)$$
$$\times\sum_{n=1}^{\infty}\frac{8n\pi\left[1+(-1)^{n+1}\exp\left(-\frac{N}{2}\right)\right]}{N^2+4n^2\pi^2}\sin(n\pi Z)\exp(-n^2\pi^2T_f) \tag{4.54}$$

The settlement of the whole layer is

$$S(T_f) = (R-B)l\left\{\frac{1-\exp(-N)}{N}\right.$$
$$\left.-\sum_{n=1}^{\infty}\frac{32n^2\pi^2\left[1+(-1)^{n+1}\exp\left(-\frac{N}{2}\right)\right]^2}{(N^2+4n^2\pi^2)^2}\exp\left(-\frac{N^2T_f}{4} - n^2\pi^2T_f\right)\right\} \tag{4.55}$$

and the average degree of consolidation is

$$U(T_f) = 1 - \frac{32\pi^2N}{1-\exp(-N)}\exp\left(-\frac{N^2T_f}{4}\right)$$
$$\times\sum_{n=1}^{\infty}\frac{n^2\left[1+(-1)^{n+1}\exp\left(-\frac{N}{2}\right)\right]^2}{(N^2+4n^2\pi^2)^2}\exp(-n^2\pi^2T_f) \tag{4.56}$$

For the limiting case of $N = 0$, (4.56) reduces via L'Hospital's rule to (4.51).

The solution chart for different N values corresponding to (4.56) and (4.51) is shown in Figure 4.4. The elapsed times for achieving an average degree of consolidation of 10, 20, 30, 40, 50, 60, 70, 80, 90, and 95 are listed in Table 4.2.

Table 4.2 Elapsed times for certain average degree of consolidation for initially normally consolidated soil layers with two-way drainage by linear finite strain theory

N	T_{f10}	T_{f20}	T_{f30}	T_{f40}	T_{f50}	T_{f60}	T_{f70}	T_{f80}	T_{f90}	T_{f95}
0	1.96E-3	7.85E-3	1.77E-2	3.14E-2	4.92E-2	7.16E-2	1.01E-1	1.42E-1	2.12E-1	2.82E-1
0.5	1.89E-3	7.61E-3	1.72E-2	3.07E-2	4.82E-2	7.05E-2	9.94E-2	1.40E-1	2.10E-1	2.80E-1
1	1.71E-3	6.94E-3	1.59E-2	2.87E-2	4.55E-2	6.71E-2	9.54E-2	1.35E-1	2.04E-1	2.72E-1
1.5	1.45E-3	6.01E-3	1.40E-2	2.57E-2	4.15E-2	6.20E-2	8.93E-2	1.28E-1	1.94E-1	2.61E-1
2	1.19E-3	5.01E-3	1.18E-2	2.22E-2	3.65E-2	5.57E-2	8.15E-2	1.18E-1	1.82E-1	2.46E-1
3	7.64E-4	3.28E-3	7.97E-3	1.54E-2	2.63E-2	4.18E-2	6.36E-2	9.60E-2	1.53E-1	2.10E-1
4	4.93E-4	2.14E-3	5.27E-3	1.03E-2	1.81E-2	2.96E-2	4.67E-2	7.34E-2	1.22E-1	1.72E-1
5	3.31E-4	1.45E-3	3.58E-3	7.09E-3	1.25E-2	2.08E-2	3.36E-2	5.45E-2	9.44E-2	1.36E-1
7	1.74E-4	7.58E-4	1.89E-3	3.75E-3	6.66E-3	1.12E-2	1.83E-2	3.04E-2	5.53E-2	8.34E-2
9	1.05E-4	4.61E-4	1.15E-3	2.28E-3	4.05E-3	6.79E-3	1.12E-2	1.86E-2	3.43E-2	5.26E-2
12	5.95E-5	2.59E-4	6.45E-4	1.28E-3	2.28E-3	3.83E-3	6.28E-3	1.05E-2	1.93E-2	2.98E-2
15	3.83E-5	1.66E-4	4.13E-4	8.21E-4	1.46E-3	2.45E-3	4.02E-3	6.72E-3	1.24E-2	1.91E-2

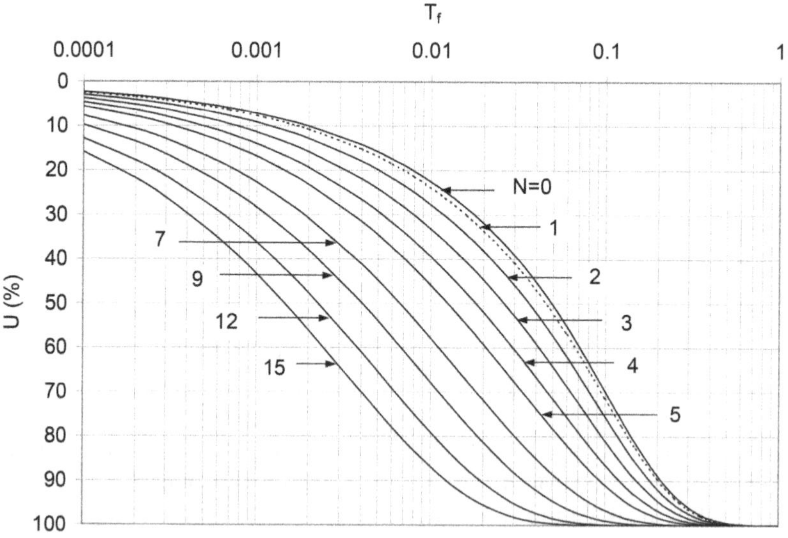

Figure 4.4 Degree of consolidation versus dimensionless time factor for initially normally consolidated soil layers with two-way drainage by linear finite strain theory.

4.5.3 Self-weight consolidation with one-way drainage

This problem is identical to the self-weight consolidation problem in Section 4.5.1 except that the soil layer is drained at the top only. The initial condition to be applied to (4.37) is (4.41). The boundary conditions are given by

$$\begin{cases} e(0,T_f) = e_{00} \\ \left[\dfrac{\partial e}{\partial Z} + N(e - e_\infty)\right]_{Z=1} = 0 \end{cases} \tag{4.57}$$

Using (4.43), (4.37) is transformed to (4.44). The initial condition becomes (4.45). The boundary conditions are

$$\begin{cases} e^*(0,T_f) = 0 \\ \left[\dfrac{\partial e^*}{\partial Z} + \dfrac{N}{2}e^*\right]_{Z=1} = 0 \end{cases} \tag{4.58}$$

The solution for e is obtained as

$$e = e_\infty + (e_{00} - e_\infty)\exp(-NZ) + (e_{00} - e_\infty)\exp\left(-\frac{NZ}{2} - \frac{N^2T_f}{4}\right) \times \sum_{n=1}^{\infty} \frac{8N\exp\left(\dfrac{N}{2}\right)\sin(\alpha_n)}{N^2 + 2N + 4\alpha_n^2}\sin(\alpha_n Z)\exp(-\alpha_n^2 T_f) \tag{4.59}$$

where α_n is the nth positive root of (4.60)

$$\alpha_n \cos(\alpha_n) + \frac{N}{2}\sin(\alpha_n) = 0 \tag{4.60}$$

The settlement of the whole layer is

$$S(T_f) = (e_{00} - e_\infty)l\left\{1 + \frac{\exp(-N) - 1}{N} - \sum_{n=1}^{\infty} \frac{32N\exp\left(\dfrac{N}{2}\right)\alpha_n \sin(\alpha_n)}{(N^2 + 2N + 4\alpha_n^2)(N^2 + 4\alpha_n^2)}\exp\left(-\frac{N^2T_f}{4} - n^2\pi^2 T_f\right)\right\} \tag{4.61}$$

and the average degree of consolidation is

$$U(T_f) = 1 - \frac{32N^2}{N + \exp(-N) - 1} \exp\left(\frac{N}{2} - \frac{N^2 T_f}{4}\right) \times \sum_{n=1}^{\infty} \frac{\alpha_n \sin(\alpha_n)}{(N^2 + 2N + 4\alpha_n^2)(N^2 + 4\alpha_n^2)} \exp(-\alpha_n^2 T_f) \tag{4.62}$$

For the limiting case of $N = 0$, (4.62) reduces to

$$U(T_f) = 1 - \frac{32}{\pi^3} \sum_{n=1,3,5,\cdots} (-1)^{1+\frac{n+1}{2}} \frac{1}{n^3} \exp\left(-\frac{n^2\pi^2}{4} T_f\right) \tag{4.63}$$

The solution chart for different N values corresponding to (4.62) and (4.63) is shown in Figure 4.5. The elapsed times for achieving an average degree of consolidation of 10, 20, 30, 40, 50, 60, 70, 80, 90, and 95 are listed in Table 4.3.

4.5.4 Initially normally consolidated soil with one-way drainage

This problem is identical to that of the consolidation of an initially normally consolidated soil layer in Section 4.5.2 except that the layer is drained at the top only. The initial and boundary conditions to be applied

Table 4.3 Elapsed times for certain average degree of consolidation for initially unconsolidated soil layers with one-way drainage by linear finite strain theory

N	T_{f10}	T_{f20}	T_{f30}	T_{f40}	T_{f50}	T_{f60}	T_{f70}	T_{f80}	T_{f90}	T_{f95}
0	5.00E-2	1.01E-1	1.57E-1	2.20E-1	2.94E-1	3.84E-1	5.01E-1	6.65E-1	9.46E-1	1.23
0.5	4.26E-2	8.59E-2	1.32E-1	1.84E-1	2.45E-1	3.20E-1	4.15E-1	5.50E-1	7.81E-1	1.01
1	3.68E-2	7.40E-2	1.13E-1	1.57E-1	2.08E-1	2.70E-1	3.49E-1	4.61E-1	6.52E-1	8.44E-1
1.5	3.21E-2	6.45E-2	9.84E-2	1.36E-1	1.79E-1	2.31E-1	2.97E-1	3.91E-1	5.51E-1	7.12E-1
2	2.84E-2	5.69E-2	8.64E-2	1.19E-1	1.55E-1	2.00E-1	2.56E-1	3.36E-1	4.71E-1	6.07E-1
3	2.28E-2	4.56E-2	6.89E-2	9.39E-2	1.22E-1	1.55E-1	1.97E-1	2.55E-1	3.55E-1	4.54E-1
4	1.89E-2	3.77E-2	5.69E-2	7.70E-2	9.92E-2	1.25E-1	1.57E-1	2.02E-1	2.78E-1	3.53E-1
5	1.60E-2	3.21E-2	4.82E-2	6.50E-2	8.32E-2	1.04E-1	1.30E-1	1.65E-1	2.24E-1	2.83E-1
7	1.22E-2	2.45E-2	3.68E-2	4.93E-2	6.26E-2	7.75E-2	9.54E-2	1.19E-1	1.58E-1	1.96E-1
9	9.88E-3	1.98E-2	2.96E-2	3.96E-2	5.01E-2	6.15E-2	7.49E-2	9.24E-2	1.20E-1	1.47E-1
12	7.64E-3	1.53E-2	2.29E-2	3.06E-2	3.85E-2	4.69E-2	5.65E-2	6.87E-2	8.77E-2	1.05E-1
15	6.22E-3	1.24E-2	1.87E-2	2.49E-2	3.12E-2	3.79E-2	4.53E-2	5.46E-2	6.85E-2	8.13E-2

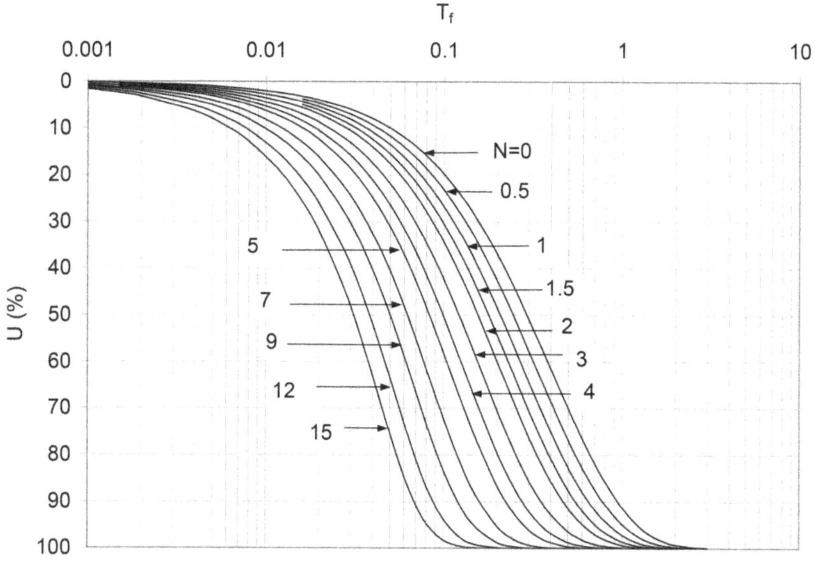

Figure 4.5 Degree of consolidation versus dimensionless time factor for initially unconsolidated soil layers with one-way drainage by linear finite strain theory.

to (4.37) are (4.38) and (4.40), respectively. Using (4.52), the governing equation, initial condition, and boundary conditions are transformed to (4.44), (4.53), and (4.58), respectively. The analysis then proceeds as for the previous cases.

The solution for e is obtained as

$$e = e_\infty + B\exp(-NZ) + (R-B)\exp\left(-\frac{NZ}{2} - \frac{N^2T_f}{4}\right) \times \sum_{n=1}^{\infty} \frac{8\alpha_n}{N^2+2N+4\alpha_n^2}\sin(\alpha_n Z)\exp(-\alpha_n^2 T_f) \tag{4.64}$$

where α_n is the nth positive root of (4.60)

The settlement of the whole layer is

$$S(T_f) = (R-B)l\left\{\frac{1-\exp(-N)}{N} - \sum_{n=1}^{\infty}\frac{32\alpha_n^2}{(N^2+2N+4\alpha_n^2)(N^2+4\alpha_n^2)}\exp\left(-\frac{N^2T_f}{4} - n^2\pi^2T_f\right)\right\} \tag{4.65}$$

and the average degree of consolidation is

$$U(T_f) = 1 - \frac{32N}{1-\exp(-N)}\exp\left(-\frac{N^2 T_f}{4}\right) \times \sum_{n=1}^{\infty} \frac{\alpha_n^2}{(N^2+2N+4\alpha_n^2)(N^2+4\alpha_n^2)}\exp(-\alpha_n^2 T_f) \tag{4.66}$$

For the limiting case of $N = 0$, (4.66) reduces to

$$U(T_f) = 1 - \frac{8}{\pi^2}\sum_{n=1,3,5,\cdots} \frac{1}{n^2}\exp\left(-\frac{n^2\pi^2}{4}T_f\right) \tag{4.67}$$

which is a special case of (2.28).

The solution chart for different N values corresponding to (4.66) and (4.67) is shown in Figure 4.6. The elapsed times for achieving an average degree of consolidation of 10, 20, 30, 40, 50, 60, 70, 80, 90, and 95 are listed in Table 4.4.

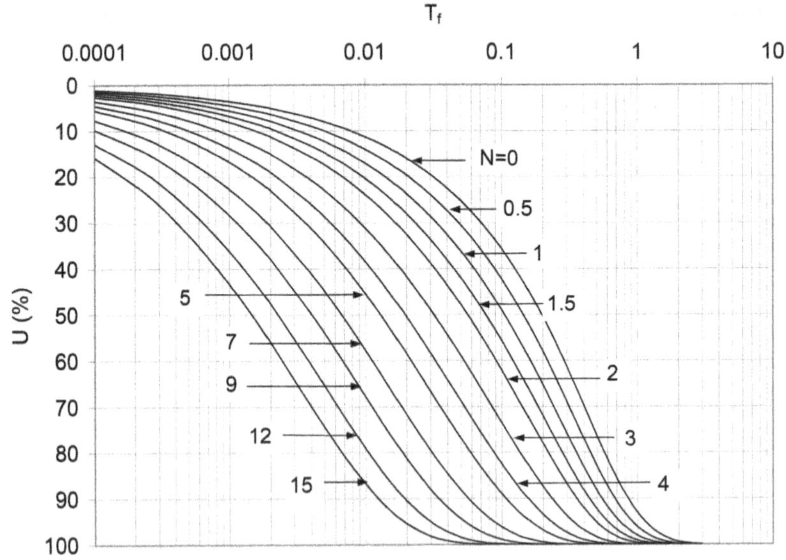

Figure 4.6 Degree of consolidation versus dimensionless time factor for initially normally consolidated soil layers with one-way drainage by linear finite strain theory.

Table 4.4 Elapsed times for certain average degree of consolidation for initially normally consolidated soil layers with one-way drainage by linear finite strain theory

N	T_{f10}	T_{f20}	T_{f30}	T_{f40}	T_{f50}	T_{f60}	T_{f70}	T_{f80}	T_{f90}	T_{f95}
0	7.85E-3	3.14E-2	7.07E-2	1.26E-1	1.97E-1	2.86E-1	4.03E-1	5.67E-1	8.48E-1	1.13
0.5	5.02E-3	2.07E-2	4.83E-2	8.89E-2	1.44E-1	2.16E-1	3.11E-1	4.46E-1	6.77E-1	9.07E-1
1	3.30E-3	1.39E-2	3.32E-2	6.28E-2	1.05E-1	1.62E-1	2.40E-1	3.51E-1	5.43E-1	7.34E-1
1.5	2.24E-3	9.61E-3	2.33E-2	4.48E-2	7.63E-2	1.21E-1	1.84E-1	2.77E-1	4.37E-1	5.97E-1
2	1.58E-3	6.80E-3	1.66E-2	3.24E-2	5.61E-2	9.08E-2	1.41E-1	2.18E-1	3.53E-1	4.88E-1
3	8.52E-4	3.71E-3	9.17E-3	1.81E-2	3.18E-2	5.26E-2	8.46E-2	1.36E-1	2.32E-1	3.31E-1
4	5.13E-4	2.24E-3	5.56E-3	1.10E-2	1.95E-2	3.26E-2	5.30E-2	8.73E-2	1.55E-1	2.29E-1
5	3.36E-4	1.47E-3	3.66E-3	7.26E-3	1.29E-2	2.16E-2	3.53E-2	5.87E-2	1.07E-1	1.61E-1
7	1.74E-4	7.60E-4	1.89E-3	3.76E-3	6.69E-3	1.12E-2	1.84E-2	3.07E-2	5.65E-2	8.68E-2
9	1.05E-4	4.61E-4	1.15E-3	2.28E-3	4.05E-3	6.80E-3	1.12E-2	1.87E-2	3.44E-2	5.29E-2
12	5.95E-5	2.59E-4	6.45E-4	1.28E-3	2.28E-3	3.83E-3	6.28E-3	1.05E-2	1.93E-2	2.98E-2
15	3.83E-5	1.66E-4	4.13E-4	8.21E-4	1.46E-3	2.45E-3	4.02E-3	6.72E-3	1.24E-2	1.91E-2

REFERENCES

Gibson, RE, Schiffman, RL, and Cargill, KW, 1981, The theory of one-dimensional consolidation of saturated clays. II. Finite nonlinear consolidation of thick homogeneous layers. *Canadian Geotechnical Journal*, **18**(2), pp. 280–293.

McNabb, A, 1960, A mathematical treatment of one-dimensional soil consolidation. *Quarterly of Applied Mathematics*, **17**(4), pp. 337–347.

Morris, PH, 2002, Analytical solutions of linear finite-strain one-dimensional consolidation. *Journal of Geotechnical and Geoenvironmental Engineering*, **128**(4), pp. 319–326.

Chapter 5

Finite element method for consolidation analysis of one-dimensional problems

5.1 INTRODUCTION

The consolidation analysis of clay strata, such as the prediction of settlement of foundation and reclamation over clay soils, is of practical importance in geotechnical engineering. The first rational approach to this problem based on the principle of effective stress was proposed by Terzaghi (1943). Since then, extensive research has been carried out in this field. The stress-strain behavior of normally consolidated clay has been clearly demonstrated to be elastoplastic and viscous in nature (Bjerrum, 1967; Schofield and Wroth, 1968; Dafalias, 1982; Adachi and Oka, 1982; Adachi *et al.*, 1987; Leroueil *et al.*, 1985, 1988; Yin and Graham, 1989, 1994). However, analytical solutions of consolidation problems considering general stress-strain behavior of soils can only be found in limited cases because of the complexity of the coupled set of partial differential equations. For general situations, numerical techniques should be utilized to obtain the consolidation solutions (Garlanger, 1972; Schiffman and Arya, 1977; Desai *et al.*, 1979; Kabbaj *et al.*, 1986; Yin and Graham, 1996; Zhu and Yin, 1999, 2000).

Although field problems are three-dimensional (3-D) in nature, application of the two-dimensional and 3-D procedures (Sandhu and Wilson, 1969; Siriwardane and Desai, 1981; Adachi and Oka, 1982; Thomas and Harb, 1990; Lewis and Schrefler, 1998) to include elasto-plastic and time-dependent behavior of soil may not be simple, and can be expensive. The one-dimensional (1-D) straining approach can be applied to a number of problems where the situations can be approximated as 1-D. In addition, the importance of this approach to geotechnical engineering practice is attested by the fact that the settlement estimation of foundations on cohesive soils is based largely on 1-D straining model. In this chapter, a 1-D finite element (FE) procedure will be derived with an emphasis on implementation of a non-linear model for the time-dependent behavior of soils. This 1-D FE procedure is then used to analyze the consolidation of clay layers.

5.2 GOVERNING EQUATIONS OF CONSOLIDATION PROBLEMS

The partial differential equations governing 1-D consolidation problems are presented in Section 2.3. For easy reference, these equations are reproduced in the following. A FE model based on these governing equations is developed for simulating the consolidation of layered soils.

a. The constitutive equations

$$\frac{\partial \varepsilon_z}{\partial t} = \frac{\partial f(\sigma_z')}{\partial t} + g(\sigma', \varepsilon_z) \tag{2.4}$$

Many 1-D models can be expressed in the form of (2.4), and $f(\sigma_z')$ and $g(\sigma_z', \varepsilon_z)$ may be different for loading and unloading for some models.

For Terzaghi's model

$$\begin{cases} f = m_v \sigma_z' \\ g = 0 \end{cases} \tag{5.1}$$

For the commonly used non-linear elastic model

$$\begin{cases} f = \dfrac{\lambda}{V} \ln \sigma_z' \\ g = 0 \end{cases} \tag{5.2}$$

where λ/V is a constant.

For Hardin's model (1989)

$$\begin{cases} f = -\dfrac{1}{a + b\sigma_z'^{p}} \\ g = 0 \end{cases} \tag{5.3}$$

where a, b, and p are constants.

For the 1-D elastic visco-plastic (EVP) model proposed by Yin and Graham (1989, 1994)

$$\begin{cases} f = \dfrac{\kappa}{V} \ln \sigma_z' \\ g = \dfrac{\psi}{V t_o} \exp\left(-\varepsilon_z \dfrac{V}{\psi}\right)\left(\dfrac{\sigma_z'}{\sigma_0}\right)^{\lambda/\psi} \end{cases} \tag{5.4}$$

where κ/V is a constant for elastic behavior (V is specific volume), λ/V is a constant for the slope of a 'reference time' line (similar to the normally consolidated compression line), ψ/V is a creep parameter (similar to the 'secondary' consolidation coefficient, but defined differently), t_o is a creep parameter in units of time, and σ'_o is a constant parameter in units of stress locating the 'reference time' line with $\varepsilon_z = 0$. The same names of the parameters κ/V and λ/V are used in the Cam-Clay model (Schofield and Wroth, 1968) for modeling the unloading/reloading behavior and the normally consolidated behavior of clays under isotropic loading.

b. The Darcy's law

$$q_z = -\frac{k}{\gamma_w}\frac{\partial u_e}{\partial z} \tag{2.6}$$

c. The continuity equation

$$\frac{\partial q_z}{\partial z} = \frac{\partial \varepsilon_z}{\partial t} \tag{2.10}$$

Equations (2.4), (2.6), and (2.10) are the governing equations for 1-D consolidation problems.

5.3 FINITE ELEMENT FORMULATION

The numerical solution of the consolidation model for a given boundary value problem is obtained here by using a Galerkin type FE weak formulation for the space variable z, coupled with a finite difference time scheme. Let the soil mass (domain Ω) be divided into n elements, with nodal points at the boundaries of each element as shown in Figure 5.1. For an element Ω_i, the nodal point coordinates are denoted as z_{i-1}, z_i and the thickness h_i of the element is z_i-z_{i-1}. The soil hydraulic conductivity in Ω_i is k_i.

Multiplying (2.10) by a virtual function $\phi(z)$, and integrating it, we have

$$\pi = \int_{\Omega}\phi\left(\frac{\partial q_z}{\partial z} - \frac{\partial \varepsilon_z}{\partial t}\right)dz = \sum_{\Omega_i}\int_{\Omega_i}\left[\frac{k}{\gamma_w}\frac{\partial u_e}{\partial z}\frac{\partial \phi}{\partial z}dz - \phi\left(\frac{\partial f}{\partial t} + g\right)\right]dz + \phi q_z\,|_{\Omega_b} = 0 \tag{5.5}$$

where Ω_b stands for the boundary. Equations (2.4) and (2.6) are used in deriving (5.5).

The standard FE approach to calculate the term $\int_{\Omega_i}\phi\left(\frac{\partial f}{\partial t} + g\right)dz$ in (5.5) is not adopted here. Otherwise, an unsymmetric stiffness matrix will be obtained.

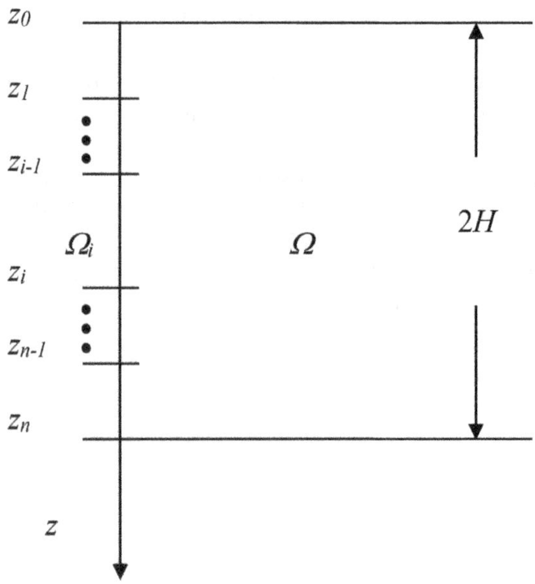

Figure 5.1 Soil profile and discretization.

In the following, this term is simply integrated using the trapezoidal formula with the same error order, that is,

$$\int_{\Omega_i} \phi\left(\frac{\partial f}{\partial t}+g\right)dz = \frac{h_i}{2}(\phi_{i-1},\phi_i)\begin{bmatrix}\left(\dfrac{\partial f}{\partial t}+g\right)_{i-1}\\ \left(\dfrac{\partial f}{\partial t}+g\right)_i\end{bmatrix} \tag{5.6}$$

where the subscripts *i*-1 and *i* denote values at z_{i-1} and z_i, respectively.

Using (5.6), some off-diagonal terms of the stiffness matrix are concentrated to the diagonal, and the positiveness of the stiffness matrix is increased. As a result, the numerical instability due to spatial oscillation discussed in Vermeer and Verruijt (1981) is avoided.

Suppose ϕ and u_e are linear in every element Ω_i, then

$$\begin{cases} u_e = N_1 u_{e_{i-1}} + N_2 u_{e_i} \\ \phi = N_1 \phi_{i-1} + N_2 \phi_i \end{cases} \tag{5.7}$$

where N_1 and N_2 are the linear shape functions.

Substituting (5.6) and (5.7) into (5.5), we get

$$\pi = \sum_{\Omega_i} (\phi_{i-1}, \phi_i) \left\{ \frac{k_i}{\gamma_w h_i} \begin{bmatrix} 1 & -1 \\ -1 & 1 \end{bmatrix} \begin{pmatrix} u_{e_{i-1}} \\ u_{e_i} \end{pmatrix} - \frac{h_i}{2} \begin{bmatrix} \left(\frac{\partial f}{\partial t} + g\right)_{i-1} \\ \left(\frac{\partial f}{\partial t} + g\right)_i \end{bmatrix} \right\} + \phi q_z \mid_{\Omega_b} = 0 \tag{5.8}$$

Integrating (5.8) with respect to time from t_j to t_{j+1} using the trapezoidal formula, we obtain

$$\begin{aligned} \pi = \sum_{\Omega_i} (\phi_{i-1}, \phi_i) \Bigg\{ & \frac{k_i \Delta t}{\gamma_w h_i} \begin{bmatrix} 1 & -1 \\ -1 & 1 \end{bmatrix} \left[\theta \begin{pmatrix} u_{e_{i-1}}^{j+1} \\ u_{e_i}^{j+1} \end{pmatrix} + (1-\theta) \begin{pmatrix} u_{e_{i-1}}^{j} \\ u_{e_i}^{j} \end{pmatrix} \right] \\ & - \frac{h_i}{2} \begin{bmatrix} f_{i-1}^{j+1} - f_{i-1}^{j} + [\theta g_{i-1}^{j+1} + (1-\theta) g_{i-1}^{j}] \Delta t \\ f_i^{j+1} - f_i^{j} + [\theta g_i^{j+1} + (1-\theta) g_i^{j}] \Delta t \end{bmatrix} \Bigg\} + \phi \int_{t_j}^{t_{j+1}} q dt \mid_{\Omega_b} = 0 \end{aligned} \tag{5.9}$$

where the superscripts j and j+1 denotes values at t_j and t_{j+1}, respectively, θ is a real number between 0 and 1, and $\Delta t = t_{j+1} - t_j$.

Differentiating π expressed in (5.9) with respect to u_e^{j+1} (a vector including all excess porewater pressures at nodes) gives

$$\delta\pi = \sum_{\Omega_i} (\phi_{i-1}, \phi_i) K_e \begin{pmatrix} \delta u_{e_{i-1}}^{j+1} \\ \delta u_{e_i}^{j+1} \end{pmatrix} \tag{5.10}$$

where

$$K_e = \frac{k_i \theta \Delta t}{\gamma_w h_i} \begin{bmatrix} 1 & -1 \\ -1 & 1 \end{bmatrix} - \frac{h_i}{2} \begin{bmatrix} \dfrac{\partial (f_{i-1}^{j+1} + \theta \Delta t g_{i-1}^{j+1})}{\partial u_{e_{i-1}}^{j+1}} & 0 \\ 0 & \dfrac{\partial (f_i^{j+1} + \theta \Delta t g_i^{j+1})}{\partial u_{e_i}^{j+1}} \end{bmatrix} \tag{5.11}$$

Therefore, the discretized (5.9) can be approximated as

$$\sum_{\Omega_i} (\phi_{i-1}, \phi_i) \left[K_e \begin{pmatrix} \delta u_{e_{i-1}}^{j+1} \\ \delta u_{e_i}^{j+1} \end{pmatrix} - R_e \right] + \phi \int_{t_j}^{t_{j+1}} q_z dt \mid_{\Omega_b} = 0 \tag{5.12}$$

where

$$R_e = \frac{k_i \Delta t}{\gamma_w h_i}\begin{bmatrix} -1 & 1 \\ 1 & -1 \end{bmatrix}\left[\theta\begin{pmatrix} u_{e_{i-1}}^{j+1} \\ u_{e_i}^{j+1} \end{pmatrix} + (1-\theta)\begin{pmatrix} u_{e_{i-1}}^{j} \\ u_{e_i}^{j} \end{pmatrix}\right]$$
$$+\frac{h_i}{2}\begin{bmatrix} f_{i-1}^{j+1} - f_{i-1}^{j} + [\theta g_{i-1}^{j+1} + (1-\theta) g_{i-1}^{j}]\Delta t \\ f_i^{j+1} - f_i^{j} + [\theta g_i^{j+1} + (1-\theta) g_i^{j}]\Delta t \end{bmatrix}$$

Following the standard FE assembly procedure, a symmetrically tri-diagonal set of equations can be derived in incremental form. This Newton form of iteration (or modified Newton form) for the non-linear equation system converges very rapidly. Generally, several iterations will produce very high accuracy.

The element stiffness matrix for (5.1) to (5.3) is straightforward. We only derive the element stiffness matrix for (5.4) here. Integrating (2.4) gives

$$(\varepsilon_z - f)^{j+1} = (\varepsilon_z - f)^{j} + [\theta' g^{j+1} + (1-\theta') g^{j}]\Delta t \tag{5.13}$$

where θ' is a real number between 0 and 1. The θ' has the same meaning as θ in (5.9), but may have a different value.

Differentiating (5.13) and $g^{j+1} = g(\sigma_z'^{j+1}, \varepsilon_z^{j+1})$ with respect to $\sigma_z'^{j+1}$ gives

$$\frac{\partial \varepsilon_z^{j+1}}{\partial \sigma_z'^{j+1}} = \left(\frac{\partial f}{\partial \sigma_z'}\right)^{j+1} + \theta' \Delta t \frac{\partial g^{j+1}}{\partial \sigma_z'^{j+1}} \tag{5.14}$$

and

$$\frac{\partial g^{j+1}}{\partial \sigma_z'^{j+1}} = \left(\frac{\partial g}{\partial \sigma_z'}\right)^{j+1} + \left(\frac{\partial g}{\partial \varepsilon_z}\right)^{j+1} \frac{\partial \varepsilon_z^{j+1}}{\partial \sigma_z'^{j+1}} \tag{5.15}$$

Substituting (5.14) into (5.15) gives

$$\frac{\partial g^{j+1}}{\partial \sigma_z'^{j+1}} = \left(\frac{\dfrac{\partial g}{\partial \sigma_z'} + \dfrac{\partial g}{\partial \varepsilon_z}\dfrac{\partial f}{\partial \sigma_z'}}{1 - \theta' \Delta t \dfrac{\partial g}{\partial \varepsilon_z}}\right)^{j+1} \tag{5.16}$$

Substituting (5.16) into (5.11) gives

$$K_e = \frac{k_i \theta \Delta t}{\gamma_w h_i}\begin{bmatrix} 1 & -1 \\ -1 & 1 \end{bmatrix} + \frac{h_i}{2}\begin{bmatrix} \left(\dfrac{\dfrac{\partial f}{\partial \sigma_z'} + \dfrac{\partial g}{\partial \sigma_z'}\theta\Delta t + (\theta - \theta')\Delta t \dfrac{\partial g}{\partial \varepsilon_z}\dfrac{\partial f}{\partial \sigma_z'}}{1 - \theta'\Delta t \dfrac{\partial g}{\partial \varepsilon_z}} \right)_{i-1}^{j+1} & 0 \\ 0 & \left(\dfrac{\dfrac{\partial f}{\partial \sigma_z'} + \dfrac{\partial g}{\partial \sigma_z'}\theta\Delta t + (\theta - \theta')\Delta t \dfrac{\partial g}{\partial \varepsilon_z}\dfrac{\partial f}{\partial \sigma_z'}}{1 - \theta'\Delta t \dfrac{\partial g}{\partial \varepsilon_z}} \right)_{i}^{j+1} \end{bmatrix} \tag{5.17}$$

Equation (5.17) is the element stiffness matrix for constitutive equations of the form of (2.4), and it may be adjusted for unloading for some models. For (5.4), the element stiffness matrix becomes

$$K_e = \frac{k_i \theta \Delta t}{\gamma_w h_i}\begin{bmatrix} 1 & -1 \\ -1 & 1 \end{bmatrix} + \frac{h_i}{2}\begin{bmatrix} \left(\dfrac{\dfrac{\kappa}{V}\dfrac{\psi}{V} + \dfrac{\lambda\theta + (\theta' - \theta)\kappa}{V} g\Delta t}{\sigma_z'\left(\dfrac{\kappa}{V} + \theta' g \Delta t\right)} \right)_{i-1}^{j+1} & 0 \\ 0 & \left(\dfrac{\dfrac{\kappa}{V}\dfrac{\psi}{V} + \dfrac{\lambda\theta + (\theta' - \theta)\kappa}{V} g\Delta t}{\sigma_z'\left(\dfrac{\kappa}{V} + \theta' g \Delta t\right)} \right)_{i}^{j+1} \end{bmatrix} \tag{5.18}$$

The positiveness of the stiffness matrix is obvious.

From (5.13), we also obtain

$$d\varepsilon_z^{j+1} = \frac{(\varepsilon_z - f)^j + [\theta' g^{j+1} + (1-\theta')g^{j+1}]\Delta t - (\varepsilon_z - f)^{j+1}}{1 - \theta'\left(\dfrac{\partial g}{\partial \varepsilon_z}\right)^{j+1}} \tag{5.19}$$

The above formulae provide a complete routine for a Newton (or Modified Newton) type solution of 1-D consolidation problem. It should be noted that if unknown variables are present in the boundary conditions, the stiffness matrix should be adjusted accordingly.

5.4 NUMERICAL CHARACTERISTICS OF THE FORMULATION

A typical consolidation problem is to determine the response of a saturated soil due to transient loading under the influence of drainage. Of special importance is the calculation of the pore pressure increments, which can be divided into two parts: one part generated by the transient loading, and another part, usually of opposite sign, due to consolidation. Normally, the accuracy of numerical solutions increases with the decrease of time increments. For many numerical algorithms, however, Vermeer and Verruijt (1981) proved that there is a lower limit for the time increments, below which spatial oscillation will occur and the results are inaccurate.

In the present formulation, the positiveness of the stiffness matrix is increased by concentrating some off-diagonal terms of the stiffness matrix to the diagonal. For simplicity, it will be proved that the numerical instability due to spatial oscillation discussed in Vermeer and Verruijt (1981) is avoided in the new formulation for Terzaghi's model (5.1) (Zhu *et al.*, 2004). Therefore, the lower limit for the time increments is algorithm-dependent. If a solution scheme is carefully constructed, the lower limit can be eliminated.

Substituting (5.1) into (5.9) gives

$$\pi = \sum_{\Omega_i} (\phi_{i-1}, \phi_i) \left\{ \frac{k_i \Delta t}{\gamma_w h_i} \begin{bmatrix} 1 & -1 \\ -1 & 1 \end{bmatrix} \left[\theta \begin{pmatrix} u_{e_{i-1}}^{j+1} \\ u_{e_i}^{j+1} \end{pmatrix} + (1-\theta) \begin{pmatrix} u_{e_{i-1}}^{j} \\ u_{e_i}^{j} \end{pmatrix} \right] - \frac{m_{vi} h_i}{2} \begin{bmatrix} \Delta\sigma - u_{e_{i-1}}^{j+1} + u_{e_{i-1}}^{j} \\ \Delta\sigma - u_{e_i}^{j+1} + u_{e_i}^{j} \end{bmatrix} \right\} + \phi \int_{t_j}^{t_{j+1}} q_z dt \,|_{\Omega_b} = 0 \tag{5.20}$$

where $\Delta\sigma = \sigma_z^{j+1} - \sigma_z^j$, and m_{vi} is the soil compressibility in Ω_i.

Letting $a_i = \dfrac{k_i \Delta t}{\gamma_w h_i}, \quad b_i = \dfrac{m_{vi} h_i}{2}, \quad u_{e_0} = u_{e_n} = 0,$

Equation (5.21) given below can be obtained following the standard FE assembly procedure from (5.20).

$$(\theta \mathbf{A} + \mathbf{M})\mathbf{u}^{j+1} + [(1-\theta)\mathbf{A} - \mathbf{M}]\mathbf{u}^j = \Delta\sigma \mathbf{b} \tag{5.21}$$

where

$$\mathbf{A} = \begin{bmatrix} a_1 + a_2 & -a_2 & & & \\ -a_2 & a_2 + a_3 & -a_3 & & \\ & \cdots & \cdots & \cdots & \\ & & -a_{n-2} & a_{n-2} + a_{n-1} & -a_{n-1} \\ & & & -a_{n-1} & a_{n-1} + a_n \end{bmatrix}, \quad \mathbf{u} = \begin{bmatrix} u_{e_1} \\ u_{e_2} \\ \cdots \\ u_{e_{n-2}} \\ u_{e_{n-1}} \end{bmatrix},$$

$$\mathbf{M} = \begin{bmatrix} b_1 + b_2 & & & & \\ & b_2 + b_3 & & & \\ & & \cdots & & \\ & & & b_{n-2} + b_{n-1} & \\ & & & & b_{n-1} + b_n \end{bmatrix}, \quad \mathbf{b} = \begin{bmatrix} b_1 + b_2 \\ b_2 + b_3 \\ \cdots \\ b_{n-2} + b_{n-1} \\ b_{n-1} + b_n \end{bmatrix}$$

The stability of time integration for consolidation problem involving the parameter θ has been examined by Booker and Small (1975) and Vermeer and Verruijt (1981). For the sake of completeness, a short proof of the stability condition is presented below.

Rewriting (5.21), we obtain

$$\mathbf{u}^{j+1} = \mathbf{Q}\mathbf{u}^{j} + \Delta\sigma(\theta\mathbf{A} + \mathbf{M})^{-1}\mathbf{b} \tag{5.22}$$

where $\mathbf{Q} = \mathbf{I} - (\theta\mathbf{A} + \mathbf{M})^{-1}\mathbf{A} = \mathbf{A}^{-1}[\mathbf{I} - (\theta\mathbf{I} + \mathbf{M}\mathbf{A}^{-1})^{-1}]\mathbf{A}$

The eigenvalues λ_i^Q of $\mathbf{Q}$ are

$$\lambda_i^Q = 1 - \frac{1}{\theta + \lambda_i^B} \tag{5.23}$$

where λ_i^B are the eigenvalues of $\mathbf{B}$ and $\mathbf{B} = \mathbf{M}\mathbf{A}^{-1}$. Since both $\mathbf{M}$ and $\mathbf{A}$ are all positive matrixes, all the eigenvalues λ_i^B are positive. Letting $|\lambda_i^Q| < 1$, the following stability condition can be derived.

$$\theta > \frac{1}{2} - \lambda_i^B \tag{5.24}$$

Vermeer and Verruijt (1981) showed that there is a lower limit for the time increments in many algorithms, below which spatial oscillation will occur and the results are inaccurate. This phenomenon is examined in detail in the following for the formulation in (5.21). It will be proved that the spatial oscillation is avoided by using (5.6).

Equation (5.21) can be rewritten as (5.25).

$$(\theta\mathbf{A}+\mathbf{M})\Delta\mathbf{u} = -\mathbf{A}\mathbf{u}^j + \Delta\sigma\mathbf{b} \tag{5.25}$$

where $\Delta\mathbf{u} = \mathbf{u}^{j+1} - \mathbf{u}^j$.

The solution $\Delta\mathbf{u}$ can be considered to consist of two parts, each part corresponding to one of the terms in the right-hand side of (5.25). The first part is determined by $\mathbf{u}^j$, the initial value in the calculation. The second part, denoted by $\Delta\mathbf{u}'$, is determined by $\Delta\sigma$. When $\mathbf{u}^j$ is zero, the solution of (5.25) represents the excess pore pressure generated by sudden loading. Since the numerical errors caused by the initial value $\mathbf{u}^j$ are controlled by (5.24), only the errors caused by the external loading are discussed in the following.

The second part of the solution in (5.25) is obtained from (5.26).

$$(\theta\mathbf{A}+\mathbf{M})\Delta\mathbf{u}' = \Delta\sigma\mathbf{b} \tag{5.26}$$

Letting

$$\Delta u'_{e_i} = \Delta\sigma(1-\varepsilon_i) \tag{5.27}$$

we will prove that all the ε_i satisfy condition (5.28) in the following.

$$0 \le \varepsilon_i < 1 \ (i = 1, 2, \ldots, n{-}1) \tag{5.28}$$

Substituting (5.27) into (5.26) gives

$$(\theta\mathbf{A}+\mathbf{M})\varepsilon = \theta\mathbf{b}^* \tag{5.29}$$

where $\mathbf{b}^* = (a_1, 0, 0, \ldots, 0, a_n)^T$

Define $\varepsilon_{i0} = \underset{i}{Min}\,\varepsilon_i$.

We will get (5.30–5.32) from (5.29).

$$\theta[a_1\varepsilon_1 + a_2(\varepsilon_1-\varepsilon_2)] + (b_1+b_2)\varepsilon_1 = \theta a_1 \quad i_0 = 1 \tag{5.30}$$

$$\theta[a_{i0}(\varepsilon_{i0}-\varepsilon_{i0-1}) + a_{i0+1}(\varepsilon_{i0}-\varepsilon_{i0+1})] + (b_{i0}+b_{i0+1})\varepsilon_{i0} = 0 \quad 1 < i_0 < n-1 \tag{5.31}$$

$$\theta[a_{n-1}(\varepsilon_{n-1}-\varepsilon_{n-2}) + a_n\varepsilon_{n-1}] + (b_{n-1}+b_n)\varepsilon_{n-1} = \theta a_n \quad i_0 = n-1 \tag{5.32}$$

Obviously, $\varepsilon_{i0} \ge 0$. Otherwise, the left-hand side of one of (5.30–5.32) will be negative, while the right-hand side is non-negative.

Define $\varepsilon_{j0} = \underset{i}{Max}\,\varepsilon_i$.

Equations (5.33–5.35) are obtained from (5.29).

$$\theta[a_1\varepsilon_1 + a_2(\varepsilon_1-\varepsilon_2)] + (b_1+b_2)\varepsilon_1 = \theta a_1 \quad j_0 = 1 \tag{5.33}$$

$$\theta[a_{j0}(\varepsilon_{j0}-\varepsilon_{j0-1})+a_{j0+1}(\varepsilon_{j0}-\varepsilon_{j0+1})]+(b_{j0}+b_{j0+1})\varepsilon_{j0}=0 \quad 1<j_0<n-1 \tag{5.34}$$

$$\theta[a_{n-1}(\varepsilon_{n-1}-\varepsilon_{n-2})+a_n\varepsilon_{n-1}]+(b_{n-1}+b_n)\varepsilon_{n-1}=\theta a_n \quad j_0=n-1 \tag{5.35}$$

Obviously, $\varepsilon_{j0}>0$ does not happen for the case of (5.34). Otherwise, the left-hand side of (5.34) will be positive, while the right-hand side is zero.

If $j_0=1$, the following relation is derived from (5.33) since $\varepsilon_1-\varepsilon_2\geq 0$.

$$\varepsilon_1\leq\frac{\theta a_1}{\theta a_1+b_1+b_2}<1 \tag{5.36}$$

Similarly, the inequality (5.37) can be obtained from (5.35).

$$\varepsilon_{n-1}\leq\frac{\theta a_n}{\theta a_n+b_{n-1}+b_n}<1 \tag{5.37}$$

In any case, $\varepsilon_{j0}<1$. Therefore, ε_i satisfies condition in (5.28), which means that the spatial oscillation phenomenon does not occur in the new FE formulation in (5.21).

The numerical characteristics of a simple FE formulation for 1-D consolidation are examined above. Numerical tests in the past also showed that the conclusion is still valid even for non-linear situation. It can be expected that the spatial oscillation due to small time increments could be avoided by carefully constructing the solution algorithms even for 3-D consolidation analysis.

5.5 ANALYSIS OF A CLAY CONSOLIDATION TEST

The FE formulation presented above has been encoded in a FE program. As an illustration of the solution and its reliability, it has been used to analyze a number of consolidation problems including those problems with closed-form solutions or measured results. One of these problems with measured results is a 1-D consolidation test with large thickness, that is, Test No. H4 published by Berre and Iversen (1972). This consolidation test was simulated by Garlanger (1972) and Yin and Graham (1996) using finite difference methods under constant loading condition.

The soil profile and location of three piezometer points of Test No. H4 are shown in Figure 5.2. The total initial thickness of Test No. H4 was 0.45 m (0.15 m for each of the three segments S_1, S_2, and S_3 as shown in Figure 5.2). The surface of the topmost segment (S_1) was freely drained and there was no drainage on the left, right, and bottom sides. The left and right sides are rigid boundary (no lateral displacement). Displacement and porewater flow are all in the vertical direction (1-D straining problem). Porewater pressures were measured at the bottom of each segment, that is, at points B, C, and D in Figure 5.2. Average strains were measured for each segment.

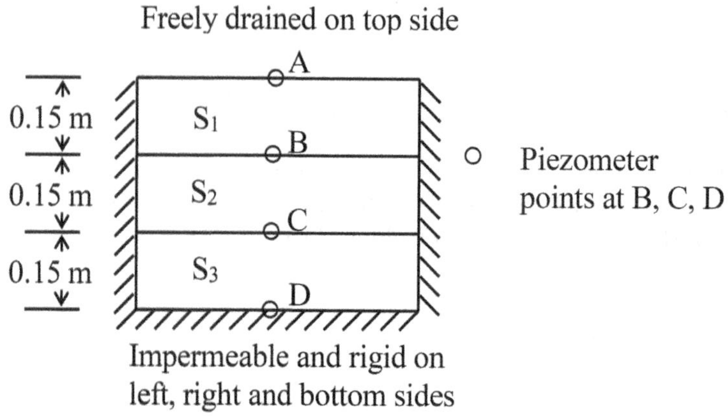

Figure 5.2 Soil profile and piezometer arrangement for Test No. H4. (After Berre and Iversen, 1972.)

The 1-D EVP model in (5.4) is used in the FE analysis of the consolidation of Test No. H4. All the material parameters used in FE modeling are listed in Table 5.1. These parameters have been determined using a different test on the same clay and published by Yin and Graham (1996).

Figure 5.3 (a) illustrates the calculated and measured pore water pressures using the EVP model under a two-step loading (from 53.4 to 89.2 kPa for the first step and from 89.2 to 134.7 kPa for the second step). Figure 5.3 (b) shows the calculated and measured strains. The computed and measured results are in good agreement.

5.6 CONSOLIDATION MODELING OF A CASE HISTORY

In this section, the FE formulation above will be used for the consolidation analysis of a case history – Berthierville Test Embankment in Quebec, Canada. The soil model suggested by Yin and Graham (1989, 1994) will be adopted in the analysis. According to Kabbaj *et al.* (1988), the stress/strain condition at the Berthierville Test Embankment was very close to 1-D straining (odometer test) condition. Therefore, the 1-D FE formulation for consolidation analysis is appropriate.

Table 5.1 Soil parameters for modeling consolidation Test No. H4 (after Yin & Graham, 1996)

k (m/min)	κ/V	λ/V	σ_0' (kPa)	ψ/V	t_0 (min)
1.0×10^{-7}	0.004	0.158	79.2	0.007	40

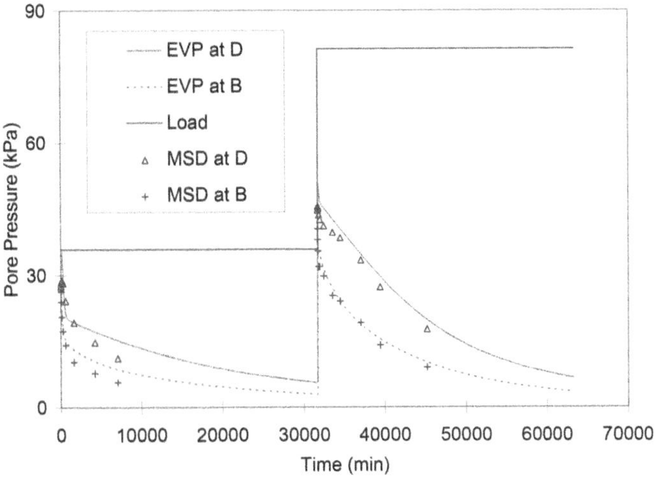

(a)

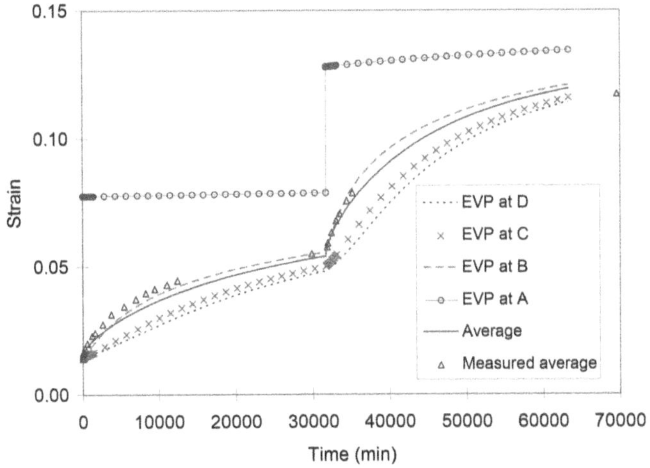

(b)

Figure 5.3 (a) Measured (MSD) and calculated (EVP) porewater pressure and (b) Measured (MSD) and calculated (EVP) strain for consolidation Test No. H4. (Test data from Berre and Iversen, 1972.)

5.6.1 Soil profile and parameters for consolidation modeling

The soil profile and measured data at Berthierville Test Embankment was reported by Kabbaj *et al.* (1988). The soil profile was identified based on the data of five vane shear test profiles, four piezocone test profiles, and three

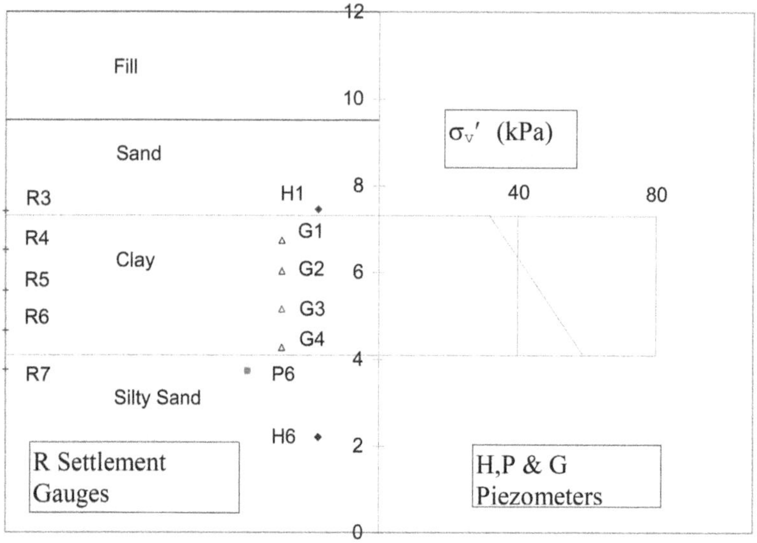

Figure 5.4 Setup of instrumentation and initial stresses. (Test data from Kabbaj et al. 1988.)

borehole records. Undisturbed clay samples were taken using both 50-mm diameter stationary piston sampler and 200-mm diameter Laval sampler. The clay samples were used for odometer testing (Leroueil *et al.*, 1988). The soil profile identified is shown in Figure 5.4.

The ground surface was essentially horizontal. Below the first 0.1 to 0.2 m of topsoil was a layer of fine to medium sand approximately 2.15-m thick. The unit weight of this material, when below the water table, was 19.7 kN/m^3, and its permeability coefficient measured in a triaxial test was 2.7×10^{-6} m/s. Underlying this fine to medium sand layer was a layer of grey silty clay about 3.2-m thick. The sand content in the silty clay was negligible and the silt content varied from 62 to 68%. The water content profiles obtained in two boreholes about 40 m apart showed the horizontal homogeneity of the site. The conventional odometer tests (MSL24) with a stress increment ratio $\Delta\sigma'/\sigma' = 0.5$ indicated an over-consolidation ratio between 1.1 and 1.3. Below this, the clay layer was a layer of silty sand. The embankment had a diameter of 24 m at the crest, a height 2.4 m, and 1:2 vertical to horizontal slope. This embankment was built in a four-day period. The stress/strain of the soils, especially in the clay layer, was very close to 1-D odometer condition (Leroueil *et al.*, 1988; Kabbaj *et al.*, 1988).

The setup of instrumentation and initial stresses are also shown in Figure 5.4. R3 to R7 are settlement gauges. H1, H6, P6, and G1 to G4 are piezometers. Except for some surface settlement plates, all the instrumentation was concentrated within a circle with a radius of 3.5 m around the central vertical axis. The water levels measured at H1, P6, and H6 are plotted in Figure 5.5. Since measured porewater data are not available after

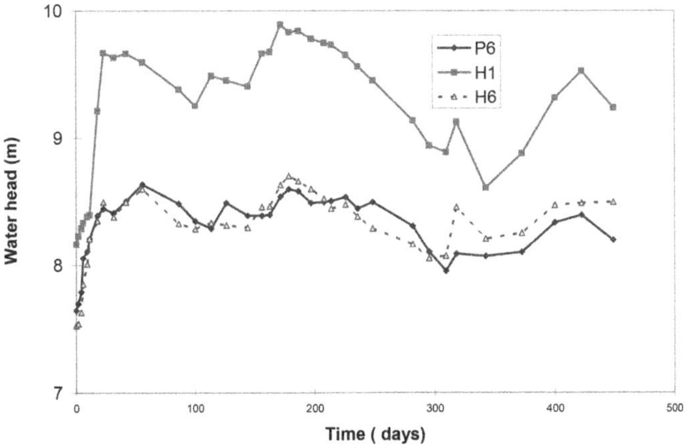

Figure 5.5 Groundwater conditions. (After Kabbaj et al. 1988.)

about 450 days from the start of construction, these data are extended horizontally in order to compare the computed settlement to the measured settlement for a period of more than 450 days. The total stress increase at the level of settlement gauge R5 is shown in Figure 5.6. This curve is the loading history used in the FE consolidation modeling.

The water heads at H1 and H6 are used as the boundary conditions. The silty sand layer underlying the clay layer is assumed to obey Terzaghi's model. The water head at P6 just before the start of construction is selected as the reference head for the calculation of the excess porewater pressure defined in Chapter 2. The element length and integration time step are constant. The model parameters used in the FE modeling are summarized in Table 5.2.

Another issue in the consolidation analysis is the determination of initial strains for the clay layer. According to the 1-D EVP model in (5.4), the creep

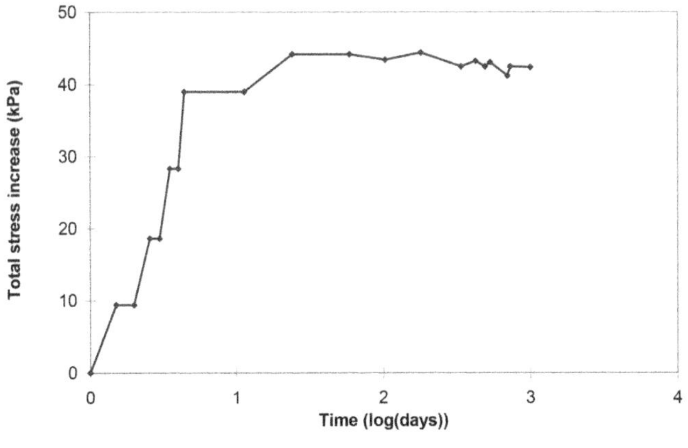

Figure 5.6 Total stress increase with time. (After Kabbaj et al. 1988.)

Table 5.2 Model parameters

Soil Layers	*Model Parameters*
Clay layer (Yin and Graham's 1-D EVP model)	$k = 1.08\times10^{-5}$ m/hour $\kappa/V = 0.01$, $\lambda/V = 0.184$, $\psi/V = 0.00911$, $\sigma'_o = 67$ kPa $t_o = 0.00805$ hour
Silty sand layer (Terzaghi's model)	For reference case: $m_v = 1.25\times10^{-4}$ /kPa $k = 2.16\times10^{-4}$ m/hour

rate is dependent on the stress-strain state. The determination of initial stress-strain states for the in-situ condition is important. The field stress and strain for the specimen tested and used for the calibration of the 1-D EVP model parameters are $\sigma' = 39$ kPa and $\varepsilon = 0$ for the specimen tested (see Figures 5.7 and 5.8). The initial strain for another stress, for example, point B in Figure 5.8, (or the same clay with higher vertical effective stress in the clay layer) can be calculated by assuming that point B has the same equivalent time as point A as shown in Figure 5.8. The dotted line in Figure 5.8 defines the initial strains for other stresses. In this way, the same 1-D EVP model calibrated using test data from one specimen can be used for modeling same soil with different initial stress-strain states.

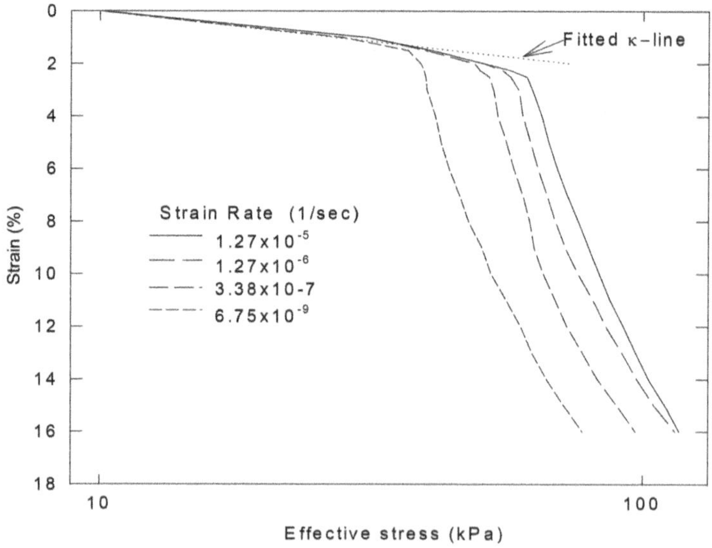

Figure 5.7 CRS tests at different strain rates and fitted κ-line. (Test data from Leroueil et al. 1988.)

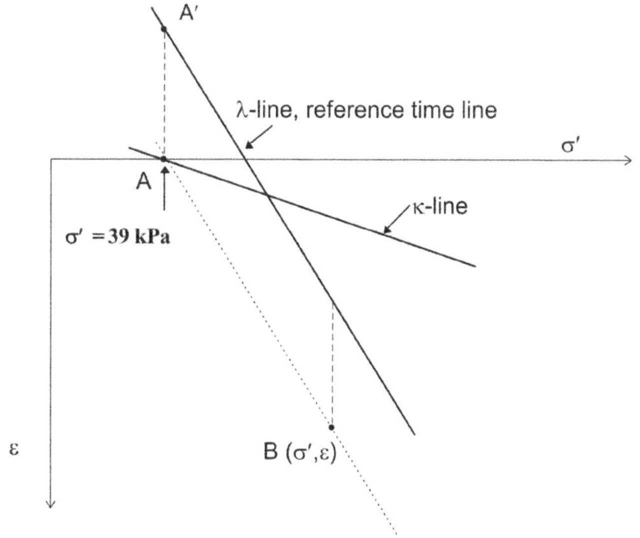

Figure 5.8 Illustration of the determination of initial strains for 1-D EVP modeling.

The equation for calculating the initial strain according to the idea above can be derived. From the definition of the 'equivalent time' (Yin and Graham, 1989, 1994) and zero strain at point A, we shall have

$$\frac{\lambda}{V}\ln\left(\frac{39}{\sigma_0'}\right)+\frac{\psi}{V}\ln\left(\frac{t_0+t_e}{t_0}\right)=0 \tag{5.38}$$

where t_e is the equivalent time for the soil creeping from point A′ to A.

For another point B, using (5.38), the initial strain can be calculated as

$$\varepsilon=\frac{\lambda}{V}\ln\left(\frac{\sigma'}{\sigma_0'}\right)+\frac{\psi}{V}\ln\left(\frac{t_0+t_e}{t_0}\right)=\frac{\lambda}{V}\ln\left(\frac{\sigma'}{39}\right) \tag{5.39}$$

In (5.39), (5.38) is used, that is, $\frac{\psi}{V}\ln\left(\frac{t_0+t_e}{t_0}\right)=\frac{\lambda}{V}\ln\left(\frac{39}{\sigma_0'}\right)$. This strain is set to be the initial strain corresponding to the initial in-situ stress in the 1-D EVP model. The increment from this initial strain after loading is the true strain for the calculation of compression of the clay layer.

5.6.2 Result of case studies

Figures 5.9 and 5.10 illustrate the isochrones of effective stresses and strains of the clay layer, respectively. As shown in Figure 5.9, the effective stress in

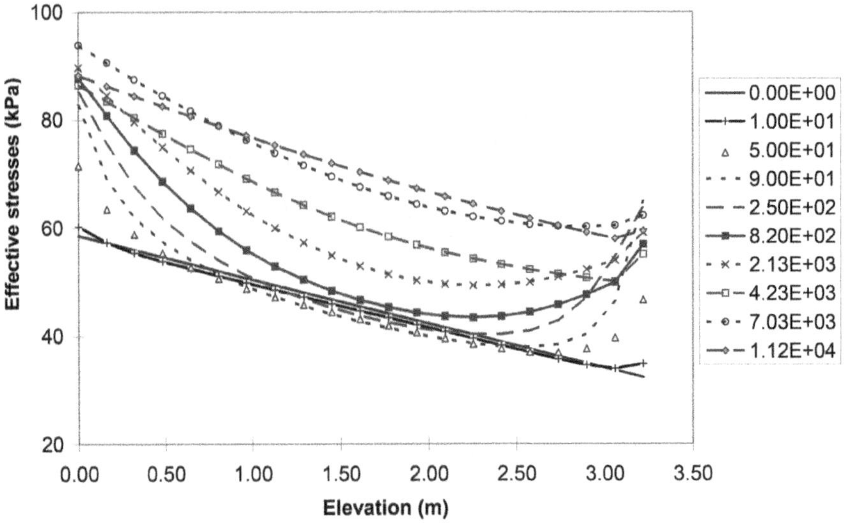

Figure 5.9 Isochrones of computed effective stresses of the clay.

the middle of the clay layer decreases a little at the initial stage of loading, which implies that the growth in the excess pore pressure is larger than the total stress increment. At the same time, the vertical strain remains almost unchanged at this stage. This porewater pressure increase phenomenon has been observed by Kabbaj *et al.* (1988). This phenomenon is due to the creep

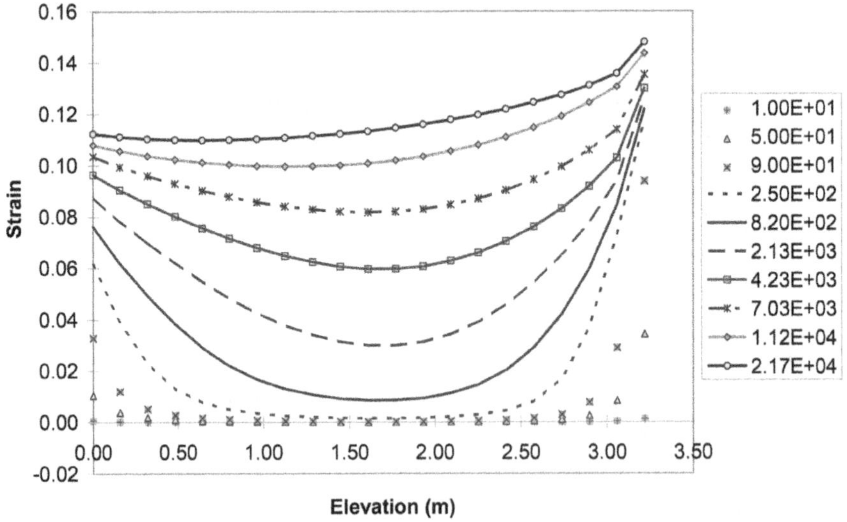

Figure 5.10 Isochrones of computed strains of the clay.

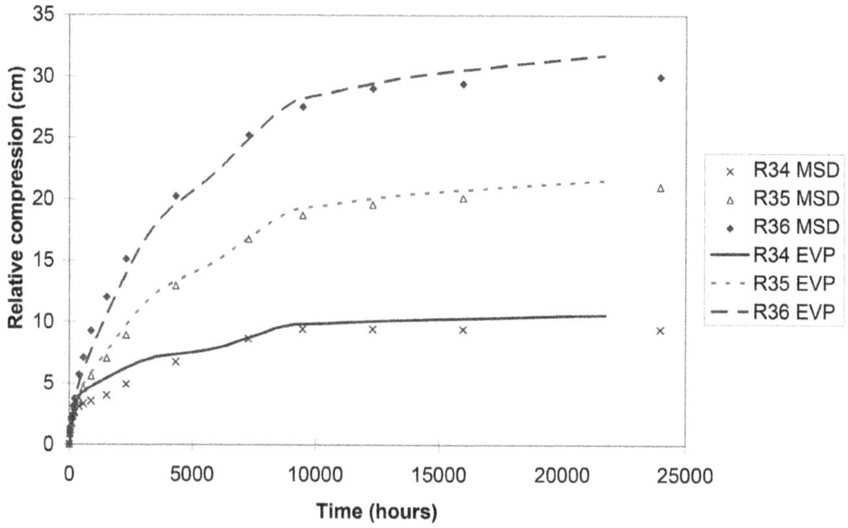

Figure 5.11 Computed and measured compression between R3 & R4, R3 & R5, and R3 & R6.

nature of clay as explained by Yin *et al.* (1994). The consolidation modeling in this chapter has simulated this phenomenon for the field condition. With the increase in time, the strain and effective stress accumulate as porewater pressure approaching steady state. Figure 5.11 displays the compression between R3 and R4, R3 and R5, R3 and R6, respectively (see Figure 5.4). As shown in Figure 5.11, the agreement between measured data and computed values is good. The computed porewater pressure values are, in general, larger than the measured values up to one and a half month after loading as shown in Figure 5.12. After this period, the computed porewater pressures are close to the measured values. The small deviation in the final stage is mainly due to extended time-history of water head in the upper sand layer and in the lower silty sand layer (extended since no data available after about 450 days). For the other time period (1.5 months to 16.5 months) when water head boundary conditions are known, the porewater pressure computed agrees quite well with the measured data.

Kabbaj *et al.* (1988) showed that the ratio of porewater pressure increase over vertical total stress increase ($\Delta u/\Delta\sigma$) was from 0.37 to 0.6 for stress increment up to 22 kPa. However, for fully saturated clay and under 1-D loading, the ratio ($\Delta u/\Delta\sigma$) shall be close to 1. The measured porewater pressure values were smaller than the values for fully saturated clay as assumed in the consolidation modeling in this chapter. The smaller values measured in the field may be due to unsaturation of the clay and/or radial porewater pressure dissipation in the clay layer.

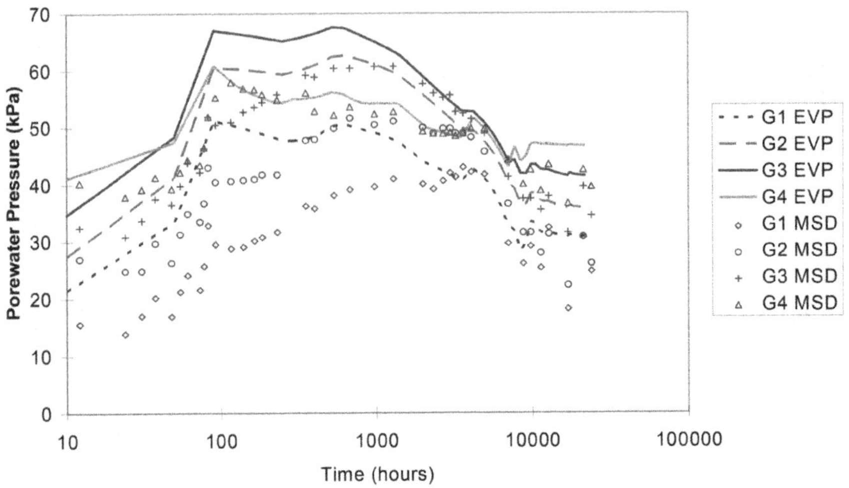

Figure 5.12 Computed and measured porewater pressures at G1, G2, G3, and G4.

REFERENCES

Adachi, T, and Oka, F, 1982, Constitutive equations for normally consolidated clay based on elasto-viscoplasticity. *Soils and Foundations*, **22**(4), pp. 57–70.

Adachi, T, Oka, F, and Mimura, M, 1987, Mathematical structure of an overstress elasto-viscoplastic model for clay. *Soils and Foundations*, **27**(3), pp. 31–42.

Berre, T, and Iversen, K, 1972, Oedometer test with different specimen heights on a clay exhibiting large secondary compression. *Géotechnique*, **22**(1), pp. 53–70.

Bjerrum, L, 1967, Engineering geology of Norwegian normally-consolidated marine clays as related to settlements of buildings. *Géotechnique*, **17**(2), pp. 83–118.

Booker, JR, and Small, JC, 1975, An investigation of the stability of numerical solutions of Biot's equations of consolidation. *International Journal of Solids and Structures*, **11**(7–8), pp. 907–917.

Dafalias, YF, 1982, Bounding surface elastoplasticity-viscoplasticity for particulate cohesive media. *Deformation and Failure of Granular Materials*, Vermeer, PA, and Luger, HJ (eds.), pp. 97–107, (Rotterdam: Balkema).

Desai, CS, Kuppusamy, T, Koutsoftas, DC, and Janardhanamm, RA, 1979, One-dimensional finite element procedure for nonlinear consolidation. *In Proceedings of the Third International Conference on Numerical Methods in Geomechanics*, pp. 143–148.

Garlanger, JE, 1972, The consolidation of soils exhibiting creep under constant effective stress. *Géotechnique*, **22**(1), pp. 71–78.

Hardin, BO, 1989, 1-D strain in normally consolidated cohesive soils. *Journal of Geotechnical Engineering Division, ASCE*, **115**(5), pp. 689–710.

Kabbaj, M, Oka, F, Leroueil, S, and Tavenas, F, 1986, Consolidation of natural clays and laboratory testing. *Consolidation of Soils: Testing and Evaluation*, Yong, RN, and Townsend, FC (eds.), ASTM Special Technical Publication 892, Philadelphia, Pa. pp. 378–404.

Kabbaj, M, Tavenas, F, and Leroueil, S, 1988, In situ and laboratory stress-strain relationships. *Géotechnique*, **38**(1), pp. 83–100.

Leroueil, S, 1988, Tenth Canadian Geotechnical Colloquium: Recent developments in consolidation of natural clays. *Canadian Geotechnical Journal*, **25**(1), pp. 85–107.

Leroueil, S, Kabbaj, M, and Tavenas, F, 1988, Study of the validity of a $\sigma_v' - \varepsilon_v - \dot{\varepsilon}_v$ model in *in situ* conditions. *Soils and Foundations*, **28**(3), pp. 13–25.

Leroueil, S, Kabbaj, M, Tavenas, F, and Bouchard, R, 1985, Stress-strain-strain rate relation for the compressibility of sensitive natural clays. *Géotechnique*, **35**(2), pp. 159–180.

Lewis, RW, and Schrefler, BA, 1998, *The Finite Element Method in the Static and Dynamic Deformation and Consolidation of Porous Media,* (Chichester: John Wiley).

Sandhu, RS, and Wilson, EL, 1969, Finite-element analysis of seepage in elastic media. *Journal of the Engineering Mechanics Division*, **95**(3), pp. 641–652.

Schiffman, RL, and Arya, SK, 1977, One-dimensional consolidation. *Numerical Methods in Geotechnical Engineering*, Desai, CS, and Christian, JT (eds.), pp. 364–398, (London: McGraw-Hall).

Schofield, A, and Wroth, P, 1968, *Critical State Soil Mechanics*, (New York: McGraw-Hill).

Siriwardane, HJ, and Desai, CS, 1981, Two numerical schemes for nonlinear consolidation. *International Journal for Numerical Methods in Engineering*, **17**(3), pp. 405–426.

Terzaghi, K, 1943, *Theoretical Soil Mechanics*, (New York: Wiley).

Thomas, HR, and Harb, HM, 1990, Analysis of normal consolidation of viscous clay. *Journal of Engineering Mechanics*, **116**(9), pp. 2035–2053.

Vermeer, PA, and Verruijt, A, 1981, An accuracy condition for consolidation by finite elements. *International Journal for Numerical and Analytical Methods in Geomechanics*, **5**(1), pp. 1–14.

Yin, JH, and Graham, J, 1989, Viscous-elastic-plastic modelling of one-dimensional time-dependent behaviour of clays. *Canadian Geotechnical Journal*, **26**(2), pp. 199–209.

Yin, JH, and Graham, J, 1994, Equivalent times and one-dimensional elastic visco-plastic modelling of time-dependent stress-strain behaviour of clays. *Canadian Geotechnical Journal*, **31**(1), pp. 42–52.

Yin, JH, and Graham, J, 1996, Elastic visco-plastic modelling of one-dimensional consolidation. *Géotechnique*, **46**(3), pp. 515–527.

Yin, J, Graham, J, Clark, JI, and Gao, L, 1994, Modelling unanticipated pore-water pressures in soft clays. *Canadian Geotechnical Journal*, **31**(5), pp. 773–778.

Zhu, GF, and Yin, J-H, 1999, Finite element analysis of consolidation of layered clay soils using an elastic visco-plastic model. *International Journal for Numerical and Analytical Methods in Geomechanics*, **23**, pp. 355–374.

Zhu, GF, and Yin, J-H, 2000, Elastic visco-plastic consolidation modelling of Berthierville test embankment. *International Journal for Numerical and Analytical Methods in Geomechanics*, **24**(5), pp. 491–508.

Zhu, GF, Yin, JH, and Luk, ST, 2004, Numerical characteristics of a simple finite element formulation for consolidation analysis. *Communications in Numerical Methods in Engineering,* **20**, pp. 767–775.

Chapter 6

Consolidation of soil with vertical drain

6.1 INTRODUCTION

Precompression of soft soils through the use of preloads and surcharge loads with vertical drains is commonly used in ground improvements. Using these techniques, the following objectives can be achieved:

a. increase the undrained shear strength of soft soil;
b. decrease the post-construction settlements;
c. minimize the differential settlements;
d. reduce the secondary consolidation settlements; and
e. shorten the time required for foundation stabilization.

Vertical drains are especially effective in stratified soils in which the horizontal permeability is generally larger than that in the vertical direction. Furthermore, for soils with isolated sand or silt lenses, vertical drains may connect these lenses to form an internal drainage network speeding the process of consolidation.

The use of precompression techniques became popular in the early 1940s in connection with the use of vertical drains for highway construction, and probably stimulated the use of preloading without vertical drains (Johnson, 1970). Although vertical sand drains and prefabricated drains were all first introduced in mid 1930s (Porter, 1936; Kjellman, 1948), the former was mainly employed until early 1970s, when the latter became more popular and most widely used for economic and other reasons.

Since the application of vertical drains, quite a few design theories have been developed (Barron, 1948; Yoshikuni and Nakanodo, 1974; Basak and Madhav, 1978; Hansbo, 1981, 1997; Onoue, 1988; Zeng and Xie, 1989; Zhu and Yin, 2001, 2004). The most well-known study on this topic was probably carried out by Barron (1948). He assumed two types of vertical strains that might occur in the subsoil: (a) 'free vertical strain' resulting from a uniform distribution of surface load and (b) 'equal vertical strain' resulting from imposing the same vertical deformation on the surface for a uniform soil.

These concepts play an important role in developing vertical drain design theories. In this chapter, the consolidation theories of soil with vertical drains are discussed for the case of linear one-dimensional (1-D) constitutive relation. The consolidation analysis for the cases of non-linear 1-D constitutive relations is discussed in Chapter 8.

6.2 SIMPLIFICATION OF THE CONSOLIDATION PROBLEM

6.2.1 Basic assumptions

The purpose of a vertical drain well is to provide an easier or shorter path for the excess water to follow as it is squeezed out of a soil layer during consolidation. Because the vertical drain is merely an auxiliary device, the consolidation of a soil layer is the fundamental subject to be considered. To obtain the governing equation for consolidation of soil with vertical drains, it is assumed that (a) the soil is fully saturated, (b) water and soil particles are incompressible, (c) Darcy's law is valid, (d) strains are small, (e) all compressive strains within the soil mass occur in the vertical direction, and (f) the coefficient of compressibility is constant (Terzaghi, 1943; Barron, 1948).

Wherever drain wells are installed, construction operations involve penetration, displacement of the soil by the mandrel, and consequently disturbance of the soil around the drain. Two principal disturbing mechanisms exist: first, displacement of the soil to create space for the mandrel; second, the dragging of soil by the sides of the mandrel. Substantial reduction in the soil permeability occurs around the drain due to this disturbance. The fast consolidation of the soil adjacent to the well periphery (Zhu and Yin, 2000) will also cause a zone of reduced permeability due to the effective stress increase. The term *smear* is used to describe these effects. These detrimental effects of smear on the efficiency of a vertical drain are grossly taken into account by assuming that there will be a uniform smear zone around the drain for application purposes.

When water flows in the vertical drains, head losses will occur due to the internal resistance of vertical drains, which is normally termed as *well resistance* (Barron, 1948; Bhide, 1979; Hansbo, 1981). The effect of well resistance is not excessive for cases where the vertical drain spacing is comparable to the half-depth of the soil layer if the vertical drains are properly installed. However, the well resistance may well have made a significant contribution to the lack of acceleration of the consolidation process if the half-depth of the drained stratum is considerably in excess of the drain spacing (Casagrande and Poulos, 1969; Atkinson and Eldred, 1981). The well resistance is described by a coefficient of permeability of the drain.

For ideal drain wells, both effects of smear and well resistance are ignored.

6.2.2 Unit cell

When vertical drains are used, generally a large number of drain wells are installed in either triangular or square patterns. An exact analysis should include the load distribution and the effect of each drain well on the rate of consolidation at any point in the foundation. However, such a solution would be extremely cumbersome and rather academic, which is not suitable for design purposes. The related consolidation problem is simplified to an axi-symmetric unit cell (Barron, 1948), as shown in Figure 6.1 in which a drain well is enclosed by a cylinder of soil. Figure 6.1 also illustrates the equivalent radius of the soil cylinder based on the same total area for different installation patterns.

For a triangular pattern, the radius, r_e, of the equivalent soil cylinder is

$$r_e = \sqrt{\frac{\sqrt{3}}{2\pi}} \times \text{Drain Spacing} = 0.525 \times \text{Drain Spacing} \tag{6.1}$$

For a square pattern, the radius of the equivalent soil cylinder is

$$r_e = \frac{1}{\sqrt{\pi}} \times \text{Drain Spacing} = 0.564 \times \text{Drain Spacing} \tag{6.2}$$

6.2.3 Equivalent drain radius of band-shaped vertical drain

The radius r_d of sand drains, or their modern derivatives such as sand wicks or plastic tube drains, can be easily determined from the size of the mandrel, which is usually circular in cross section. The most common radius of sand drains is in the range from 15 to 25 cm, which is required to guarantee a well-constructed continuous drain. For prefabricated drains, however, the situation is different. Owing to the band shape of prefabricated drains, the flow pattern around the drain is considerably altered from the cylindrical case. Therefore, an equivalent drain radius ought to be calculated.

Kjellman (1948) first suggested that the equivalent radius could be estimated from a consideration of the drain surface area (circumference of its cross section). Hansbo (1979) showed that the average degrees of consolidation from a band-shaped drain and from a circular drain are practically the same when the equivalent radius is calculated from (6.3).

$$r_d = \frac{b+h}{\pi} \tag{6.3}$$

where b is the width of prefabricated drain, h is the thickness of prefabricated drain, and r_d is the equivalent radius of the prefabricated drain.

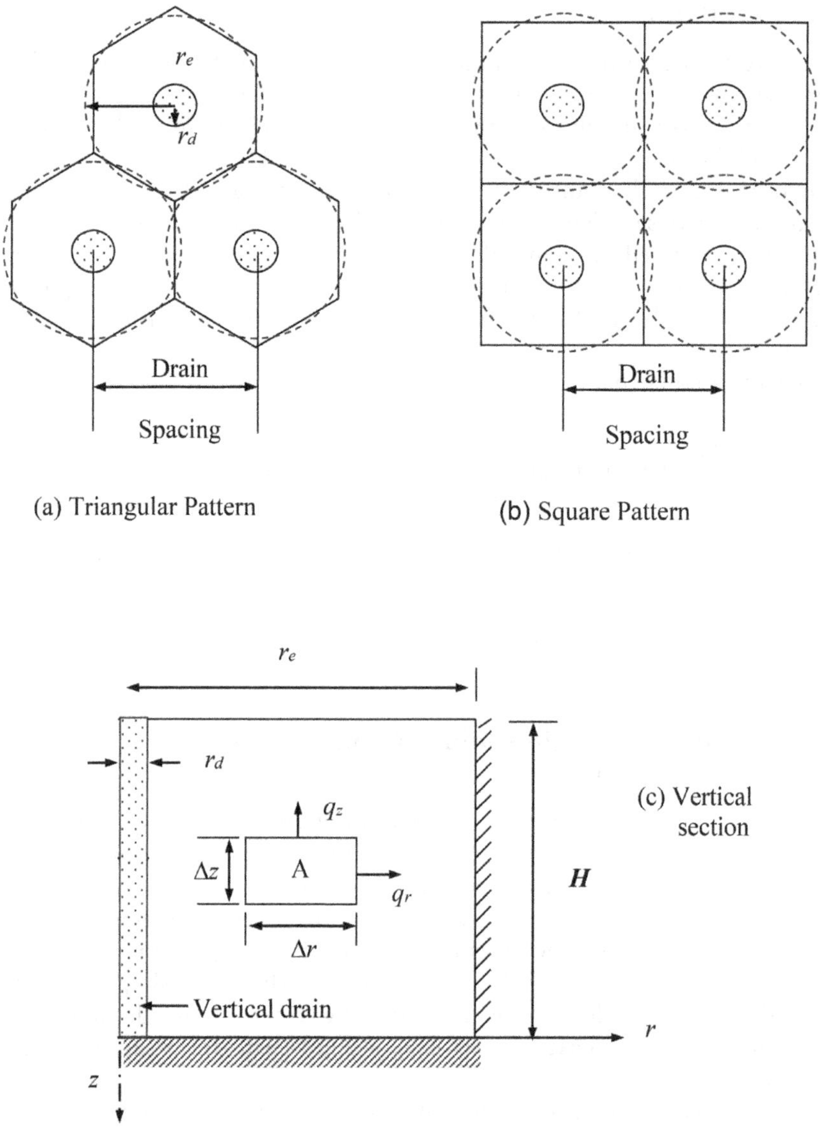

Figure 6.1 Typical drain patterns and the radiusof equivalent soil cylinder (a) triangular pattern, (b) square pattern, and (c) vertical section.

Atkinson and Eldred (1981) proposed that a factor of $\pi/4$ should be applied to (6.3) to take account of the throttle, which must occur close to the drain corner. That is, r_d should be calculated using (6.4).

$$r_d = \frac{b+h}{4} \tag{6.4}$$

Other studies (Long and Alvaro, 1994) list several equations that might be used in computing the equivalent radius, but made no recommendation regarding selection. The reported equations are

$$r_d = \sqrt{\frac{bh}{\pi}} \tag{6.5}$$

$$r_d = \sqrt{\frac{b'h'}{\pi}} \tag{6.6}$$

$$r_d = \frac{b'+h'}{4} \tag{6.7}$$

where $2(b'+h')$ = free or open circumference of the drain; and $b'h'$ = free or open cross section of the drain. The reduced dimensions, b' and h' take into account some clogging of the drain surface, but are difficult to evaluate.

Long and Alvaro (1994) suggested that (6.8) be used to compute the equivalent radius. Equation (6.8) was determined using an electrical analogue field plotter.

$$r_d = \frac{b+h}{4} + \frac{h}{10} \tag{6.8}$$

Considering the assumptions in deriving these equations, (6.4) or (6.8) is normally adopted in calculating the equivalent radius of the band-shaped drain.

6.3 BASIC EQUATIONS AND SOLUTIONS FOR THE CASE OF FREE STRAIN

In addition to the assumptions in Section 6.2, it is further assumed that (a) the load is uniform over the circular zone of influence for each well and that the differential settlements occurring over such a zone, as consolidation progresses, have no effect on redistribution of stresses by arching of the fill; and (b) shear strains developed in the foundation by differential settlements will have no effect on the consolidation process.

The pore fluid moves through the soil skeleton in accordance with Darcy's law and is expressed as

$$\begin{cases} q_r = -\dfrac{k_r}{\gamma_w}\dfrac{\partial u_e}{\partial r} \\ q_z = -\dfrac{k_z}{\gamma_w}\dfrac{\partial u_e}{\partial z} \end{cases} \tag{6.9}$$

where r is radial co-ordinate; z is vertical co-ordinate; q_r and q_z are the radial (horizontal) and vertical flow rates, respectively; and k_r and k_z are radial (horizontal) and vertical coefficients of permeability, respectively.

Consider an infinitesimal (ring) element A as shown in Figure 6.1. The rate of outflow of water from element A is

$$\text{Net outflow rate of water} = 2\pi \frac{\partial (rq_r)}{\partial r} drdz + 2\pi r \frac{\partial q_z}{\partial z} drdz \tag{6.10}$$

which must equal the volume compression rate of the element

$$\text{Volume compression rate} = 2\pi r \frac{\partial \varepsilon_v}{\partial t} drdz \tag{6.11}$$

so that

$$\frac{\partial q_r}{\partial r} + \frac{q_r}{r} + \frac{\partial q_z}{\partial z} = \frac{\partial \varepsilon_v}{\partial t} = \frac{\partial \varepsilon_z}{\partial t} \tag{6.12}$$

where ε_v and ε_z are the volume and vertical strains (positive for compression), respectively. Since all compressive strains within the soil mass occur in the vertical direction, these two strains are equal.

Substituting $\varepsilon_z = m_v(\sigma - u_e)$ and (6.9) into (6.12) gives

$$\frac{\partial u_e}{\partial t} = c_r \left(\frac{\partial^2 u_e}{\partial r^2} + \frac{1}{r} \frac{\partial u_e}{\partial r} \right) + c_v \frac{\partial^2 u_e}{\partial z^2} + \frac{\partial \sigma}{\partial t} \tag{6.13}$$

where σ is vertical total stress increase; $c_r = \frac{k_r}{\gamma_w m_v}$ is horizontal (or radial) consolidation coefficient; $c_v = \frac{k_z}{\gamma_w m_v}$ is vertical consolidation coefficient. Equation (6.13) is similar to the equation derived by Barron (1948) except for the non-homogenous term $\frac{\partial \sigma}{\partial t}$.

The vertical total stress increase is assumed to vary linearly with time and remain unchanged after time t_c (see Figure 6.2) in the following sections, that is,

$$\sigma(r,z,t) = \sigma_0 \min\left(1, \frac{t}{t_c}\right) \tag{6.14}$$

where σ_o is a constant (final vertical total stress increment). The 'min' means taking the minimum value between 1 and $\frac{t}{t_c}$.

6.3.1 Solutions for an ideal drain

The boundary conditions for the soil layer in Figure 6.1 are expressed in (6.15).

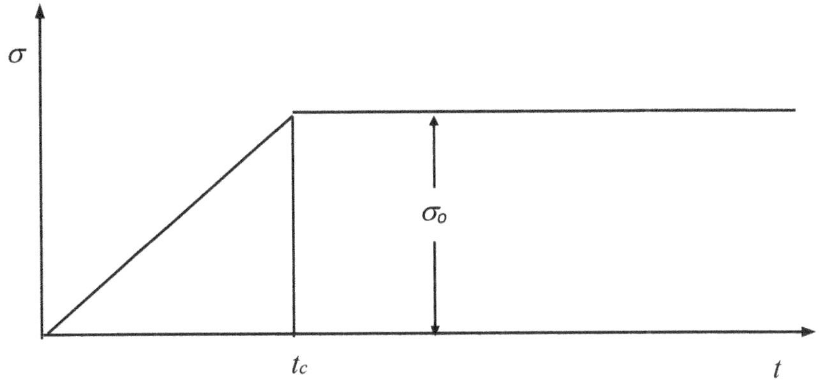

Figure 6.2 Variation of the vertical total stress.

$$u_e\, |_{r=r_d} = 0, \left.\frac{\partial u_e}{\partial r}\right|_{r=r_e} = 0 \tag{6.15a}$$

$$u_e\, |_{z=0} = 0, \left.\frac{\partial u_e}{\partial z}\right|_{z=H} = 0 \tag{6.15b}$$

Equation (6.15) means that water is freely drained at top and along well ($r = r_d$), but impermeable at the bottom and at $r = r_e$. If drainage is available at the bottom of the soil, the drainage path H is halved. The initial excess porewater pressure is assumed to be zero.

Using the method of separation of variables, the consolidation (6.13) under the given loading condition in (6.14) and the boundary conditions in (6.15) can be solved. The solutions are

$$u_e = \sum_{m,n=1}^{\infty} A_{mn}(T) R_m(r) \sin\left(\lambda_n \frac{z}{H}\right) \tag{6.16}$$

$$R_m(r) = Y_1(N\mu_m) J_0\left(\mu_m \frac{r}{r_d}\right) - J_1(N\mu_m) Y_0\left(\mu_m \frac{r}{r_d}\right) \tag{6.17}$$

$$A_{mn}(T) =$$

$$\begin{cases} \dfrac{B_m(\mu_1^2 + \lambda_1^2 L)}{\lambda_n(\mu_m^2 + \lambda_n^2 L)T_c}\left[1 - \exp\left(-\dfrac{\mu_m^2 + \lambda_n^2 L}{\mu_1^2 + \lambda_1^2 L} T\right)\right] & T \le T_c \\ \dfrac{B_m(\mu_1^2 + \lambda_1^2 L)}{\lambda_n(\mu_m^2 + \lambda_n^2 L)T_c}\left[1 - \exp\left(-\dfrac{\mu_m^2 + \lambda_n^2 L}{\mu_1^2 + \lambda_1^2 L} T_c\right)\right]\exp\left[-\dfrac{\mu_m^2 + \lambda_n^2 L}{\mu_1^2 + \lambda_1^2 L}(T - T_c)\right] & T \ge T_c \end{cases} \tag{6.18}$$

$$T = \left(c_r \frac{\mu_1^2}{r_d^2} + \frac{c_v \pi^2}{4H^2} \right) t = \left(\mu_1^2 + \frac{\pi^2}{4} L \right) \frac{c_r t}{r_d^2} \tag{6.19}$$

$$T_c = \left(c_r \frac{\mu_1^2}{r_d^2} + \frac{c_v \pi^2}{4H^2} \right) t_c = \left(\mu_1^2 + \frac{\pi^2}{4} L \right) \frac{c_r t_c}{r_d^2} \tag{6.20}$$

$$\lambda_n = n\pi - \frac{1}{2}\pi \tag{6.21}$$

$$B_m = \frac{4\sigma_0 r_d R_m(r_d)\pi^2}{4 - [\pi r_d R_m(r_d)]^2} \tag{6.22}$$

$$r_d R'_m(r_d) = \mu_m [J_1(N\mu_m) Y_1(\mu_m) - Y_1(N\mu_m) J_1(\mu_m)] \tag{6.23}$$

$$N = \frac{r_e}{r_d} \tag{6.24}$$

$$L = \frac{c_v r_d^2}{c_r H^2} \tag{6.25}$$

where N is the ratio of the equivalent radius over well radius; L is a dimensionless parameter; T is a (dimensionless) time factor; T_c is the time factor corresponding to construction time t_c; J_0 and J_1 are first-order and second-order Bessel functions of the first kind, respectively; and Y_0 and Y_1 are first-order and second-order Bessel functions of the second kind, respectively. The quantity μ_m is the m-th positive root of the following (6.26):

$$Y_1(N\mu_m) J_0(\mu_m) - J_1(N\mu_m) Y_0(\mu_m) = 0 \tag{6.26}$$

The first five roots (eigenvalues μ_1 to μ_5) of (6.26) are listed in Table 6.1 for different N values.

The average degree of consolidation $U(T)$ is defined as

$$U(T) = \frac{\overline{S(t)}}{\overline{S_f}} = \frac{\int_{r_d}^{r_e} r\,dr \int_0^H m_v(\sigma - u_e)\,dz}{\int_{r_d}^{r_e} r\,dr \int_0^H m_v \sigma_{t=\infty}\,dz} \tag{6.27}$$

where $\overline{S_f}$ and $\overline{S(t)}$ are average final settlement and settlement at t. Substituting (6.14) and (6.16) into (6.27) gives

Table 6.1 Eigenvalues μ_1 to μ_5 for different N

N	μ_1	μ_2	μ_3	μ_4	μ_5
5	0.2823580	1.1392100	1.9391800	2.7312100	3.5204000
6	0.2180720	0.9074940	1.5485600	2.1828000	2.8145600
7	0.1765100	0.7534490	1.2883800	1.8173400	2.3441000
8	0.1475980	0.6437050	1.1027000	1.5564000	2.0081200
9	0.1264120	0.5615990	0.9635600	1.3607800	1.7561900
10	0.1102690	0.4978840	0.8554290	1.2086800	1.5602900
11	0.0975920	0.4470220	0.7689930	1.0870500	1.4036000
12	0.0873916	0.4054910	0.6983290	0.9875770	1.2754300
13	0.0790199	0.3709470	0.6394870	0.9047130	1.1686500
14	0.0720346	0.3417700	0.5897340	0.8346240	1.0783100
15	0.0661241	0.3168040	0.5471200	0.7745720	1.0009000
16	0.0610626	0.2952020	0.5102130	0.7225450	0.9338180
17	0.0566830	0.2763290	0.4779420	0.6770390	0.8751390
18	0.0528588	0.2597000	0.4494860	0.6369010	0.8233750
19	0.0494927	0.2449390	0.4242070	0.6012360	0.7773730
20	0.0465086	0.2317500	0.4016030	0.5693350	0.7362220
21	0.0438462	0.2198950	0.3812720	0.5406350	0.6991950
22	0.0414571	0.2091820	0.3628870	0.5146770	0.6657010
23	0.0393022	0.1994540	0.3461840	0.4910860	0.6352580
24	0.0373493	0.1905820	0.3309410	0.4695540	0.6074680
25	0.0355718	0.1824580	0.3169750	0.4498220	0.5820000
26	0.0339475	0.1749920	0.3041340	0.4316750	0.5585740
28	0.0310871	0.1617380	0.2813210	0.3994260	0.5169390
30	0.0286505	0.1503330	0.2616730	0.3716420	0.4810610
32	0.0265515	0.1404170	0.2445760	0.3474570	0.4498250
34	0.0247258	0.1317180	0.2295640	0.3262140	0.4223840
36	0.0231241	0.1240240	0.2162780	0.3074090	0.3980880
38	0.0217083	0.1171730	0.2044380	0.2906450	0.3764270
40	0.0204483	0.1110320	0.1938200	0.2756070	0.3569930
50	0.0157890	0.0879276	0.1538070	0.2189020	0.2836850
60	0.0128038	0.0727350	0.1274340	0.1814910	0.2352950
70	0.0107363	0.0619918	0.1087500	0.1549680	0.2009720
80	0.0092239	0.0539969	0.0948257	0.1351880	0.1753660
90	0.0080720	0.0478175	0.0840496	0.1198720	0.1555330
100	0.0071669	0.0428995	0.0754640	0.1076630	0.1397200
200	0.0033038	0.0210817	0.0372540	0.0532549	0.0691909

$$U(T)=\begin{cases} \dfrac{T}{T_c}-\displaystyle\sum_{m=1}^{\infty}\sum_{n=1,3,5,\dots}\frac{32C_m\left(\mu_1^2+\dfrac{\pi^2}{4}L\right)}{n^2\pi^2\left(\mu_m^2+\dfrac{n^2\pi^2}{4}L\right)T_c}\left[1-\exp\left(-\frac{\mu_m^2+\dfrac{n^2\pi^2}{4}L}{\mu_1^2+\dfrac{\pi^2}{4}L}T\right)\right] & T\le T_c \\ 1-\displaystyle\sum_{m=1}^{\infty}\sum_{n=1,3,5,\dots}\frac{32C_m\left(\mu_1^2+\dfrac{\pi^2}{4}L\right)}{n^2\pi^2\left(\mu_m^2+\dfrac{n^2\pi^2}{4}L\right)T_c}\left[1-\exp\left(-\frac{\mu_m^2+\dfrac{n^2\pi^2}{4}L}{\mu_1^2+\dfrac{\pi^2}{4}L}T_c\right)\right] & \\ \times\exp\left[-\dfrac{\mu_m^2+\dfrac{n^2\pi^2}{4}L}{\mu_1^2+\dfrac{\pi^2}{4}L}(T-T_c)\right] & T\ge T_c \end{cases} \tag{6.28}$$

where

$$C_m=\frac{[\pi r_d R_m'(r_d)]^2}{(N^2-1)\mu_m^2\{4-[\pi r_d R_m'(r_d)]^2\}} \tag{6.29}$$

Equation (6.28) can be used to calculate the average degree of consolidation of soils with vertical drains. If the vertical total stress σ is composed of several ramp loads, the superposition method may be used to calculate the excess porewater pressure and the average degree of consolidation. It should be noted that (6.28) is also valid if the final total stress σ_0 is linear with depth and water is freely drained at the bottom of the soil.

6.3.1.1 Special case of a suddenly applied loading

For the situation of a suddenly applied loading, (6.13) becomes a homogeneous equation, and the above problem can be solved by combining the solution for radial flow and the solution for vertical flow as the following theorem states:

Theorem 6.1 (Carrillo, 1942). If $u_1(r, t)$ satisfies (6.15a) and (6.30)

$$\frac{\partial u_1}{\partial t}=c_r\left(\frac{\partial^2 u_1}{\partial r^2}+\frac{1}{r}\frac{\partial u_1}{\partial r}\right)\quad r_d<r<r_e \tag{6.30}$$

and the initial condition $u_1|_{t=0}=\sigma_0$, and $u_2(z, t)$ satisfies (6.15b) and (6.31)

$$\frac{\partial u_2}{\partial t}=c_v\frac{\partial^2 u_2}{\partial z^2} \tag{6.31}$$

and the initial condition $u_2 \mid_{t=0} = \sigma_0$, then $u = \frac{u_1 u_2}{\sigma_0}$ is necessarily a solution of (6.15) and (6.32)

$$\frac{\partial u}{\partial t} = c_r \left(\frac{\partial^2 u}{\partial r^2} + \frac{1}{r}\frac{\partial u}{\partial r} \right) + c_v \frac{\partial^2 u}{\partial z^2} \tag{6.32}$$

and the initial condition $u \mid_{t=0} = \sigma_0$.

To prove the theorem, let us substitute the function $u = \frac{u_1 u_2}{\sigma_0}$ in (6.32):

$$\frac{\partial}{\partial t}\left(\frac{u_1 u_2}{\sigma_0} \right) = c_r \left[\frac{\partial^2}{\partial r^2}\left(\frac{u_1 u_2}{\sigma_0} \right) + \frac{1}{r}\frac{\partial}{\partial r}\left(\frac{u_1 u_2}{\sigma_0} \right) \right] + c_v \frac{\partial^2}{\partial z^2}\left(\frac{u_1 u_2}{\sigma_0} \right) \tag{6.33}$$

or

$$u_2 \frac{\partial u_1}{\partial t} + u_1 \frac{\partial u_2}{\partial t} = u_2 c_r \left(\frac{\partial^2 u_1}{\partial r^2} + \frac{1}{r}\frac{\partial u_1}{\partial r} \right) + u_1 c_v \frac{\partial^2 u_2}{\partial z^2} \tag{6.34}$$

since u_1 does not depend on z, and u_2 does not depend on r.

Inserting in the left side of (6.34) the values given in (6.30) and (6.31), the same right side is obtained. Since u satisfies the initial condition $u \mid_{t=0} = \sigma_0$ and the boundary condition (6.15), the theorem is proved.

Substituting $u = \frac{u_1 u_2}{\sigma_0}$ in (6.27) gives

$$U(T) = 1 - \frac{\int_{r_d}^{r_e} r\,dr \int_0^H m_v \frac{u_1 u_2}{\sigma_0} dz}{\int_{r_d}^{r_e} r\,dr \int_0^H m_v \sigma_0\, dz} = 1 - \frac{\int_{r_d}^{r_e} m_v u_1 r\,dr}{\int_{r_d}^{r_e} m_v \sigma_0 r\,dr} \times \frac{\int_0^H m_v u_2\, dz}{\int_0^H m_v \sigma_0\, dz} \tag{6.35}$$

Thus

$$U(T) = 1 - (1 - U_r)(1 - U_v) \tag{6.36}$$

where U_r is the average degree of consolidation for radial flow only; U_v is the average degree of consolidation for vertical flow only. This formula is called Carrillo's (1942) formula.

6.3.1.2 Influence of N and L on average degree of consolidation

Normally, the value of parameter N in (6.24) is in the range of 5 to 80, and the value of parameter L in (6.25) is in the range of 0 to 0.01. These two parameters obviously have a very large influence on the average degree of consolidation. However, if the time factor T is normalized using (6.19), the average degree of consolidation shows very good normalized feature with respect to N. Figures 6.3–6.4 show some typical relations between normalized time

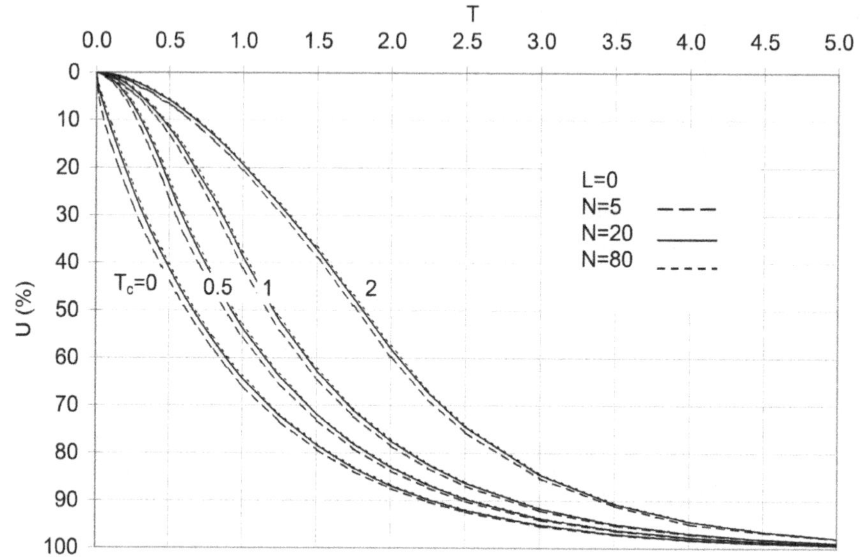

Figure 6.3 Normalized time factor *T* vs. the average degree of consolidation *U* for *L* = 0.

factor *T* and average degree of consolidation *U* for different *N*, *L*, and T_c. Clearly, the difference is very small for different *N*-values. The parameter *L* has a little bit larger influence on the normalized relations than *N*, as shown in Figure 6.5. However, the difference is still small. For the whole range calculated ($N = 5$ to 80 and $L = 0$ to 0.01), the difference of average degree of

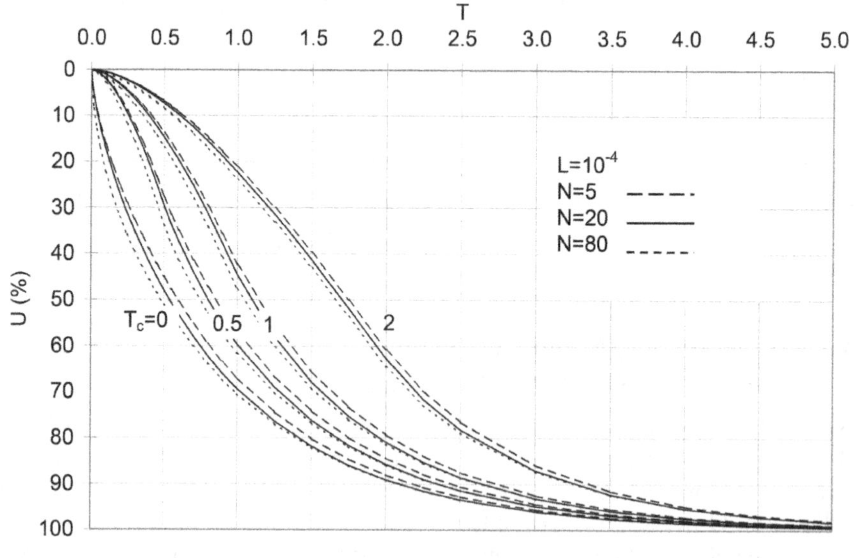

Figure 6.4 Normalized time factor *T* vs. the average degree of consolidation *U* for $L = 1\times10^{-4}$.

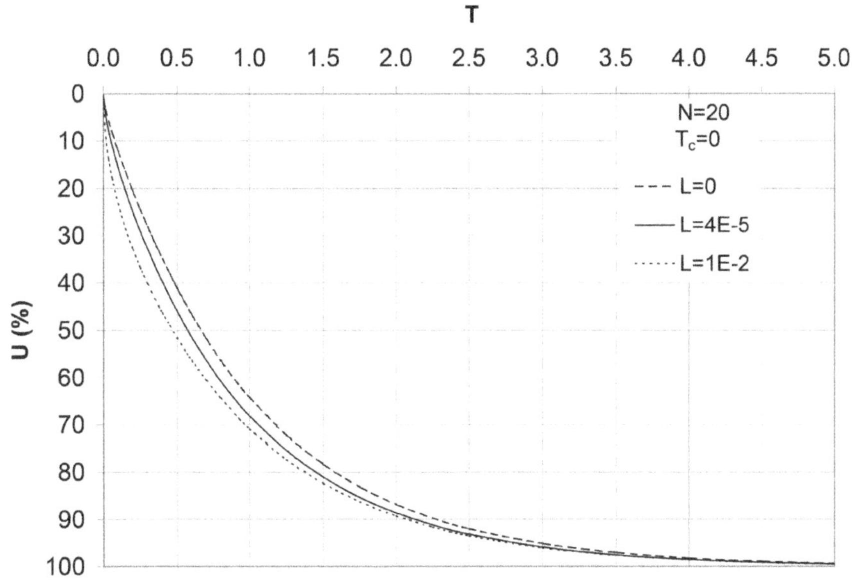

Figure 6.5 Influence of L for $T_c = 0$ and $N = 20$.

consolidation falls in a band with width less than 13%. This indicates that the relationship of average degree of consolidation U vs. time factor T is approximately independent of the dimensionless parameters N and L. It is noted that the influence of N and L is included in the normalized time factor T in (6.19) since μ_1 is related to N in Table 6.1.

The normalized time T makes that the relationship of U vs. T is approximately dependent on the construction time factor T_c only. In this way, the total number of charts of U vs. T is largely reduced. For practical applications, the relationships of U vs. T for a wide of ranges of T_c-values ($N = 10$, 40 and $L = 0$, 2×10^{-5}, 5×10^{-4}, 1×10^{-2}) are calculated and presented in Figures 6.6–6.9.

6.3.1.3 Design charts for vertical drains considering construction time

Vertical drains are usually spaced in either a square or a triangular pattern. As the object of vertical drain installation is to reduce the length of the drainage path, the spacing of the drain is the most important design consideration. Most design charts available use a dimensionless time factor (Barron, 1942; Hansbo, 1979; Yeung, 1994) obtained by normalizing the product of the coefficient of radial consolidation and elapsed time by the square of the radius of influence zone of vertical drain, namely

$$T_r = \frac{c_r t}{r_e^2} \tag{6.37}$$

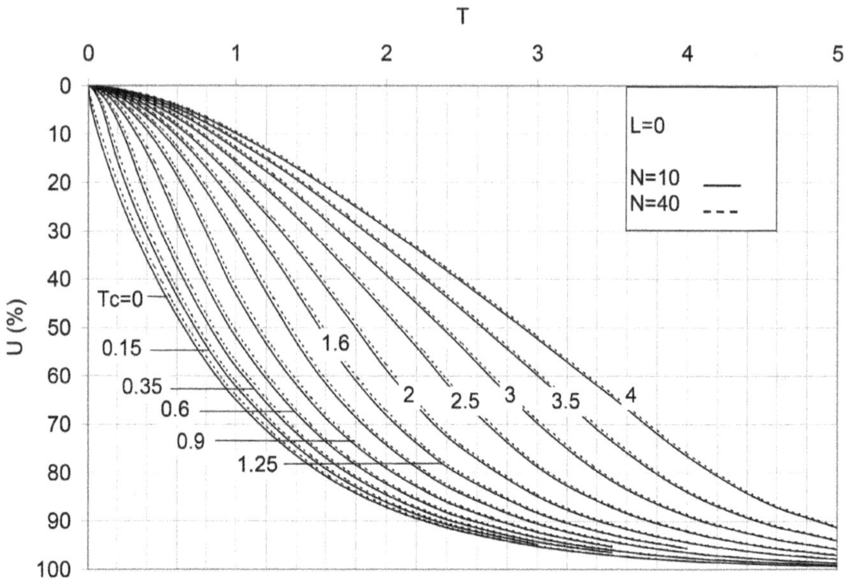

Figure 6.6 Normalized time factor *T* vs. the average degree of consolidation *U* for $L = 0$.

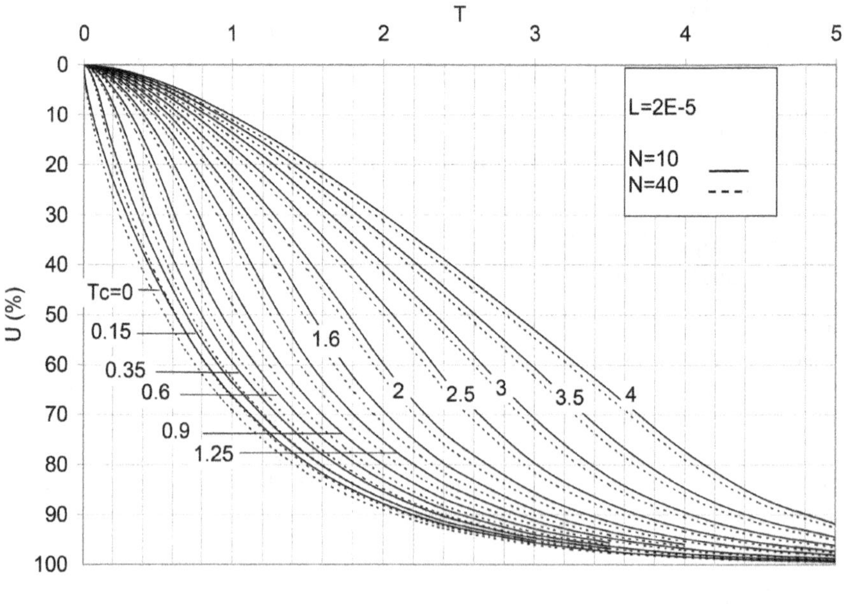

Figure 6.7 Normalized time factor *T* vs. the average degree of consolidation *U* for $L = 2\times10^{-5}$.

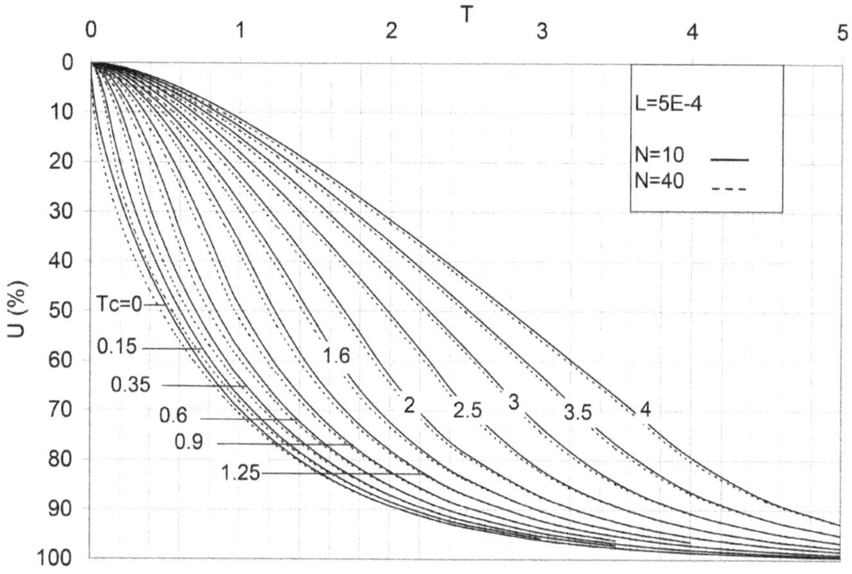

Figure 6.8 Normalized time factor T vs. the average degree of consolidation U for $L = 5\times10^{-4}$.

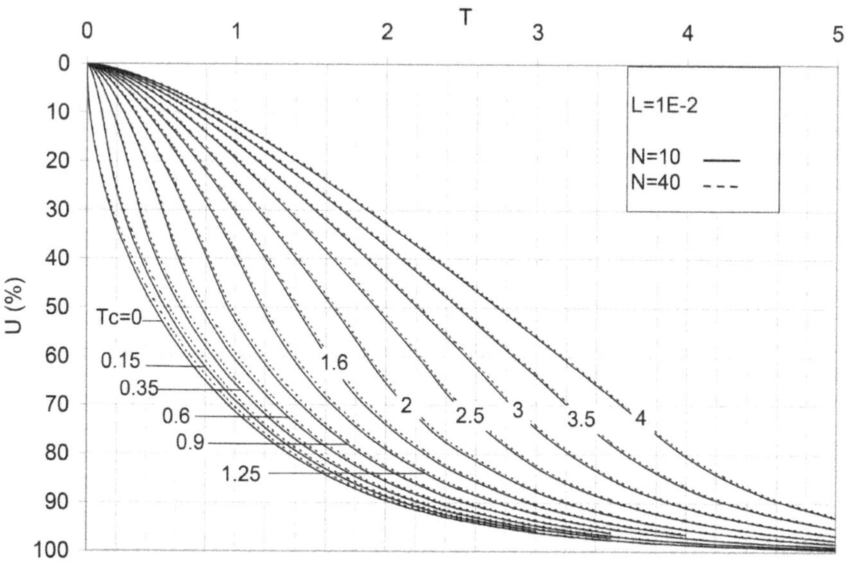

Figure 6.9 Normalized time factor T vs. the average degree of consolidation U for $L = 1\times10^{-2}$.

where T_r = dimensionless time factor for consolidation due to radial drainage. When using these charts, the drain spacing has to be assumed at the onset of the design process to calculate the dimensionless time factor using (6.37). Then the drain radius, which is needed to achieve the degree of consolidation required within the time available, is determined as a dependent variable. For a prefabricated drain type, the design procedure has to be performed repeatedly using different values of equivalent radius of cylinder of soil around drain until the drain size is met.

Since the normalized time factor T makes that the relationship of U *vs.* T is primarily dependent on the construction time factor T_c only, the analytical solution in (6.28) is used to produce design charts for determining the required drain spacing explicitly taking into consideration of time-dependent loading as shown in Figure 6.2. The design charts are prepared by selecting $N = 30$ and $L = 0$, 5×10^{-6}, 3×10^{-5}, 10^{-4}, and 10^{-2} for different ratio of t_c over t, and plotted in Figures 6.10–6.14. The values of L selected shall be adequate for normal design conditions.

6.3.1.4 Design procedure

The procedure for determining the drain spacing can be summarized as follows:

1. Determine the soil parameters c_r, c_v and the thickness H of the soil layer (see Figure 6.1).
2. Determine the radius of vertical drain. If a prefabricated drain is selected, the equivalent radius of vertical drain can be calculated using $r_d = \frac{b+h}{\pi}$ (Hansbo, 1979), or $r_d = \frac{b+h}{4} + \frac{h}{10}$ (Long and Alvaro, 1994).
3. Calculate L using (6.25).

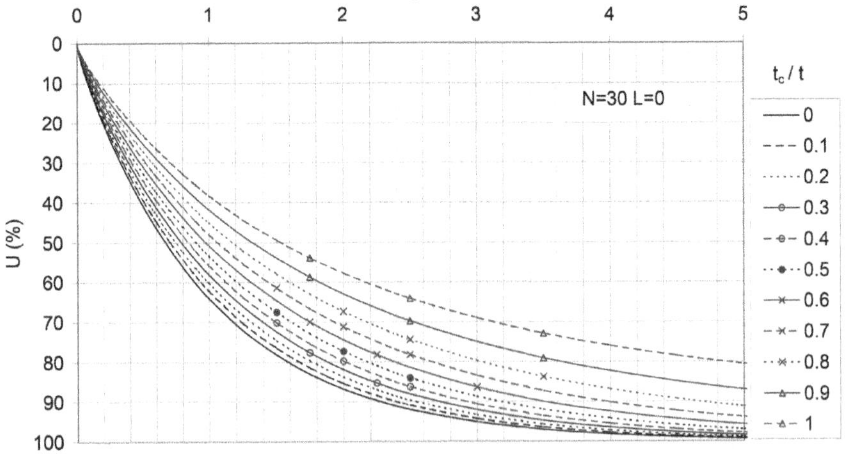

Figure 6.10 Average degree of consolidation U vs. T for $L = 0$.

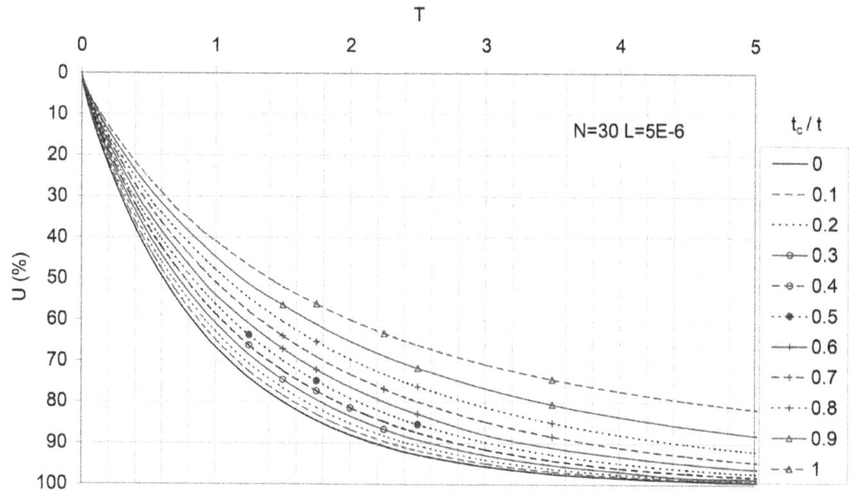

Figure 6.11 Average degree of consolidation U vs. T for $L = 5\times10^{-6}$.

4. Determine the construction time t_c, the average degree of consolidation required and the corresponding time t available.
5. Determine the time factor T using L, U, and t_c/t from Figures 6.10–6.14.
6. Determine μ_1 using (6.19), or $\mu_1 = \sqrt{\frac{r_d^2 T}{c_r t} - \frac{\pi^2}{4} L}$.
7. Determine N from Table 6.1 or $\lg N = 0.202 - 0.8318 \lg \mu_1$ since the relationship of μ_1 *vs.* N is an approximately straight line when plotted in double logarithmic scale.

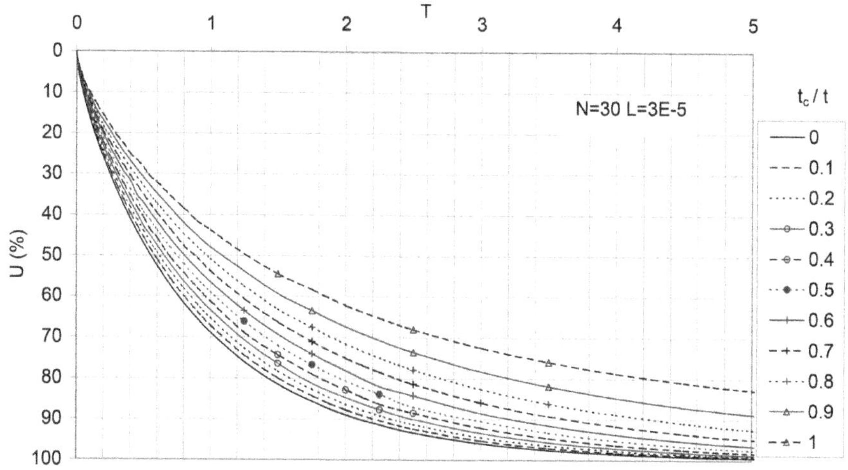

Figure 6.12 Average degree of consolidation U vs. T for $L = 3\times10^{-5}$.

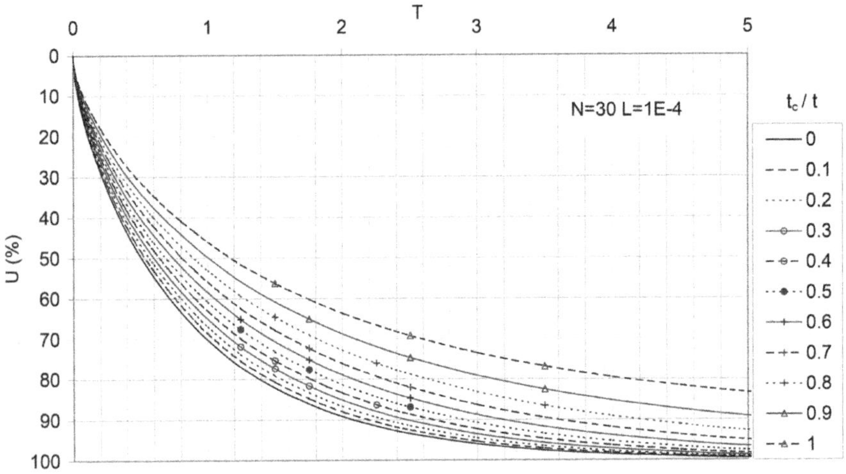

Figure 6.13 Average degree of consolidation *U* vs. *T* for $L = 10^{-4}$.

8. Determine the radius of the equivalent cylindrical block of soil using $r_e = N\, r_d$.
9. Determine the drain spacing *S*. The *S* equals $1.905r_e$ for a triangular pattern or $1.772r_e$ for a square grid pattern.

6.3.1.5 Example of application

As an example of calculation, let us consider the case where $H = 2.5$ m, $r_d = 25$ mm, $c_v = c_r = 1.5$ m²/year, subjected to a single ramp loading with a construction time of $t_c = 0.15$ year. It is required that an average degree

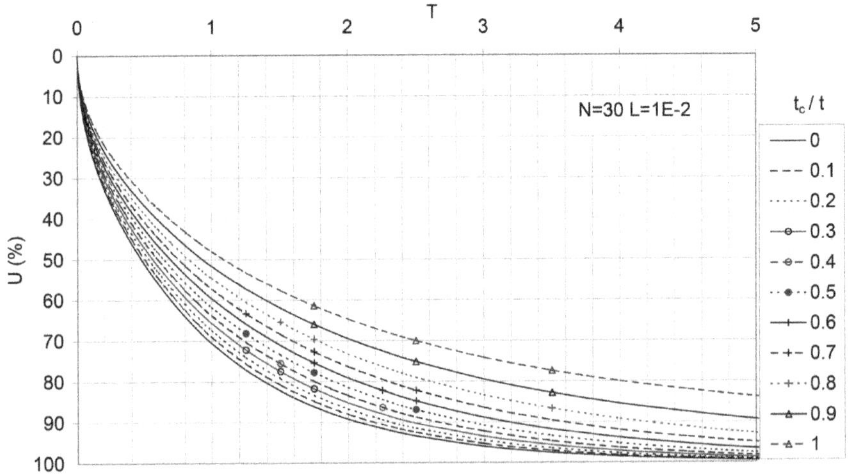

Figure 6.14 Average degree of consolidation *U* vs. *T* for $L = 10^{-2}$.

of consolidation of 70% would be achieved at time $t = 0.3$ year since the start of construction. From Figure 6.13, the corresponding time factor T is 1.35 for the parameters $L = \frac{c_v r_d^2}{c_r H^2} = \frac{1.5 \times 0.025^2}{1.5 \times 2.5^2} = 1 \times 10^{-4}$, $t_c/t = 0.15/0.3 = 0.5$ and $U = 0.7$. Substituting the parameters into (6.19), μ_1 is found to be $\mu_1 = \sqrt{\frac{r_d^2 T}{c_r t} - \frac{\pi^2}{4} L} = \sqrt{\frac{0.025^2 \times 1.35}{1.5 \times 0.3} - \frac{\pi^2}{4} 1 \times 10^{-4}} = 0.04$. From Table 6.1, the corresponding N is 22.5. Thus, the radius of the equivalent cylindrical block of soil is $r_e = N r_d = 22.5 \times 0.025 = 0.5625$ m. The spacing is $S = 1.905 r_e = 1.905 \times 0.5625 = 1.07$ m for a triangular pattern or $S = 1.772 r_e = 1.772 \times 0.5625 = 0.997$ m for a square grid pattern.

6.3.2 Solutions for a vertical drain with smear zone

If a smear zone is considered, the unit cell is shown in Figure 6.15. The differential equation for the dissipation of excess porewater pressure using a free strain assumption is

$$\frac{\partial u_e}{\partial t} = \begin{cases} c_{rs}\left(\frac{\partial^2 u_e}{\partial r^2} + \frac{1}{r}\frac{\partial u_e}{\partial r}\right) + c_v \frac{\partial^2 u_e}{\partial z^2} + \frac{\partial \sigma}{\partial t} & r_d < r < r_s \\ c_r\left(\frac{\partial^2 u_e}{\partial r^2} + \frac{1}{r}\frac{\partial u_e}{\partial r}\right) + c_v \frac{\partial^2 u_e}{\partial z^2} + \frac{\partial \sigma}{\partial t} & r_s < r < r_e \end{cases} \tag{6.38}$$

where r_s is radius of the smear zone; $c_{rs} = \frac{k_s}{\gamma_w m_v}$ is horizontal (or radial) consolidation coefficient in smear zone; and k_s is horizontal hydraulic conductivity in smear zone.

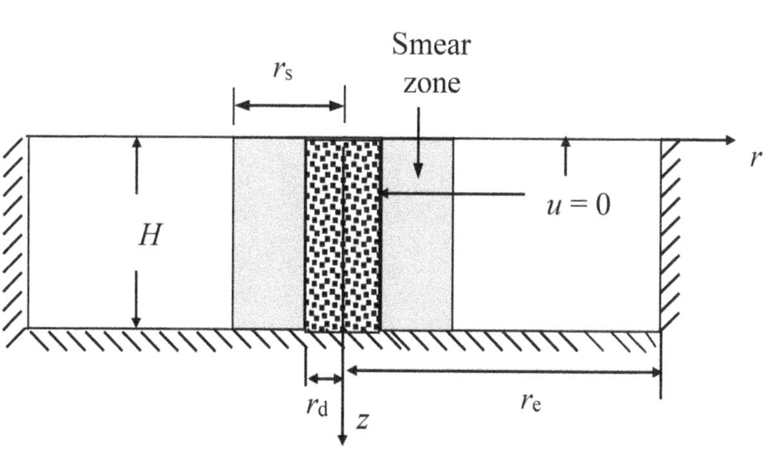

Figure 6.15 Co-ordinates and boundary conditions for a vertical drain with smear zone.

At the interface of the smeared zone and undisturbed zone, the excess pore pressure is the same and the rate of flow out of the undisturbed zone must be equal to that into the smeared zone, so that:

$$\begin{cases} u_e\Big|_{r=r_s^-} = u_e\Big|_{r=r_s^+} \\ k_s \dfrac{\partial u_e}{\partial r}\Big|_{r=r_s^-} = k_r \dfrac{\partial u_e}{\partial r}\Big|_{r=r_s^+} \end{cases} \tag{6.39}$$

The solution to (6.38) under the given loading condition in (6.14) and the boundary conditions in (6.15) and (6.39) becomes

$$u_e = \sum_{m,n=1}^{\infty} A'_{mn}(T) R'_m(r) \sin\left(\lambda_n \frac{z}{H}\right) \tag{6.40}$$

$$R'_m(r) = \begin{cases} V_0^m\left(\eta\mu_m \dfrac{r}{r_d}\right) & r_d < r < r_s \\ W_0^m\left(\mu_m \dfrac{r}{r_d}\right) & r_s < r < r_e \end{cases} \tag{6.41}$$

$$V_v^m(x) = \frac{Y_0(\eta\mu_m)J_v(x) - J_0(\eta\mu_m)Y_v(x)}{Y_0(\eta\mu_m)J_0(\eta s\mu_m) - J_0(\eta\mu_m)Y_0(\eta s\mu_m)} \tag{6.42}$$

$$W_v^m(x) = \frac{Y_1(N\mu_m)J_v(x) - J_1(N\mu_m)Y_v(x)}{Y_1(N\mu_m)J_0(s\mu_m) - J_1(N\mu_m)Y_0(s\mu_m)} \tag{6.43}$$

$$A'_{mn}(T) =$$

$$\begin{cases} \dfrac{2B'_m\sigma_0(\mu_1^2 + \lambda_1^2 L)}{\lambda_n(\mu_m^2 + \lambda_n^2 L)T_c}\left[1 - \exp\left(-\dfrac{\mu_m^2 + \lambda_n^2 L}{\mu_1^2 + \lambda_1^2 L}T\right)\right] & T \le T_c \\ \dfrac{2B'_m\sigma_0(\mu_1^2 + \lambda_1^2 L)}{\lambda_n(\mu_m^2 + \lambda_n^2 L)T_c}\left[1 - \exp\left(-\dfrac{\mu_m^2 + \lambda_n^2 L}{\mu_1^2 + \lambda_1^2 L}T_c\right)\right]\exp\left[-\dfrac{\mu_m^2 + \lambda_n^2 L}{\mu_1^2 + \lambda_1^2 L}(T - T_c)\right] & T \ge T_c \end{cases} \tag{6.44}$$

$$B'_m = \frac{\displaystyle\int_{r_w}^{r_e} rR'_m dr}{\displaystyle\int_{r_w}^{r_e} rR'^2_m dr} \tag{6.45}$$

$$s = \frac{r_s}{r_w} \tag{6.46}$$

$$\eta = \sqrt{\frac{c_r}{c_{rs}}} \tag{6.47}$$

where J_v, Y_v denote, respectively, Bessel functions of the first and second kind of order v; s is the ratio of the smeared zone radius over well radius; η expresses the difference in the coefficients of consolidation between the undisturbed zone and the smeared zone; and T, T_c, and λ_n are defined in (6.19–6.21). However, the μ_l in (6.19–6.20) and the μ_m in (6.41–6.44) are the m-th positive root of the following (6.48), which is a generalized form of (6.26) to consider the smear effects.

$$\begin{aligned}&[Y_0(\eta\mu_m)J_1(\eta s\mu_m) - J_0(\eta\mu_m)Y_1(\eta s\mu_m)][Y_1(N\mu_m)J_0(s\mu_m) - J_1(N\mu_m)Y_0(s\mu_m)]\\&-\eta[Y_0(\eta\mu_m)J_0(\eta s\mu_m) - J_0(\eta\mu_m)Y_0(\eta s\mu_m)][Y_1(N\mu_m)J_1(s\mu_m) - J_1(N\mu_m)Y_1(s\mu_m)] = 0\end{aligned} \tag{6.48}$$

The first eigenvalue μ_l of (6.48) is listed in Table 6.2 for different N, s, and η values.

Substituting (6.40) into (6.27), the average degree of consolidation $U(T)$ is obtained as

$$U(T) = \begin{cases} \dfrac{T}{T_c} - \displaystyle\sum_{m=1}^{\infty}\sum_{n=1,3,5,\dots} \frac{32C'_m\left(\mu_1^2 + \frac{\pi^2}{4}L\right)}{n^2\pi^2\left(\mu_m^2 + \frac{n^2\pi^2}{4}L\right)T_c}\left[1 - \exp\left(-\frac{\mu_m^2 + \frac{n^2\pi^2}{4}L}{\mu_1^2 + \frac{\pi^2}{4}L}T\right)\right] & T \le T_c \\ 1 - \displaystyle\sum_{m=1}^{\infty}\sum_{n=1,3,5,\dots} \frac{32C'_m\left(\mu_1^2 + \frac{\pi^2}{4}L\right)}{n^2\pi^2\left(\mu_m^2 + \frac{n^2\pi^2}{4}L\right)T_c}\left[1 - \exp\left(-\frac{\mu_m^2 + \frac{n^2\pi^2}{4}L}{\mu_1^2 + \frac{\pi^2}{4}L}T_c\right)\right] & \\ \times \exp\left[-\dfrac{\mu_m^2 + \frac{n^2\pi^2}{4}L}{\mu_1^2 + \frac{\pi^2}{4}L}(T - T_c)\right] & T \ge T_c \end{cases} \tag{6.49}$$

Table 6.2 Eigenvalue μ_l for different N, s, and η

	$\eta^2 = 1$						$\eta^2 = 1.4$					
N	s = 1.2	1.4	1.6	2	2.8	4.5	s = 1.2	1.4	1.6	2	2.8	4.5
5	0.282	0.282	0.282	0.282	0.282	0.282	0.274	0.267	0.261	0.253	0.244	0.239
6	0.218	0.218	0.218	0.218	0.218	0.218	0.212	0.207	0.204	0.198	0.191	0.185
7	0.177	0.177	0.177	0.177	0.177	0.177	0.172	0.169	0.166	0.161	0.156	0.151
8	0.148	0.148	0.148	0.148	0.148	0.148	0.144	0.142	0.139	0.136	0.131	0.127
10	0.11	0.11	0.11	0.11	0.11	0.11	0.108	0.106	0.105	0.102	0.0992	0.0956
12	0.0874	0.0874	0.0874	0.0874	0.0874	0.0874	0.0858	0.0844	0.0834	0.0816	0.0793	0.0764
14	0.072	0.072	0.072	0.072	0.072	0.072	0.0708	0.0698	0.0689	0.0676	0.0657	0.0634
16	0.0611	0.0611	0.0611	0.0611	0.0611	0.0611	0.0601	0.0592	0.0586	0.0575	0.056	0.0541
18	0.0529	0.0529	0.0529	0.0529	0.0529	0.0529	0.052	0.0514	0.0508	0.0499	0.0486	0.0471
20	0.0465	0.0465	0.0465	0.0465	0.0465	0.0465	0.0458	0.0452	0.0448	0.044	0.0429	0.0416
23	0.0393	0.0393	0.0393	0.0393	0.0393	0.0393	0.0387	0.0383	0.0379	0.0373	0.0364	0.0353
26	0.0339	0.0339	0.0339	0.0339	0.0339	0.0339	0.0335	0.0331	0.0328	0.0323	0.0316	0.0306
30	0.0287	0.0287	0.0287	0.0287	0.0287	0.0287	0.0283	0.028	0.0277	0.0273	0.0267	0.026
35	0.0239	0.0239	0.0239	0.0239	0.0239	0.0239	0.0236	0.0234	0.0232	0.0228	0.0224	0.0218
40	0.0204	0.0204	0.0204	0.0204	0.0204	0.0204	0.0202	0.02	0.0198	0.0196	0.0192	0.0187
45	0.0178	0.0178	0.0178	0.0178	0.0178	0.0178	0.0176	0.0175	0.0173	0.0171	0.0168	0.0163
50	0.0158	0.0158	0.0158	0.0158	0.0158	0.0158	0.0156	0.0155	0.0154	0.0152	0.0149	0.0145
	$\eta^2 = 2$						$\eta^2 = 2.5$					
N	s = 1.2	1.4	1.6	2	2.8	4.5	s = 1.2	1.4	1.6	2	2.8	4.5
5	0.262	0.247	0.237	0.223	0.208	0.2	0.253	0.234	0.221	0.204	0.187	0.179
6	0.204	0.194	0.186	0.176	0.164	0.155	0.198	0.184	0.175	0.162	0.149	0.139
7	0.166	0.159	0.153	0.145	0.135	0.127	0.161	0.151	0.144	0.134	0.123	0.114
8	0.139	0.134	0.129	0.122	0.115	0.107	0.136	0.128	0.122	0.114	0.105	0.0963
10	0.105	0.101	0.0977	0.0931	0.0874	0.0816	0.102	0.097	0.0929	0.0871	0.0802	0.0736
12	0.0835	0.0805	0.0782	0.0748	0.0704	0.0657	0.0817	0.0776	0.0746	0.0702	0.0649	0.0595
14	0.069	0.0667	0.0649	0.0622	0.0587	0.0549	0.0676	0.0645	0.0621	0.0586	0.0543	0.0499
16	0.0586	0.0568	0.0553	0.0531	0.0503	0.047	0.0575	0.055	0.053	0.0501	0.0466	0.0428
18	0.0509	0.0493	0.0481	0.0463	0.0438	0.0411	0.0499	0.0478	0.0462	0.0438	0.0408	0.0375
20	0.0448	0.0435	0.0425	0.0409	0.0388	0.0364	0.044	0.0422	0.0408	0.0388	0.0362	0.0333
23	0.0379	0.0369	0.0361	0.0348	0.0331	0.0311	0.0373	0.0359	0.0347	0.033	0.0309	0.0285
26	0.0328	0.032	0.0313	0.0302	0.0288	0.0271	0.0323	0.0311	0.0301	0.0287	0.0269	0.0249
30	0.0277	0.027	0.0265	0.0256	0.0244	0.0231	0.0273	0.0263	0.0256	0.0244	0.0229	0.0212
35	0.0232	0.0226	0.0222	0.0215	0.0205	0.0194	0.0228	0.0221	0.0214	0.0205	0.0193	0.0179
40	0.0199	0.0194	0.019	0.0184	0.0177	0.0167	0.0196	0.0189	0.0184	0.0176	0.0166	0.0155
45	0.0173	0.017	0.0166	0.0161	0.0155	0.0147	0.0171	0.0166	0.0161	0.0155	0.0146	0.0136
50	0.0154	0.015	0.0148	0.0143	0.0138	0.0131	0.0152	0.0147	0.0143	0.0137	0.013	0.0121
	$\eta^2 = 3.5$						$\eta^2 = 5$					
N	s = 1.2	1.4	1.6	2	2.8	4.5	s = 1.2	1.4	1.6	2	2.8	4.5
5	0.238	0.213	0.197	0.178	0.16	0.151	0.219	0.189	0.172	0.152	0.135	0.126
6	0.187	0.169	0.157	0.142	0.127	0.118	0.173	0.151	0.138	0.122	0.108	0.0985
7	0.153	0.139	0.13	0.118	0.106	0.0965	0.143	0.126	0.115	0.102	0.0899	0.0809
8	0.129	0.118	0.111	0.101	0.0905	0.0819	0.121	0.107	0.0984	0.0877	0.0771	0.0688

Table 6.2 (Continued) Eigenvalue μ_l for different N, s, and η

10	0.098	0.0903	0.0849	0.0778	0.07	0.0629	0.0923	0.0824	0.076	0.0681	0.06	0.0531
12	0.0784	0.0726	0.0685	0.063	0.0569	0.0511	0.0742	0.0667	0.0617	0.0555	0.049	0.0432
14	0.0651	0.0605	0.0573	0.0529	0.0478	0.0429	0.0618	0.0558	0.0518	0.0467	0.0414	0.0364
16	0.0555	0.0518	0.0491	0.0454	0.0412	0.037	0.0528	0.0479	0.0446	0.0403	0.0357	0.0315
18	0.0482	0.0451	0.0429	0.0398	0.0361	0.0325	0.046	0.0418	0.039	0.0354	0.0314	0.0277
20	0.0426	0.0399	0.038	0.0353	0.0322	0.029	0.0407	0.0371	0.0347	0.0315	0.028	0.0247
23	0.0361	0.034	0.0324	0.0302	0.0276	0.0249	0.0346	0.0317	0.0297	0.027	0.0241	0.0213
26	0.0313	0.0295	0.0282	0.0263	0.0241	0.0217	0.03	0.0276	0.0259	0.0236	0.0211	0.0187
30	0.0265	0.0251	0.024	0.0224	0.0206	0.0186	0.0255	0.0235	0.0221	0.0202	0.0181	0.016
35	0.0222	0.021	0.0202	0.0189	0.0174	0.0158	0.0214	0.0197	0.0186	0.0171	0.0153	0.0136
40	0.0191	0.0181	0.0174	0.0163	0.015	0.0137	0.0184	0.017	0.0161	0.0148	0.0133	0.0118
45	0.0167	0.0158	0.0152	0.0143	0.0132	0.012	0.0161	0.0149	0.0141	0.013	0.0117	0.0104
50	0.0148	0.0141	0.0135	0.0127	0.0118	0.0107	0.0143	0.0133	0.0126	0.0116	0.0105	0.00934

	$\eta^2 = 7$						$\eta^2 = 10$					
N	s = 1.2	1.4	1.6	2	2.8	4.5	s = 1.2	1.4	1.6	2	2.8	4.5
5	0.199	0.167	0.15	0.13	0.114	0.107	0.178	0.145	0.128	0.11	0.0961	0.0894
6	0.159	0.135	0.121	0.105	0.0918	0.0833	0.143	0.117	0.104	0.0896	0.0773	0.0697
7	0.132	0.112	0.101	0.0885	0.0768	0.0685	0.119	0.0986	0.0875	0.0754	0.0648	0.0574
8	0.112	0.0963	0.0869	0.0761	0.066	0.0583	0.102	0.0848	0.0754	0.0651	0.0558	0.0489
10	0.0861	0.0745	0.0676	0.0594	0.0515	0.0451	0.0787	0.066	0.059	0.051	0.0437	0.0379
12	0.0695	0.0606	0.0551	0.0487	0.0423	0.0368	0.0638	0.0539	0.0483	0.0419	0.0359	0.031
14	0.058	0.0509	0.0465	0.0411	0.0358	0.0311	0.0535	0.0455	0.0408	0.0355	0.0305	0.0262
16	0.0497	0.0438	0.0401	0.0356	0.031	0.0269	0.0459	0.0393	0.0353	0.0308	0.0264	0.0227
18	0.0434	0.0384	0.0352	0.0313	0.0273	0.0237	0.0402	0.0345	0.0311	0.0272	0.0234	0.0201
20	0.0384	0.0341	0.0313	0.0279	0.0244	0.0212	0.0357	0.0307	0.0278	0.0243	0.0209	0.018
23	0.0328	0.0292	0.0269	0.024	0.021	0.0183	0.0305	0.0264	0.0239	0.0209	0.0181	0.0155
26	0.0285	0.0255	0.0235	0.021	0.0185	0.0161	0.0266	0.0231	0.0209	0.0184	0.0159	0.0136
30	0.0242	0.0217	0.0201	0.018	0.0159	0.0138	0.0227	0.0197	0.0179	0.0158	0.0137	0.0118
35	0.0204	0.0183	0.017	0.0153	0.0135	0.0118	0.0191	0.0167	0.0152	0.0134	0.0116	0.01
40	0.0175	0.0158	0.0147	0.0133	0.0117	0.0102	0.0165	0.0144	0.0132	0.0117	0.0101	0.00873
45	0.0154	0.0139	0.0129	0.0117	0.0103	0.00905	0.0145	0.0127	0.0116	0.0103	0.00895	0.00773
50	0.0137	0.0124	0.0115	0.0104	0.00925	0.00811	0.0129	0.0113	0.0104	0.00922	0.00802	0.00693

where

$$C'_m = \frac{[V_1^m(\eta\mu_m)]^2}{(N^2-1)\eta^2\mu_m^2\left\{N^2[W_0^m(N\mu_m)]^2 - [V_1^m(\eta\mu_m)]^2 + \left(1-\frac{c_{rs}}{c_r}\right)s^2[V_1^m(s\eta\mu_m)]^2\right\}} \tag{6.50}$$

6.3.2.1 *Special case of a suddenly applied loading*

For the situation of a suddenly applied loading, (6.38) becomes a homogeneous equation. As in Section 6.3.1, the above problem can be solved by combining the solution for radial flow and the solution for vertical flow as the following theorem states:

Theorem 6.2. If $u_1(r, t)$ satisfies (6.15a), (6.39), and (6.51)

$$\frac{\partial u_1}{\partial t}=\begin{cases} c_{rs}\left(\dfrac{\partial^2 u_1}{\partial r^2}+\dfrac{1}{r}\dfrac{\partial u_1}{\partial r}\right) & r_d < r < r_s \\ c_r\left(\dfrac{\partial^2 u_1}{\partial r^2}+\dfrac{1}{r}\dfrac{\partial u_1}{\partial r}\right) & r_s < r < r_e \end{cases} \tag{6.51}$$

and the initial condition $u_1 |_{t=0}=\sigma_0$, and $u_2(z, t)$ satisfies (6.15b) and (6.31) and the initial condition $u_2 |_{t=0}=\sigma_0$, then $u_e = \frac{u_1 u_2}{\sigma_0}$ is necessarily a solution of (6.15), (6.39), and (6.52)

$$\frac{\partial u_e}{\partial t}=\begin{cases} c_{rs}\left(\dfrac{\partial^2 u_e}{\partial r^2}+\dfrac{1}{r}\dfrac{\partial u_e}{\partial r}\right)+c_v\dfrac{\partial^2 u_e}{\partial z^2} & r_d < r < r_s \\ c_r\left(\dfrac{\partial^2 u_e}{\partial r^2}+\dfrac{1}{r}\dfrac{\partial u_e}{\partial r}\right)+c_v\dfrac{\partial^2 u_e}{\partial z^2} & r_s < r < r_e \end{cases} \tag{6.52}$$

and the initial condition $u_e |_{t=0}=\sigma_0$.

Similarly, (6.36) still holds for the case of a vertical drain with smear zone.

6.3.2.2 Influence of N, s, η, and L on average degree of consolidation

As in Section 6.3.1, the average degree of consolidation shows very little sensitivity to N, s, η, and L if the time factor T is defined by (6.19). Figure 6.16 shows a typical result of the effect of η ($=\sqrt{c_r/c_{rs}}$) on the relations between normalized time factor T and the average degree of consolidation U for $N=20$, $s=2$, $L=10^{-4}$, and $T_c=0.3$. Obviously, the difference is very small for different η-values. This conclusion is also true for N, s, and L. Using this property, the average degree of consolidation is prepared for $\frac{c_r}{c_{rs}}=2.5$, $s=2.5$ and for $\frac{c_r}{c_{rs}}=2.5$, $s=4.5$ as shown in Figures 6.17–6.24 for practical applications. If s is in the range of 1.2 to 2.5, the solution charts for $\frac{c_r}{c_{rs}}=2.5$, $s=2.5$ can be used. If s is in the range of 2.5 to 4.5, the solution charts for $\frac{c_r}{c_{rs}}=2.5$, $s=4.5$ can be adopted. The maximum error in average degree of consolidation induced in the substitution of $\frac{c_r}{c_{rs}}$ by 2.5 and s by 2.5 or 4.5 is less than 3% for the whole range calculated ($N=5$ to 80, $s=1.2$ to 4.5, $\frac{c_r}{c_{rs}}=1$ to 10, and $L=$ 0 to 0.01).

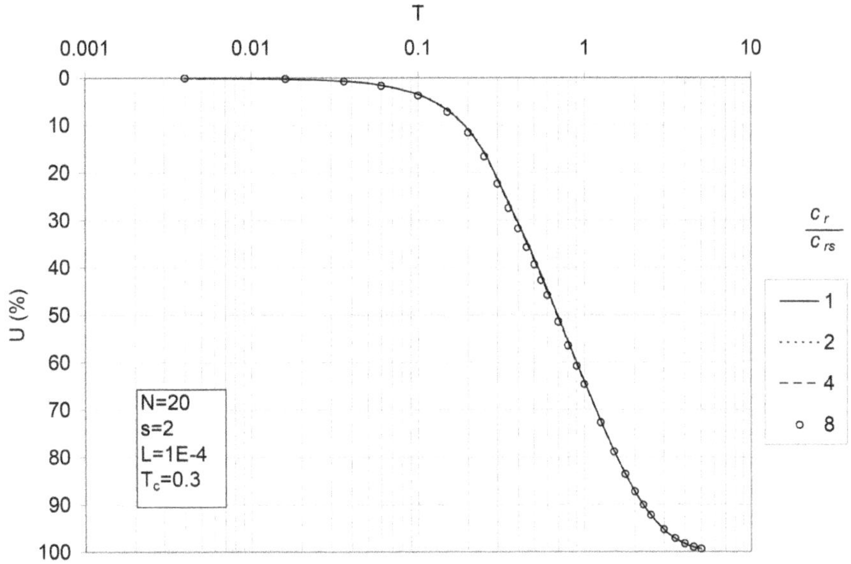

Figure 6.16 Normalized time vs. degree of consolidation for different c_r/c_{rs}.

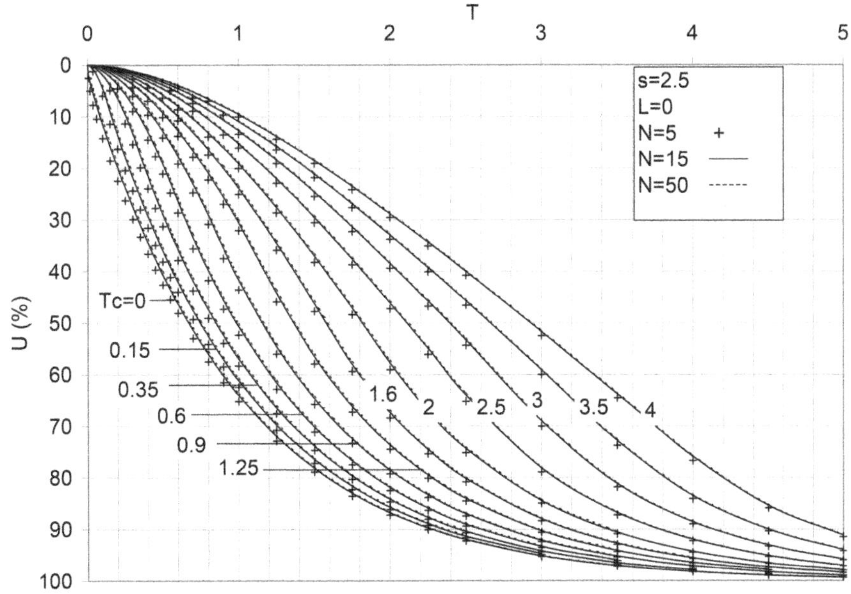

Figure 6.17 Degree of consolidation for s = 2.5 and $L = 0$.

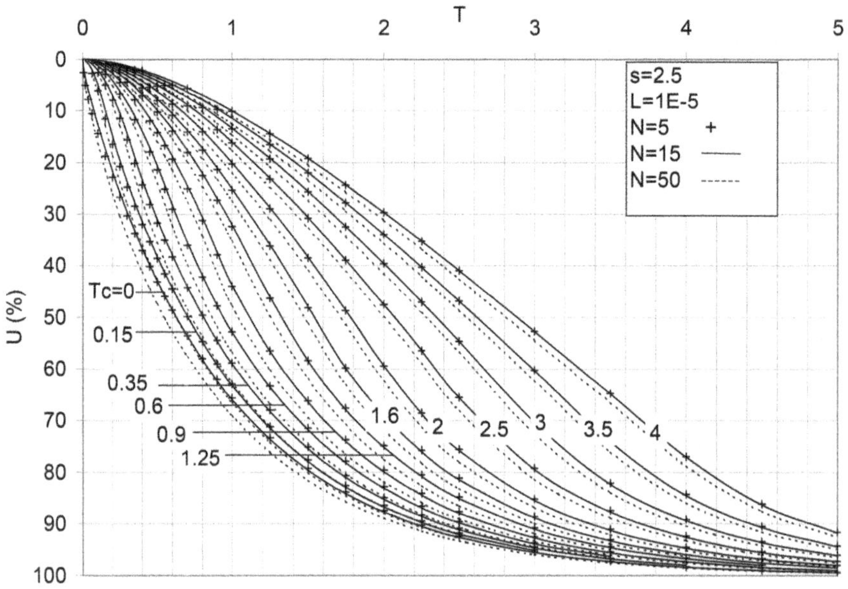

Figure 6.18 Degree of consolidation for $s = 2.5$ and $L = 1\times10^{-5}$.

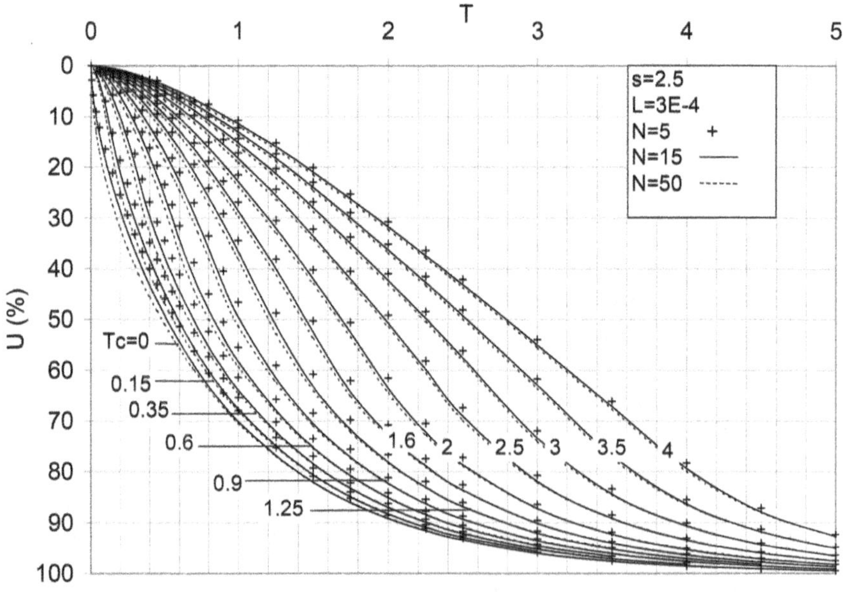

Figure 6.19 Degree of consolidation for $s = 2.5$ and $L = 3\times10^{-4}$.

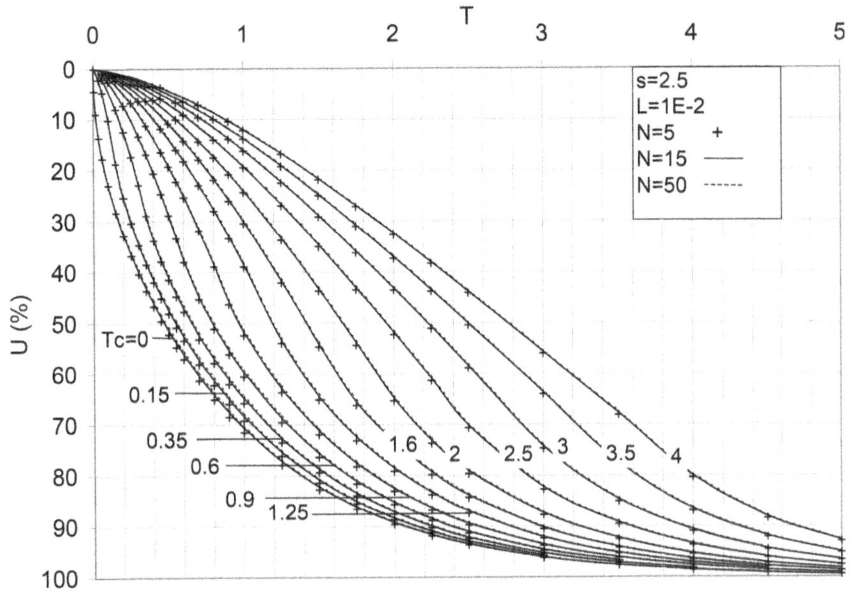

Figure 6.20 Degree of consolidation for s = 2.5 and $L = 1 \times 10^{-2}$.

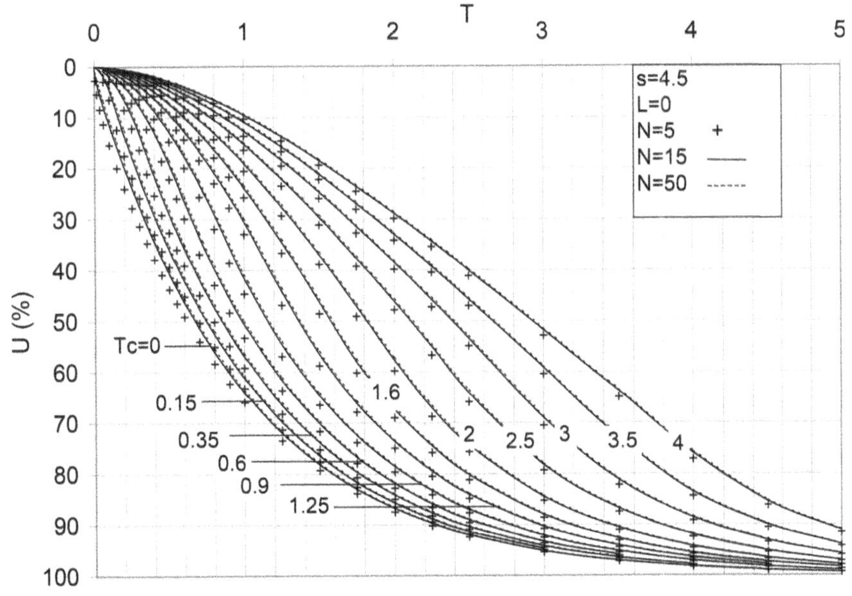

Figure 6.21 Degree of consolidation for s = 4.5 and $L = 0$.

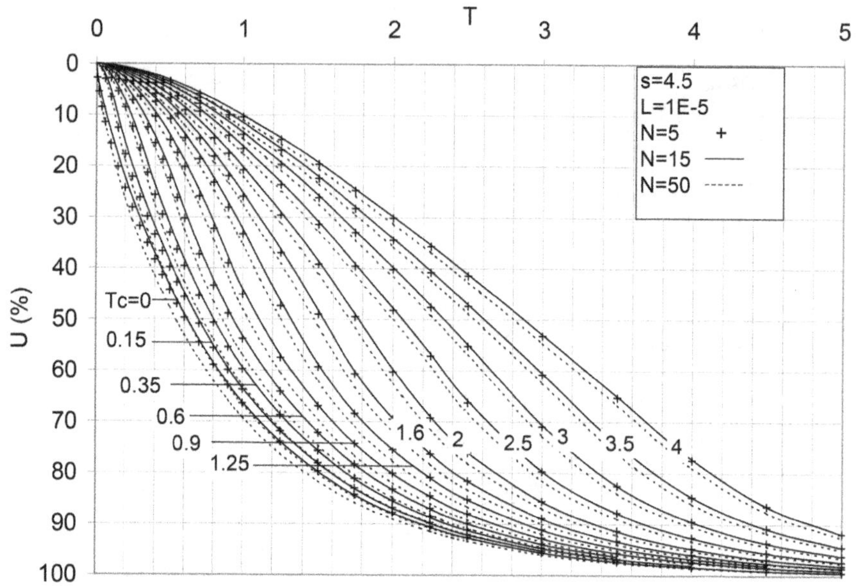

Figure 6.22 Degree of consolidation for s = 4.5 and $L = 1 \times 10^{-5}$.

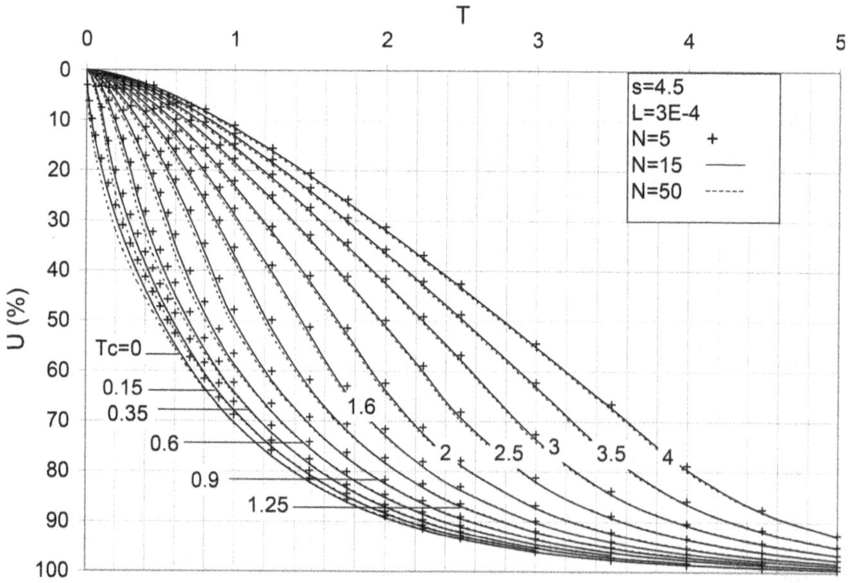

Figure 6.23 Degree of consolidation for s = 4.5 and $L = 3 \times 10^{-4}$.

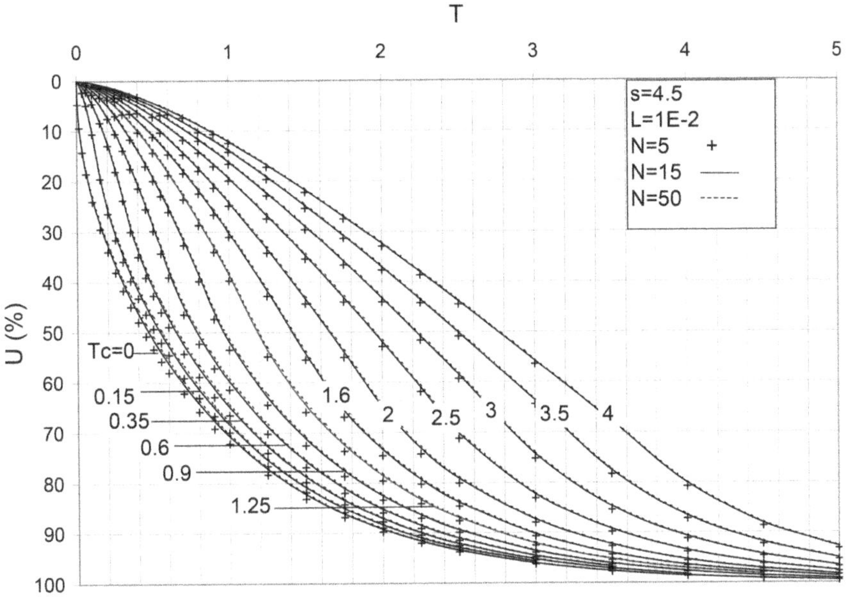

Figure 6.24 Degree of consolidation for s = 4.5 and $L = 1\times10^{-2}$.

6.4 BASIC EQUATIONS AND SOLUTIONS FOR THE CASE OF EQUAL STRAIN

For consolidation of soil with vertical drain with free strains permitted, the soil adjacent to the well consolidates and compresses faster than soil farther away from the drain. This difference in the rate of consolidation develops differential settlement of the upper surface of the soil mass and shear strains within the mass. For free strains, it is assumed that these effects do not influence the redistribution of load to the soil nor the rate of consolidation. It is obvious, however, that these effects will redistribute the load to some extent, depending on the amount of arching developed in the material above the compressible soil. The extreme case would be where the arching process redistributes the load to the consolidating soil so that all vertical strains at any depth z are equal and no differential settlement develops. This condition of *equal vertical strain* is rather severe. It can be obtained in the laboratory by use of a rigid loading platform, and probably is approached in the field if the ratio of H to r_e is large.

The governing equation assuming equal strain condition can be derived from the continuity equation (6.12). Substituting $\varepsilon_z = m_v[\sigma(t) - \bar{u}_e]$ and (6.9) into (6.12) gives

$$\frac{\partial \bar{u}_e}{\partial t} = c_r\left(\frac{\partial^2 u_e}{\partial r^2} + \frac{1}{r}\frac{\partial u_e}{\partial r}\right) + c_v\frac{\partial^2 u_e}{\partial z^2} + \frac{\partial \sigma}{\partial t} \tag{6.53}$$

where

$$\bar{u}_e = \frac{2}{r_e^2 - r_d^2} \int_{r_d}^{r_e} r u_e dr \tag{6.54}$$

For radial flow only, (6.53) becomes

$$\frac{\partial \bar{u}_e}{\partial t} = c_r \left(\frac{\partial^2 u_e}{\partial r^2} + \frac{1}{r} \frac{\partial u_e}{\partial r} \right) + \frac{\partial \sigma}{\partial t} \tag{6.55}$$

It is difficult to solve (6.53) considering both radial and the vertical flow. Normally, (6.55) is used to analyze the radial flow. Then, the influence of vertical flow is taken into account by using Theorems 6.1–6.2 for free strain assumption.

6.4.1 Solutions for a vertical drain with smear zone

For the consolidation of the unit cell in Figure 6.15 due to radial flow, the differential equation can be expressed as

$$\frac{\partial \bar{u}_e}{\partial t} = \begin{cases} c_{rs} \left(\dfrac{\partial^2 u_e}{\partial r^2} + \dfrac{1}{r} \dfrac{\partial u_e}{\partial r} \right) + \dfrac{\partial \sigma}{\partial t} & r_d < r < r_s \\ c_r \left(\dfrac{\partial^2 u_e}{\partial r^2} + \dfrac{1}{r} \dfrac{\partial u_e}{\partial r} \right) + \dfrac{\partial \sigma}{\partial t} & r_s < r < r_e \end{cases} \tag{6.56}$$

Integrating (6.56) with respect to r and introducing (6.14), (6.15a), and (6.39), we obtain

$$u_e = \begin{cases} \dfrac{\gamma_w r_e^2}{2k_s} \dfrac{\partial \varepsilon_z}{\partial t} \left[\ln\left(\dfrac{r}{r_d} \right) - \dfrac{r^2 - r_d^2}{2r_e^2} \right] & r < r_s \\ \dfrac{\gamma_w r_e^2}{2k_r} \dfrac{\partial \varepsilon_z}{\partial t} \left[\ln\left(\dfrac{r}{r_s} \right) - \dfrac{r^2 - r_s^2}{2r_e^2} + \dfrac{k_r}{k_s} \left(\ln s - \dfrac{s^2 - 1}{2N^2} \right) \right] & r \geq r_s \end{cases} \tag{6.57}$$

$$\frac{\partial \varepsilon_z}{\partial t} = \begin{cases} \dfrac{m_v \sigma_0}{t_c} \left[1 - \exp\left(-\dfrac{2T_r}{\mu_s} \right) \right] & T_r < T_{rc} \\ \dfrac{m_v \sigma_0}{t_c} \left[1 - \exp\left(-\dfrac{2T_{rc}}{\mu_s} \right) \right] \exp\left[-\dfrac{2(T_r - T_{rc})}{\mu_s} \right] & T_r \geq T_{rc} \end{cases} \tag{6.58}$$

$$\bar{u}_e = \begin{cases} \dfrac{\mu_s \sigma_0}{2T_{rc}}\left[1-\exp\left(-\dfrac{2T_r}{\mu_s}\right)\right] & T_r < T_{rc} \\ \dfrac{\mu_s \sigma_0}{2T_{rc}}\left[1-\exp\left(-\dfrac{2T_{rc}}{\mu_s}\right)\right]\exp\left[-\dfrac{2(T_r - T_{rc})}{\mu_s}\right] & T_r \geq T_{rc} \end{cases} \tag{6.59}$$

$$T_r = \frac{c_r t}{r_e^2} \tag{6.60}$$

$$T_{rc} = \frac{c_r t_c}{r_e^2} \tag{6.61}$$

$$\mu_s = \frac{1}{(N^2-1)}\left[N^2 \ln\left(\frac{N}{s}\right) + s^2 - \frac{3}{4}N^2 - \frac{s^4}{4N^2} + \frac{k_r}{k_s}\left(N^2 \ln s + \frac{s^4-1}{4N^2} + 1 - s^2\right)\right] \tag{6.62}$$

$$U_r(T_r) = \begin{cases} \dfrac{T_r}{T_{rc}} - \dfrac{\mu_s}{2T_{rc}}\left[1-\exp\left(-\dfrac{2T_r}{\mu_s}\right)\right] & T_r < T_{rc} \\ 1 - \dfrac{\mu_s}{2T_{rc}}\left[1-\exp\left(-\dfrac{2T_{rc}}{\mu_s}\right)\right]\exp\left[-\dfrac{2(T_r - T_{rc})}{\mu_s}\right] & T_r \geq T_{rc} \end{cases} \tag{6.63}$$

For a suddenly applied loading $d\sigma(\tau)$ at time τ, the generated excess pore pressure due to radial flow only at time t from (6.57–6.58) will be

$$du_e = \begin{cases} \dfrac{k_r}{k_s}\dfrac{d\sigma(\tau)}{\mu_s}\exp\left[-\dfrac{2c_r(t-\tau)}{\mu_s r_e^2}\right]\left[\ln\left(\dfrac{r}{r_d}\right) - \dfrac{r^2 - r_d^2}{2r_e^2}\right] & r < r_s \\ \dfrac{d\sigma(\tau)}{\mu_s}\exp\left[-\dfrac{2c_r(t-\tau)}{\mu_s r_e^2}\right]\left[\ln\left(\dfrac{r}{r_s}\right) - \dfrac{r^2 - r_s^2}{2r_e^2} + \dfrac{k_r}{k_s}\left(\ln s - \dfrac{s^2-1}{2N^2}\right)\right] & r \geq r_s \end{cases} \tag{6.64}$$

Using (2.20), (6.64) and Theorem 6.2, the excess pore pressure generated by $\sigma(t)$ considering both radial and the vertical flow becomes

$$u_e = \begin{cases} \dfrac{k_r}{k_s}\left[\ln\left(\dfrac{r}{r_d}\right) - \dfrac{r^2 - r_d^2}{2r_e^2}\right]\displaystyle\sum_{n=1}^{\infty} T_n^*(t)\sin\left(\lambda_n \dfrac{z}{H}\right) & r < r_s \\ \left[\ln\left(\dfrac{r}{r_s}\right) - \dfrac{r^2 - r_s^2}{2r_e^2} + \dfrac{k_r}{k_s}\left(\ln s - \dfrac{s^2-1}{2N^2}\right)\right]\displaystyle\sum_{n=1}^{\infty} T_n^*(t)\sin\left(\lambda_n \dfrac{z}{H}\right) & r \geq r_s \end{cases} \tag{6.65}$$

where

$$T_n^*(t) = \int_0^t \frac{2}{\mu_s \lambda_n} \exp\left[-\left(\frac{c_v \lambda_n^2}{H^2} + \frac{2c_r}{\mu_s r_e^2}\right)(t-\tau)\right] d\sigma(\tau) \tag{6.66}$$

For the ramp loading (6.14)

$$T_n^*(t) = \begin{cases} \dfrac{2\sigma_0}{\mu_s \lambda_n t_c \left(\dfrac{c_v \lambda_n^2}{H^2} + \dfrac{2c_r}{\mu_s r_e^2}\right)} \left\{1 - \exp\left[-\left(\dfrac{c_v \lambda_n^2}{H^2} + \dfrac{2c_r}{\mu_s r_e^2}\right)t\right]\right\} & t < t_c \\ \dfrac{2\sigma_0 \left\{1 - \exp\left[-\left(\dfrac{c_v \lambda_n^2}{H^2} + \dfrac{2c_r}{\mu_s r_e^2}\right)t_c\right]\right\} \exp\left[-\left(\dfrac{c_v \lambda_n^2}{H^2} + \dfrac{2c_r}{\mu_s r_e^2}\right)(t - t_c)\right]}{\mu_s \lambda_n t_c \left(\dfrac{c_v \lambda_n^2}{H^2} + \dfrac{2c_r}{\mu_s r_e^2}\right)} & t \geq t_c \end{cases} \tag{6.67}$$

Substituting (6.65) in (6.27), the average degree of consolidation is obtained as

$$U = \begin{cases} \dfrac{t}{t_c} - \sum_{n=1}^{\infty} \dfrac{2}{\lambda_n^2 t_c \left(\dfrac{c_v \lambda_n^2}{H^2} + \dfrac{2c_r}{\mu_s r_e^2}\right)} \left\{1 - \exp\left[-\left(\dfrac{c_v \lambda_n^2}{H^2} + \dfrac{2c_r}{\mu_s r_e^2}\right)t\right]\right\} & t < t_c \\ 1 - \sum_{n=1}^{\infty} \dfrac{2\left\{1 - \exp\left[-\left(\dfrac{c_v \lambda_n^2}{H^2} + \dfrac{2c_r}{\mu_s r_e^2}\right)t_c\right]\right\} \exp\left[-\left(\dfrac{c_v \lambda_n^2}{H^2} + \dfrac{2c_r}{\mu_s r_e^2}\right)(t - t_c)\right]}{\lambda_n^2 t_c \left(\dfrac{c_v \lambda_n^2}{H^2} + \dfrac{2c_r}{\mu_s r_e^2}\right)} & t \geq t_c \end{cases} \tag{6.68}$$

6.4.2 Cases of equal vertical strain with well resistance

In reality, the discharge capacity of a drain is limited. Therefore, well resistance will delay the consolidation process and the consolidation degree according to (6.63) will be overestimated. Hansbo (1981) derived a formula that the effect of well resistance was included by further assuming that the strain rate was independent of the depth. Hansbo's (1981) approach is adopted in this section.

Rewriting (6.12) by neglecting the vertical flow in the soil gives

$$\frac{\partial}{\partial r}(rq_r) = r\frac{\partial \varepsilon_z}{\partial t} \tag{6.69}$$

Integrating (6.69) with respect to r, we obtain

$$q_r = \frac{1}{2}\left(r - \frac{r_e^2}{r}\right)\frac{\partial \varepsilon_z}{\partial t} \tag{6.70}$$

All flow in the vertical drain can be assumed to be vertical. The radial flow from the soil into the well at depth, z, is equal to the increase in flow up the well:

$$\left[\pi r_d^2 \frac{k_d}{\gamma_w}\frac{\partial^2 u_e}{\partial z^2} - 2\pi r_d q_r\right]_{r=r_d} = 0 \tag{6.71}$$

where k_d is the coefficient of permeability of the vertical drain.

Substituting (6.70) into (6.71) gives

$$\left.\frac{\partial^2 u_e}{\partial z^2}\right|_{r=r_d} = -\frac{(N^2-1)\gamma_w}{k_d}\frac{\partial \varepsilon_z}{\partial t} \tag{6.72}$$

Integrating (6.72) with respect to z by assuming that the strain rate is independent of z and introducing (6.15b), we obtain

$$u_e\,|_{r=r_d} = \frac{(N^2-1)\gamma_w}{2k_d}\frac{\partial \varepsilon_z}{\partial t}(2Hz - z^2) \tag{6.73}$$

Substituting the Darcy's law in (6.70) gives

$$q_r = \begin{cases} -\dfrac{k_s}{\gamma_w}\dfrac{\partial u_e}{\partial r} = \dfrac{1}{2}\left(r - \dfrac{r_e^2}{r}\right)\dfrac{\partial \varepsilon_z}{\partial t} & r < r_s \\ -\dfrac{k_r}{\gamma_w}\dfrac{\partial u_e}{\partial r} = \dfrac{1}{2}\left(r - \dfrac{r_e^2}{r}\right)\dfrac{\partial \varepsilon_z}{\partial t} & r \geq r_s \end{cases} \tag{6.74}$$

Integrating (6.74) with respect to r and introducing (6.14) and (6.73), we obtain

$$u_e = \begin{cases} \dfrac{\gamma_w r_e^2}{2k_r}\dfrac{\partial \varepsilon_z}{\partial t}\left\{\dfrac{k_r}{k_s}\left[\ln\left(\dfrac{r}{r_d}\right) - \dfrac{r^2 - r_d^2}{2r_e^2}\right] + \dfrac{(N^2-1)k_r}{k_d r_e^2}(2Hz - z^2)\right\} & r < r_s \\ \dfrac{\gamma_w r_e^2}{2k_r}\dfrac{\partial \varepsilon_z}{\partial t}\left[\ln\left(\dfrac{r}{r_s}\right) - \dfrac{r^2 - r_s^2}{2r_e^2} + \dfrac{k_r}{k_s}\left(\ln s - \dfrac{s^2-1}{2N^2}\right) + \dfrac{(N^2-1)k_r}{k_d r_e^2}(2Hz - z^2)\right] & r \geq r_s \end{cases} \tag{6.75}$$

$$\frac{\partial \varepsilon_z}{\partial t} = \begin{cases} \dfrac{m_v \sigma_0}{t_c}\left[1-\exp\left(-\dfrac{2T_r}{\mu_r}\right)\right] & T_r < T_{rc} \\ \dfrac{m_v \sigma_0}{t_c}\left[1-\exp\left(-\dfrac{2T_{rc}}{\mu_r}\right)\right]\exp\left[-\dfrac{2(T_r - T_{rc})}{\mu_r}\right] & T_r \geq T_{rc} \end{cases} \tag{6.76}$$

$$\bar{u}_e = \begin{cases} \dfrac{\mu_r \sigma_0}{2T_{rc}}\left[1-\exp\left(-\dfrac{2T_r}{\mu_r}\right)\right] & T_r < T_{rc} \\ \dfrac{\mu_r \sigma_0}{2T_{rc}}\left[1-\exp\left(-\dfrac{2T_{rc}}{\mu_r}\right)\right]\exp\left[-\dfrac{2(T_r - T_{rc})}{\mu_r}\right] & T_r \geq T_{rc} \end{cases} \tag{6.77}$$

$$\mu_r = \mu_s + \frac{(N^2-1)k_r}{k_d r_e^2}(2Hz - z^2) \tag{6.78}$$

$$U_r(T_r) = \begin{cases} \dfrac{T_r}{T_{rc}} - \dfrac{\mu_r}{2T_{rc}}\left[1-\exp\left(-\dfrac{2T_r}{\mu_r}\right)\right] & T_r < T_{rc} \\ 1-\dfrac{\mu_r}{2T_{rc}}\left[1-\exp\left(-\dfrac{2T_{rc}}{\mu_r}\right)\right]\exp\left[-\dfrac{2(T_r - T_{rc})}{\mu_r}\right] & T_r \geq T_{rc} \end{cases} \tag{6.79}$$

6.5 ACCURACY OF CARRILLO'S FORMULA

As proved in Section 6.3, (6.36) holds for homogeneous differential equations where the load is assumed to be applied instantaneously. For non-homogeneous equations (loading is gradually applied), (6.36) can only be regarded as an approximate relationship. Since this formula is widely used in applications even for the cases of time-dependent loading (Olson, 1977), its accuracy is examined against (6.49) for ramp loading in this section.

6.5.1 Cases when U_r is calculated using free strain assumption

The average degree of consolidation U_r based on free strain assumption for radial flow only can be obtained by substituting $L = 0$ into (6.49) and is expressed in (6.80).

$$U_r(T_r) = \begin{cases} \dfrac{T_r}{T_{rc}} - \displaystyle\sum_{m=1}^{\infty}\dfrac{4C'_m}{N^2 T_{rc}\mu_m^2}[1-\exp(-N^2\mu_m^2 T_r)] & T_r < T_{rc} \\ 1-\displaystyle\sum_{m=1}^{\infty}\dfrac{4C'_m}{N^2 T_{rc}\mu_m^2}[1-\exp(-N^2\mu_m^2 T_{rc})]\exp[-N^2\mu_m^2(T_r - T_{rc})] & T_r \geq T_{rc} \end{cases} \tag{6.80}$$

The average degree of consolidation U_v for vertical flow only is (2.28) and expressed as follows:

$$U_v(T_v)=\begin{cases}\dfrac{T_v}{T_{vc}}-\sum\limits_{n=1,3,5,\ldots}\dfrac{32}{n^4\pi^4T_{vc}}\left[1-\exp\left(-\dfrac{n^2\pi^2}{4}T_v\right)\right] & T_v<T_{vc}\\ 1-\sum\limits_{n=1,3,5,\ldots}\dfrac{32}{n^4\pi^4T_{vc}}\left[1-\exp\left(-\dfrac{n^2\pi^2}{4}T_{vc}\right)\right]\exp\left[-\dfrac{n^2\pi^2}{4}(T_v-T_{vc})\right] & T_v\geq T_{vc}\end{cases} \tag{6.81}$$

The overall average degree of consolidation calculated using Carrillo's formula (6.36) is denoted as U_c, and $U_c = 1-(1-U_r)(1-U_v)$. In order to examine the accuracy of Carrillo's formula, the difference ΔU between U_c and the overall average degree of consolidation U using the rigorous solution in (6.49) is used:

$$\Delta U = U_c - U = [1-(1-U_r)(1-U_v)] - U \tag{6.82}$$

For the case of suddenly applied loading, it has been verified that $\Delta U = 0$. That is, Carrillo's formula holds for the situation without well resistance.

Figure 6.25 shows the influence of T_c on ΔU for a typical case of $N = 20$, $s = 1.6$, $\eta^2 = 2$, and $L = 3\times10^{-4}$. It can be seen that Carrillo's formula overestimates the average degree of consolidation and the maximum difference increases with the increase of T_c. However, ΔU reaches its maximum for $T< T_c$.

Figure 6.26 shows the influence of N on ΔU for a typical case of $s = 2$, $\eta^2 = 2$, $L = 3\times10^{-4}$, and $T_c = 2$. ΔU first increases and then decreases with the increase of N. This is consistent with the fact that the radial flow will dominate the drainage

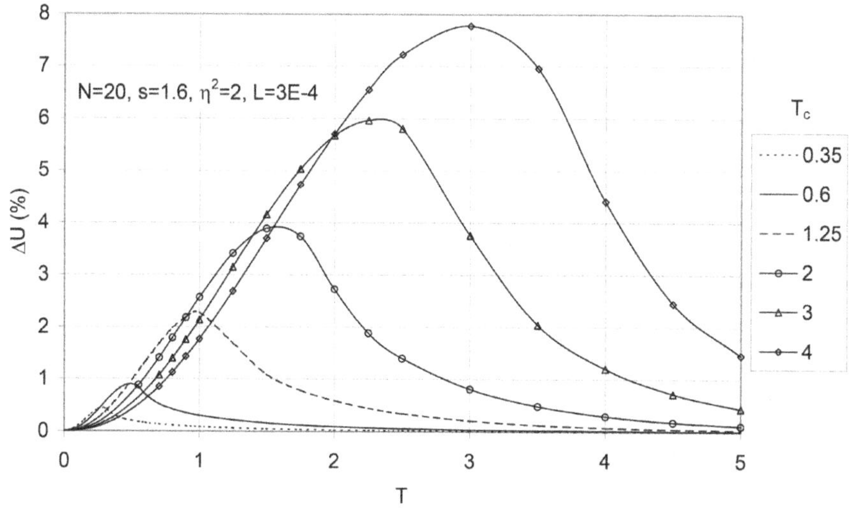

Figure 6.25 ΔU vs. T for different T_c for free strain situation.

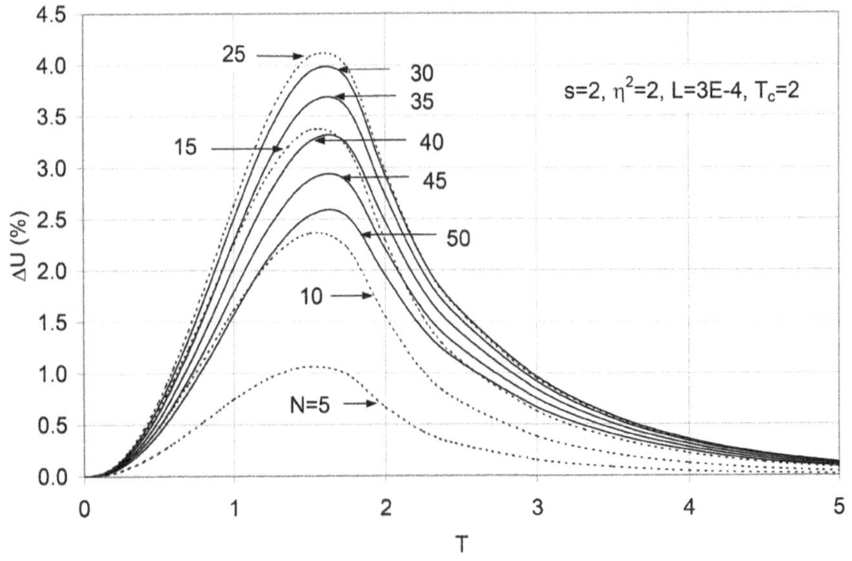

Figure 6.26 ΔU vs. *T* for different *N* for free strain situation.

for smaller N and the vertical flow will play an important role for larger N. For these two extreme situations of water flow, Carrillo's formula holds.

Figure 6.27 shows the influence of L on ΔU for a typical case of $N=20$, $\eta^2=2$, $s=2$, and $T_c=4$. ΔU first increases and then decreases with the increase of L. The maximum difference is about 8.5%. The influence of L is similar to that of N since L is an index of the two drainage paths.

The influences of s and η on ΔU are shown in Figures 6.28 and 6.29, respectively. Relative small errors are generated by these two parameters.

Thus, the difference in average degree of consolidation in the application of Carrillo's formula is caused mainly by T_c, N, and L. However, for the range analyzed ($N=5$ to 80, $s=1.2$ to 4.5, $\frac{c_r}{c_{rs}}=1$ to 10, $L=0$ to 0.01, and $T_c=0$ to 4), the overestimate is less than 9%. If $T_c \leq 1.25$, the overestimate is less than 3%.

6.5.2 Cases when U_r is calculated using equal strain assumption

In Carrillo's formula $U_c=1-(1-U_r)(1-U_v)$, the U_r is calculated using (6.63) for the case of equal strain, while the vertical U_v is still calculated using (6.81). The rigorous U is calculated using the (6.49). The difference ΔU, that is, U_c-U where applying Carrillo's formula using the equal strain assumption for U_r is then calculated using (6.82).

Figure 6.30 shows the influence of T_c on ΔU for a typical case of $N=20$, $s=1.6$, $\eta^2=2$, and $L=3\times10^{-4}$. Unlike the free strain situation, Carrillo's formula (1942) underestimates the average degree of consolidation at the

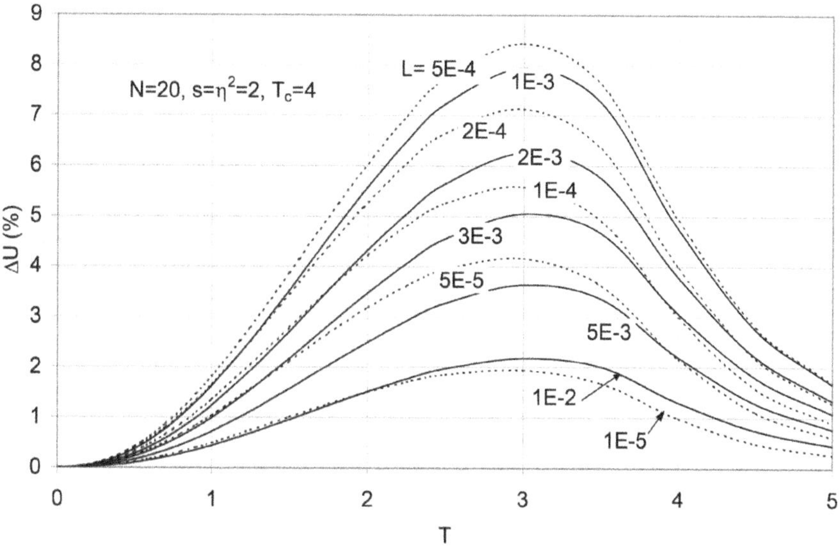

Figure 6.27 ΔU vs. T for different L for free strain situation.

initial loading stage for smaller construction time factor T_c and overestimates the average degree of consolidation for larger construction time factor. The maximum difference also increases with the increase of T_c and ΔU reaches its maximum before the construction time for larger construction time factor T_c.

Figure 6.31 shows the influence of N on ΔU for a typical case of $s = 2$, $\eta^2 = 2$, $L = 3\times10^{-4}$, and $T_c = 0$, and 2. For the case of suddenly applied loading $T_c = 0$,

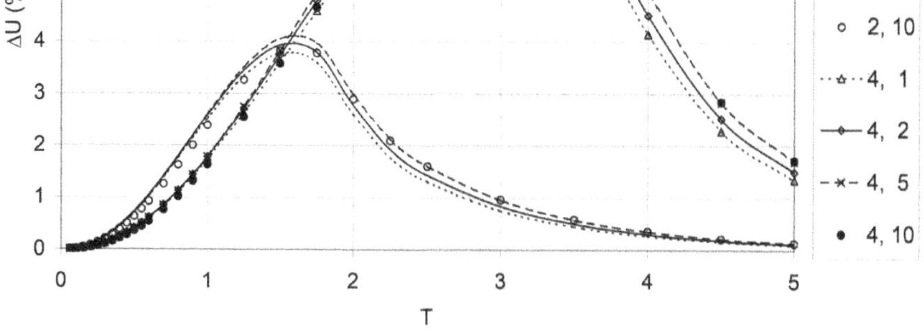

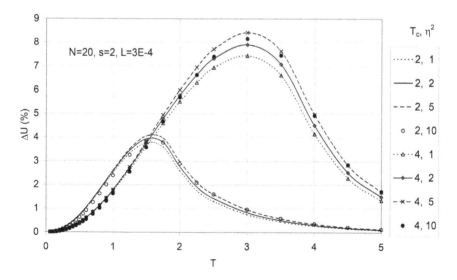

Figure 6.28 ΔU vs. T for different η for free strain situation.

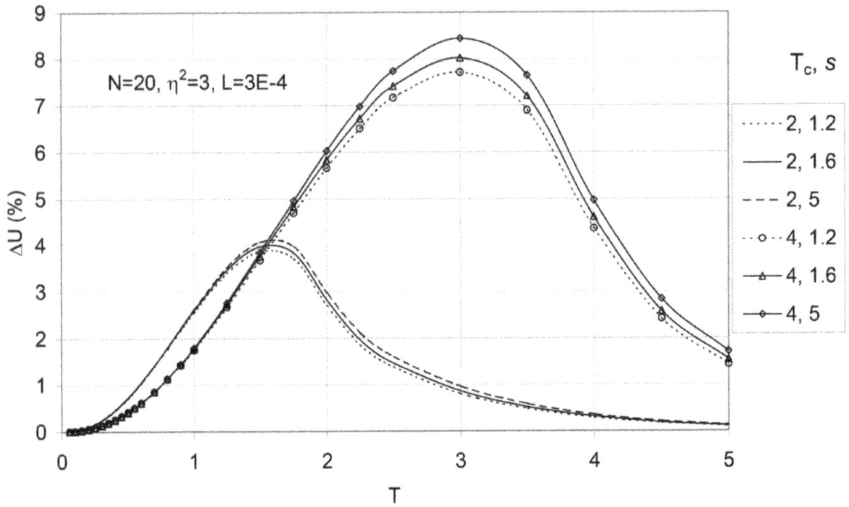

Figure 6.29 ΔU vs. T for different s for free strain situation.

ΔU is the difference between the equal strain solution and the free strain solution. Except for the initial loading stage, the difference is generally small. For larger time factor T_c, ΔU first increases and then decreases with the increase of N similar to the free strain situation.

Figure 6.32 shows the influence of L on ΔU for a typical case of $N = 20$, $\eta^2 = 2$, $s = 2$, and $T_c = 4$. Like the free strain situation, ΔU first increases and then decreases with the increase of L. The maximum difference is about 8.5%.

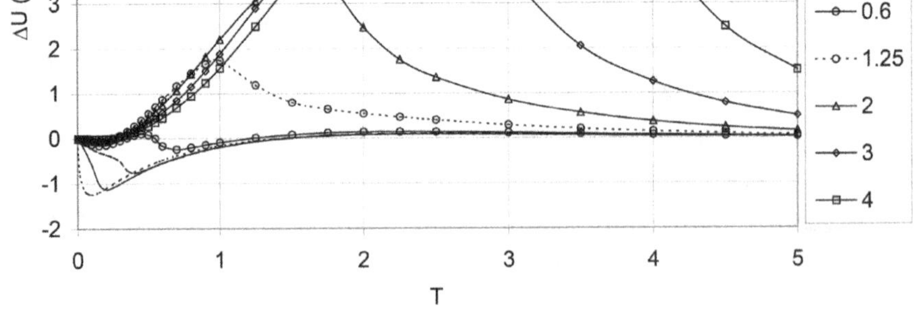

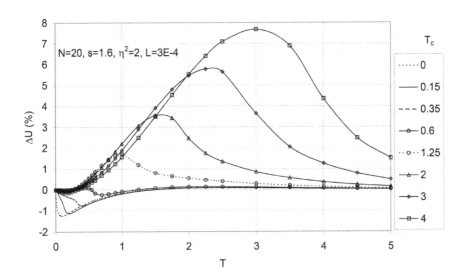

Figure 6.30 ΔU vs. T for different T_c for equal strain situation.

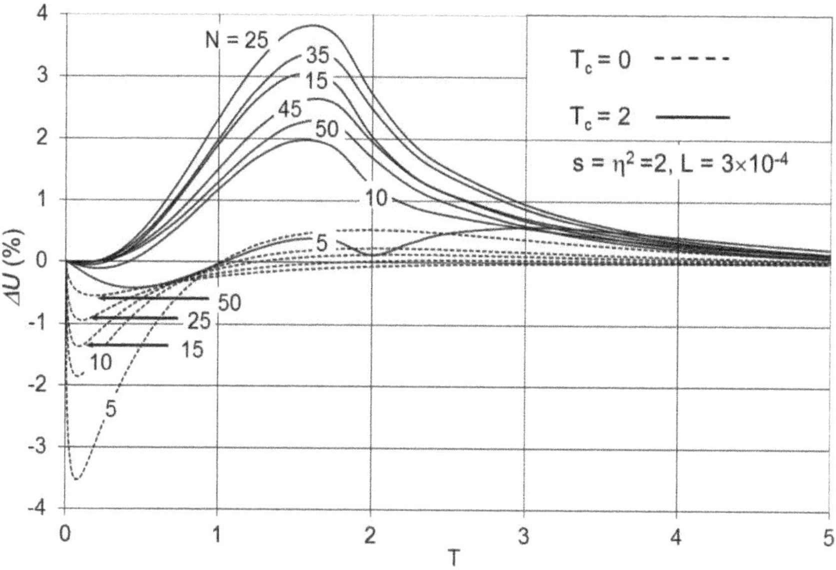

Figure 6.31 ΔU vs. T for different N for equal strain situation.

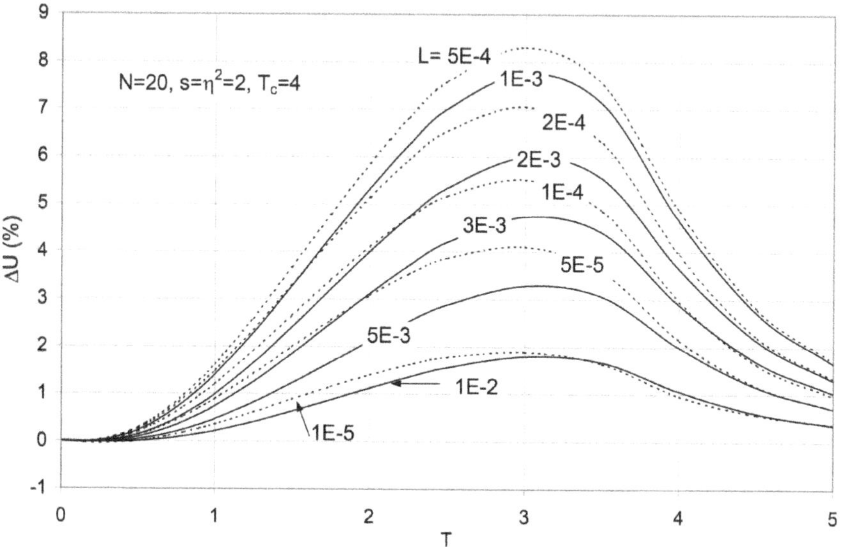

Figure 6.32 ΔU vs. T for different L for equal strain situation.

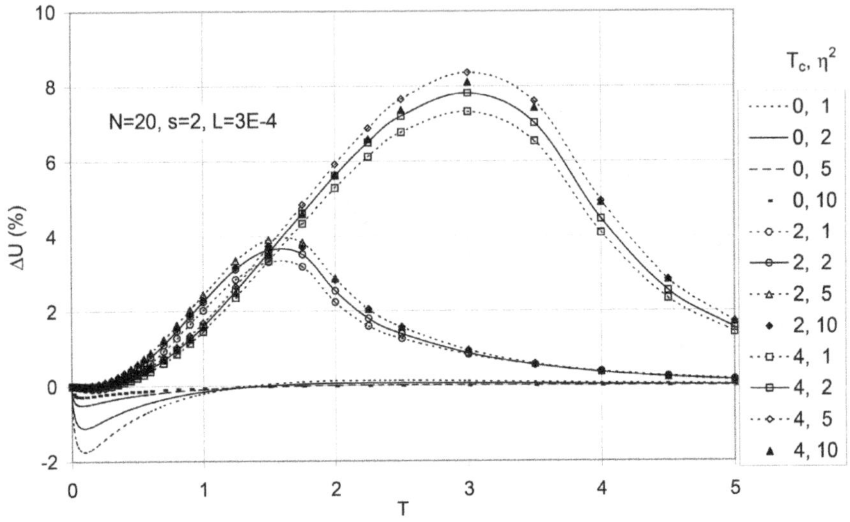

Figure 6.33 Δ*U* vs. *T* for different *η* for equal strain situation.

The influences of η and s on ΔU are shown in Figures 6.33 and 6.34, respectively. Relative small errors are generated by these two parameters similar to the free strain solution.

From the results above, it is clear that the difference in the average degree of consolidation in applying Carrillo's (1942) formula is also caused mainly by T_c, N, and L. The results from equal strain assumption will underestimate the average degree of consolidation at the initial loading stage, but smaller

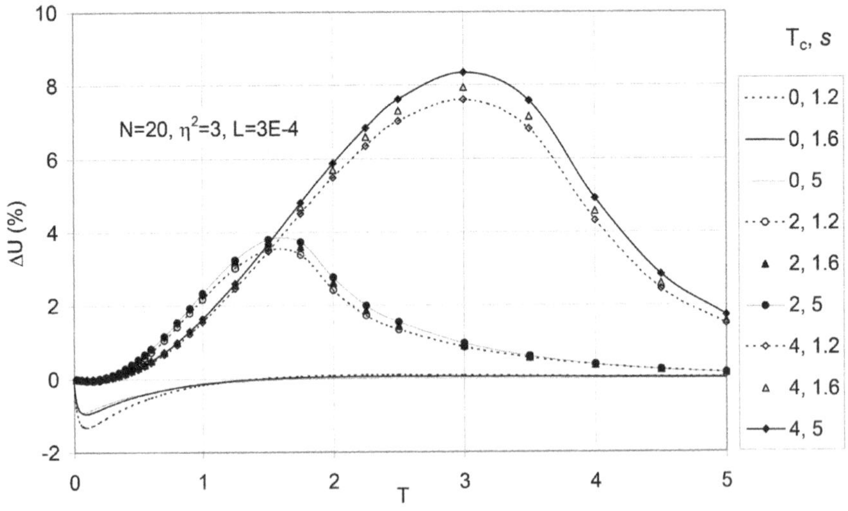

Figure 6.34 Δ*U* vs. *T* for different s for equal strain situation.

than 5% for $N \geq 5$ and smaller than 3% for $N \geq 10$. The results also overestimate the average degree of consolidation for larger construction time factor. Nevertheless, the overestimate for the range analyzed ($N = 5$ to 80, $s = 1.2$ to 4.5, $\frac{c_r}{c_{rs}} = 1$ to 10, $L = 0$ to 0.01, and $T_c = 0$ to 4) is less than 9%. If $T_c \leq 1.25$, the overestimate is less than 3%.

6.6 COMPARISON OF FREE STRAIN AND EQUAL STRAIN SOLUTIONS

Richart (1959) compared the average degrees of consolidation between free strain and equal strain solutions for ideal drain wells due to radial flow only. The differences in overall average degrees of consolidation between free strain and equal strain solutions are analyzed for time-dependent loading in this section. As in the previous sections, the following difference ΔU^* is used:

$$\Delta U^* = U^* - U \tag{6.83}$$

where U^* is calculated using (6.68) and U is calculated using (6.49).

Figure 6.35 illustrates the influence of T_c on ΔU^* for a typical case of $N = 15$, $s = 1.6$, $\eta^2 = 2$, and $L = 10^{-4}$. At the initial stage, the average degree of consolidation using the equal strain assumption is smaller than the value using the free strain assumption. In the late stage, the equal strain solution is slightly larger than the free strain solution. The maximum difference decreases with the increase of T_c.

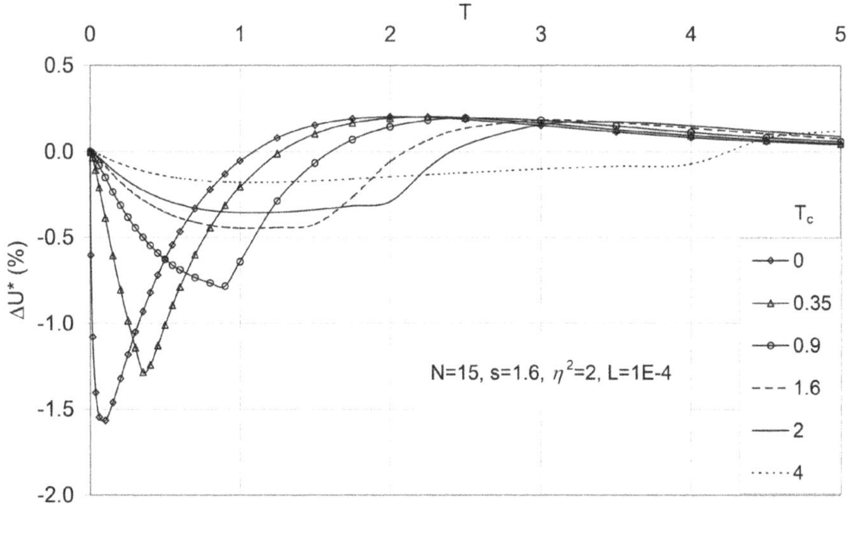

Figure 6.35 ΔU^* vs. T for different T_c.

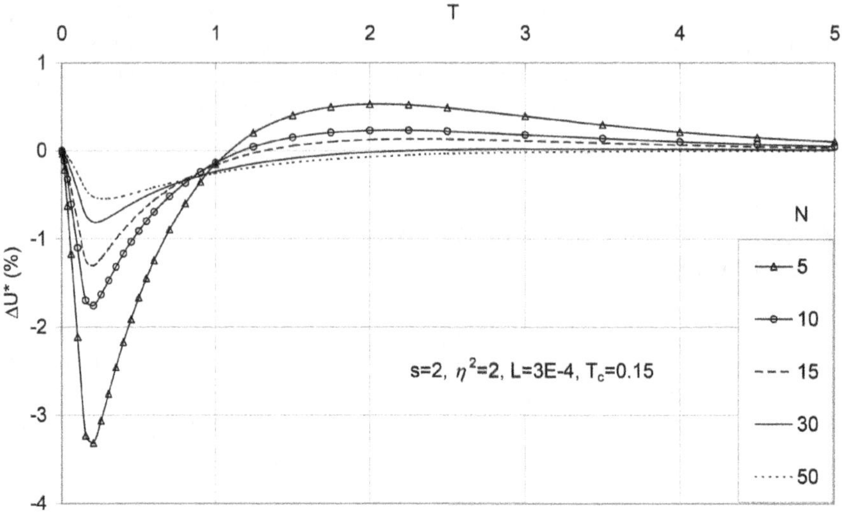

Figure 6.36 ΔU* vs. *T* for different *N*.

Figure 6.36 shows the influence of N on ΔU^* for a typical case of $s = 2$, $\eta^2 = 2$, $L = 3\times10^{-4}$, and $T_c = 0.15$. As shown in Figure 6.36, the maximum difference decreases with the increases of N.

Figure 6.37 shows the influence of η on ΔU^* for a typical case of $N = 15$, $s = 2$, $L = 10^{-4}$, and $T_c = 0$ and 0.35. The maximum difference decreases with the increases of η.

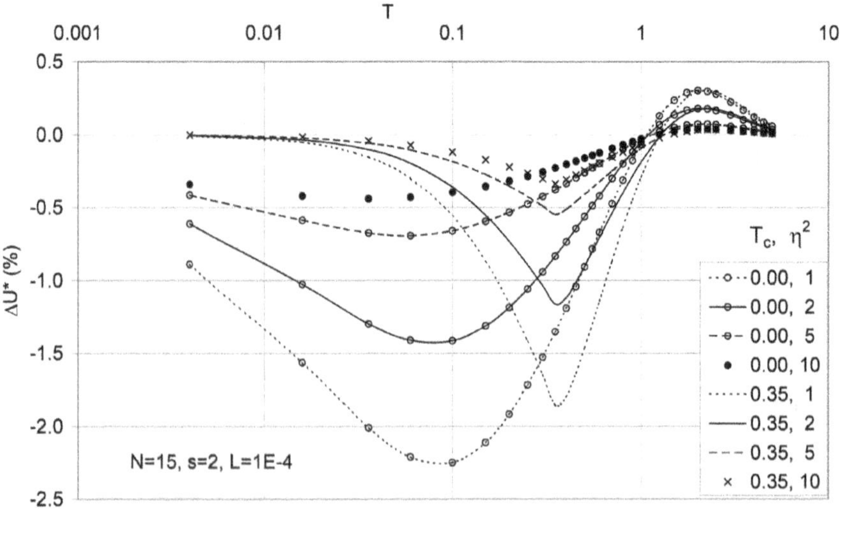

Figure 6.37 ΔU* vs. *T* for different *η*.

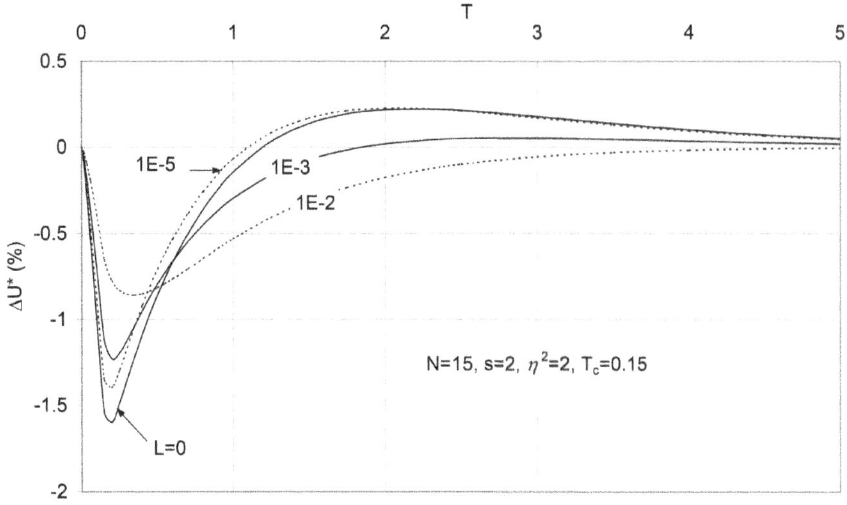

Figure 6.38 ΔU^* vs. *T* for different *L*.

The influences of L and s on ΔU^* are shown in Figures 6.38 and 6.39, respectively. The differences generated by these two parameters are within 2%.

From these results, it is clear that there are some small differences in free strain and equal strain solutions caused mainly by T_c, N and η. The results from equal strain assumption are smaller in the initial stage, but slightly larger in the late stage than the free strain values. For the range analyzed ($N = 5$ to 80, $s = 1.2$ to 4.5, $\frac{c_r}{c_{rs}}$=1 to 10, $L = 0$ to 0.01, and $T_c = 0$ to 4) ΔU^* is within −5.5

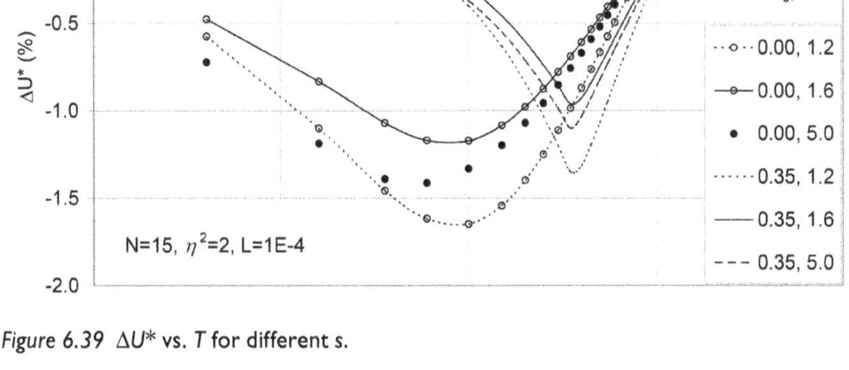

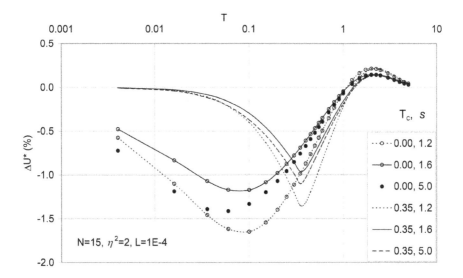

Figure 6.39 ΔU^* vs. *T* for different s.

to 1% and ΔU^* falls in the range of –2.5 to 1% for $T_c \geq 1.25$. If $N \geq 10$, ΔU^* is within –3.5 to 0.5%. This is in agreement with Richart's (1959) conclusion for ideal drain wells due to radial flow only.

6.7 CALCULATION FOR PARTIALLY PENETRATING VERTICAL DRAIN

Vertical drains sometimes may only partially penetrate the soft soil stratum. In this situation, the solutions developed for the fully penetrating case above cannot be applied directly, and analytical solutions are not available so far owing to the complex boundary conditions. Hart *et al.* (1958) suggested that the overall average degree of consolidation of the stratum could be estimated by (6.84).

$$U = U_{r,z}\frac{l}{H} + U_z\left(1 - \frac{l}{H}\right) \tag{6.84}$$

where $U_{r,z}$ is the combined average degree of consolidation of the soil above the elevation of the drain tip; U_z is the average degree of consolidation of the soil below the drain tip by vertical drainage only, assuming the drain tip as a free drainage boundary; and l is the length of vertical drain.

REFERENCES

Atkinson, MS, and Eldred, PJL, 1981, Consolidation of soil using vertical drains. *Géotechnique*, 31(1), pp. 33–43.

Barron, RA, 1948, Consolidation of fine-grained soils by drain wells. *Transactions*, ASCE, 113(1), No. 2346, pp. 718–742.

Basak, P, and Madhav, MR, 1978, Analytical solutions of sand drain problems. *Journal of the Geotechnical Engineering Division*, 104(1), pp. 129–135.

Bhide, SB, 1979, Sand drains: Additional design criteria. *Journal of the Geotechnical Engineering Division*, 105(4), pp. 559–563.

Carrillo, N, 1942, Simple two- and three-dimensional cases in the theory of consolidation of soils. *Journal of Mathematics and Physics*, 21, pp. 1–5.

Casagrande, L, and Poulos, S, 1969, On the effectiveness of sand drains. *Canadian Geotechnical Journal*, 6(3), pp. 287–326.

Hansbo, S, 1979, Consolidation of clay by band-shaped prefabricated drains. *Ground Engineering*, 12(5), pp. 16–25.

Hansbo, S, 1981, Consolidation of fine-grained soils by prefabricated drains. *In Proceedings of the 10th International Conference on Soil Mechanics and Foundation Engineering*, 3, pp. 667–682.

Hansbo, S, 1997, Aspects of vertical drain design: Darcian or non-Darcian flow. *Géotechnique*, 47(5), pp. 983–992.

Hart, EG, Kondner, RL, and Boyer, WC, 1958, Analysis for partially penetrating sand drains. *Journal of the Soil Mechanics and Foundations Division,* ASCE, 84(4), pp. 1–15.

Johnson, SJ, 1970, Foundation precompression with vertical sand drains. *Journal of the Soil Mechanics and Foundations Division*, 96(1), pp. 145–175.

Kjellman, W, 1948, Accelerating construction of fine-grained soils by means of card board wicks. *In Proceedings of the 2nd International Conference on Soil Mechanics and Foundation Engineering*, 2, pp. 302–305.

Long, RP, and Alvaro, C, 1994, Equivalent diameter of vertical drains with an oblong cross section. *Journal of Geotechnical Engineering*, ASCE, 120(9), pp. 1625–1629.

Olson, RE, 1977, Consolidation under time-dependent loading. *Journal of the Geotechnical Engineering Division*, 103(1), pp. 55–60.

Onoue, A, 1988, Consolidation by vertical drains taking well resistance and smear into consideration. *Soils and Foundations*, 28(4), pp. 165–174.

Porter, OJ, 1936, Studies of fill construction over mud flats including a description of experimental construction using vertical sand drains to hasten stabilisation. *In Proceedings of the 1st International Conference on Soil Mechanics and Foundation Engineering*, 1, pp. 229–235.

Richart, FE, 1959, Review of the theories for sand drains. *Transaction*, ASCE, 124, pp. 709–739.

Terzaghi, K, 1943, *Theoretical Soil Mechanics*, (New York: Wiley).

Yeung, AT, 1994, Design curves for prefabricated vertical drains. *Journal of Geotechnical and Geoenvironmental Engineering*, 123(8), pp. 755–759.

Yoshikuni, H, and Nakanodo, H, 1974, Consolidation of soils by vertical drain wells with finite permeability. *Soils and Foundations*, 14(2), pp. 35–46.

Zeng, GX, and Xie, KH, 1989, New development of the vertical drain theories. *In Proceedings of the 12th International Conference on Soil Mechanics and Foundation Engineering*, 2, pp. 1435–1438.

Zhu, GF, and Yin, J-H, 2000, Finite element consolidation analysis of soils with vertical drain. *International Journal for Numerical and Analytical Methods in Geomechanics*, 24(4), pp. 337–366.

Zhu, GF, and Yin, J-H, 2001, Consolidation of soil with vertical and horizontal drainage under ramp load. *Géotechnique*, 51(4), pp. 361–367.

Zhu, GF, and Yin, J-H, 2004, Consolidation analysis of soil with vertical and horizontal drainage under ramp loading considering smear effects. *Geotextiles and Geomembranes*, 22(1–2), pp. 63–74.

Zhu, GF, Yin, J-H, 2004, Accuracy of the Carrillo's formula for consolidation of soil with vertical and horizontal drainage under time-dependent loading. *Communications in Numerical Methods in Engineering*, 20, pp. 721–735.

Chapter 7

Finite difference consolidation analysis of one-dimensional problems

7.1 INTRODUCTION

Chapter 5 introduces a finite element (FE) method for the numerical consolidation analysis of saturated soils in one-dimensional (1-D) compression. The 1-D consolidation problem can also be solved using a finite difference method. Since there is only one dimension in vertical direction, the finite difference method is very effective for such 1-D consolidation analysis of liner or non-linear problems. In this chapter, the six common assumptions on consolidation analysis, discussed in **Section 2.3.1**, are still adopted.

The assumptions and the condition of continuity require that

$$\frac{\partial}{\partial z}\left(\frac{k}{\gamma_w}\frac{\partial u_e}{\partial z}\right)=-\frac{\partial \varepsilon_z}{\partial t} \tag{7.1}$$

where k is the coefficient of hydraulic conductivity which may change with depth and time, γ_w is the unit weight of water, which is a constant, u_e is the excess porewater pressure (PWP), z is the vertical co-ordinate axis, ε_z is the vertical strain, and t is time.

As summarized in **Chapter 1**, all 1-D constitutive relationships for the stress-strain behavior of soil skeleton can be written in a general equation in (1.59) in terms of effective stress σ_z' and is re-written here in a partial differential form for convenience of analysis:

$$\frac{\partial \varepsilon_z}{\partial t}=\frac{\partial f(\sigma_z')}{\partial t}+g(\sigma_z',\varepsilon_z) \tag{7.2}$$

The vertical effective stress σ_z' is equal to the total stress σ_z minus the total PWP u_w:

$$\sigma_z'=\sigma_z-u_w=\sigma_z-u_s-u_e \tag{7.3}$$

Note that the total water pressure u_w is equal to the static PWP u_s plus the excess PWP u_e, that is, $u_w=u_s+u_e$ where u_s does not change with time. The total stress σ_z is known for a given load on the boundary but may change with depth and time. Substituting σ_z' in (7.3) into (7.2), we have:

$$\frac{\partial \varepsilon_z}{\partial t} = \frac{\partial f[(\sigma_z - u_s - u_e)]}{\partial t} + g[(\sigma_z - u_s - u_e),\varepsilon_z] \tag{7.4}$$

Equation (7.4) is then substituted into (7.1), we have

$$\frac{\partial}{\partial z}\left(\frac{k}{\gamma_w}\frac{\partial u_e}{\partial z}\right) = -\frac{\partial f[(\sigma_z - u_s - u_e)]}{\partial t} - g[(\sigma_z - u_s - u_e),\varepsilon_z] \tag{7.5}$$

Equations (7.4) and (7.5) can be used to solved two unknowns, that is, the excess PWP u_e and vertical strain ε_z as a function of time t and depth z under given initial and boundary conditions.

7.2 GENERAL FINITE DIFFERENCE METHOD FOR 1-D CONSOLIDATION ANALYSIS OF MULTIPLE SOIL LAYERS

This section presents a general finite difference method and scheme for solving (7.4) with a general constitutive equation in (7.5). The two partial differential equations are highly non-linear and can be solved using the Crank-Nicholson finite difference procedure. A soil profile with N horizontal layers is shown in Figure 7.1. The top layer is subjected to a uniform surcharge of $q(t)$. For convenience, the direction of the vertical coordinate z is positive downward with origin zero at the top boundary. The thickness of each layer is h^L. The total thickness of the multi-layered system is $H = \Sigma_{L=1}^{L=N} h^L$. Equations (7.4) and (7.5) are valid for all N layers in Figure 7.1:

$$\begin{cases} \dfrac{\partial}{\partial z}\left(\dfrac{k^L}{\gamma_w}\dfrac{\partial u_e^L}{\partial z}\right) = -\dfrac{\partial f[(\sigma_z^L - u_s^L - u_e^L)]}{\partial t} - g[(\sigma_z^L - u_s^L - u_e^L),\varepsilon_z^L] \\ \dfrac{\partial \varepsilon_z^L}{\partial t} = \dfrac{\partial f[(\sigma_z^L - u_s^L - u_e^L)]}{\partial t} + g[(\sigma_z^L - u_s^L - u_e^L),\varepsilon_z^L] \end{cases} \quad L = 1,2,\ldots,N \tag{7.6}$$

Two common boundary conditions are:

a. Top boundary with $z = 0$

$$\begin{cases} u_e^1(0,t) = a^1 \\ or \\ \dfrac{\partial u_e^1(0,t)}{\partial z} = b^1 \end{cases} \tag{7.7}$$

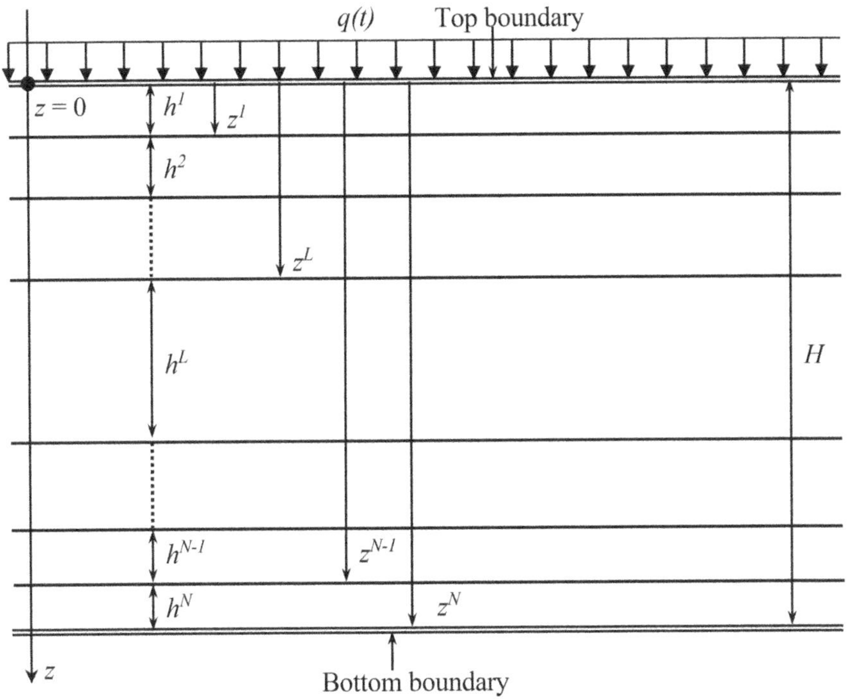

Figure 7.1 A soil profile with *N* number of horizontal layers subject a uniform surcharge *q*(*t*).

b. Bottom boundary with $z = H$

$$\begin{cases} u_e^N(H,t) = a^N \\ or \\ \dfrac{\partial u_e^N(H,t)}{\partial z} = b^N \end{cases} \tag{7.8}$$

In (7.7) and (7.8), a^1, b^1, a^N, b^N can be zero or a known function of time.

The interface location between layer 'L-1' with thickness h^{L-1} and layer 'L' with thickness h^L is at depth z^L as shown in Figure 7.1. At all interfaces, a full continuity condition must be satisfied, that is, the excess PWPs and the flow velocities in adjacent layers must be equal. These conditions are expressed as follows:

$$\begin{cases} u_e^{L-1}(z^L,t) = u_e^L(z^L,t) \\ k^{L-1}\dfrac{\partial u_e^{L-1}(z^L,t)}{\partial z} = k^L \dfrac{\partial u_e^L(z^L,t)}{\partial z} \end{cases} \tag{7.9}$$

where $L = 2, 3, \ldots N$.

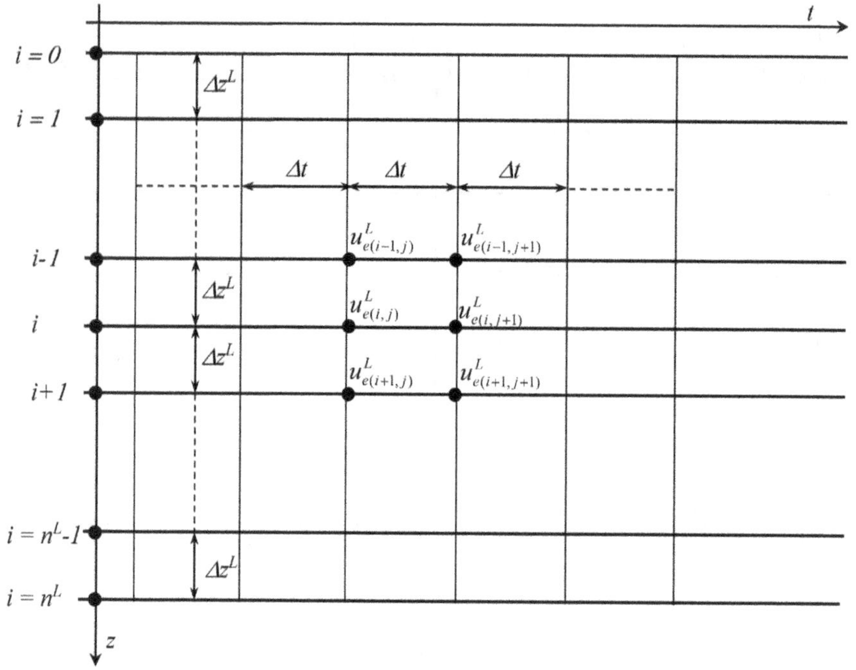

Figure 7.2 A finite difference scheme for layer L with node number $i = 0, 1, 2, \ldots n^L$.

Referring to Figure 7.2, layer L is divided into a number of sub-layers in the time direction with the same thickness Δz^L which may be different for different layers. The time increment is Δt which is assumed to be the same for all layers.

The following finite difference is used:

$$\left(\frac{k^L}{\gamma_w}\frac{\partial u_e^L}{\partial z}\right)_{(i,j)} \approx \frac{k_{i,j}^L}{\gamma_w}\frac{u_{e(i,j)}^L - u_{e(i-1,j)}^L}{\Delta z^L} \tag{7.10a}$$

$$\left(\frac{k^L}{\gamma_w}\frac{\partial u_e^L}{\partial z}\right)_{(i,j+1)} \approx \frac{k_{i,j}^L}{\gamma_w}\frac{u_{e(i,j+1)}^L - u_{e(i-1,j+1)}^L}{\Delta z^L} \tag{7.10b}$$

$$\left(\frac{k^L}{\gamma_w}\frac{\partial u_e^L}{\partial z}\right)_{(i+1,j)} \approx \frac{k_{i+1,j}^L}{\gamma_w}\frac{u_{e(i+1,j)}^L - u_{e(i,j)}^L}{\Delta z^L} \tag{7.10c}$$

$$\left(\frac{k^L}{\gamma_w}\frac{\partial u_e^L}{\partial z}\right)_{(i+1,j+1)} \approx \frac{k_{i+1,j}^L}{\gamma_w}\frac{u_{e(i+1,j+1)}^L - u_{e(i,j+1)}^L}{\Delta z^L} \tag{7.10d}$$

At point (i, j) in Figure 7.2, we have the following approximation:

$$\frac{\partial}{\partial z}\left(\frac{k^L}{\gamma_w}\frac{\partial u_e^L}{\partial z}\right) \approx \frac{\left(\frac{k^L}{\gamma_w}\frac{\partial u_e^L}{\partial z}\right)_{(i+1,j+1)} - \left(\frac{k^L}{\gamma_w}\frac{\partial u_e^L}{\partial z}\right)_{(i,j+1)}}{2\Delta z^L} + \frac{\left(\frac{k^L}{\gamma_w}\frac{\partial u_e^L}{\partial z}\right)_{(i+1,j)} - \left(\frac{k^L}{\gamma_w}\frac{\partial u_e^L}{\partial z}\right)_{(i,j)}}{2\Delta z^L} \tag{7.11}$$

Substituting (7.10) into (7.11), we have

$$\begin{aligned}\frac{\partial}{\partial z}\left(\frac{k^L}{\gamma_w}\frac{\partial u_e^L}{\partial z}\right) \approx \frac{1}{2\gamma_w(\Delta z^L)^2}&[k_{i+1,j}^L(u_{e(i+1,j+1)}^L - u_{e(i,j+1)}^L) - k_{i,j}^L(u_{e(i,j+1)}^L - u_{e(i-1,j+1)}^L)\\ &+ k_{i+1,j}^L(u_{e(i+1,j)}^L - u_{e(i,j)}^L) - k_{i,j}^L(u_{e(i,j)}^L - u_{e(i-1,j)}^L)]\end{aligned} \tag{7.12}$$

The finite difference approximation for the right side of the first equation in (7.6) is:

$$\begin{aligned}&-\frac{\partial f[(\sigma_z^L - u_s^L - u_e^L)]}{\partial t} - g[(\sigma_z^L - u_s^L - u_e^L),\varepsilon_z^L] = -\frac{\partial f}{\partial \sigma_z^L}\frac{\partial \sigma_z^L}{\partial t} - \frac{\partial f}{\partial u_s^L}\frac{\partial u_s^L}{\partial t} - \frac{\partial f}{\partial u_e^L}\frac{\partial u_e^L}{\partial t} - g^L\\ &\approx -\frac{1}{\Delta t}\left(\frac{\partial f}{\partial u_e^L}\right)_{i,j}[u_{e,(i,j+1)}^L - u_{e(i,j)}^L] - \left(\frac{\partial f}{\partial \sigma_z^L}\frac{\partial \sigma_z^L}{\partial t}\right)_{i,j} - \left(\frac{\partial f}{\partial u_s^L}\frac{\partial u_s^L}{\partial t}\right)_{i,j} - g_{i,j}^L\end{aligned} \tag{7.13}$$

Using (7.12) and (7.13), the first equation in (7.6) is

$$\begin{aligned}&\frac{1}{2\gamma_w(\Delta z^L)^2}[k_{i+1,j}^L(u_{e(i+1,j+1)}^L - u_{e(i,j+1)}^L) - k_{i,j}^L(u_{e(i,j+1)}^L - u_{e(i-1,j+1)}^L)\\ &+ k_{i+1,j}^L(u_{e(i+1,j)}^L - u_{e(i,j)}^L) - k_{i,j}^L(u_{e(i,j)}^L - u_{e(i-1,j)}^L)] =\\ &-\frac{1}{\Delta t}\left(\frac{\partial f}{\partial u_e^L}\right)_{i,j}[u_{e,(i,j+1)}^L - u_{e(i,j)}^L] - \left(\frac{\partial f}{\partial \sigma_z^L}\frac{\partial \sigma_z^L}{\partial t}\right)_{i,j} - \left(\frac{\partial f}{\partial u_s^L}\frac{\partial u_s^L}{\partial t}\right)_{i,j} - g_{i,j}^L\end{aligned} \tag{7.14}$$

The finite difference approximation for the second equation in (7.6) is

$$\varepsilon_{z(i,j+1)}^L = \varepsilon_{z(i,j)}^L + [f_{i,j+1}^L - f_{i,j}^L] + \Delta t \times g_{i,j}^L \tag{7.15}$$

It is noted that $g_{i,j}^L$ in (7.14) is $g[(\sigma_z^L - u_s^L - u_e^L)_{i,j}, \varepsilon_{z(i,j)}^L]$, which is a function of the total vertical strain $\varepsilon_{z(i,j)}^L$. $\varepsilon_{z(i,j)}^L$ shall be updated for the next time increment as $\varepsilon_{z(i,j+1)}^L$ from (7.15) which is used in the calculation of the next $g_{i,j+1}^L$.

Defining $a = \Delta t/[\gamma_w(\Delta z^L)^2]$, (7.14) can be written as:

$$-0.5ak_{i,j}^L u_{e(i-1,j+1)}^L + 0.5a\left[k_{i,j}^L + k_{i+1,j}^L + 2\left(\frac{\partial f}{\partial u_e^L}\right)_{i,j}\right]u_{e(i,j+1)}^L - 0.5ak_{i+1,j}^L u_{e(i+1,j+1)}^L$$
$$= 0.5ak_{i,j}^L u_{e(i-1,j)}^L + 0.5a\left[k_{i,j}^L + k_{i+1,j}^L - 2\left(\frac{\partial f}{\partial u_e^L}\right)_{i,j}\right]u_{e(i,j)}^L + 0.5ak_{i+1,j}^L u_{e(i+1,j)}^L$$
$$+\Delta t\left(\frac{\partial f}{\partial \sigma_z^L}\frac{\partial \sigma_z^L}{\partial t}\right)_{i,j} + \Delta t\left(\frac{\partial f}{\partial u_s^L}\frac{\partial u_s^L}{\partial t}\right)_{i,j} + \Delta t g_{i,j}^L \tag{7.16}$$

where $i = 1, 2, 3, \ldots n^{L-1}$; $j = 0, 1, 2, \ldots m$; $L = 1, 2, 3, \ldots N$.

The finite difference for the layer interface condition in (7.9) is:

$$\begin{cases} u_{e,(n^{L-1},j)}^{L-1}(z^L,t) = u_{e,(0,j)}^L(z^L,t) \\ k_{n^{L-1},j}^{L-1}\dfrac{u_{e,(n^{L-1},j)}^{L-1} - u_{e,(n^{L-1},j)}^{L-1}}{\Delta z^{L-1}} = k_{0,j}^L\dfrac{u_{e,(1,j)}^L - u_{e,(0,j)}^L}{\Delta z^L} \end{cases} \tag{7.17}$$

where $j = 0, 1, 2, \ldots m$; $L = 2, 3, \ldots N\text{-}1$.

The finite difference for the boundary conditions in (7.7) and (7.8) is:

$$\begin{cases} u_{e,(0,j)}^1(0,t) = a^1 \\ or \\ \dfrac{u_{e,(1,j)}^1(0,t) - u_{e,(0,j)}^1(0,t)}{\Delta z^1} = b^1 \end{cases} \tag{7.18}$$

$$\begin{cases} u_{e,(n^N,j)}^N(0,t) = a^N \\ or \\ \dfrac{u_{e,(n^N,j)}^N(0,t) - u_{e,(n^N-1,j)}^N(0,t)}{\Delta z^N} = b^N \end{cases} \tag{7.19}$$

where $j = 0, 1, 2, \ldots m$.

Equations (7.15), (7.16), (7.17), (7.18), and (7.19) are a set of equations for solving two unknowns u_e and ε_z. The settlement or total compression of soil layers is an integration of the vertical strain ε_z.

7.3 FINITE DIFFERENCE SCHEME FOR 1-D CONSOLIDATION ANALYSIS OF ONE CLAY LAYER USING AN ELASTIC VISCO-PLASTIC MODEL

As presented in Chapter 1, a general elastic visco-plastic (EVP) constitutive model for 1-D straining (Yin and Graham, 1989, 1994) has been presented with the basic idea of 'equivalent time,' which describes creep strains in both the over-consolidated and normally consolidated ranges using constant creep parameters ψ and t_o. And the general 1-D EVP model for 1-D straining is expressed in (1.29), which is re-written considering the effective stress $\sigma'_z = \sigma_z - u$ with known total stress σ_z during consolidation:

$$\frac{\partial \varepsilon_z}{\partial t} = \frac{\kappa / V}{\sigma_z - u}\left(\frac{\partial \sigma_z}{\partial t} - \frac{\partial u}{\partial t}\right) + \frac{\psi / V}{t_o}\exp\left(-(\varepsilon_z - \varepsilon^r_{zo})\frac{V}{\psi}\right)\left(\frac{\sigma_z - u}{\sigma'_{zo}}\right)^{\lambda/\psi} \quad (7.20)$$

Equations (7.1) and (7.20) are a general differential system for solving 1-D consolidation problems under any load condition and load history. The finite difference method is adopted in this section to solve the consolidation analysis of one clay layer using the 1-D EVP model on (7.20).

In this chapter, the simple case of constant external loading with time is considered. Using the condition $\partial \sigma_z / \partial t = 0$, (7.20) becomes:

$$\frac{\partial \varepsilon_z}{\partial t} = \frac{-\kappa / V}{\sigma_z - u}\frac{\partial u}{\partial t} + \frac{\psi / V}{t_o}\exp\left(-(\varepsilon_z - \varepsilon^r_{zo})\frac{V}{\psi}\right)\left(\frac{\sigma_z - u}{\sigma'_{zo}}\right)^{\lambda/\psi} \quad (7.21)$$

Substituting (7.21) into (7.1):

$$\frac{k}{\gamma_w}\frac{\partial^2 u}{\partial z^2} = \frac{\kappa / V}{\sigma_z - u}\frac{\partial u}{\partial t} - \frac{\psi / V}{t_o}\exp\left(-(\varepsilon_z - \varepsilon^r_{zo})\frac{V}{\psi}\right)\left(\frac{\sigma_z - u}{\sigma'_{zo}}\right)^{\lambda/\psi} \quad (7.22)$$

The vertical strain ε_z can be solved by combing (7.21) and (7.22). In Terzaghi's theory, *coefficient of volume compressibility* (m_v) and *coefficient of consolidation* (c_v) are usually used. In this chapter, $m_v = \partial \varepsilon_z / \partial \sigma'_z = (\kappa / V)(\sigma_z - u)$ and $c_v = k/(m_v \gamma_w)$ can also be introduced for the EVP model. It should be noted that in Terzaghi's theory these parameters are constant, but in the consolidation analysis using 1-D EVP model, they depend on time t and depth z. Equations (7.21) and (7.22) can then be re-written:

$$c_v \frac{\partial^2 u}{\partial z^2} = \frac{\partial u}{\partial t} - \frac{1}{m_v} g(u, \varepsilon_z) \quad (7.23a)$$

$$\frac{\partial \varepsilon_z}{\partial t} = -m_v \frac{\partial u}{\partial t} + g(u, \varepsilon_z) \quad (7.23b)$$

where

$$g(u,\varepsilon_z)=\frac{\psi/V}{t_o}\exp\left(-(\varepsilon_z-\varepsilon_{zo}^r)\frac{V}{\psi}\right)\left(\frac{\sigma_z-u}{\sigma'_{zo}}\right)^{\lambda/\psi} \tag{7.23c}$$

Equation (7.23) can be solved by using the Crank-Nicholson finite difference procedure to determine consolidation problems with respect to time (Yin and Graham, 1996). With the terminology shown in Figure 7.2, (7.23a) becomes:

$$(c_v)_{i,j}\frac{[(u_{i+1,j+1}-2u_{i,j+1}+u_{i-1,j+1})+(u_{i,j+1}-2u_{i,j}+u_{i-1,j})]}{2(\Delta z)^2}=\frac{1}{\Delta t}(u_{i,j+1}-u_{i,j})$$

$$-\left(\frac{1}{m_v}g(u,\varepsilon_z)\right)_{i,j} \tag{7.24}$$

The depth increment $\Delta z=z_{i+1}-z_i$ is kept constant, but the time increment $\Delta t=t_{j+1}-t_j$ is allowed to increase as consolidation proceeds. c_v, m_v, and $g(u,\varepsilon_z)$ in (7.23) may vary with depth and time. As a simple approximation, their values at (i, j) are used in (7.24).

Defining $r=[c_v\Delta t/(\Delta z)^2]_{i,j}$, (7.24) is allowed to be written:

$$-0.5r\times u_{i-1,j+1}+(1+r)u_{i,j+1}-0.5r\times u_{i+1,j+1}$$

$$=0.5r\times u_{i-1,j}+(1-r)u_{i,j}+0.5r\times u_{i+1,j}+\Delta t\left(\frac{1}{m_v}g(u,\varepsilon_z)\right)_{i,j} \tag{7.25}$$

where (i = 1,2,3...n–1; j = 0,1,2,...m–1). Similarly, (7.23b) becomes:

$$(\varepsilon_z)_{i,j+1}=(\varepsilon_z)_{i,j}-(m_v)_{i,j}(u_{i,j+1}-u_{i,j})+\Delta t[g(u,\varepsilon_z)]_{i,j} \tag{7.26}$$

The boundary conditions are shown in Figure 7.2 with L = 1. The initial thickness of the soil layer before consolidation is H_o. The top of the soil layer (z = 0) is freely draining, $u_e=0$. At the bottom of the layer ($z=H_o$), the bottom of the soil layer is impermeable, which is represented as $\partial u/\partial z=0$. In finite difference form, the two boundary conditions are written:

a. $u_{o,j}=0$ (where j = 0, 1, 2, ... m)
b. $u_{n,j}-u_{n-1,j}=0$ (where j = 0, 1, 2, ... m)

It is shown in Chapter 1 that total compression strain ε_z of a viscous clay is uniquely related to (σ'_z,t_e) state in (1.24) and the creep strain rate is uniquely related to $(\sigma'_z,\varepsilon_z)$ state in (1.28). Therefore, deformation during consolidation depends not only on the initial PWPs but also on the initial state $(\sigma'_z,\varepsilon_z)_{i,0}$.

Three sets of initial values, at time $t = +0$, are required before computation begins:

a. $(\sigma_z')_{i,o}$ $(i = 0, 1, 2, ...n)$
b. $(\varepsilon_z)_{i,o}$ $(i = 0, 1, 2,...n)$
c. $u_{i,o}$ $(i = 0, 1, 2, ... n)$

Conditions (a) and (c) show that the initial effective stresses can be written as:

$$(\sigma_z')_{i,o} = (\sigma_z)_{i,o} - u_{i,o} \; (i = 0, 1, 2, ... n)$$

Using these boundary conditions and initial conditions, (7.23c), (7.25), and (7.26) form a linear tridiagonal equation system for unknowns $u_{i,j}, (\varepsilon_z)_{i,j}$ $(i = 1, 2, 3...n{-}1; j = 1, 2, 3 ... m)$. The time-dependent settlement S_t at the top of the soil layer ($z = 0$) is given by the sum:

$$S_t = \int_{z=0}^{z=H_o} \varepsilon_z(t,z)dz \tag{7.27}$$

where the strain ε_z is a function of time and depth. An approximate numerical solution of (7.27) can be written as:

$$S_t = (0.5(\varepsilon_z)_{o,j} + \sum_{i=1}^{i=n-1} (\varepsilon_z)_{i,j} + 0.5(\varepsilon_z)_{n,j})\Delta z \tag{7.28}$$

Numerical solutions of (7.25), (7.26), and (7.28) can be programmed for solution to the consolidation problem using the 1-D EVP model (Yin and Graham, 1989, 1994).

7.4 SIMULATION OF CONSOLIDATION IN A CLAY LAYER WITH DIFFERENT THICKNESSES AND COMPARISON WITH TEST DATA

Laboratory modified odometer model data of strain and PWP presented by Berre and Iversen (1972) are used to evaluate the finite different method for consolidation of a soil using the 1-D EVP model. In the consolidation model, four different initial thicknesses of soil specimens were tested: 18.8 mm (test 7), 75.7 mm (test 6), 150.1 mm (test H6), and 450.1 mm (test H4). The soil material was a carefully sampled, lightly over-consolidated, marine post-glacial clay from 5.2 to 6.7 m depth. Its clay fraction (<2 μm) was 45%, natural water content 57–60%, liquid limit 54–60%, plastic limit 28–34%, OCR ~1.35, and sensitivity 10–12. The tests involved multi-stage loading with

Table 7.1 Soil parameters in the modeling of 1-D consolidation

κ/V	λ/V	ε^r_{zo}	σ'_{zo} (kPa)	ψ/V	t_o (min)	k (m/min)	c_v (m^2/min)	m_v (m^2/kN)
0.004	0.158	0	79.2	0.007	40	1.0×10^{-7}	1.51×10^{-4}	6.75×10^{-5}

various time durations. They allowed examination of any creep straining that might occur during primary consolidation.

Permeability k and coefficient of volume compressibility m_v are determined using data in test 7 and listed in Table 7.1. Using values of k and m_v, the coefficient of consolidation c_v is calculated using (2.14) and listed in Table 7.1. Six parameters in the 1-D EVP model (Yin and Graham, 1989, 1994) were determined using the rigorous method in Chapter 1 and listed in Table 7.1. Conditions of four odometer model tests are summarized in Table 7.2.

Results of relationships of S/H_o and log(time) and (b) u_e and log(time) are shown in Figures 7.3, 7.4, 7.5, and 7.6 for test 7, test 6, test H6, and test H4, respectively (Yin and Graham, 1996). S and H_o are the settlement and initial thickness of each specimen. S/H_o is a ratio or called average strain of each specimen. Specimen H4 was instrumented in three separate segments with free drainage at the top of the topmost segment and no drainage at the bottom of final segment as shown in Figure 7.6. Average compression of each segment in H4 was measured (Berre and Iversen, 1972). Total compression S(or settlement) of H4 is sum of compressions of the three segments in Figure 7.6.

In Table 7.2, $(\varepsilon_z)_{i,o}$, which is an initial strain at the start of a new load increment, has been accumulated up to the end of the previous total stress

Table 7.2 Thickness, initial stress, strain, PWP, and duration times from odometer tests on Drammen clay (Berre and Iversen, 1972)

Test	Increment	H_o (m)	$(\varepsilon_z)_{i,o}$ (%)	σ_{zo} (kPa)	$(\sigma_z)_{i,o}$ (kPa)	$u_{i,o} = \Delta\sigma_z$ (kPa)	t (min)
7	4	0.0188	2.25	55.3	92.5	37.2	7055
	5	0.0188	6.08	92.5	140.2	47.7	10000
6	4	0.0757	1.25	55.9	93.3	37.4	8060
	5	0.0757	5.51	90.3	140.5	47.7	5964
H6	4	0.150	1.20	55.2	92.5	37.3	11230
	5	0.150	4.53	92.5	140.2	47.7	10000
H4	4						
	Top	0.450	1.15	53.4	89.2	35.8	30440
	Middle	0.450	1.25	53.4	89.2	35.8	30440
	Lower	0.450	1.50	53.4	89.2	35.8	30440
H4	5						
	Top	0.450	4.25	89.2	134.7	45.5	61450
	Middle	0.450	5.21	89.2	134.7	45.5	61450
	Lower	0.450	6.30	89.2	134.7	45.5	61450

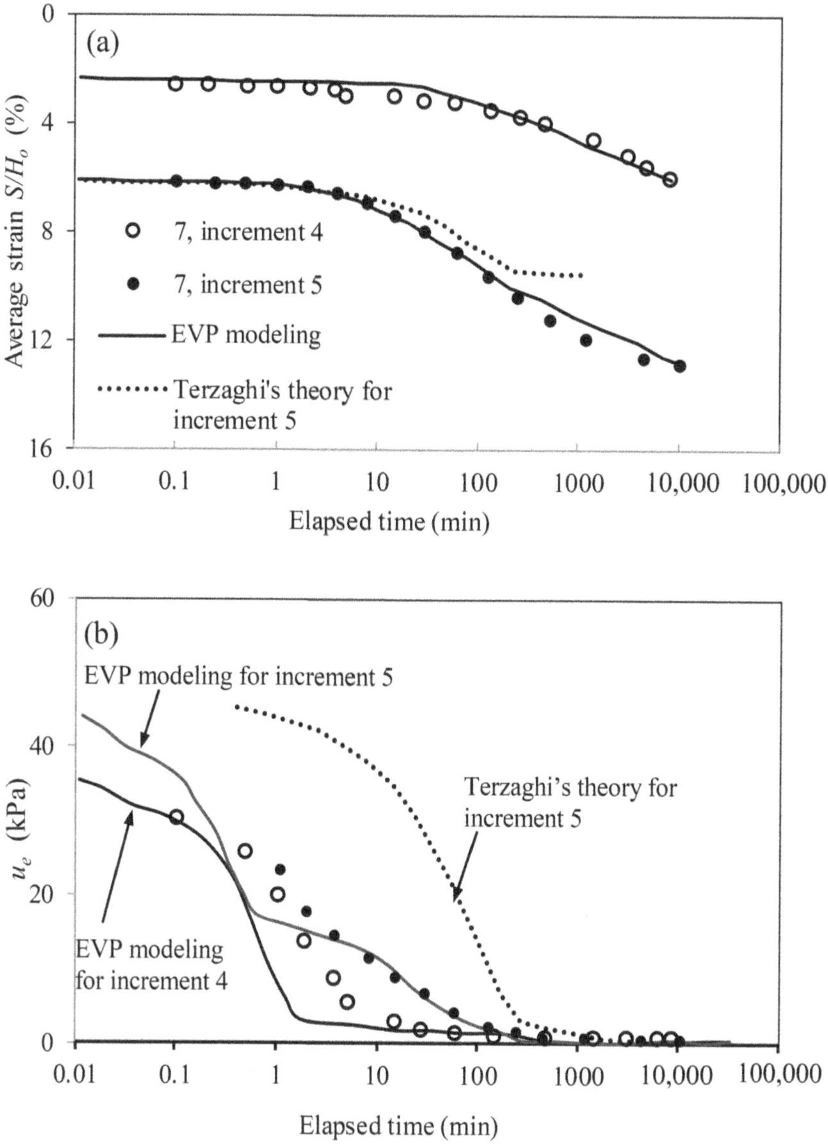

Figure 7.3 Comparison of measured and calculated results using the 1-D EVP model and Terzaghi's theory: (a) S/H_o versus time and (b) u_e versus time at increments 4 and 5 in test 7.

σ_{zo}. σ_{zo} is the previous stress before the new loading is applied. Since $u = 0$ at equilibrium, $\sigma_{zo} = \sigma'_{zo}$. The initial total stress $(\sigma_z)_{i,o}$ is the new value of σ_z immediately after new loading, i.e., at $t = 0^+$. The initial PWP $u_{i,o}$ at $t = 0^+$ is equal to the total stress increment $\Delta\sigma_z$. The time t is the duration of each load increment which is keep unchanged.

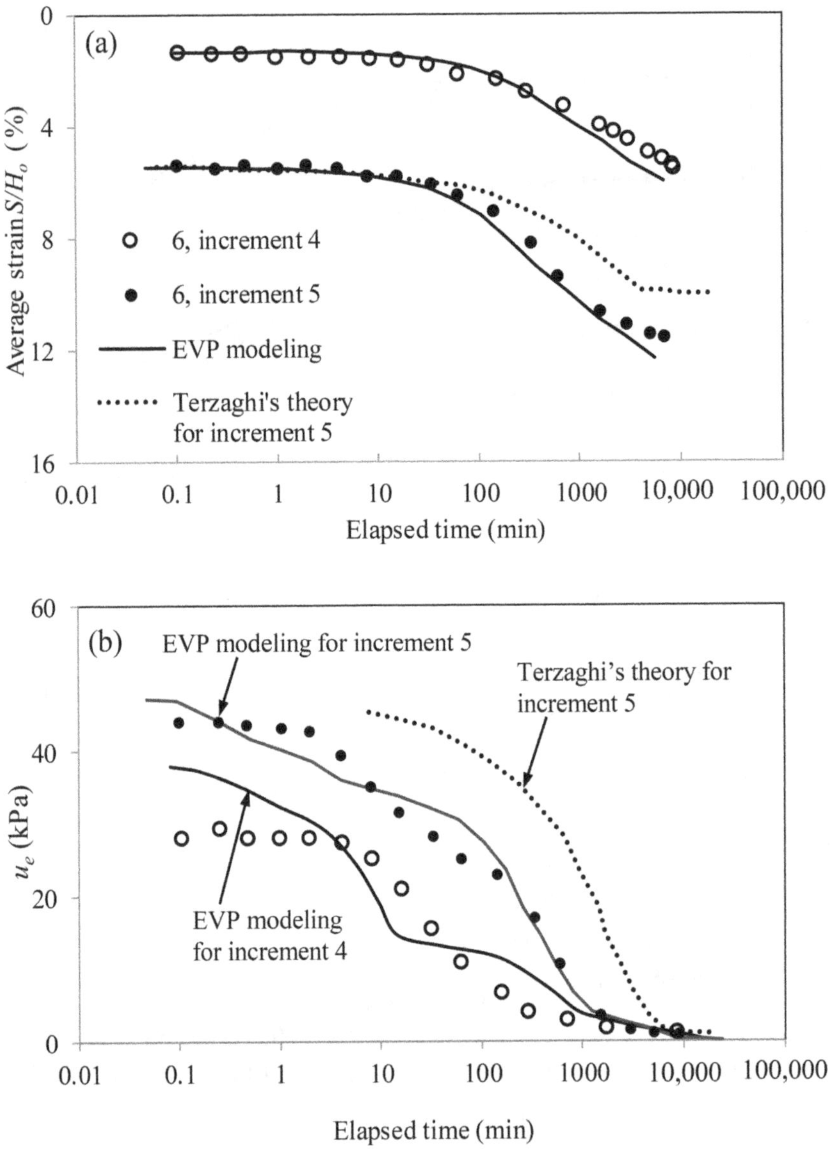

Figure 7.4 Comparison of measured and calculated results using the EVP model and Terzaghi's theory: (a) S/H_o versus time and (b) u_e versus time at increments 4 and 5 in test 6.

With the finite difference procedure outlined above, (7.23a) and (7.23b) have been solved for the conditions in tests 6, H6, and H4 (Berre and Iversen, 1972). All model parameters have been determined using data when $u_e \approx 0$ from test 7 and listed in Table 7.1. Modeling results are compared with measured values to examine the validity of the 1-D EVP model in Figures 7.3, 7.4, 7.5, and 7.6.

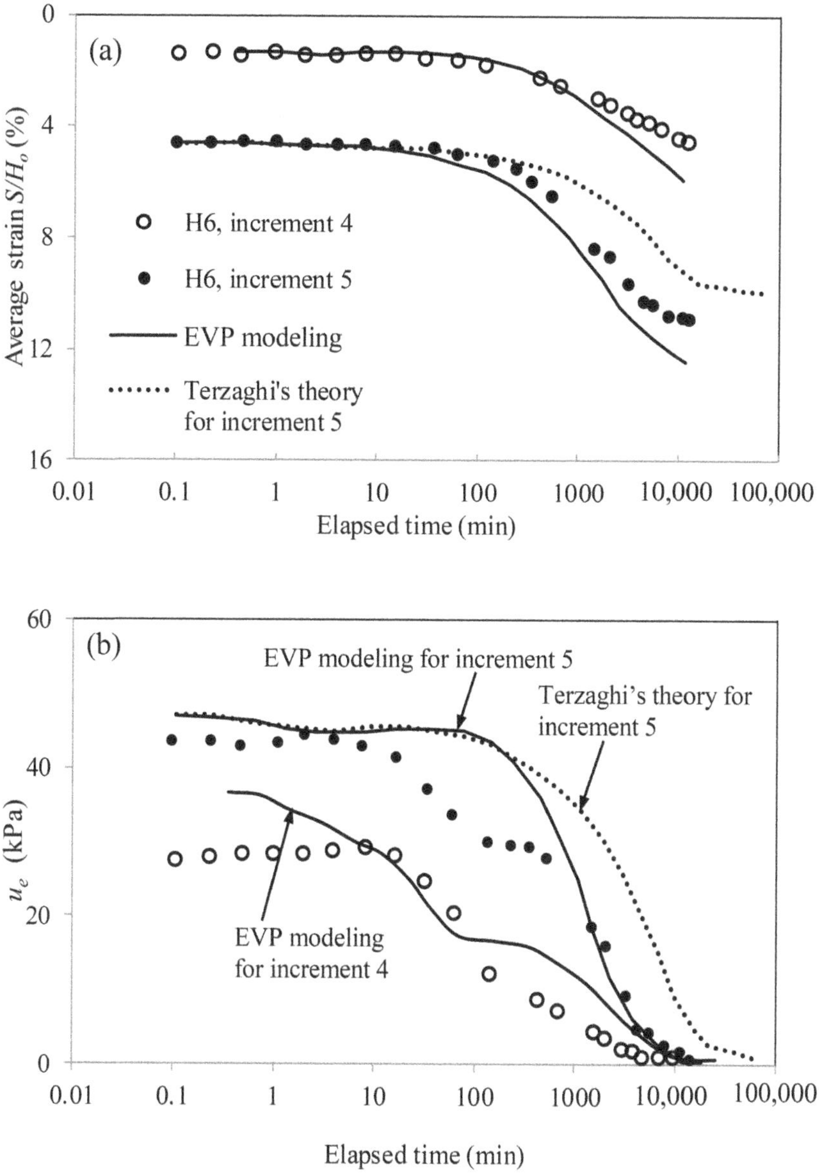

Figure 7.5 Comparison of measured and calculated results using the EVP model and Terzaghi's theory: (a) S/H_o versus time and (b) u_e versus time at increments 4 and 5 in test H6.

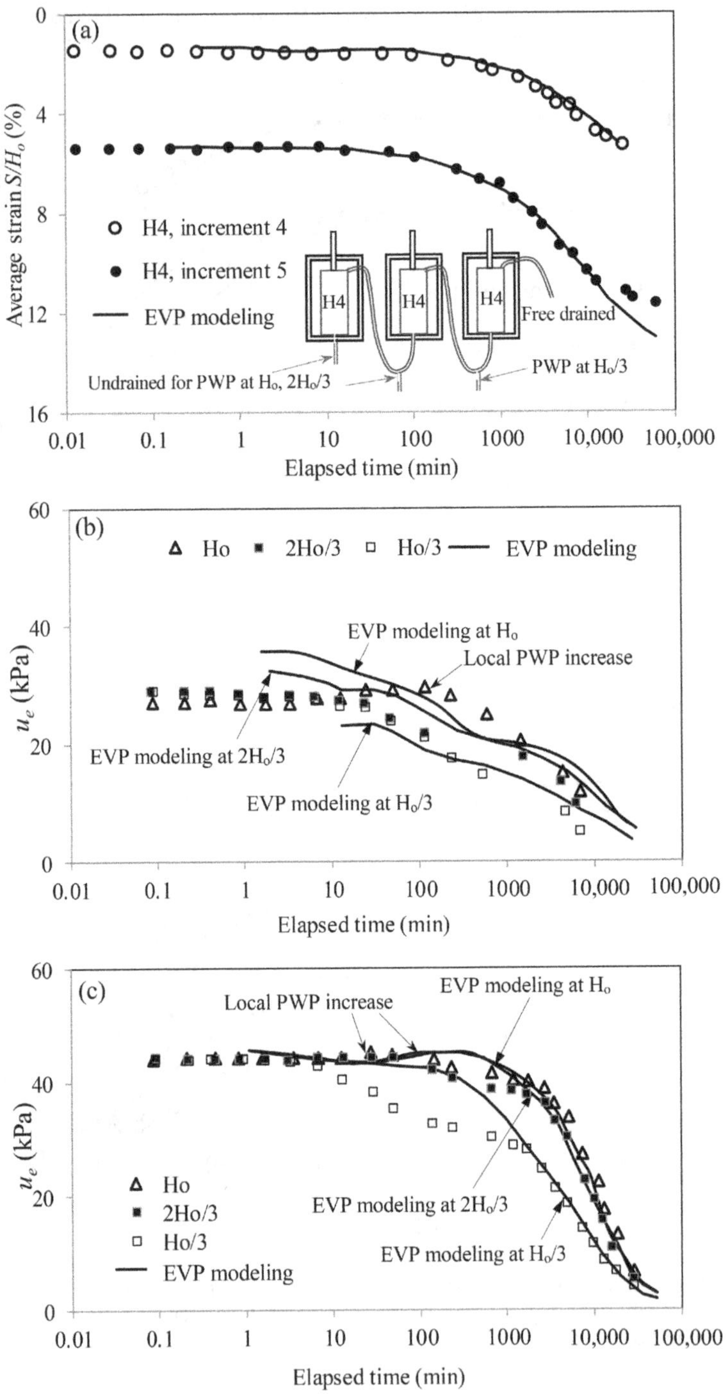

Figure 7.6 Comparison of measured and calculated results using the EVP model and Terzaghi's theory: (a) S/H_o versus time, (b) u_e versus time at increment 4, and (c) u_e versus time at increment 5 in test H4.

For comparison, calculated values using Terzaghi's theory are also shown in these figures. If S_f is the settlement at the end of primary consolidation, the ratio S/H_o at any time is given by:

$$S/H_o = (S_f/H_o)\times U_v \quad for\, 0 < t \le t_{EOP} \tag{7.29}$$

where U_v is the average degree of consolidation in vertical direction obtained from traditional solutions of Terzaghi's theory.

Figure 7.3 shows comparison of measured results and values from 1-D EVP modeling and Terzaghi's theory for increments 4 and 5 of test 7 (Berre and Iversen, 1972). It is found that Terzaghi's theory gradually underestimates the average strain (or settlement) at increment 5, and since test 7 was used to determine the model parameters in the EVP constitutive equation, the good agreement is expected for the average strain in Figure 7.3(a). The PWP from Terzaghi's theory is delayed with the time of consolidation while the EVP model gives predictions closed to test data in Figure 7.3(b).

Figure 7.4 shows comparisons of measured data and predictions from the EVP model and Terzaghi's theory for test 6. Again, Terzaghi's theory underestimates the average strain in the specimen, while the EVP model agrees well with the test data. The relationship of excess PWP for Terzaghi's theory is largely different from the measured data at increment 5. Values from the EVP modeling are closer to measured PWPs at increments 4 and 5.

Figure 7.5 shows a good agreement between measured and predicted average strains up to about 1000 min of increment load duration. However, thereafter, the calculated values of EVP modeling are larger than the measured data. The calculated excess PWPs of EVP measured agree largely with measured data for the two increments with more differences for time less than 10 min at increment 4. Again, Terzaghi's theory underestimates compression and predicts delayed response of excess PWP with more errors than that of the EVP modeling at increment 5.

For test H4 data, the predicted average strains of EVP modeling agree well with measured data at increment 4 and increment 5 within first 20,000 min as shown in Figure 7.6(a). It is noted that the excess PWPs at $H_o/3, 2H_o/3, H_o$ were measured at the end of the first segment at location $z = \frac{1}{3}H_o$, the end of the second segment at location $z = \frac{2}{3}H_o$, and the end of the third segment at location $z = H_o$ of H4 specimen as shown in Figure 7.6(a). As shown in Figure 7.6(b), the predicted PWP values of EVP modeling are larger than the measured values for time less than 100 min and are in good agreement with measured data afterwards at increment 4. Figure 7.6(c) shows that the predicted PWPs of EVP modeling agree largely well with the measured data at increment 5. It is interesting to note that the measured PWP data at local middle periods show an increasing phenomenon in Figure 7.6(b) and (c). This increasing phenomenon of local PWP during the consolidation process can be predicted with EVP modeling as shown in Figure 7.6(c). The mechanism of this increasing phenomenon of local PWP will be discussed later in this chapter.

The linear rheological model of Gibson and Lo (1961) and the non-linear rheological model of Barden (1965) are used to calculate compression and PWP and compared with measured data in Figures 7.7 and 7.8. It is noted that the average strains in Figures 7.7 and 7.8 are the incremental average

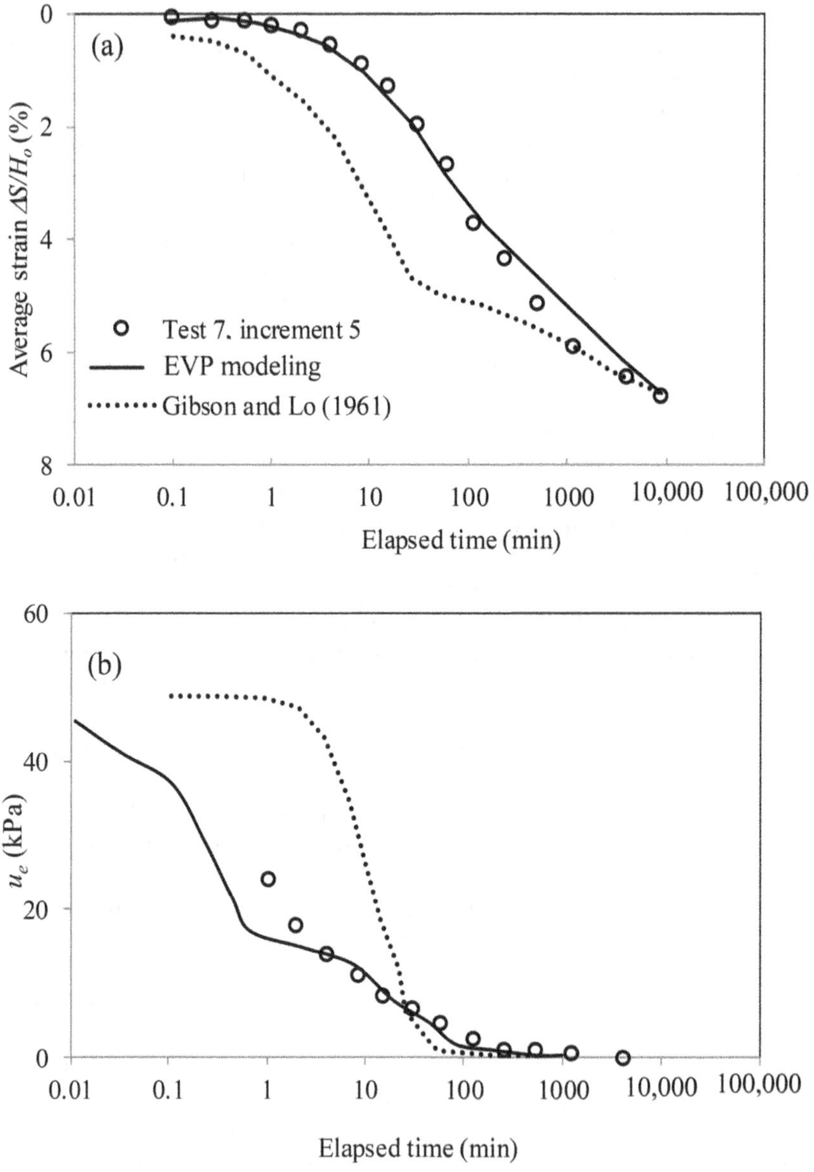

Figure 7.7 Comparison of measured and calculated results using the EVP model and Gibson and Lo's (1961) model: (a) $\Delta S/H_o$ versus time and (b) u_e versus time at increment 5 in test 7.

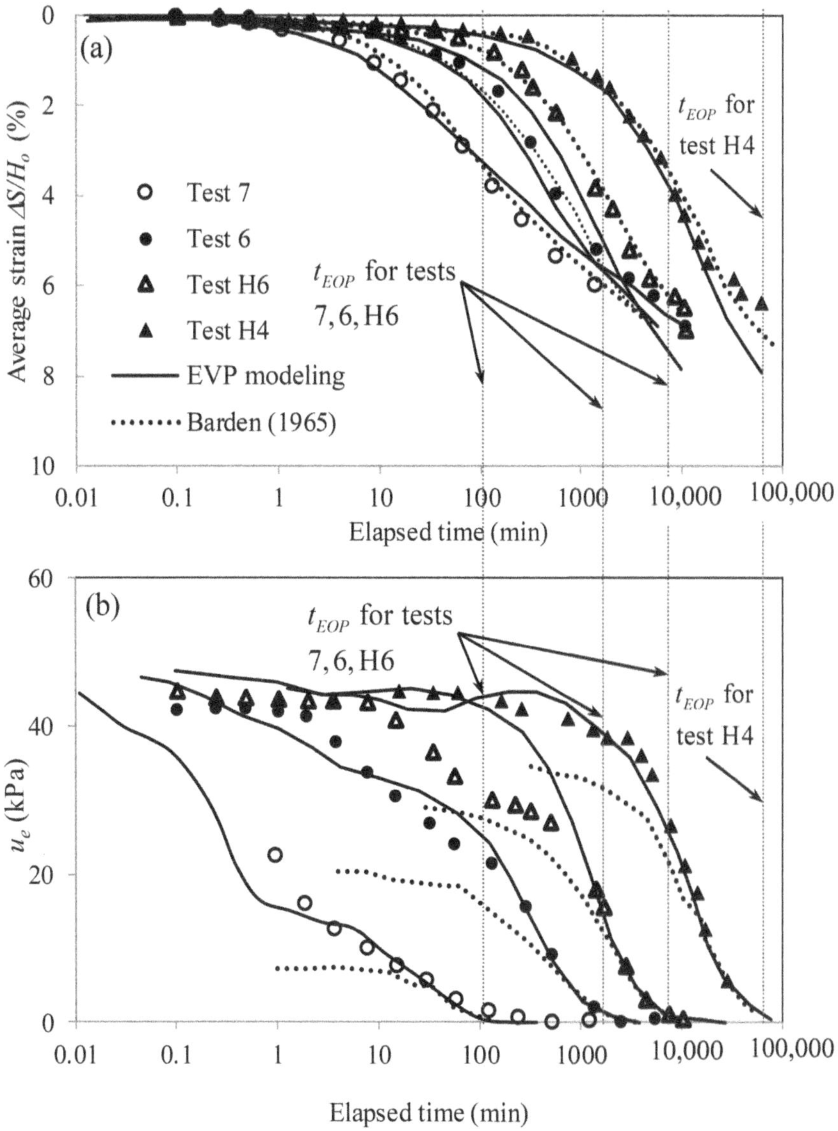

Figure 7.8 Comparison of measured and calculated results using the EVP model and Barden's (1965) model: (a) $\Delta S/H_o$ versus time and (b) u_e versus time at increment 5 in test 7, 6, H6, and H4.

strains $\Delta S/H_o$, where ΔS is the incremental compression (or settlement) from increment 4 to increment 5. It is seen from Figure 7.7 that calculated curves of $\Delta S/H_o$ and u_e with log(time) agree well with measured data. Gibson and Lo's model underestimates the average strains in the middle stage of the testing and overestimates the PWPs except at the end of the increment. Similarly, the comparison of measured and calculated results using EVP model and Barden's model is shown in Figure 7.8 for all four test conditions. It can be observed that Barden's model (Barden, 1965) predicts the average strain quite well at increment 5, however, predicts a lower PWP than the measured data. Values from the 1-D EVP model are in good agreement with measured data in all four tests for both the average strain and excess PWP. There are discrepancies on average strains, when time is large, from the EVP modeling. This may be caused by a few reasons: (a) hydraulic conductivity k is assumed to be constant and (b) soil specimens were taken from less undisturbed samples from the field and the composition and degree of disturbance of all for specimens might not be the same.

7.5 INFLUENCES OF THICKNESS ON CONSOLIDATION OF A SOIL LAYER AND METHODS OF HYPOTHESIS A AND HYPOTHESIS B

7.5.1 Influences of thickness on consolidation of a soil layer by EVP consolidation modeling

According to Terzaghi's theory (Terzaghi, 1925, 1943), the time factor T_v is defined in (2.18) as $T_v = c_v t / H^2$ or $t = T_v H^2 / c_v$. This means that for the value of T_v, the consolidation time is inversely proportional to the maximum drainage path H^2. The larger the thickness of a soil layer, the longer the consolidation process. It is known that Terzaghi's theory does consider viscous behavior or creep of a clayey soil. How does the layer thickness of a clayey soil exhibiting creep layer influence the consolidation process including settlement and excess PWP responses needs further study.

The 1-D EVP model (Yin and Graham, 1989, 1994) described in Chapter 1 and in (7.20) is used in the consolidation equation in (7.1) for the analysis of consolidation of a soil layer with four different thicknesses $H_o = 0.0188$, 0.0752, 0.3008, and 1.203 m. The conditions for $H_o = 0.0188$ m are the same as for test 7, increment 5 in Table 7.1. The top of the layer is freely drained, and the bottom is completely undrained. Values of all parameters in the 1-D EVP modeling are those in Table 7.1. The initial conditions of the layer with four thicknesses are assumed to be the same: $(\varepsilon_z)_{i,o} = 6.08\%$, $\sigma'_{zo} = 92.5$ kPa (the previous effective stress level), $u_{i,o} = 47.7$ kPa. The total stress $(\sigma_z)_{i,o}$ under the new loading is 140.2 kPa.

Figure 7.9 shows curves of average strains S/H_o and excess PWPs u_e with log(time) for four different clay layer thicknesses that are assumed to be freely drained at the top and completely undrained at the bottom. Figure 7.9(a) shows that the average strain curves with log(time) from by EVP consolidation modeling shift upper right and finally converges to the curve with thickness of 0.0188 m. Figure 7.9(b) shows that the curves of excess PWP with log(time), in which PWP dissipation is delayed more as the layer thickness increases from EVP consolidation modeling. The PWP in Figure 7.9(b) is u_e at the

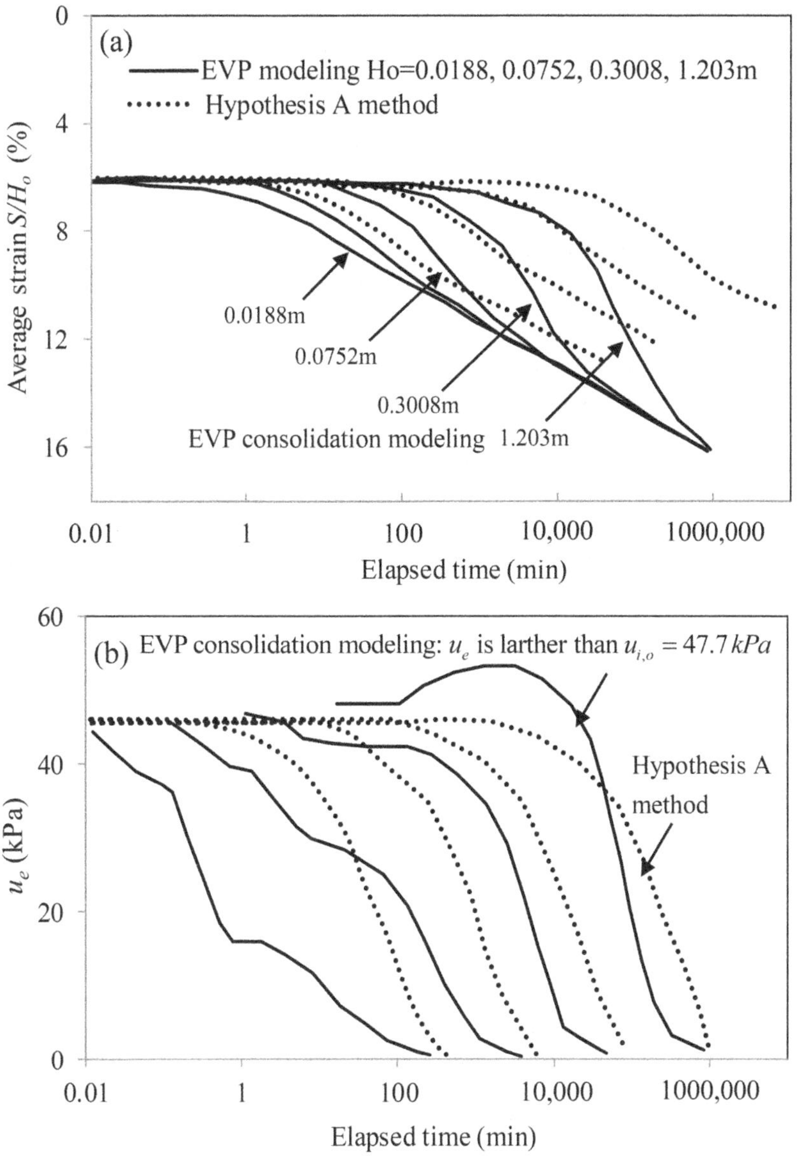

Figure 7.9 Comparison of results calculated using the EVP model and Hypothesis A method for clay layers $H_o = 0.0188$, 0.0752, 0.3008, and 1.203 m: (a) S/H_o versus time and (b) u_e versus time.

undrained bottom of the clay layer. It is noted that for thickness $H_o = 1.203$ m, the excess PWP u_e becomes larger than the initial excess $u_{i,o} = 47.7$ kPa in the initial period and is finally approaching zero as time increases. This phenomenon of the excess PWP u_e larger than the initial excess $u_{i,o}$ does not occur according to Terzaghi's 1-D consolidation theory. More explanation on this phenomenon is presented in Section 7.6.

7.5.2 Consolidation analyses of a soil layer using Hypothesis A method and Hypothesis B method

Since Terzaghi's consolidation theory cannot consider the creep or viscous compression of a soil layer, methods based on Hypothesis A method and Hypothesis B method were proposed to calculate consolidation settlement of a clayey soil exhibiting creep. Hypothesis A method assumes that creep or 'secondary' compression only after the End-Of-Primary (EOP) consolidation (Ladd *et al.*, 1977; Mesri and Godlewski, 1977; Mesri and Choi, 1985). This means that there is no creep (or viscous) compression during 'primary' consolidation. On contrary, Hypothesis B method considers that creep or viscous compression occurs during and after the EOP consolidation (Gibson and Lo, 1961; Barden, 1965; Bjerrum, 1967; Garlanger, 1972; Leroueil *et al.*, 1985).

Based on Hypothesis A assumption, a mathematical equation of Hypothesis A method for calculation of the total consolidation settlement S_{totalA} in the field is:

$$S_{totalA} = S_{primary} + S_{\text{'}secondary\text{'}} = \begin{cases} U_v S_f & for \quad t < t_{EOP,field} \\ U_v S_f + \dfrac{C_{\alpha e}}{1+e_o}\log\left(\dfrac{t}{t_{EOP,field}}\right)H & for \quad t \geq t_{EOP,field} \end{cases} \tag{7.30}$$

where $S_{primary}$ is the 'primary' consolidation settlement at time t and is equal to $U_v S_f$ in which U_v is the average degree of consolidation and S_f is the final 'primary' consolidation settlement. U_v can be calculated using Terzaghi's 1-D consolidation theory. In (7.30), e_o is the initial void ratio corresponding to initial zero strain. The 'secondary' compression settlement is $S_{\text{'}secondary\text{'}} = \frac{C_{\alpha e}}{1+e_o}\log\left(\frac{t}{t_{EOP,field}}\right)H$ and is calculated for $t \geq t_{EOP,field}$ which is the time at 'EOP' in the field condition. The $t_{EOP,field}$ is dependent on the thickness of the soil layer and hydraulic permeability of the soil. $C_{\alpha e}$ is the coefficient of 'secondary' consolidation or compression. The two main problems in (7.30) are as follows:

a. The separation of the 'primary' consolidation period and 'secondary' consolidation period is subjective. According to Terzaghi's 1-D consolidation theory, the time corresponding to excess PWP $u_e = 0$, that is, $U_v = 100\%$, is infinite, that is, $t_{EOP,field}$ shall be infinite. To get around this problem, it is often assumed that the time corresponding to $U_v \geq 98\%$ is the time $t_{EOP,field}$ at 'EOP.'

b. There is no creep when time $t < t_{EOP,field}$ as shown in (7.30), that is, no creep in the 'primary' consolidation and creep occurs right after $t_{EOP,field}$ as 'secondary' compression.

In fact, under the action of the effective stress, the skeleton of a clayey soil exhibits viscous deformation, or ongoing settlement in 1-D straining case. Even in the 'primary' consolidation period, there are effective stresses acting on soil skeleton, magnitude of which may vary with time. The rate of creep compression depends on the state of consolidation (normal consolidation or over-consolidation state) as explained in Chapter 1. Therefore, due to exclusion of creep compression in the 'primary' consolidation period, the method based on Hypothesis A normally underestimates the total consolidation settlement.

Hypothesis B method does not need those assumptions in Hypothesis A method. The method based on Hypothesis B is a coupled consolidation analysis using a proper constitutive relationship for the time-dependent stress-strain behavior of clayey soils. In this approach, the time-dependent compression or creep of the clay skeleton in the 'primary' consolidation' is naturally included in the coupled consolidation analysis. In fact, the consolidation analysis using (7.1) and (7.2) in a general case or using (7.22) or (7.23) with 1-D EVP model belongs to Hypothesis B method.

The conditions for $H_o = 0.0188$ m are the same as those in increment 5 of test 7 in Table 7.2. As shown in Figure 7.9(a) for comparison of consolidation analyses of four thicknesses of the same clay, Hypothesis A method gives smaller settlements than those from EVP consolidation modeling. The larger the thickness, the bigger the differences, or the smaller the settlements from Hypothesis A method. Figure 7.9(b) shows comparison of excess PWP dissipation with time from Terzaghi's 1-D consolidation theory adopted in Hypothesis A method and EVP consolidation modeling. It is seen that the PWP decreases with time from the initial maximum value to zero value, which is delayed as the thickness increases. EVP consolidation modeling shows the same trend but shows an increase in the initial period for thickness of 1.203 m.

7.5.3 Proof of viscous compression in 'primary' consolidation

There has been an argument whether or not viscous compression occurs in 'primary' consolidation, from which there are Hypothesis A and Hypothesis B. The proof of viscous compression occurring in 'primary' consolidation is presented below in logical reasoning and test evidence:

a. Viscous compression is time-dependent compression of the skeleton of clay particles. The vertical effective stress σ'_z is the average vertical contact forces $\Sigma N'$ of particles per area A, that is, $\sigma'_z = \Sigma N'/A$. This is the well-known effective stress principal (Terzaghi, 1925, 1943; Knappett, 2019). As explained in Chapter 1 and in Figure 1.5, action of effective stress results in viscous compression or creep. If the effective stress is constant, resulted compression is called creep compression. If the effective stress is changed with time, for example, in 'primary' consolidation, the resulted compression is called viscous compression. The creep (or viscous) strain rate is uniquely dependent on the state of effective stress and strain $(\sigma'_z, \varepsilon_z)$. From this fundamental understanding and logic, viscous compression of soil skeleton under action of effective stress always occurs in 'primary' consolidation period with positive effective values in soil skeleton.
b. Figure 7.8 shows the real data of the incremental average strain $\Delta S/H_o$ and excess PWP u_e with log(time) for Tests 7, 6, H6, and H4 with H_o = 0.0188, 0.0757, 0.150, and 0.450 m, respectively. The lines of time t_{EOP} at the EOP are also drawn in Figures 7.8(a) and (b). If we use test H4 as a reference, t_{EOP} has a value of 98,000 min. According to Hypothesis A, there is no creep compression for time smaller than t_{EOP} = 98,000 min. However, Figure 7.8(a) shows that three curves of $\Delta S/H_o$ and log(time) from tests 7, 6, and H6 with small thickness are all under the curve of test H4. This means that the compressions of tests 7, 6, and H6 are larger than that of test H4. In particular, for test 7, t_{EOP} is only 115 min as shown in Figure 7.8. But compression of test 7 from time 105 min to time of 98,000 min shall be creep compression under constant vertical effective stress since u_e is nearly zero or zero. This period from time 115 min to time of 98,000 min is, in fact, within 'primary' consolidation period of test H4. Keeping in mind that the four specimens tested were for the same clay. This proves that creep compression, or more generally speaking, viscous compression does occur in 'primary' consolidation period.
c. Figure 7.9 shows comparisons of average strain S/H_o and excess PWP u_e with log(time) for four tests with H_o = 0.0188, 0.0757, 0.3008, and 1.203 m, respectively. If we use test H_o = 1.203 m as a reference, t_{EOP} has a value of 10×10^6 min. According to Hypothesis A, there is no creep compression for time smaller than $t_{EOP} = 10 \times 10^6$ min. However, Figure 7.9(a) shows that three curves of S/H_o and log(time) from other three tests with small thickness are all under the curve of test with H_o = 1.203 m. This means that the compressions of the three tests are larger than that of test with H_o = 1.203 m. In particular, for test with H_o = 0.0188 m, t_{EOP} is only 115 min as shown in Figure 7.9. Compression from time 115 min to time of 10×10^6 min shall be creep compression under constant vertical effective stress since u_e is nearly zero or zero. This period from time 115 min to time of 10×10^6 min is within 'primary' consolidation

period of test with $H_o = 1.203$ m. This again proves that creep compression, or more generally speaking, viscous compression does occur in 'primary' consolidation period.

7.6 EXCESS POREWATER PRESSURE INCREASES DUE TO CREEP, CYCLIC LOADING, AND MANDEL-CRYER EFFECTS

7.6.1 Increase of excess porewater pressure due to creep compression

Figure 7.10 shows isochrones of vertical strain ε_z and excess PWPs u_e respectively for the clay with initial thickness $H_o = 1.203$ m. It is noted that the top of the layer is freely drained and the bottom is undrained. It is seen from Figure 7.10(a) that in the early stages of consolidation, say at time of 1.69×10^3 min, vertical strain ε_z near the undrained side is nearly unchanged, which means nearly no volume change. It is seen from Figure 7.10(b) that in the early stages of consolidation, say at time of 1.69×10^3 min, excess PWPu_e near the undrained side is larger than the initial excess PWP $u_{i,o} = 47.7\,\text{kPa}$.

Why is the excess PWP increased in clays under constant vertical stress in 1-D straining condition? Yin *et al.* (1994) were among the first who simulated and explained the phenomenon of the excess PWP increase in the initial consolidation period in 1-D straining and the creep mechanism causing such increase. In this study, Yin *et al.* (1994) used three constitutive models of (a) linear isotropic elastic model, (b) the modified Cam-Clay model (Roscoe and Burland, 1968), and (c) 1-D EVP model (Yin and Graham, 1989, 1994) to simulate the consolidation of a marine clay from Halifax, Canada. The layer is 1 m in thickness, freely drained in the top, and impermeable at the bottom. They found that when linear isotropic elastic model and the modified Cam-Clay model were used, the excess PWP at the bottom would decrease monotonically with time from the maximum value of the initial excess PWP to zero and there is no PWP increase at all. This excess PWP dissipation process is consistent with that by using Terzaghi's 1-D consolidation theory (Terzaghi, 1925, 1943). However, the consideration simulation using 1-D EVP model (Yin and Graham 1989, 1994) shows that the excess PWP dissipation process increases in the initial period for a time period of $0 < t < 20{,}000$ min, larger than the initial excess PWP $u_{i,0} = 30\,\text{kPa}$. They explained that the creep mechanism for causing the excess PWP to increase in clay. Becker *et al.* (1985) reported the site conditions, soil properties, and measured PWPs in the clay underneath artificial Tarsiut Island in the Canadian Beaufort Sea. Excess PWPs at piezometer Y in the marine clay at 2 m below the base of the sane fill and 28 m from the central symmetric line the of the artificial island were measured using electrical piezometers starting at 2 months after completion of the island. Becker *et al.* (1985) found that the observed excess PWP increased for

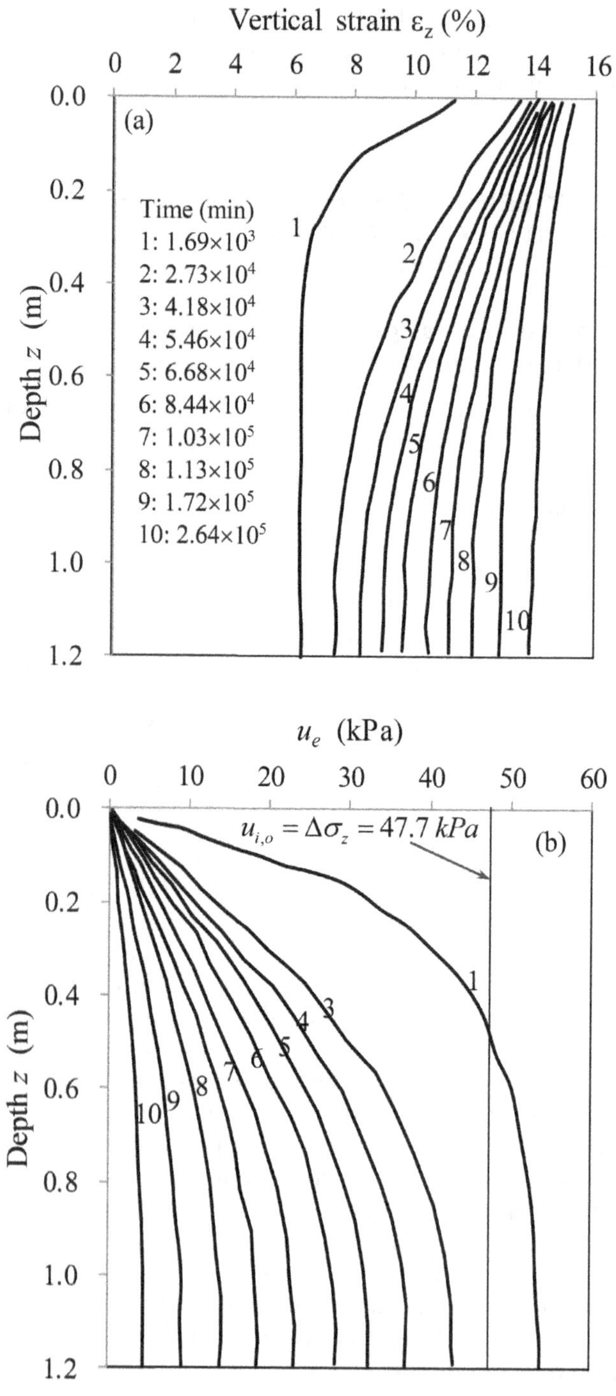

Figure 7.10 (a) Strain isochrones and (b) excess porewater pressure isochrones in a clay layer with H_o = 1.203 m from EVP consolidation modeling.

approximately 4 months (120 days) after the piezometer installation from the initial value of 170 kPa to a peak value of 190 kPa at 120 days. After completion of the island, the total load on the foundation clay was unchanged. Why was the excess PWP still increased? To answer this question, Conlin *et al.* (1985) did fully coupled FE consolidation analysis of the foundation clay underneath Tarsiut Island using the modified Cam-Clay model (Roscoe and Burland, 1968). They found that their FE modeling could not simulate the phenomenon of the excess PWP increase in the clay (Conlin *et al.,* 1985). They could not explain the reason causing such excess PWP increase.

Yin and Zhu (1999) did fully coupled FE analysis of deformation and PWP responses in the clay underneath Tarsiut Island using a three-dimensional (3-D) EVP model (Yin, 1990; Yin and Graham, 1999). Their FE model successfully simulated the phenomenon of the excess PWP increase in the clay (Yin and Zhu, 1999). In addition, they explained the reason causing such excess PWP increase in the clay at Tarsiut Island.

Yin *et al.* (1994) and Yin and Zhu (1999) discovered that the creep compression of clay skeleton is an independent factor causing the increase of excess PWP in clay. They called this causing factor as creep mechanism, which is different from the Mande-Cryer effects, but more like the reason causing PWP increase in sandy soils due to seismic loading.

Figure 7.11 shows the causes of PWP increase due (a) mechanism of creep compression, (b) cyclic shear volume compression, and (c) Mandel-Cryer effects. Figure 7.11(a) shows that for 1-D straining, the vertical effective stress is σ'_z, vertical strain is ε_z. Let us consider a physical point A at or near the undrained bottom side of a clay layer. Under loading of a constant total stress σ_z on the top drained side of the layer, the initial stress-strain state at the physical point A is at point 1 of $(\sigma'_{z1},\varepsilon_{z1})$. The corresponding vertical strain is ε_{z1} on *a-a* line in Figure 7.11(a). Noting for 1-D straining, vertical strain is equal to volume strain, $\varepsilon_z = \varepsilon_v$. Since point A is far from the top drainage side, the water at point A cannot come out in the initial period so that the total volume at point A is unchanged. However, as explained in Chapter 1, under action of constant vertical effective stress, the soil skeleton has the potential to have creep compression from point 1 to point 2 of $(\sigma'_{z2},\varepsilon_{z2})$. Since the total volume at point A is unchanged, as a result, point 2 will move, in unloading, to point 3 of $(\sigma'_{z3},\varepsilon_{z3})$ to maintain a constant volume. With time increasing, the volume at point 1 may be reduced a little, say to *b-b* line. In Figure 7.11(a), point 2 is on to *b-b* line as a general case, but there is still an unloading from point 2 to point 3.

According to the effective stress principle (Terzaghi, 1925, 1943), we have:

$$\because \sigma'_z = \sigma_z - u \quad \text{and} \quad u = \sigma_z - \sigma'_z$$
$$\therefore \Delta u = -\Delta\sigma'_z = \sigma'_{z1} - \sigma'_{z3} \quad (\because \sigma_z = \text{constant})$$

This means that the PWP is increased by the difference $(\sigma'_{z1} - \sigma'_{z3})$.

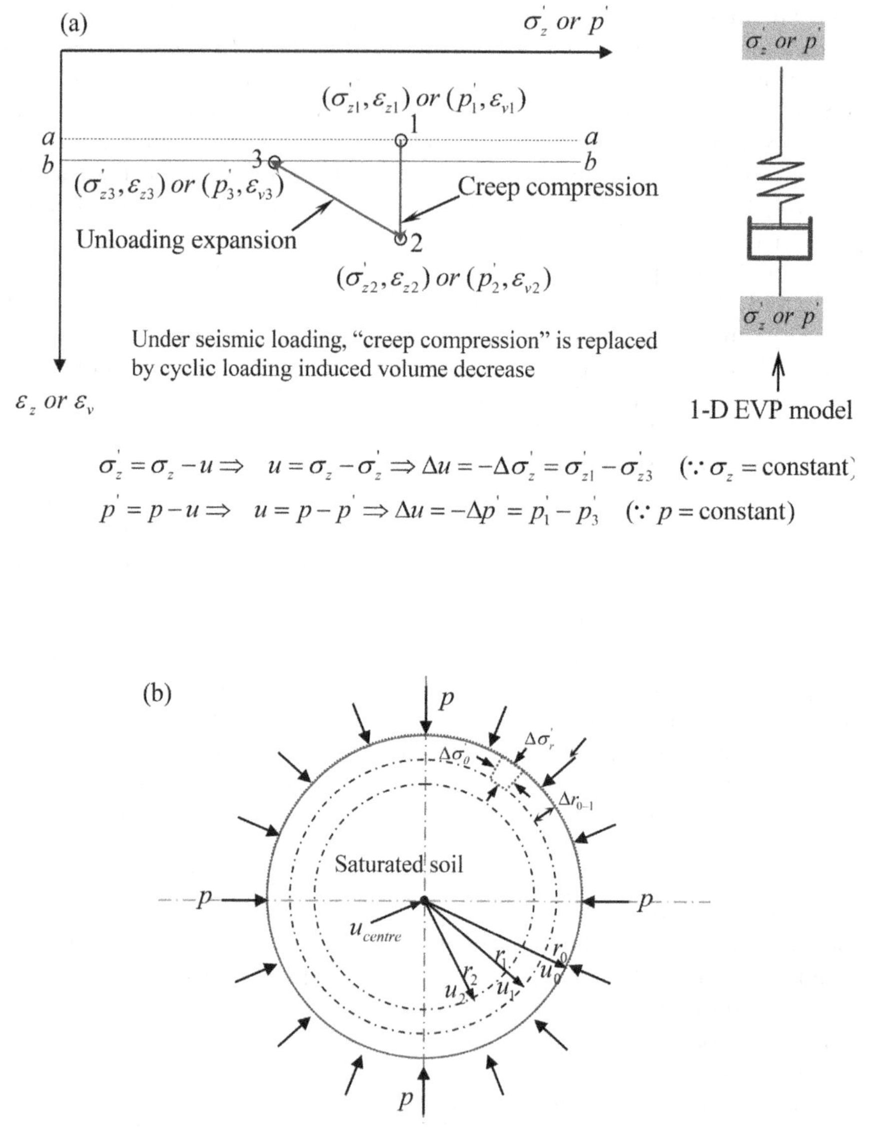

Figure 7.11 Causes of porewater pressure increase: (a) mechanism of creep compression or cyclic loading induced volume decrease and (b) Mandel-Cryer effects.

For a general 3-D stress state, Figure 7.11(a) also shows the coordinates of mean effective stress p' and volume strain ε_v. The three points will be (p'_1, ε_{v1}), (p'_2, ε_{v2}), and (p'_3, ε_{v3}). In the same explanation, the increased PWP is $\Delta u = -\Delta p' = p'_1 - p'_3 \quad (\because p = \text{constant})$.

It is noted that with time increases more, the water at point A will come out so that the volume of the soil is compressed due to the increase of effective stress, no longer kept unchanged (or a small change). In this case, there is no need to have an unloading process to compensate the creep compression of soil skeleton. Therefore, no PWP is generated. The volume compression due to the increase of effective stress is a fully coupled consolidation process, which is affected by permeability of the soil and boundary conditions.

7.6.2 Increase of excess porewater pressure due to cyclic loading

Casagrande (1936) firstly and qualitatively examined the mechanism underlying the liquefaction of saturated sands subjected to cyclic loading. Seed and Lee (1966) firstly carried out undrained cyclic triaxial tests on saturated sands and observed the PWP generated under cyclic loading resulting in liquefaction. Martin *et al.* (1975) further explained fundamentals of liquefaction under cyclic loading. It was found that the application of cyclic loading to sands results in a progressive decrease in volume. For a saturated sand, if drainage is unable to occur during the time of cyclic loading, then the tendency for volume reduction during cyclic loading results in a corresponding progressive increase in PWP. If the PWP builds up to a value equals to the confining pressure, the effective stress becomes zero. In this case, the sand lost its strength and this phenomenon is called liquefaction.

The key factor causing the PWP building up is the progress decrease in volume under cyclic loading. Figure 7.11(a) can still be used to explain this mechanism with slightly different terms. The tendency of volume decrease is due to cyclic loading rather than creep compression from point 1 to point 2. Since the sand is undrained in the initial period under quick cyclic loading, the total volume could be regarded as nearly unchanged. To balance this nearly unchanged volume, point 2 is moved to point 3 in unloading. Finally, the PWP in sand is increased by $\Delta u = -\Delta p' = p_1' - p_3'$ $(\because p = \text{constant})$.

7.6.3 Excess porewater pressure increases due to Mandel-Cryer effects

Mandel (1953) and Cryer (1963) did fully coupled Biot's consolidation analysis of soils using the simplest linear isotopic elastic model. They discovered that the PWP in a soil at locations away from drainage boundary might become larger than its initial value in 2-D and 3-D cases. This increase in PWP u is dependent on Poisson's ratio μ with a range of $0 \leq \mu \leq 0.5$. The smaller Poisson's ratio, the larger the increase in PWP. When Poisson's ratio is 0.5, there is no increase in PWP u. It is well known that when Poisson's ratio is equal to 0.5, there is no volume change under loading. The mechanism of such PWP increase is called Mandel-Cryer effects (Mandel, 1953; Cryer, 1963).

Figure 7.11(b) can be used to explain Mandel-Cryer effects. Figure 7.11(b) shows a cylinder of saturated soil, which is similar to a triaxial soil specimen.

If the soil is subjected to a pressure loading of p suddenly, the PWP u inside the soil shall have the same magnitude as the applied pressure, that is, $u = p$ if the soil surface with radius $r = r_0$ is impermeable. However, if this soil surface boundary suddenly becomes freely drained, then PWP at $r = r_0$ is zero, that is, $u_0 = 0$. With time increment Δt, the PWPs at $r = r_1, r_2$ are decreased to u_1, u_2. The average PWP decreases in the circular strips of $\Delta r_{0-1} = r_0 - r_1$ and $\Delta r_{1-2} = r_1 - r_2$ are $\Delta u_{0-1} = u_1 - u_0$ and $\Delta u_{1-2} = u_2 - u_1$. Total stresses in the saturated soil are unchanged and uniform, that is, $\sigma_r = \sigma_\theta = p$. The decrease in the PWP results in an increase in effective stresses, that is, $\Delta\sigma'_{r,0-1} = \Delta\sigma'_{\theta,0-1} = \Delta u_{0-1}$ and $\Delta\sigma'_{r,1-2} = \Delta\sigma'_{\theta,1-2} = \Delta u_{1-2}$ in the two circular strips of Δr_{0-1} and Δr_{1-2}. Under the increase of the effective stresses of $\Delta\sigma'_{r,0-1} = \Delta\sigma'_{\theta,0-1}$ and $\Delta\sigma'_{r,1-2} = \Delta\sigma'_{\theta,1-2}$, the volume of the two strips is decreased or values of radius of r_0, r_1, r_2 are decreased. This decrease will generate additional total radial and circumferential stresses Δp on the soil mass inside. Due to this Δp, the PWP inside the soil mass, say at the center of the cylinder of soil mass, is increased by Δu. If the soil has no volume change under loading, say Poisson's ratio is equal to 0.5 in an ideal linear isotropic elastic material, then the two circular strips of Δr_{0-1} and Δr_{1-2} have neither volume change nor reduction in radius. In this case, there is no increase of total radial and circumferential stresses ($\Delta p = 0$) and then no PWP increase ($\Delta u = 0$).

It is noted that if $r = \infty$, this cylinder of soil becomes a 1-D straining problem (odometer test condition) and radius in Figure 7.11(b) is regarded as the depth. In this case, the decrease in volume and depth will not generate additional stress Δp, that is, $\Delta p = 0$, then there is no increase in PWP inside the drainage boundary. This is why in Terzaghi's 1-D consolidation analysis, there is no Mandel-Cryer effects, that is, no PWP increases if the soil skeleton is not viscous. Therefore, Mandel-Cryer effects are different from the creep compression mechanism mentioned above (Yin *et al.*, 1994; Yin and Graham, 1996; Yin and Zhu, 1999).

REFERENCES

Barden, L, 1965, Consolidation of clay with non-linear viscosity. *Geotechnique*, **15**(4), pp. 345–362.

Becker, DE, Jefferies, MG, Shinder, SB, and Crooks, JHA, 1985, Porewater pressure in clays below caisson islands. *In Proceedings of the American Society of Civil Engineers Arctic 85 Conference*, San Francisco, March 1985, pp. 75–83.

Berre, T, and Iversen, K, 1972, Oedometer tests with different specimen heights on a clay exhibiting large secondary compression. *Geotechnique*, **22**(1), pp. 53–70.

Bjerrum, L, 1967, Seventh Rankine Lecture. Engineering geology of Norwegian normally-consolidated marine clays as related to settlement of buildings. *Geotechnique*, **17**(2), pp. 81–118.

Casagrande, A, 1936, Characteristics of cohesionless soils affecting the stability of slopes and earth fills. *Journal of the Boston Society of Civil Engineers*, **23**, pp. 257–276.

Conlin, BH, Jefferies, MG, and Maddock, WP, 1985, An assessment of the behaviour of foundation clay at Tarsiut N-44 caisson retained island. *In Proceedings of the* 17th *Offshore Technology Conference*, Houston, Texas, 6–9 May 1985, pp. 379–387.

Cryer, CW, 1963, A comparison of the three-dimensional consolidation of Biot and Terzaghi. *Quarterly Journal of Mechanics and Applied Mathematics*, **16**, pp. 401–412.

Mandel, J, 1953, Consolidation des sols (etude mathematique). *Geotechnique*, **3**(7), pp. 287–299.

Garlanger, JE, 1972, The consolidation of soils exhibiting creep under constant effective stress. *Geotechnique*, **22**(1), pp. 71–78.

Gibson, RE, and Lo, KY, 1961, *A Theory of Consolidation for Soils Exhibiting Secondary Compression*. Publication 41, pp. 1–16, (Oslo: Norwegian Geotechnical Institute).

Knappett, J, 2019, *Craig's Soil Mechanics*, 9th ed., (Oxford: Taylor & Francis).

Ladd, CC, Foott, R, Ishihara, K, Schlosser, F, and Poulos, HJ, 1977, Stress-deformation strength characteristics. *In Proceedings of the 9th International Conference on Soil Mechanics and Foundation Engineering, Tokyo*, pp. 421–494.

Leroueil, S, Kabbaj, M, Tavenas, F, and Bouchard, R, 1985, Stress-strain-strain rate relationship for the compressibility of sensitive natural clays. *Geotechnique*, **35**(2), pp. 159–180.

Martin, GR, Finn, WDL, and Seed, HB, 1975, Fundamentals of liquefaction under cyclic loading. *Journal of the Geotechnical Engineering Division*, ASCE, GT5, pp. 423–438.

Mesri, G, and Choi, YK, 1985, The uniqueness of end of primary (EOP) void ratio effective stress relationship. *In Proceedings of the 11th international conference on soil mechanics, San Francisco*, **2**, pp. 587–590.

Mesri, G, and Godelewski, PM, 1977, Time and stress-compressibility interrelationship. *Journal of the Geotechnical Engineering Division*, ASCE, **105**(1), pp. 106–113.

Roscoe, KH, and Burland, JB, 1968, *On the generalised stress-strain behaviour of wet clay, Engineering Plasticity*. Heyman, J, and Leckie, FA (eds.), pp. 535–609, (Cambridge: Cambridge at the University Press).

Seed, HB, and Lee, KL, 1966, Liquefaction of saturated sands during cyclic loading. *Journal of the Soil Mechanics and Foundation Division*, ASCE, **92**(6), pp. 105–134.

Terzaghi, K, 1925, *Erdbaumechanik auf Boden-physicalischen Grundlagen*, (Vienna: Deuticke).

Terzaghi, K, 1943, *Theoretical Soil Mechanics*, (New York: Wiley).

Yin, JH, 1990, *Constitutive modelling of time-dependent stress-strain behaviour of soils*. Ph.D. Thesis, The University of Manitoba, 314 pages, pp. 161–230.

Yin, JH, and Graham, J, 1989, Viscous-elastic-plastic modeling of one-dimensional time-dependent behaviour of clays. *Canadian Geotechnical Journal*, **26**, pp. 199–209.

Yin, JH, and Graham, J, 1994, Equivalent times and one-dimensional elastic visco-plastic modeling of time-dependent stress-strain behaviour of clays. *Canadian Geotechnical Journal*, **31**, pp. 42–52.

Yin, JH, and Graham, J, 1996, Elastic visco-plastic modelling of one-dimensional consolidation. *Geotechnique*, **46**(3), pp. 515–527.

Yin, JH, and Graham, J, 1999, Elastic visco-plastic modelling of the time-dependent stress-strain behavior of soils. *Canadian Geotechnical Journal*, **36**(4), pp. 736–745.

Yin, JH, Graham, J, Clark, JI, and Gao, L, 1994, Modelling unanticipated pore water pressures in soft clays. *Canadian Geotechnical Journal*, **31**, pp. 773–778.

Yin, JH, and Zhu, JG, 1999, Elastic visco-plastic consolidation modelling and interpretation of porewater pressure responses in clay underneath Tarsiut island. *Canadian Geotechnical Journal*, **36**(4), pp. 708–717.

Chapter 8

Simplified methods for calculating consolidation settlements of soils exhibiting creep

8.1 INTRODUCTION

When the stress-strain behavior of soils is time-dependent, calculation of consolidation settlements shall take into consideration of the creep compression during and after 'primary' consolidation. As discussed in Chapter 7, there are two methods for calculating consolidation settlements of soils exhibiting creep, that is, Hypothesis A (Ladd et al. 1977; Mesri and Godlewski 1977) and Hypothesis B (Gibson and Lo 1961; Barden 1965, 1969; Bjerrum 1967; Garlanger 1972; Leroueil et al. 1985; Yin and Graham 1996). It has been pointed that Hypothesis A underestimates consolidation settlements since creep (or viscous) compression in 'primary' consolidation is ignored. Hypothesis B is a rigorous method by fully coupled consolidation analysis or simulation using a suitable time-dependent constitutive relationship for soil skeleton. The consolidation analysis using one-dimensional elastic visco-plastic (1-D EVP) model (Yin and Graham 1989, 1994) belongs to Hypothesis B method as presented in Chapter 7. Nash and Ryde (2000, 2001) used Hypothesis B method with Yin and Graham's the 1-D EVP model to analyze the 1-D consolidation settlement of an embankment on soft ground with vertical drains. They used a finite difference method (Yin and Graham 1996) to solve the coupled consolidation equations. Their computed settlement values were in good agreement with the observed values. Hypothesis B method normally gives the accurate calculation of consolidation settlements and porewater pressure (Yin and Graham 1996). One limitation of Hypothesis B method is that a numerical method must be used to solve a set of non-linear partial different equations with a computer program. However, such computer program is still not readily available to engineers. Development of a simple method with good approximation is very much needed by engineers. This chapter presents firstly a formulation of a general simplified Hypothesis B method for calculating consolidation settlement of multi-layered viscous soils with or without vertical drains under any loading condition. After this, a few special simplified Hypothesis B methods (Yin and Feng 2017; Feng and Yin 2017) are introduced. Verifications and examples of these methods are presented.

8.2 FORMULATION OF A GENERAL SIMPLIFIED HYPOTHESIS B METHOD FOR CALCULATING CONSOLIDATION SETTLEMENT OF MULTI-LAYERED SOILS EXHIBITING CREEP

8.2.1 Formulation of a general simplified Hypothesis B method

Formulation of a general simplified Hypothesis B method for calculating consolidation settlement of multi-layered viscous soils with or without vertical drains under any loading condition is presented for a soil profile under uniform surcharge $q(t)$ as shown in Figure 8.1. Figure 8.1 shows a soil profile with n-layers with corresponding thicknesses $(H_1, H_2, ...H_n)$ and depths $(z_1, z_2, ...z_n)$. The total thickness is H. A vertical drain with smear zone is shown in Figure 8.1, where $d_d = 2r_d$ is diameter of a drain equal to twice radius r_d, $d_s = 2r_s$ is diameter of a smear zone equal to twice radius r_s, $d_e = 2r_e$ is diameter of an equivalent unit cell equal to twice radius r_e. Illustration of these parameters is shown in Figure 6.1. It is noted that vertical drains are installed all in the same triangular pattern or the same square pattern and are subjected to a uniform surcharge overall vertical drains, then deformation of soils in all unit cells is approximately in vertical direction. Therefore, we can

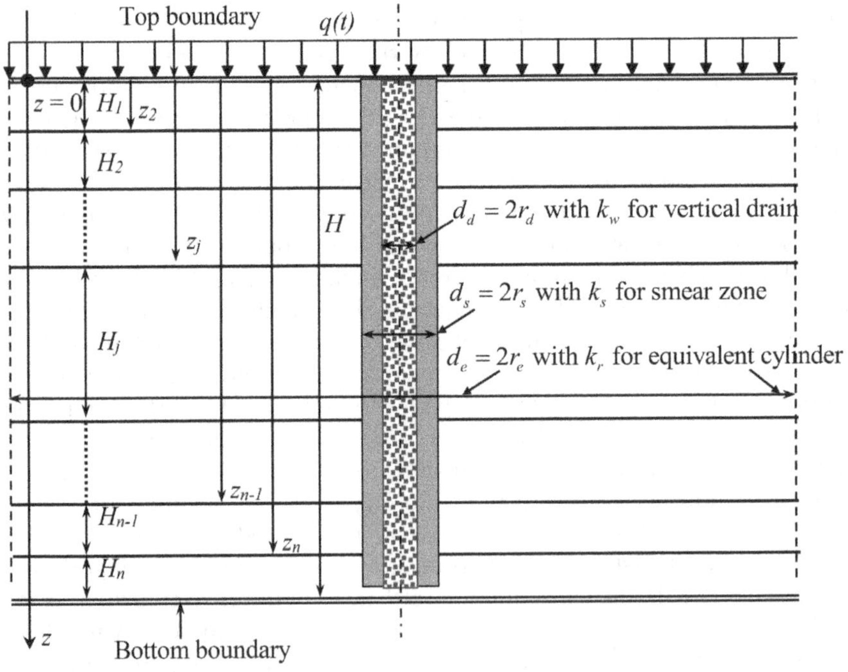

Figure 8.1 A soil profile with n-layers with vertical drain subjected to uniform surcharge $q(t)$ with time.

assume that soils in each unit cell are in 1-D straining. 1-D straining constitutive models can be used, for example, 1-D EVP model (Yin and Graham 1989, 1994). If a horizontal soil profile has no vertical drains, then $d_d = d_s = 0$ and $d_e = \infty$ in Figure 8.1 which is also suitable for multi-layered soils without vertical drains.

Referring to Figure 8.1, the formulation of a general simplified Hypothesis B method is presented below for calculating consolidation settlement of multi-layered viscous soils with or without vertical drains under any loading condition:

$$
\begin{aligned}
S_{totalB} &= S_{primary} + S_{creep} = \sum_{j=1}^{j=n} U_j S_{fj} + \sum_{j=1}^{j=n} S_{creepj} \\
&= U \sum_{j=1}^{j=n} S_{fj} + \sum_{j=1}^{j=n} [\alpha U_j^{\beta} S_{creep,fj} + (1-\alpha U_j^{\beta}) S_{creep,dj}] \\
&\quad for \;\; all \;\; t \geq t_{\mathrm{EOP},lab} \quad (t \geq t_{\mathrm{EOP},field} \; for \; S_{creep,dj})
\end{aligned}
\tag{8.1}
$$

The formulation in (8.1) is a de-coupled simplified Hypothesis B method. The 'de-coupled' means that 'primary' consolidation settlement $S_{primary}$ is separated from creep settlement S_{creep}. The separation of 'primary' consolidation from 'secondary' compression is shown in Figure 8.2. As explained in Chapter 1, value of $t_{\mathrm{EOP},lab}$ is small only tens of minutes for a standard soil specimen of 20 mm with double drainage, compared to $t_{\mathrm{EOP},field}$ from a few years to tens of years depending on the thickness and permeability of soils in the field. t_{24hrs} is the time with duration of 24 hours in an oedometer test, normally larger than $t_{\mathrm{EOP},lab}$.

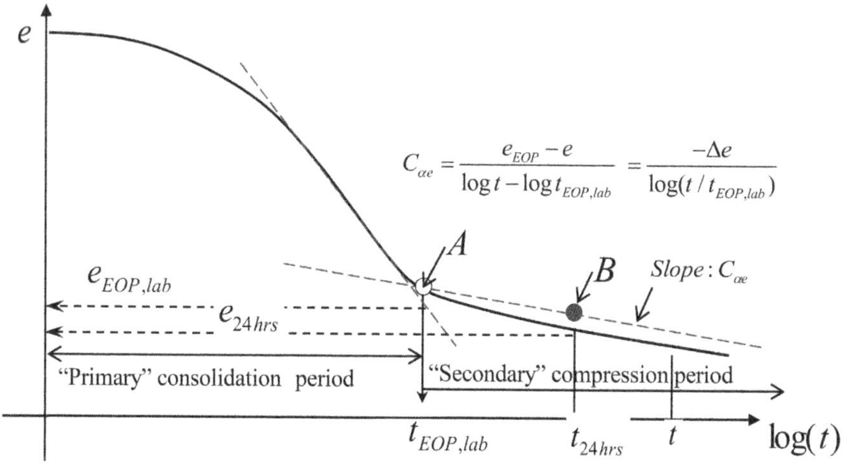

Figure 8.2 Curve of void ratio versus log(time) and 'secondary' compression coefficient.

Equation (8.1) can consider creep compression in ‘primary’ consolidation and still belongs to Hypothesis B approach. In (8.1), ‘primary’ consolidation settlement $S_{primary}$ shall be calculated for multiple soil layers with or without a vertical drain by assuming the soil skeleton is elastic or at least elastic for the applied load in consideration. Therefore, many existing analytical solutions to various consolidation problems can be utilized. In general, we have:

$$S_{\text{‘}primary\text{’}} = \sum_{j=1}^{j=n} U_j S_{fj} = U \sum_{j=1}^{j=n} S_{fj} \tag{8.2}$$

where U_j is combined average degree of consolidation for j-layer and U is combined average degree of consolidation for all multiple soil layers with or without a vertical drain:

$$U_j = 1 - (1 - U_{vj})(1 - U_{rj}) \tag{8.3a}$$

$$U = 1 - (1 - U_v)(1 - U_r) \tag{8.3b}$$

Equation (8.3) is called Carrillo’s (1942) formula where U_v and U_r are average degree of vertical consolidation and radial consolidation for j-layer or multiple soil layers. If there is no vertical drain, $U_r = 0$, from (8.3), $U = U_v$. For multiple soil layers, the superposition of the average degree of consolidation for each layer is not valid since the continuation condition for each interface of two layers in (7.9) must be satisfied. S_{fj} is the final ‘primary’ consolidation settlement at the end-of-primary (EOP) consolidation for j-layer. S_{fj} can be calculated using the coefficient of volume compressibility m_v or compression indexes C_c, C_r of j-layer in Chapter 1. More details on calculations of S_{fj} and U are presented in Section 8.2.2.

In (8.1), S_{creepj} is creep settlement of soil skeleton in j-layer and is equal to:

$$S_{creepj} = \alpha U_j^{\beta} S_{creep,fj} + (1 - \alpha U_j^{\beta}) S_{creep,dj} \tag{8.4}$$

where U_j is from (8.4) with value from 0 to 1 only and β is a power index with value from 0 to 1. Yin (2011) used a parameter $\alpha = 1$ without U_j^{β}. But this over-predicted total consolidation settlement. Yin and Feng (2017) and Feng and Yin (2017) used $\alpha = 0.8$ without U_j^{β} and gave results in close agreement with measured data and values from rigorous fully coupled consolidation modeling. In this chapter, a general term of αU_j^{β} is suggested. See more examples later in this chapter on more accurate prediction results.

$S_{creep,fj}$ is creep settlement of j-layer under the ‘final’ vertical effective stress ignoring the excess porewater pressure. $S_{creep,dj}$ is calculated for $t \geq t_{EOP,field}$. $S_{creep,dj}$ is ‘delayed’ creep settlement of j-layer under the ‘final’ vertical effective stress ignoring the excess porewater pressure. $S_{creep,dj}$ starts for $t \geq t_{EOP,field}$, in other words, is 'delayed' by time of $t_{EOP,field}$ to occur. $t_{EOP,field}$ is the time at EOP for field condition of j-later. More discussion on $S_{creep,fj}$ and $S_{creep,dj}$ are in Section 8.3.

8.2.2 Calculation of S_{fj}

In (8.2), the total primary consolidation settlement $S_{primary}$ is sum of settlements S_{fj} of all sub-layers multiplied by an overall average degree of consolidation U. This section presents methods and solutions for calculating S_{fj}. In the following calculations, in order make all equations and text in following paragraphs concise, the layer index 'j' is removed, keeping in minds that these equations are for one soil layer.

If the coefficient of volume compressibility m_v is used, vertical effective stress increment $\Delta\sigma'_z$ and thickness H are is known for a soil j-layer. S_f for j-layer is:

$$S_f = m_v \Delta\sigma'_z H \tag{8.5}$$

As pointed out in Chapter 1, m_v is not a constant, depending on vertical effective stress, and shall be used with care.

For clayey soils or soft soils, it is better to use C_c and C_r to calculate S_f for higher accuracy. An oedometer test is normally done on the same specimen in multi-stages. According to British Standard 1377 (1990), the standard duration for each load shall normally last for 24 hours. In this chapter, the indexes C_r, C_c and pre-consolidation stress point $(\sigma'_{zp}, \varepsilon_{zp})$ are all determined from the standard oedometer test with duration of 24 hours (1 day) for each load and for each layer. The idealized relationship between the vertical strain and the log(effective stress) is shown in Figure 8.3 with loading, unloading, and reloading states.

In Chapter 1, limitations of using a logarithmic function in (1.21) for stress are pointed out. The main limitation is that when the vertical effective stress σ'_z is zero or near zero, for example, σ'_z at surface or near surface of seabed soils or soil ground, the strain is infinite or to large. To overcome this problem, a unit stress σ'_{unit} is added to the logarithmic function in (1.41a) and also in Yin's a non-linear logarithmic-hyperbolic function in (1.41b). In this chapter, σ'_{unit} is added in linear logarithmic stress function. This is particularly necessary for very soft soils in a soil ground with initial effective stress zero at the top of the surface. For example, the initial vertical effective stress at the top surface in soft Hong Kong Marine Clay (HKMC) in seedbed is zero.

Based on Figure 8.3 and assuming stresses in each layer are uniform, the final settlements S_f for j-layer in (8.2) for six cases are calculated as follows by adding σ'_{unit} in logarithmic stress function.

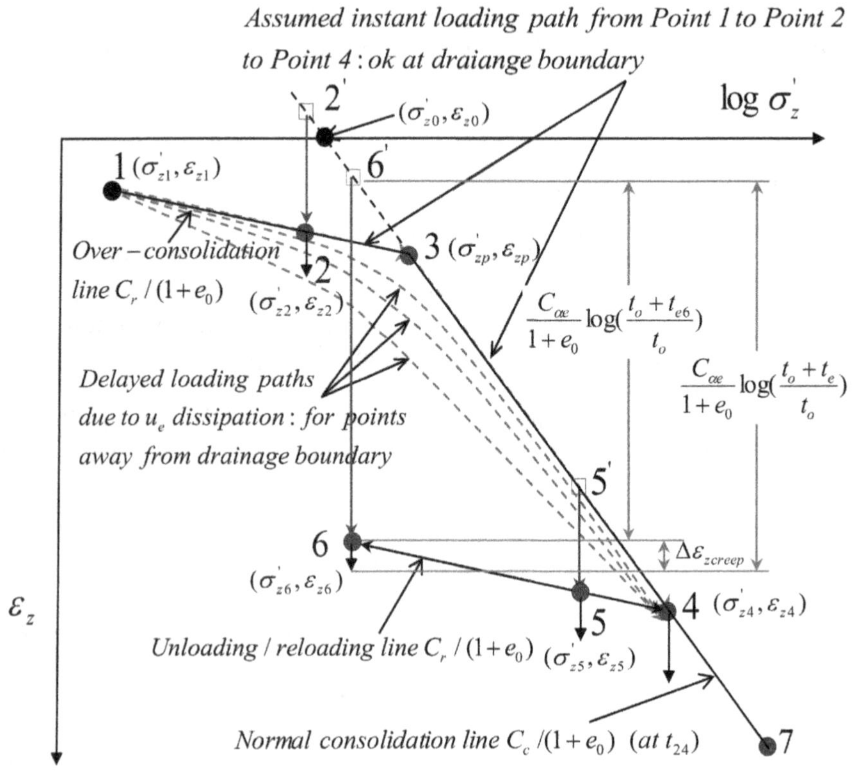

Figure 8.3 Relationship of void ratio (or strain) and log(effective stress) with different consolidation states.

i. Loading from point 1 to point 2 with $OCR = \sigma'_{zp}/\sigma'_{z1}$ and point 2 in over-consolidation line (OCL):

$$S_{f,1\text{-}2} = \varepsilon_{z,1\text{-}2}H = \frac{C_r}{1+e_o}\log\left(\frac{\sigma'_{z2}+\sigma'_{unit}}{\sigma'_{z1}+\sigma'_{unit}}\right)H \tag{8.6a}$$

The $\varepsilon_{z,1\text{-}2}$ is the vertical strain increase due to stress increases from σ'_{z1} to σ'_{z2}. The OCR is over-consolidation ratio and OCL is an over-consolidation line. If σ'_{unit} is zero, (8.6a) goes back to conventional logarithmic stress function. The value of σ'_{unit} is from 0.01 to 1 kPa. For very soft soils, σ'_{unit} takes values close 0.01 kPa. Similar strain increase symbols are used in the following equations.

ii. Loading from point 1 to point 4 with $OCR = \sigma'_{zp}/\sigma'_{z1} > 1$ and point 4 in normal consolidation line (NCL):

$$S_{f,1\text{-}4} = \varepsilon_{z,1\text{-}4}H = \left[\frac{C_r}{1+e_o}\log\left(\frac{\sigma'_{zp}+\sigma'_{unit}}{\sigma'_{z1}+\sigma'_{unit}}\right)+\frac{C_c}{1+e_o}\log\left(\frac{\sigma'_{z4}+\sigma'_{unit}}{\sigma'_{zp}+\sigma'_{unit}}\right)\right]H \tag{8.6b}$$

NCL is a normal consolidation line.

iii. Loading from point 3 to point 4 with $OCR = \sigma'_{zp}/\sigma'_{z3} = 1$ and point 4 in NCL:

$$S_{f,3-4} = \varepsilon_{z,3-4}H = \frac{C_c}{1+e_o}\log\left(\frac{\sigma'_{z4}+\sigma'_{unit}}{\sigma'_{zp}+\sigma'_{unit}}\right)H \tag{8.6c}$$

iv. Unloading from point 4 to point 6:

$$S_{f,4-6} = \varepsilon_{z,4-6}H = \frac{C_r}{1+e_o}\log\left(\frac{\sigma'_{z6}+\sigma'_{unit}}{\sigma'_{z4}+\sigma'_{unit}}\right)H \tag{8.6d}$$

v. Reloading from point 6 to point 5:

$$S_{f,6-5} = \varepsilon_{z,6-5}h = \frac{C_r}{1+e_o}\log\left(\frac{\sigma'_{z5}+\sigma'_{unit}}{\sigma'_{z6}+\sigma'_{unit}}\right)h \tag{8.6e}$$

vi. Reloading from point 6 to point 7:

$$S_{f,6-7} = \varepsilon_{z,6-7}H = \left[\frac{C_r}{1+e_o}\log\left(\frac{\sigma'_{z4}+\sigma'_{unit}}{\sigma'_{z6+\sigma'_{unit}}}\right)+\frac{C_c}{1+e_o}\log\left(\frac{\sigma'_{z7}+\sigma'_{unit}}{\sigma'_{z4}+\sigma'_{unit}}\right)\right]H \tag{8.6f}$$

However, the initial stresses and stress increments in a clayey soil layer are not uniform, (8.6) cannot be used. There are two approaches to consider this.

a. Dividing *j*-layer into sub-layers
A general method is to divide this soil layer into sub-layers with smaller thickness, say, 0.25 to 0.5 m. The stresses and parameters in each sub-layer are considered uniform and constant. The final settlement S_f for *j*-layer is sum of settlements of all sub-layers (Yin and Feng 2017; Feng and Yin 2017). For each sub-layer with uniform stresses, equations in (8.6a) to (8.6f) can be used depending on the initial and final stress points. This method is flexible and valid for complicated cases in which vertical stress, pre-consolidation pressure, and even permeability may not be uniform.
b. Special case of constant parameters C_c, C_r and linear changes of initial stresses, stress increments, and pre-consolidation pressure for *j*-layer
For a clayey soil layer of thickness H, C_c, C_r are often constant, but stresses may vary with depth z. Figure 8.4 shows linear changes of initial vertical effective stress, total vertical effective stress, and vertical

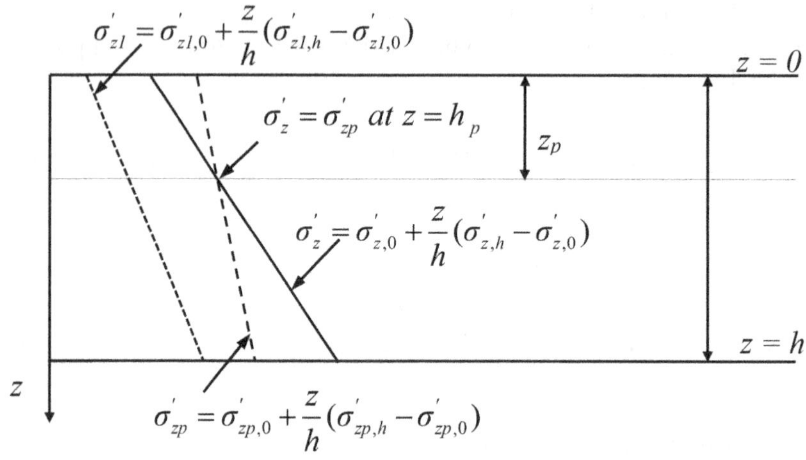

Figure 8.4 Linear changes of initial vertical effective stress, total vertical effective stress, and vertical pre-consolidation stress for a soil layer.

pre-consolidation stress for a soil layer. Linear changes are in following equations:

$$\sigma'_{z1} = \sigma'_{z1,0} + \frac{z}{H}(\sigma'_{z1,H} - \sigma'_{z1,0}) \quad (8.7a)$$

$$\sigma'_{zp} = \sigma'_{zp,0} + \frac{z}{H}(\sigma'_{zp,H} - \sigma'_{zp,0}) \quad (8.7b)$$

$$\sigma'_{z} = \sigma'_{z4,0} + \frac{z}{H}(\sigma'_{z4,H} - \sigma'_{z4,0}) \quad (8.7c)$$

σ'_{z1} is the initial vertical effective stress. It is noted that the increase of pre-consolidation stress (or pressure) σ'_{zp} may not be as fast as the total vertical effective stress σ'_z as shown in Figure 8.4. Therefore, there is a point which $\sigma'_{zp} = \sigma'_z$ at depth z_p.

Let us consider a general case of loading form point 1 to point 4, the calculation of S_f of j-layer for four different cases are in followings.

i. Normal consolidation case: $OCR = \sigma'_{zp} / \sigma'_{z1} = 1$
   In this case, initial effective stress σ'_{z1} and pre-consolidation stress σ'_{zp} are the same, after the stress increase, $\sigma'_z > \sigma'_{zp} = \sigma'_{z1}$. In this case, $S_{f,1-4}$ is:

$$S_{f,1-4} = \int_{z=0}^{z=H} \varepsilon_{z,1-4} dz = \int_{z=0}^{z=H} \frac{C_c}{1+e_o} \log\left(\frac{\sigma_z + \sigma'_{unit}}{\sigma'_{zp} + \sigma'_{unit}}\right) dz \quad (8.8a)$$

Substituting (8.6) into above equation:

$$S_{f,1-4} = \int_{z=0}^{z=H} \frac{C_c}{(1+e_o)\ln(10)} \ln\left[\frac{\sigma'_{z4,0} + \frac{z}{H}(\sigma'_{z4,H} - \sigma'_{z4,0}) + \sigma'_{unit}}{\sigma'_{z1,0} + \frac{z}{H}(\sigma'_{z1,H} - \sigma'_{z1,0}) + \sigma'_{unit}}\right] dz = \frac{C_c}{(1+e_o)\ln(10)}$$

$$\left\{\int_{z=0}^{z=H} \ln[\sigma'_{z4,0} + \frac{z}{H}(\sigma'_{z4,H} - \sigma'_{z4,0}) + \sigma'_{unit}] dz\right.$$

$$\left. - \int_{z=0}^{z=H} \ln[\sigma'_{z1,0} + \frac{z}{H}(\sigma'_{z1,H} - \sigma'_{z1,0}) + \sigma'_{unit}] dz\right\}$$

Let us introduce a new variable $x = \sigma'_{z4,0} + \frac{z}{H}(\sigma'_{z4,H} - \sigma'_{z4,0}) + \sigma'_{unit}$ and $y = \sigma'_{z1,0} + \frac{z}{H}(\sigma'_{z1,H} - \sigma'_{z1,0}) + \sigma'_{unit}$, we have $dz = [H/(\sigma'_{z4,H} - \sigma'_{z4,0})]dx$ and $dz = [H/(\sigma'_{z1,H} - \sigma'_{z1,0})]dy$. Note that for $z = 0$ and H, we have $x_{z=0} = \sigma'_{z4,0} + \sigma'_{unit}$ and $x_{z=H} = \sigma'_{z4,H} + \sigma'_{unit}$; $y_{z=0} = \sigma'_{z1,0} + \sigma'_{unit}$ and $y_{z=H} = \sigma'_{z1,H} + \sigma'_{unit}$. The above equation can be written as:

$$S_{f,1-4} = \frac{C_c}{(1+e_o)\ln(10)} \left\{\int_{x=\sigma'_{z4,0}+\sigma'_{unit}}^{x=\sigma'_{z4,H}+\sigma'_{unit}} \frac{H}{(\sigma'_{z4,H} - \sigma'_{z4,0})} \ln x dx\right.$$

$$\left. - \int_{y=\sigma'_{z1,0}+\sigma'_{unit}}^{y=\sigma'_{z1,H}+\sigma'_{unit}} \frac{H}{(\sigma'_{z1,H} - \sigma'_{z1,0})} \ln y dy\right\}$$

Since $\int \ln x dx = x \ln x - x$ and $\int \ln y dy = y \ln y - y$, the above equation becomes:

$$S_{f,1-4} = \frac{C_c}{(1+e_o)\ln(10)} \left\{\frac{H}{(\sigma'_{z4,H} - \sigma'_{z4,0})}[x \ln x - x]_{x=\sigma'_{z4,0}+\sigma'_{unit}}^{x=\sigma'_{z4,H}+\sigma'_{unit}}\right.$$

$$\left. - \frac{H}{(\sigma'_{z1,H} - \sigma'_{z1,0})}[y \ln y - y]_{y=\sigma'_{z1,0}+\sigma'_{unit}}^{y=\sigma'_{z1,H}+\sigma'_{unit}}\right\}$$

From above, we have:

$$S_{f,1-4}=\frac{C_c}{(1+e_o)\ln(10)}\left\{\frac{H}{(\sigma'_{z4,H}-\sigma'_{z4,0})}[(\sigma'_{z4,H}+\sigma'_{unit})\ln(\sigma'_{z4,H}+\sigma'_{unit})\right.$$

$$-(\sigma'_{z4,H}+\sigma'_{unit})-((\sigma'_{z4,0}+\sigma'_{unit})\ln(\sigma'_{z4,0}+\sigma'_{unit})-(\sigma'_{z4,0}+\sigma'_{unit}))]$$

$$-\frac{H}{(\sigma'_{z1,H}-\sigma'_{z1,0})}[(\sigma'_{z1,H}+\sigma'_{unit})\ln(\sigma'_{z1,H}+\sigma'_{unit})-(\sigma'_{z1,H}+\sigma'_{unit})$$

$$\left.-((\sigma'_{z1,0}+\sigma'_{unit})\ln(\sigma'_{z1,0}+\sigma'_{unit})-(\sigma'_{z1,0}+\sigma'_{unit}))]\right\} \tag{8.8b}$$

ii. Over-consolidation case: $OCR=\sigma'_{zp}/\sigma'_{z1}>1$ and $\sigma'_z\geq\sigma'_{zp}$ $for\ 0\leq z\leq H$
Figure 8.4 shows a case commonly encountered in the field. Initially, the soil is over-consolidated with $OCR=\sigma'_{zp}/\sigma'_{z1}>1$. After increased loading $\Delta\sigma'_z=\sigma'_z-\sigma'_{z1}$, we have and $\sigma'_z\geq\sigma'_{zp}$ $for\ 0\leq z\leq H$. In this case, we have:

$$S_{f,1-4}=\int_{z=0}^{z=h}\varepsilon_{z,1-4}dz=\int_{z=z_p}^{z=h}\left[\frac{C_r}{1+e_o}\log\left(\frac{\sigma'_{zp}+\sigma'_{unit}}{\sigma'_{z1}+\sigma'_{unit}}\right)+\frac{C_c}{1+e_o}\log\left(\frac{\sigma'_z+\sigma'_{unit}}{\sigma'_{zp}+\sigma'_{unit}}\right)\right]dz \tag{8.8c}$$

Substituting equations in (8.6) for $\sigma'_{z1},\sigma'_{zp},\sigma'_z$ into (8.7c):

$$S_{f,1-4}=\int_{z=z_p}^{z=H}\left[\frac{C_r}{1+e_o}\log\left(\frac{\sigma'_{zp,0}+\frac{z}{H}(\sigma'_{zp,H}-\sigma'_{zp,0})+\sigma'_{unit}}{\sigma'_{z1,0}+\frac{z}{H}(\sigma'_{z1,H}-\sigma'_{z1,0})+\sigma'_{unit}}\right)+\right.$$

$$\left.\frac{C_c}{1+e_o}\log\left(\frac{\sigma'_{z4,0}+\frac{z}{H}(\sigma'_{z4,H}-\sigma'_{z4,0})+\sigma'_{unit}}{\sigma'_{zp,0}+\frac{z}{H}(\sigma'_{zp,H}-\sigma'_{zp,0})+\sigma'_{unit}}\right)\right]dz$$

Using the same method in (i), the integration of above equation is:

$$S_{f,1-4}=\frac{C_c}{(1+e_o)\ln(10)}\left\{\frac{H}{(\sigma'_{zp,H}-\sigma'_{zp,0})}[(\sigma'_{zp,H}+\sigma'_{unit})\ln(\sigma'_{z4,p}+\sigma'_{unit})\right.$$

$$-(\sigma'_{zp,h}+\sigma'_{unit})-((\sigma'_{zp,0}+\sigma'_{unit})\ln(\sigma'_{zp,0}+\sigma'_{unit})-(\sigma'_{zp,0}+\sigma'_{unit}))]$$

$$-\frac{H}{(\sigma'_{z1,H}-\sigma'_{z1,0})}[(\sigma'_{z1,H}+\sigma'_{unit})\ln(\sigma'_{z1,H}+\sigma'_{unit})-(\sigma'_{z1,H}+\sigma'_{unit})$$

$$\left.-((\sigma'_{z1,0}+\sigma'_{unit})\ln(\sigma'_{z1,0}+\sigma'_{unit})-(\sigma'_{z1,0}+\sigma'_{unit}))]\right\}$$

$$+\frac{C_c}{(1+e_o)\ln(10)}\left\{\frac{H}{(\sigma'_{z4,H}-\sigma'_{z4,0})}[(\sigma'_{z4,H}+\sigma'_{unit})\ln(\sigma'_{z4,H}+\sigma'_{unit})\right.$$

$$-(\sigma'_{z4,0}+\sigma'_{unit})-((\sigma'_{z4,0}+\sigma'_{unit})\ln(\sigma'_{z4,0}+\sigma'_{unit})-(\sigma'_{z4,0}+\sigma'_{unit}))]$$

$$-\frac{H}{(\sigma'_{zp,H}-\sigma'_{zp,0})}[(\sigma'_{zp,H}+\sigma'_{unit})\ln(\sigma'_{zp,H}+\sigma'_{unit})-(\sigma'_{zp,H}+\sigma'_{unit})$$

$$\left.-((\sigma'_{zp,0}+\sigma'_{unit})\ln(\sigma'_{zp,0}+\sigma'_{unit})-(\sigma'_{zp,0}+\sigma'_{unit}))]\right\} \tag{8.8d}$$

iii. Over-consolidation case: $OCR = \sigma'_{zp}/\sigma'_{z1} > 1$ and $\sigma'_z < \sigma'_{zp}\ for\ 0 \le z \le z_p$
Figure 8.4 shows a case in which $OCR = \sigma'_{zp}/\sigma'_{z1} > 1$, but $\sigma'_z < \sigma'_{zp}\ for\ 0 \le z \le z_p$ and $\sigma'_z \ge \sigma'_{zp}\ for\ z_p \le z \le H$. In this case, the settlement calculation shall consider depth z_p:

$$S_{f,1-4} = \int_{z=0}^{z=H} \varepsilon_{z,1-4} dz = \begin{cases} \displaystyle\int_{z=0}^{z=z_p} \frac{C_r}{1+e_o}\log\left(\frac{\sigma'_z+\sigma'_{unit}}{\sigma'_{z1}+\sigma'_{unit}}\right)dz \quad for\ 0 \le z \le z_p \\ \displaystyle\int_{z=z_p}^{z=H}\left[\frac{C_r}{1+e_o}\log\left(\frac{\sigma'_{zp}+\sigma'_{unit}}{\sigma'_{z1}+\sigma'_{unit}}\right)+\frac{C_c}{1+e_o}\log\left(\frac{\sigma'_z+\sigma'_{unit}}{\sigma'_{zp}+\sigma'_{unit}}\right)\right]dz \\ for\ z_p \le z \le H \end{cases} \tag{8.8e}$$

Linear equations in (8.6) for $\sigma'_{z1}, \sigma'_{zp}, \sigma'_z$ can be substituted into (8.7e). Analytical integration solution can be obtained using the same method in (i) and is not presented here. Equations like (8.7) can be obtained for other loading, unloading, and reloading cases with linear changes of stresses and are not discussed here.

In many calculations, m_v is needed, for example in (8.5) and $c_v = k_v/(m_v\gamma_w)$ and $c_r = k_r/(\gamma_w m_v)$ in order to calculate U_v and U_r. If indexes C_r, C_c and preconsolidation stress point $(\sigma'_{zp}, \varepsilon_{zp})$ are used to calculate final settlements in (8.6), (8.7), and (8.8), the coefficient of vertical volume compressibility m_v can be back-calculated as

i. for the case of (8.6b) in normal loading:

$$m_{v,1-4} = \frac{S_{f,1-4}}{\sigma'_{z4}-\sigma'_{z1}} \tag{8.9a}$$

ii. for the case of (8.6d) in unloading:

$$m_{v,4-6} = \frac{S_{f,4-6}}{\sigma'_{z4} - \sigma'_{z6}} \tag{8.9b}$$

In (8.9a) and (8.9b), settlements and stress increments are known so that m_v corresponding to the same stress increment can be calculated. In (8.9b), $S_{f,4-6}$ and $(\sigma'_{z4} - \sigma'_{z6})$ are both negative so that $m_{v,4-6}$ is positive. The above method for m_v can be applied to all other loading cases.

8.2.3 Calculation of U_j and U

In (8.1) and (8.2), an average degree of consolidation U_j for j-layer or over-all average degree of consolidation U is needed. The basic definition of U_j for j-layer is:

$$U_j = \frac{S_j(t)}{S_{fj}} = \frac{\int_{z=0}^{z=H_j} m_{vj}\Delta\sigma'_{zj}(t)\,dz}{\int_{z=0}^{z=H_j} m_{vj}\Delta\sigma'_{zjf}\,dz} = \frac{\int_{z=0}^{z=H_j} [u_{eij} - u_{ej}(t)]dz}{\int_{z=0}^{z=H_j} u_{eij}\,dz} = 1 - \frac{\int_{z=0}^{z=H_j} u_{ej}(t)\,dz}{\int_{z=0}^{z=H_j} u_{eij}\,dz} \tag{8.10a}$$

S_{fj} is the final settlement for j-layer using m_{vj} and $\Delta\sigma'_{zjf}$, calculated using (8.5). It is noted that the final vertical effective stress increment $\Delta\sigma'_{zjf}$ is equal to the initial excess porewater pressure u_{eij} for j-layer. $u_{ej}(t)$ is the excess porewater pressure at time t for j-layer. Equation (8.10a) can be written:

$$U_j = 1 - \frac{\frac{1}{H_j}\int_{z=0}^{z=H_j} u_{ej}(t)\,dz}{\frac{1}{H_j}\int_{z=0}^{z=H_j} u_{eij}\,dz} = 1 - \frac{\bar{u}_{ej}(t)}{\bar{u}_{eij}} \tag{8.10b}$$

$\bar{u}_{eij}$ and $\bar{u}_{ej}$ are the average initial and current excess porewater pressures, respectively.

The overall average degree of consolidation U is:

$$U = \frac{S_{primary}}{S_f} = \frac{\sum_{j=1}^{j=n} S_j(t)}{\sum_{j=1}^{j=n} S_{fj}} = \frac{\sum_{j=1}^{j=n}\int_{z=0}^{z=H_j} m_{vj}\Delta\sigma'_{zj}(t)\,dz}{\sum_{j=1}^{j=n}\int_{z=0}^{z=H_j} m_{vj}\Delta\sigma'_{zjf}\,dz} = \frac{\sum_{j=1}^{j=n}\int_{z=0}^{z=H_j} m_{vj}\Delta\sigma'_{zj}(t)\,dz}{\sum_{j=1}^{j=n}\int_{z=0}^{z=H_j} m_{vj}\Delta\sigma'_{zjf}\,dz} =$$

$$= \frac{\sum_{j=1}^{j=n} m_{vj}\int_{z=0}^{z=H_j} [u_{eij} - u_{ej}(t)]dz}{\sum_{j=1}^{j=n} m_{vj}\int_{z=0}^{z=H_j} u_{eij}\,dz} = 1 - \frac{\sum_{j=1}^{j=n} m_{vj}\int_{z=0}^{z=H_j} u_{ej}(t)\,dz}{\sum_{j=1}^{j=n} m_{vj}\int_{z=0}^{z=H_j} u_{eij}\,dz} \tag{8.11a}$$

From (8.10b), $\bar{u}_{ej}(t)=(1-U_j)\bar{u}_{eij}$. Using this relation, (8.11a) can be written:

$$U=1-\frac{\sum_{j=1}^{j=n} m_{vj}\frac{H_j}{H_j}\int_{z=0}^{z=H_j} u_{ej}(t)\,dz}{\sum_{j=1}^{j=n} m_{vj}\frac{H_j}{H_j}\int_{z=0}^{z=H_j} u_{eij}\,dz}=1-\frac{\sum_{j=1}^{j=n} m_{vj}H_j\bar{u}_{ej}(t)}{\sum_{j=1}^{j=n} m_{vj}H_j\bar{u}_{eij}}=1-\frac{\sum_{j=1}^{j=n} m_{vj}H_j(1-U_j)\bar{u}_{eij})}{\sum_{j=1}^{j=n} m_{vj}H_j\bar{u}_{eij}}=$$

$$=1-\frac{\sum_{j=1}^{j=n} m_{vj}H_j\bar{u}_{eij}-\sum_{j=1}^{j=n} m_{vj}H_jU_j\bar{u}_{eij}}{\sum_{j=1}^{j=n} m_{vj}H_j\bar{u}_{eij}}=\frac{\sum_{j=1}^{j=n} m_{vj}H_j\bar{u}_{eij}U_j}{\sum_{j=1}^{j=n} m_{vj}H_j\bar{u}_{eij}} \tag{8.11b}$$

If $\bar{u}_{eij}$ is the same for all layers, (8.11b) becomes:

$$U=\frac{\bar{u}_{eij}\sum_{j=1}^{j=n} m_{vj}H_jU_j}{\bar{u}_{eij}\sum_{j=1}^{j=n} m_{vj}H_j}=\frac{\sum_{j=1}^{j=n} m_{vj}H_jU_j}{\sum_{j=1}^{j=n} m_{vj}H_j} \tag{8.11c}$$

If m_{vj} and $\bar{u}_{eij}$ are the same for all layers, (8.11b) becomes:

$$U=\frac{\sum_{j=1}^{j=n} H_jU_j}{\sum_{j=1}^{j=n} H_j}=\frac{1}{H}\sum_{j=1}^{j=n} H_jU_j \tag{8.11d}$$

Note that the total thickness is $H=\Sigma_{j=1}^{j=n}H_j$.

Attention shall be paid to the definition and differences of U_j and U. The following paragraphs summarize existing solutions for U_v, U_r, U_j, and U.

a. A single soil layer without vertical drains
i. Classic solutions and correction of construction period

The early analytical solutions were obtained by Terzaghi (1925, 1943) for a single soil layer with thickness H under suddenly applied load for 1-D straining. Charts of these solutions can be found in Craig's Soil Mechanics (Knappett 2019). For double drainage with linear excess porewater pressure u_e distribution or one way drainage with uniform u_e distribution, the following approximate equation is good and simple to calculating U_v:

$$\begin{cases} \text{For } U_v<0.6: \quad T_v=\frac{\pi}{4}U_v^2, \quad U_v=\sqrt{\frac{4T_v}{\pi}} \\ \text{For } U_v\geq 0.6: \quad T_v=-0.944\log(1-U_v)-0.085, \quad U_v=1-10^{-\frac{T_v+0.085}{0.933}} \end{cases} \tag{8.12a}$$

If we assume that then $U_v = 98\%$, $u_e \approx 0$, this is the time at EOP in the field $t_{EOP,field}$. We have:

$$T_v = -0.944\log(1-U_v) - 0.085 = 0.150 \tag{8.12b}$$

$$t_{EOP,field} = \frac{T_v d^2}{c_v} = \frac{1.500 d^2}{c_v} \tag{8.12c}$$

where d is the maximum drainage path of a soil layer, if double drainage $d = H/2$; c_v is the coefficient of vertical consolidation and defined in (2.14).

To consider ramp loading as shown in Figure 2.4, a simple correction method for U_v was proposed by Terzaghi (1925, 1943) and is presented here for easy use:

$$\begin{cases} For \quad 0 < t \leq t_c: \quad S_{t,corr}(t) = S_t\left(\frac{t}{2}\right)\frac{\sigma_{top}(t)}{\sigma_0} = S_t\left(\frac{t}{2}\right)\frac{\sigma_{bottom}(t)}{\sigma_1} \\ where\ S_t\left(\frac{t}{2}\right) = U\left(\frac{t}{2}\right)S_f \\ For \quad t > t_c: \quad S_{t,corr}(t) = S_t\left(t - \frac{t_c}{2}\right) \\ where\ S_t\left(t - \frac{t_c}{2}\right) = U\left(t - \frac{t_c}{2}\right)S_f \end{cases} \tag{8.13}$$

The above correction method can also be applied to the case with vertical drains and multiple layers.

iii. Solutions to 1-D consolidation under depth-dependent ramp load and to special 1-D consolidation problems (Zhu and Yin 1998, 2012)

For 1-D consolidation of one-layered soil profile with depth-dependent ramp load as shown in Figure 2.4, analytical solutions are provided for U_v in (2.26) and (2.27) in Section 2.41 of Chapter 2. Analytical solutions to several special 1-D consolidation problems of one soil layer are presented for U_v in Chapter 3.

b. Double soil layers without vertical drains under ramp load (Zhu and Yin 1999b)

Figure 2.13 shows a double-layered soil profile. Analytical solutions are obtained and presented in Section 2.4.2 of Chapter 2 for this soil profile under ramp loading. For boundary conditions in (2.16) and (2.17), the analytical solutions for U_v are in (2.42) and (2.43). If the vertical total stress increment is uniform with depth, the average degree of consolidation U_v is in

(2.44). Charts and tables of these solutions are presented in Section 2.4.2 of Chapter 2 including special case for abruptly applied loading with $t_c = 0$. It is noted that U_v obtained is the overall average degree of vertical consolidation of two layers.

c. A single soil layer with vertical drains
i. Solutions to two-dimensional (2-D) consolidation of a single soil layer with vertical drains under ramp load by Zhu and Yin (2001a, b, 2004a, b)

Section 6.3 in Chapter 6 presents basic equations and solutions to a single layer of a soil with vertical drains in case of free strain, which assumes the top surface is a flexible boundary under uniform loading. Section 6.3.1 presents a solution for an ideal drain (no smear zone and no well resistance) as shown in Figure 6.1 for ramp loading. Analytic solution is presented in (6.28). Charts of this solution for U are also produced for suddenly applied load or ramp load.

Section 6.3.2 presents a solution for one soil layer with vertical drains considering smear zone (no well resistance). Solution for U is presented in (6.49) under ramp loading. Solution and charts for a special case of suddenly applied loading are also obtained.

Section 6.4 in Chapter 6 presents basic equations and solutions to a single layer of a soil with vertical drains (smear zone and well resistance) in case of equal strain, which assumes the top surface is compressed with equal displacement under uniform loading. Section 6.4.1 presents solutions for a soil layer with vertical drain considering smear zone in case of equal strain. Solution for U under free strain assumption is presented in (6.68) under ramp loading. Section 6.4.2 presents a solution to cases of a single soil layer with vertical drains considering both smear zone and well resistance in case of equal strain. Solution for U_r under equal vertical strain assumption is presented in (6.79) under well resistance and ramp loading. Zhu and Yin (2004b) discussed accuracy of the Carrillo's formula (Carrillo 1942) for consolidation of soil with vertical and horizontal drainage under time-dependent loading.

ii. Solutions to 2-D consolidation of a single soil layer with vertical drains without well resistance under suddenly applied load by Barron (1948)

Under equal vertical strain assumption, Barron (1948) presented firstly analytical solution to consolidation problem of a soil with drain wells without smear zone and well resistance under suddenly applied load. Equation of his solution is, with a referenced to Figure 6.1:

$$U_r = 1 - \exp\left(\frac{-2T_r}{F(n)}\right) \tag{8.14a}$$

$$F(n) = \frac{n^2}{n^2 - 1}\ln n - \frac{3}{4} + \frac{1}{4n^2} \tag{8.14b}$$

Noting that $T_r = \frac{c_r t}{r_e^2}$, $n = \frac{r_e}{r_d}$. Neither smear zone nor well resistance is considered in (8.14). In the same paper, still under equal vertical strain assumption, Barron (1948) presented analytical solution to consolidation problem of a soil with drain wells with smear zone and without well resistance under suddenly applied load. Equation of his solution is, with a reference to Figure 6.1:

$$U_r = 1 - \exp\left(\frac{-2T_r}{F(n,s,k_r,k_s)}\right) \tag{8.15a}$$

$$F(n,s,k_r,k_s) = \frac{n^2}{n^2 - s^2}\ln\frac{n}{s} - \frac{3}{4} + \frac{s^2}{4n^2} + \frac{k_r}{k_s}\frac{n^2 - s^2}{n^2}\ln s \tag{8.15b}$$

r_s is radius of smear zone, and $s = r_s / r_d$. If there is no smear zone, $r_s = r_d$ and $s = 1$. Equation (8.15a) is reduced to (8.14b).

iii. Solutions to 2-D consolidation of a single soil layer with vertical drains considering smear zone and well resistance under suddenly applied load by Hansbo (1981)

Hansbo (1981) presented analytical solution to consolidation problem of a soil with vertical drains considering both smear zone and well resistance under suddenly applied load. Under equal vertical strain assumption, equation of his solution is, with a reference to Figure 8.1 for a single soil layer:

$$U_r = 1 - \exp\left(\frac{-2T_r}{\mu}\right) \tag{8.16a}$$

$$\begin{aligned}\mu = &\frac{n^2}{n^2-1}\left(\ln\frac{n}{s} - \frac{3}{4} + \frac{k_r}{k_s}\ln s\right) + \frac{s^2}{n^2-1}\left(1 - \frac{s^2}{4n^2}\right) + \\ &+ \frac{k_r}{k_s}\frac{1}{n^2-1}\left(\frac{s^4-1}{4n^2} - s^2 + 1\right) + \pi z(2l - z)\frac{k_r}{q_w}\left(1 - \frac{1}{n^2}\right)\end{aligned} \tag{8.16b}$$

In (8.16a), $T_r = \frac{c_r t}{r_e^2}$, $n = \frac{r_e}{r_d}$, $s = \frac{r_s}{r_d}$. $q_w = k_w \pi r_w^2$ is specific discharge capacity of drain (vertical hydraulic gradient i=1). z is the vertical coordinate in Figure 8.1 and l is length of drain when closed at bottom or a half of drain when bottom is open. If hydraulic resistance of vertical drains is zero, this means $q_w = k_w \pi r_d^2 \Rightarrow \infty$. Equation (8.16b) can be simplified:

$$\mu = \frac{n^2}{n^2-1}\left(\ln\frac{n}{s} - \frac{3}{4} + \frac{k_r}{k_s}\ln s\right) + \frac{s^2}{n^2-1}\left(1 - \frac{s^2}{4n^2}\right) + \frac{k_r}{k_s}\frac{1}{n^2-1}\left(\frac{s^4-1}{4n^2} - s^2 + 1\right) \tag{8.16c}$$

A standard band drain has $b = 100$ mm, $h = 3$ mm. According to (6.3), $r_d = (b + h)/\pi = (0.1 + 0.003)/\pi = 0.0328$ m. Equivalent radius r_e is normally 1 to 2 m, much bigger radius of well r_d to that $n = r_e/r_w = 24 \sim 48$, which is much bigger than 1. In addition, radius of smear zone r_s is normally one to four times of well r_d so that $s = r_s/r_d = 1 \sim 4$. If $s = r_s/r_d = 1$, there is no smear zone. It is found that $s/n = 1/48 \sim 4/24 \Rightarrow 0.0204 \sim 0.167$ and $s^2/n^2 = 0.000416 \sim 0.0279$. Considering all these, (8.11c) can be simplified:

$$\mu = \ln\frac{n}{s} - \frac{3}{4} + \frac{k_r}{k_s}\ln s \tag{8.16d}$$

If well resistance is included, (8.16d) is written as (Hansbo 1981):

$$\mu = \ln\frac{n}{s} - \frac{3}{4} + \frac{k_r}{k_s}\ln s + \pi z(2l - z)\frac{k_r}{q_w} \tag{8.16e}$$

If there is neither smear zone nor well resistance, (8.16c) is reduced to:

$$\begin{aligned}
\mu &= \frac{n^2}{n^2-1}\left(\ln n - \frac{3}{4}\right) + \frac{1}{n^2-1}\left(1 - \frac{1}{4n^2}\right) = \frac{n^2}{n^2-1}\ln n + \frac{1}{n^2-1}\left(-\frac{3n^2}{4} + 1 - \frac{1}{4n^2}\right)\\
&= \frac{n^2}{n^2-1}\ln n + \frac{1}{4(n^2-1)n^2}(-3n^4 + 4n^2 - 1)\\
&= \frac{n^2}{n^2-1}\ln n + \frac{-3}{4(n^2-1)n^2}(n^2-1)\left(n^2 - \frac{1}{3}\right)
\end{aligned}$$

The above equation can be further simplified:

$$\mu = \frac{n^2}{n^2-1}\ln n - \frac{3}{4} + \frac{1}{4n^2} \tag{8.16f}$$

Equation (8.16f) is the same as (8.14b) obtained by Barren (1948).

a. Multiple soil layers without or with vertical drains
i. Solutions for consolidation of n-layer soil profile

Chapter 2.4.3 presents analytical solutions for calculating excess porewater pressure u_e in (2.50) and the overall average degree of vertical consolidation $U = U_v$ in (2.67) for n-layers. However, these solutions are expressed in forms of series and integration, which cannot be used easily to calculate values of u_e and U_v.

ii. Solutions for a stratified soil with vertical and horizontal drainage using the spectral method (Walker and Indraratna 2009; Walker *et al.* 2009)

Walker and Indraratna (2009) and Walker *et al.* (2009) used a spectral method to find analytical solutions for multi-layered soils with vertical and horizontal drainage under ramp loading. The main partial differential equation for the average excess porewater pressure $\bar{u}$ using spectral method is:

$$\frac{m_v}{\bar{m}_v}\frac{\partial \bar{u}}{\partial t} = -\left[dT_r \frac{\eta}{\bar{\eta}}\bar{u} - dT_v \left(\frac{\partial}{\partial Z}\left(\frac{k_v}{\bar{k}_v}\right)\frac{\partial \bar{u}}{\partial Z} + \frac{k_v}{\bar{k}_v}\frac{\partial^2 \bar{u}}{\partial Z^2} \right) \right] + \frac{m_v}{\bar{m}_v}\frac{\partial \bar{\sigma}}{\partial t} + dT_r \frac{\eta}{\bar{\eta}} w \tag{8.17a}$$

where, $\eta = \frac{k_r}{r_e^2 \mu}$, $\mathrm{d}T_v = \frac{\bar{c}_v}{H^2}$, $\mathrm{d}T_r = \frac{2\bar{\eta}}{\gamma_w \bar{m}_v}$, $\bar{c}_v = \frac{\bar{k}_v}{\gamma_w \bar{m}_v}$, $Z = \frac{z}{H}$. Vertical and horizontal drainages are considered simultaneously in (8.17a). All parameters are explained below:

- $\bar{u}$: averaged excess porewater pressure (averaged along radial coordinate r) at depth Z, a function of time t and Z.
- $\bar{\sigma}$: average total stress (averaged along r) at depth Z, a function of time t and Z.
- w: water pressure applied on the vertical drains, varying with depth Z, which is zero without vacuum pre-loading pressure.
- r_w: unit weight of water.
- k_r: the horizontal permeability coefficient of the undisturbed soil, a function of Z.
- m_v: coefficient of volume compressibility (assumed the same in smear and undisturbed zone), calculated using total incremental strain resulted from primary consolidation under total stress increment, and a function of Z.

Parameters k_v, m_v and η can be depth-dependent in a piecewise linear way or kept constant within each layer. $\bar{k}_v$, $\bar{m}_v$, and $\bar{\eta}$ are convenient reference values at certain depth, for example, values $k_{v,j=1}$, $m_{v,j=1}$, and $\eta_{j=1}$ of layer 1. If so, $\bar{c}_v = \bar{k}_v / \gamma_w \bar{m}_v = k_{v,j=1}/\gamma_w m_{v,j=1} = c_{v,j=1}$. $\bar{\eta} = k_r / r_e^2 \mu = k_{r,j=1} / r_e^2 \mu = \eta_{j=1}$. All the parameters in (8.17a) have been normalized and may be different for different soil layers (no layer index is used here to make presentation concise). Normalized parameters in (8.17a) are: $m_v / \bar{m}_v$, $\eta / \bar{\eta}$, $k / \bar{k}_v$.

The parameter $\eta = k_r / (r_e^2 \mu)$ is related to radial permeability k_r, equivalent radius r_e of cylinder cell, and μ. If there is no horizontal drainage in a soil layer, $k_r = 0$ to that $\eta = 0$. This is useful for consolidation analysis of soils with partially penetrating vertical drains. All soil layers below vertical drains have $\eta = 0$. Walker and Indraratna (2009) and Walker *et al.* (2009) discussed that their method can also simulate the effect of using long and short drains in union. For example, in lower soil layers where only long drains are installed, η shall have smaller value than that of upper soil layers where both short and long drains are present.

μ inside η is a dimensionless drain geometry/smear zone parameter. Expressions for μ can be taken as those in (8.16b, c, d, e, f) by considering effects of smear zone, well resistance, or approximation. Walker and Indraratna (2006) also provided an expression for μ considering parabolic smear zone permeability but ignoring smear zone:

$$\mu = \ln\frac{n}{s} - \frac{3}{4} + \frac{\kappa(s-1)^2}{(s^2 - 2\kappa s + \kappa)}\ln\frac{s}{\sqrt{\kappa}} - \frac{s(s-1)\sqrt{\kappa(\kappa-1)}}{2(s^2 - 2\kappa s + \kappa)}\ln\left(\frac{\sqrt{\kappa}+\sqrt{\kappa-1}}{\sqrt{\kappa}-\sqrt{\kappa-1}}\right) \quad (8.17b)$$

κ is ratio of undisturbed horizontal permeability k_r to smear zone permeability k_{so} at the drain/soil interface, (at $r = r_d$, $k_s = k_{s0}$). At $r = r_s$, $k_s = k_r$.

Walker and Indraratna (2009) provided an excel spreadsheet implemented with a 'Visual Basic for Applications' (VBA) program named SPECCON to enable convenient adoption of this method for consolidation analysis of multiple soils layers with or without vertical drains. With inputted all parameters and load $\bar{\sigma}$, this program gives excess porewater pressure at time t $\bar{u}_{ej}(t)$ for j-layer and $\bar{u}_e(t)$ for all layers together. The combined average degree of consolidation U_j for each j-layer is calculated using (8.10a). The overall combined average degree of consolidation U for all layers shall be calculated using (8.11a) or (8.11b). Once U_j and U with time t are known, total 'primary' consolidation settlement $S_{primary}$ can be calculated using (8.2).

iii. Approximate solutions for consolidation of n-layer soil profile without vertical drains

US Department of the Navy (1982) proposed a simplified procedure for consolidation analysis of multiple soil layers in Figure 8.1. For n-layers, we use layer 1 with H_1 and c_{v1} as a reference layer. The thicknesses $(H_2..., H_j, j = 2...n)$ of all other layers are calculated using:

$$H_2' = H_2\left(c_{v1}/c_{v2}\right)^{1/2}, ..., H_j' = H_j\left(c_{v1}/c_{vj}\right)^{1/2}, ..., H_n' = H_j\left(c_{v1}/c_{vn}\right)^{1/2}$$

The equivalent new total thickness is:

$$H' = \sum_{j=1}^{j=n} H_j' = \sum_{j=1}^{j=n} H_j\left(c_{v1}/c_{vj}\right)^{1/2} \quad (8.18a)$$

$$T_v = \frac{c_{v1}t}{H'^2} \quad (8.18b)$$

The n-layers are converted to an equivalent single layer with thickness H' in (8.18a) and time factor T_v in (8.18b). All existing solutions for a single layer can be utilized.

8.2.4 Calculation of S_{creep}, S_{creepj}, $S_{creep,fj}$, and $S_{creep,dj}$

In (8.1) or (8.4), the total creep settlement S_{creep} of all layers together is sum of S_{creepj} for all layers. This is a simple superposition. The key items for calculating S_{creepj} are $S_{creep,fj}$ and $S_{creep,dj}$.

a. Calculation of $S_{creep,fj}$ for different stress-strain states

Creep is the compression under a constant vertical effectible stress. The real case is that the vertical total stress (or pressure) is increased from an initial stress-strain point suddenly or in ramp to the final stress-strain point and then is kept constant. Creep settlement $S_{creep,fj}$ of j-layer is calculated as creep compression under the 'final' vertical effective stress ignoring coupling of excess porewater pressure nor any ramp loading process. This is an ideal case in order to decouple this consolidation problem.

To consider creep compression occurred in 'primary' consolidation, (1.14) shall be re-written as, in the same logic in (1.22):

$$e = e_o - C_{\alpha e} \log \frac{t_o + t_e}{t_o} \tag{8.19}$$

The parameter $C_{\alpha e}$ above has the same symbol as $C_{\alpha e}$ in (1.14), but is re-defined in (8.19). Equation (8.19) is definite when $t_e = 0$, while (1.14) is not when $t = 0$. The $C_{\alpha e}$ in (1.14) is called 'secondary' consolidation coefficient; while $C_{\alpha e}$ in (8.19) is called creep coefficient. The physical meanings of t_o and t_e are explained in Chapter 1 before. In fact, the value of the newly defined $C_{\alpha e}$ in (8.19) is nearly equal to the value of the old $C_{\alpha e}$ in (1.14). The relationship between $C_{\alpha e}$ in (8.19) and ψ, and other parameters can be found in (1.33) in Chapter 1. These relations are: $C_{\alpha e} = \ln(10)\psi$, $t_o = t_{24} = 1440\,\text{min}$; $C_{r,24} = \ln(10)\kappa$; $C_{c,24} = \ln(10)\lambda$, $\varepsilon^r_{zo} = \varepsilon_{zc,24}$, $\sigma'_{zo} = \sigma'_{zc,24}$. These relations assume that all test data for determining $C_r = C_{r,24}$, $C_c = C_{c,24}$ and pre-consolidation pressure point (ε_{zc}, $\sigma'_{zc,24}$) have a standard duration of 1 day or 24 hours.

According to the 'equivalent time' concept (Yin and Graham 1989; Yin 1990, 2011, 2015), the total strain ε_z at any stress-strain state in Figure 8.3 can be calculated by the following equation:

$$\varepsilon_z = \varepsilon_{zp} + \frac{C_c}{V} \log \frac{\sigma'_z}{\sigma'_{zp}} + \frac{C_{\alpha e}}{V} \log \frac{t_o + t_e}{t_o} \tag{8.20}$$

where $\varepsilon_{zp} + \frac{C_c}{V} \log \frac{\sigma'_z}{\sigma'_{zp}}$ is the strain on the NCL under stress σ'_z (also called 'reference time line' and noting specific volume $V = 1 + e_o$); and $\frac{C_{\alpha e}}{V} \log \frac{t_o + t_e}{t_o}$ is the creep strain occurring from the NCL under the same stress σ'_z. The above equation is valid for any 1-D loading path.

The calculation of $S_{creep,fj}$ is dependent on the final stress-strain state $(\sigma_z, \varepsilon_z)$. To make presentation concise, in the following equations, layer index j is removed.

i. The final stress-strain point is on an NCL line, for example, at point 4

The final creep settlement for any point on NCL line is:

$$S_{creep,f} = \frac{C_{\alpha e}}{1+e_o}\log\left(\frac{t_o+t_e}{t_o}\right)H \quad for\ \ t_e \geq 0 \tag{8.21a}$$

For a suddenly applied load kept for a duration time t, we have:

$$t_e = t - t_o$$

Submitting the above relation into (8.21a), we have:

$$S_{creep,f} = \frac{C_{\alpha e}}{1+e_o}\log\left(\frac{t}{t_o}\right)H \quad for\ \ t \geq 1\,day \tag{8.21b}$$

Note $t_o = 1$ day since C_r and C_c are determined using data with 1 day duration. In (8.21a), if $t = 1$ day, $t_e = 0$. This means that at time $t = 1$ day, creep settlement $S_{creep,f}$ on NCL is zero. According to the EVP model theory (Yin and Graham 1989, 1994), the compression strain rate is sum of elastic strain rate and visco-plastic strain rate. The NCL line in Figure 8.3 or in (8.20), in fact, has included both elastic strain and visco-plastic strain (or creep strain). The creep settlement in (8.21a) is additional creep compression starting from 1 day or below NCL.

ii. The final stress-strain point is on an OCL line, for example at point 2

Consider a sudden load increase from point 1 to point 2, which is kept unchanged with a duration time t. The final creep settlement for any point, for example point 2, on OCL line is:

$$S_{creep,f} = \frac{C_{\alpha e}}{1+e_o}\log\left(\frac{t_o+t_e}{t_o+t_{e2}}\right)H \quad for\ \ t_e \geq t_{e2} \tag{8.21c}$$

Equation (8.21c) can be written:

$$S_{creep,f} = \left[\frac{C_{\alpha e}}{1+e_o}\log\left(\frac{t_o+t_e}{t_o}\right) - \frac{C_{\alpha e}}{1+e_o}\log\left(\frac{t_o+t_{e2}}{t_o}\right)\right]H = \Delta\varepsilon_{zcreep}H$$

$$\Delta\varepsilon_{zcreep} = \frac{C_{\alpha e}}{1+e_o}\log\left(\frac{t_o+t_e}{t_o}\right) - \frac{C_{\alpha e}}{1+e_o}\log\left(\frac{t_o+t_{e2}}{t_o}\right)$$

Referring to Figure 8.3, it is seen that $\frac{C_{\alpha e}}{1+e_o}\log\left(\frac{t_o+t_{e2}}{t_o}\right)$ is the strain from point 2′ to point 2; while $\frac{C_{\alpha e}}{1+e_o}\log\left(\frac{t_o+t_e}{t_o}\right)$ is the strain from point 2′ to a point below point 2 downward. The increased strain for further creep done from point 2 is $\Delta\varepsilon_{zcreep}$, which is what we want to use it to calculate creep settlement $S_{creep,f}$ under loading at point 2. It is noted that the relationship between t_e and the creep duration time t under the stress σ'_{z2} is:

$$t_e = t_{e2} + t - t_o$$

t_{e2} above or in (8.21b) can be calculated below. Using (8.20), at point 2 with stress-strain $(\sigma_{z2}, \varepsilon_{z2})$:

$$\varepsilon_{z2} = \varepsilon_{zp} + \frac{C_c}{V}\log\frac{\sigma_{z2}}{\sigma_{zp}} + \frac{C_{\alpha e}}{V}\log\frac{t_o + t_{e2}}{t_o}$$

From the above, we have:

$$\log\frac{t_o + t_{e2}}{t_o} = (\varepsilon_{z2} - \varepsilon_{zp})\frac{V}{C_{\alpha e}} - \frac{C_c}{C_{\alpha e}}\log\frac{\sigma'_{z2}}{\sigma'_{zp}}$$

$$\therefore \frac{t_o + t_{e2}}{t_o} = 10^{\left[(\varepsilon_{z2}-\varepsilon_{zp})\frac{V}{C_{\alpha e}} + \log\left(\frac{\sigma'_{z2}}{\sigma'_{zp}}\right)^{-\frac{C_c}{C_{\alpha e}}}\right]} = 10^{\left[(\varepsilon_{z2}-\varepsilon_{zp})\frac{V}{C_{\alpha e}}\right]} \times \left(\frac{\sigma'_{z2}}{\sigma'_{zp}}\right)^{-\frac{C_c}{C_{\alpha e}}}$$

From which, we obtain:

$$t_{e2} = t_o \times 10^{(\varepsilon_{z2}-\varepsilon_{zp})\frac{V}{C_{\alpha e}}}\left(\frac{\sigma'_{z2}}{\sigma'_{zp}}\right)^{-\frac{C_c}{C_{\alpha e}}} - t_o \qquad (8.21d)$$

It is seen from (8.21d) that the equivalent time t_{e2} at point 2 is uniquely related to the stress-strain point $(\sigma'_{z2}, \varepsilon_{z2})$. Substituting $t_e = t_{e2} + t - t_o$ into (8.21c), we have:

$$S_{creep,f} = \frac{C_{\alpha e}}{1+e_o}\log\left(\frac{t + t_{e2}}{t_o + t_{e2}}\right)H \qquad t \geq 1\text{ day} \qquad (8.21e)$$

If we consider unloading from point 4 to point 6 in Figure 8.3, using the same approach above, we can derive the following equations:

$$t_{e6} = t_o \times 10^{(\varepsilon_{z6}-\varepsilon_{zp})\frac{V}{C_{\alpha e}}}\left(\frac{\sigma'_{z6}}{\sigma'_{zp}}\right)^{-\frac{C_c}{C_{\alpha e}}} - t_o \qquad (8.21f)$$

$$S_{creep,f} = \frac{C_{\alpha e}}{1+e_o}\log\left(\frac{t_o+t_e}{t_o+t_{e6}}\right)H = \frac{C_{\alpha e}}{1+e_o}\log\left(\frac{t+t_{e6}}{t_o+t_{e6}}\right)H \qquad t \geq 1\text{ day} \quad (8.21g)$$

Reloading from point 6 to point 5:

$$t_{e5} = t_o \times 10^{(\varepsilon_{z5}-\varepsilon_{zp})\frac{V}{C_{\alpha e}}}\left(\frac{\sigma'_{z5}}{\sigma'_{zp}}\right)^{-\frac{C_c}{C_{\alpha e}}} - t_o \qquad (8.21h)$$

$$S_{creep,f} = \frac{C_{\alpha e}}{1+e_o}\log\left(\frac{t_o+t_e}{t_o+t_{e5}}\right)H = \frac{C_{\alpha e}}{1+e_o}\log\left(\frac{t+t_{e5}}{t_o+t_{e5}}\right)H \qquad t \geq 1\text{ day} \quad (8.21i)$$

b. Calculation of $S_{creep,dj}$ for different stress-strain states

$S_{creep,dj}$ is called 'delayed' creep settlement of j-layer under the 'final' vertical effective stress ignoring the excess porewater pressure. $S_{creep,dj}$ starts for $t \geq t_{EOP,field}$, that is, is 'delayed' by time of $t_{EOP,field}$. The selection of time at EOP is subjective since the separation of 'primary' consolidation from 'secondary' compressions is not scientific and subjective. In all simplified methods in this chapter, the time at $U_j = 98\%$ is considered time at EOP, that is, $t_{EOP,field}$ for field condition for j-layer. Equations (8.12b) and (8.12c) can be used to calculate $t_{EOP,field}$ for a single layer case. Equations for calculating $S_{creep,dj}$ for different 'final' stress-strain state are presented below. The layer index j is removed in following equations.

i. The final stress-strain point is on an NCL line, for example at point 4

Equation (8.21a) is the final creep settlement for any point on NCL line for $t_e \geq 0$ or $t \geq 1$ day:

$$S_{creep,f} = \frac{C_{\alpha e}}{1+e_o}\log\left(\frac{t_o+t_e}{t_o}\right)H \quad t_e \geq 0$$

$S_{creep,d}$ is delayed by $t_{EOP,field}$:

$$\begin{aligned} S_{creep,d} &= \frac{C_{\alpha e}}{1+e_o}\log\left(\frac{t_o+t_e}{t_o}\right)H - \frac{C_{\alpha e}}{1+e_o}\log\left(\frac{t_o+t_{e,EOP,field}}{t_o}\right)H \\ &= \frac{C_{\alpha e}}{1+e_o}\log\left(\frac{t_o+t_e}{t_o+t_{e,EOP,field}}\right)H \qquad for\ \ t_e \geq t_{e,EOP,field} \end{aligned} \qquad (8.22a)$$

Note $\because t_e = t - t_o \quad \therefore t_{e,EOP,field} = t_{EOP,field} - t_o$. Substituting these time relations into (8.22a):

$$S_{creep,d} = \frac{C_{\alpha e}}{1+e_o}\log\left(\frac{t}{t_{EOP,field}}\right)H \qquad for\ \ t \geq t_{EOP,field} \tag{8.22b}$$

In (8.22b) $S_{creep,d}$ is calculated for $t \geq t_{EOP,field}$, that is, 'delayed' by time $t_{EOP,field}$. Equation (8.22b) is the same as the equation for calculating 'secondary' compression of soil in (7.30).

ii. The final stress-strain point is on an OCL line, for example at point 2

The final creep settlement at point 2 is:

$$S_{creep,f} = \frac{C_{\alpha e}}{1+e_o}\log\left(\frac{t+t_{e2}}{t_o+t_{e2}}\right)H \quad for\ \ t_e \geq t_{e2}\ or\ t \geq t_o = 1\ \text{day}$$

$S_{creep,d}$ is delayed by $t_{EOP,field}$:

$$\begin{aligned} S_{creep,d} &= \frac{C_{\alpha e}}{1+e_o}\log\left(\frac{t+t_{e2}}{t_o+t_{e2}}\right)H - \frac{C_{\alpha e}}{1+e_o}\log\left(\frac{t_{EOP,field}+t_{e2}}{t_o+t_{e2}}\right)H \\ S_{creep,d} &= \frac{C_{\alpha e}}{1+e_o}\log\left(\frac{t+t_{e2}}{t_{EOP,field}+t_{e2}}\right)H \qquad for\ \ t \geq t_{EOP,field} \end{aligned} \tag{8.23a}$$

When $t \leq t_{EOP,field}$, $S_{creep,d}$ in (8.23a) is zero.

Using the same approach, at point 6:

$$S_{creep,d} = \frac{C_{\alpha e}}{1+e_o}\log\left(\frac{t+t_{e6}}{t_{EOP,field}+t_{e6}}\right)H \quad for\ \ t \geq t_{EOP,field} \tag{8.23b}$$

at point 5:

$$S_{creep,d} = \frac{C_{\alpha e}}{1+e_o}\log\left(\frac{t+t_{e5}}{t_{EOP,field}+t_{e5}}\right)H \quad for\ \ t \geq t_{EOP,field} \tag{8.23c}$$

8.3 CONSOLIDATION SETTLEMENTS OF A CLAY LAYER WITH UNIFORM STRESSES

A clay layer has thickness of 1 m, is free drained at the top, and impermeable at bottom. Values of parameters (C_r, C_c, $C_{\alpha e}$, t_o, e_o, k) are listed in Table 8.1. Referring to Figure 8.3, point 1 is assumed to be the initial effective stress-strain point ($\sigma'_{z1}, \varepsilon_{z1}$). This layer is subjected to suddenly applied uniform surcharge in different stages, for example, from point 1 to point 4 ($\sigma'_{z4}, \varepsilon_{z4}$), or loading from point 1 to point 2 ($\sigma'_{z2}, \varepsilon_{z2}$), unloading from point 4 to point 6

Table 8.1 Values of parameters and stress-strain states for a single soil layer

$C_r = 0.07$	$C_c = 0.5$	$C_{\alpha e} = 0.018$	$t_o = 1$ day	$e_o = 1$	$k = 10^{-6}$ m/sec
$\sigma'_{z1} = 30$ kPa	$\sigma'_{z2} = 40$ kPa	$\sigma'_{zp} = 60$ kPa	$\sigma'_{z4} = 120$ kPa	$\sigma'_{z6} = 50$ kPa	$\sigma'_{z7} = 200$ kPa
$\varepsilon_{z1} = 0.001$	$\varepsilon_{z2} = 0.0054$	$\varepsilon_{zp} = 0.0115$	$\varepsilon_{z4} = 0.0868$	$\varepsilon_{z6} = 0.0735$	$\varepsilon_{z7} = 0.1423$

$(\sigma'_{z6}, \varepsilon_{z6})$, and reloading from point 6 to point 7 $(\sigma'_{z7}, \varepsilon_{z7})$. Values of these points are listed in Table 8.1. All stresses in this layer are considered uniform. A thick layer can be divided into a few sub-layers with thickness 0.25 to 1 m. Stresses and strains in each sub-layer can be assumed uniform. Therefore, uniform stress assumption is practical. The calculation of consolidation settlements for different cases using the simplified Hypothesis B method in (8.1) is explained below.

a. Settlement-time curve of the layer for sudden loading from point 1 to point 4

Using (8.6b) with $\sigma'_{unit} = 0$:

$$S_{f,1-4} = \varepsilon_{z,1-4} H = \left[\frac{C_r}{1+e_o}\log\left(\frac{\sigma'_{zp}}{\sigma'_{z1}}\right) + \frac{C_c}{1+e_o}\log\left(\frac{\sigma'_{z4}}{\sigma'_{zp}}\right)\right] H$$

$$= \left[\frac{0.07}{1+1}\log\frac{60}{30} + \frac{0.5}{1+1}\log\left(\frac{120}{60}\right)\right] \times 1 = 0.0858 \text{ m}$$

$$\because \Delta\sigma'_{z,1-4} = 120 - 30 = 90 \text{ kPa}; \Delta\varepsilon_{z,1-4} = S_{f,1-4}/H = 0.0867/1 = 0.0867$$

$$\therefore m_{v,1-4} = \Delta\varepsilon_{z,1-4} / \Delta\sigma'_{z,1-4} = 0.0858/90 \text{ (1/kPa)} = 0.953 \text{ (1/MPa)}$$

$$c_{v,1-4} = \frac{k}{\gamma_w m_{v,1-4}} = \frac{10^{-10}\text{(m/s)}}{9.81(\text{kN/m}^3) \times 9.533 \times 10^{-4}(\text{1/kPa})} = 0.337 \text{ (m}^2\text{/year)}$$

For $U_v = 98\%$:

$$t_{EOP,field} = \frac{T_v d^2}{c_v} = \frac{1.500 d^2}{c_v} = \frac{1.5 \times 1^2}{0.337} = 4.451 \text{ (years)}$$

Using the simplified Hypothesis B method, (8.1) for one layer is written:

$$S_{totalB} = U_v S_f + [\alpha U^\beta S_{creep,f} + (1 - \alpha U^\beta) S_{creep,d}]$$
$$for \ \ all \ \ t \geq 1 \, day \ (t \geq t_{EOP,field} = 4.451 \text{ years for } S_{creep,d})$$

It is assumed that $\alpha = 0.8$ and $\beta = 0$ so that $U^{\beta} = 1$. This was the case studied by Yin and Feng (2017). Equation (8.21b) is for $S_{creep,f}$ and (8.22b) is for $S_{creep,d}$. Equation (8.1) is written:

$$S_{totalB} = U_v S_f + [0.8 S_{creep,f} + (1-0.8) S_{creep,d}] \;\; for \;\; t \geq 1 \, day \, (t \geq t_{EOP,field}$$

$$= 4.451 \text{ years for } S_{creep,d}) = U_v S_f + \left[0.8 \frac{C_{\alpha e}}{1+e_o} \log\left(\frac{t}{t_o}\right) H \right.$$

$$\left. + \; (1-0.8) \frac{C_{\alpha e}}{1+e_o} \log\left(\frac{t}{t_{EOP,field}}\right) H \right] \tag{8.24a}$$

Substituting values of parameters into above equation:

$$S_{totalB} = 0.0858 U_v + \left[0.8 \frac{0.018}{2} \log(t) \times 1 + (1-0.8) \frac{0.018}{2} \log\left(\frac{t}{4.451}\right) \times 1 \right] \tag{8.24b}$$

Note $\frac{0.018}{2} \log\left(\frac{t}{4.451}\right)$ will start at $t \geq 4.451$ years.

If Hypothesis A is used, from (7.30):

$$S_{totalA} = S_{primary} + S_{\text{`}secondary\text{'}}$$

$$= \begin{cases} 0.0858 U_v & for \quad t < t_{EOP,field} = 4.451 \text{ years} \\ 0.0858 U_v + \dfrac{0.018}{1+1} \log\left(\dfrac{t}{4.451 \times 365}\right) \times 1 & for \quad t \geq t_{EOP,field} = 4.451 \text{ years} \end{cases} \tag{8.25}$$

Equations (8.24) and (8.25) are used to calculate settlements of this layer up to 50 years. Results are plotted in Figure 8.5(a). It is seen that at time to 50 years $S_{totalB} = 0.1184$ m, $S_{totalA} = 0.09525$. Difference is 0.02315 m, and relative difference is $|S_{totalB} - S_{totalA}| / S_{totalB} = 19.55\%$ for a layer of only 1-m thickness.

b. Settlement-time curve of the layer for sudden loading from point 1 to point 2

Using (8.6a) with $\sigma'_{unit} = 0$:

$$S_{f,1\text{-}2} = \varepsilon_{z,1\text{-}2} H = \frac{C_r}{1+e_o} \log\left(\frac{\sigma'_{z2}}{\sigma'_{z1}}\right) H = \frac{0.07}{1+1} \log\left(\frac{40}{30}\right) \times 1 = 0.00437 \text{ m}$$

$$\because \Delta\sigma'_{z,1\text{-}2} = 40 - 30 = 10 \text{ kPa}; \; \Delta\varepsilon_{z,1\text{-}2} = S_{f,1\text{-}2} / H = 0.00437/1 = 0.00437$$

$$\therefore m_{v,1\text{-}2} = \Delta\varepsilon_{z,1\text{-}2} / \Delta\sigma'_{z,1\text{-}2} = 0.00437 / 10 \; (1/kPa) = 0.437 \; (1/\text{MPa})$$

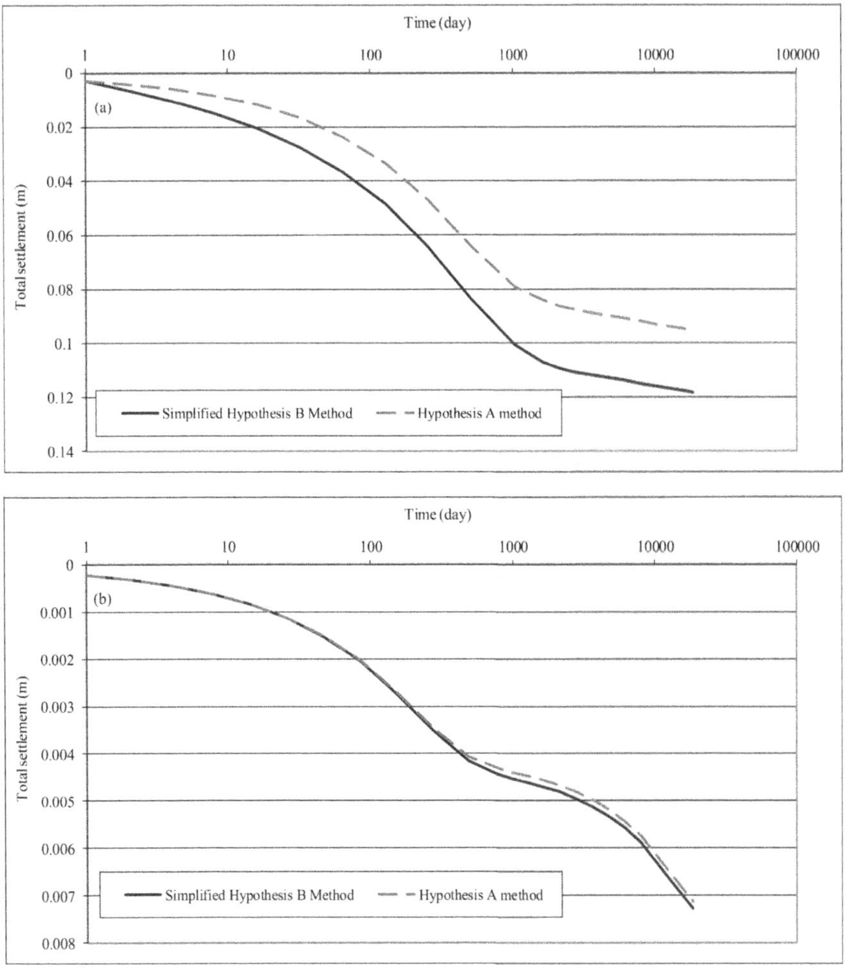

Figure 8.5 Comparison of curves of settlement and log(time) from simplified Hypothesis B method and Hypothesis A method for a 1-m soil layer – (a) for loading from point 1 to point 4 and (b) for loading from point 1 to point 2.

$$c_{v,1-2} = \frac{k}{\gamma_w m_{v,1-2}} = \frac{10^{-10}(\text{m/s})}{9.81(\text{kN/m}^3)\times 4.373\times 10^{-4}(1/\text{kPa})} = 0.735\ (\text{m}^2/\text{year})$$

For $U_v = 98\%$:

$$t_{EOP,field} = \frac{T_v d^2}{c_v} = \frac{1.500 d^2}{c_v} = \frac{1.5\times 1^2}{0.735} = 2.041\ (\text{years})$$

Using the simplified Hypothesis B method, (8.1) for one layer is written:

$$S_{totalB} = U_v S_f + [\alpha U^{\beta} S_{creep,f} + (1-\alpha U^{\beta}) S_{creep,d}]$$
$$\text{for all } t \geq 1 \text{ day } (t \geq t_{EOP,field} = 2.041 \text{ years for } S_{creep,d})$$

$\alpha = 0.8$ and $\beta = 0$ so that $U^{\beta} = 1$ (Yin and Feng 2017). Equation (8.21e) is for $S_{creep,f}$ and (8.23a) is for $S_{creep,d}$. Equation (8.1) is written:

$$S_{totalB} = U_v S_f + [0.8 S_{creep,f} + (1-0.8) S_{creep,d}] \text{ for } t \geq 1 \text{ day } (t \geq t_{EOP,field}$$
$$= 2.041 \text{ years for } S_{creep,d}) = U_v S_f + \left[0.8 \frac{C_{\alpha e}}{1+e_o} \log\left(\frac{t+t_{e2}}{t_o + t_{e2}} \right) H \right.$$
$$\left. + (1-0.8) \frac{C_{\alpha e}}{1+e_o} \log\left(\frac{t+t_{e2}}{t_{EOP,field} + t_{e2}} \right) H \right] \quad (8.26a)$$

In (8.26a), $t \geq t_{EOP,field} = 2.041$ years for $S_{creep,d}$. t_{e2} can be calculated using (8.21d)

$$t_{e2} = t_o \times 10^{(\varepsilon_{z2} - \varepsilon_{zp}) \frac{V}{C_{\alpha e}}} \left(\frac{\sigma'_{z2}}{\sigma'_{zp}} \right)^{-\frac{C_c}{C_{\alpha e}}} - t_o = 1 \times 10^{(0.0054 - 0.0115) \frac{2}{0.018}} \left(\frac{40}{60} \right)^{-\frac{0.5}{0.018}} - 1$$
$$= 16354 \text{ days}$$

Here t_{e2} is 16,354 days or 44.8055 years. This means that creep from point 2′ to point 2 in Figure 8.3 takes 44.8055 years. Substituting values of t_{e2} and other parameters into above equation:

$$S_{totalB} = 0.00437 U_v + \left[0.8 \frac{0.018}{2} \log\left(\frac{t+16354}{1+16354} \right) \times 1 \right.$$
$$\left. + (1-0.8) \frac{0.018}{2} \log\left(\frac{t+16354}{2.041 \times 365 + 16354} \right) \times 1 \right] \quad (8.26b)$$

Note $\frac{0.018}{2} \log\left(\frac{t+16354}{2.041 \times 365 + 16354} \right)$ will start at $t \geq 2.041$ years.

If Hypothesis A is used, from (7.30):

$$S_{totalA} = S_{primary} + S_{secondary},$$
$$= \begin{cases} 0.00436 U_v & \text{for } t < t_{EOP,field} = 2.041 \text{ years} \\ 0.00437 U_v + \frac{0.018}{1+1} \log\left(\frac{t}{2.041 \times 365} \right) \times 1 & \text{for } t \geq t_{EOP,field} = 2.041 \text{ years} \end{cases} \quad (8.27)$$

Equations (8.26) and (8.27) are used to calculate settlements of this layer up to 50 years. Results are plotted in Figure 8.5(b). It is seen that at time to 50 years $S_{totalB} = 0.00727$ m, $S_{totalA} = 0.00713$ m. Difference is 0.00014 m, and relative difference is $|S_{totalB} - S_{totalA}|/S_{totalB} = 1.926\%$ for a layer of only 1-m thickness. The difference of values from Hypothesis B and Hypothesis A is small because creep settlement at point 2, which is in heavily over-consolidated state, is small.

c. Consolidation settlements at 50 years for unloading from point 4 to point 6 and re-loading from point 6 to point 7
i. Unloading from point 4 to point 6

Using (8.6d) with $\sigma'_{unit} = 0$, the final primary consolidation settlement for unloading from point 4 to point 6 is:

$$S_{f,4-6} = \varepsilon_{z,4-6}H = \frac{C_r}{1+e_o}\log\left(\frac{\sigma'_{z6}}{\sigma'_{z4}}\right)H = \frac{0.07}{1+1}\log\left(\frac{50}{120}\right)\times 1 = -0.0133 \text{ m}$$

$$\because \Delta\sigma'_{z,4-6} = 50 - 120 = -70 \text{ kPa}; \Delta\varepsilon_{z,4-6} = S_{f,4-6}/H = -0.0133/1 = -0.0133$$

$$\therefore m_{v,4-6} = \Delta\varepsilon_{z,4-6}/\Delta\sigma'_{z,4-6} = -0.0133/(-70)\ (1/kPa) = 0.190\ (1/\text{MPa})$$

$$c_{v,4-6} = \frac{k}{\gamma_w m_{v,4-6}} = \frac{10^{-10}(\text{m/s})}{9.81(\text{kN/m}^3)\times 1.901\times 10^{-4}(1/\text{kPa})} = 1.691\ (\text{m}^2/\text{year})$$

For $U_v = 98\%$:

$$t_{EOP,field} = \frac{T_v d^2}{c_v} = \frac{1.500d^2}{c_v} = \frac{1.5\times 1^2}{1.169} = 1.283 \text{ (years)}$$

t_{e6} can be calculated using (8.21f)

$$t_{e6} = t_o \times 10^{(\varepsilon_{z6}-\varepsilon_{zp})\frac{V}{C_{\alpha e}}}\left(\frac{\sigma'_{z6}}{\sigma'_{zp}}\right)^{-\frac{C_c}{C_{\alpha e}}} - t_o = 1\times 10^{(0.0735-0.0115)\frac{2}{0.018}}\left(\frac{50}{60}\right)^{-\frac{0.5}{0.018}} - 1$$

$$= 1.226\times 10^9 \text{ days}$$

From (8.21g), final creep settlement at 50 years is:

$$S_{creep,f} = \frac{C_{\alpha e}}{1+e_o}\log\left(\frac{t+t_{e6}}{t_o+t_{e6}}\right)H = \frac{0.018}{1+1}\log\left(\frac{50\times 365+1.226\times 10^9}{1+1.226\times 10^9}\right)\times 1 \approx 0$$

$$S_{creep,d} = \frac{C_{\alpha e}}{1+e_o}\log\left(\frac{t+t_{e6}}{t_{EOP,field}+t_{e6}}\right)H = \frac{0.018}{2}\log\left(\frac{50\times 365+1.226\times 10^9}{1.283\times 365+1.226\times 10^9}\right)\times 1 \approx 0$$

Therefore:

$$S_{totalB} = U_v S_f + [0.8 S_{creep,f} + (1-0.8) S_{creep,d}] \ \ for \ \ t \geq 1 \, day \, (t \geq t_{EOP,field}$$

$$= 1.283 \text{ years for } S_{creep,d}) = U_v S_f + \left[0.8 \frac{C_{\alpha e}}{1+e_o} \log\left(\frac{t + t_{e6}}{t_o + t_{e6}} \right) H \right.$$

$$\left. + (1-0.8) \frac{C_{\alpha e}}{1+e_o} \log\left(\frac{t + t_{e6}}{t_{EOP,field} + t_{e6}} \right) H \right] = \ \approx 1 \times (-0.0133)$$

$$+ [0.8 \times 0 + (1-0.8) \times 0] = -0.0133 \text{ m}$$

ii. Loading from point 6 to point 7

Using (8.6f) with $\sigma'_{unit} = 0$, the final primary consolidation settlement is:

$$S_{f,6-7} = \varepsilon_{z,6-7} H = \left[\frac{C_r}{1+e_o} \log\left(\frac{\sigma'_{z4}}{\sigma'_{z6}} \right) + \frac{C_c}{1+e_o} \log\left(\frac{\sigma'_{z7}}{\sigma'_{z4}} \right) \right] H =$$

$$= \left[\frac{0.07}{1+1} \log\left(\frac{120}{50} \right) + \frac{0.5}{1+1} \log\left(\frac{200}{120} \right) \right] \times 1 = 0.0688 \text{ m}$$

$$\varepsilon_{z,7} = \varepsilon_{z,4} + \frac{C_c}{1+e_o} \log\left(\frac{\sigma'_{z7}}{\sigma'_{z4}} \right) = 0.0868 + \frac{0.5}{1+1} \log\left(\frac{200}{120} \right) = 0.1423$$

$$\because \Delta\sigma'_{z,6-7} = 200 - 50 = 150 \text{ kPa}; \ \Delta\varepsilon_{z,6-7} = S_{f,6-7}/H = 0.0688/1 = 0.0688$$

$$\therefore m_{v,6-7} = \Delta\varepsilon_{z,6-7} / \Delta\sigma'_{z,6-7} = 0.0688/150 \text{ (1/kPa)} = 0.459 \text{ (1/MPa)}$$

$$c_{v,6-7} = \frac{k}{\gamma_w m_{v,6-7}} = \frac{10^{-10} \text{(m/s)}}{9.81 \text{(kN/m}^3) \times 4.585 \times 10^{-4} \text{(1/kPa)}} = 0.701 \text{ (m}^2\text{/year)}$$

For $U_v = 98\%$:

$$t_{EOP,field} = \frac{T_v d^2}{c_v} = \frac{1.500 d^2}{c_v} = \frac{1.5 \times 1^2}{0.701} = 2.140 \text{ (years)}$$

Since point 7 is on the NCL, final creep settlement at 50 years is:

$$S_{creep,f} = \frac{C_{\alpha e}}{1+e_o} \log\left(\frac{t}{t_o} \right) H = \frac{0.018}{1+1} \log\left(\frac{50 \times 365}{1} \right) \times 1 = 0.0384 \text{ m}$$

$$S_{creep,d} = \frac{C_{\alpha e}}{1+e_o} \log\left(\frac{t}{t_{EOP,field}} \right) H = \frac{0.018}{2} \log\left(\frac{50 \times 365}{2.140 \times 365} \right) \times 1 = 0.0123 \text{ m}$$

Therefore:

$$S_{totalB} = U_v S_f + [0.8S_{creep,f} + (1-0.8)S_{creep,d}] \ for \ t \geq 1 \, day \, (t \geq t_{EOP,field}$$

$$= 2.140 \text{ years for } S_{creep,d}) = U_v S_f + \left[0.8\frac{C_{\alpha e}}{1+e_o}\log\left(\frac{t}{t_o}\right)H\right.$$

$$\left.+(1-0.8)\frac{C_{\alpha e}}{1+e_o}\log\left(\frac{t}{t_{EOP,field}}\right)H\right] = 1\times 0.0688 + [0.8\times 0.0384$$

$$+(1-0.8)\times 0.0123] = 0.102 \text{ m}$$

If needed, curves of settlement vs. log(time) can be calculated using above equations and values of parameters.

8.4 CONSOLIDATION SETTLEMENTS OF A CLAY LAYER WITH OCR OF 1 AND 1.5 FROM SIMPLIFIED HYPOTHESIS B METHOD AND FULLY COUPLED CONSOLIDATION ANALYSES

In this section, consolidation settlements of an idealized horizontal layer of HKMC are calculated using the simplified Hypothesis B method and two fully coupled finite element (FE) consolidation models. This HKMC layer has 4 m in thickness and is free drained on the top surface and impermeable at the bottom. OCR is 1 or 1.5. Two FE programs are used for fully coupled consolidation analysis of the HKMC layer: one is software 'Consol' developed by Zhu and Yin (1999a, 2000), and the other one is Plaxis software (2D 2015 version). In the 'Consol' analysis, the 1-D EVP model (Yin and Graham 1989, 1994) presented in Chapter 1 is used for the consolidation modeling. In Plaxis software (2D 2015 version), a soft soil creep (SSC) model, is adopted in the FE simulations. The 1-D EVP model was applied by Zhu and Yin (1999, 2000, 2001) for consolidation analyses of different soils. The description of the SSC model is referred to Vermeer and Neher (1999) and Plaxis user's manual (2015).

Values of all parameters used in FE consolidation simulation are listed in Table 8.2. In all FE simulations, a vertical stress of 20 kPa is assumed to be instantly applied on the top surface and kept constant for a period of 18,250 days (50 years). Since HKMC layer is in seabed, the initial vertical effective stress is zero at the top of the HKMC layer surface. Therefore, the unit stress σ'_{unit} in (8.6) cannot be zero. The best value of σ'_{unit} shall be determined by oedometer compression test data at very small vertical effective stress. Here we may assume that σ'_{unit} takes values from 0.01 to 1 kPa and discuss difference of calculated settlement values.

Table 8.2 Values of parameters for the upper marine clay of Hong Kong

(a) Values of basic property

$V=1+e_o$	γ_i (kN/m³)	OCR	w_i (%)
3.65	15	1 or 1.5	100

(b) Values of parameters used in Consol software

κ/V	λ/V	ψ/V	t_o (day)	k_v (m/day)	σ'_{zo}* (kPa)
0.01086	0.174	0.0076	1	1.90×10^{-4}	1

*: σ'_{zo} is the value of the effective vertical stress when the vertical strain of the reference time line is zero ($\varepsilon_{zo}=0$). Further details can be found in Zhu and Yin (2000).

(c) Values of parameters used in PLAXIS

κ^*	λ^*	μ^*	t_o (day)	k_v (m/day)	OCR	c' (kPa	ϕ' (deg)
0.02172	0.174	0.0076	1	1.90×10^{-4}	1 or 1.5	0.1	30

(d) Values of parameters used in the simplified Hypothesis B method

C_e [$C_e=\kappa/\ln(10)$]	C_c [$C_c=\lambda/\ln(10)$]	$C_{\alpha e}$ [$C_{\alpha e}=\psi/\ln(10)$]	V	t_o (day)	k_v (m/day)
0.0913	1.4624	0.0639	3.65	1	1.90×10^{-4}

a. Normally consolidated HKMC layer with $H = 4$ m and OCR = 1

Equation (8.8b) is used to calculate the final 'primary' settlement $S_{f,1-4}$. The values of all parameters are listed in Table 8.2. The values of all stresses are $\sigma'_{z1,0}=0, \sigma'_{z1,H}=20.76kPa,\ \sigma'_{z4,0}=20, \sigma'_{z4,H}=40.76kPa$. $S_{f,1-4}$ is:

$$S_{f,1-4}=\frac{1.4624}{3.65\ln(10)}\left\{\frac{4}{(40.76-20)}[(40.76+\sigma'_{unit})\ln(40.76+\sigma'_{unit})\right.$$
$$-(40.76+\sigma'_{unit})-((20+\sigma'_{unit})\ln(20+\sigma'_{unit})-(20+\sigma'_{unit}))]$$
$$-\frac{4}{(20.76-0)}[(20.76+\sigma'_{unit})\ln(20.76+\sigma'_{unit})-(20.76+\sigma'_{unit})$$
$$\left.-((0+\sigma'_{unit})\ln(0+\sigma'_{unit})-(0+\sigma'_{unit}))]\right\}$$

Using above equation, for σ'_{unit} = 0.01 kPa, using above equation, $S_{f,1-4}$ = 0.944 m; if σ'_{unit} = 0.1 kPa, $S_{f,1-4}$ = 0.928 m; σ'_{unit} = 0.5 kPa, $S_{f,1-4}$ = 0.879 m; if σ'_{unit} = 1 kPa, $S_{f,1-4}$ = 0.834 m. This means that the final 'primary' settlement $S_{f,1-4}$ is sensitive to the value of σ'_{unit}. In this example, we select σ'_{unit} = 0.1 kPa so that the final 'primary' settlement $S_{f,1-4}$ is 0.928 m.

The calculation of average m_v and c_v is below:

$$\Delta\varepsilon_{z,1-4} = S_{f,1-4}/H = 0.928/4 = 0.232$$
$$m_v = \Delta\varepsilon_{z,1-4}/\Delta\sigma'_{z,1-4} = 0.232/20 = 0.0116 \quad (1/\text{kPa})$$

$$c_v = k/(\gamma_w m_v) = 1.9\times10^{-4}/(9.81\times0.0116) = 1.670\times10^{-3}\ (\text{m}^2/\text{day})$$
$$= 0.610\ (\text{m}^2/\text{year})$$

As explained in this chapter before, a thick layer can be divided into small sub-layers. The stresses and values of soil parameters in each sub-layer are assumed be constant. In this case, simple equations in (8.6c) can be used to calculate that the final 'primary' settlement $S_{f,1-4}$ for each sub-layer. This layer of 4 m can be divided into 2, 4, or 8 sub-layers with thickness of 2, 1, and 0.5 m, respectively. The final 'primary' settlement $S_{f,1-4}$ calculated is 0.743, 0.831, 0.881, and 0.910 m sub-layers with thickness of 4, 2, 1, and 0.5 m, respectively. Values of S_f, m_v, and c_v for sub-layer thickness of 0.5 m for OCR = 1 are listed in Table 8.3. ε_{zp} in Table 8.3 is the strain from the initial effective stress to pre-consolidation pressure. Since OCR=1, $\sigma'_{zp} = \sigma'_{zi}$, the strain ε_{zp} is zero. Summary of values of S_f, m_v, and c_v for different number of sub-layers for OCR=1 are listed in Table 8.4 including S_f obtained by more accurate integration method. Figure 8.6 shows values of S_f, ε_z, m_v, and c_v for different number of sub-layer for OCR=1.

In this example, the simplified Hypothesis B method in (8.1) together with other equations on relevant parameters is used to calculate the total settlement S_{totalB} using $\alpha = 0.8$ and $\beta = 0$ (denoted B Method 1), $\beta = 0.3$ (denoted B Method 2), and $\beta = 1$ (denoted B Method 3). B Method 1 using $\alpha = 0.8$ and

Table 8.3 Calculation of S_f, m_v, and c_v for sub-layer thickness of 0.5 m for OCR=1

Mid sub-layer depth (m)	σ'_{zi} *(kPa)*	$\sigma'_z = \sigma'_{zi} + \Delta\sigma'_z$ *(kPa)*	$\sigma'_{zp} = \sigma'_{zi}$ *(kPa)*	ε_{zp}	$\Delta\varepsilon_z$	m_v*(1/kPa)*	$c_v = \frac{k_v}{\gamma_w m_v}$ *(m²/day)*
0.25	1.2975	21.2975	1.2975	0	0.4748	0.01137	1.704E-03
0.75	3.8925	23.8925	3.8925	0	0.3120		
1.25	6.4875	26.4875	6.4875	0	0.2428		
1.75	9.0825	29.0825	9.0825	0	0.2012		
2.25	11.6775	31.6775	11.6775	0	0.1727		
2.75	14.2725	34.2725	14.2725	0	0.1517		
3.25	16.8675	36.8675	16.8675	0	0.1355		
3.5	18.1650	38.1650	18.1650	0	0.1287		
				0	0.2274		
				Total strain:	0.2274		
				Settlement:	0.9097	(m)	

Table 8.4 Summary of S_f, m_v, and c_v for different number of sub-layers for OCR=1

Number sub-layers	*Strain ε_z after loading*	*m_v (1/kPa)*	*$c_v = \frac{k_v}{\gamma_w m_v}$ (m²/day)*	*$S_f = \varepsilon_z \times H$ (m)*	*S_f in integration (m)*
1	0.1858	0.0093	2.086E-03	0.743	0.928
2	0.2077	0.0104	1.866E-03	0.831	0.928
4	0.2202	0.0110	1.760E-03	0.881	0.928
8	0.2274	0.0114	1.704E-03	0.910	0.928

$\beta = 0$ is in fact the method published by Yin and Feng (2017). The calculated curves of settlements with log(time) from the simplified Hypothesis B method are shown in Figure 8.7(a) for time up to 100 years. At the same time, Hypothesis A method and two fully coupled FE models are used to calculate the curves of settlements with log(time) which are also shown in Figure 8.7(a) for comparison. Table 8.5 lists values of $S_{creep,f}$, $S_{creep,d}$, S_{totalA}, and S_{totalB} from the simplified Hypothesis B method and Hypothesis A method for OCR=1. It is seen from Figure 8.7(a) that when $\alpha = 0.8$ and $\beta = 0.3$m B Method 2

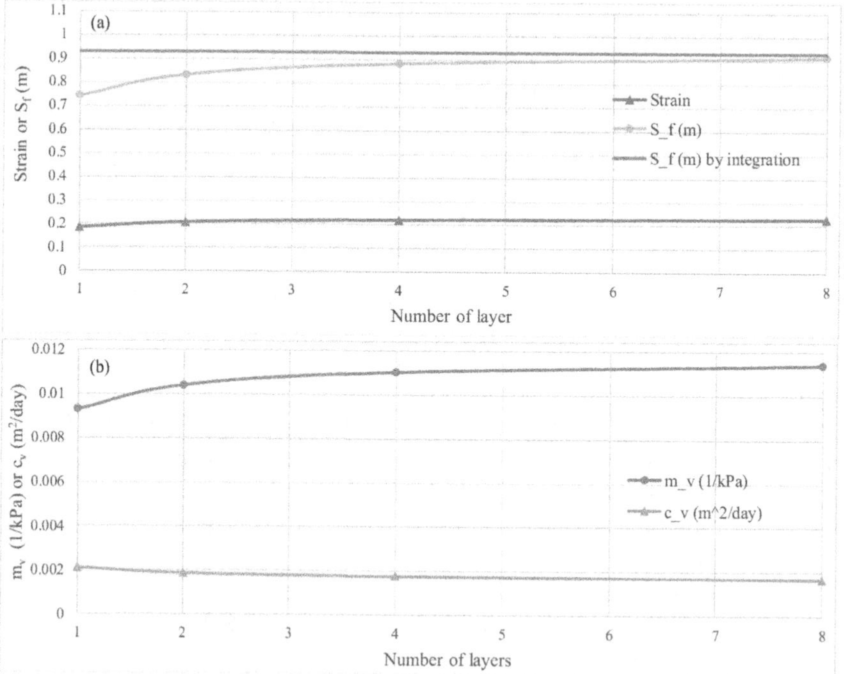

Figure 8.6 fx1, fx2, fx3, and fx4 for different number of sub-layer for OCR = 1 (a) strain and final settlement and (b) the coefficient of volume compressibility and coefficient of consolidation (a) strain and final settlement and (b) coefficient of volume compressibility and coefficient of consolidation.

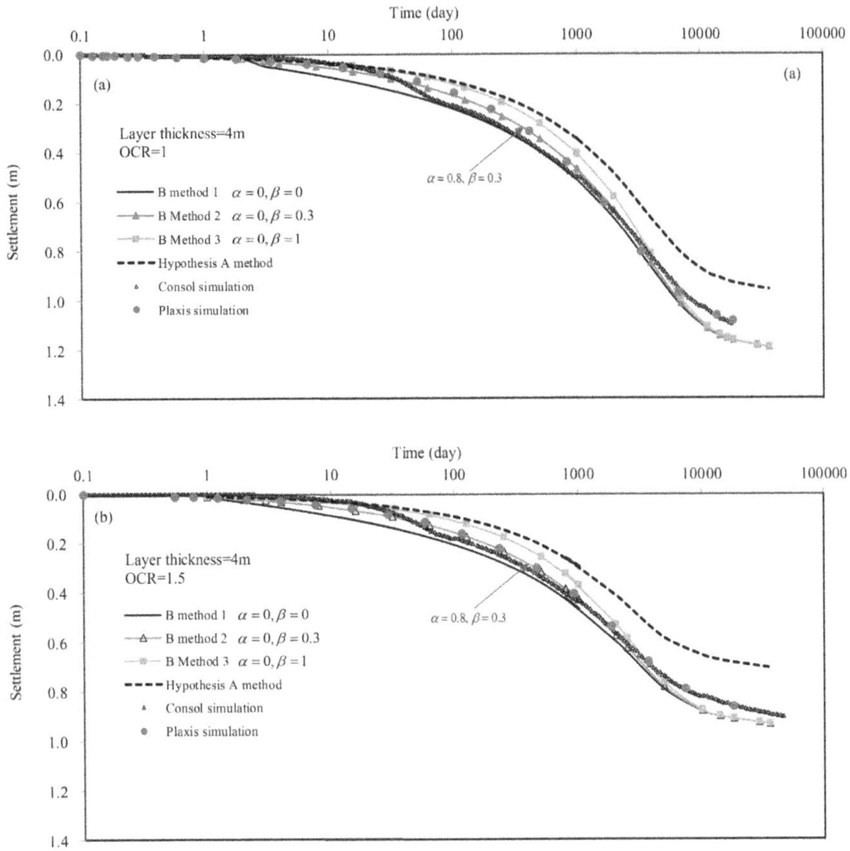

Figure 8.7 Comparisons of curves of settlements with log(time) from the simplified Hypothesis B method, Hypothesis A method, and two fully coupled finite element models (a) Over-Consolidation Ratio (OCR) = 1 and (b) OCR = 1.5.

gives curves much closer to the curves from the two FE models of 'Consol' by Zhu and Yin (1999a, 2000) and Plaxis software (2D 2015 version). Values of parameters used in Consol software are listed in Table 8.2(b) and those of Plaxis in Table 8.2(c). As shown in Figure 8.7(a), again, Hypothesis A method underestimates the total settlement for the time period.

b. Over-consolidated HKMC layer with $H = 4$ m and OCR = 1.5

Equation (8.8d) is used to calculate the final 'primary' settlement $S_{f,1\text{-}4}$. The values of all parameters are listed in Table 8.2. The values of all stresses are $\sigma'_{z1,0} = 0, \sigma'_{z1,H} = 20.76$ kPa, $\sigma'_{zp,0} = 0, \sigma'_{zp,H} = 31.14$ kPa $\sigma'_{z4,0} = 20, \sigma'_{z4,H} = 40.76$ kPa. $S_{f,1\text{-}4}$ is:

Table 8.5 Summary of $S_{creep,f}$ · $S_{creep,d}$, S_{totalA}, S_{totalB} from simplified Hypothesis B method and Hypothesis A method for OCR=1

Time (day)	$T_v = \frac{c_v t}{d^2}$	U_v	$S_{primary} = U_v S_f$ (m)	$S_{creep,f}$ (m)	$S_{creep,d}$ (m)	A Method S_{totalA} (m)	B Method 1 S_{totalB} (m)	B Method 2 S_{totalB} (m)	B Method 3 S_{totalB} (m)
1	0.0001	0.0117	0.0108	0	0	0.0108	0.0108	0.01081	0.0108
2	0.0002	0.0165	0.0153	0	0	0.0153	0.0153	0.0153	0.0153
3	0.0003	0.0202	0.0187	0.0334	0	0.0187	0.0454	0.0270	0.0193
4	0.0004	0.0233	0.0216	0.0421	0	0.0216	0.0553	0.0325	0.0224
8	0.0009	0.0329	0.0306	0.0632	0	0.0306	0.0811	0.0487	0.0322
16	0.0017	0.0466	0.0432	0.0843	0	0.0432	0.1107	0.0701	0.0464
32	0.0034	0.0659	0.0611	0.1054	0	0.0611	0.1454	0.0984	0.0667
64	0.0068	0.0932	0.0865	0.1264	0	0.0865	0.1876	0.1361	0.0959
128	0.0136	0.1317	0.1223	0.1475	0	0.1223	0.2403	0.1865	0.1378
256	0.0273	0.1863	0.1729	0.1686	0	0.1729	0.3078	0.2544	0.1980
512	0.0545	0.2635	0.2445	0.1897	0	0.2445	0.3963	0.3462	0.2845
1024	0.1091	0.3726	0.3458	0.2107	0	0.3458	0.5144	0.4712	0.4087
1000	0.1065	0.3682	0.3418	0.2100	0	0.3418	0.5098	0.4662	0.4036
2000	0.2130	0.5208	0.4833	0.2311	0	0.4833	0.6682	0.6353	0.5796
4000	0.4260	0.7167	0.6651	0.2521	0	0.6651	0.8668	0.8477	0.8097
7000	0.7455	0.8712	0.8086	0.2692	0	0.8088	1.0239	1.0152	0.9962
11300	1.2034	0.9584	0.8895	0.2837	0	0.8895	1.1165	1.1136	1.1070
14255	1.5181	0.9809	0.9103	0.2908	0	0.9103	1.1430	1.1416	1.1385
16425	1.7492	0.9892	0.9181	0.2951	0.0043	0.9224	1.1550	1.1542	1.1525
18250	1.9436	0.9933	0.9219	0.2983	0.0075	0.9294	1.1620	1.1615	1.1605
29200	3.1097	0.9996	0.9278	0.3126	0.0218	0.9496	1.1821	1.1821	1.1821
36500	3.8872	0.9999	0.9280	0.3194	0.0286	0.9566	1.1892	1.1892	1.1892

$$
\begin{aligned}
S_{f,1-4} = \frac{0.0913}{3.656\ln(10)} \Bigg\{ & \frac{4}{(31.14-0)}[(31.14+\sigma'_{unit})\ln(31.14+\sigma'_{unit}) - (31.14+\sigma'_{unit}) \\
& - ((0+\sigma'_{unit})\ln(0+\sigma'_{unit}) - (0+\sigma'_{unit}))] - \frac{4}{(20.76-0)}[(20.76+\sigma'_{unit}) \\
& \ln(20.76+\sigma'_{unit}) - (20.76+\sigma'_{unit}) - ((0+\sigma'_{unit})\ln(0+\sigma'_{unit}) - (0+\sigma'_{unit}))] \Bigg\} \\
+ \frac{1.4624}{3.65\ln(10)} \Bigg\{ & \frac{4}{(40.76-20)}[(40.76+\sigma'_{unit})\ln(40.76+\sigma'_{unit}) - (40.76+\sigma'_{unit}) \\
& - ((20+\sigma'_{unit})\ln(20+\sigma'_{unit}) - (20+\sigma'_{unit}))] - \frac{4}{(40.76-0)}[(31.14+\sigma'_{unit}) \\
& \ln(31.14+\sigma'_{unit}) - (31.14+\sigma'_{unit}) - ((0+\sigma'_{unit})\ln(0+\sigma'_{unit}) - (0+\sigma'_{unit}))] \Bigg\}
\end{aligned}
$$

Using above equation, for σ'_{unit} = 0.01 kPa, using above equation, $S_{f,1-4}$ = 0.681 m; if σ'_{unit}=0.1 kPa, $S_{f,1-4}$ = 0.669 m; σ'_{unit} = 0.5 kPa, $S_{f,1-4}$ = 0.635 m; if σ'_{unit} = 1 kPa, $S_{f,1-4}$ = 0.604 m. This means that the final 'primary' settlement $S_{f,1-4}$ is sensitive to the value of σ'_{unit}. In this example, we select σ'_{unit} = 0.1 kPa so that the final 'primary' settlement $S_{f,1-4}$ is 0.669 m.

The calculation of average m_v and c_v is below:

$$\Delta\varepsilon_{z,1-4} = S_{f,1-4}/H = 0.669/4 = 0.167$$

$$m_v = \Delta\varepsilon_{z,1-4}/\Delta\sigma'_{z,1-4} = 0.167/20 = 0.00837 \quad (1/\text{kPa})$$

$$c_v = k/(\gamma_w m_v) = 1.9\times10^{-4}/(9.81\times0.00837) = 2.316\times10^{-3}\ (\text{m}^2/\text{day})$$

$$= 0.845\ (\text{m}^2/\text{year})$$

This 4-m thick layer can be divided into small sub-layers. The stresses and values of soil parameters in each sub-layer are assumed be constant. In this case, simple equations in (8.6b) can be used to calculate that the final 'primary' settlement $S_{f,1-4}$ in each sub-layer. This layer of 4 m can be divided into 2, 4, or 8 sub-layers with thickness of 2, 1, and 0.5 m, respectively. The final 'primary' settlement $S_{f,1-4}$ calculated is 0.487, 0.573, 0.625, and 0.646 m sub-layers with thickness of 4, 2, 1, and 0.5 m, respectively. Values of S_f, m_v and c_v for sub-layer thickness of 0.5 m for OCR=1.5 are listed in Table 8.6. Summary of values of S_f, m_v, and c_v for different number of sub-layers for OCR=1.5 are listed in Table 8.7 including S_f obtained by more accurate integration method. Figure 8.8 shows values of S_f, ε_z, m_v, and c_v for different number of sub-layer for OCR=1.5.

Table 8.6 Calculation of S_f, m_v, and c_v for sub-layer thickness of 0.5 m for OCR=1.5

Mid sub-layer depth (m)	σ'_{zi} *(kPa)*	$\sigma'_z = \sigma'_{zi} + \Delta\sigma'_z$ *(kPa)*	$\sigma'_{zp} = \sigma'_{zi}$ *(kPa)*	ε_{zp}	$\Delta\varepsilon_z$	m_v*(1/kPa)*	$c_v = \dfrac{k_v}{\gamma_w m_v}$ *(m²/day)*
0.25	1.2975	21.2975	1.94625	0.004141	0.4084	0.00808	2.399E-03
0.75	3.8925	23.8925	5.83875	0.004312	0.2429		
1.25	6.4875	26.4875	9.73125	0.004348	0.1731		
1.75	9.0825	29.0825	13.62375	0.004364	0.1313		
2.25	11.6775	31.6775	17.51625	0.004373	0.1026		
2.75	14.2725	34.2725	21.40875	0.004378	0.0816		
3.25	16.8675	36.8675	25.30125	0.004382	0.0653		
3.75	19.4625	39.4625	29.19375	0.004385	0.0523		
				0.004335	0.1572		
				Total strain:	0.1615		
				Settlement:	0.6461	(m)	

Table 8.7 Summary of S_f, m_v, and c_v for different number of sub-layers for OCR=1.5

Number sub-layers	*Strain ε_z after loading*	*m_v (1/kPa)*	$c_v = \dfrac{k_v}{\gamma_w m_v}$ *(m²/day)*	$S_f = \varepsilon_z \times H$ *(m)*	S_f *in integration (m)*
1	0.1207	0.006036	3.210E-03	0.483	0.669
2	0.1432	0.007158	2.707E-03	0.573	0.669
4	0.1562	0.00781	2.481E-03	0.625	0.669
8	0.1615	0.008076	2.399E-03	0.646	0.669

The simplified Hypothesis B method in (8.1) together with other equations on relevant parameters is used to calculate the total settlement S_{totalB} using $\alpha = 0.8$ and $\beta = 0$ (denoted B Method 1), $\beta = 0.3$ (denoted B Method 2), and $\beta = 1$ (denoted B Method 3) for OCR=1.5. The calculated curves of settlements with log(time) from the simplified Hypothesis B method are shown in Figure 8.7(b) for time up to 100 years. At the same time, Hypothesis A method and two fully coupled FE models are used to calculate the curves of settlements with log(time) which are also shown in Figure 8.7(b) for comparison. Table 8.8 lists values of $S_{creep,f}$,$S_{creep,d}$, S_{totalA}, and S_{totalB} from the simplified Hypothesis B method and Hypothesis A method for OCR=1.5. It is seen from

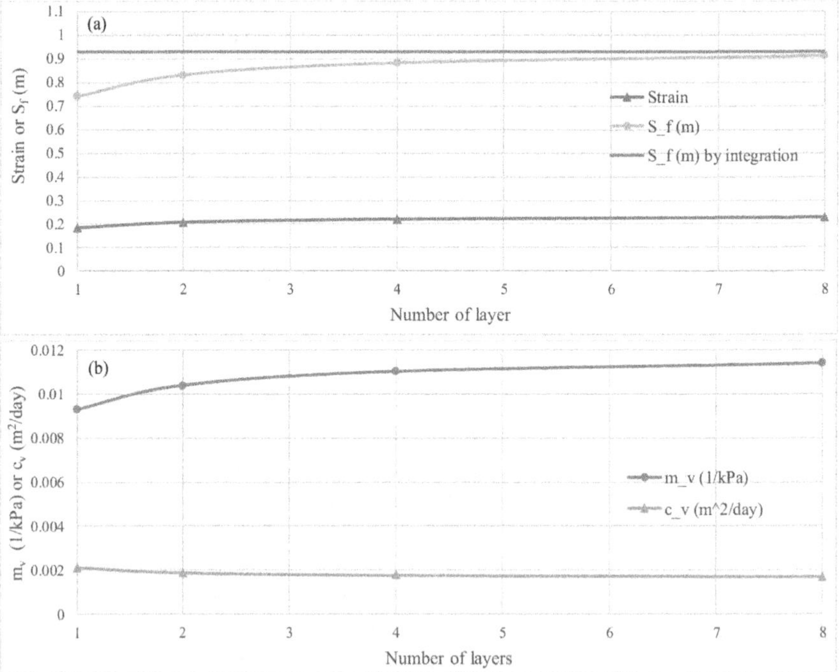

Figure 8.8 fx5, fx6, fx7, and fx8 for different number of sub-layer for OCR = 1.5 (a) strain and final settlement and (b) the coefficient of volume compressibility and coefficient of consolidation.

Table 8.8 Summary of $S_{creep,f}$, $S_{creep,d}$, S_{totalA}, S_{totalB} from simplified Hypothesis B method and Hypothesis A method for OCR=1.5

Time (day)	$T_v = \frac{c_v t}{d^2}$	U_v	$S_{primary} = U_v S_f$ (m)	$S_{creep,f}$ (m)	$S_{creep,d}$ (m)	A Method S_{totalA} (m)	B Method 1 S_{totalB} (m)	B Method 2 S_{totalB} (m)	B Method 3 S_{totalB} (m)
1	0.0001	0.0138	0.0092	0.0000	0	0.0092	0.0092	0.0092	0.0092
2	0.0003	0.0195	0.0131	0.0211	0	0.0131	0.0299	0.0182	0.0134
3	0.0004	0.0239	0.0160	0.0334	0	0.0160	0.0427	0.0247	0.0166
4	0.0006	0.0276	0.0185	0.0421	0	0.0185	0.0522	0.0300	0.0194
8	0.0012	0.0391	0.0261	0.0632	0	0.0261	0.0767	0.0453	0.0281
16	0.0024	0.0553	0.0370	0.0843	0	0.0370	0.1044	0.0653	0.0407
32	0.0048	0.0782	0.0523	0.1054	0	0.0523	0.1366	0.0915	0.0589
64	0.0096	0.1105	0.0739	0.1264	0	0.0739	0.1751	0.1262	0.0851
128	0.0192	0.1563	0.1046	0.1475	0	0.1046	0.2226	0.1722	0.1230
256	0.0384	0.2211	0.1479	0.1686	0	0.1479	0.2828	0.2336	0.1777
512	0.0768	0.3126	0.2092	0.1896	0	0.2092	0.3609	0.3162	0.2566
1024	0.1535	0.4421	0.2958	0.2107	0	0.2958	0.4644	0.4278	0.3703
800	0.1200	0.3908	0.2615	0.2032	0	0.2615	0.4240	0.3841	0.3250
2048	0.3071	0.6200	0.4148	0.2318	0	0.4148	0.6002	0.5755	0.5298
2500	0.3749	0.6785	0.4539	0.2379	0	0.4539	0.6442	0.6233	0.5831
5000	0.7497	0.8725	0.5837	0.2589	0	0.5837	0.7909	0.7826	0.7645
10110	1.5159	0.9808	0.6561	0.2803	0	0.6561	0.8804	0.8791	0.8761
14000	2.0992	0.9954	0.6659	0.2902	0.0099	0.6758	0.9001	0.8998	0.8991
18250	2.7365	0.9991	0.6684	0.2983	0.0180	0.6863	0.9106	0.9105	0.9104
29200	4.3783	1.0000	0.6690	0.3126	0.0322	0.7012	0.9255	0.9255	0.9255
36500	5.4729	1.0000	0.6690	0.3194	0.0390	0.7080	0.9323	0.9323	0.9323

Figure 8.7(b) that when $\alpha = 0.8$ and $\beta = 0.3$ m, B Method 2 gives curves much closer to the curves from the two FE models of 'Consol' by Zhu and Yin (1999, 2000) and Plaxis software (2D 2015 version). Again, Hypothesis A method underestimates the total settlement for the time period.

8.5 CONSOLIDATION SETTLEMENTS OF DOUBLE CLAY LAYERS FROM SIMPLIFIED HYPOTHESIS B METHOD AND FULLY COUPLED CONSOLIDATION ANALYSIS

Feng and Yin (2017) studied double clay layers with total thickness of 4 and 8 m with OCR of 1, 1.5, and 2. Here the cases of 4-m thick double clay layers with OCR of 1 and 1.5 are presented here. Feng and Yin (2017) selected $\alpha = 0.6$ for OCR=1 and $\alpha = 0.7$ for OCR=1.5. They did not have the item of U_v^β. In fact, their approach was equivalent to $\beta = 0$ so that $U_v^\beta = 1$. This is discussed in Section 8.4, denoted as B Method 1. In this case, (8.1) can be written:

$$S_{totalB} = \sum_{j=1}^{j=2} U_{vj} S_{fj} + \sum_{j=1}^{j=2} S_{creepj} = U_v \sum_{j=1}^{j=2} S_{fj} + \sum_{j=1}^{j=2} [\alpha S_{creep,fj} + (1-\alpha) S_{creep,dj}]$$
$$for \quad all \quad t \geq 1\,day \quad (t \geq t_{EOP,field} \text{ for } S_{creep,dj}) \tag{8.28}$$

Two methods are used to calculate the average degree of consolidation for the double soil layers. The first one is from the analytical solutions and charts by Zhu and Yin (1999b, 2005), called Zhu and Yin Method, is presented in Section 2.4.2 of Chapter 2. The second method is an approximate method in (8.18) by US Department of the Navy (1982). For double soil layers, we can convert soil layer 2 to an equivalent thickness of soil layer 1, using:

$$H_2' = H_2 \left(c_{v1} / c_{v2} \right)^{1/2}$$
$$T = \frac{c_{v1} t}{(H_1 + H_2')^2} \tag{8.29}$$

where H_2 is the height of the soil layer 2, H_2' is the equivalent thickness of soil layer 2 as if it is made up of soil layer 1, c_{v1} and c_{v2} are the coefficients of consolidation for layers 1 and 2, respectively. T is the overall time factor of the whole deposit. After the conversion, the average degree of consolidation, U_v, can be determined as one single soil layer. This method is named as US Navy method.

Figure 8.9 shows the profile of two cases of double soil layers. The total thickness of *CaseI* (4*m*) and *CaseII* (4*m*) is 4 m with 2 m 'Upper Marine

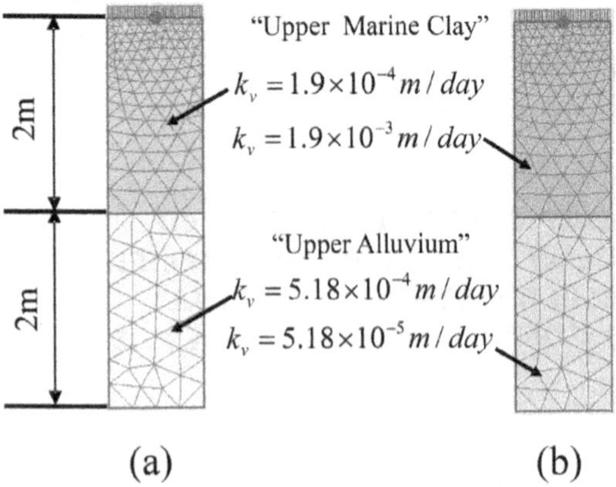

Figure 8.9 Profiles of double soil layers for Plaxis simulation with free drainage in the top and impermeable in the bottom: (a) Case I (4 m) and (b) Case II (4 m).

Clay' in the top followed by 2 m 'Upper Alluvium,' the bottom of which is impermeable. Compared to *CaseI* (4*m*), the difference of *CaseII* (4*m*) is that the permeability value of 'Upper Marine Clay' is increased by one order and the permeability value of 'Upper Alluvium' is decreased by one order. Table 8.9 presents values of all parameters of two cases of double soil layers used for consolidation analysis using the simplified Hypothesis B method and Finite Element Modeling (FEM) using Plaxis (2D version 2015). A vertical stress of 20*kPa* is assumed suddenly applied on the two layers in Figure 8.9.

In order to verify the accuracy of the new simplified Hypothesis B method for double soil layers, the FE software Plaxis (2D version 2015) is used for the numerical simulation adopting the soft soil creep (SSC) model (Vermeer and Neher 1999; Waterman and Broere 2011), which is, in fact,

Table 8.9 Values of parameters used in the simplified Hypothesis B method and Plaxis for double soil layers

(a) Values of parameters in the simplified Hypothesis B method

Case	*Soil layer*	e_o	γ_{soil} (kN/m^3)	*OCR*	C_e	C_c	$C_{\alpha e}$	k_v (*m/day*)	t_o (*day*)
Case I (4*m*)	'Upper Marine Clay'	2.65	15	1, 1.5	0.0913	1.4624	0.0639	1.9×10^{-4}	1
	'Upper Alluvium'	1	19.5	1, 1.5	0.05	0.2993	0.016	5.18×10^{-4}	1
Case II (4*m*)	'Upper Marine Clay'	2.65	15	1, 1.5	0.0913	1.4624	0.0639	1.9×10^{-3}	1
	'Upper Alluvium'	1	19.5	1, 1.5	0.05	0.2993	0.016	5.18×10^{-5}	1

(b) Values of parameters used in Plaxis

Case	*Soil layer*	γ_{soil} (kN/m^3)	*OCR*	κ^*	λ^*	μ^*	k_v (*m/day*)	c' (*kPa*)	ϕ' (°)
Case I (4*m*)	'Upper Marine Clay'	15	1, 1.5	0.0217	0.174	0.0076	1.9×10^{-4}	0.1	25
	'Upper Alluvium'	19.5	1, 1.5	0.0217	0.065	0.00347	5.18×10^{-4}	0.1	30
Case II (4*m*)	'Upper Marine Clay'	15	1, 1.5	0.0217	0.174	0.0076	1.9×10^{-3}	0.1	25
	'Upper Alluvium'	19.5	1, 1.5	0.0217	0.065	0.00347	5.18×10^{-5}	0.1	30

Note λ^* is the modified compression index, $\lambda^* = \frac{C_c}{2.3(1+e_0)}$, κ^* is the modified swelling index, $\kappa^* \approx \frac{2C_e}{2.3(1+e_0)}$, μ^* is the modified creep index, $\mu^* = \frac{C_{\alpha e}}{2.3(1+e_0)}$, all used in Plaxis.

an non-linear EVP constitutive model (Yin and Graham 1999; Yin 2015). A 2-D plane strain FE mesh with 15-node triangular elements is used in Plaxis simulation. As illustrated in Figure 8.9, the top elements of the soil have free drainage and the bottom elements are impermeable when conducting the consolidation analysis. The left and right vertical boundaries of the models are impermeable and are confined to have vertical movements only. A vertical stress of $20kPa$ is instantly applied on the top of all FE simulation models and the loading period is up to 100,000 days to make sure that consolidation is totally completed in all simulation cases. The monitoring point of red dot for the settlement is at the top surface of the FE model, as illustrated in Figure 8.9. The definition of the SSC parameters can be found in the Plaxis manual (2D version 2015), and values of parameters used in Plaxis are listed in Table 8.9(b).

The initial pre-consolidation stress plays an important role in the ground settlement prediction when adopting the SSC model. When considering OCR value effects, OCR values of 'Upper Marine Clay' and 'Upper Alluvium' are set to be 1 or 1.5, respectively. When establishing an FE model in Plaxis, initial stress-strain states before adding the vertical loading and consolidation is generated with the in-situ at-rest K_o condition.

When the simplified Hypothesis B method is used, the total thickness of 4 m is divided into a number of sub-layers with 0.5-m thickness in order to calculate the final primary settlement $S_{f,j}$ more accurately for each soil type layer. The initial effective stress $\sigma'_{z1,j}$, the pre-consolidation stress $\sigma'_{zp,j} = OCR \times \sigma'_{z1,j}$, corresponding strain $\varepsilon_{zp,j}$, final 'primary' consolidation settlement $S_{f,j}$, final 'primary' consolidation strain $\varepsilon_{z,j}$, coefficient of volume compressibility $m_{v,j}$, and coefficient of vertical consolidation $c_{v,j}$ for j-layer are calculated in the same way as that in Section 8.4 for a single layer case. However, in Feng and Yin (2017), the unit stress σ'_{unit} was zero.

The factors for double soil layers, p and q, can be calculated with (2.35) and (2.36) in Chapter 2 by substituting the values of m_v and c_v, and the corresponding values are also listed in Table 8.10. Take $CaseI(4m)$ with $OCR = 1$ as an example: $H_1 = H_2 = 2m$, $c_{v1} = 0.00122$ m²/day and $c_{v2} = 0.02208$ m²/day in Table 2, the time factor, T, after a loading time of 100 days with one-way drainage condition could be determined as follows:

$$T = \frac{c_{v1}c_{v2}t}{\left(H_1\sqrt{c_{v2}} + H_2\sqrt{c_{v1}}\right)^2} = \frac{0.0012 \times 0.02208 \times 100}{4(\sqrt{0.02208} + \sqrt{0.0012})^2} = 0.02 \tag{8.30}$$

From the solution charts for one-way drainage condition (Zhu and Yin 2005), the average degree of consolidation U_v is 18% for $p = -0.3$ and $q = 0.62$ (from solution charts in Zhu and Yin (1999b, 2005) and is 13% for $p = 0.3$ and $q = 0.62$. It is noted that $T_c = 0$ since the loading is suddenly applied. With the help of the interpolation method for $p = -0.22$ in Table 2, the average degree

Table 8.10 Summary of calculated values of parameters used in the new simplified Hypothesis B method for 4-m thick double soil layers

		'Upper Marine Clay'			*'Upper Alluvium'*					
Case	*OCR*	S_{f1} *(m)*	m_{v1} *(kPa^{-1})*	c_{v1} *(m^2/day)*	S_{f2}*(m)*	m_{v2} *(kPa^{-1})*	c_{v2}*(m^2/day)*	S_{f1} + S_{f2}(m)	p	q
Case I (4m)	1	0.635	0.01588	0.00122	0.096	0.00239	0.02208	0.731	0.22	0.62
	1.5	0.500	0.01258	0.00154	0.052	0.00130	0.04080	0.552	0.31	0.67
Case II (4m)	1	0.635	0.01588	0.0122	0.096	0.00239	0.00221	0.731	0.88	0.40
	1.5	0.500	0.01258	0.0154	0.052	0.00130	0.00408	0.552	0.90	0.32

of consolidation U_v at time of 100 days and for $p = -0.22$ and $q = 0.62$ could be obtained:

$$U_v = \frac{[0.3-(-0.22)]\times 18\% + [(-0.22)-(-0.3)]\times 13\%}{[0.3-(-0.3)]} = 17.3\% \tag{8.31}$$

Similarly, the average degree of consolidation, U_v, for double soil layers in other different times or other conditions can also be determined.

In order to compare with the US Navy method (1982), the average degree of consolidation, U_v, is also calculated by transferring the 'Upper Alluvium' into 'Upper Marine Clay' soil considering the difference of coefficient of consolidation, c_v, with Equation (8.29). Then, the average degree of consolidation, U_v, could be easily determined as one equivalent single layer.

For 'Upper Marine Clay,' the creep compression $\varepsilon_{creep,f,j}$ is calculated by adopting (8.21b) since the final effective stress state is in a normal consolidation state. Equation (8.21c) shall be used for calculating the creep compression $\varepsilon_{creep,f,j}$ when the final effective stress state of some sub-layers is in over-consolidation state. Values of $t_{EOP,field}$ are determined to be the time when the average degree of consolidation is 98% for double soil layers.

In addition, Hypothesis A method is also used to calculate curves of settlements with time for the two cases. Figure 8.10 shows a comparison of settlement-log(time) curves from FE simulation, the simplified Hypothesis B method, and Hypothesis A method for double layers with OCR=1: (a) Case I (4 m) and (b) Case II (4 m). It is seen that the curves from the simplified Hypothesis B method using U_v by Zhu and Yin (1999b, 2005) are closer to the curves from FE simulations. The simplified Hypothesis B method using U_v by the US Navy method (1982) gives settlements close to those from the FE simulations in Case I, but not in Case II. This means that the U_v by the US Navy method (1982) is not accurate for the case with large differences in the permeability k_v. In Case II, the permeability is $k_v = 1.9\times10^{-3}$ (m/day) for 'Upper Marine Clay,' but $k_v = 5.18\times10^{-5}$ (m/day) for 'Upper Alluvium.' The settlements from Hypothesis A method are always much smaller than those from FE simulations for the two cases.

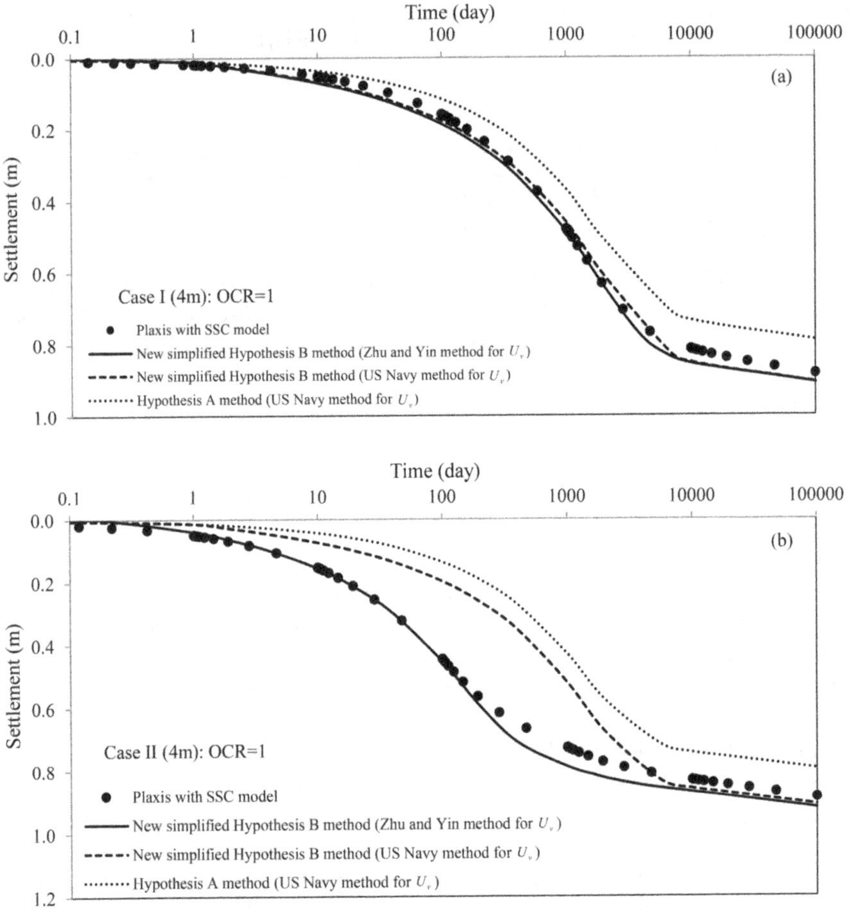

Figure 8.10 **Settlement-log(time) curves from FE simulation, the simplified Hypothesis B method, and Hypothesis A method for 4-m double layers with OCR = 1: (a) Case I (4 m) and (b) Case II (4 m).**

Figure 8.11 shows a comparison of settlement-log(time) curves from FE simulation, the simplified Hypothesis B method, and Hypothesis A method 4 m for double layers with OCR=1.5: (a) Case I (4 m) and (b) Case II (4 m). It is seen that the curves from the simplified Hypothesis B method using U_v by Zhu and Yin (1999b, 2005) are closer to the curves from FE simulations. The simplified Hypothesis B method using U_v by the US Navy method (1982) gives settlements close to those from the FE simulations in Case I, but not in Case II. This means that the U_v by the US Navy method (1982) is not accurate for the case with large differences in the permeability k_v. In Case II, the permeability is $k_v = 1.9 \times 10^{-3}$ (m/day) for 'Upper Marine Clay', but $k_v = 5.18 \times 10^{-5}$ (m/day) for 'Upper Alluvium'. The settlements from Hypothesis A method is always much smaller than those from FE simulations for the two cases.

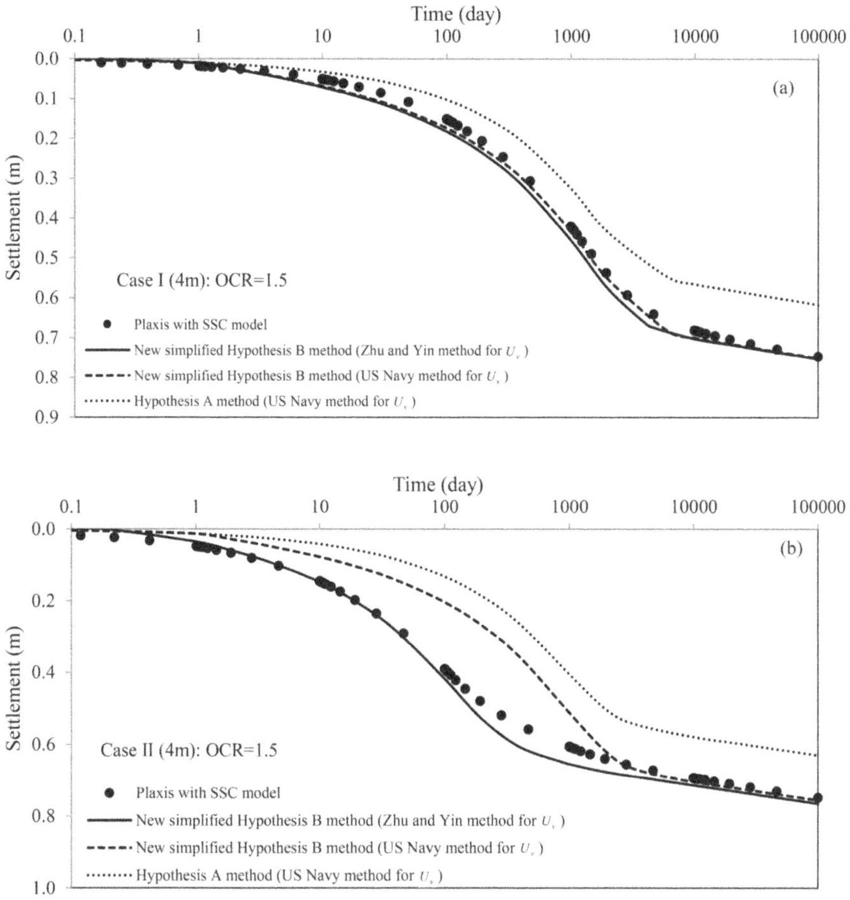

Figure 8.11 Settlement-log(time) curves from FE simulation, the simplified Hypothesis B method, and Hypothesis A method for 4-m double layers with OCR = 1.5: (a) Case I (4 m) and (b) Case II (4 m).

8.6 CONSOLIDATION SETTLEMENTS OF MULTIPLE SOIL LAYERS WITH VERTICAL DRAINS UNDER STAGED LOADING FROM SIMPLIFIED HYPOTHESIS B METHOD AND FULLY COUPLED CONSOLIDATION ANALYSIS

8.6.1 General equations for calculating consolidation settlements of multiple soil layers using simplified hypothesis b method

If the average degree of combined consolidation U_j for a single j-layer is known, (8.1) can be written:

$$S_{totalB} = S_{primary} + S_{creep} = \sum_{j=1}^{j=n} U_j S_{fj} + \sum_{j=1}^{j=n} [\alpha U_j^{\beta} S_{creep,fj} + (1-\alpha U_j^{\beta}) S_{creep,dj}] \quad (8.32)$$
$$for \;\; all \;\; t \geq t_{\mathrm{EOP},lab} \quad (t \geq t_{\mathrm{EOP},field} \; for \; S_{creep,dj})$$

The calculation of either $S_{primary}$ or S_{creep} is a simple supervision of the 'primary' consolidation settlement or creep consolidation settlement for each layer:

$$S_{primary} = \sum_{j=1}^{j=n} S_{primaryj} = \sum_{j=1}^{j=n} U_j S_{fj} \quad (8.33a)$$

$$S_{primaryj} = U_j S_{fj} \quad (8.33b)$$

$$S_{creep} = \sum_{j=1}^{j=n} S_{creepj} = \sum_{j=1}^{j=n} [\alpha U_j^{\beta} S_{creep,fj} + (1-\alpha U_j^{\beta}) S_{creep,dj}] \quad (8.34a)$$
$$for \;\; all \;\; t \geq t_{\mathrm{EOP},lab} \quad (t \geq t_{\mathrm{EOP},field} \; for \; S_{creep,dj})$$

$$S_{creepj} = \alpha U_j^{\beta} S_{creep,fj} + (1-\alpha U_j^{\beta}) S_{creep,dj} \quad (8.34b)$$
$$for \;\; all \;\; t \geq t_{\mathrm{EOP},lab} \quad (t \geq t_{\mathrm{EOP},field} \; for \; S_{creep,dj})$$

According to (8.33) and (8.34), under multi-staged loading, only $S_{primaryj}$ and S_{creepj} for each layer need to be calculated.

In this section, Test Embankment at Chek Lap Kok for Hong Kong International Airport (HKIA) project in 1980s is used as an example to demonstrate the new simplified Hypothesis B method in (8.32) to (8.34). Consolidation settlements of this HKIA Chek Lap Kok Test Embankment are calculated using the new simplified Hypothesis B method in (8.32) to (8.34) and are compared with measured data and values from the simplified FE method described in Chapter 11 and reported by (Zhu *et al.* 2001). Details of the site conditions, properties of soils, parameters of vertical drains, construction process, and parameters used in the FE model can be found in Chapter 11.

Figure 8.12 shows soil profile and points of settlements measured and calculated for comparison of Chek Lap Kok Test Embankment (Handfelt *et al.* 1987; Koutsoftas *et al.* 1987). Elevation in mPD (meter in Principal Datum), depth coordinate, thickness values of four major layers, 8 settlement monitoring points by Sondex anchors, and 9 porewater pressure measurement points are all shown in Figure 8.12. In this section, only 4 points at depths 0, 3, 6, and 14.5 m are selected to calculate settlements with comparison to measured data at the same depth.

Figure 8.13 shows soil profile and vertical drain with smear zone. It is noted that the vertical drain penetrated only 5.1 m into 'Lower marine clay.'

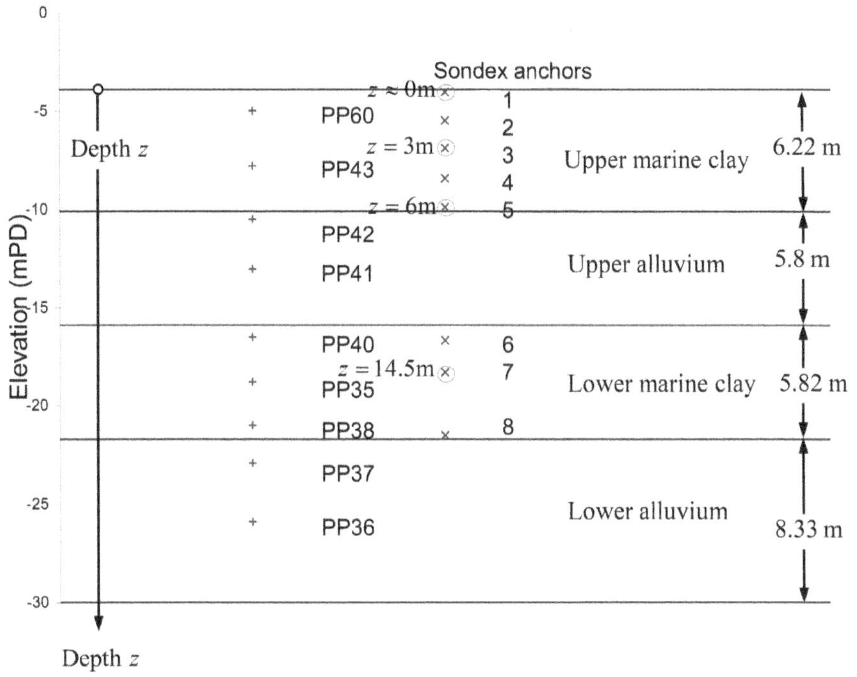

Figure 8.12 Soil profile and points of settlements measured and calculated for comparison.

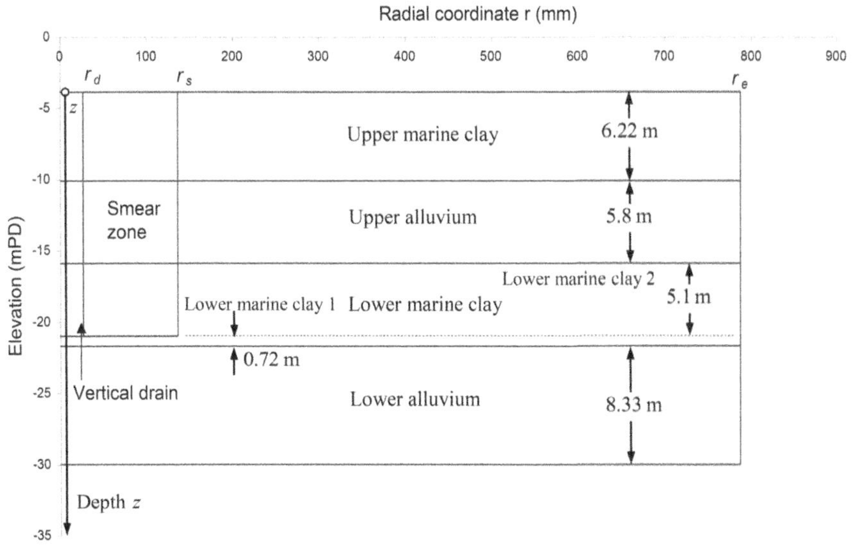

Figure 8.13 Soil profile and vertical drain with smear zone.

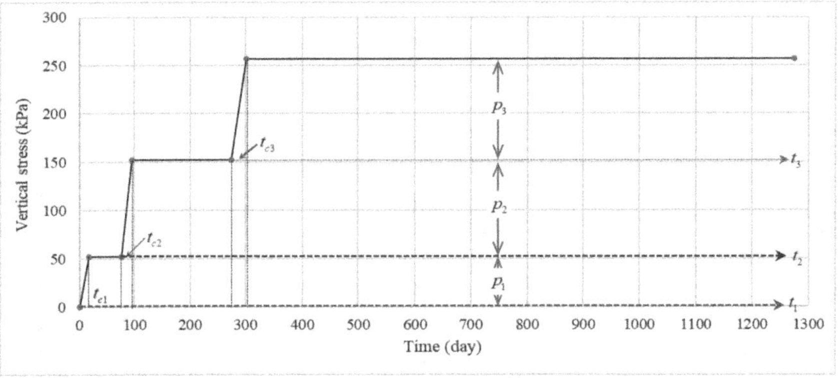

	Stage 1	Stage 1	Stage 1	Stage 2	Stage 2	Stage 3	Stage 3
Time (day)	0	17	77	96	275	301	1275
Total pressure (kPa)	0	52	52	152	152	257	257
t_c and stage ending time (day)	$t_{c1}=17, t_1=77$			$t_{c2}=19, t_2=198$		$t_{c3}=26, t_3=1000$	

Figure 8.14 Construction time, stage time, and vertical pressures of three staged loadings.

Therefor, 'Lower marine clay' is divided into two layers: 'Lower marine clay 1' with thickness 5.1 m and 'Lower marine clay 2' with thickness 0.72 m in order to calculate the average degree of consolidation of each layer better.

Figure 8.14 shows construction time (t_{c1}, t_{c2}, or t_{c3}), loading stage time (t_1, t_2, or t_3), and vertical pressure (p_1, p_2, or p_3) for each of three staged loadings. It is noted that the construction time, loading stage time, and vertical pressure are defined for each stage loading for easy calculation like a single-staged loading.

Values of parameters of soils and data of vertical drains for HKIA Chek Lap Kok Test Embankment are listed in Table 8.11. For more accurate calculate of settlements and the average degree of consolidation, 'Upper marine clay' is divided into two main layers of $H_j = 3.01$ m and 3.21 m, 'Lower marine clay 1' is divided into $H_j = 2.47$ m and 2.63 m, 'Lower alluvium' is divided into two layers with $H_j = 4.165$ m each. There are a total of 8 layers (H_j, j=1…8). 'POP' in Table 8.11 is called pre-over-consolidation pressure (after Zhu *et al.* 2001) and is $\text{POP} = \sigma'_{zp} - \sigma'_{zi}$. σ'_{zi} is the initial vertical effective stress and σ'_{zp} is over-consolidation pressure.

The average degree of combined consolidation U_j is calculated using (8.17a) with a spreadsheet with VBA program developed by (Walker and Indraratna 2009; Walker *et al.* 2009) for each H_j (j=1…8). In fact, when the spreadsheet excel program is used to calculate S_f, ε_z, m_v, and c_v, each j-layer H_j is further divided into sub-layers with thickness of 0.1 to 0.5 m as explained in Sections 8.4 and 8.5. In this way, more accurate values of S_f, ε_z, m_v, and c_v

Table 8.11 Parameters of soils and vertical drains for HKIA Chek Lap Kok Test Embankment

Main layer	*Upper marine clay*		*Upper alluvium*	*Lower marine clay 1*		*Lower marine clay 2*	*Lower alluvium*	
Layer	1	2	3	4	5	6	7	8
H_j (m)	3.01	3.21	5.8	2.47	2.63	0.72	4.165	4.165
POP (kPa)	17		350	150		150	200	
γ (kN/m³)	14.22		19.13	18.15		18.15	19.72	
$V = 1 + e_o$	3.65		2.06	2.325		2.325	2.06	
κ	0.0396		0.0224	0.0353		0.0353	0.0020	
λ	0.5081		0.1339	0.3030		0.3030	0.0087	
ψ	0.0078		0.0035	0.0061		0.0061	0.0000	
$C_r = C_e$	0.0913		0.0515	0.0814		0.0814	0.0046	
C_c	1.1699		0.3083	0.6977		0.6977	0.0200	
$C_{\alpha e}$	0.0180		0.0080	0.0140		0.0140	0.0000	
t_o (year)	0.0027		0.0027	0.0027		0.0027	0.0027	
k_v(m/yr)	0.03469		0.09461	0.00394		0.00394	0.1577	
$k_r = k_h$ (m/yr)	0.06307		0.18922	0.00978		0.00978	0.3155	
Drain spacing S (m)	1.5		1.5	1.5		No drain	No drain	
Drain pattern	Triangular		Triangular	Triangular		No drain	No drain	
$r_d = r_w$ (m)	0.0275		0.0275	0.0275		No drain	No drain	
r_s / r_d	5		5	5		No drain	No drain	
k_r / k_s	1.82		2.00	2.48		No drain	No drain	

are obtained for j-layer H_j. From these, values of U_j are calculated for H_j (j=1…8). To make presentation concise, the sub-index 'j' is not used.

Under multi-staged loading as shown in Figure 8.14, how to calculate S_{fj}, $S_{creep,fj}$, and $S_{creep,dj}$ is important. It is noted that the j-layer H_j can be further divided into a few sub-layers. Below are details of equations for calculating ΔS_f, $\Delta S_{creep,f}$, and $\Delta S_{creep,d}$ of each sub-layer with thickness ΔH_j, in which, the initial stress and strain are assumed be uniform. The S_{fj}, $S_{creep,fj}$, and $S_{creep,dj}$ of the j-layer H_j are summation of ΔS_f, $\Delta S_{creep,f}$, and $\Delta S_{creep,d}$ of all each sub-layers (no sub-index is used). This approach of calculation for each sub-layer is very flexible for general cases.

8.6.2 Calculation of primary consolidation settlements of soil layers under staged loadings

According to (8.32), primary consolidation settlement $S_{primary}$ is calculated separately (de-coupled). For j-layer, $S_{primaryj} = U_j S_{fj}$ in (8.33b). The question is how to calculate final settlements S_{fj} and combined average degree of consolidation under staged loadings.

1. ***Stage 1***

Refer to Stage 1 with vertical load $p_1 = 52$ kPa and construction time $t_{c1} = 17$ days and time coordinate t_1 for Stage 1 in Figure 8.14. For a sub-layer of j-layer H_j with thickness ΔH, the initial stress-strain state is assumed to at point 1 with $(\sigma'_{z1}, \varepsilon_{z1})$ as shown in Figure 8.15. Here $\sigma'_{z1} = \sigma'_{zi}$ and $\varepsilon_{z1} = \varepsilon_{zi}$ as the initial state. With the increase of vertical pressure p_1, the final stress $\sigma'_{z1} + p_1$ may be at point 2 or point 4.

i. The final stress-strain state is in over-consolidated state with $\sigma'_{z1} + p_1 \le \sigma'_{zp}$.

In this case, ΔS_f of each sub-layer is calculated using (8.6a), for example, from point 1 to point 2.

$$\Delta S_f = \varepsilon_{z,1+p_1} \Delta H = \frac{C_r}{1+e_o} \log\left(\frac{\sigma'_{z1} + \sigma'_{unit} + p_1}{\sigma'_{z1} + \sigma'_{unit}}\right) \Delta H \tag{8.35a}$$

At point 2, $\sigma'_{z2} = \sigma'_{z1} + p_1$. The strain at point 2 is:

$$\varepsilon_{z,1+p_1} = \frac{C_r}{1+e_o} \log\left(\frac{\sigma'_{z1} + \sigma'_{unit} + p_1}{\sigma'_{z1} + \sigma'_{unit}}\right) \tag{8.35b}$$

In (8.35a), ΔH is the thickness of this sub-layer. In all calculations of settlements in this section, the σ'_{unit} is 0.1 kPa.

ii. The final stress-strain state is in a normal consolidation state with $\sigma'_{z1} + p_1 \ge \sigma'_{zp}$.

In this case, ΔS_f of each sub-layer is calculated using (8.6b), for example at point 4.

$$\Delta S_f = \varepsilon_{z,1+p_1} \Delta H = \left[\frac{C_r}{1+e_o} \log\left(\frac{\sigma'_{zp} + \sigma'_{unit}}{\sigma'_{z1} + \sigma'_{unit}}\right) + \frac{C_c}{1+e_o} \log\left(\frac{\sigma'_{z1} + \sigma'_{unit} + p_1}{\sigma'_{zp} + \sigma'_{unit}}\right)\right] \Delta H \tag{8.35c}$$

At point 4, $\sigma'_{z4} = \sigma'_{z1} + p_1$. The strain at point 4 is:

$$\varepsilon_{z,1+p_1} = \frac{C_r}{1+e_o} \log\left(\frac{\sigma'_{zp} + \sigma'_{unit}}{\sigma'_{z1} + \sigma'_{unit}}\right) + \frac{C_c}{1+e_o} \log\left(\frac{\sigma'_{z1} + \sigma'_{unit} + p_1}{\sigma'_{zp} + \sigma'_{unit}}\right) \tag{8.35d}$$

Final 'primary' consolidation settlement S_{fj} and strain $\Delta\varepsilon$ for j-layer are summation of all ΔS_f and strains $\varepsilon_{z,1+p_1}$, respectively, for all sub-layers with

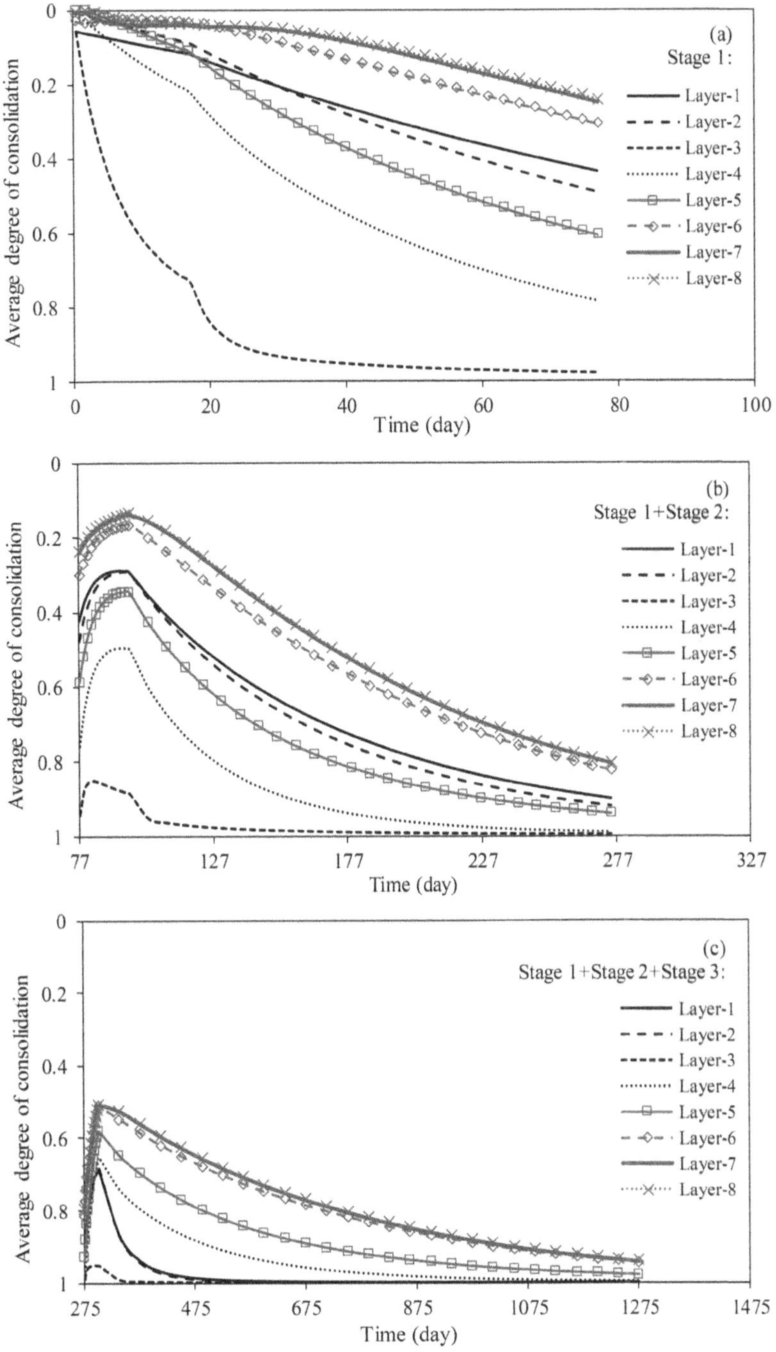

Figure 8.15 Multi-staged average degree of consolidation of eight layers for three stages of loading with accumulated time (a) Stage 1, (b) Stage 1 + Stage 2, and (c) Stage 1 + Stage 2 + Stage 3.

thickness ΔH. Now the strain and stress increment for j-layer are known, the coefficient of volume compressibility m_v for j-layer can be calculated as:

$$m_v = \frac{\Delta\varepsilon}{\Delta\sigma'_z} \tag{8.35e}$$

In (8.35e), $\Delta\sigma'_z = p_1$. From (8.35e) and known permeability k_v and k_r, the coefficients of vertical c_v and radial c_r (horizontal) consolidation for j-layer are:

$$\begin{aligned} c_v &= \frac{k_v}{\gamma_w m_v} \\ c_r &= \frac{k_r}{\gamma_w m_v} \end{aligned} \tag{8.35f}$$

Using c_v and c_r, U_v and U_r or combined U_j directly can be calculated for j-layer. Values of S_{fj}, $\Delta\varepsilon$, $\Delta\sigma'_z$, m_v, c_v, and c_r for j-layers are listed in Table 8.12.

The combined U_j of Stage 1 is calculated for j-layer using the solutions by Walker and Indraratna (2009) and Walker *et al.* (2009). The combined U_j of Stage 1 with time is plotted in Figure 8.15(a).

2. ***Stage 2***

Refer to Stage 2 with vertical load $p_2 = (152 - 52) = 100$ kPa and construction time $t_{c2} = 19$ days and time coordinate t_2 for Stage 2 in Figure 8.14. Stage 2 starts at total accumulated time of 77 days from beginning construction. At beginning of Stage 2 loading, t_2 is zero. With the increase of 100 kPa in vertical pressure in all soil layers, the final stress is $\sigma'_{z1} + p_1 + p_2$. When we calculate ΔS_f for each sub-layer, the creep is ignored. The final stress-strain state at Stage 2 can be still in an over-consolidation state or into normal consolidation state. In either case, ΔS_f under Stage 2 loading with $p_2 = 100$ kPa can be calculated as an incremental settlement:

i. The final stress-strain state is an over-consolidation state $\sigma'_{z1} + p_1 + p_2 \le \sigma'_{zp}$. If the previous stress-strain state is in an over-consolidation state, say, at point 2, ΔS_f is calculated below:

$$\Delta S_f = \varepsilon_{z,1+p_1+p_2}\Delta H = \frac{C_r}{1+e_o}\log\left(\frac{\sigma'_{z1} + \sigma'_{unit} + p_1 + p_2}{\sigma_{z1} + \sigma'_{unit} + p_1}\right)\Delta H \tag{8.36a}$$

The strain at point $\sigma'_{z1} + p_1 + p_2 + p_2$ is:

$$\varepsilon_{z,1+p_1+p_2} = \frac{C_r}{1+e_o}\log\left(\frac{\sigma'_{z1} + \sigma'_{unit} + p_1 + p_2}{\sigma'_{z1} + \sigma'_{unit} + p_1}\right) \tag{8.36b}$$

Table 8.12 Calculated values of parameters of *j*-layers for HKIA Chek Lap Kok Test Embankment

	Stage	Layer H_j							
		1	2	3	4	5	6	7	8
S_{fj} (m)	1	0.4291	0.3035	0.0415	0.0160	0.0143	0.0035	0.0011	0.0009
S_{fj} (m)	2	0.4157	0.3796	0.0409	0.0214	0.0204	0.0052	0.0015	0.0013
S_{fj} (m)	3	0.2123	0.2101	0.0255	0.1128	0.1113	0.0290	0.0031	0.0028
$\Delta\sigma'_z$ (kPa)	1	52	52	52	52	52	52	52	52
$\Delta\sigma'_z$ (kPa)	2	100	100	100	100	100	100	100	100
$\Delta\sigma'_z$ (kPa)	3	105	105	105	105	105	105	105	105
$\Delta\varepsilon_z$	1	0.1426	0.0946	0.0072	0.0065	0.0054	0.0049	0.0003	0.0002
$\Delta\varepsilon_z$	2	0.1381	0.1182	0.0070	0.0087	0.0078	0.0072	0.0004	0.0003
$\Delta\varepsilon_z$	3	0.0705	0.0654	0.0044	0.0456	0.0423	0.0403	0.0008	0.0007
m_v (1/kPa)	1	2.741×10^{-3}	1.818×10^{-3}	1.375×10^{-4}	1.245×10^{-4}	1.043×10^{-4}	9.386×10^{-5}	5.182×10^{-6}	4.202×10^{-6}
m_v (1/kPa)	2	1.381×10^{-3}	1.182×10^{-3}	7.045×10^{-5}	8.683×10^{-5}	7.765×10^{-5}	7.242×10^{-5}	3.170×10^{-6}	3.178×10^{-6}
m_v (1/kPa)	3	6.716×10^{-4}	6.233×10^{-4}	4.187×10^{-5}	4.348×10^{-4}	4.031×10^{-4}	3.840×10^{-4}	7.165×10^{-6}	6.431×10^{-6}
c_v (m^2/yr)	1	1.27	1.91	68.78	3.17	3.78	4.20	3042.97	3752.62
c_v (m^2/yr)	2	2.51	2.93	134.30	4.54	5.08	5.44	4250.04	4962.32
c_v (m^2/yr)	3	5.17	5.57	225.95	0.91	0.98	1.03	2200.71	2451.95
c_r (m^2/yr)	1	2.30	3.47	137.56	7.85	9.37	10.42	6085.94	7505.24
c_r (m^2/yr)	2	4.57	5.33	268.60	11.26	12.59	13.50	8500.09	9924.63
c_r (m^2/yr)	3	9.39	10.12	451.91	2.25	2.43	2.55	4401.41	4903.91

ii. The final stress-strain state is a normal consolidation state $\sigma'_{z1} + p_1 + p_2 \geq \sigma'_{zp}$. If the previous stress-strain state is in an over-consolidation state, say, at point 2, ΔS_f of each sub-layer is calculated using (8.6b) below:

$$\Delta S_f = \varepsilon_{z,1+p_1+p_2}\Delta H = \left[\frac{C_r}{1+e_o}\log\left(\frac{\sigma'_{zp}+\sigma'_{unit}}{\sigma'_{z1}+\sigma'_{unit}+p_1}\right) + \frac{C_c}{1+e_o}\log\left(\frac{\sigma'_{z1}+\sigma'_{unit}+p_1+p_2}{\sigma'_{zp}+\sigma'_{unit}}\right)\right]\Delta H \tag{8.36c}$$

The strain at point $\sigma'_{z1} + p_1 + p_2$ is:

$$\varepsilon_{z,1+p_1+p_2} = \frac{C_r}{1+e_o}\log\left(\frac{\sigma'_{zp} + \sigma'_{unit}}{\sigma'_{z1} + \sigma'_{unit} + p_1}\right) + \frac{C_c}{1+e_o}\log\left(\frac{\sigma'_{z1} + \sigma'_{unit} + p_1 + p_2}{\sigma'_{zp} + \sigma'_{unit}}\right) \quad (8.36d)$$

Final 'primary' consolidation settlement S_{fi} and strain $\Delta\varepsilon$ for j-layer are summation of all ΔS_f and strains $\varepsilon_{z,1+p_1}$, respectively, for all sub-layers with thickness ΔH. The strain $\Delta\varepsilon$ under Stage 2 loading p_2 for j-layer is summation of strains $\varepsilon_{z,1+p_1+p_2}$ for all sub-layers with thickness ΔH. Now the strain and stress increment for j-layer are known, the coefficient of volume compressibility m_v for j-layer can be calculated as:

$$m_v = \frac{\Delta\varepsilon}{\Delta\sigma'_z} \quad (8.36e)$$

In (8.36e), $\Delta\sigma'_z = p_2$. From (8.36e) and known permeability k_v and k_r, the coefficients of vertical c_v and radial c_r (horizontal) consolidation for j-layer are:

$$c_v = \frac{k_v}{\gamma_w m_v}$$
$$c_r = \frac{k_r}{\gamma_w m_v} \quad (8.36f)$$

Using c_v and c_r, U_v and U_r or combined U_j directly can be calculated for j-layer under Stage 2 loading p_2. Values of S_{fi}, $\Delta\varepsilon$, $\Delta\sigma'_z$, m_v c_v, and c_r for j-layers for Stage 2 are listed in Table 8.12. The combined U_j for both Stage 1 and Stage 2, that is, $U_{multi,j}$ for multi-staged loadings is calculated for j-layer using the solutions by Walker and Indraratna (2009) and Walker *et al.* (2009).

The total average degree of consolidation $U_{multi,j}$ for multi-staged loadings for j-layer can be calculated below. We refer to j-layer with thickness and the same property m_v. The final settlements $S_{1f}, S_{2f}, ..., S_{nf}$, respectively, under $p_1, p_2, p_3, ..., p_n$, that is, n stages of loading is:

$$S_{1f} = m_v H_j p_1, S_{2f} = m_v H_j p_2, S_{3f} = m_v H_j p_3, ..., S_{nf} = m_v H_j p_n$$

The time-dependent consolidation settlement is:

$$S_{1t} = U_1 S_{1f}, S_{2t} = U_2 S_{2f}, S_{3t} = U_3 S_{3f}, ..., S_{nt} = U_n S_{nf}$$

$U_1, U_2, U_3, ..., U_n$ are average degrees of consolidation for j-layer for $p_1, p_2, p_3, ..., p_n$, respectively, and shall be calculated at corresponding time coordinates $t_1, t_2, t_3, ..., t_n$.

For ramp loading, the correction can be applied to each $S_{1t}, S_{2t}, S_{3t}, ..., S_{nt}$. The total consolidation settlement for n-stages of loading for j-layer is:

$$\begin{aligned} S_{multi,j}(t) &= S_{1t} + S_{2t} + S_{3t} + ... + S_{nt} = U_1 S_{1f} + U_2 S_{2f} + U_3 S_{3f} + ... + U_n S_{nf} \\ &= U_1 m_v H_j p_1 + U_2 m_v H_j p_2 + U_3 m_v H_j p_3 + ... + U_n m_v H_j p_n \end{aligned} \quad (8.37a)$$

The above equation is re-written:

$$\begin{aligned} S_{multi,j}(t) &= \left(U_1 \frac{p_1}{p_0} + U_2 \frac{p_2}{p_0} + U_3 \frac{p_3}{p_0} + ... + U_n \frac{p_n}{p_0} \right) m_v H_j p_0 \\ &= \left(U_1 \frac{p_1}{p_0} + U_2 \frac{p_2}{p_0} + U_3 \frac{p_3}{p_0} + ... + U_n \frac{p_n}{p_0} \right) S_{multif,j} = U_{multi,j} \times S_{multif,j} \end{aligned} \quad (8.37b)$$

where, p_0 is the total load of all staged loads and $p_0 = p_1 + p_2 + p_3 + ... + p_n$. The total final settlement due to n-stages of loading is:

$$S_{multif,j} = m_v H_j p_0 \quad (8.37c)$$

The total average degree of consolidation $U_{multi,j}$ for multi-staged loading is:

$$U_{multi,j} = \left(U_1 \frac{p_1}{p_0} + U_2 \frac{p_2}{p_0} + U_3 \frac{p_3}{p_0} + ... + U_n \frac{p_n}{p_0} \right) \quad (8.37d)$$

For multi-staged loading, $U_{multi,j}$ in (8.37d) will replace U_j in (8.1). It is noted that at Stage 2, $p_0 = p_1 + p_2$, we have:

$$S_{multif,j} = m_v H_j p_0 = m_v H_j (p_1 + p_2) \quad (8.37e)$$

$$U_{multi,j} = \left(U_1 \frac{p_1}{p_0} + U_2 \frac{p_2}{p_0} \right) \quad (8.37f)$$

The combined multi-staged $U_{multi,j}$ for Stage 1 and Stage 2 together with accumulated time is plotted in Figure 8.15(b).

3. *Stage 3*

Refer to Stage 3 with vertical load $p_3 = (257 - 152) = 103$ kPa and construction time $t_{c3} = 26$ days and time coordinate t_3 for Stage 3 in Figure 8.14. Stage 3 starts at time of 275 days from beginning construction. The final stress is $\sigma'_{z1} + p_1 + p_2 + p_3$. The final stress-strain state at Stage 3 can be still in an overconsolidation state or into normal consolidation state. In either case, ΔS_f

under Stage 3 loading with $p_3 = 103$ kPa can be calculated as an incremental settlement:

i. The final stress-strain state is an over-consolidation state $\sigma'_{z1} + p_1 + p_2 + p_3 \le \sigma'_{zp}$. If the previous stress-strain state is in an over-consolidation state, ΔS_f of each sub-layer is calculated below:

$$\Delta S_f = \varepsilon_{z,1+p_1+p_2+p_3}\Delta H = \frac{C_r}{1+e_o}\log\left(\frac{\sigma'_{z1} + \sigma'_{unit} + p_1 + p_2 + p_3}{\sigma'_{z1} + \sigma'_{unit} + p_1 + p_2}\right)\Delta H \quad (8.38a)$$

The strain at point $\sigma'_{z1} + p_1 + p_2 + p_3$ is:

$$\varepsilon_{z,1+p_1+p_2+p_3} = \frac{C_r}{1+e_o}\log\left(\frac{\sigma'_{z1} + \sigma'_{unit} + p_1 + p_2 + p_3}{\sigma'_{z1} + \sigma'_{unit} + p_1 + p_2}\right) \quad (8.38b)$$

ii. The final stress-strain state is a normal consolidation state $\sigma'_{z1} + p_1 + p_2 + p_3 \ge \sigma'_{zp}$. If the previous stress-strain state is in an over-consolidation state, ΔS_f of each sub-layer is calculated using (8.6b) below:

$$\Delta S_f = \varepsilon_{z,1+p_1+p_2+p_3}\Delta H = \left[\frac{C_r}{1+e_o}\log\left(\frac{\sigma'_{zp} + \sigma'_{unit}}{\sigma'_{z1} + \sigma'_{unit} + p_1 + p_2}\right) + \frac{C_c}{1+e_o}\log\left(\frac{\sigma'_{z1} + \sigma'_{unit} + p_1 + p_2 + p_3}{\sigma'_{zp} + \sigma'_{unit}}\right)\right]\Delta H \quad (8.38c)$$

The strain at point $\sigma'_{z1} + p_1 + p_2 + p_34$ is:

$$\varepsilon_{z,1+p_1+p_2+p_3} = \frac{C_r}{1+e_o}\log\left(\frac{\sigma'_{zp} + \sigma'_{unit}}{\sigma'_{z1} + \sigma'_{unit} + p_1 + p_2}\right) + \frac{C_c}{1+e_o}\log\left(\frac{\sigma'_{z1} + \sigma'_{unit} + p_1 + p_2 + p_3}{\sigma'_{zp} + \sigma'_{unit}}\right) \quad (8.38d)$$

The strain $\Delta\varepsilon$ under Stage 3 loading p_3 for j-layer is summation of strains $\varepsilon_{z,1+p_1+p_2+p_3}$ for all sub-layers with thickness ΔH. Now the strain and stress increment for j-layer are known, the coefficient of volume compressibility m_v for j-layer can be calculated as:

$$m_v = \frac{\Delta\varepsilon}{\Delta\sigma'_z} \quad (8.38e)$$

In (8.38e), $\Delta\sigma_z' = p_3$. From (8.38e) and known permeability k_v and k_r, the coefficients of vertical c_v and radial c_r (horizontal) consolidation for j-layer are:

$$c_v = \frac{k_v}{\gamma_w m_v}$$
$$c_r = \frac{k_r}{\gamma_w m_v} \quad (8.38f)$$

Using c_v and c_r, U_v and U_r or combined U_j directly can be calculated for j-layer under Stage 3 loading p_3. Values of S_{fj} , $\Delta\varepsilon$, $\Delta\sigma_z'$, m_v c_v, and c_r for j-layers for Stage 2 are listed in Table 8.12. The combined U_j for both Stages 1, 2, 3, that is, $U_{multi,j}$ for multi-staged loadings is calculated for j-layer using the solutions by Walker and Indraratna (2009) and Walker *et al.* (2009).

The total average degree of consolidation $U_{multi,j}$ for multi-staged loadings for j-layer can be calculated below. It is noted that at Stage 3, $p_0 = p_1 + p_2 + p_3$, we have:

$$S_{multif,j} = m_v H_j p_0 = m_v H_j (p_1 + p_2 + p_3) \quad (8.38g)$$

$$U_{multi,j} = \left(U_1 \frac{p_1}{p_0} + U_2 \frac{p_2}{p_0} + U_3 \frac{p_3}{p_0} \right) \quad (8.38h)$$

The combined multi-staged $U_{multi,j}$ for Stages 1, 2, and 3 together with accumulated time is plotted in Figure 8.15(c).

U_v and U_r can be calculated separately. In this case, the combined U can be calculated using (8.3), that is, $U = 1-(1-U_v)(1-U_r)$. If solutions or charts for U_v and U_r are for suddenly applied, the correction for construction period (or ramp loading) can be done using Terzaghi's method. Referring to Figure 8.14, t_{c1}, t_{c2}, and t_{c3} is the construction time for Stage 1, 2, and 3 loadings p_1, p_2, and p_3, respectively. The correction for construction period is done for each stage separately. For example, for Stage 1, the corrected settlement $S_{1t,corr}(t)$ is:

For $0 < t \le t_{c1}$:

$$S_{1t,corr}(t) = S_{1t}\left(\frac{t}{2}\right)\frac{p_1(t)}{p_1}$$
$$\text{where } S_{1t}\left(\frac{t}{2}\right) = U_1\left(\frac{t}{2}\right)S_{1f} \quad (8.39a)$$

For $t > t_{c1}$:

$$S_{1t,corr}(t) = S_{1t}\left(t - \frac{t_{c1}}{2}\right)$$
$$where\ S_{1t}\left(t - \frac{t_c}{2}\right) = U_1\left(t - \frac{t_{c1}}{2}\right)S_{1f} \tag{8.39b}$$

$S_{1t}\left(\frac{t}{2}\right)$ is un-corrected settlement using $U_1\left(\frac{t}{2}\right)$ from solutions under suddenly applied load. S_{1f} is the final 'primary' consolidation settlement under final vertical pressure p_1. $p_1(t)$ is the vertical pressure, changing linearly from zero to p_1 (ramp loading). It shall be pointed out that the time t follows coordinate t_1 in Figure 8.14.

For Stage 2 the corrected settlement $S_{2t,corr}(t)$ is:

For $0 < t \le t_{c2}$:

$$S_{2t,corr}(t) = S_{2t}\left(\frac{t}{2}\right)\frac{p_2(t)}{p_2}$$
$$\text{where } S_{2t}\left(\frac{t}{2}\right) = U_2\left(\frac{t}{2}\right)S_{2f} \tag{8.39c}$$

For $t > t_{c2}$:

$$S_{2t,corr}(t) = S_{2t}\left(t - \frac{t_{c2}}{2}\right)$$
$$where\ S_{2t}\left(t - \frac{t_c}{2}\right) = U_2\left(t - \frac{t_{c2}}{2}\right)S_{2f} \tag{8.39d}$$

In some solutions, the combined U under ramp loading is calculated directly (Zhu and Yin 2001a, b, 2004a; Walker and Indraratna 2009; Walker *et al.* 2009). In this example, combined U under ramp loading is calculated using the method and excel program developed by Walker and Indraratna (2009) and Walker *et al.* (2009).

8.6.3 Calculation of creep settlements of soil layers under staged loadings

To calculate creep settlement S_{creep} in (8.32), we need to calculate S_{creepj} for each layer. For S_{creepj}, the creep settlement $S_{creep,fj}$ at the final vertical effective stress and the creep settlement $S_{creep,dj}$ delayed by $t_{EOP,field}$ must be calculated first. The following are details how to calculate $S_{creep,fj}$ and $S_{creep,dj}$ for j-layer under staged loading.

1. ***Stage 1***

Refer to Stage 1 with vertical load $p_1 = 52$ kPa and construction time $t_{c1} = 17$ days and time coordinate t_1 for Stage 1 in Figure 8.14. For a sub-layer of j-layer H_j, the initial stress-strain state is assumed to at point 1 with $(\sigma'_{z1}, \varepsilon_{z1})$ as shown in Figure 8.16. Here $\sigma'_{z1} = \sigma'_{zi}$ and $\varepsilon_{z1} = \varepsilon_{zi}$ as the initial state. With the increase of vertical pressure p_1, the final vertical effective stress $\sigma'_{z1} + p_1$ may be at point 2 or point 4.

i. The final stress-strain state is at point 2, in over-consolidated state with $\sigma'_{z1} + p_1 \le \sigma'_{zp}$.

In this case, ΔS_f and $\varepsilon_{z,1+p_1}$ are calculated in (8.35a) and (8.35b) for a sub-layer of this j-layer. $\Delta S_{creep,f}$ and $\Delta S_{creep,d}$ are final creep settlement and delayed settlement for this sub-layer with thickness ΔH of this j-layer (sub-index 'j' is not included in following equations). Since the final point is at point 2, (8.21c) is used for $\Delta S_{creep,f}$ at ending time $t_1 = 77$ days. At time of $t_1 = 77$ days, the stress-strain state is at 2″ with $(\sigma'_{z2''}, \varepsilon_{z2''})$. The stress at point 2″ is

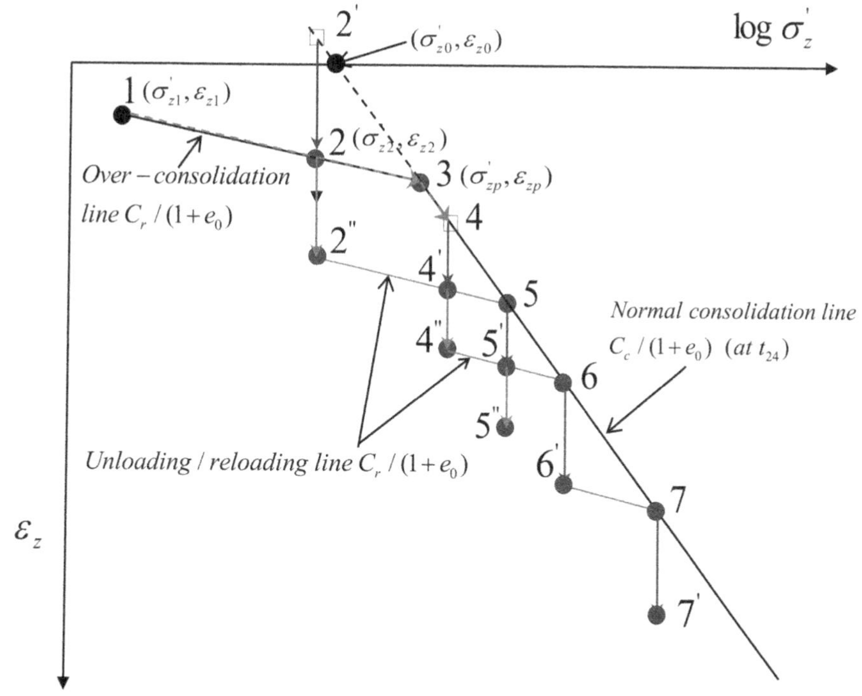

Figure 8.16 Calculation of fx9 and fx10 at different stress-strain states under multi-staged loadings.

$\sigma'_{z2''} = \sigma'_{z1} + p_1$. The creep settlement $\Delta S_{creep,f}$ and creep strain $\Delta\varepsilon_{z,creep}$ are from (8.21c) or (8.21e):

$$\Delta S_{creep,f} = \frac{C_{\alpha e}}{1+e_o}\log\left(\frac{t+t_{e2}}{t_o+t_{e2}}\right)\Delta H \qquad 1\text{ day} \le t \le t_1 = 77\text{ days} \qquad (8.40a)$$

$$\Delta\varepsilon_{z,creep} = \frac{C_{\alpha e}}{1+e_o}\log\left(\frac{t+t_{e2}}{t_o+t_{e2}}\right) \qquad 1\text{ day} \le t \le t_1 = 77\text{ days} \qquad (8.40b)$$

The strain at point 2″ is

$$\varepsilon_{z2''} = \varepsilon_{z2} + \Delta\varepsilon_{zcreep2''}$$

where:

$$\varepsilon_{z2} = \varepsilon_{z1} + \frac{C_r}{1+e_o}log\frac{\sigma'_{z2}+\sigma'_{unit}}{\sigma'_{z1}+\sigma'_{unit}}$$

$$\Delta\varepsilon_{z,creep,2''} = \frac{C_{\alpha e}}{1+e_o}\log\left(\frac{t_1+t_{e2}}{t_o+t_{e2}}\right) \qquad (t_1 = 77\text{ days})$$

t_{e2} is calculated using (8.21d):

$$t_{e2} = t_o \times 10^{(\varepsilon_{z2}-\varepsilon_{zp})\frac{V}{C_{\alpha e}}}\left(\frac{\sigma'_{z2}}{\sigma'_{zp}}\right)^{-\frac{C_c}{C_{\alpha e}}} - t_o$$

Equation (8.23a) is used to calculate $\Delta S_{creep,d}$ for sub-layer with ΔH of this j-layer:

$$\Delta S_{creep,d} = \frac{C_{\alpha e}}{1+e_o}\log\left(\frac{t+t_{e2}}{t_{EOP,field}+t_{e2}}\right)\Delta H \qquad for\ \ t \ge t_{EOP,field} \qquad (8.40c)$$

The creep settlement $S_{creep,fj}$ at the final vertical effective stress is summation of all $\Delta S_{creep,f}$ of all sub-layers with ΔH of this j-layer. The delayed creep settlement $S_{creep,dj}$ is summation of all $\Delta S_{creep,d}$ of all sub-layers with ΔH of this j-layer. It is noted that $S_{creep,dj}$ is delayed by $t_{EOP,field}$. $t_{EOP,field}$ is determined when multi-staged $U_{multi,j} = 98\%$.

ii. The final stress-strain state is at point 4, in a normal consolidation state with $\sigma'_{z1} + p_1 \ge \sigma'_{zp}$.

In this case, ΔS_f and $\varepsilon_{z,1+p_1}$ are calculated in (8.35c) and (8.35d) for a sub-layer of this *j*-layer. At time of $t_1 = 77$ days, the stress-strain state is at 4′ with $(\sigma'_{z4}, \varepsilon_{z4'})$. The stress at point 4′ is $\sigma'_{z4} = \sigma'_{z4'} = \sigma'_{z1} + p_1$. The creep settlement $\Delta S_{creep,f}$ and creep strain $\Delta\varepsilon_{z,creep}$ are from (8.21a) or (8.21b):

$$\Delta S_{creep,f} = \frac{C_{\alpha e}}{1+e_o}\log\left(\frac{t}{t_o}\right)\Delta H \qquad 1\text{ day} \le t \le t_1 = 77\text{ days} \tag{8.40d}$$

$$\Delta\varepsilon_{z,creep} = \frac{C_{\alpha e}}{1+e_o}\log\left(\frac{t}{t_o}\right) \qquad 1\text{ day} \le t \le t_1 = 77\text{ days} \tag{8.40e}$$

The strain at point 4′ is

$$\varepsilon_{z4'} = \varepsilon_{z4} + \Delta\varepsilon_{zcreep4'}$$

where:

$$\varepsilon_{z4} = \varepsilon_{z1} + \frac{C_r}{1+e_o}\log\frac{\sigma'_{zp} + \sigma'_{unit}}{\sigma'_{z1} + \sigma'_{unit}} + \frac{C_c}{1+e_o}\log\frac{\sigma'_{z4} + \sigma'_{unit}}{\sigma'_{zp} + \sigma'_{unit}}$$

$$\Delta\varepsilon_{z,creep,4'} = \frac{C_{\alpha e}}{1+e_o}\log\left(\frac{t_1}{t_o}\right) \qquad (t_1 = 77\text{ days})$$

Equation (8.22b) is used to calculate $\Delta S_{creep,d}$ for sub-layer with ΔH of this *j*-layer:

$$\Delta S_{creep,d} = \frac{C_{\alpha e}}{1+e_o}\log\left(\frac{t}{t_{\text{EOP},field}}\right)\Delta H \qquad \textit{for } t \ge t_{\text{EOP},field} \tag{8.40f}$$

The creep settlement $S_{creep,fj}$ and delayed creep settlement $S_{creep,dj}$ are summation of all $\Delta S_{creep,f}$ and $\Delta S_{creep,d}$ of all sub-layers with ΔH of this *j*-layer. $S_{creep,dj}$ is delayed by $t_{\text{EOP},field}$, which is determined when multi-staged $U_{multi,j} = 98\%$.

2. ***Stage 2***

Refer to Stage 2 with added vertical load $p_2 = (152 - 52) = 100$ kPa and construction time $t_{c2} = 19$ days and time coordinate t_2 for Stage 2 in Figure 8.14. Stage 2 starts at time of 77 days from beginning construction. With the increase of 100 kPa in vertical pressure in all soil layers, the final stress is $\sigma'_{z1} + p_1 + p_2$. There are two possible cases for the final stress-strain state from Stage 1 loading and consolidation:

i. The final vertical stress and strain state is in over-consolidated state, say, at point 4′ from point 2″ with $\sigma'_{z1} + p_1 + p_2 \le \sigma'_{z5}$.

σ'_{z5} is called 'apparent pre-consolidation pressure' due to previous creep history or 'aging' (Bjerrum 1967). The stress and strain at this point 5 shall be calculated first. For loading from point 2″ to point 5, we have:

$$\varepsilon_{z5} = \varepsilon_{z2''} + \frac{C_r}{1+e_o}\log\frac{\sigma'_{z5}+\sigma'_{unit}}{\sigma'_{z2}+\sigma'_{unit}} \tag{8.41a}$$

Since point 5 is also in the NCL line, we have:

$$\varepsilon_{z5} = \varepsilon_{zp} + \frac{C_c}{1+e_o}\log\frac{\sigma'_{z5}+\sigma'_{unit}}{\sigma'_{zp}+\sigma'_{unit}} \tag{8.41b}$$

From the above two equations, (8.41b) – (8.41a):

$$\varepsilon_{zp} + \frac{C_c}{1+e_o}\log\frac{\sigma'_{z5}+\sigma'_{unit}}{\sigma'_{zp}+\sigma'_{unit}} - \varepsilon_{z2''} - \frac{C_r}{1+e_o}\log\frac{\sigma'_{z5}+\sigma'_{unit}}{\sigma'_{z2}+\sigma'_{unit}} = 0$$

$$(\varepsilon_{zp} - \varepsilon_{z2''}) + \frac{C_c}{1+e_o}\log(\sigma'_{z5}+\sigma'_{unit}) - \frac{C_c}{1+e_o}\log(\sigma'_{zp}+\sigma'_{unit}) -$$
$$\frac{C_r}{1+e_o}\log(\sigma'_{z5}+\sigma'_{unit}) + \frac{C_r}{1+e_o}\log(\sigma'_{z2}+\sigma'_{unit}) = 0$$

$$\frac{C_c - C_r}{1+e_o}\log(\sigma'_{z5}+\sigma'_{unit}) = -(\varepsilon_{zp} - \varepsilon_{z2''}) + \frac{C_c}{1+e_o}\log(\sigma'_{zp}+\sigma'_{unit})$$
$$-\frac{C_r}{1+e_o}\log(\sigma'_{z2}+\sigma'_{unit})$$

$$\frac{C_c - C_r}{1+e_o}\log(\sigma'_{z5}+\sigma'_{unit}) = \log(\sigma'_{zp}+\sigma'_{unit})^{\frac{C_c}{1+e_o}} - \log(\sigma'_{z2}+\sigma'_{unit})^{\frac{C_r}{1+e_o}} - (\varepsilon_{zp} - \varepsilon_{z2''})$$

$$\frac{C_c - C_r}{1+e_o}\log(\sigma'_{z5}+\sigma'_{unit}) = \log\frac{(\sigma'_{zp}+\sigma'_{unit})^{\frac{C_c}{1+e_o}}}{(\sigma'_{z2}+\sigma'_{unit})^{\frac{C_r}{1+e_o}}} - (\varepsilon_{zp} - \varepsilon_{z2''})$$

$$\log(\sigma'_{z5}+\sigma'_{unit}) = \frac{1+e_o}{C_c - C_r}\log\frac{(\sigma'_{zp}+\sigma'_{unit})^{\frac{C_c}{1+e_o}}}{(\sigma'_{z2}+\sigma'_{unit})^{\frac{C_r}{1+e_o}}} - (\varepsilon_{zp} - \varepsilon_{z2''})\frac{1+e_o}{C_c - C_r}$$

$$\log(\sigma'_{z5}+\sigma'_{unit}) = \log\frac{(\sigma'_{zp}+\sigma_{unit})^{\frac{C_c}{C_c-C_r}}}{(\sigma'_{z2}+\sigma'_{unit})^{\frac{C_r}{C_c-C_r}}} - (\varepsilon_{zp} - \varepsilon_{z2''})\frac{1+e_o}{C_c - C_r}$$

From the above, we have:

$$\sigma'_{z5} = \frac{(\sigma'_{zp} + \sigma'_{unit})^{\frac{C_c}{C_c - C_r}}}{(\sigma'_{z2} + \sigma'_{unit})^{\frac{C_r}{C_c - C_r}}} \times 10^{-(\varepsilon_{zp} - \varepsilon_{z2''})\frac{1+e_o}{C_c - C_r}} - \sigma'_{unit} \qquad (8.41c)$$

Substituting σ'_{z5} into (8.41a) or (8.41b), ε_{z5} is calculated. When we know σ'_{z5}, the condition $\sigma'_{z1} + p_1 + p_2 \leq \sigma'_{z5}$ can be checked.

Point 2″ is the final point after Stage 1 loading and creep. The creep settlement is due to creep strain from point 4′ to point 4″ with time t_2 of 198 days. The strain $\varepsilon_{z4'}$ at 4′ is

$$\varepsilon_{z4'} = \varepsilon_{z2''} + \frac{C_r}{1+e_o} log \frac{\sigma'_{z4} + \sigma'_{unit}}{\sigma'_{z2} + \sigma'_{unit}} \qquad (8.42a)$$

The stress and strain at point 4′ are known as $(\varepsilon_{z4'}, \sigma'_{z4})$, the creep settlement $\Delta S_{creep,f}$ and strain $\Delta\varepsilon_{z,creep}$ from point 4′ to point 4″ are from (8.21c) or (8.21e):

$$\Delta S_{creep,f} = \frac{C_{\alpha e}}{1+e_o} \log\left(\frac{t + t_{e4'}}{t_o + t_{e4'}}\right) \Delta H \qquad 1\text{ day} \leq t \leq t_2 = 198\text{ days} \qquad (8.42b)$$

$$\Delta\varepsilon_{z,creep} = \frac{C_{\alpha e}}{1+e_o} \log\left(\frac{t + t_{e4'}}{t_o + t_{e4'}}\right) \qquad 1\text{ day} \leq t \leq t_2 = 198\text{ days} \qquad (8.42c)$$

The strain at point 4″ is

$$\varepsilon_{z4''} = \varepsilon_{z4'} + \Delta\varepsilon_{zcreep4''}$$

where:

$$\Delta\varepsilon_{z,creep,4''} = \frac{C_{\alpha e}}{1+e_o} \log\left(\frac{t_2 + t_{e4'}}{t_o + t_{e4'}}\right) \qquad (t_2 = 198\text{ days})$$

$t_{e4'}$ is calculated using (8.21d):

$$t_{e4'} = t_o \times 10^{(\varepsilon_{z4'} - \varepsilon_{zp})\frac{V}{C_{\alpha e}}} \left(\frac{\sigma'_{z4}}{\sigma'_{zp}}\right)^{-\frac{C_c}{C_{\alpha e}}} - t_o$$

Equation (8.23a) is used to calculate $\Delta S_{creep,d}$ for sub-layer with ΔH of this j-layer:

$$\Delta S_{creep,d} = \frac{C_{\alpha e}}{1+e_o} \log\left(\frac{t + t_{e4'}}{t_{EOP,field} + t_{e4'}}\right) \Delta H \qquad for\ t \geq t_{EOP,field} \qquad (8.42d)$$

The accumulated creep settlement $S_{creep,fj}$ and accumulated delayed creep settlement $S_{creep,dj}$ up to Stage 2 are summation of all $\Delta S_{creep,f}$ and $\Delta S_{creep,d}$ of all sub-layers with ΔH of this j-layer. $S_{creep,dj}$ is delayed by $t_{EOP,field}$, which is determined when multi-staged $U_{multi,j} = 98\%$.

ii. The final vertical stress and strain state is normal consolidation state, say, at point 6 from point 2″ to point 5, then to point 6 with $\sigma'_{z1} + p_1 + p_2 \geq \sigma'_{z5}$.

Since we know σ'_{z5}, the condition $\sigma'_{z1} + p_1 + p_2 \geq \sigma'_{z5}$ can be checked. Point 2″ is the final point after Stage 1 loading and creep. The creep settlement is due to creep strain from point 6 to 6′ with time t_2 of 198 days. The strain ε_{z6} at 6 is

$$\varepsilon_{z6} = \varepsilon_{z2''} + \frac{C_r}{1+e_o}\log\frac{\sigma'_{z5} + \sigma'_{unit}}{\sigma'_{z2} + \sigma'_{unit}} + \frac{C_c}{1+e_o}\log\frac{\sigma'_{z6} + \sigma'_{unit}}{\sigma'_{z5} + \sigma'_{unit}} \tag{8.43a}$$

The creep settlement $\Delta S_{creep,f}$ and strain $\Delta\varepsilon_{z,creep}$ from point 6 to point 6′ are from (8.21a) or (8.21b):

$$\Delta S_{creep,f} = \frac{C_{\alpha e}}{1+e_o}\log\left(\frac{t}{t_o}\right)\Delta H \qquad 1\text{ day} \leq t \leq t_2 = 198\text{ days} \tag{8.43b}$$

$$\Delta\varepsilon_{z,creep} = \frac{C_{\alpha e}}{1+e_o}\log\left(\frac{t}{t_o}\right) \qquad 1\text{ day} \leq t \leq t_2 = 198\text{ days} \tag{8.43c}$$

The strain at point 6′ is

$$\varepsilon_{z4'} = \varepsilon_{z,1+p_1} + \Delta\varepsilon_{zcreep4'}$$

where:

$$\Delta\varepsilon_{z,creep,6'} = \frac{C_{\alpha e}}{1+e_o}\log\left(\frac{t_2}{t_o}\right) \qquad (t_2 = 198\text{ days})$$

Equation (8.22b) is used to calculate $\Delta S_{creep,d}$ for sub-layer with ΔH of this j-layer:

$$\Delta S_{creep,d} = \frac{C_{\alpha e}}{1+e_o}\log\left(\frac{t}{t_{EOP,field}}\right)\Delta H \qquad \text{for } t \geq t_{EOP,field} \tag{8.43d}$$

Again, the accumulated creep settlement $S_{creep,fj}$ and accumulated delayed creep settlement $S_{creep,dj}$ up to Stage 2 are summation of all $\Delta S_{creep,f}$ and $\Delta S_{creep,d}$ of all sub-layers with ΔH of this j-layer. $S_{creep,dj}$ is delayed by $t_{EOP,field}$, which is determined when multi-staged $U_{multi,j} = 98\%$.

3. *Stage 3*

Refer to Stage 3 with vertical load $p_3 = (255 - 152) = 103$ kPa and construction time $t_{c3} = 26$ days and time coordinate t_3 for Stage 3 in Figure 8.14. Stage 3 starts at time of 275 days from beginning construction. With the increase of 103 kPa in vertical pressure in all soil layers, the final stress is $\sigma'_{z1} + p_1 + p_2 + p_3$. There are two possible cases for the final stress-strain state from Stage 2 loading and consolidation:

i. The final vertical stress and strain state is in over-consolidated state, say, at point 5′ from point 4″ with $\sigma'_{z1} + p_1 + p_2 + p_3 \le \sigma'_{z6}$.

σ'_{z6} is called 'apparent pre-consolidation pressure' due to previous creep history or 'aging' (Bjerrum 1967). The stress and strain at this point 6 shall be calculated first. For loading from point 4″ to point 6, we have:

$$\varepsilon_{z6} = \varepsilon_{z4''} + \frac{C_r}{1+e_o} log \frac{\sigma'_{z6} + \sigma'_{unit}}{\sigma'_{z4} + \sigma'_{unit}} \tag{8.44a}$$

Since point 6 is also in the NCL line, we have:

$$\varepsilon_{z6} = \varepsilon_{zp} + \frac{C_c}{1+e_o} log \frac{\sigma'_{z6} + \sigma'_{unit}}{\sigma'_{zp} + \sigma'_{unit}} \tag{8.44b}$$

From the above two equations, (8.44b) – (8.44a):

$$\varepsilon_{zp} + \frac{C_c}{1+e_o} log \frac{\sigma'_{z6} + \sigma'_{unit}}{\sigma'_{zp} + \sigma'_{unit}} - \varepsilon_{z4''} - \frac{C_r}{1+e_o} log \frac{\sigma'_{z6} + \sigma''_{unit}}{\sigma'_{z4} + \sigma''_{unit}} = 0$$

From the above, we have:

$$\sigma'_{z6} = \frac{(\sigma'_{zp} + \sigma'_{unit})^{\frac{C_c}{C_c - C_r}}}{(\sigma'_{z4} + \sigma'_{unit})^{\frac{C_r}{C_c - C_r}}} \times 10^{-(\varepsilon_{zp} - \varepsilon_{z4''})\frac{1+e_o}{C_c - C_r}} - \sigma'_{unit} \tag{8.44c}$$

Substituting σ'_{z6} into (8.44a) or (8.44b), ε_{z6} is calculated. When we know σ'_{z6}, the condition $\sigma'_{z1} + p_1 + p_2 + p_3 \le \sigma'_{z6}$ can be checked.

Point 4″ is the final point after Stage 2 loading and creep. The creep settlement is due to creep strain from point 5′ to point 5″ with time t_3 of 1000 days. The strain $\varepsilon_{z5'}$ at 5′ is

$$\varepsilon_{z5'} = \varepsilon_{z4''} + \frac{C_r}{1+e_o} \log \frac{\sigma'_{z5} + \sigma'_{unit}}{\sigma'_{z4} + \sigma'_{unit}} \tag{8.45a}$$

The stress and strain at point 5′ are known as $(\varepsilon_{z5'}, \sigma'_{z5})$, the creep settlement $\Delta S_{creep,f}$ and strain $\Delta\varepsilon_{z,creep}$ from point 5′ to point 5″ are from (8.21c) or (8.21e):

$$\Delta S_{creep,f} = \frac{C_{\alpha e}}{1+e_o}\log\left(\frac{t+t_{e5'}}{t_o+t_{e5'}}\right)\Delta H \qquad 1\text{ day} \le t \le t_3 = 1000\text{ days} \quad (8.45b)$$

$$\Delta\varepsilon_{z,creep} = \frac{C_{\alpha e}}{1+e_o}\log\left(\frac{t+t_{e5'}}{t_o+t_{e5'}}\right) \qquad 1\text{ day} \le t \le t_3 = 1000\text{ days} \quad (8.45c)$$

The strain at point 5″ is

$$\varepsilon_{z5''} = \varepsilon_{z5'} + \Delta\varepsilon_{zcreep5''}$$

where:

$$\Delta\varepsilon_{z,creep,5''} = \frac{C_{\alpha e}}{1+e_o}\log\left(\frac{t_3+t_{e5'}}{t_o+t_{e5'}}\right) \qquad (t_3 = 1000\text{ days})$$

$t_{e5'}$ is calculated using (8.21d):

$$t_{e5'} = t_o \times 10^{(\varepsilon_{z5'}-\varepsilon_{zp})\frac{V}{C_{\alpha e}}}\left(\frac{\sigma'_{z5}}{\sigma'_{zp}}\right)^{-\frac{C_c}{C_{\alpha e}}} - t_o$$

Equation (8.23a) is used to calculate $\Delta S_{creep,d}$ for sub-layer with ΔH of this j-layer:

$$\Delta S_{creep,d} = \frac{C_{\alpha e}}{1+e_o}\log\left(\frac{t+t_{e5'}}{t_{EOP,field}+t_{e5'}}\right)\Delta H \qquad for\ \ t \ge t_{EOP,field} \quad (8.45d)$$

The accumulated creep settlement $S_{creep,fj}$ and accumulated delayed creep settlement $S_{creep,dj}$ up to Stage 2 are summation of all $\Delta S_{creep,f}$ and $\Delta S_{creep,d}$ of all sub-layers with ΔH of this j-layer. $S_{creep,dj}$ is delayed by $t_{EOP,field}$, which is determined when multi-staged $U_{multi,j} = 98\%$.

ii. The final vertical stress and strain state is normal consolidation state, say, at point 7 from point 4″ to point 6, then to point 7 with $\sigma'_{z1} + p_1 + p_2 + p_3 \ge \sigma'_{z6}$.

Since we know σ'_{z5}, the condition $\sigma'_{z1} + p_1 + p_2 \ge \sigma'_{z5}$ can be checked. Point 4″ is the final point after Stage 2 loading and creep. The creep settlement is due to creep strain from point 7 to 7′ with time t_3 of 1000 days. The strain ε_{z7} at 7 is

$$\varepsilon_{z7} = \varepsilon_{z4''} + \frac{C_r}{1+e_o}\log\frac{\sigma'_{z6}+\sigma'_{unit}}{\sigma'_{z4}+\sigma'_{unit}} + \frac{C_c}{1+e_o}\log\frac{\sigma'_{z7}+\sigma'_{unit}}{\sigma'_{z6}+\sigma'_{unit}} \tag{8.46a}$$

The creep settlement $\Delta S_{creep,f}$ and strain $\Delta\varepsilon_{z,creep}$ from point 7 to point 7′ are from (8.21a) or (8.21b):

$$\Delta S_{creep,f} = \frac{C_{\alpha e}}{1+e_o}\log\left(\frac{t}{t_o}\right)\Delta H \qquad 1\text{ day} \le t \le t_3 = 1000\text{ days} \tag{8.46b}$$

$$\Delta\varepsilon_{z,creep} = \frac{C_{\alpha e}}{1+e_o}\log\left(\frac{t}{t_o}\right) \qquad 1\text{ day} \le t \le t_3 = 1000\text{ days} \tag{8.46c}$$

The strain at point 7′ is

$$\varepsilon_{z7'} = \varepsilon_{z7} + \Delta\varepsilon_{zcreep7'}$$

where:

$$\Delta\varepsilon_{z,creep,7'} = \frac{C_{\alpha e}}{1+e_o}\log\left(\frac{t_3}{t_o}\right) \qquad (t_3 = 1000\text{ days})$$

Equation (8.22b) is used to calculate $\Delta S_{creep,d}$ for sub-layer with ΔH of this j-layer:

$$\Delta S_{creep,d} = \frac{C_{\alpha e}}{1+e_o}\log\left(\frac{t}{t_{EOP,field}}\right)\Delta H \qquad for\ \ t \ge t_{EOP,field} \tag{8.46d}$$

Again, the accumulated creep settlement $S_{creep,fj}$ and accumulated delayed creep settlement $S_{creep,dj}$ up to Stage 3 are summation of all $\Delta S_{creep,f}$ and $\Delta S_{creep,d}$ of all sub-layers with ΔH of this j-layer. $S_{creep,dj}$ is delayed by $t_{EOP,field}$, which is determined when multi-staged $U_{multi,j} = 98\%$.

Based on the above calculations, cures of $S_{creep,f}$ and time for each j-layer under three stages of loading are plotted in Figure 8.17. Cures of $S_{creep,d}$ and time for each j-layer under three stages of loading are plotted in Figure 8.18.

8.6.4 Calculation of total consolidation settlements of soil layers under staged loadings

Using all calculation results above, the total consolidation settlement S_{totalB} can be calculated using (8.1) and the multi-staged average degree of combined

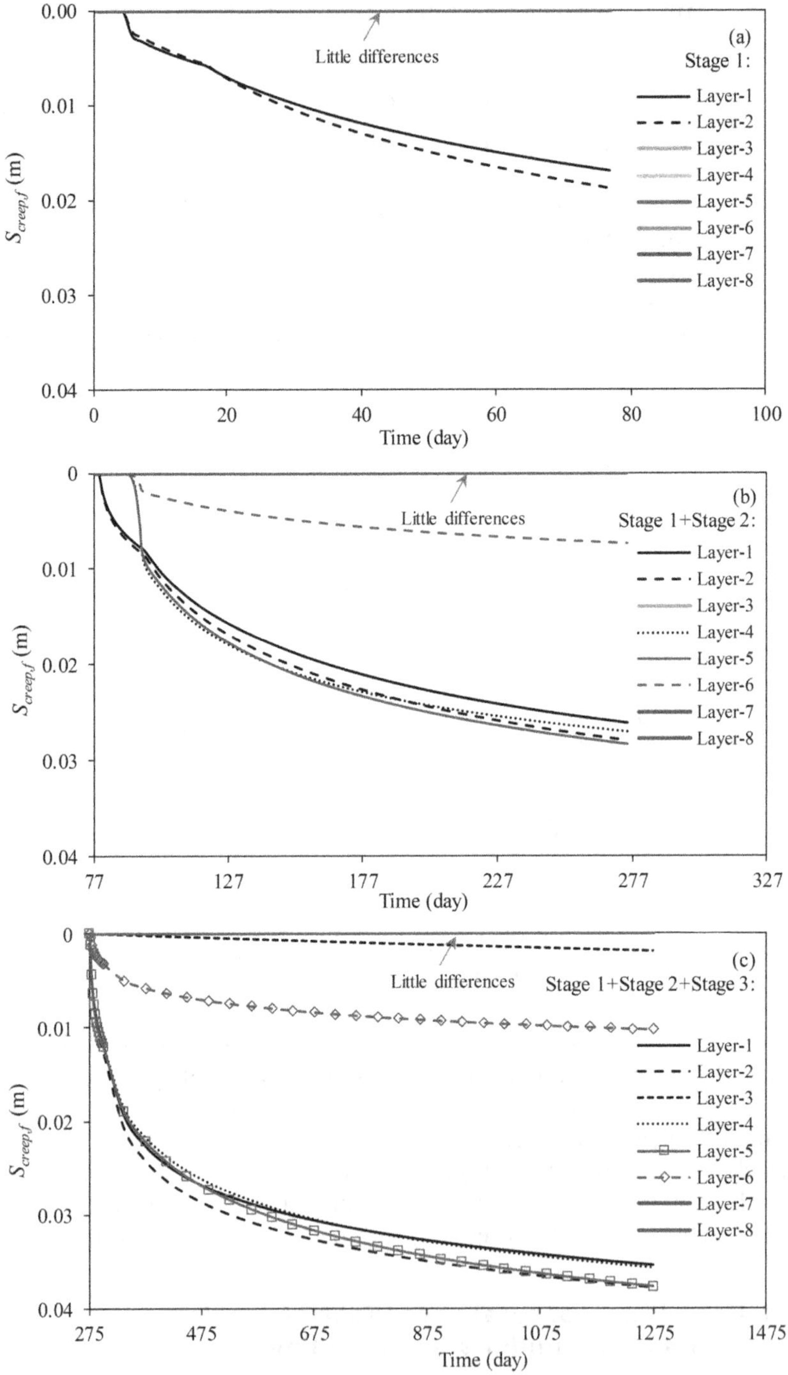

Figure 8.17 Calculation of fx11 for each j-layer under three stages of loading (a) Stage 1, (b) Stage 1 + Stage 2, and (c) Stage 1 + Stage 2 + Stage 3.

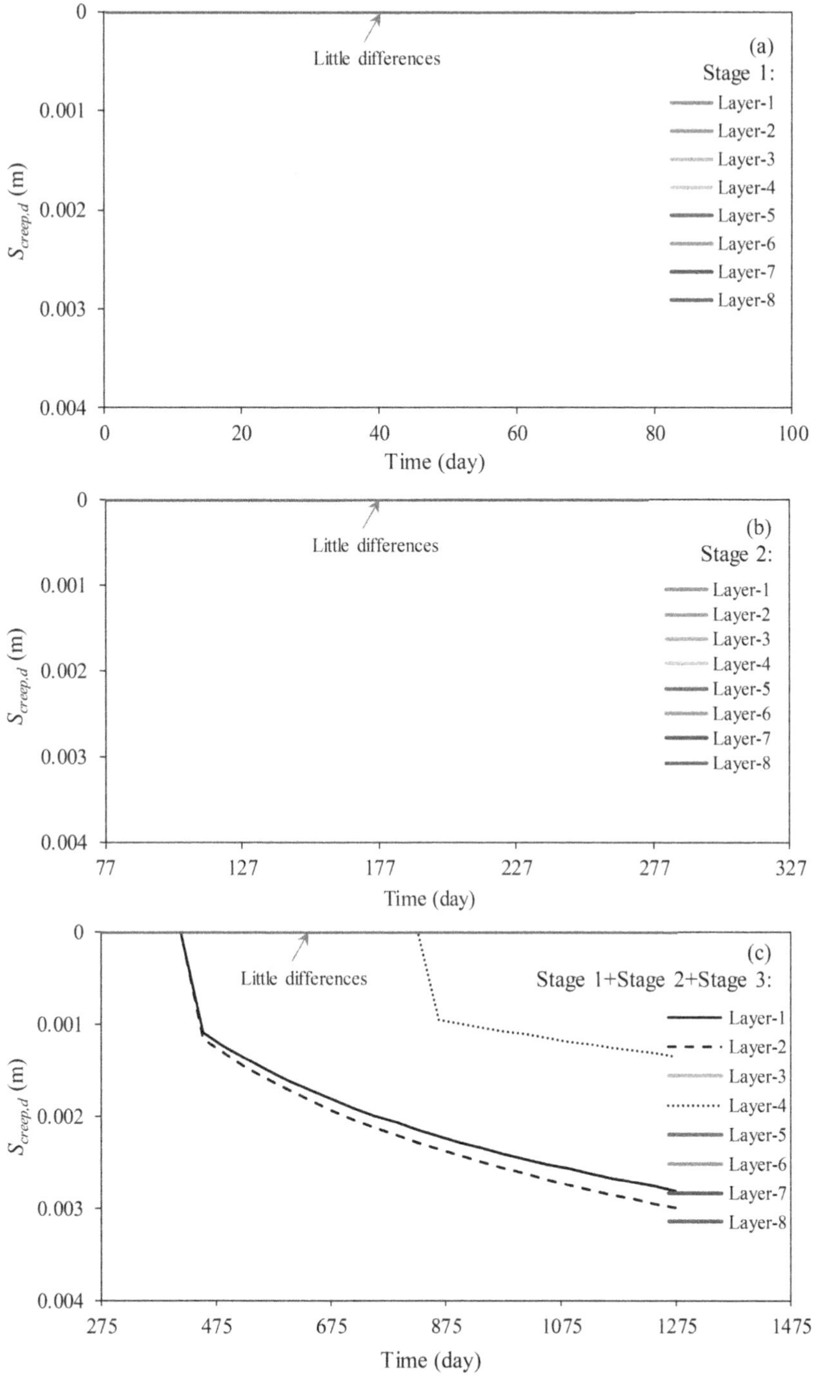

Figure 8.18 Calculation of fx12 for each j-layer under three stages of loading.

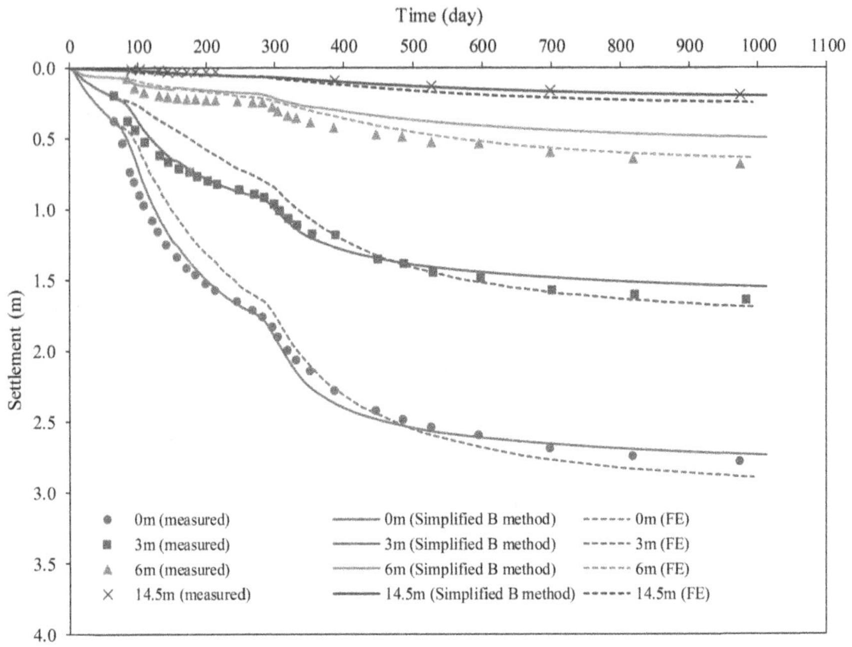

Figure 8.19 Comparison of curves of settlements with accumulated time at depths 0, 3, 6, and 14.5 m from the simplified Hypothesis B method, fully coupled finite element modeling and measurement.

consolidation $U_{multi,j}$ for a single j-layer. In this case, $\alpha = 0.8$ and $\beta = 0.3$. Equation (8.32) can be written as:

$$S_{totalB} = \sum_{j=1}^{j=n} U_{multi,j} S_{fj} + \sum_{j=1}^{j=n} [\alpha U^{\beta}_{multi,j} S_{creep,fj} + (1 - \alpha U^{\beta}_{multi,j}) S_{creep,dj}] \quad (8.47)$$

$$for \quad all \quad t \geq t_{EOP,lab} \quad (t \geq t_{EOP,field} \, for \, S_{creep,dj})$$

Using (8.46), the total consolidation settlements S_{totalB} at depths of 0, 3, 6, and 14.5 m are calculated using the simplified Hypothesis B method for three stages of loading. Comparison of curves of settlements with accumulated time at depths 0, 3, 6, and 14.5 m from the simplified Hypothesis B method, fully coupled FE modeling and measurement are shown in Figure 8.19. It is found that the values from the simplified Hypothesis B method are in good agreement with measured data and results from fully coupled FE modeling (Zhu *et al.* 2001) using a 1-D EVP model (Yin and Graham 1989, 1994).

REFERENCES

Barden, L, 1965, Consolidation of clay with non-linear viscosity. *Géotechnique*, **15**(4), pp. 345–362.

Barden, L, 1969, Time-dependent deformation of normally consolidated clays and peats. *Journal of the Soil Mechanics and Foundations Division*, ASCE, **95**, SM1, pp. 1–31.

Barron, RA, 1948, Consolidation of fine-grained soils by drain wells. *Transactions*, ASCE, **113**(2346), pp. 718–742.

Bjerrum, L, 1967, Engineering geology of Norwegian normally consolidated marine clays as related to the settlements of buildings. *Géotechnique*, **17**(2), pp. 83–118.

British Standard 1377, 1990, *Methods of Test for Soils for Civil Engineering Purposes* (Part 5), (London: British Standards Institution).

Carrillo, N, 1942, Simple two- and three-dimensional cases in the theory of consolidation of soils. *Journal of Mathematical Physics*, **21**, pp. 1–5.

Feng, WQ, and Yin, JH, 2017, A new simplified Hypothesis B method for calculating consolidation settlements of double soil layers exhibiting creep. *International Journal for Numerical and Analytical Methods in Geomechanics*, **41**, pp. 899–917.

Garlanger, JE, 1972, The consolidation of soils exhibiting creep under constant effective stress. *Géotechnique*, **22**(1), pp. 71–78.

Gibson, RE, and Lo, KY, 1961, A theory of consolidation for soils exhibiting secondary compression. *Publication*, **41**, pp. 1–16, (Oslo: Norwegian Geotechnical Institute).

Handfelt, LD, Koutsoftas, DC, and Foott, R, 1987, Instrumentation for test fill in Hong Kong. *Journal of Geotechnical Engineering*, ASCE, **113**(GT2), pp. 127–146.

Hansbo, S, 1981, Consolidation of fine-grained soils by prefabricated drains. *In Proceedings of the Tenth International Conference on Soil Mechanics and Foundation Engineering*, Stochholm, Sweden, **3**, pp. 667–682.

Knappett, J, 2019, *Craig's Soil Mechanics*, 9th ed., (Oxford: Taylor & Francis).

Koutsoftas, DC, Foott, R, and Handfelt, LD, 1987, Geotechnical investigations offshore Hong Kong. *Journal of Geotechnical Engineering*, ASCE, **96**(SM1), pp. 145–175.

Ladd, CC, Foott, R, Ishihara, K, Schlosser, F, and Poulos, HJ, 1977, Stress-deformation and strength characteristics. *In Proceedings of the 9th International Conference on Soil Mechanics and Foundation Engineering, Tokyo, 4210494. Estimating settlements of structures supported on cohesive soils.* Special summer program.

Leroueil, S, Kabbaj, M, Tavenas, F, and Bouchard, R, 1985, Stress-strain-time rate relation for the compressibility of sensitive natural clays. *Géotechnique*, **35**(2), pp. 159–180.

Mesri, G, and Godlewski, PM, 1977, Time- and stress-compressibility interrelationship. *Journal of Geotechnical Engineering*, ASCE, **103** (GT5), pp. 417–430.

Nash, DFT, and Ryde, SJ, 2000, Modelling the effects of surcharge to reduce long term settlement of reclamations over soft clays. *In Proceedings of Soft Soil Engineering Conference*, Japan, 2000.

Nash, DFT, and Ryde, SJ, 2001, Modelling consolidation accelerated by vertical drains in soils subject to creep. *Géotechnique*, **51**(3), pp. 257–273.

Plaxis, 2015, See https://www.plaxis.com/news/software-update/update-pack-plaxis-2d-2015-02/

US Department of the Navy, 1982, *Soil Mechanics Design Manual* 7.1. NAVFAC DM-7.1., (Arlington: US Department of the Navy).

Vermeer, PA, and Neher, HP, 1999, A soft soil model that accounts for creep. *In Proceedings of 'Beyond 2000 in Computational Geotechnics 10 Years of Plaxis International,'* Balkema, pp. 249–261.

Walker, R, and Indraratna, B, 2009a, Consolidation analysis of a stratified soil with vertical and horizontal drainage using the spectral method. *Géotechnique*, **59**, pp. 439–449. https://doi.org/10.1680/geot.2007.00019.

Walker, R, Indraratna, B, and Sivakugan, N, 2009b, Vertical and radial consolidation analysis of multilayered soil using the spectral method. *Journal of Geotechnical and Geoenvironmental Engineering*, **135**, pp. 657–663. https://doi.org/10.1061/(asce)gt.1943-5606.0000075.

Walker, R, and Indraratna, B, 2006, Vertical drain consolidation with parabolic distribution of permeability in smear zone. *Journal of Geotechnical and Geoenvironmental Engineering*, **132**, pp. 937–941. https://doi.org/10.1061/(asce)1090-0241(2006)132:7(937).

Waterman, D, and Broere, W, 2011, Practical application of the soft soil creep model. *Plaxis Bulletin*, **15**, pp. 1–15.

Terzaghi, K, 1925, *Erdbaumechanik auf Boden-physicalischen Grundlagen*. (Vienna: Deuticke).

Terzaghi, K, 1943, *Theoretical Soil Mechanics*, (New York: Wiley).

Yin, JH, 1990, Constitutive modelling of time-dependent stress-strain behaviour of soils. Ph.D. thesis, Univ. of Manitoba, Winnipeg, Canada, March, 1990, 314 pages.

Yin, JH, 2011, From constitutive modeling to development of laboratory testing and optical fiber sensor monitoring technologies. *Chinese Journal of Geotechnical Engineering*, **33**(1), pp. 1–15. (14th 'Huang Wen-Xi Lecture' in China)

Yin, JH, 2015, Fundamental issues on constitutive modelling of the time-dependent stress-strain behaviour of geomaterials. *International Journal of Geomechanics*, **15**(5), A4015002, pp. 1–9.

Yin, JH, and Feng, WQ, 2017, A new simplified method and its verification for calculation of consolidation settlement of a clayey soil with creep. *Canadian Geotechnical Journal*, **54**(3), pp. 333–347.

Yin, JH, and Graham, J, 1989, Visco-elastic-plastic modeling of one-dimensional time-dependent behaviour of clays. *Canadian Geotechnical Journal*, **26**, pp. 199–209.

Yin, JH, and Graham, J, 1994, Equivalent times and one-dimensional elastic visco-plastic modeling of time-dependent stress-strain behavior of clays. *Canadian Geotechnical Journal*, **31**, pp. 42–52.

Yin, JH, and Graham, J, 1996, Elastic visco-plastic modelling of one-dimensional consolidation. *Géotechnique*, **46**(3), pp. 515–527.

Yin, JH, and Graham, J, 1999, Elastic visco-plastic modelling of the time-dependent stress-strain behavior of soils. *Canadian Geotechnical Journal*, **36**(4), pp. 736–745.

Zhu, GF, and Yin, JH, 1998, Consolidation of soil under depth-dependent ramp load. *Canadian Geotechnical Journal*, **35**(2), pp. 344–350.

Zhu, GF, and Yin, JH, 1999a, Finite element analysis of consolidation of layered clay soils using an elastic visco-plastic model. *International Journal for Numerical and Analytical Methods in Geomechanics*, **23**, pp. 355–374.

Zhu, GF, and Yin, JH, 1999b, Consolidation of double soil layers under depth-dependent ramp load. *Géotechnique*, **49**(3), pp. 415–421.

Zhu, GF, and Yin, JH, 2000, Elastic visco-plastic finite element consolidation modeling of Berthierville test embankment. *International Journal for Numerical and Analytical Methods in Geomechanics,* **24**, pp. 491–508.

Zhu, GF, and Yin, JH, 2001a, Consolidation of soil with vertical and horizontal drainage under ramp load. *Géotechnique*, **51**(2), pp. 361–367.

Zhu, GF, and Yin, JH, 2001b, Design charts for vertical drains considering construction time. *Canadian Geotechnical Journal*, **38**(5), pp. 1142–1148.

Zhu, GF, and Yin, JH, 2004a, Consolidation analysis of soil with vertical and horizontal drainage under ramp loading considering smear effects. *Geotextiles and Geomembranes*, **22**(1 &2), pp. 63–74.

Zhu, GF, and Yin, JH, 2004b, Accuracy of the Carrillo's formula for consolidation of soil with vertical and horizontal drainage under time-dependent Loading. *Communications in Numerical Methods in Engineering*, **20**, pp. 721–735

Zhu, GF, and Yin, JH, 2012, Analysis and mathematical solutions for consolidation of a soil layer with depth-dependent parameters under confined compression. *International Journal of Geomechanics*, **12**(4), pp. 451–461.

Zhu, GF, Yin, JH, and Graham, J, 2001, Consolidation modelling of soils under the Test Embankment at Chek Lap Kok International Airport in Hong Kong using a simplified finite element method. *Canadian Geotechnical Journal*, **38**(2), pp. 349–363.

Zhu, GF, and Yin, JH, 2005, Solution charts for the consolidation of double soil layers. *Canadian geotechnical journal*, 42(3), pp. 949–956.

Chapter 9

Three-dimensional consolidation equations

9.1 INTRODUCTION

The mathematical theory for one-dimensional (1-D) consolidation was discussed in Chapter 2. Published field evidence of the rate at which foundations on clay settle suggests that the actual rates are generally faster than those predicted by the 1-D theory. It is obvious that in many practical cases, the geometric conditions are far from 1-D and that horizontal dissipation of porewater pressure should be taken into account. A logical extension of Terzaghi's 1-D consolidation theory to three-dimensional (3-D) situation is due to Biot's fully coupled poroelastic formulation (1941). Since the stress-strain behavior of normally consolidated clay has been clearly demonstrated to be elastoplastic and viscous in nature, equations governing the consolidation of a soil with an elasto-plastic or visco-plastic skeleton have been developed (Small *et al.*, 1976; Zienkiewicz *et al.*, 1977). This chapter describes the mathematical model of 3-D consolidation.

9.2 BASIC ASSUMPTIONS FOR 3-D CONSOLIDATION THEORY

The 3-D consolidation theory is based on the underlying assumptions:

1. The soil is completely saturated with water.
2. The porewater and soil particles are incompressible.
3. Darcy's law is valid.
4. Strains and displacements are small.
5. Inertial force can be neglected.

9.3 FORCE EQUILIBRIUM EQUATIONS

9.3.1 Cartesian co-ordinates

Figure 9.1 shows an element of a larger continuous body. Forces applied on the element can be described in terms of the stress tensor $\boldsymbol{\sigma}$ in Cartesian co-ordinates.

$$\sigma = \begin{bmatrix} \sigma_x & \tau_{xy} & \tau_{xz} \\ \tau_{yx} & \sigma_y & \tau_{yz} \\ \tau_{zx} & \tau_{zy} & \sigma_z \end{bmatrix} \tag{9.1}$$

Each component of stress tensor $\boldsymbol{\sigma}$ represents a force acting in a specific co-ordinate direction on a unit area oriented in a particular way. Thus, τ_{xy} is the force in the negative x direction acting on a unit area whose outward normal is the positive y direction. σ_x is the force in the negative x direction acting on a unit area whose outward normal is the positive x direction. The terms σ_x, σ_y, and σ_x are normal stresses, and the rest are shear stresses. Rotational equilibrium requires that the stress tensor be symmetric so that $\tau_{xy} = \tau_{yx}, \tau_{xz} = \tau_{zx}$ and $\tau_{yz} = \tau_{zy}$. The stress tensor is also denoted in vector form as

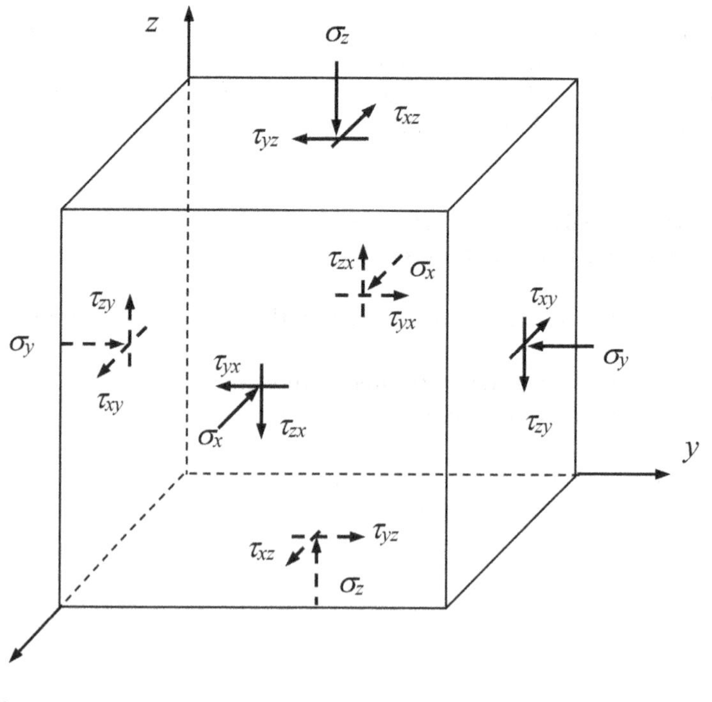

Figure 9.1 Sign convention for stress, compression positive.

$$\vec{\sigma}^T = [\,\sigma_x\ \sigma_y\ \sigma_z\ \tau_{xy}\ \tau_{yz}\ \tau_{zx}\,] \tag{9.2}$$

where the superscript T stands for 'transpose.'

The equilibrium of local forces will result in

$$\frac{\partial \sigma_x}{\partial x} + \frac{\partial \tau_{xy}}{\partial y} + \frac{\partial \tau_{xz}}{\partial z} = f_x \tag{9.3a}$$

$$\frac{\partial \tau_{yx}}{\partial x} + \frac{\partial \sigma_y}{\partial y} + \frac{\partial \tau_{yz}}{\partial z} = f_y \tag{9.3b}$$

$$\frac{\partial \tau_{xz}}{\partial x} + \frac{\partial \tau_{zy}}{\partial y} + \frac{\partial \sigma_z}{\partial z} = f_z \tag{9.3c}$$

where f_x, f_y, and f_z are the body force per unit bulk volume in the x, y, and z directions, respectively.

9.3.2 Cylindrical co-ordinates

In the cylindrical co-ordinates (r, θ, z) (Figure 9.2), where $r = \sqrt{x^2 + y^2}$ and $\theta = \arctan(y/x)$, the stress tensor $\boldsymbol{\sigma}$ is expressed as

$$\sigma = \begin{bmatrix} \sigma_r & \tau_{r\theta} & \tau_{rz} \\ \tau_{\theta r} & \sigma_\theta & \tau_{\theta z} \\ \tau_{zr} & \tau_{z\theta} & \sigma_z \end{bmatrix} \tag{9.4}$$

$$\vec{\sigma}^T = [\,\sigma_r\ \sigma_\theta\ \sigma_z\ \tau_{r\theta}\ \tau_{\theta z}\ \tau_{zr}\,] \tag{9.5}$$

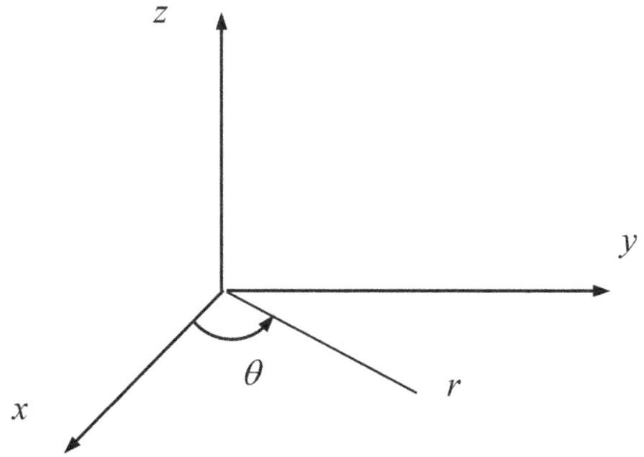

Figure 9.2 Cylindrical co-ordinates.

The equilibrium of local forces becomes

$$\frac{\partial \sigma_r}{\partial r} + \frac{1}{r}\frac{\partial \tau_{r\theta}}{\partial \theta} + \frac{\partial \tau_{rz}}{\partial z} + \frac{\sigma_r - \sigma_\theta}{r} = f_r \tag{9.6a}$$

$$\frac{\partial \tau_{r\theta}}{\partial r} + \frac{1}{r}\frac{\partial \sigma_\theta}{\partial \theta} + \frac{\partial \tau_{\theta z}}{\partial z} + \frac{2\tau_{r\theta}}{r} = f_\theta \tag{9.6b}$$

$$\frac{\partial \tau_{rz}}{\partial r} + \frac{1}{r}\frac{\partial \tau_{\theta z}}{\partial \theta} + \frac{\partial \sigma_z}{\partial z} + \frac{\tau_{rz}}{r} = f_z \tag{9.6c}$$

where f_r, f_θ, and f_z are the body force per unit bulk volume in the r, θ, and z directions, respectively.

9.3.3 Spherical co-ordinates

The spherical co-ordinates (r, θ, ϕ) (Figure 9.3) are defined by $r = \sqrt{x^2 + y^2 + z^2}$, $\theta = \arctan(y/x)$ and $\phi = \cot^{-1}(z/\sqrt{x^2 + y^2})$. The stress tensor $\boldsymbol{\sigma}$ is

$$\sigma = \begin{bmatrix} \sigma_r & \tau_{r\theta} & \tau_{r\phi} \\ \tau_{\theta r} & \sigma_\theta & \tau_{\theta\phi} \\ \tau_{\phi r} & \tau_{\phi\theta} & \sigma_\phi \end{bmatrix} \tag{9.7}$$

$$\vec{\sigma}^T = [\,\sigma_r \ \sigma_\theta \ \sigma_\phi \ \tau_{r\theta} \ \tau_{\theta\phi} \ \tau_{\phi r}\,] \tag{9.8}$$

The equilibrium of local forces is

$$\frac{\partial \sigma_r}{\partial r} + \frac{1}{r\sin\phi}\frac{\partial \tau_{r\theta}}{\partial \theta} + \frac{1}{r}\frac{\partial \tau_{r\phi}}{\partial \phi} + \frac{2\sigma_r - \sigma_\theta - \sigma_\phi + \tau_{r\phi}\cot\phi}{r} = f_r \tag{9.9a}$$

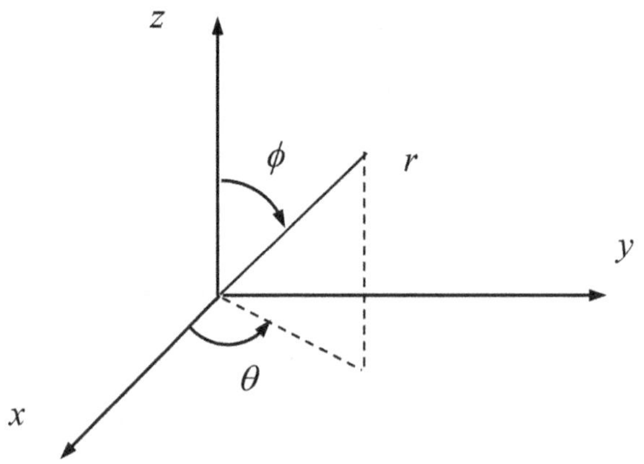

Figure 9.3 Spherical co-ordinates.

$$\frac{\partial \tau_{r\theta}}{\partial r} + \frac{1}{r\sin\phi}\frac{\partial \sigma_\theta}{\partial \theta} + \frac{1}{r}\frac{\partial \tau_{\theta\phi}}{\partial \phi} + \frac{3\tau_{r\theta} + 2\tau_{\theta\phi}\cot\phi}{r} = f_\theta \tag{9.9b}$$

$$\frac{\partial \tau_{r\phi}}{\partial r} + \frac{1}{r\sin\phi}\frac{\partial \tau_{\theta\phi}}{\partial \theta} + \frac{1}{r}\frac{\partial \sigma_\phi}{\partial \phi} + \frac{3\tau_{r\phi} + (\sigma_\phi - \sigma_\theta)\cot\phi}{r} = f_\phi \tag{9.9c}$$

where f_r, f_θ, and f_ϕ are the body force per unit bulk volume in the r, θ, and ϕ directions, respectively.

9.4 DEFORMATION OF SOIL SKELETON

9.4.1 Cartesian co-ordinates

Suppose that a representative particle of the soil skeleton occupies the point defined by the vector $\mathbf{X}^T = (x, y, z)$ as shown in Figure 9.4. When the soil is displaced, the same particle will occupy the point $\mathbf{X}^{*T} = (x^*, y^*, z^*)$. The difference $\mathbf{X}^* - \mathbf{X}$ is the displacement of the particle and will be denoted as $\mathbf{w}(\mathbf{X}) = (w_x, w_y, w_z)^T$.

Consider a neighboring particle labeled by $\mathbf{X} + \Delta\mathbf{X}$. In the displaced soil skeleton, the position of this particle will be

$$\mathbf{X}^* + \Delta\mathbf{X}^* = \mathbf{X} + \Delta\mathbf{X} + \mathbf{w}(\mathbf{X} + \Delta\mathbf{X}) \tag{9.10}$$

so that

$$\Delta\mathbf{X}^* = \Delta\mathbf{X} + \mathbf{w}(\mathbf{X} + \Delta\mathbf{X}) - \mathbf{w}(\mathbf{X}) \tag{9.11}$$

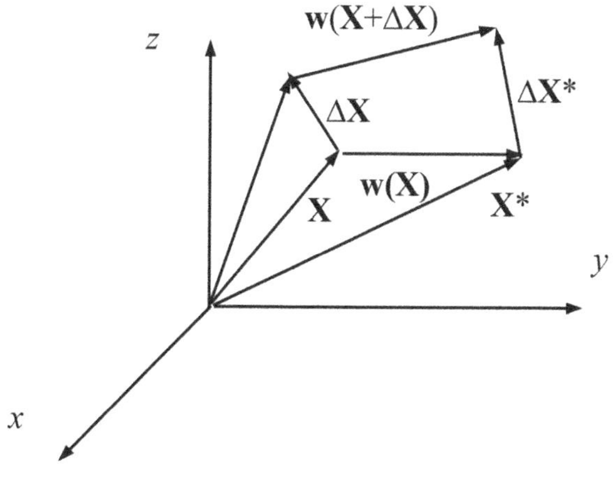

Figure 9.4 Displacement of soil skeleton.

If $\Delta \mathbf{X}$ is sufficiently small, (9.11) can be replaced by (9.12).

$$d\mathbf{X}^* = \left[\mathbf{I} + \frac{\partial \mathbf{w}}{\partial \mathbf{X}}\right] d\mathbf{X} \tag{9.12}$$

where **I** is the unit matrix.

The deformation of an infinitesimal neighborhood of the particle labeled by **X** may be measured by the extent to which the lengths of the infinitesimal vectors $d\mathbf{X}$ emanating from **X** change in the course of the displacement. The square of the length of $d\mathbf{X}$ is

$$d\mathbf{X}^{*T} d\mathbf{X}^* = d\mathbf{X}^T \left[\mathbf{I} + \left(\frac{\partial \mathbf{w}}{\partial \mathbf{X}}\right)^T\right]\left[\mathbf{I} + \frac{\partial \mathbf{w}}{\partial \mathbf{X}}\right] d\mathbf{X} = d\mathbf{X}^T [\mathbf{I} - 2\boldsymbol{\varepsilon}] d\mathbf{X} \tag{9.13}$$

where

$$\boldsymbol{\varepsilon} = -\frac{1}{2}\left[\frac{\partial \mathbf{w}}{\partial \mathbf{X}} + \left(\frac{\partial \mathbf{w}}{\partial \mathbf{X}}\right)^T\right] = \begin{bmatrix} \varepsilon_x & \varepsilon_{xy} & \varepsilon_{xz} \\ \varepsilon_{yx} & \varepsilon_y & \varepsilon_{yz} \\ \varepsilon_{zx} & \varepsilon_{zy} & \varepsilon_z \end{bmatrix} \tag{9.14a}$$

$$\varepsilon_x = -\frac{\partial w_x}{\partial x} \tag{9.14b}$$

$$\varepsilon_y = -\frac{\partial w_y}{\partial y} \tag{9.14c}$$

$$\varepsilon_z = -\frac{\partial w_z}{\partial z} \tag{9.14d}$$

$$\varepsilon_{xy} = -\frac{1}{2}\left(\frac{\partial w_x}{\partial y} + \frac{\partial w_y}{\partial x}\right) \tag{9.14e}$$

$$\varepsilon_{yz} = -\frac{1}{2}\left(\frac{\partial w_y}{\partial z} + \frac{\partial w_z}{\partial y}\right) \tag{9.14f}$$

$$\varepsilon_{zx} = -\frac{1}{2}\left(\frac{\partial w_z}{\partial x} + \frac{\partial w_x}{\partial z}\right) \tag{9.14g}$$

$\boldsymbol{\varepsilon}$ is a symmetric second-rank tensor for infinitesimal strains, called the infinitesimal *strain tensor.* Clearly, $\boldsymbol{\varepsilon}$ describes the deformation of the infinitesimal neighborhood of **X**. The strain tensor is also denoted in vector form as

$$\vec{\varepsilon}^T = [\,\varepsilon_x\ \varepsilon_y\ \varepsilon_z\ 2\varepsilon_{xy}\ 2\varepsilon_{yz}\ 2\varepsilon_{zx}\,] = [\,\varepsilon_x\ \varepsilon_y\ \varepsilon_z\ \gamma_{xy}\ \gamma_{yz}\ \gamma_{zx}\,] \tag{9.15}$$

The trace of the strain tensor has a special geometric significance: it is the infinitesimal *volumetric strain* ε_v (compression positive), defined as $-\Delta V/V_0$, where ΔV is the volume change and V_0 is the initial volume of a small neighborhood. An easy way to show this is to look at an infinitesimal cube ($V_0 = dxdydz$) whose edges parallel the co-ordinates. When the cube is infinitesimally deformed, the lengths of the edges change to $(1-\varepsilon_x)dx, (1-\varepsilon_y)dy$ and $(1-\varepsilon_z)dz$, respectively, making the volume change $\Delta V = -(\varepsilon_x + \varepsilon_y + \varepsilon_z)dxdydz$. Thus,

$$\varepsilon_v = \varepsilon_x + \varepsilon_y + \varepsilon_z \tag{9.16}$$

9.4.2 Cylindrical co-ordinates

In the cylindrical co-ordinates (r, θ, z) (Figure 9.2), the strain tensor $\boldsymbol{\varepsilon}$ is expressed as

$$\boldsymbol{\varepsilon} = \begin{bmatrix} \varepsilon_r & \varepsilon_{r\theta} & \varepsilon_{rz} \\ \varepsilon_{\theta r} & \varepsilon_\theta & \varepsilon_{\theta z} \\ \varepsilon_{zr} & \varepsilon_{z\theta} & \varepsilon_z \end{bmatrix} \tag{9.17a}$$

$$\varepsilon_r = -\frac{\partial w_r}{\partial r} \tag{9.17b}$$

$$\varepsilon_\theta = -\left(\frac{1}{r}\frac{\partial w_\theta}{\partial \theta} + \frac{w_r}{r}\right) \tag{9.17c}$$

$$\varepsilon_z = -\frac{\partial w_z}{\partial z} \tag{9.17d}$$

$$\varepsilon_{r\theta} = \frac{\gamma_{r\theta}}{2} = -\frac{1}{2}\left(\frac{1}{r}\frac{\partial w_r}{\partial \theta} + \frac{\partial w_\theta}{\partial r} - \frac{w_\theta}{r}\right) \tag{9.17e}$$

$$\varepsilon_{\theta z} = \frac{\gamma_{\theta z}}{2} = -\frac{1}{2}\left(\frac{\partial w_\theta}{\partial z} + \frac{1}{r}\frac{\partial w_z}{\partial \theta}\right) \tag{9.17f}$$

$$\varepsilon_{zr} = \frac{\gamma_{zr}}{2} = -\frac{1}{2}\left(\frac{\partial w_z}{\partial r} + \frac{\partial w_r}{\partial z}\right) \tag{9.17g}$$

$$\vec{\varepsilon}^T = [\, \varepsilon_r \; \varepsilon_\theta \; \varepsilon_z \; \gamma_{r\theta} \; \gamma_{\theta z} \; \gamma_{zr} \,] \tag{9.18}$$

$$\varepsilon_v = \varepsilon_r + \varepsilon_\theta + \varepsilon_z \tag{9.19}$$

where $\mathbf{w} = (w_r, w_\theta, w_z)^T$.

9.4.3 Spherical co-ordinates

In the spherical co-ordinates (r, θ, ϕ) (Figure 9.3), the strain tensor $\boldsymbol{\varepsilon}$ is expressed as

$$\varepsilon = \begin{bmatrix} \varepsilon_r & \varepsilon_{r\theta} & \varepsilon_{r\phi} \\ \varepsilon_{\theta r} & \varepsilon_\theta & \varepsilon_{\theta\phi} \\ \varepsilon_{\phi r} & \varepsilon_{\phi\theta} & \varepsilon_\phi \end{bmatrix} \tag{9.20a}$$

$$\varepsilon_r = -\frac{\partial w_r}{\partial r} \tag{9.20b}$$

$$\varepsilon_\theta = -\left(\frac{1}{r\sin\phi}\frac{\partial w_\theta}{\partial\theta} + \frac{w_r + w_\phi \cot\phi}{r}\right) \tag{9.20c}$$

$$\varepsilon_\phi = -\left(\frac{1}{r}\frac{\partial w_\phi}{\partial\phi} + \frac{w_r}{r}\right) \tag{9.20d}$$

$$\varepsilon_{r\theta} = \frac{\gamma_{r\theta}}{2} = -\frac{1}{2}\left(\frac{1}{r\sin\phi}\frac{\partial w_r}{\partial\theta} + \frac{\partial w_\theta}{\partial r} - \frac{w_\theta}{r}\right) \tag{9.20e}$$

$$\varepsilon_{\theta\phi} = \frac{\gamma_{\theta\phi}}{2} = -\frac{1}{2}\left(\frac{1}{r}\frac{\partial w_\theta}{\partial\phi} + \frac{1}{r\sin\phi}\frac{\partial w_\phi}{\partial\theta} - \frac{w_\theta\cot\phi}{r}\right) \tag{9.20f}$$

$$\varepsilon_{\phi r} = \frac{\gamma_{\phi r}}{2} = -\frac{1}{2}\left(\frac{\partial w_\phi}{\partial r} + \frac{1}{r}\frac{\partial w_r}{\partial\phi} - \frac{w_\phi}{r}\right) \tag{9.20g}$$

$$\vec{\varepsilon}^T = [\,\varepsilon_r\ \varepsilon_\theta\ \varepsilon_\phi\ \gamma_{r\theta}\ \gamma_{\theta\phi}\ \gamma_{\phi r}\,] \tag{9.21}$$

$$\varepsilon_v = \varepsilon_r + \varepsilon_\theta + \varepsilon_\phi \tag{9.22}$$

where $\mathbf{w} = (w_r, w_\theta, w_\phi)^T$.

9.5 EFFECTIVE STRESS

The stresses defined in Section 9.3 are total stresses. The total stresses are supported in part by the skeleton of the soil particles and in part by the interstitial fluid since soil is a two-phase material. The state of stress of the porewater can be described by a simple porewater pressure u_w, and even if the porewater is in motion, the departure from this hydrostatic state is insignificant. It is also observed that a uniform increase of the hydrostatic porewater pressure causes only insignificant strains in the soil skeleton. Thus, any straining of the soil is due to the difference of the total stress and the porewater pressure. The difference between total stress and porewater pressure is called the effective stress (see Section 2.3.2). For 3-D stress states, the effective stress equation is expressed as

$$\sigma' = \sigma - u_w \mathbf{I} \tag{9.23}$$

where σ' is the effective stress tensor and **I** is the unit tensor. The effective stress tensor is also denoted in vector form as

$$\vec{\sigma}' = \vec{\sigma} - u_w \mathbf{m} \tag{9.24}$$

where $\mathbf{m}^T = [\,1\;1\;1\;0\;0\;0\,]$.

The *effective stress principle* for 3-D cases is the same as in Section 2.3.2, and listed as follows:

1. The *effective stress* is equal to the total stress minus the porewater pressure.
2. The effective stress controls certain aspects of soil behavior, notably compression, and strength.

9.6 CONSTITUTIVE EQUATIONS

The constitutive equations of soil for 3-D cases are expressed in a quite general form in this section. Suppose the total strain rate $\dot{\vec{\varepsilon}}$ is composed of an elastic component $\dot{\vec{\varepsilon}}^e$ and an inelastic component $\dot{\vec{\varepsilon}}^{vp}$, as follows:

$$\dot{\vec{\varepsilon}} = \dot{\vec{\varepsilon}}^e + \dot{\vec{\varepsilon}}^{vp} \tag{9.25}$$

$$\dot{\vec{\varepsilon}}^e = \mathbf{D}_e^{-1} \dot{\vec{\sigma}}' \tag{9.26}$$

where the dot denotes differentiation with respect to time and $\mathbf{D}_e$ is an elastic matrix. The limit between elastic recoverable behavior and inelastic behavior is some combination of effective stresses and material constants:

$$F = F(\vec{\sigma}', \vec{\varepsilon}^{vp}) = 0 \tag{9.27}$$

when $F < 0$, the inelastic strain rate becomes zero. F is called a *yield surface*.

Suppose the inelastic strain rate satisfies the flow rule

$$\dot{\vec{\varepsilon}}^{vp} = \gamma_F \frac{\partial Q}{\partial \vec{\sigma}'} \tag{9.28}$$

where γ_F is a one-signed multiplier which takes the value of zero if the soil skeleton is in an elastic state, and

$$Q = Q(\vec{\sigma}, \vec{\varepsilon}^{vp}) \tag{9.29}$$

is called a *flow potential*. If the function F itself is a flow potential, that is $Q = F$, the flow rule in (9.28) is called an *associated* flow rule. A flow rule derivable from a flow potential Q that is distinct from F is accordingly called a *nonassociated* flow rule. The flow potential and yield surface define the inelastic strain rate, and the stress-strain relation in incremental form is derived as follows.

Equations (9.25), (9.26), and (9.28) can be combined to give

$$\dot{\vec{\sigma}}' = \mathbf{D}_e \dot{\vec{\varepsilon}} - \gamma_F \mathbf{D}_e \frac{\partial Q}{\partial \dot{\vec{\sigma}}'} \tag{9.30}$$

Differentiating (9.27) with respect to time, we have

$$\left[\frac{\partial F}{\partial \vec{\sigma}'}\right]^T \dot{\vec{\sigma}}' + \gamma_F \left[\frac{\partial F}{\partial \vec{\varepsilon}^{vp}}\right]^T \frac{\partial Q}{\partial \vec{\sigma}'} = 0 \tag{9.31}$$

Eliminating the multiplier γ_F in (9.30–9.31) gives

$$\dot{\vec{\sigma}} = \mathbf{D}_{evp} \dot{\vec{\varepsilon}} \tag{9.32}$$

where

$$\mathbf{D}_{evp} = \mathbf{D}_e - \frac{\mathbf{D}_e \dfrac{\partial Q}{\partial \vec{\sigma}'} \left[\dfrac{\partial F}{\partial \vec{\sigma}'}\right]^T \mathbf{D}_e}{\left[\dfrac{\partial F}{\partial \vec{\sigma}'}\right]^T \mathbf{D}_e \dfrac{\partial Q}{\partial \vec{\sigma}'} - \left[\dfrac{\partial F}{\partial \vec{\varepsilon}^{vp}}\right]^T \dfrac{\partial Q}{\partial \vec{\sigma}'}} \tag{9.33}$$

Thus,

$$\dot{\vec{\sigma}}' = \mathbf{D}\dot{\vec{\varepsilon}} = \begin{cases} \mathbf{D}_e \dot{\vec{\varepsilon}} & F < 0 \\ \mathbf{D}_{evp} \dot{\vec{\varepsilon}} & F = 0 \end{cases} \tag{9.34}$$

9.7 DARCY'S LAW

9.7.1 Cartesian co-ordinates

With the water head defined in (2.1), the Darcy's empirical law for a fully saturated soil is expressed as:

$$q = -k\nabla h = -k\nabla\left(z + \frac{u_w}{\gamma_w}\right) \tag{9.35}$$

where $\mathbf{q}^T = (\, q_x \; q_y \; q_z)$ is the water flow rate, $\nabla^T = \left(\frac{\partial}{\partial x} \; \frac{\partial}{\partial y} \; \frac{\partial}{\partial z} \right)$ is the gradient operator, and **k** is a symmetric second-rank hydraulic conductivity tensor:

$$\mathbf{k} = \begin{bmatrix} k_x & k_{xy} & k_{xz} \\ k_{yx} & k_y & k_{yz} \\ k_{zx} & k_{zy} & k_z \end{bmatrix} \tag{9.36}$$

The hydraulic conductivity tensor will in general be strongly dependent on the total volumetric strain ε_v reached. In isotropic situations, **k** can be replaced by a scalar value k.

When expressed using excess porewater pressure, (9.35) becomes

$$q = -k\nabla h = -\frac{\mathbf{k}}{\gamma_w}\nabla u_e \tag{9.37}$$

9.7.2 Cylindrical co-ordinates

In the cylindrical co-ordinates (r, θ, z) (Figure 9.2), we have

$$\mathbf{k} = \begin{bmatrix} k_r & k_{r\theta} & k_{rz} \\ k_{\theta r} & k_\theta & k_{\theta z} \\ k_{zr} & k_{z\theta} & k_z \end{bmatrix} \tag{9.38}$$

$$\mathbf{q}^T = (\, q_r \; q_\theta \; q_z) \tag{9.39}$$

and

$$\nabla^T = \left(\frac{\partial}{\partial r} \; \frac{1}{r}\frac{\partial}{\partial \theta} \; \frac{\partial}{\partial z} \right) \tag{9.40}$$

9.7.3 Spherical co-ordinates

In the spherical co-ordinates (r, θ, ϕ) (Figure 9.3), we have

$$\mathbf{k} = \begin{bmatrix} k_r & k_{r\theta} & k_{r\phi} \\ k_{\theta r} & k_\theta & k_{\theta\phi} \\ k_{\phi r} & k_{\phi\theta} & k_\phi \end{bmatrix} \tag{9.41}$$

$$\mathbf{q}^T = (\,q_r \;\; q_\theta \;\; q_\phi) \tag{9.42}$$

and

$$\nabla^T = \left(\frac{\partial}{\partial r} \;\; \frac{1}{r\sin\phi}\frac{\partial}{\partial \theta} \;\; \frac{1}{r}\frac{\partial}{\partial \phi} \right) \tag{9.43}$$

9.8 CONTINUITY EQUATION

9.8.1 Cartesian co-ordinates

Consider an infinitesimal cube as shown in Figure 9.5. In a differential time period dt, the net outflow of water from the cube is

$$\text{Net outflow of water} = \left(\frac{\partial q_x}{\partial x} + \frac{\partial q_y}{\partial y} + \frac{\partial q_z}{\partial z} - q_s \right) dxdydzdt \tag{9.44}$$

where q_s, volume of water per unit bulk volume per unit time, is the water source (positive when water is added) within the cube. Since the water and soil

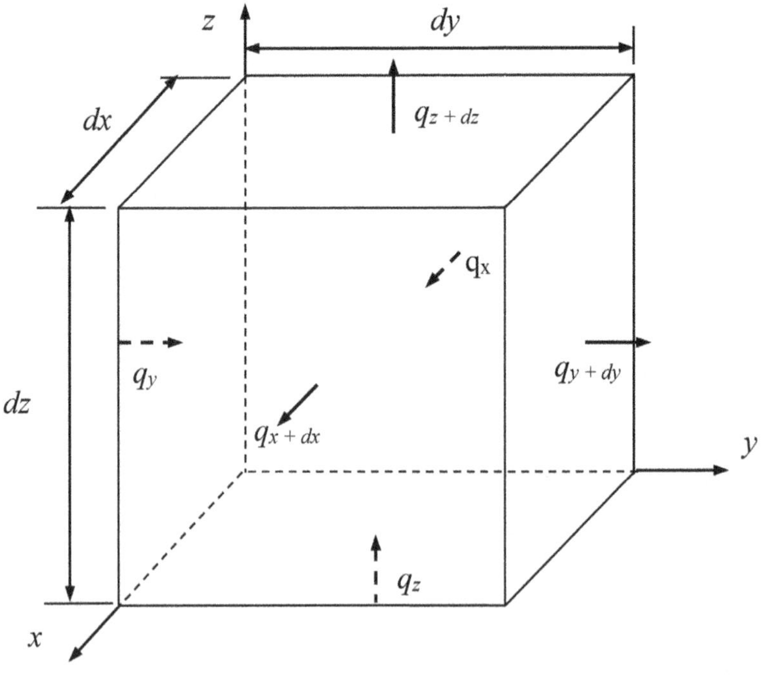

Figure 9.5 Water conservation in an infinitesimal element.

particles are incompressible, the volume of net outflow of water must be equal to the volume change of the infinitesimal element after a differential time dt

$$\left(\frac{\partial q_x}{\partial x}+\frac{\partial q_y}{\partial y}+\frac{\partial q_z}{\partial z}-q_s\right)dxdydzdt=\frac{\partial \varepsilon_v}{\partial t}dxdydzdt \tag{9.45}$$

Therefore, we obtain the following continuity equation

$$\frac{\partial \varepsilon_v}{\partial t}=\frac{\partial q_x}{\partial x}+\frac{\partial q_y}{\partial y}+\frac{\partial q_z}{\partial z}-q_s=\nabla\cdot\mathbf{q}-q_s \tag{9.46}$$

9.8.2 Cylindrical co-ordinates

In the cylindrical co-ordinates (r, θ, z) (Figure 9.2), we have

$$\frac{\partial \varepsilon_v}{\partial t}=\nabla\cdot\mathbf{q}-q_s=\frac{\partial q_r}{\partial r}+\frac{1}{r}\frac{\partial q_\theta}{\partial \theta}+\frac{\partial q_z}{\partial z}+\frac{q_r}{r}-q_s \tag{9.47}$$

9.8.3 Spherical co-ordinates

In the spherical co-ordinates (r, θ, ϕ) (Figure 9.3), we have

$$\frac{\partial \varepsilon_v}{\partial t}=\nabla\cdot\mathbf{q}-q_s=\frac{\partial q_r}{\partial r}+\frac{1}{r\sin\phi}\frac{\partial q_\theta}{\partial \theta}+\frac{1}{r}\frac{\partial q_\phi}{\partial \phi}+\frac{2q_r+q_\phi\cot\phi}{r}-q_s \tag{9.48}$$

9.9 BOUNDARY AND INITIAL CONDITIONS

The governing equations for the consolidation of saturated soil are presented in Sections 9.3–9.8. The displacement **w**, strain $\boldsymbol{\varepsilon}$, and excess porewater pressure u_e are quantities measured with respect to the initial configuration of soil skeleton (a given initial time $t = 0\_$ before the change or sudden change of boundary conditions). Other quantities such as stress $\boldsymbol{\sigma}$, water pressure u_w, and flow rate **q** are current values. In order to solve the governing equations, the initial and boundary conditions should be prescribed.

Assume the soil occupy a region Ω, part of its surface S_σ is subjected to applied tractions $\mathbf{p}_{ns}$, while the reminder of its surface S_w is subjected to displacements $\mathbf{w}_s$. The surface may also be divided into a portion S_q which is subjected to prescribed flow rate q_{ns}, while the reminder S_{u_e} is subjected to prescribed excess porewater pressure u_{es} (or porewater pressure) (see Figure 9.6).

Thus, the boundary conditions are

$$\boldsymbol{\sigma}\mathbf{n}\,|_{S_\sigma}=-\mathbf{p}_{ns} \tag{9.49a}$$

$$\mathbf{w}\,|_{S_w}=\mathbf{w}_s \tag{9.49b}$$

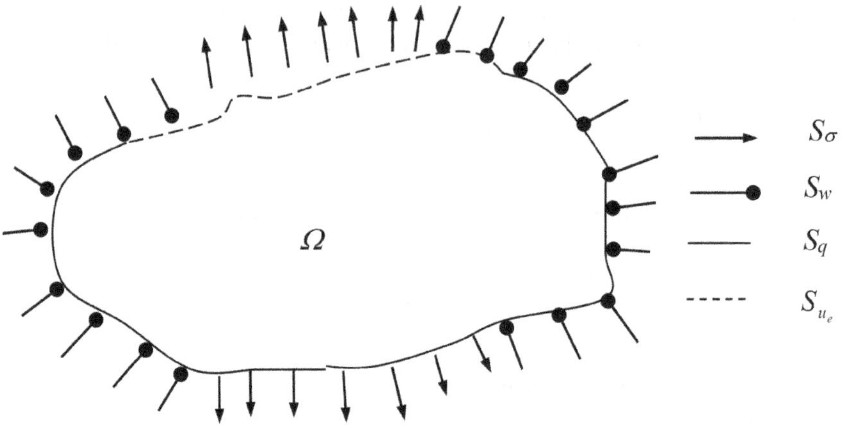

Figure 9.6 Prescribed boundary conditions.

$$\mathbf{q}\mathbf{n}\,|_{S_q} = q_{ns} \tag{9.49c}$$

$$u_e\,|_{S_{u_e}} = u_{es} \tag{9.49d}$$

where **n** denotes the outward normal to the surface.

The initial conditions are

$$\boldsymbol{\sigma}\,|_{t=0\_} = \boldsymbol{\sigma}_0 \tag{9.50a}$$

$$\mathbf{w}\,|_{t=0\_} = 0 \tag{9.50b}$$

$$u_e\,|_{t=0\_} = 0 \tag{9.50c}$$

The initial stress $\boldsymbol{\sigma}_0$ and excess porewater pressure are in equilibrium with the initial external force applied on the soil. Since the initial displacement is zero, the initial strain must be zero. Therefore, in the constitutive relation, any disagreement with this initial state should be adjusted.

9.10 EQUATIONS FOR AN ELASTIC SKELETON AND CONSTANT HYDRAULIC CONDUCTIVITY

When a soil has an elastic soil skeleton and constant hydraulic conductivity, the governing equations in Sections 9.3–9.9 can be simplified using *quantities in excess of the initial values* at $t = 0\_$. We use the *same symbols* to denote the quantities in excess of the initial values at $t = 0\_$. The strain-displacement relation in Section 9.4 and the continuity relation in Section 9.8 are the same. The others are listed in the following.

9.10.1 Force equilibrium equations

1. Cartesian co-ordinates

$$\frac{\partial \sigma'_x}{\partial x}+\frac{\partial \tau_{xy}}{\partial y}+\frac{\partial \tau_{xz}}{\partial z}+\frac{\partial u_e}{\partial x}=0 \tag{9.51a}$$

$$\frac{\partial \tau_{yx}}{\partial x}+\frac{\partial \sigma'_y}{\partial y}+\frac{\partial \tau_{yz}}{\partial z}+\frac{\partial u_e}{\partial y}=0 \tag{9.51b}$$

$$\frac{\partial \tau_{xz}}{\partial x}+\frac{\partial \tau_{zy}}{\partial y}+\frac{\partial \sigma'_z}{\partial z}+\frac{\partial u_e}{\partial z}=0 \tag{9.51c}$$

where

$$\sigma' = \sigma - u_e \mathrm{I} \tag{9.52}$$

$\boldsymbol{\sigma}$ and $\boldsymbol{\sigma}'$ represent excesses over the initial values.

2. Cylindrical co-ordinates

$$\frac{\partial \sigma'_r}{\partial r}+\frac{1}{r}\frac{\partial \tau_{r\theta}}{\partial \theta}+\frac{\partial \tau_{rz}}{\partial z}+\frac{\partial u_e}{\partial r}+\frac{\sigma'_r-\sigma'_\theta}{r}=0 \tag{9.53a}$$

$$\frac{\partial \tau_{r\theta}}{\partial r}+\frac{1}{r}\frac{\partial \sigma'_\theta}{\partial \theta}+\frac{\partial \tau_{\theta z}}{\partial z}+\frac{1}{r}\frac{\partial u_e}{\partial \theta}+\frac{2\tau_{r\theta}}{r}=0 \tag{9.53b}$$

$$\frac{\partial \tau_{rz}}{\partial r}+\frac{1}{r}\frac{\partial \tau_{\theta z}}{\partial \theta}+\frac{\partial \sigma'_z}{\partial z}+\frac{\partial u_e}{\partial z}+\frac{\tau_{rz}}{r}=0 \tag{9.53c}$$

3. Spherical co-ordinates

$$\frac{\partial \sigma'_r}{\partial r}+\frac{1}{r\sin\phi}\frac{\partial \tau_{r\theta}}{\partial \theta}+\frac{1}{r}\frac{\partial \tau_{r\phi}}{\partial \phi}+\frac{\partial u_e}{\partial r}+\frac{2\sigma'_r-\sigma'_\theta-\sigma'_\phi+\tau_{r\phi}\cot\phi}{r}=0 \tag{9.54a}$$

$$\frac{\partial \tau_{r\theta}}{\partial r}+\frac{1}{r\sin\phi}\frac{\partial \sigma'_\theta}{\partial \theta}+\frac{1}{r}\frac{\partial \tau_{\theta\phi}}{\partial \phi}+\frac{1}{r\sin\phi}\frac{\partial u_e}{\partial \theta}+\frac{3\tau_{r\theta}+2\tau_{\theta\phi}\cot\phi}{r}=0 \tag{9.54b}$$

$$\frac{\partial \tau_{r\phi}}{\partial r}+\frac{1}{r\sin\phi}\frac{\partial \tau_{\theta\phi}}{\partial \theta}+\frac{1}{r}\frac{\partial \sigma'_\phi}{\partial \phi}+\frac{1}{r}\frac{\partial u_e}{\partial \phi}+\frac{3\tau_{r\phi}+(\sigma'_\phi-\sigma'_\theta)\cot\phi}{r}=0 \tag{9.54c}$$

9.10.2 Constitutive equations

The constitutive equation of soil that first adopted (Biot, 1941) for 3-D situation assumed an isotropic elastic material. Since most soils exhibit evidence of stratification due to deposition, a transversely isotropic elastic model is presented for the more general consolidation problem. (Transverse isotropy means there is some plane, usually horizontal, in which all stress-train relations are isotropic.) Taking the z-axis normal to the planes of isotropy, the constitutive relation has the form

$$\begin{bmatrix} \sigma'_x \\ \sigma'_y \\ \sigma'_z \\ \tau_{xy} \\ \tau_{yz} \\ \tau_{zx} \end{bmatrix} = \begin{bmatrix} D_{11} & D_{12} & D_{13} & 0 & 0 & 0 \\ D_{21} & D_{22} & D_{23} & 0 & 0 & 0 \\ D_{31} & D_{32} & D_{33} & 0 & 0 & 0 \\ 0 & 0 & 0 & D_{44} & 0 & 0 \\ 0 & 0 & 0 & 0 & D_{55} & 0 \\ 0 & 0 & 0 & 0 & 0 & D_{66} \end{bmatrix} \begin{bmatrix} \varepsilon_x \\ \varepsilon_y \\ \varepsilon_z \\ \gamma_{xy} \\ \gamma_{yz} \\ \gamma_{zx} \end{bmatrix} \tag{9.55}$$

where

$$D_{11} = D_{22} = \frac{E_h(E_h/E_v - \nu_{hv}^2)}{(1+\nu_h)[(1-\nu_h)E_h/E_v - 2\nu_{hv}^2]} \tag{9.56a}$$

$$D_{33} = \frac{E_h(1-\nu_h^2)}{(1+\nu_h)[(1-\nu_h)E_h/E_v - 2\nu_{hv}^2]} \tag{9.56b}$$

$$D_{12} = D_{21} = \frac{E_h(\nu_h E_h/E_v + \nu_{hv}^2)}{(1+\nu_h)[(1-\nu_h)E_h/E_v - 2\nu_{hv}^2]} \tag{9.56c}$$

$$D_{13} = D_{31} = D_{23} = D_{32} = \frac{E_h(1+\nu_h)\nu_{hv}}{(1+\nu_h)[(1-\nu_h)E_h/E_v - 2\nu_{hv}^2]} \tag{9.56d}$$

$$D_{44} = \frac{E_h}{2(1+\nu_h)} \quad D_{55} = D_{66} = G_v \tag{9.56e}$$

E_h is the modulus of elasticity in the horizontal direction; E_v is the modulus of elasticity in the vertical direction; ν_h is Poisson's ratio for the effect of horizontal strain on complementary horizontal strain; ν_{hv} is Poisson's ratio for the effect of horizontal strain on vertical strain; and G_v is the shear modulus of elasticity in the vertical direction.

The transversely isotropic elastic model (9.56) reduces to an isotropic elastic model if properties are rotationally symmetric about an arbitrarily chosen axis. For an isotropic elastic stress-strain relation, only two independent material constants, Young's modulus E and Poisson's ratio ν, exist:

$$E = E_h = E_v \quad \nu = \nu_h = \nu_{hv} \quad G = G_v = \frac{E}{2(1+\nu)} \tag{9.57a}$$

$$D_{11} = D_{22} = D_{33} = \frac{E(1-\nu)}{(1+\nu)(1-2\nu)} \tag{9.57b}$$

$$D_{12} = D_{21} = D_{13} = D_{31} = D_{23} = D_{32} = \frac{E\nu}{(1+\nu)(1-2\nu)} \tag{9.57c}$$

$$D_{44} = D_{55} = D_{66} = G \tag{9.57d}$$

9.10.3 Darcy's law

The hydraulic conductivity tensor k in Darcy's empirical law (9.37) for a transversely isotropic soil is:

$$\mathrm{k} = \begin{bmatrix} k_h & 0 & 0 \\ 0 & k_h & 0 \\ 0 & 0 & k_v \end{bmatrix} \tag{9.58}$$

where k_h and k_v denote the horizontal and vertical hydraulic conductivity, respectively. For an isotropic soil in hydraulic conductivity, we have

$$k = k_h = k_v \tag{9.59}$$

9.10.4 Boundary and initial conditions

The boundary conditions are the same as (9.49). The initial conditions are given in (9.60)

$$\sigma \,|_{t=0_-} = 0 \tag{9.60a}$$

$$\mathrm{w} \,|_{t=0_-} = 0 \tag{9.60b}$$

$$u_e \,|_{t=0_-} = 0 \tag{9.60c}$$

9.11 BIOT'S CONSOLIDATION EQUATIONS AND SIMPLE DIFFUSION THEORY

9.11.1 Biot's theory

For an isotropic elastic soil, substituting (9.14) and (9.55) in (9.51) gives

$$-G\nabla^2 w_x + \frac{G}{1-2\nu}\frac{\partial \varepsilon_v}{\partial x} + \frac{\partial u_e}{\partial x} = 0 \tag{9.61a}$$

$$-G\nabla^2 w_y + \frac{G}{1-2\nu}\frac{\partial \varepsilon_v}{\partial y} + \frac{\partial u_e}{\partial y} = 0 \tag{9.61b}$$

$$-G\nabla^2 w_z + \frac{G}{1-2\nu}\frac{\partial \varepsilon_v}{\partial z} + \frac{\partial u_e}{\partial z} = 0 \tag{9.61c}$$

where

$\nabla^2 = \frac{\partial^2}{\partial x^2} + \frac{\partial^2}{\partial y^2} + \frac{\partial^2}{\partial z^2}$ is the Laplace operator.

Substituting (9.37) and (9.58) in (9.46), we have

$$\frac{\partial \varepsilon_v}{\partial t} + \frac{k}{\gamma_w}\nabla^2 u_e + q_s = 0 \tag{9.62}$$

Equations (9.61) and (9.62) are Biot's (1941) equations of consolidation. The displacement and excess porewater pressure are coupled. Although these equations satisfy not only Darcy's law for water flow through the soil but also the requirement of displacement compatibility, application of Biot's equations of consolidation should be very careful. In many situations, foundation soils display significant anisotropy with regard to mechanical and flow properties. In addition, soils behave quite differently in compression and in swelling.

Under 3-D strain and drainage conditions, (9.62) can be written as (9.63) using the isotropic stress strain relation (9.55)

$$\frac{\partial u_e}{\partial t} = c_{v3}\nabla^2 u_e + \frac{\partial \sigma_m}{\partial t} + \frac{E}{3(1-2\nu)}q_s \tag{9.63}$$

where $\sigma_m = (\sigma_x + \sigma_y + \sigma_z)/3$ is the mean total stress increment,

$$c_{v3} = \frac{kE}{3(1-2\nu)\gamma_w} \tag{9.64}$$

c_{v3} is the coefficient of consolidation under 3-D strain conditions.

Under two-dimensional (2-D) strain and drainage conditions $\varepsilon_y = q_y = 0$

$$\frac{\partial u_e}{\partial t} = c_{v2}\left(\frac{\partial^2 u_e}{\partial x^2} + \frac{\partial^2 u_e}{\partial z^2}\right) + \frac{1}{2}\frac{\partial(\sigma_x + \sigma_z)}{\partial t} + \frac{E}{2(1+\nu)(1-2\nu)}q_s \tag{9.65}$$

where

$$c_{v2} = \frac{kE}{2(1+\nu)(1-2\nu)\gamma_w} \tag{9.66}$$

c_{v2} is the coefficient of consolidation under 2-D strain conditions.

Under 1-D strain and drainage conditions $\varepsilon_x = \varepsilon_y = q_x = q_y = 0$

$$\frac{\partial u_e}{\partial t} = c_{v1}\frac{\partial^2 u_e}{\partial z^2} + \frac{\partial \sigma_z}{\partial t} + \frac{E(1-\nu)}{(1+\nu)(1-2\nu)}q_s \tag{9.67}$$

where

$$c_{v1} = \frac{kE(1-\nu)}{(1+\nu)(1-2\nu)\gamma_w} \tag{9.68}$$

c_{v1} is the usual coefficient of consolidation under 1-D strain conditions. Equation (9.67) is the same as (2.13) if $q_s = 0$.

9.11.2 Simple diffusion theory

The term $\partial\sigma_m/\partial t$ in (9.63), or its equivalent in (9.65), is generally not zero, even when the foundation load remains constant, since stress redistribution will generally take place within a soil mass during consolidation. The distribution of total stresses due to a suddenly applied foundation load corresponds to a Poisson's ratio of 0.5 at the start but corresponds to a Poisson's ratio of the soil skeleton at the completion of consolidation. However, if the term $\partial\sigma_m/\partial t$ in (9.63) is assumed to be zero as an approximation and there is no water source, the equation becomes

$$\frac{\partial u_e}{\partial t} = c_{v3}\nabla^2 u_e \tag{9.69}$$

Correspondingly, (9.65) becomes

$$\frac{\partial u_e}{\partial t} = c_{v2}\left(\frac{\partial^2 u_e}{\partial x^2} + \frac{\partial^2 u_e}{\partial z^2}\right) \tag{9.70}$$

This approximation was implicitly made by Terzaghi (1925) and Rendulic (1937). The derived equation is a much simpler diffusion equation. When ν tends to 0.5, the term $\partial\sigma_m/\partial t$ does in fact become zero, and (9.69) and (9.70) therefore give the same porewater pressure as in Biot's theory. This can be seen by writing (9.63) as (9.71) for $q_s = 0$.

$$\frac{\partial u_e}{\partial[t(1-2\nu)]} = \frac{kE}{3\gamma_w}\nabla^2 u_e + (1-2\nu)\frac{\partial \sigma_m}{\partial t} \tag{9.71}$$

As ν tends to 0.5, the influence of the term $\partial\sigma_m/\partial t$ vanishes.

To determine the excess porewater pressure using (9.69) or (9.70), the initial excess porewater pressure should be the excess porewater pressure just after a load is suddenly applied. The suddenly applied load will remain constant thereafter.

REFERENCES

Biot, MA, 1941, General theory of three-dimensional consolidation. *Journal of Applied Physics*, **12**(2), pp. 155–164.

Rendulic, L, 1937, Das grundgesetz der tonmechanik und sein experimenteller beweis. *Bauingenieur*, **18**, pp. Heft-31.

Small, JC, Booker, JR, and Davis, EH, 1976, Elasto-plastic consolidation of soil. *International Journal of Solids and Structures*, **12**(6), pp. 431–448.

Terzaghi, K, 1925, Principles of soil mechanics, IV-settlement and consolidation of clay. *Engineering News-Record*, **95**(3), pp. 874–878.

Zienkiewicz, OC, Humpheson, C, and Lewis, RW, 1977, A unified approach to soil mechanics problems (including plasticity and visco-plasticity). *In Finite Elements in Geomechanics*, Gudehus, G. (ed.), pp. 151–177. (Chichester: Wiley).

Chapter 10

Solutions of typical three-dimensional consolidation problems

10.1 INTRODUCTION

The mathematical theory of three-dimensional (3-D) consolidation was presented in Chapter 9. Because of the complexity of the equations, relatively few solutions have been obtained. This chapter presents several typical solutions for linear 3-D consolidation problems. The general numerical methods will be discussed in Chapters 11 and 12.

10.2 IMMEDIATE SETTLEMENT AND LONG-TERM SETTLEMENT

At the time of the application of a load on saturated soil, water cannot be expelled of the voids immediately. Since the compressibility of water and soil grains is negligible, the volume of a soil element remains constant. This does not imply that the *immediate deformation*, the deformation at the time of the load application, is null, but only that it comprises a null cubic expansion. A soil element can also undergo a distortion but not immediate variation of volume. The distortion of the soil skeleton is accompanied by movements of water; but, in these movements, the mean relative velocity of water with respect to the skeleton and the speed of filtration remain null (more exactly negligible with respect to speeds of the soil skeleton).

Figure 10.1 shows the case of a suddenly applied point load P that is normal to the surface of an isotropic saturated elastic semi-infinite space. The point load remains constant. The solution of Boussinesq gives the stresses and displacements in A at the time of loading as

$$\sigma_r = \frac{3P}{2\pi}\frac{zr^2}{(r^2+z^2)^{5/2}} \tag{10.1a}$$

$$\sigma_\theta = 0 \tag{10.1b}$$

$$\sigma_z = \frac{3P}{2\pi}\frac{z^3}{(r^2+z^2)^{5/2}} \tag{10.1c}$$

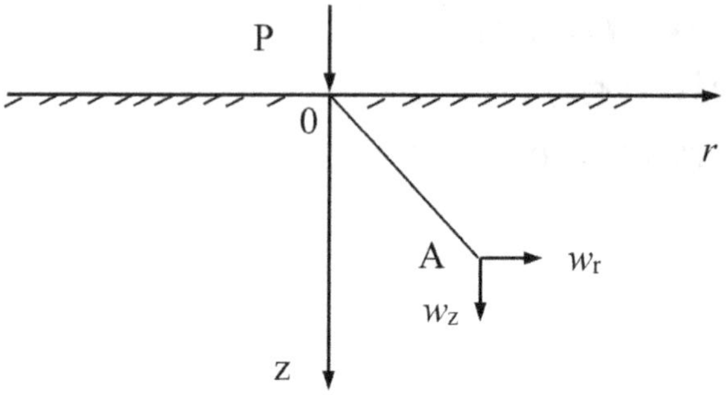

Figure 10.1 Point load on a semi-infinite space (cylindrical co-ordinates).

$$\tau_{rz} = \frac{3P}{2\pi}\frac{z^2 r}{(r^2+z^2)^{5/2}} \tag{10.1d}$$

$$\sigma_m = \frac{P}{2\pi}\frac{z}{(r^2+z^2)^{3/2}} \tag{10.1e}$$

$$w_r = \frac{P}{4\pi G}\frac{zr}{(r^2+z^2)^{3/2}} \tag{10.1f}$$

$$w_z = \frac{P}{4\pi G}\left[\frac{z^2}{(r^2+z^2)^{3/2}} + \frac{1}{(r^2+z^2)^{1/2}}\right] \tag{10.1g}$$

After the excess porewater pressure completely dissipates, the stresses and displacements in A become

$$\sigma_r = \frac{P}{2\pi}\left[\frac{3zr^2}{(r^2+z^2)^{5/2}} - \frac{1-2v}{\sqrt{r^2+z^2}\,(z+\sqrt{r^2+z^2})}\right] \tag{10.2a}$$

$$\sigma_\theta = \frac{P(1-2v)}{2\pi}\left[\frac{1}{\sqrt{r^2+z^2}\,(z+\sqrt{r^2+z^2})} - \frac{z}{(r^2+z^2)^{3/2}}\right] \tag{10.2b}$$

$$\sigma_z = \frac{3P}{2\pi}\frac{z^3}{(r^2+z^2)^{5/2}} \tag{10.2c}$$

$$\tau_{rz} = \frac{3P}{2\pi}\frac{z^2 r}{(r^2+z^2)^{5/2}} \tag{10.2d}$$

$$\sigma_m = \frac{P(1+v)}{3\pi}\frac{z}{(r^2+z^2)^{3/2}} \tag{10.2e}$$

$$w_r = \frac{P}{4\pi G}\left[\frac{zr}{(r^2+z^2)^{3/2}} - \frac{(1-2\nu)r}{\sqrt{r^2+z^2}\,(\sqrt{r^2+z^2}+z)}\right] \tag{10.2f}$$

$$w_z = \frac{P}{4\pi G}\left[\frac{z^2}{(r^2+z^2)^{3/2}} + \frac{2(1-\nu)}{(r^2+z^2)^{1/2}}\right] \tag{10.2g}$$

The mean total stress increment σ_m is clearly quite different at the time of load application and after the excess porewater pressure completely dissipates. The ratio of immediate settlement to final at an unspecified point of the surface of the semi-infinite space is

$$\frac{w_z\,|_{t=0^+}}{w_z\,|_{t=\infty}} = \frac{1}{2(1-\nu)} \tag{10.3}$$

After an external load is applied, a soil element will continue to deform as the water flows out of the element even though the load remains constant. For thick soil layers, the deformation of a foundation soil can last for a long time. This time-dependent deformation is caused partially by the dissipation of excess porewater pressure and partially by the creep nature of the soil skeleton. The deformation corresponds to $t = \infty$ is called *long-term deformation.* If the creep nature of the soil skeleton can be neglected, the total settlement S_t of a foundation at time t is given by

$$S_t = S_u + U(S_f - S_u) \tag{10.4}$$

where S_u is the immediate or undrained settlement, U is the average degree of consolidation, and S_f is the final (or long-term) settlement.

Under true one-dimensional (1-D) conditions $S_u = 0$, U is the average degree of consolidation given by the 1-D theory.

10.3 MANDEL'S PROBLEM

Mandel (1953) presented one of the first solutions for the 3-D consolidation theory of Biot (1941). In Mandel's (1953) original solution, the material was isotropic. Abousleiman *et al.* (1996) extended the solution to include material transverse isotropy.

Mandel's problem involves an infinitely long rectangular specimen sandwiched at the top and the bottom by two rigid frictionless plates as shown in Figure 10.2. The lateral sides are free from surface tractions and pore pressure. A compressive force $2p_0a$ is suddenly applied to the rigid plates at $t = 0$ and then held constant, where a is the half width of the specimen and p_0 is a constant having a unit of stress. The specimen is a transversely isotropic elastic material with the z-axis normal to the planes of isotropy. Since the

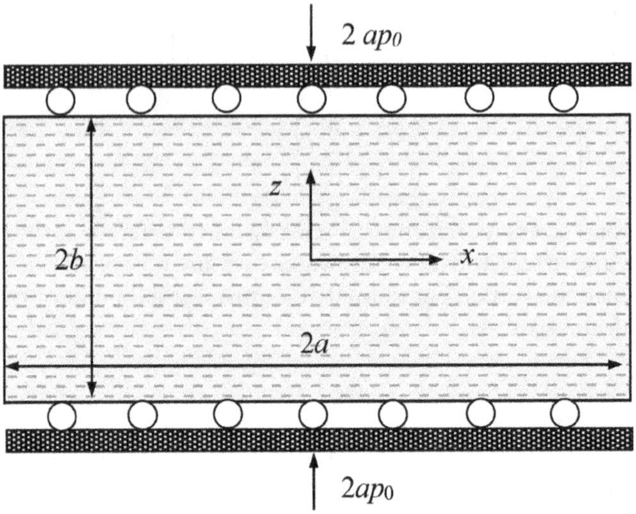

Figure 10.2 Mandel's problem.

specimen is long, the plane strain condition $\varepsilon_y = \gamma_{xy} = \gamma_{yz} = q_y = 0$ prevails. The boundary conditions are

$$\sigma_x = \tau_{xz} = u_e = 0 \ at \ x = \pm a \tag{10.5a}$$

$$\tau_{xz} = q_z = 0 \ at \ z = \pm b \tag{10.5b}$$

$$w_z = constant \ at \ z = \pm b \tag{10.5c}$$

$$\int_{-a}^{a} \sigma_z dx = 2ap_0 \ at \ z = \pm b \tag{10.5d}$$

where b is the half height of the specimen.

Restricted by the geometry and boundary conditions, every horizontal plane is a plane of folding symmetry and fluid flows only in the horizontal direction, thus

$$q_z = \tau_{xz} = 0 \tag{10.6}$$

$$\varepsilon_z = \varepsilon_z(t) \tag{10.7}$$

The following conditions can be obtained from (9.51) and Darcy's law

$$\sigma_x = 0 \tag{10.8}$$

$$\sigma_z = \sigma_z(x,t) \tag{10.9}$$

$$u_e = u_e(x,t) \tag{10.10}$$

and the continuity equation becomes

$$\frac{\partial \varepsilon_v}{\partial t} = -\frac{k_h}{\gamma_w}\frac{\partial^2 u_e}{\partial x^2} \tag{10.11}$$

Rewriting the effective stress-strain equation (9.55) gives

$$E_h\varepsilon_x = -\nu_{hv}(1+\nu_h)\sigma_z - [1-\nu_h^2 - \nu_{hv}(1+\nu_h)]u_e \tag{10.12a}$$

$$E_h\varepsilon_z = \left[\frac{E_h}{E_v} - \nu_{hv}^2\right]\sigma_z - \left[\frac{E_h}{E_v} - \nu_{hv}^2 - \nu_{hv}(1+\nu_h)\right]u_e \tag{10.12b}$$

$$E_h\varepsilon_v = \left[\frac{E_h}{E_v} - \nu_{hv}^2 - \nu_{hv}(1+\nu_h)\right](\sigma_z - A_1u_e) \tag{10.12c}$$

where

$$A_1 = 1 + \frac{1-\nu_h^2 - \nu_{hv}(1+\nu_h)}{E_h/E_v - \nu_{hv}^2 - \nu_{hv}(1+\nu_h)} \tag{10.13}$$

Utilizing (10.7) and (10.12b), (10.14) is obtained

$$\sigma_z - A_2u_e = \sigma_z(a,t) \tag{10.14}$$

where

$$A_2 = 1 - \frac{\nu_{hv}(1+\nu_h)}{E_h/E_v - \nu_{hv}^2} \tag{10.15}$$

Using (10.12c) and (10.14), (10.11) becomes

$$\frac{\partial(\sigma_z - A_1u_e)}{\partial t} = c_0\frac{\partial^2(\sigma_z - A_1u_e)}{\partial x^2} \tag{10.16}$$

where

$$c_0 = \frac{E_hk_h(E_h/E_v - \nu_{hv}^2)}{\gamma_w[E_h/E_v(1-\nu_h^2) - 2(1+\nu_h)\nu_{hv}^2]} \tag{10.17}$$

Applying the Laplace transform to (10.16) gives

$$\frac{d^2}{dx^2}(\bar{\sigma}_z - A_1\bar{u}_e) = \frac{s}{c_0}(\bar{\sigma}_z - A_1\bar{u}_e) \tag{10.18}$$

where the overbar denotes the Laplace transform of the function, that is,

$$\bar{\sigma}_z(x,s) = \int_0^{\infty} e^{-st}\sigma_z(x,t)dt \tag{10.19}$$

Thus,

$$\bar{\sigma}_z - A_1\bar{u}_e = C(s)\cosh\left(\sqrt{\frac{s}{c_0}}x\right) \tag{10.20}$$

Taking the Laplace transform of (10.5d) and (10.14) leads to

$$\int_{-a}^{a} \bar{\sigma}_z dx = \frac{2ap_0}{s} \tag{10.21}$$

and

$$\bar{\sigma}_z - A_2\bar{u}_e = \bar{\sigma}_z(a,s) = C(s)\cosh\left(\sqrt{\frac{s}{c_0}}a\right) \tag{10.22}$$

Solutions of (10.20–10.22) are

$$C(s) = \frac{(A_1 - A_2)p_0}{s}\frac{1}{\left[A_1\cosh\left(\sqrt{\frac{s}{c_0}}a\right) - \frac{A_2}{a}\sqrt{\frac{c_0}{s}}\sinh\left(\sqrt{\frac{s}{c_0}}a\right)\right]} \tag{10.23}$$

$$\bar{\sigma}_z = \frac{p_0}{s}\frac{A_1\cosh\left(\sqrt{\frac{s}{c_0}}a\right) - A_2\cosh\left(\sqrt{\frac{s}{c_0}}x\right)}{\left[A_1\cosh\left(\sqrt{\frac{s}{c_0}}a\right) - \frac{A_2}{a}\sqrt{\frac{c_0}{s}}\sinh\left(\sqrt{\frac{s}{c_0}}a\right)\right]} \tag{10.24}$$

$$\bar{u}_e = \frac{p_0}{s}\frac{\cosh\left(\sqrt{\frac{s}{c_0}}a\right) - \cosh\left(\sqrt{\frac{s}{c_0}}x\right)}{\left[A_1\cosh\left(\sqrt{\frac{s}{c_0}}a\right) - \frac{A_2}{a}\sqrt{\frac{c_0}{s}}\sinh\left(\sqrt{\frac{s}{c_0}}a\right)\right]} \tag{10.25}$$

The singularities of $\bar{\sigma}_z$ and $\bar{u}_e$ are

$$s_n = -\frac{c_0\lambda_n^2}{a^2} \tag{10.26}$$

where λ_n is the nth positive root of the following equation

$$\cos\lambda - \frac{A_2}{A_1}\frac{\sin\lambda}{\lambda} = 0 \tag{10.27}$$

Therefore, using the contour integral for the inverse Laplace transform as shown in Figure 10.3, we have

$$\begin{aligned}\sigma_z(x,t) &= \frac{1}{2\pi i}\int_{c'-i\infty}^{c'+i\infty} \exp(st)\bar{\sigma}_z ds \\ &= p_0 + 2p_0\sum_{n=1}^{\infty}\frac{\sin\lambda_n\left[A_2\cos\left(\lambda_n \frac{x}{a}\right) - A_1\cos\lambda_n\right]}{A_1(\lambda_n - \sin\lambda_n\cos\lambda_n)}\exp\left(-\frac{\lambda_n^2 c_0 t}{a^2}\right)\end{aligned} \tag{10.28}$$

and

$$\begin{aligned}u_e(x,t) &= \frac{1}{2\pi i}\int_{c'-i\infty}^{c'+i\infty} \exp(st)\bar{u}_e ds \\ &= 2p_0\sum_{n=1}^{\infty}\frac{\sin\lambda_n\left[\cos\left(\lambda_n \frac{x}{a}\right) - \cos\lambda_n\right]}{A_1(\lambda_n - \sin\lambda_n\cos\lambda_n)}\exp\left(-\frac{\lambda_n^2 c_0 t}{a^2}\right)\end{aligned} \tag{10.29}$$

where c' is a constant greater than the real part of all singularities.

Finally, the displacements are obtained from (10.12) as follows:

$$\begin{aligned}w_x(x,t) &= \frac{p_0(E_h/E_v - \nu_{hv}^2)}{E_h}\left\{(1-A_2)x + \frac{2(A_1 - A_2)}{A_1}\right. \\ &\left.\times\sum_{n=1}^{\infty}\frac{\sin\lambda_n\left[\frac{A_2 a}{\lambda_n}\sin\left(\lambda_n \frac{x}{a}\right) - x\cos\lambda_n\right]}{\lambda_n - \sin\lambda_n\cos\lambda_n}\exp\left(-\frac{\lambda_n^2 c_0 t}{a^2}\right)\right\}\end{aligned} \tag{10.30}$$

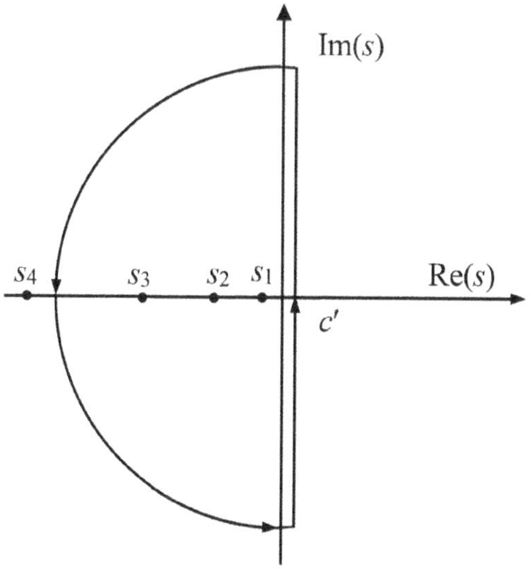

Figure 10.3 Contour integral for the inverse Laplace transform.

$$w_z(x,t) = -\frac{p_0\left(\frac{E_h}{E_v} - v_{hv}^2\right)z}{E_h}\left[1 + \frac{2(A_2 - A_1)}{A_1}\sum_{n=1}^{\infty}\frac{\sin\lambda_n \cos\lambda_n}{\lambda_n - \sin\lambda_n \cos\lambda_n}\exp\left(-\frac{\lambda_n^2 c_0 t}{a^2}\right)\right] \tag{10.31}$$

The stress and strain states just after loading and when the excess porewater pressure completely dissipates are examined in the following. At $t = 0^+$, $\varepsilon_v = 0$. Utilizing (10.12) and (10.5d), (10.32) can be established

$$\sigma_z = p_0 \tag{10.32a}$$

$$u_e = \frac{p_0}{A_1} \tag{10.32b}$$

$$\varepsilon_x = -\frac{E_h/E_v - v_{hv}^2}{E_h}\frac{A_1 - A_2}{A_1}p_0 \tag{10.32c}$$

$$\varepsilon_z = \frac{E_h/E_v - v_{hv}^2}{E_h}\frac{A_1 - A_2}{A_1}p_0 \tag{10.32d}$$

At $t = \infty$, (10.33) is obtained since $u_e = 0$.

$$\sigma_z = p_0 \tag{10.33a}$$

$$\varepsilon_x = -\frac{v_{hv}(1 + v_h)}{E_h}p_0 \tag{10.33b}$$

$$\varepsilon_z = \frac{E_h/E_v - v_{hv}^2}{E_h}p_0 \tag{10.33c}$$

Clearly, uniform stress-strain states exist both at $t = 0^+$ and at $t = \infty$. However, the total force is partially supported by the soil skeleton at $t = 0^+$ if $A_1 \neq 1$.

The average degree of consolidation defined in (10.4) is derived using (10.31) as

$$U = 1 + 2\sum_{n=1}^{\infty}\frac{\sin\lambda_n(\lambda_n \cos\lambda_n - \sin\lambda_n)}{\lambda_n(\lambda_n - \sin\lambda_n \cos\lambda_n)}\exp\left(-\frac{\lambda_n^2 c_0 t}{a^2}\right) \tag{10.34}$$

For an isotropic material, we have

$$A_1 = 2 \tag{10.35a}$$

$$A_2 = \frac{1 - 2v}{1 - v} \tag{10.35b}$$

$$c_0 = \frac{Ek(1 - v)}{\gamma_w(1 + v)(1 - 2v)} \tag{10.35c}$$

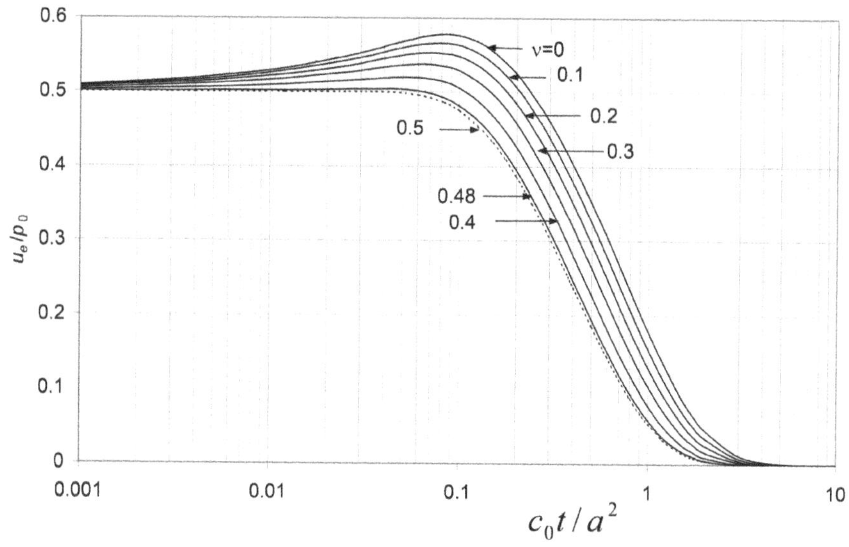

Figure 10.4 The excess porewater pressure at $x = z = 0$ for isotropic soil.

Substituting (10.35) to the analytical solutions for Mandel's problem, the consolidation behavior of an isotropic material is obtained. Figure 10.4 shows the excess porewater pressure at $x = z = 0$ for different Poisson's ratio. The excess porewater pressure first increases and then decreases with time. This non-monotonic response is different from the monotonic behavior in the solution of diffusion equation. Mandel (1953) first demonstrated this type of non-monotonic porewater pressure response in Biot's consolidation theory. Later, Cryer (1963) obtained similar results at the center of a sphere. This type of non-monotonic response is normally referred to as the Mandel-Cryer effect. The physical phenomenon has been confirmed in the laboratory (Gibson *et al.* 1963) as well as in the field (Rodrigues 1983). As the Poisson's ratio tends to 0.5, the excess porewater pressure approaches the solution of diffusion equation. The excess porewater pressure distribution along x-axis is shown in Figure 10.5. The excess porewater pressure is $0.5p_0$ just after the load is suddenly applied. As time progresses, the porewater pressure near the lateral edges dissipates faster, and thus the specimen becomes more compliant near the sides with the reduction of excess porewater pressure. To support the horizontal rigid plates, there is a load transfer of total stress towards the stiffer center region. Therefore, the excess porewater pressure in the center region continues to rise after the load is applied since the water in this region cannot dissipate easily. At large times, the excess porewater pressure vanishes. Hence, the excess porewater pressure behavior is non-monotonic.

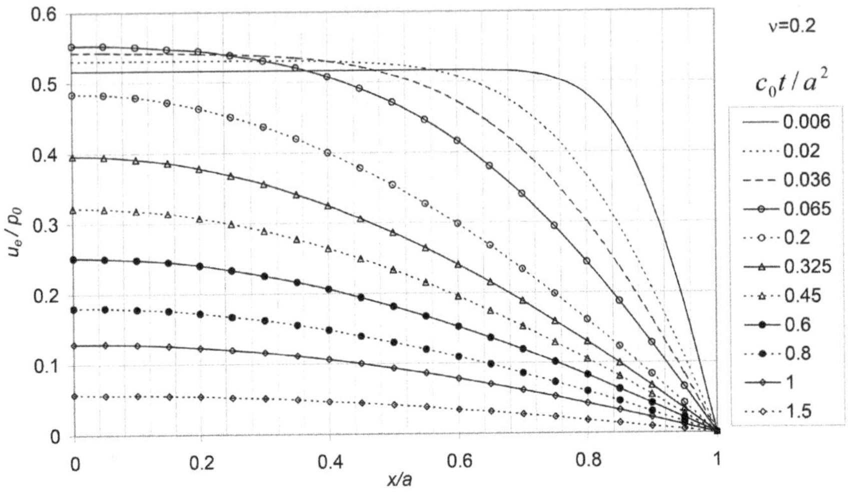

Figure 10.5 Excess porewater pressure distribution along *x*-axis for isotropic soil.

Figure 10.6 shows the horizontal displacements of the specimen edge for different Poisson's ratios. Clearly, the specimen is driven out. The largest horizontal displacement takes place at $t = 0^+$, and then the soil skeleton shrinks with the excess porewater pressure dissipation. The figure also implies that a load on a saturated soil foundation may result in damages to neighboring structures.

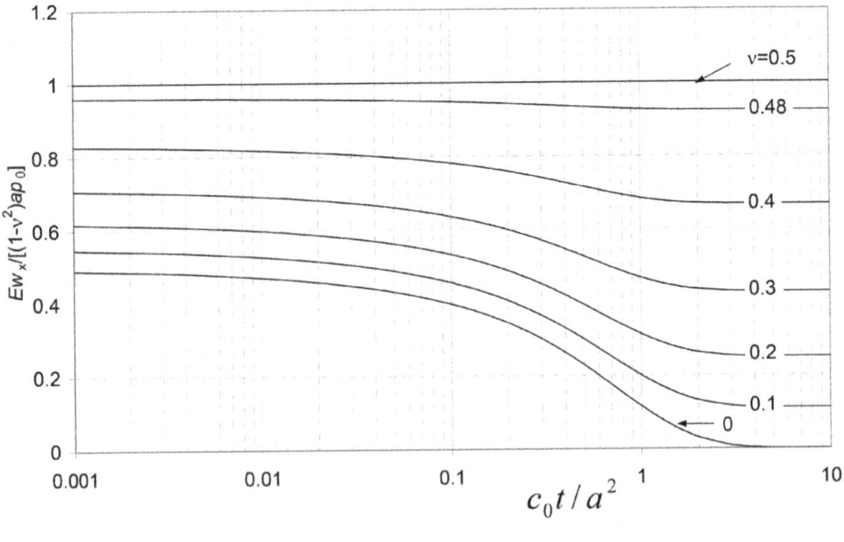

Figure 10.6 The horizontal displacements of the specimen edge for isotropic soil.

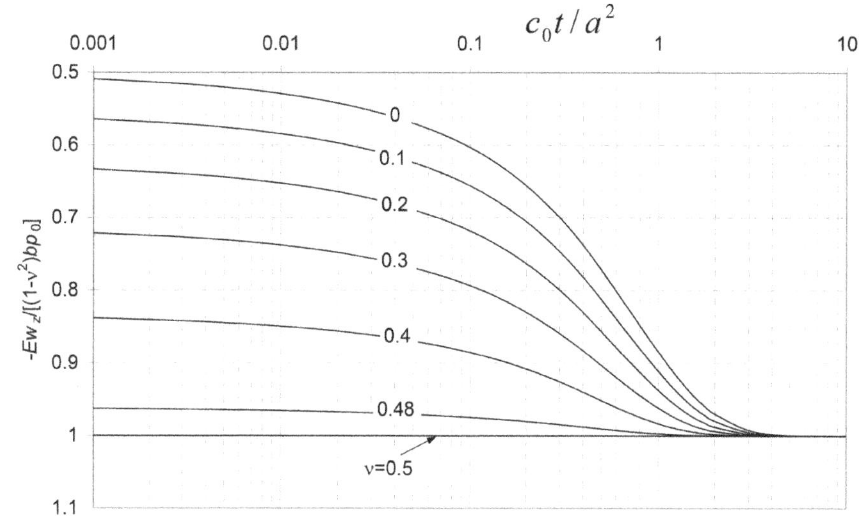

Figure 10.7 The vertical displacement of the upper rigid plate for isotropic soil.

Figure 10.7 illustrates the vertical displacement of the upper rigid plate for different Poisson's ratios. The immediate settlement takes up a large portion of the total settlement. As time progresses, the rigid plate continues settling until the excess porewater pressure completely dissipates.

The average degree of consolidation is shown in Figure 10.8. The effect of Poisson's ratio on the consolidation behavior is clearly expressed in

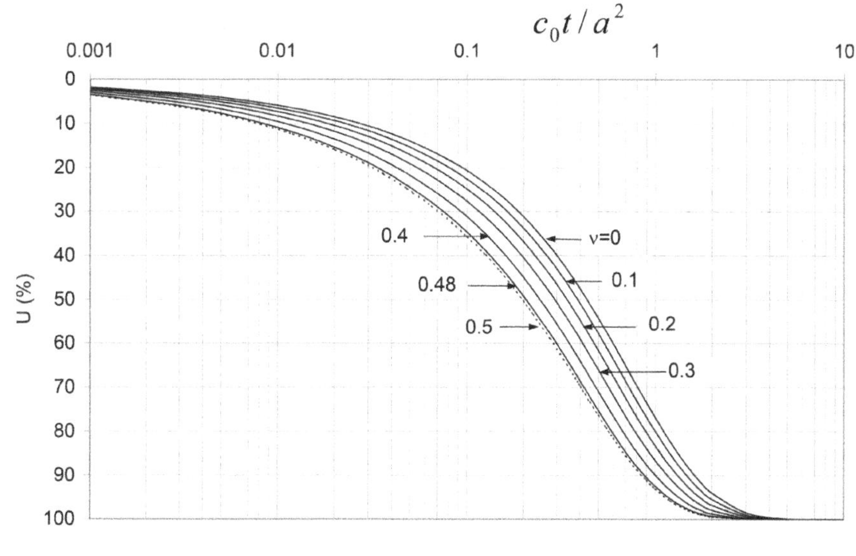

Figure 10.8 The average degree of consolidation for isotropic soil.

this figure. As the Poisson's ratio tends to 0.5, the average degree of consolidation approaches the results of uncoupled consolidation theory.

10.4 CRYER'S PROBLEM

Cryer (1963) presented solutions of Biot and Terzaghi consolidation theories for a saturated sphere of soil subjected to an isotropic pressure change. As in Mandel (1953), the non-monotonic excess porewater pressure response was found at the center of the sphere. In this section, the consolidation of a saturated sphere of isotropic elastic soil is examined.

As shown in Figure 10.9, a sphere of isotropic elastic soil is suddenly applied at $t = 0$ by a normal stress p_0 at the surface. The radius of the sphere is a. The surface is a permeable boundary. Since the consolidation problem is spherically symmetric, the spherical co-ordinates (Figure 9.3) are adopted. The governing equations in Chapter 9 are simplified as follows:

$$\frac{\partial \sigma'_r}{\partial r} + \frac{2(\sigma'_r - \sigma'_\theta)}{r} + \frac{\partial u_e}{\partial r} = 0 \tag{10.36}$$

$$\sigma'_r = 2G\left[\varepsilon_r + \frac{\nu}{1-2\nu}\varepsilon_v\right] \tag{10.37a}$$

$$\sigma'_\theta = 2G\left[\varepsilon_\theta + \frac{\nu}{1-2\nu}\varepsilon_v\right] \tag{10.37b}$$

$$\varepsilon_r = -\frac{\partial w_r}{\partial r} \tag{10.38a}$$

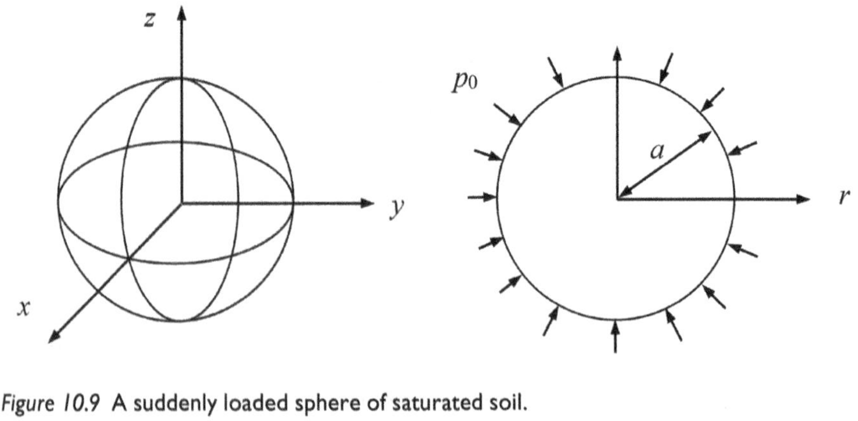

Figure 10.9 A suddenly loaded sphere of saturated soil.

$$\varepsilon_\theta = -\frac{w_r}{r} \tag{10.38b}$$

$$\varepsilon_v = -\left[\frac{\partial w_r}{\partial r} + \frac{2w_r}{r}\right] = -\frac{1}{r^2}\frac{\partial}{\partial r}(r^2 w_r) \tag{10.38c}$$

$$\frac{\partial \varepsilon_v}{\partial t} = c_0 \nabla^2 \varepsilon_v = c_0\left(\frac{\partial^2 \varepsilon_v}{\partial r^2} + \frac{2}{r}\frac{\partial \varepsilon_v}{\partial r}\right) \tag{10.39}$$

$$\sigma_r'(a,t) = 2G\left(\varepsilon_r + \frac{\nu}{1-2\nu}\varepsilon_v\right)\bigg|_{r=a} = p_0 \tag{10.40}$$

where c_0 is the same as in (10.35c).

Substituting (10.37) and (10.38) in (10.36) gives

$$2G\eta\frac{\partial \varepsilon_v}{\partial r} + \frac{\partial u_e}{\partial r} = 0 \tag{10.41}$$

where

$$\eta = \frac{1-\nu}{1-2\nu} \tag{10.42}$$

Integrating (10.41), we have

$$u_e = 2G\eta[\varepsilon_v(a,t) - \varepsilon_v(r,t)] \tag{10.43}$$

Rewriting (10.38c) gives

$$w_r = -\frac{1}{r^2}\int_0^r \zeta^2 \varepsilon_v(\zeta,t)d\zeta \tag{10.44}$$

Substituting (10.44) in (10.38a) gives

$$\varepsilon_r = \varepsilon_v - \frac{2}{r^3}\int_0^r \zeta^2 \varepsilon_v(\zeta,t)d\zeta \tag{10.45}$$

Substituting (10.45) in (10.40), we obtain

$$2G\left[\eta\varepsilon_v(a,t) - \frac{2}{a^3}\int_0^a \zeta^2 \varepsilon_v(\zeta,t)d\zeta\right] = p_0 \tag{10.46}$$

In order to solve the above equations, let us introduce an auxiliary variable, ψ

$$\psi(r,t) = r\left[\varepsilon_v - \frac{p_0}{2G(\eta - 2/3)}\right] \tag{10.47}$$

in terms of which (10.39) and (10.46) become

$$\frac{\partial \psi}{\partial t} = c_0 \frac{\partial^2 \psi}{\partial r^2} \tag{10.48}$$

$$\eta\psi(a,t) - \frac{2}{a^2}\int_0^a \zeta\psi(\zeta,t)d\zeta = 0 \tag{10.49}$$

with initial condition

$$\psi(r,0) = -\frac{p_0 r}{2G(\eta - 2/3)} \tag{10.50}$$

Applying the Laplace transform to (10.48) and (10.49) leads to

$$\frac{d^2\overline{\psi}}{dr^2} = \frac{s}{c_0}\overline{\psi} + \frac{p_0 r}{2G(\eta - 2/3)s} \tag{10.51}$$

and

$$\eta\overline{\psi}(a,s) - \frac{2}{a^2}\int_0^a \zeta\overline{\psi}(\zeta,s)d\zeta = 0 \tag{10.52}$$

Solution of (10.51–10.52) is

$$\overline{\psi} = \frac{ap_0}{2Gs}\frac{\sinh\left(\sqrt{\frac{s}{c_0}}r\right)}{\left[\left(\eta + \frac{2c_0}{a^2 s}\right)\sinh\left(\sqrt{\frac{s}{c_0}}a\right) - \frac{2}{a}\sqrt{\frac{c_0}{s}}\cosh\left(\sqrt{\frac{s}{c_0}}a\right)\right]} - \frac{p_0 r}{2G(\eta - 2/3)s} \tag{10.53}$$

The singularities of $\overline{\psi}$ are 0 and

$$s_n = -\frac{c_0\mu_n^2}{a^2} \tag{10.54}$$

where μ_n is the nth positive root of the following equation

$$\left(\frac{\mu^2}{2} - \frac{1}{\eta}\right)\sin\mu + \frac{\mu}{\eta}\cos\mu = 0 \tag{10.55}$$

Thus, using the contour integral for the inverse Laplace transform as shown in Figure 10.3, we obtain

$$\begin{aligned}\psi(r,t) &= \frac{1}{2\pi i}\int_{c'-i\infty}^{c'+i\infty} \exp(st)\bar{\psi}ds \\ &= \frac{ap_0}{2G\eta}\sum_{n=1}^{\infty}\frac{\sin\left(\mu_n \frac{r}{a}\right)}{\frac{\mu_n}{2}\cos\mu_n + \left(1-\frac{1}{\eta}\right)\sin\mu_n}\exp\left(-\frac{\mu_n^2 c_0 t}{a^2}\right)\end{aligned} \tag{10.56}$$

The displacement, excess porewater pressure, and average degree of consolidation are obtained accordingly as follows:

$$w_r(r,t) = -\frac{p_0 r}{6G\left(\eta-\frac{2}{3}\right)} - \sum_{n=1}^{\infty}\frac{p_0\frac{a^3}{r^2}\left[\sin\left(\mu_n\frac{r}{a}\right)-\mu_n\frac{r}{a}\cos\left(\mu_n\frac{r}{a}\right)\right]}{2G\eta\mu_n^2\left[\frac{\mu_n}{2}\cos\mu_n+\left(1-\frac{1}{\eta}\right)\sin\mu_n\right]}\exp\left(-\frac{\mu_n^2 c_0 t}{a^2}\right) \tag{10.57}$$

$$u_e(r,t) = p_0\sum_{n=1}^{\infty}\frac{\sin\mu_n - \frac{a}{r}\sin\left(\mu_n\frac{r}{a}\right)}{\frac{\mu_n}{2}\cos\mu_n+\left(1-\frac{1}{\eta}\right)\sin\mu_n}\exp\left(-\frac{\mu_n^2 c_0 t}{a^2}\right) \tag{10.58}$$

$$U = 1+\sum_{n=1}^{\infty}\frac{\left(3-\frac{2}{\eta}\right)(\sin\mu_n-\mu_n\cos\mu_n)}{\left[\frac{\mu_n}{2}\cos\mu_n+\left(1-\frac{1}{\eta}\right)\sin\mu_n\right]\mu_n^2}\exp\left(-\frac{\mu_n^2 c_0 t}{a^2}\right) \tag{10.59}$$

Figure 10.10 shows the excess porewater pressure at the sphere center. Figure 10.11 shows the excess porewater pressure distribution along the

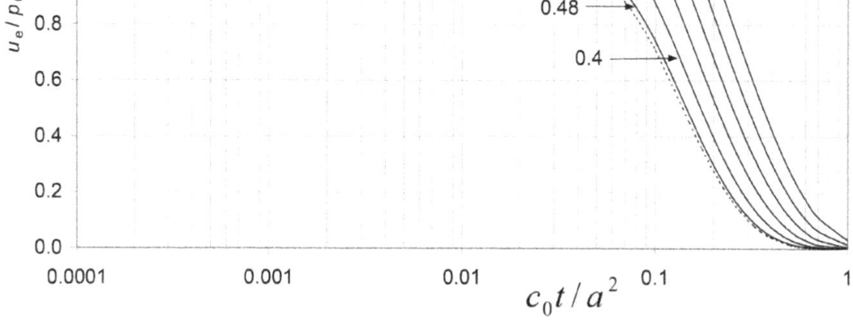

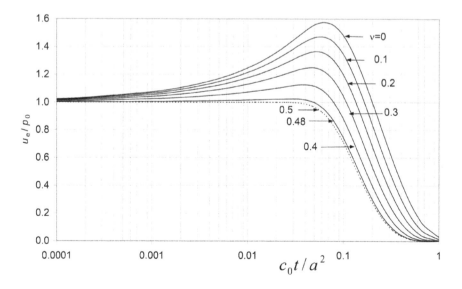

Figure 10.10 Excess pore pressure at the sphere center.

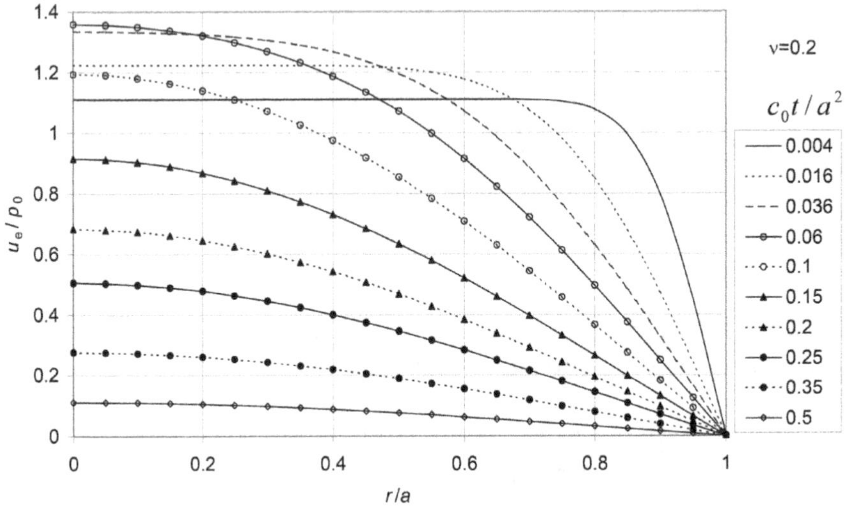

Figure 10.11 Excess pore pressure distribution along the radius of the sphere.

radius of the sphere. Figure 10.12 presents the average degree of consolidation. It is clear that the excess porewater pressure response is similar to the one in Mandel's problem.

10.5 CONSOLIDATION OF A SEMI-INFINITE LAYER SUBJECT TO NORMAL CIRCULAR LOADING

Problems in soil mechanics frequently relate to the semi-infinite body or the infinite layer. In this section, we will examine the consolidation of a semi-infinite layer subjected to a normal surface loading.

As shown in Figure 10.13, a uniform circular normal stress p_0 is suddenly applied on the surface of a semi-infinite space. The radius of the circle is r_0. The soil in the semi-infinite space is a linear isotropic material. We consider two boundary conditions on the excess pore pressure here, namely, (a) that the excess pore pressure vanishes at $z = 0$, and (b) that the normal gradient of the excess pore pressure vanishes at $z = 0$. This consolidation problem is axially symmetric, so the cylindrical co-ordinates (Figure 9.2) are adopted. The governing equations in Chapter 9 are simplified as follows:

$$\frac{\partial \sigma_r'}{\partial r} + \frac{\partial \tau_{rz}}{\partial z} + \frac{\partial u_e}{\partial r} + \frac{\sigma_r' - \sigma_\theta'}{r} = 0 \tag{10.60a}$$

$$\frac{\partial \tau_{rz}}{\partial r} + \frac{\partial \sigma_z'}{\partial z} + \frac{\partial u_e}{\partial z} + \frac{\tau_{rz}}{r} = 0 \tag{10.60b}$$

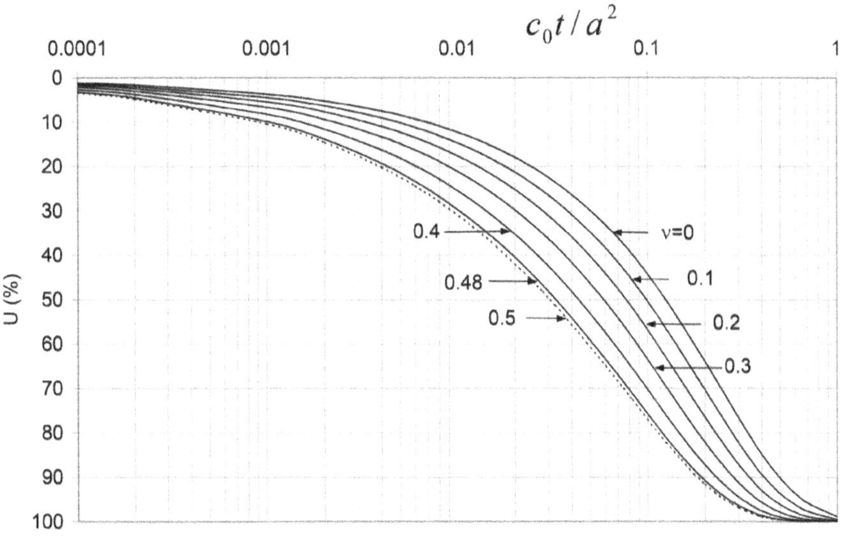

Figure 10.12 The average degree of consolidation of the sphere.

$$\sigma'_r = 2G\left[\varepsilon_r + \frac{v}{1-2v}\varepsilon_v\right] \tag{10.61a}$$

$$\sigma'_\theta = 2G\left[\varepsilon_\theta + \frac{v}{1-2v}\varepsilon_v\right] \tag{10.61b}$$

$$\sigma'_z = 2G\left[\varepsilon_z + \frac{v}{1-2v}\varepsilon_v\right] \tag{10.61c}$$

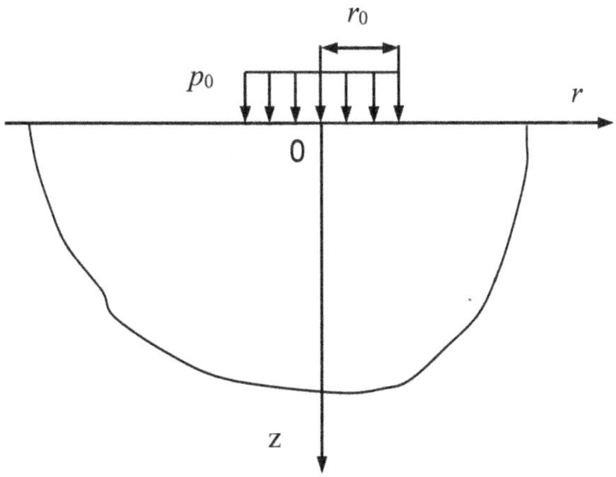

Figure 10.13 A semi-infinite isotropic space subjected normal circular loading.

$$\tau_{rz} = G\gamma_{rz} \tag{10.61d}$$

$$\varepsilon_r = -\frac{\partial w_r}{\partial r} \tag{10.62a}$$

$$\varepsilon_\theta = -\frac{w_r}{r} \tag{10.62b}$$

$$\varepsilon_z = -\frac{\partial w_z}{\partial z} \tag{10.62c}$$

$$\varepsilon_{zr} = \frac{\gamma_{zr}}{2} = -\frac{1}{2}\left(\frac{\partial w_z}{\partial r} + \frac{\partial w_r}{\partial z}\right) \tag{10.62d}$$

$$\varepsilon_v = -\left[\frac{\partial w_r}{\partial r} + \frac{w_r}{r} + \frac{\partial w_z}{\partial z}\right] = -\left[\frac{1}{r}\frac{\partial}{\partial r}(rw_r) + \frac{\partial w_z}{\partial z}\right] \tag{10.62e}$$

$$\frac{\partial \varepsilon_v}{\partial t} = c_0 \nabla^2 \varepsilon_v \tag{10.63}$$

where $\nabla^2 = \frac{\partial^2}{\partial r^2} + \frac{1}{r}\frac{\partial}{\partial r} + \frac{\partial^2}{\partial z^2}$ is the Laplace operator in cylindrical co-ordinates and c_0 is defined in (10.35c).

Substituting (10.61–10.62) in (10.60) gives

$$-G\left(\nabla^2 - \frac{1}{r^2}\right)w_r + \frac{G}{1-2\nu}\frac{\partial \varepsilon_v}{\partial r} + \frac{\partial u_e}{\partial r} = 0 \tag{10.64a}$$

$$-G\nabla^2 w_z + \frac{G}{1-2\nu}\frac{\partial \varepsilon_v}{\partial z} + \frac{\partial u_e}{\partial z} = 0 \tag{10.64b}$$

In (10.63–10.64), there are three independent variables (the excess pore pressure and displacements). These independent variables are expressed in terms of two displacement functions Φ and Ψ (McNamee and Gibson 1960a, b) in the following:

$$w_r = -\frac{\partial \Phi}{\partial r} + z\frac{\partial \Psi}{\partial r} \tag{10.65a}$$

$$w_z = -\frac{\partial \Phi}{\partial z} + z\frac{\partial \Psi}{\partial z} - \Psi \tag{10.65b}$$

$$u_e = 2G\left(\frac{\partial \Psi}{\partial z} - \eta\varepsilon_v\right) \tag{10.65c}$$

Substituting (10.65) in (10.63–10.64) results in

$$\frac{\partial \nabla^2 \Phi}{\partial t} = c_0 \nabla^4 \Phi \tag{10.66}$$

$$\nabla^2 \Psi = 0 \tag{10.67}$$

The total stresses are obtained from the displacement functions

$$\sigma_r = 2G\left[\left(\frac{\partial^2}{\partial r^2} - \nabla^2\right)\Phi - z\frac{\partial^2 \Psi}{\partial r^2} + \frac{\partial \Psi}{\partial z}\right] \tag{10.68a}$$

$$\sigma_z = 2G\left[\left(\frac{\partial^2}{\partial z^2} - \nabla^2\right)\Phi - z\frac{\partial^2 \Psi}{\partial z^2} + \frac{\partial \Psi}{\partial z}\right] \tag{10.68b}$$

$$\sigma_\theta = 2G\left[\left(\frac{1}{r}\frac{\partial}{\partial r} - \nabla^2\right)\Phi - \frac{z}{r}\frac{\partial \Psi}{\partial r} + \frac{\partial \Psi}{\partial z}\right] \tag{10.68c}$$

$$\tau_{rz} = 2G\left(\frac{\partial^2 \Phi}{\partial r\, \partial z} - z\frac{\partial^2 \Psi}{\partial r\, \partial z}\right) \tag{10.68d}$$

The partial differential equations (10.66–10.67) can be transformed into ordinary differential equations (10.69–10.70) by means of a Laplace transform and a Hankel transform of order zero:

$$\left(\frac{d^2}{dz^2} - \rho^2\right)\left(\frac{d^2}{dz^2} - \rho^2 - \frac{s}{c_0}\right)\tilde{\Phi} = 0 \tag{10.69}$$

$$\left(\frac{d^2}{dz^2} - \rho^2\right)\tilde{\Psi} = 0 \tag{10.70}$$

where

$$\tilde{\Phi} = \int_0^{+\infty}\left[\int_0^{+\infty} rJ_0(r\rho)\Phi dr\right]e^{-st}dt \tag{10.71}$$

$$\tilde{\Psi} = \int_0^{+\infty}\left[\int_0^{+\infty} rJ_0(r\rho)\Psi dr\right]e^{-st}dt \tag{10.72}$$

The over tilde in (10.71–10.72) represents taking both integral transforms. J_0 is the zero-order Bessel function of the first kind. The solutions of (10.69–10.70) are straightforward. Choosing the solutions, which vanish as z approaches infinity, we may write

$$\tilde{\Phi} = a_1 e^{-z\rho} + a_2 e^{-z\rho\sqrt{1+\xi}} \tag{10.73}$$

$$\tilde{\Psi} = b_1 e^{-z\rho} \tag{10.74}$$

$$s = c_0 \rho^2 \xi \tag{10.75}$$

where a_1, a_2, and b_1 are functions of ρ and ξ which are determinable from the boundary conditions.

Since τ_{rz} is zero on $z = 0$ for all r and for $t > 0$, applying the Laplace transform and the Hankel transform of order one to (10.68d) gives

$$a_1 + \sqrt{1+\xi}\, a_2 = 0 \tag{10.76}$$

The compressive stress σ_z is defined by (10.68b). The boundary conditions require that

$$\sigma_z \big|_{z=0,t>0} = \begin{cases} p_0 & r < r_0 \\ 0 & r > r_0 \end{cases} \tag{10.77}$$

Taking the Laplace transform and the Hankel transform of order zero to (10.68b) and (10.77), we obtain

$$(a_1 + a_2)\rho - b_1 = \frac{p_0 r_0}{2Gc_0 \xi \rho^4} J_1(\rho r_0) \tag{10.78}$$

where J_1 is the Bessel function of the first kind of order 1. To determine the functions a_1, a_2, and b_1, we need one additional equation. We will discuss the permeable and impermeable boundaries in the following, respectively.

10.5.1 Solution when the boundary z = 0 is permeable

The excess pore pressure u_e is expressed as (10.65c′) (see (10.65c)) using the displacement functions

$$u_e = 2G\left(\frac{\partial \Psi}{\partial z} - \eta \nabla^2 \Phi\right) \tag{10.65c′}$$

Taking the repeated integral transforms, (10.65c′) becomes

$$\tilde{u}_e = -2G\rho\left(b_1 e^{-z\rho} + a_2 \eta \rho \xi\, e^{-z\rho\sqrt{1+\xi}}\right) \tag{10.79}$$

When the boundary $z = 0$ is permeable, we must have

$$b_1 + a_2\eta\rho\xi = 0 \tag{10.80}$$

The functions a_1, a_2, and b_1 are now given by

$$a_1 = -\frac{p_0 r_0\sqrt{1+\xi}\,J_1(r_0\rho)}{2Gc_0\xi\rho^5[1+\eta\xi-\sqrt{1+\xi}\,]} \tag{10.81a}$$

$$a_2 = \frac{p_0 r_0 J_1(r_0\rho)}{2Gc_0\xi\rho^5[1+\eta\xi-\sqrt{1+\xi}\,]} \tag{10.81b}$$

$$b_1 = -\frac{p_0 r_0 \eta J_1(r_0\rho)}{2Gc_0\rho^4[1+\eta\xi-\sqrt{1+\xi}\,]} \tag{10.81c}$$

Thus,

$$\tilde{\Phi} = \frac{p_0 r_0 J_1(r_0\rho)}{2Gc_0\xi\rho^5[1+\eta\xi-\sqrt{1+\xi}\,]}[e^{-z\rho\sqrt{1+\xi}} - \sqrt{1+\xi}\,e^{-z\rho}] \tag{10.82}$$

$$\tilde{\Psi} = -\frac{p_0 r_0 \eta J_1(r_0\rho)}{2Gc_0\rho^4[1+\eta\xi-\sqrt{1+\xi}\,]}e^{-z\rho} \tag{10.83}$$

Then the displacement functions can be obtained by inverting the integral transforms as

$$\Phi = \frac{p_0 r_0}{2G}\int_0^\infty\left[\frac{J_0(r\rho)J_1(r_0\rho)}{\rho^2}\frac{1}{2\pi i}\int_{B_\xi}\frac{e^{c_0\rho^2\xi t}(e^{-z\rho\sqrt{1+\xi}} - \sqrt{1+\xi}\,e^{-z\rho})}{\xi(1+\eta\xi-\sqrt{1+\xi}\,)}d\xi\right]d\rho \tag{10.84}$$

$$\Psi = -\frac{p_0 r_0 \eta}{2G}\int_0^\infty\left[\frac{J_0(r\rho)J_1(r_0\rho)}{\rho}\frac{1}{2\pi i}\int_{B_\xi}\frac{e^{c_0\rho^2\xi t - z\rho}}{1+\eta\xi-\sqrt{1+\xi}}d\xi\right]d\rho \tag{10.85}$$

where B_ξ is used to signify the Bromwich-Wagner contour $c' - i\infty$ to $c' + i\infty$ in the ξ–plane (see Figure 10.14).

The displacement w_z of the medium is given by (10.65b). On the surface $z = 0$, the first two terms vanish and we have

$$w_z(r,0,t) = -\Psi(r,0,t) = \frac{p_0 r_0 \eta}{2G}\int_0^\infty\left[\frac{J_0(r\rho)J_1(r_0\rho)}{\rho}\frac{1}{2\pi i}\int_{B_\xi}\frac{e^{c_0\rho^2\xi t}}{1+\eta\xi-\sqrt{1+\xi}}d\xi\right]d\rho \tag{10.86}$$

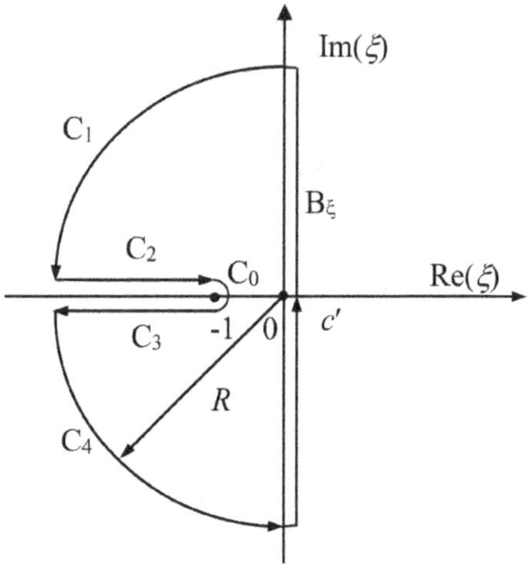

Figure 10.14 Contour integral for the inverse Laplace transform.

The contour integral in (10.86) is calculated using the residue theorem here. Let us introduce a function $\Lambda(\xi)$ as

$$\Lambda(\xi) = \frac{e^{c_0 \rho^2 \xi t}}{1 + \eta\xi - \sqrt{1+\xi}} \tag{10.87}$$

It is clear that $\Lambda(\xi)$ has a pole at $\xi = 0$ and a branch point at $\xi = -1$. Hence, we integrate $\Lambda(\xi)$ around the contour shown in Figure 10.14. The integrals along C_0, C_1, and C_4 all tend to zero as $R \to \infty$ (Jordan's lemma) and C_0 approaches $\xi = -1$. Along C_2 and C_3, we find

$$\int_{C_2+C_3} \Lambda(\xi) d\xi = -\frac{2\pi i}{2\eta - 1}\left[\frac{\eta - 1}{\eta} erfc\left(\frac{\eta-1}{\eta}\rho\sqrt{c_0 t}\right)\exp\left(-\frac{2\eta-1}{\eta^2}\rho^2 c_0 t\right) - erfc(\rho\sqrt{c_0 t})\right] \tag{10.88}$$

by using the known integral (Gradshteyn and Ryzhik 1994)

$$\int_0^\infty \frac{e^{-\alpha x^2}}{x^2 + \beta^2} dx = \frac{\pi}{2\beta} e^{\alpha\beta^2} erfc(\beta\sqrt{\alpha}) \quad (\beta \neq 0, \quad \alpha \geq 0) \tag{10.89}$$

Calculating the residue at $\xi = 0$ yields

Residue$\{\Lambda(\xi); \xi = 0\} = \frac{2}{2\eta - 1}$

Hence, it follows that

$$\frac{1}{2\pi i}\int_{B_\xi} \frac{e^{c_0 \rho^2 \xi t}}{1 + \eta\xi - \sqrt{1+\xi}} d\xi = \frac{1}{2\eta - 1}\varphi_1(\rho\sqrt{c_0 t}) \tag{10.90}$$

$$w_z(r,0,t) = \frac{p_0 r_0 \eta}{2G(2\eta - 1)} \int_0^\infty \left[\frac{J_0(r\rho) J_1(r_0\rho)}{\rho} \varphi_1\left(r_0 \rho \sqrt{\frac{c_0 t}{r_0^2}} \right) \right] d\rho \qquad (10.91)$$

where

$$\varphi_1(x) = 1 + erf(x) + \frac{\eta - 1}{\eta} erfc\left(\frac{\eta - 1}{\eta} x \right) \exp\left(-\frac{2\eta - 1}{\eta^2} x^2 \right) \qquad (10.92)$$

From (10.91–10.92), it will be seen that the sudden application of the surface loads at $t = 0$ leads to an immediate surface settlement of magnitude

$$w_z(r,0,0^+) = \frac{p_0 r_0}{2G} \int_0^\infty \left[\frac{J_0(r\rho) J_1(r_0\rho)}{\rho} \right] d\rho \qquad (10.93)$$

and a final surface settlement

$$w_z(r,0,\infty) = \frac{p_0 r_0 \eta}{G(2\eta - 1)} \int_0^\infty \left[\frac{J_0(r\rho) J_1(r_0\rho)}{\rho} \right] d\rho \qquad (10.94)$$

The degree of consolidation at the circular center thus becomes

$$U = \frac{w_z(0,0,t) - w_z(0,0,0^+)}{w_z(0,0,\infty) - w_z(0,0,0^+)} = \eta \int_0^\infty \frac{J_1(\rho)}{\rho} \left[\varphi_1\left(\rho \sqrt{\frac{c_0 t}{r_0^2}} \right) - \frac{2\eta - 1}{\eta} \right] d\rho \qquad (10.95)$$

The results for different Poisson's ratios are shown in Figure 10.15.

The average settlement of the uniformly loaded circular area may be of interest, and of some use to engineers. Using (10.91), the average settlement can be easily established:

$$\bar{w}_z(t) = \frac{2}{r_0^2} \int_0^{r_0} w_z(r,0,t) r dr = \frac{p_0 r_0 \eta}{G(2\eta - 1)} \int_0^\infty \frac{J_1^2(\rho)}{\rho^2} \varphi_1\left(\rho \sqrt{\frac{c_0 t}{r_0^2}} \right) d\rho \qquad (10.96)$$

and, in particular, the average immediate and final surface settlements are

$$\bar{w}_z(0^+) = \frac{4}{3\pi G} \qquad (10.97)$$

$$\bar{w}_z(\infty) = \frac{8\eta}{3\pi G(2\eta - 1)} \qquad (10.98)$$

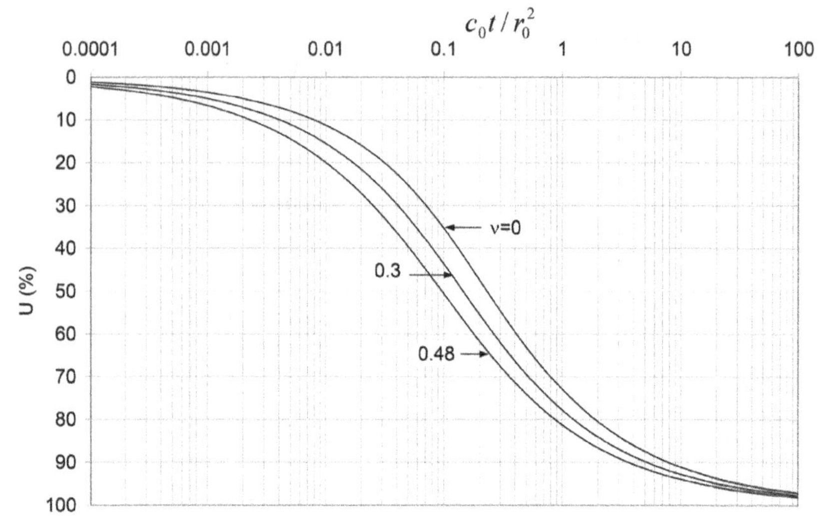

Figure 10.15 **Degree of consolidation of center of circular load (permeable surface).**

by using the known integral (Gradshteyn and Ryzhik 1994)

$$\int_0^\infty \frac{J_1^2(\rho)}{\rho^2}\,d\rho = \frac{4}{3\pi} \tag{10.99}$$

The rate at which the average final settlement is attained can be defined in terms of the average degree of consolidation

$$\bar{U} = \frac{\bar{w}_z(t) - \bar{w}_z(0^+)}{\bar{w}_z(\infty) - \bar{w}_z(0^+)} = \frac{3\pi\eta}{4}\int_0^\infty \frac{J_1^2(\rho)}{\rho^2}\left[\varphi_1\left(\rho\sqrt{\frac{c_0 t}{r_0^2}}\right) - \frac{2\eta-1}{\eta}\right]d\rho \tag{10.100}$$

which is given in Figure 10.16 for different Poisson's ratios. The results in Figure 10.15 are also plotted in Figure 10.16. Clearly, the difference between U and $\bar{U}$ is small.

10.5.2 Solution when the boundary z = 0 is impermeable

When the boundary $z = 0$ is impermeable, we must have

$$\left.\frac{\partial \tilde{u}_e}{\partial z}\right|_{z=0} = 0 \tag{10.101}$$

Substituting (10.79) in (10.101) gives

$$b_1 + a_2\eta\rho\xi\sqrt{1+\xi} = 0 \tag{10.102}$$

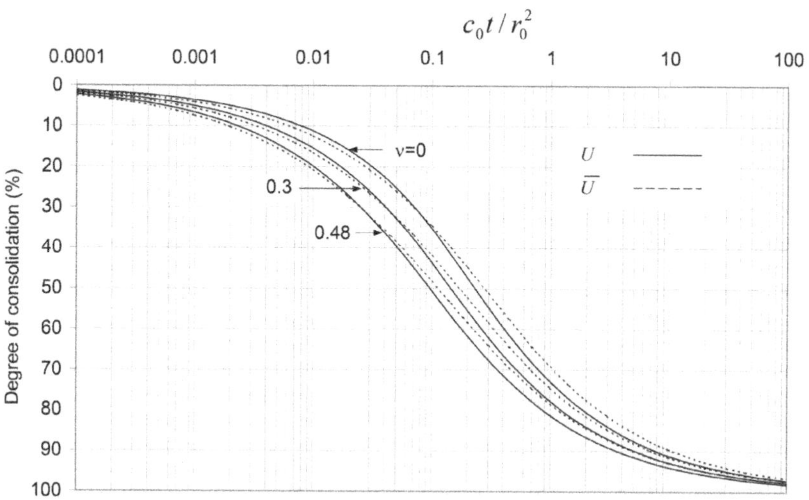

Figure 10.16 U and $\bar{U}$ for the circular normal loading (permeable surface).

The functions a_1, a_2, and b_1 in (10.73–10.74) are now given by

$$a_1 = -\frac{p_0 r_0 \sqrt{1+\xi}\, J_1(r_0\rho)}{2Gc_0\xi\rho^5\left[1-\sqrt{1+\xi}+\eta\xi\sqrt{1+\xi}\right]} \tag{10.103a}$$

$$a_2 = \frac{p_0 r_0 J_1(r_0\rho)}{2Gc_0\xi\rho^5\left[1-\sqrt{1+\xi}+\eta\xi\sqrt{1+\xi}\right]} \tag{10.103b}$$

$$b_1 = -\frac{p_0 r_0 \eta\sqrt{1+\xi}\, J_1(r_0\rho)}{2Gc_0\rho^4\left[1-\sqrt{1+\xi}+\eta\xi\sqrt{1+\xi}\right]} \tag{10.103c}$$

Thus,

$$\tilde{\Phi} = \frac{p_0 r_0 J_1(r_0\rho)}{2Gc_0\xi\rho^5\left[1-\sqrt{1+\xi}+\eta\xi\sqrt{1+\xi}\right]}\left[e^{-z\rho\sqrt{1+\xi}} - \sqrt{1+\xi}\,e^{-z\rho}\right] \tag{10.104}$$

$$\tilde{\Psi} = -\frac{p_0 r_0 \eta\sqrt{1+\xi}\, J_1(r_0\rho)}{2Gc_0\rho^4\left[1-\sqrt{1+\xi}+\eta\xi\sqrt{1+\xi}\right]}e^{-z\rho} \tag{10.105}$$

The displacement functions can be obtained by inverting the integral transforms as

$$\Phi = \frac{p_0 r_0}{2G}\int_0^\infty\left[\frac{J_0(r\rho)J_1(r_0\rho)}{\rho^2}\frac{1}{2\pi i}\int_{B_\xi}\frac{\Lambda_1(\xi)(e^{-z\rho\sqrt{1+\xi}} - \sqrt{1+\xi}\,e^{-z\rho})}{\xi\sqrt{1+\xi}}d\xi\right]d\rho \tag{10.106}$$

$$\Psi = -\frac{p_0 r_0 \eta}{2G}\int_0^\infty \left[\frac{J_0(r\rho)J_1(r_0\rho)}{\rho}\frac{1}{2\pi i}\int_{B_\xi}\Lambda_1(\xi)e^{-z\rho}d\xi\right]d\rho \tag{10.107}$$

where

$$\Lambda_1(\xi) = \frac{\sqrt{1+\xi}\,e^{c_0\rho^2\xi t}}{1-\sqrt{1+\xi}+\eta\xi\sqrt{1+\xi}} \tag{10.108}$$

On the surface $z = 0$, the displacement w_z of the medium is

$$w_z(r,0,t) = \frac{p_0 r_0 \eta}{2G}\int_0^\infty \left[\frac{J_0(r\rho)J_1(r_0\rho)}{\rho}\frac{1}{2\pi i}\int_{B_\xi}\Lambda_1(\xi)d\xi\right]d\rho \tag{10.109}$$

As in Section 10.5.1, the contour integral in (10.109) is calculated using the residue theorem. The function $\Lambda_1(\xi)$ has a branch point at $\xi = -1$ and two poles at $\xi = 0$ and at $\xi = \xi_2$ where

$$\xi_2 = -\frac{\sqrt{\eta^2+4\eta}+\eta-2}{2\eta} \tag{10.110}$$

which satisfies $-1 < \xi_2 < 0$. Calculating the residues at $\xi = 0$ and at $\xi = \xi_2$ yields

Residue$\{\Lambda_1(\xi);\ \xi = 0\} = \frac{2}{2\eta-1}$

Residue$\{\Lambda_1(\xi);\ \xi = \xi_2\} = \frac{2(1+\xi_2)}{3\eta\xi_2+2\eta-1}e^{c_0\rho^2\xi_2 t}$

We integrate $\Lambda_1(\xi)$ around the contour shown in Figure 10.14. The integrals along C_0, C_1, and C_4 all tend to zero as $R \to \infty$ (Jordan's lemma) and C_0 approaches $\xi = -1$. Along C_2 and C_3, we have

$$\begin{aligned}\int_{C_2+C_3}\Lambda_1(\xi)d\xi = \frac{2\pi i}{2\eta-1}\Big\{ & erfc\left(\rho\sqrt{c_0 t}\right) - B_1 erfc\left[\rho\sqrt{(1+\xi_2)c_0 t}\right]\exp(\xi_2\rho^2 c_0 t) \\ & - B_2 erfc\left[\rho\sqrt{\left(\frac{2}{\eta}-\xi_2\right)c_0 t}\right]\exp\left[c_0 t\rho^2\left(\frac{2}{\eta}-\xi_2-1\right)\right]\Big\}\end{aligned} \tag{10.111}$$

where

$$B_1 = \frac{1+\xi_2-\dfrac{1}{\eta^2}}{\left(1+2\xi_2-\frac{2}{\eta}\right)\sqrt{1+\xi_2}} \tag{10.112}$$

$$B_2 = \frac{\xi_2+\frac{1}{\eta^2}-\frac{2}{\eta}}{\left(1+2\xi_2-\frac{2}{\eta}\right)\sqrt{\frac{2}{\eta}-\xi_2}} \tag{10.113}$$

Hence, it follows that

$$\frac{1}{2\pi i}\int_{B_\xi}\Lambda_1(\xi)d\xi = \frac{1}{2\eta-1}\varphi_2\left(\rho\sqrt{c_0 t}\right) \tag{10.114}$$

$$w_z(r,0,t) = \frac{p_0 r_0 \eta}{2G(2\eta-1)}\int_0^\infty\left[\frac{J_0(r\rho)J_1(r_0\rho)}{\rho}\varphi_2\left(r_0\rho\sqrt{\frac{c_0 t}{r_0^2}}\right)\right]d\rho \tag{10.115}$$

where

$$\varphi_2(x) = 1 + erf(x) + \frac{2(1+\xi_2)(2\eta-1)}{3\eta\xi_2+2\eta-1}\exp(\xi_2 x^2) + B_1 erfc\left(x\sqrt{1+\xi_2}\right)\exp(\xi_2 x^2)$$
$$+ B_2 erfc\left(x\sqrt{\frac{2}{\eta}-\xi_2}\right)\exp\left[\left(\frac{2}{\eta}-\xi_2-1\right)x^2\right] \tag{10.116}$$

The immediate and final surface settlements are the same as those in (10.93) and (10.94). From (10.115–10.116), the degree of consolidation at the circular center then becomes

$$U = \frac{w_z(0,0,t)-w_z(0,0,0^+)}{w_z(0,0,\infty)-w_z(0,0,0^+)} = \eta\int_0^\infty\frac{J_1(\rho)}{\rho}\left[\varphi_2\left(\rho\sqrt{\frac{c_0 t}{r_0^2}}\right)-\frac{2\eta-1}{\eta}\right]d\rho \tag{10.117}$$

The results for different Poisson's ratios are shown in Figure 10.17.

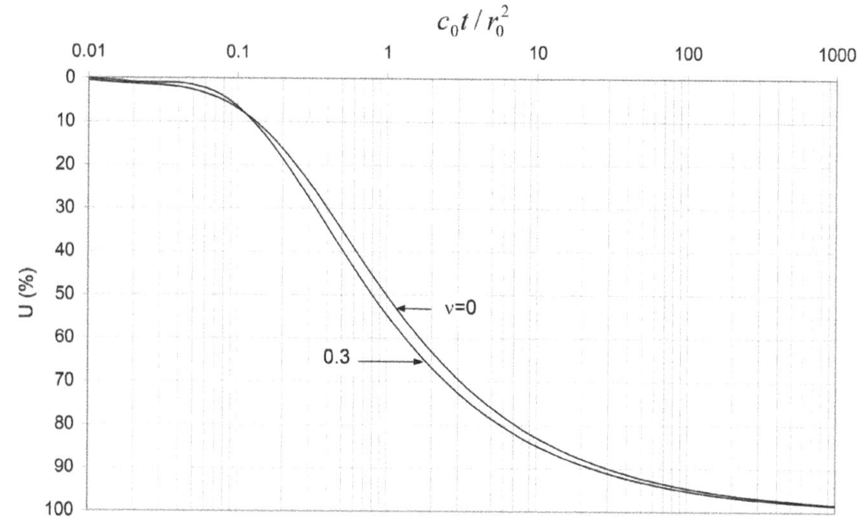

Figure 10.17 U for the circular normal loading (impermeable surface).

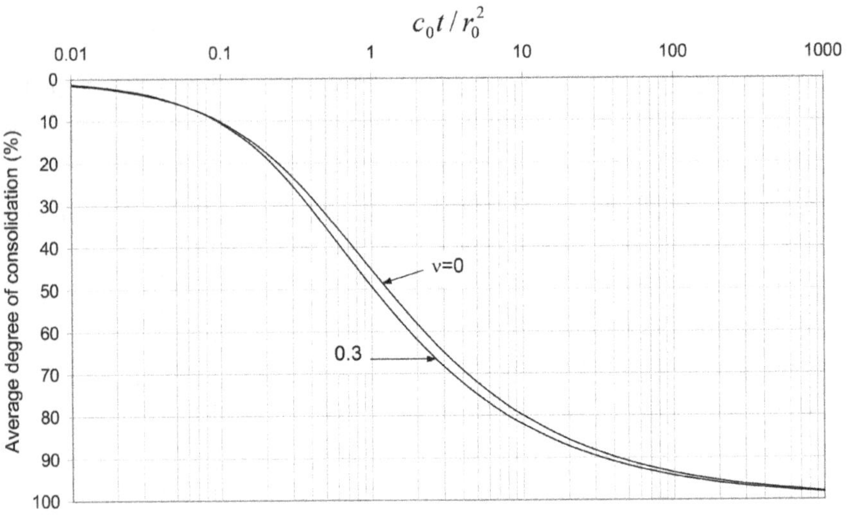

Figure 10.18 $\bar{U}$ for the circular normal loading (impermeable surface).

The average settlement of the uniformly loaded circular area is

$$\bar{w}_z(t) = \frac{2}{r_0^2}\int_0^{r_0} w_z(r,0,t)r\,dr = \frac{p_0 r_0 \eta}{G(2\eta-1)}\int_0^{\infty}\frac{J_1^2(\rho)}{\rho^2}\varphi_2\left(\rho\sqrt{\frac{c_0 t}{r_0^2}}\right)d\rho \tag{10.118}$$

and, in particular, the average immediate and final surface settlements are the same as those in (10.97) and (10.98). The rate at which the average final settlement is attained can again be defined in terms of the average degree of consolidation

$$\bar{U} = \frac{\bar{w}_z(t)-\bar{w}_z(0^+)}{\bar{w}_z(\infty)-\bar{w}_z(0^+)} = \frac{3\pi\eta}{4}\int_0^{\infty}\frac{J_1^2(\rho)}{\rho^2}\left[\varphi_2\left(\rho\sqrt{\frac{c_0 t}{r_0^2}}\right)-\frac{2\eta-1}{\eta}\right]d\rho \tag{10.119}$$

which is given in Figure 10.18 for different Poisson's ratios.

10.6 CONSOLIDATION OF A SEMI-INFINITE LAYER SUBJECT TO NORMAL RECTANGULAR LOADING

In Section 10.5, the solution for the consolidation of a semi-infinite layer subjected to a normal circular surface loading is presented. For the foundation engineer, the rectangular area is a convenient unit from which more complicated load distributions may be built. In this section, the consolidation behavior of a semi-infinite layer subjected to a normal rectangular surface loading is analyzed.

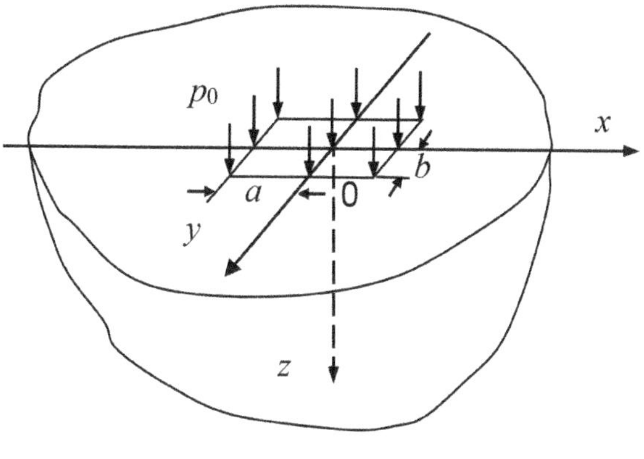

Figure 10.19 A semi-infinite isotropic space subjected normal rectangular loading.

As shown in Figure 10.19, a uniform rectangular normal stress p_0 is suddenly applied on the surface of a semi-infinite space. The length of the longer edge is $2a$. The width is $2b$. The soil in the semi-infinite space is a linear isotropic material. The surface of the semi-infinite space is permeable.

The displacements and excess pore pressure in (9.61–9.62) are expressed in terms of two displacement functions Φ and Ψ (Gibson and McNamee 1963) as in Section 10.5 in the following:

$$w_x = -\frac{\partial \Phi}{\partial x} + z\frac{\partial \Psi}{\partial x} \tag{10.120a}$$

$$w_y = -\frac{\partial \Phi}{\partial y} + z\frac{\partial \Psi}{\partial y} \tag{10.120b}$$

$$w_z = -\frac{\partial \Phi}{\partial z} + z\frac{\partial \Psi}{\partial z} - \Psi \tag{10.120c}$$

$$u_e = 2G\left(\frac{\partial \Psi}{\partial z} - \eta\varepsilon_v\right) \tag{10.120d}$$

Substituting (10.120) in (9.61–9.62) results in

$$\frac{\partial \nabla^2 \Phi}{\partial t} = c_0 \nabla^4 \Phi \tag{10.121}$$

$$\nabla^2 \Psi = 0 \tag{10.122}$$

where $\nabla^2 = \frac{\partial^2}{\partial x^2} + \frac{\partial^2}{\partial y^2} + \frac{\partial^2}{\partial z^2}$ is the Laplace operator in Cartesian co-ordinates.

The total stresses are obtained from the displacement functions

$$\sigma_x = 2G\left[\left(\frac{\partial^2}{\partial x^2} - \nabla^2\right)\Phi - z\frac{\partial^2 \Psi}{\partial x^2} + \frac{\partial \Psi}{\partial z}\right] \tag{10.123a}$$

$$\sigma_y = 2G\left[\left(\frac{\partial^2}{\partial y^2} - \nabla^2\right)\Phi - z\frac{\partial^2 \Psi}{\partial y^2} + \frac{\partial \Psi}{\partial z}\right] \tag{10.123b}$$

$$\sigma_z = 2G\left[\left(\frac{\partial^2}{\partial z^2} - \nabla^2\right)\Phi - z\frac{\partial^2 \Psi}{\partial z^2} + \frac{\partial \Psi}{\partial z}\right] \tag{10.123c}$$

$$\tau_{xy} = 2G\left(\frac{\partial^2 \Phi}{\partial x \partial y} - z\frac{\partial^2 \Psi}{\partial x \partial y}\right) \tag{10.123d}$$

$$\tau_{xz} = 2G\left(\frac{\partial^2 \Phi}{\partial x \partial z} - z\frac{\partial^2 \Psi}{\partial x \partial z}\right) \tag{10.123e}$$

$$\tau_{yz} = 2G\left(\frac{\partial^2 \Phi}{\partial y \partial z} - z\frac{\partial^2 \Psi}{\partial y \partial z}\right) \tag{10.123f}$$

$$u_e = 2G\left(\frac{\partial \Psi}{\partial z} - \eta \nabla^2 \Phi\right) \tag{10.123g}$$

The partial differential equations (10.121–10.122) are reduced to ordinary differential equations (10.124–10.125) by the means of repeated complex Fourier and Laplace transforms:

$$\left(\frac{d^2}{dz^2} - \rho^2\right)\left(\frac{d^2}{dz^2} - \rho^2 - \frac{s}{c_0}\right)\Phi^* = 0 \tag{10.124}$$

$$\left(\frac{d^2}{dz^2} - \rho^2\right)\Psi^* = 0 \tag{10.125}$$

where

$$\Phi^* = \frac{1}{2\pi}\int_0^{+\infty}\left[\int_{-\infty}^{+\infty} dx \int_{-\infty}^{+\infty} e^{-i(\alpha x + \beta y)}\Phi dy\right] e^{-st} dt \tag{10.126}$$

$$\Psi^* = \frac{1}{2\pi}\int_0^{+\infty}\left[\int_{-\infty}^{+\infty} dx \int_{-\infty}^{+\infty} e^{-i(\alpha x + \beta y)}\Psi dy\right] e^{-st} dt \tag{10.127}$$

$$\rho = \sqrt{\alpha^2 + \beta^2} \tag{10.128}$$

The superscript star in (10.126–10.127) represents taking both integral transforms. Choosing the solutions of (10.126–10.127) which vanish as z tends infinity, we have

$$\Phi^* = a_1 e^{-z\rho} + a_2 e^{-z\rho\sqrt{1+\xi}} \tag{10.129}$$

$$\Psi^* = b_1 e^{-z\rho} \tag{10.130}$$

$$s = c_0 \rho^2 \xi \tag{10.131}$$

where a_1, a_2, and b_1 are functions of α, β, and ξ which are determinable from the boundary conditions.

Since the surface is free of shear stress and permeable, applying the repeated complex Fourier and Laplace transforms to (10.123e–10.123g) gives

$$a_1 + \sqrt{1+\xi}\, a_2 = 0 \tag{10.132}$$

$$b_1 + a_2 \eta \rho \xi = 0 \tag{10.133}$$

The compressive stress σ_z is defined by (10.123c). The boundary conditions require that

$$\sigma_z \big|_{z=0,t>0} = \begin{cases} p_0 & |x| < a,\ |y| < b \\ 0 & \text{otherwise} \end{cases} \tag{10.134}$$

Taking the repeated complex Fourier and Laplace transforms to (10.123c) and (10.134) gives

$$(a_1 + a_2)\rho - b_1 = \frac{p_0 \sin(a\alpha)\sin(b\beta)}{\pi G c_0 \xi \rho^3 \alpha\beta} \tag{10.135}$$

Thus, the functions a_1, a_2, and b_1 are given by

$$a_1 = -\frac{p_0 \sin(a\alpha)\sin(b\beta)\sqrt{1+\xi}}{\pi G c_0 \xi \rho^4 \alpha\beta[1+\eta\xi - \sqrt{1+\xi}\,]} \tag{10.136a}$$

$$a_2 = \frac{p_0 \sin(a\alpha)\sin(b\beta)}{\pi G c_0 \xi \rho^4 \alpha\beta[1+\eta\xi - \sqrt{1+\xi}\,]} \tag{10.136b}$$

$$b_1 = -\frac{p_0 \eta \sin(a\alpha)\sin(b\beta)}{\pi G c_0 \rho^3 \alpha\beta[1+\eta\xi - \sqrt{1+\xi}\,]} \tag{10.136c}$$

Therefore,

$$\Phi^* = \frac{p_0 \sin(a\alpha)\sin(b\beta)}{\pi G c_0 \xi \rho^4 \alpha\beta[1+\eta\xi - \sqrt{1+\xi}\,]}\left[e^{-z\rho\sqrt{1+\xi}} - \sqrt{1+\xi}\, e^{-z\rho}\right] \tag{10.137}$$

$$\Psi^* = -\frac{p_0\eta\sin(a\alpha)\sin(b\beta)}{\pi G c_0 \rho^3\alpha\beta[1+\eta\xi-\sqrt{1+\xi}]}e^{-z\rho} \tag{10.138}$$

Then the displacement functions can be obtained by inverting the integral transforms as

$$\Phi = \frac{p_0}{2G\pi^2}\int_{-\infty}^{+\infty}d\alpha\int_{-\infty}^{+\infty}\left[e^{i(\alpha x+\beta y)}\frac{\sin(a\alpha)\sin(b\beta)}{\alpha\beta\rho^2}\frac{1}{2\pi i}\int_{B_\xi}\frac{\Lambda(\xi)(e^{-z\rho\sqrt{1+\xi}}-\sqrt{1+\xi}e^{-z\rho})}{\xi}d\xi\right]d\beta \tag{10.139}$$

$$\Psi = -\frac{p_0\eta}{2G\pi^2}\int_{-\infty}^{+\infty}d\alpha\int_{-\infty}^{+\infty}\left[e^{i(\alpha x+\beta y)}\frac{\sin(a\alpha)\sin(b\beta)}{\alpha\beta\rho}\frac{1}{2\pi i}\int_{B_\xi}\Lambda(\xi)e^{-z\rho}d\xi\right]d\beta \tag{10.140}$$

The function $\Lambda(\xi)$ is defined in (10.87).

Using (10.90), the displacement w_z of the medium on the surface $z = 0$ is

$$w_z(x,y,0,t) = \frac{p_0\eta}{2G\pi^2(2\eta-1)}\int_{-\infty}^{+\infty}d\alpha\int_{-\infty}^{+\infty}\left[e^{i(\alpha x+\beta y)}\frac{\sin(a\alpha)\sin(b\beta)}{\alpha\beta\rho}\varphi_1(\rho\sqrt{c_0 t})\right]d\beta \tag{10.141}$$

The settlement at the center of the rectangle can now be obtained by setting $x = y = 0$ and the corner settlement w_{zc} of a rectangle with sides a and b subjected to a normal stress p_0 is clearly one quarter of this value.

$$w_{zc}(t) = \frac{ap_0\eta}{8G\pi^2(2\eta-1)}\int_{-\infty}^{+\infty}d\alpha\int_{-\infty}^{+\infty}\left[\frac{\sin(\alpha)\sin(\lambda\beta)}{\alpha\beta\rho}\varphi_1\left(\rho\sqrt{\frac{c_0 t}{a^2}}\right)\right]d\beta \tag{10.142}$$

where $\lambda = b/a$.

Thus, we have the immediate surface settlement of the corner

$$w_{zc}(0^+) = \frac{ap_0}{8G\pi^2}\int_{-\infty}^{+\infty}\int_{-\infty}^{+\infty}\left[\frac{\sin(\alpha)\sin(\lambda\beta)}{\alpha\beta\rho}\right]d\alpha d\beta \tag{10.143}$$

and its final surface settlement

$$\begin{aligned} w_{zc}(\infty) &= \frac{ap_0\eta}{4G\pi^2(2\eta-1)}\int_{-\infty}^{+\infty}\int_{-\infty}^{+\infty}\left[\frac{\sin(\alpha)\sin(\lambda\beta)}{\alpha\beta\rho}\right]d\alpha d\beta \\ &= \frac{ap_0\eta}{2G\pi(2\eta-1)}\left[\sinh^{-1}\lambda+\lambda\sinh^{-1}\left(\frac{1}{\lambda}\right)\right] \end{aligned} \tag{10.144}$$

The degree of consolidation at the corner of a rectangle with sides a and b thus becomes

$$U = \frac{w_{zc}(t) - w_{zc}(0^+)}{w_{zc}(\infty) - w_{zc}(0^+)} \tag{10.145}$$

Using the relation

$$\begin{aligned} &\int_{-\infty}^{+\infty} d\alpha \int_{-\infty}^{+\infty} \left[\frac{\sin(\alpha)\sin(\lambda\beta)}{\alpha\beta\rho} \exp(-A\rho^2 T) erfc(B\rho\sqrt{T}) \right] d\beta \\ &= 2\pi\sqrt{\pi T} \int_{B}^{+\infty} erf\left[\frac{1}{2\sqrt{(A+\beta^2)T}} \right] erf\left[\frac{\lambda}{2\sqrt{(A+\beta^2)T}} \right] d\beta \end{aligned} \tag{10.146}$$

where A and B are any constants, (10.145) may now be expressed in the form:

$$U = \frac{\int_0^T \psi_1(\tau,\lambda)\,d\tau - (\eta - 1)\int_T^{\infty} \left(1 - \sqrt{\frac{\tau}{\tau - \mu_0 T}}\right) \psi_1(\tau,\lambda)\,d\tau}{\int_0^{\infty} \psi_1(\tau,\lambda)\,d\tau} \tag{10.147}$$

where

$$\begin{aligned} &\psi_1(\tau,\lambda) = \frac{1}{\sqrt{\tau}} erf\left(\frac{1}{2\sqrt{\tau}}\right) erf\left(\frac{\lambda}{2\sqrt{\tau}}\right); \\ &\mu_0 = \frac{2\eta - 1}{\eta^2}; \qquad T = \frac{c_0 t}{a^2} \end{aligned} \tag{10.148}$$

The degrees of consolidation for different λ and Poisson's ratios are shown in Figure 10.20.

The average settlement of the uniformly loaded rectangular area is

$$\begin{aligned} \bar{w}_z(t) &= \int_{-a}^{a}\int_{-b}^{b} \frac{w_z(x,y,0,t)}{4ab} dxdy \\ &= \frac{p_0 a\eta}{2G\pi^2(2\eta - 1)\lambda} \int_{-\infty}^{\infty}\int_{-\infty}^{\infty} \frac{\sin^2\alpha \sin^2(\lambda\beta)}{\alpha^2\beta^2\rho} \varphi_2\left(\rho\sqrt{\frac{c_0 t}{a^2}}\right) d\alpha d\beta \end{aligned} \tag{10.149}$$

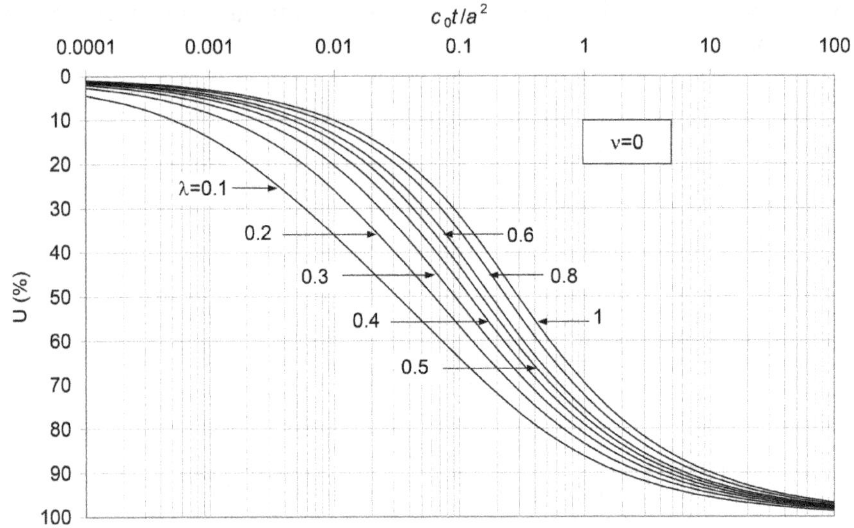

Figure 10.20(a) U for rectangular loading corner with permeable surface $\nu = 0$.

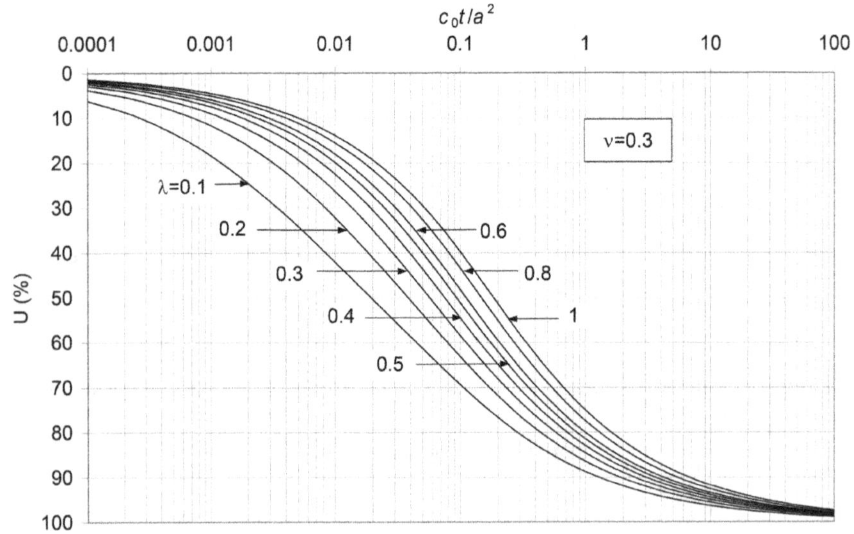

Figure 10.20(b) U for rectangular loading corner with permeable surface $\nu = 0.3$.

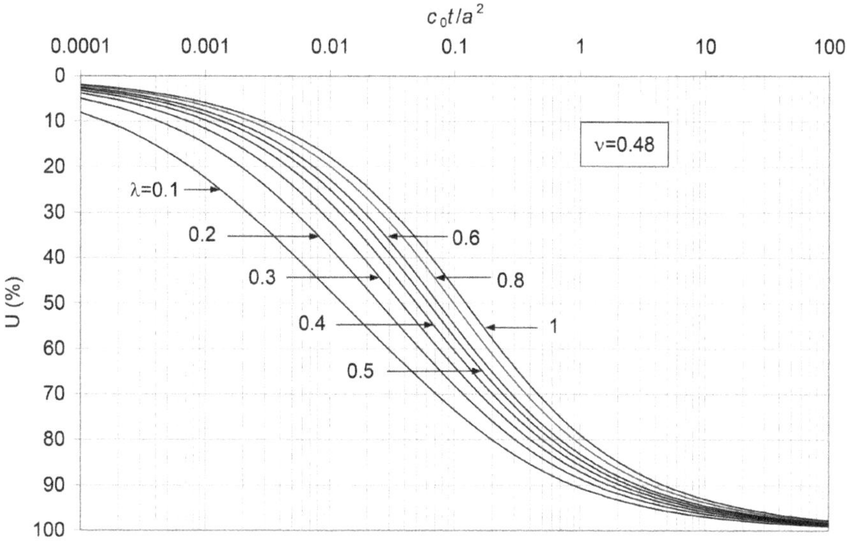

Figure 10.20(c) U for rectangular loading corner with permeable surface $\nu = 0.48$.

The double integration may be reduced to a single integration by using the following relation

$$\int_{-\infty}^{\infty}\int_{-\infty}^{\infty}\frac{\sin^2\alpha\sin^2(\lambda\beta)}{\alpha^2\beta^2\rho}\exp(-A\rho^2T)erfc(B\rho\sqrt{T})d\alpha d\beta = \lambda\sqrt{\pi}\int_{(A+B^2)T}^{\infty}\psi_2(\tau,\lambda)\psi_2(\tau,1)\sqrt{\frac{\tau}{\tau-AT}}d\tau \tag{10.150}$$

where

$$\psi_2(\tau,\lambda) = \pi^{\frac{1}{2}}\tau^{-\frac{1}{4}}erf\left(\lambda\tau^{-\frac{1}{2}}\right) - \tau^{\frac{1}{4}}\lambda^{-1}[1-\exp(-\lambda^2\tau^{-1})] \tag{10.151}$$

It follows that

$$\bar{w}_z(t) = \frac{p_0 a\eta}{2G\pi^{\frac{3}{2}}(2\eta-1)}\left[\int_0^{\infty}\psi_2(\tau,\lambda)\psi_2(\tau,1)d\tau + \int_0^{T}\psi_2(\tau,\lambda)\psi_2(\tau,1)d\tau + \frac{\eta-1}{\eta}\int_T^{\infty}\sqrt{\frac{\tau}{\tau-\mu_0 T}}\psi_2(\tau,\lambda)\psi_2(\tau,1)d\tau\right] \tag{10.152}$$

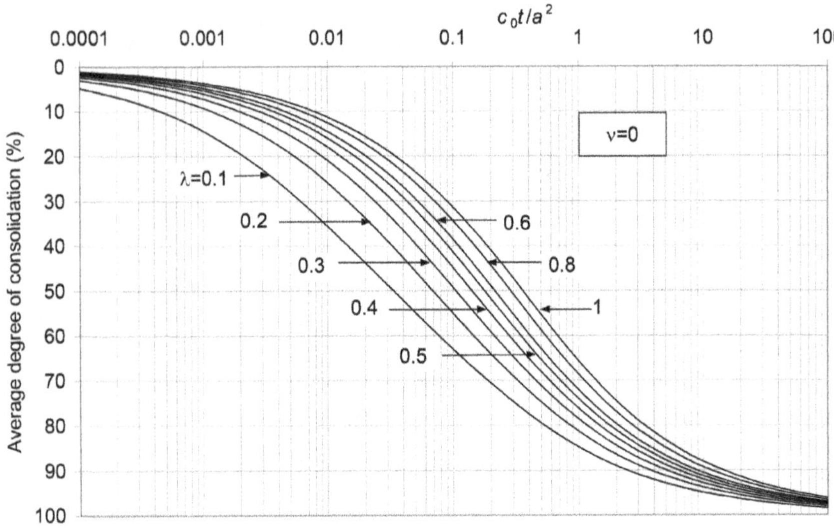

Figure 10.21(a) $\bar{U}$ for rectangular loading with permeable surface $\nu = 0.0$.

and, in particular, the final average settlement is

$$\bar{w}_z(\infty) = \frac{2\eta}{(2\eta-1)}\bar{w}_z(0^+) = \frac{2p_0 a\eta}{\pi G(2\eta-1)}\left[\sinh^{-1}\lambda + \lambda\sinh^{-1}\left(\frac{1}{\lambda}\right) + \frac{1+\lambda^3-(1+\lambda^2)^{\frac{3}{2}}}{3\lambda}\right] \tag{10.153}$$

The average degree of consolidation defined as in (10.119) is illustrated in Figure 10.21.

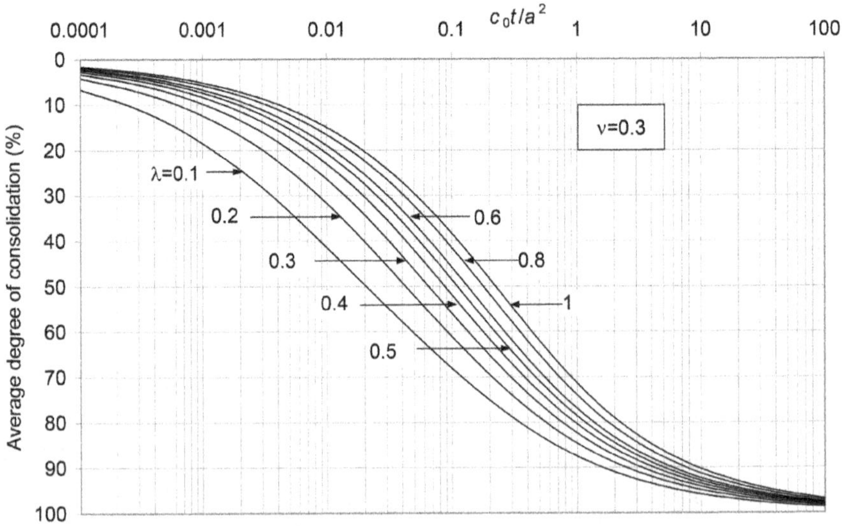

Figure 10.21(b) $\bar{U}$ for rectangular loading with permeable surface $\nu = 0.3$.

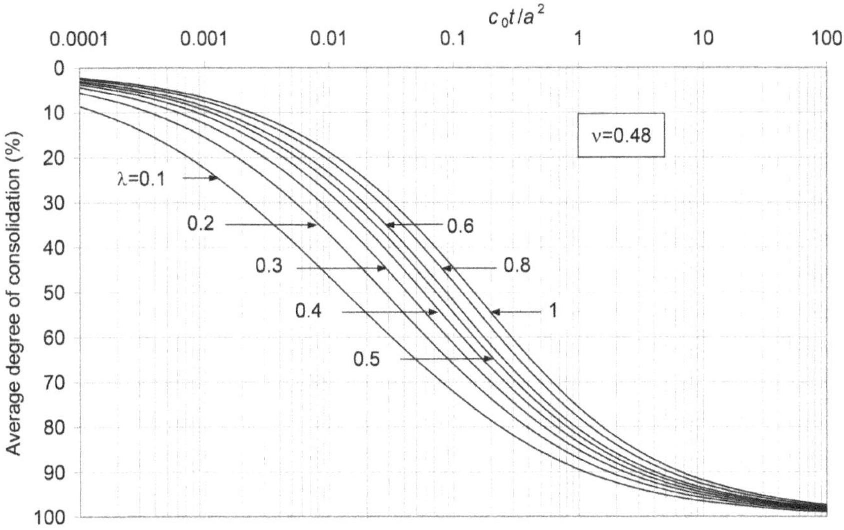

Figure 10.21(c) $\bar{U}$ for rectangular loading with permeable surface $\nu = 0.48$.

10.7 SUBSIDENCE DUE TO A POINT SINK IN A CROSS-ANISOTROPIC POROUS ELASTIC HALF SPACE

When porewater is withdrawn from saturated ground by pumping, there is a reduction of pore pressure in the neighborhood of the region of withdrawal. This leads to an increase in effective stress and consequently some surface subsidence. In some cases, the settlement could cause problems for structures founded on or near the surface. The well-known examples of this phenomenon occur in Bangkok, Venice, and Mexico City where widespread subsidence has been caused by withdrawal of water from aquifers for industrial and domestic purposes. Recorded settlements in Mexico City have been as large as 8 m and in Venice they have reached rates of 5 to 6 cm per year (Scott 1978). Problems with subsidence due to fluid extraction have also been reported in a number of other regions of the world.

Biot's 3-D consolidation theory is generally regarded as the fundamental theory for modeling land subsidence. Based on Biot's theory, Booker and Carter (1986a, b) presented closed-form solutions for the long-term settlement caused by withdrawal of fluid at a constant rate from a point sink at finite depth below the surface of a homogeneous, isotropic, porous, and elastic half space. Tarn and Lu (1991) extended the results to a cross-anisotropic porous elastic half space. These solutions are discussed in the following.

10.7.1 Problem definition and governing equations

As shown in Figure 10.22, a point sink of strength Q is placed at the point $(0, 0, h)$ below the surface of a semi-infinite space. The soil is a saturated

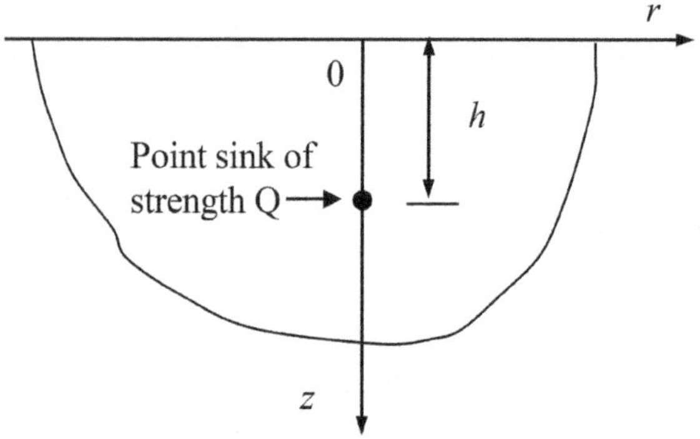

Figure 10.22 A point sink embedded in a cross-anisotropic half space (cylindrical co-ordinates).

cross-anisotropic elastic material. The co-ordinate directions are the principle directions. The stress-strain relation can be expressed using (9.55). The Darcy's law can be expressed using (9.58). The surface is permeable. Since the problem is cylindrically symmetric, the cylindrical co-ordinates are adopted.

Suppose that water is being pumped out of the saturated soil. It will be assumed that there has been sufficient rainfall or inflow of groundwater and that no lowering of the water table has occurred. After a long period of time the soil will reach a steady state. The following equations are obtained from the governing equations in Chapter 9.

$$D_{11}\left(\frac{\partial^2 w_r}{\partial r^2}+\frac{1}{r}\frac{\partial w_r}{\partial r}-\frac{w_r}{r^2}\right)+G_v\frac{\partial^2 w_r}{\partial z^2}+(D_{13}+G_v)\frac{\partial^2 w_z}{\partial r\,\partial z}-\frac{\partial u_e}{\partial r}=0 \quad (10.154)$$

$$(D_{13}+G_v)\left(\frac{\partial^2 w_r}{\partial r\,\partial z}+\frac{1}{r}\frac{\partial w_r}{\partial z}\right)+G_v\left(\frac{\partial^2 w_z}{\partial r^2}+\frac{1}{r}\frac{\partial w_z}{\partial r}\right)$$
$$+D_{33}\frac{\partial^2 w_z}{\partial z^2}-\frac{\partial u_e}{\partial z}=0 \quad (10.155)$$

$$-\frac{k_h}{\gamma_w}\left(\frac{\partial^2 u_e}{\partial r^2}+\frac{1}{r}\frac{\partial u_e}{\partial r}\right)-\frac{k_v}{\gamma_w}\frac{\partial^2 u_e}{\partial z^2}+\frac{Q}{\pi r}\delta(r)\delta(z-h)=0 \quad (10.156)$$

where $\delta(r)$ is the Dirac delta function.

Since the half space surface is a traction-free pervious boundary, the boundary conditions to (10.154–10.156) are

$$G_v\left(\frac{\partial w_r}{\partial z}+\frac{\partial w_z}{\partial r}\right)\bigg|_{z=0}=0 \quad (10.157a)$$

$$\left[D_{13}\left(\frac{\partial w_r}{\partial r}+\frac{w_r}{r}\right)+D_{33}\frac{\partial w_z}{\partial z}\right]\bigg|_{z=0}=0 \tag{10.157b}$$

$$u_e\,|_{z=0}=0 \tag{10.157c}$$

10.7.2 Solutions

The governing (10.154–10.156) can be reduced to ordinary differential equations by performing Hankel transforms of first, zeroth, and zeroth order with respect to the radial co-ordinate r, respectively. In doing so, we obtain

$$-\rho^2 D_{11}\tilde{w}_r+G_v\frac{d^2\tilde{w}_r}{dz^2}-\rho(D_{13}+G_v)\frac{d\tilde{w}_z}{dz}+\rho\tilde{u}_e=0 \tag{10.158a}$$

$$\rho(D_{13}+G_v)\frac{d\tilde{w}_r}{dz}-\rho^2 G_v\tilde{w}_z+D_{33}\frac{d^2\tilde{w}_z}{dz^2}-\frac{d\tilde{u}_e}{dz}=0 \tag{10.158b}$$

$$-\rho^2\frac{k_h}{\gamma_w}\tilde{u}_e+\frac{k_v}{\gamma_w}\frac{d^2\tilde{u}_e}{dz^2}=\frac{Q}{2\pi}\delta(z-h) \tag{10.158c}$$

where

$$\tilde{w}_r(\rho,z)=\int_0^{+\infty} rJ_1(r\rho)w_r dr \tag{10.159a}$$

$$\tilde{w}_z(\rho,z)=\int_0^{+\infty} rJ_0(r\rho)w_z dr \tag{10.159b}$$

$$\tilde{u}_e(\rho,z)=\int_0^{+\infty} rJ_0(r\rho)u_e dr \tag{10.159c}$$

Equation (10.158) can be written in matrix form as

$$\frac{d\mathbf{V}}{dz}=\mathbf{BV}+\frac{Q\gamma_w\delta(z-h)}{2\pi k_v}\mathbf{R} \tag{10.160a}$$

where

$$\mathbf{V}^T=\left[\begin{matrix}\tilde{w}_r & \dfrac{d\tilde{w}_r}{dz} & \tilde{w}_z & \dfrac{d\tilde{w}_z}{dz} & \tilde{u}_e & \dfrac{d\tilde{u}_e}{dz}\end{matrix}\right] \tag{10.160b}$$

$$\mathbf{R}^T = [\,0 \;\; 0 \;\; 0 \;\; 0 \;\; 0 \;\; 1\,] \tag{10.160c}$$

$$\mathbf{B} = \begin{bmatrix} 0 & 1 & 0 & 0 & 0 & 0 \\ \dfrac{\rho^2 D_{11}}{G_v} & 0 & 0 & \dfrac{\rho(D_{13}+G_v)}{G_v} & -\dfrac{\rho}{G_v} & 0 \\ 0 & 0 & 0 & 1 & 0 & 0 \\ 0 & -\dfrac{\rho(D_{13}+G_v)}{D_{33}} & \dfrac{\rho^2 G_v}{D_{33}} & 0 & 0 & \dfrac{1}{D_{33}} \\ 0 & 0 & 0 & 0 & 0 & 1 \\ 0 & 0 & 0 & 0 & \rho^2\mu_3^2 & 0 \end{bmatrix} \tag{10.160d}$$

$$\mu_3 = \sqrt{\frac{k_h}{k_v}} \tag{10.160e}$$

The eigenvalues $\zeta_j\rho$ of $\mathbf{B}$ are

$$\mu_1\rho,\ \mu_2\rho,\ -\mu_1\rho,\ -\mu_2\rho,\ \mu_3\rho,\ -\mu_3\rho$$

where μ_1 and μ_2 are the roots of the characteristic equation:

$$D_{33}G_v\mu^4 - [D_{11}D_{33} - D_{13}(D_{13}+2G_v)]\mu^2 + D_{11}G_v = 0 \tag{10.161}$$

The roots of (10.161) cannot be purely imaginary. In fact, consider a uniform stress field and assume $u_e = 0$ as follows:

$$\sigma_r = \beta^2 \quad \sigma_z = 1 \quad \sigma_\theta = \nu_h\beta^2 + \nu_{hv} \quad \tau_{rz} = \beta \quad \tau_{r\theta} = \tau_{\theta z} = 0 \tag{10.162}$$

where β is an arbitrary real number. The corresponding elastic potential requires that

$$D_{33}G_v\beta^4 + [D_{11}D_{33} - D_{13}(D_{13}+2G_v)]\beta^2 + D_{11}G_v > 0 \tag{10.163}$$

If one root of (10.161) is purely imaginary, then substituting $\mu = i\beta$ into (10.161) gives

$$D_{33}G_v\beta^4 + [D_{11}D_{33} - D_{13}(D_{13}+2G_v)]\beta^2 + D_{11}G_v = 0 \tag{10.164}$$

which contradicts (10.163). Thus, μ_1 and μ_2 can be chosen such that the real part is positive. For isotropic soil mass, the roots μ_1 and μ_2 are equal to one.

The eigenvectors Y_j of $\mathbf{B}$ corresponding the eigenvalues $\zeta_j\rho$ are jth column of $\mathbf{Y}$:

$$\mathbf{Y} = \begin{bmatrix} 1 & 1 & 1 & 1 & 1 & 1 \\ \mu_1\rho & \mu_2\rho & -\mu_1\rho & -\mu_2\rho & \mu_3\rho & -\mu_3\rho \\ -S_1 & -S_2 & S_1 & S_2 & -S_3 & S_3 \\ -S_1\mu_1\rho & -S_2\mu_2\rho & -S_1\mu_1\rho & -S_2\mu_2\rho & -S_3\mu_3\rho & -S_3\mu_3\rho \\ 0 & 0 & 0 & 0 & S_4\rho & S_4\rho \\ 0 & 0 & 0 & 0 & S_4\mu_3\rho^2 & -S_4\mu_3\rho^2 \end{bmatrix} \tag{10.165a}$$

where

$$S_1 = -\frac{G_v\mu_1^2 - D_{11}}{(G_v + D_{13})\mu_1}, \quad S_2 = -\frac{G_v\mu_2^2 - D_{11}}{(G_v + D_{13})\mu_2} \tag{10.165b}$$

$$S_3 = -\frac{G_v\mu_3^3 + (D_{13} + G_v - D_{11})\mu_3}{(G_v + D_{13} - D_{33})\mu_3^2 + G_v} \tag{10.165c}$$

$$S_4 = \frac{D_{33}G_v\mu_3^4 - [D_{11}D_{33} - D_{13}(D_{13} + 2G_v)]\mu_3^2 + D_{11}G_v}{(G_v + D_{13} - D_{33})\mu_3^2 + G_v} \tag{10.165d}$$

Thus the general solutions of (10.160) are

$$\mathbf{V} = \begin{cases} \sum_{j=1}^{6} c_j^1 \mathbf{Y}_j \exp(\varsigma_j z\rho) & z < h \\ \sum_{j=1}^{6} c_j^2 \mathbf{Y}_j \exp(\varsigma_j z\rho) & z > h \end{cases} \tag{10.166}$$

Integrating (10.160a) gives

$$\int_{h^-}^{h^+} \frac{d\mathbf{V}}{dz} dz = \mathbf{V}\,|_{z=h^+} - \mathbf{V}\,|_{z=h^-} = \frac{Q\gamma_w}{2\pi k_v}\mathbf{R} \tag{10.167}$$

Taking Hankel transform of first order for τ_{rz}, zeroth order for σ_z and for u_e, (10.157) becomes

$$\left[\frac{d\tilde{w}_r}{dz} - \rho\tilde{w}_r\right]\Big|_{z=0} = 0 \tag{10.168a}$$

$$\left[D_{13}\rho\tilde{w}_r + D_{33}\frac{d\tilde{w}_z}{dz}\right]\Big|_{z=0} = 0 \tag{10.168b}$$

$$\tilde{u}_e\,|_{z=0} = 0 \tag{10.168c}$$

The constants in (10.166) can be determined from (10.167–10.168) and the conditions as z → ∞, where the effect of the point sink must vanish. Finally, the desired quantities w_r, w_z, and u_e can be obtained by applying appropriate inverse Hankel transforms as follows:

$$
\begin{aligned}
w_r = \frac{Q\gamma_w}{4\pi k_v}\Bigg\{ & -\frac{a_1 r}{\sqrt{\mu_1^2(z-h)^2+r^2}+\mu_1|z-h|}+\frac{a_2 r}{\sqrt{\mu_2^2(z-h)^2+r^2}+\mu_2|z-h|} \\
& -\frac{r}{\mu_3 S_4[\sqrt{\mu_3^2(z-h)^2+r^2}+\mu_3|z-h|]}+\frac{r}{\mu_3 S_4[\sqrt{\mu_3^2(z+h)^2+r^2}+\mu_3(z+h)]} \\
& \frac{\mu_2+\mu_1}{\mu_2-\mu_1}\left[\frac{a_1 r}{\sqrt{\mu_1^2(z+h)^2+r^2}+\mu_1(z+h)}+\frac{a_2 r}{\sqrt{\mu_2^2(z+h)^2+r^2}+\mu_2(z+h)}\right] \\
& -\frac{2}{\mu_2-\mu_1}\left[\frac{\mu_2 a_2 b_2 r}{b_1[\sqrt{(\mu_1 z+\mu_2 h)^2+r^2}+\mu_1 z+\mu_2 h]}\right. \\
& \left.+\frac{\mu_1 a_1 b_1 r}{b_2[\sqrt{(\mu_2 z+\mu_1 h)^2+r^2}+\mu_2 z+\mu_1 h]}\right] \\
& \frac{2(\mu_3+S_3)}{(\mu_2-\mu_1)S_4\mu_3}\left[\frac{r}{b_1[\sqrt{(\mu_1 z+\mu_3 h)^2+r^2}+\mu_1 z+\mu_3 h]}\right. \\
& \left.-\frac{r}{b_2[\sqrt{(\mu_2 z+\mu_3 h)^2+r^2}+\mu_2 z+\mu_3 h]}\right]\Bigg\}
\end{aligned}
\tag{10.169a}
$$

$$
\begin{aligned}
w_z = \frac{Q\gamma_w}{4\pi k_v}\Bigg\{ & S_1 a_1 \sinh^{-1}\left[\frac{\mu_1(z-h)}{r}\right]-S_2 a_2 \sinh^{-1}\left[\frac{\mu_2(z-h)}{r}\right] \\
& +\frac{S_3}{\mu_3 S_4}\sinh^{-1}\left[\frac{\mu_3(z-h)}{r}\right]-\frac{S_3}{\mu_3 S_4}\sinh^{-1}\left[\frac{\mu_3(z+h)}{r}\right] \\
& -\frac{\mu_2+\mu_1}{\mu_2-\mu_1}\left[S_1 a_1 \sinh^{-1}\frac{\mu_1(z+h)}{r}+S_2 a_2 \sinh^{-1}\frac{\mu_2(z+h)}{r}\right] \\
& +\frac{2}{\mu_2-\mu_1}\left[\frac{S_1\mu_2 a_2 b_2}{b_1}\sinh^{-1}\left(\frac{\mu_1 z+\mu_2 h}{r}\right)+\frac{S_2\mu_1 a_1 b_1}{b_2}\sinh^{-1}\left(\frac{\mu_2 z+\mu_1 h}{r}\right)\right] \\
& -\frac{2(\mu_3+S_3)}{(\mu_2-\mu_1)S_4\mu_3}\left[\frac{S_1}{b_1}\sinh^{-1}\left(\frac{\mu_1 z+\mu_3 h}{r}\right)-\frac{S_2}{b_2}\sinh^{-1}\left(\frac{\mu_2 z+\mu_3 h}{r}\right)\right]\Bigg\}
\end{aligned}
\tag{10.169b}
$$

$$u_e = -\frac{Q\gamma_w}{4\pi k_v \mu_3}\left[\frac{1}{\sqrt{\mu_3^2(z-h)^2+r^2}} - \frac{1}{\sqrt{\mu_3^2(z+h)^2+r^2}}\right] \tag{10.169c}$$

where

$$a_j = \frac{(D_{13}+G_v-D_{33})\mu_j^2+G_v}{(\mu_2^2-\mu_1^2)(\mu_3^2-\mu_j^2)\mu_j D_{33}G_v}, \quad (j=1,\ 2) \tag{10.169d}$$

$$b_1 = \frac{D_{13}+D_{33}\mu_2^2}{D_{13}+G_v}, \quad b_2 = \frac{D_{13}+D_{33}\mu_1^2}{D_{13}+G_v} \tag{10.169e}$$

Solutions of an isotropic soil mass with cross-anisotropic permeability can be obtained from (10.169) by taking an appropriate limit $\mu_1 = \mu_2 = 1$ and using l'Hôpital's rule. Carrying out the procedure gives

$$\begin{aligned} w_r = \frac{Q\gamma_w}{8\pi G\eta k_v}\frac{1}{\mu_3^2-1}\Bigg\{ & -\frac{r}{\sqrt{(z-h)^2+r^2}+|z-h|} + \frac{r}{\mu_3\sqrt{\mu_3^2(z-h)^2+r^2}+\mu_3^2|z-h|} \\ & -\frac{2\eta+1}{2\eta-1}\frac{r}{\sqrt{(z+h)^2+r^2}+z+h} - \frac{r}{\mu_3\sqrt{\mu_3^2(z+h)^2+r^2}+\mu_3^2(z+h)} \\ & +\frac{2rz}{[\sqrt{(z+h)^2+r^2}+z+h]\sqrt{(z+h)^2+r^2}} + \frac{4\eta}{2\eta-1}\frac{r}{\sqrt{(z+\mu_3 h)^2+r^2}+z+\mu_3 h} \\ & -\frac{2rz}{[\sqrt{(z+\mu_3 h)^2+r^2}+z+\mu_3 h]\sqrt{(z+\mu_3 h)^2+r^2}}\Bigg\} \end{aligned} \tag{10.170a}$$

$$\begin{aligned} w_z = \frac{Q\gamma_w}{8\pi G\eta k_v}\frac{1}{\mu_3^2-1}\Bigg\{ & \sinh^{-1}\frac{(z-h)}{r} - \sinh^{-1}\frac{\mu_3(z-h)}{r} - \frac{2\eta+1}{2\eta-1}\sinh^{-1}\frac{(z+h)}{r} \\ & +\frac{2}{2\eta-1}\sinh^{-1}\frac{(z+\mu_3 h)}{r} + \sinh^{-1}\frac{\mu_3(z+h)}{r} + \frac{2z}{\sqrt{(z+h)^2+r^2}} \\ & -\frac{2z}{\sqrt{(z+\mu_3 h)^2+r^2}}\Bigg\} \end{aligned} \tag{10.170b}$$

$$u_e = -\frac{Q\gamma_w}{4\pi k_v \mu_3}\left[\frac{1}{\sqrt{\mu_3^2(z-h)^2+r^2}} - \frac{1}{\sqrt{\mu_3^2(z+h)^2+r^2}}\right] \tag{10.170c}$$

If the hydraulic conductivity is also isotropic, the (10.170) can be reduced further by letting $k_h = k_v = k$. Using l'Hôpital's rule, we obtain

$$w_r = \frac{Q\gamma_w}{16\pi G\eta k}\left\{-\frac{r}{\sqrt{(z-h)^2+r^2}}+\frac{r}{\sqrt{(z+h)^2+r^2}}+\frac{2hrz}{[\sqrt{(z+h)^2+r^2}\,]^3}\right.$$
$$\left.-\frac{4\eta}{2\eta-1}\frac{rh}{[\sqrt{(z+h)^2+r^2}+z+h]\sqrt{(z+h)^2+r^2}}\right\} \tag{10.171a}$$

$$w_z = \frac{Q\gamma_w}{16\pi G\eta k}\left\{-\frac{z-h}{\sqrt{(z-h)^2+r^2}}+\frac{z}{\sqrt{(z+h)^2+r^2}}+\frac{2\eta+1}{2\eta-1}\frac{h}{\sqrt{(z+h)^2+r^2}}\right.$$
$$\left.+\frac{2hz(z+h)}{[\sqrt{(z+h)^2+r^2}\,]^3}\right\} \tag{10.171b}$$

$$u_e = -\frac{Q\gamma_w}{4\pi k}\left[\frac{1}{\sqrt{(z-h)^2+r^2}}-\frac{1}{\sqrt{(z+h)^2+r^2}}\right] \tag{10.171c}$$

10.7.3 Results

The solutions in (10.169–10.171) have been evaluated for a number of cases. The results are discussed in the following.

Figure 10.23 shows the distribution of excess pore pressure along the z-axis. The value of u_e decreases with depth from zero at the surface and is unbound at the location of the sink. Beneath the sink, the excess pore pressure asymptotically approaches zero again. As can be seen from (10.169c), the magnitude of excess pore pressure is proportional to the strength of the sink and inversely proportional to the permeability of the soil for a designated μ_3.

The distributions of displacement for a soil skeleton with isotropic permeability and elasticity are plotted in Figures 10.24–10.26. Figure 10.24 shows the horizontal displacement w_r on the surface. The maximum displacement takes place between one to two times the sink depth. However, at a distance seven times the sink depth, the value of horizontal displacement is still as large as half the maximum.

Figure 10.25 indicates the surface settlement profile. Of course, the greatest settlement occurs directly above the sink but the settlement bowl is quite extensive, e.g. at $r = 7h$ the settlement is still about 14% of its maximum value.

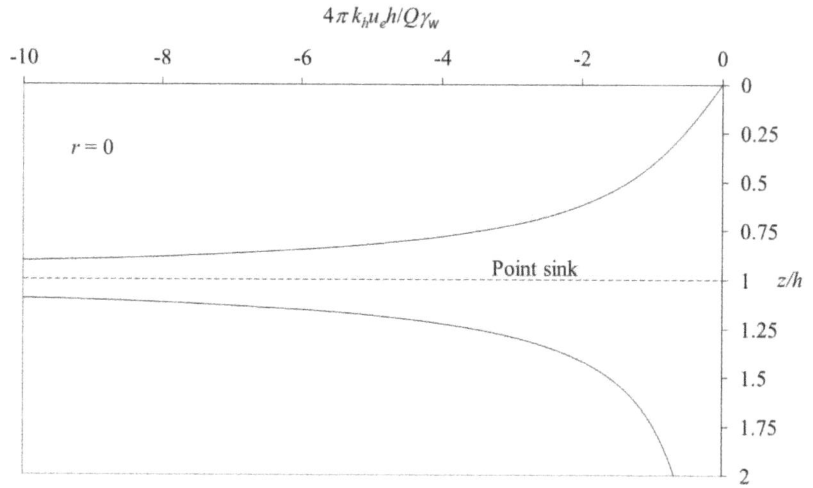

Figure 10.23 Distribution of excess pore pressure along the z-axis.

Figure 10.26 illustrates the vertical displacement w_z along the z-axis. It is clear that the changes in vertical movement are small both above and below the sink. The surface settlement at $r = 0$ is given by the expression

$$\frac{Q\gamma_w}{4\pi Gk(2\eta - 1)}$$

It is noted that the magnitude of this settlement is independent of the depth of the embedment. There is a discontinuity in displacement at the location of the sink but at no other point. This discontinuity is due to considering the

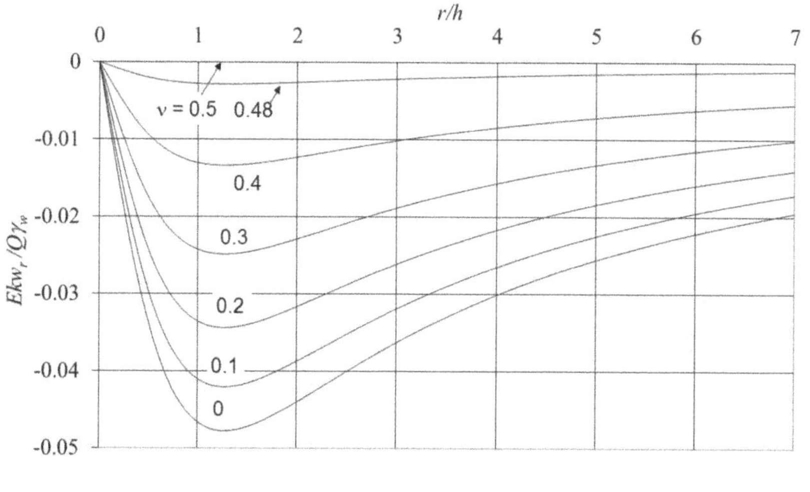

Figure 10.24 Horizontal displacement on the surface z = 0 for an isotropic soil.

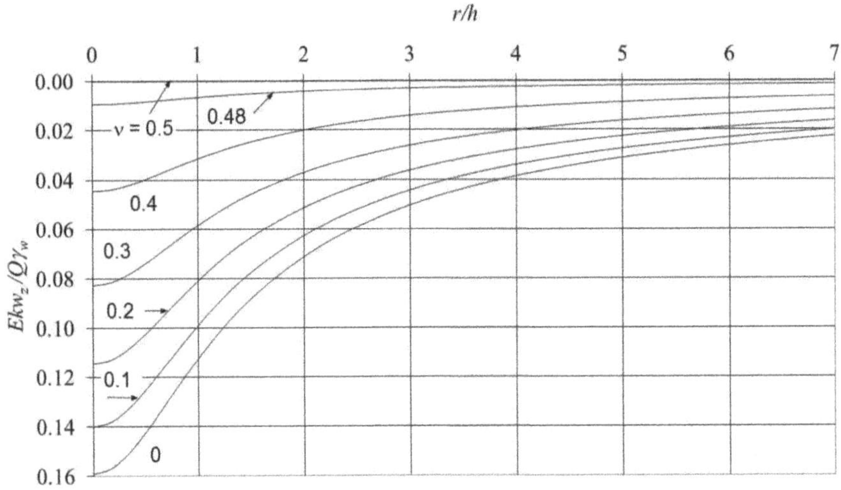

Figure 10.25 Vertical displacement on the surface z = 0 for an isotropic soil.

physically artificial but mathematical convenient case of a point sink. If the sink were distributed over a finite region, no such discontinuity would exist.

To examine the effect of anisotropy on subsidence, the elastic constants in Table 10.1 for five different cases are used. Case 0 in Table 10.1 is isotropic. Figure 10.27 indicates the influence of anisotropy on the surface settlement. Clearly, the anisotropy of soil has a significant effect on land subsidence. For example, the maximum vertical displacement of case 1 is reduced to about 45% of the corresponding value for the isotropic case.

Figure 10.28 shows the effect of G_v/E_v. As can be seen in this figure, the effect of G_v/E_v on land subsidence is secondary. Figure 10.29 illustrates that

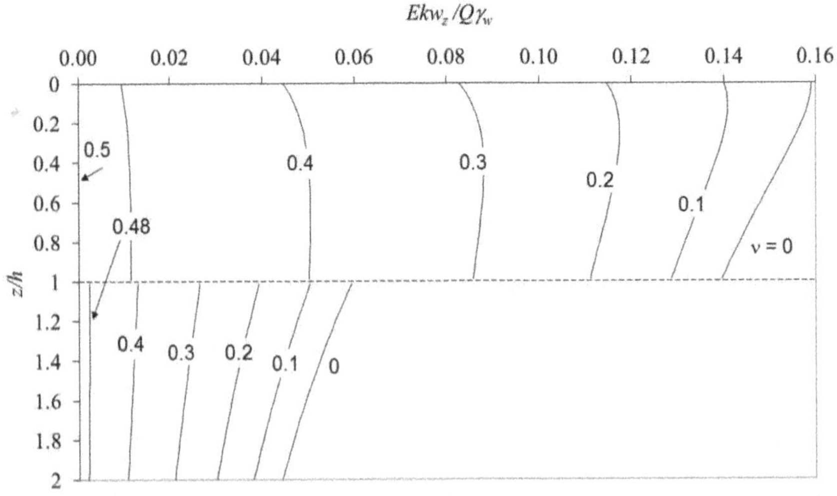

Figure 10.26 Distribution of settlement along the z-axis for isotropic soil.

Table 10.1 Material properties of cross-anisotropic soils (after Tarn and Lu 1991)

Case	ν_h	ν_{hv}	G_v/E_v	E_h/E_v	k_h/k_v
0	0.25	0.25	0.4	1	1
1	0.125	0.75	0.445	2	1
2	0.125	0.75	0.64	3	1
3	0.125	0.75	0.64	4	1
4	0.0	0.38	0.38	1.84	1

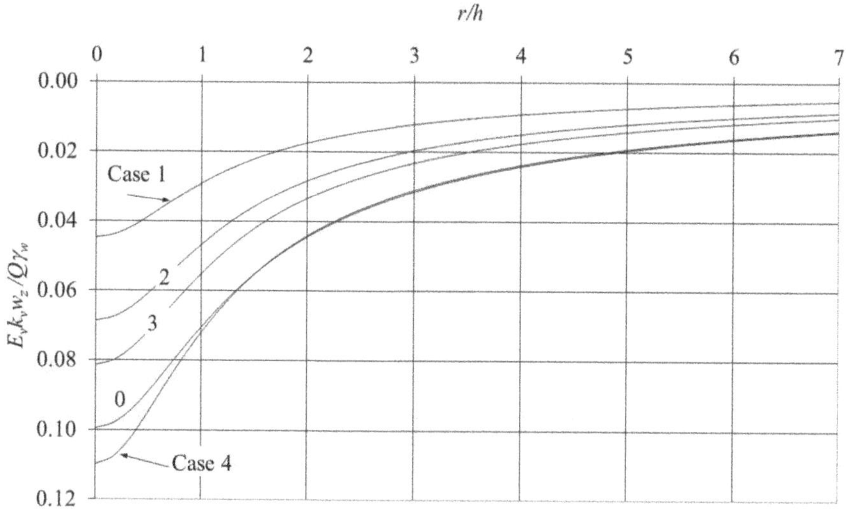

Figure 10.27 Influence of anisotropy on the settlement.

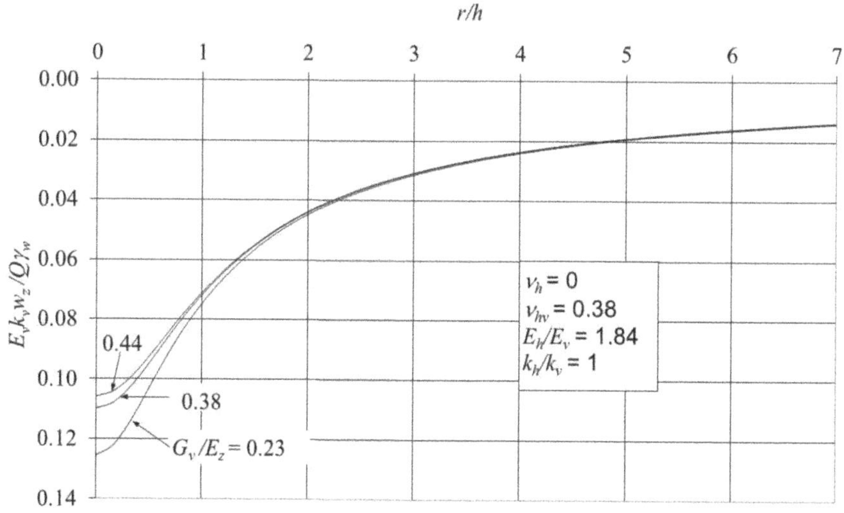

Figure 10.28 Influence of the degree of anisotropy G_v/E_v on settlement.

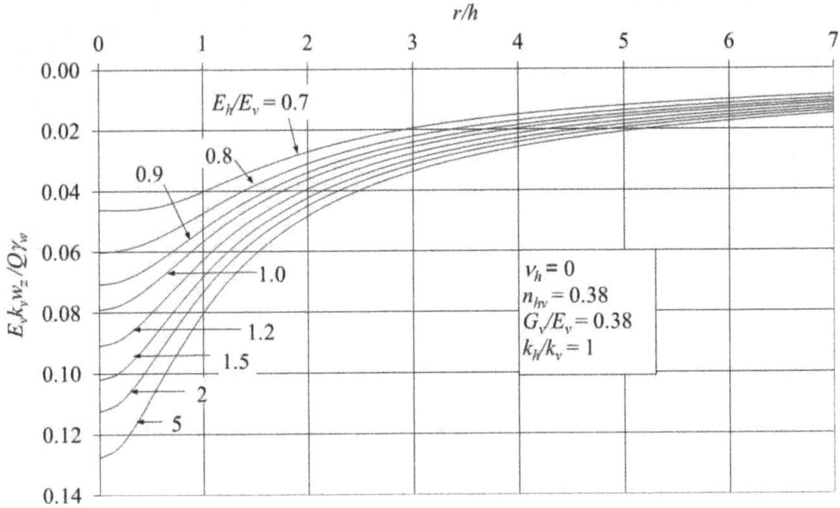

Figure 10.29 Influence of the degree of anisotropy E_h/E_v on settlement.

the ratio of E_h/E_v has a marked influence on surface settlement. Figure 10.30 demonstrates that the degree of anisotropic flow properties, k_h/k_v has the most significant effect on land subsidence. The maximum surface settlement of the anisotropic case $k_h/k_v = 10$ is only approximately a quarter of the isotropic case $k_h/k_v = 1$. The larger magnitude of subsidence in the isotropic case is due to a wider region influenced by the pumping.

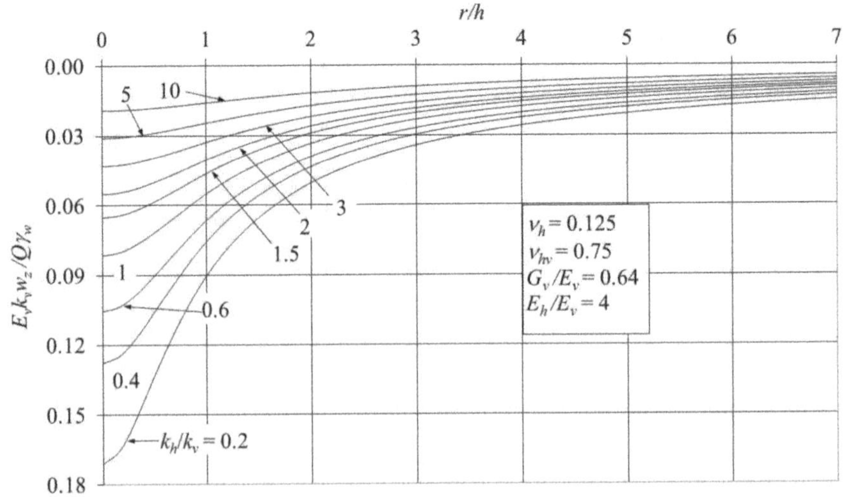

Figure 10.30 Influence of the degree of anisotropy k_h/k_v on settlement.

10.8 CONSOLIDATION OF A FINITE LAYER SUBJECT TO NORMAL STRIP LOADING

In Sections 10.5–10.7, the consolidation behaviors of a semi-infinite layer are examined. These solutions have applications to problems in engineering practice where the thickness of the clay stratum is great compared to the dimensions of the loaded area. When this is not so, the influence of the thickness should be taken into account (Gibson *et al.* 1970; Booker 1974). This section presents the results of a finite layer subject to normal strip loading.

As shown in Figure 10.31, a uniform normal strip stress p_0 is suddenly applied on the surface of a clay layer ($0 < z < H$) of infinite lateral extent resting on a rigid medium. The width of the strip load is $2b$. The soil in the stratum is a linear isotropic material. At the interface $z = H$, the soil completely adheres to the rigid base. The excess pore pressure vanishes at $z = 0$. At $z = H$, we consider two boundary conditions on the excess pore pressure here, namely, (a) that the excess pore pressure vanishes at $z = H$, and (b) that the normal gradient of the excess pore pressure vanishes at $z = H$. Since closed-form solutions are difficult to obtain, this consolidation problem is solved using finite element method.

For a suddenly applied strip stress p_0, the immediate settlement S_u and the final settlement S_f at the loading center can be expressed as

$$S_u = \frac{2p_0 b}{E} I_{us} \tag{10.172}$$

$$S_f = \frac{2p_0 b}{E} I_{fs} \tag{10.173}$$

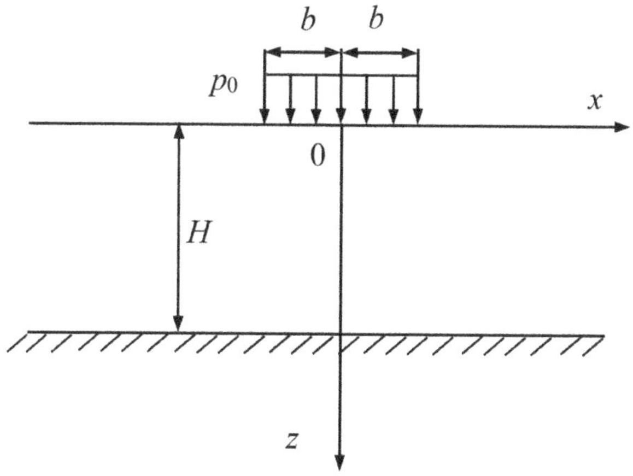

Figure 10.31 A finite layer underlain by a rough rigid base subjected to strip loading.

Table 10.2 I_{us} and I_{fs} for different *H*/*b* and Poisson's ratios

	I_{us}			I_{fs}		
H/*b*	$\nu = 0$	$\nu = 0.3$	$\nu = 0.48$	$\nu = 0$	$\nu = 0.3$	$\nu = 0.48$
0.5	0.0204	0.0265	0.0302	0.2535	0.1889	0.0536
1	0.0973	0.1265	0.1439	0.5049	0.3914	0.1803
1.5	0.1819	0.2365	0.2692	0.7095	0.5676	0.3136
2	0.2553	0.3319	0.3778	0.8714	0.7106	0.4278
3	0.3702	0.4813	0.5479	1.1132	0.9273	0.6058
4	0.4564	0.5934	0.6756	1.2904	1.0873	0.7389
5	0.5244	0.6817	0.7761	1.43	1.2133	0.844
7	0.6259	0.8138	0.9265	1.6406	1.4053	1.0016
10	0.7276	0.9459	1.0768	1.8642	1.6081	1.1614

where I_{us} and I_{fs} are influence factors for the vertical displacements at the loading center ($x = z = 0$). I_{us} and I_{fs} are listed in Table 10.2 for different *H*/*b* and Poisson's ratios.

The average degrees of consolidation *U* for the loading center are presented in Figures 10.32–10.33 for a permeable base and an impermeable base as follows, respectively.

10.8.1 Permeable rigid base

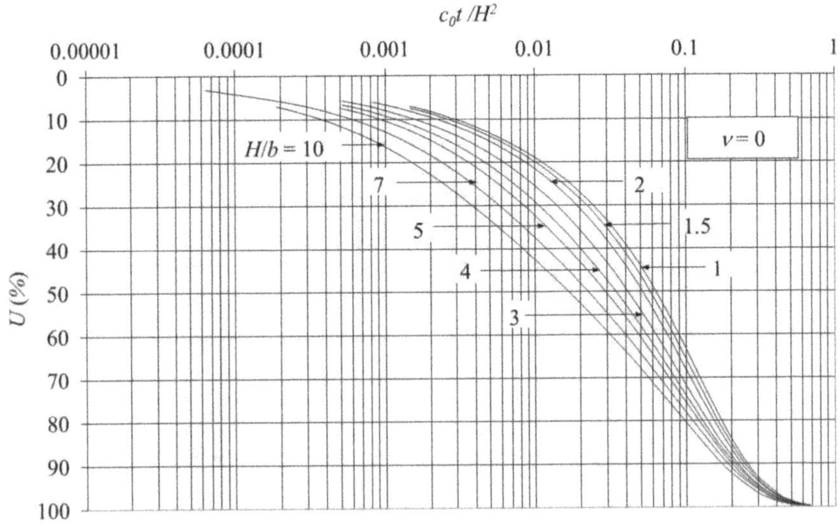

Figure 10.32(a) *U* for strip loading with permeable rigid base $\nu = 0.0$.

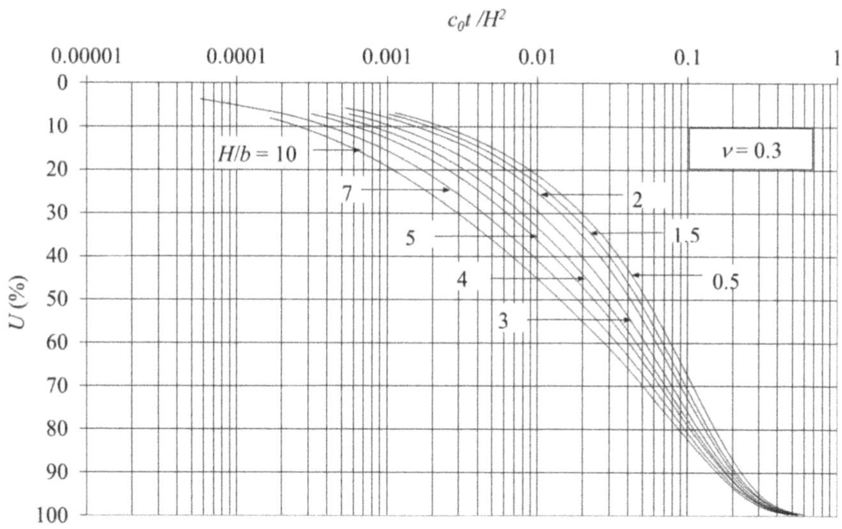

Figure 10.32(b) U for strip loading with permeable rigid base $\nu = 0.3$.

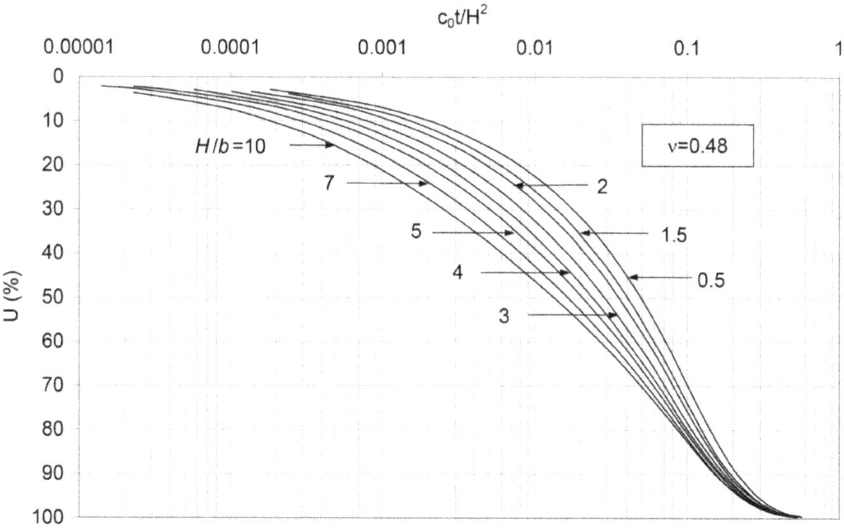

Figure 10.32(c) U for strip loading with permeable rigid base $\nu = 0.48$.

10.8.2 Impermeable rigid base

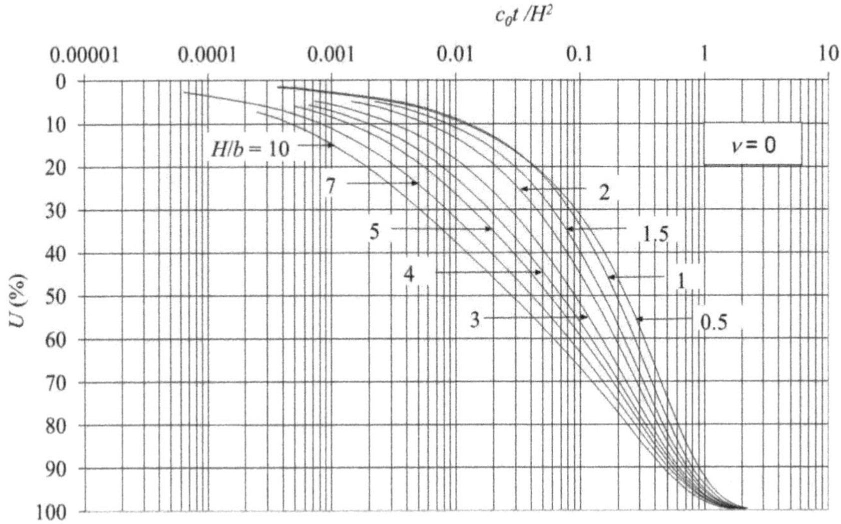

Figure 10.33(a) U for strip loading with impermeable rigid base $\nu = 0.0$.

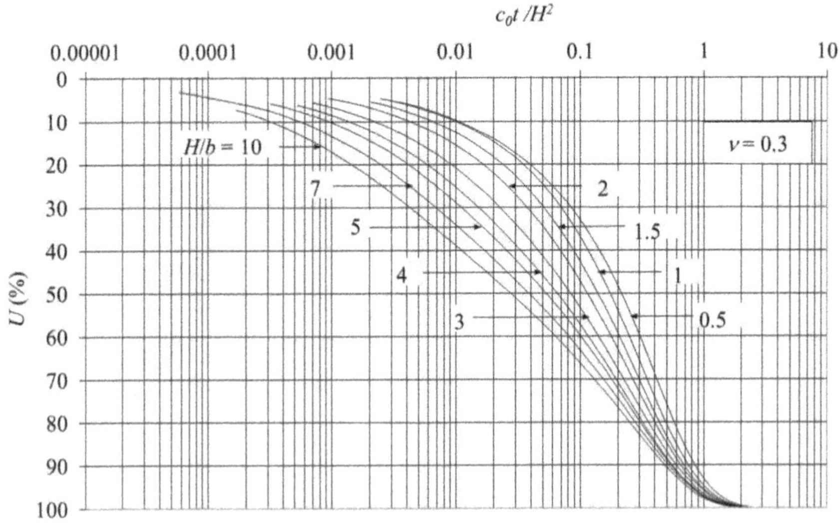

Figure 10.33(b) U for strip loading with impermeable rigid base $\nu = 0.3$.

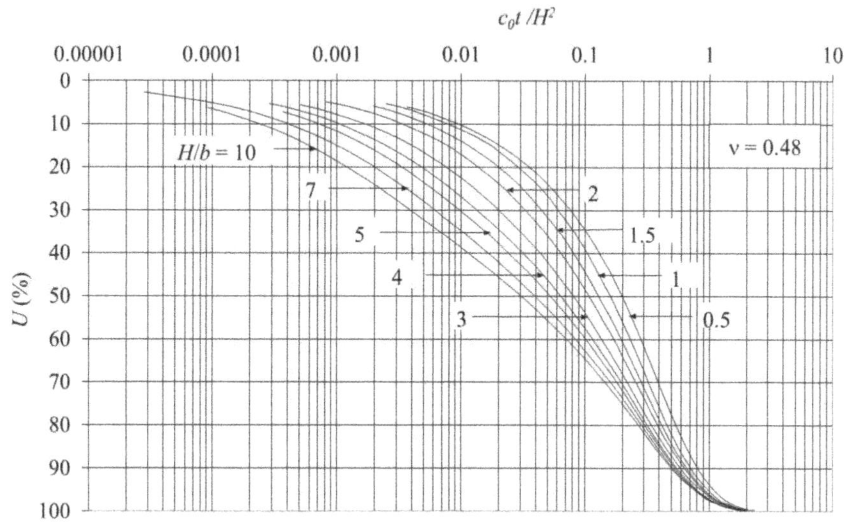

Figure 10.33(c) U for strip loading with impermeable rigid base $\nu = 0.48$.

10.9 CONSOLIDATION OF A FINITE LAYER SUBJECT TO NORMAL LOADING ON A CIRCULAR AREA

As shown in Figure 10.34, a uniform normal stress p_0 is suddenly applied on a circular area of the surface of a clay layer $(0 < z < H)$ of infinite lateral extent resting on a rigid medium. The radius of the circle is r_0.

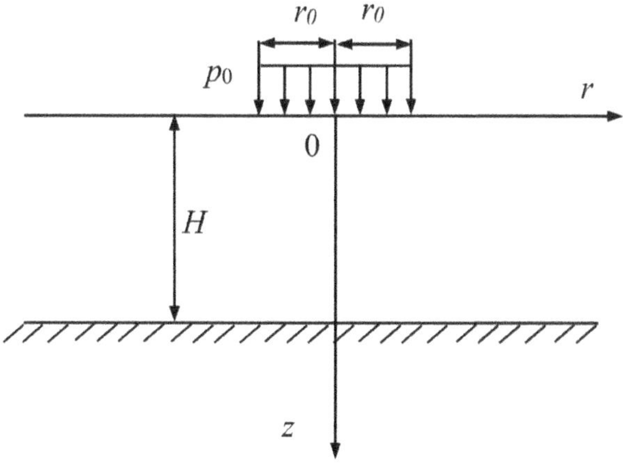

Figure 10.34 A finite layer underlain by a rough rigid base subjected to normal loading on a circular area (cylindrical co-ordinates).

Table 10.3 I_{uc} and I_{fc} for different H/r_0 and Poisson's ratios

	I_{uc}			I_{fc}		
H/r_0	$\nu = 0$	$\nu = 0.3$	$\nu = 0.48$	$\nu = 0$	$\nu = 0.3$	$\nu = 0.48$
0.5	0.0388	0.0505	0.0575	0.2583	0.1938	0.077
1	0.1525	0.1984	0.2258	0.4946	0.4009	0.2517
1.5	0.2408	0.3129	0.3562	0.6354	0.5372	0.384
2	0.2973	0.3865	0.4399	0.7193	0.6209	0.4685
3	0.3615	0.4699	0.5348	0.8108	0.714	0.5639
4	0.3957	0.5144	0.5855	0.8589	0.7631	0.6148
5	0.4166	0.5416	0.6166	0.8883	0.7931	0.646
7	0.4406	0.5726	0.652	0.9222	0.828	0.6816
10	0.4577	0.5951	0.6774	0.9477	0.8541	0.7075

The soil in the stratum is a linear isotropic material. At the interface $z = H$, the soil completely adheres to the rigid base. The excess pore pressure vanishes at $z = 0$. At $z = H$, we consider two boundary conditions on the excess pore pressure, that is, (a) that the excess pore pressure vanishes at $z = H$, and (b) that the normal gradient of the excess pore pressure vanishes at $z = H$. This consolidation problem is solved using finite element method.

For a suddenly applied normal stress p_0, the immediate settlement S_u and the final settlement S_f at the circle center can be expressed as

$$S_u = \frac{2p_0 r_0}{E} I_{uc} \tag{10.174}$$

$$S_f = \frac{2p_0 r_0}{E} I_{fc} \tag{10.175}$$

where I_{uc} and I_{fc} are influence factors for the vertical displacements at the circle center ($r = z = 0$). I_{uc} and I_{fc} are listed in Table 10.3 for different H/r_0 and Poisson's ratios.

The average degrees of consolidation U for the loading center are presented in Figures 10.35–10.36 for a permeable base and an impermeable base as follows, respectively.

10.9.1 Permeable rigid base

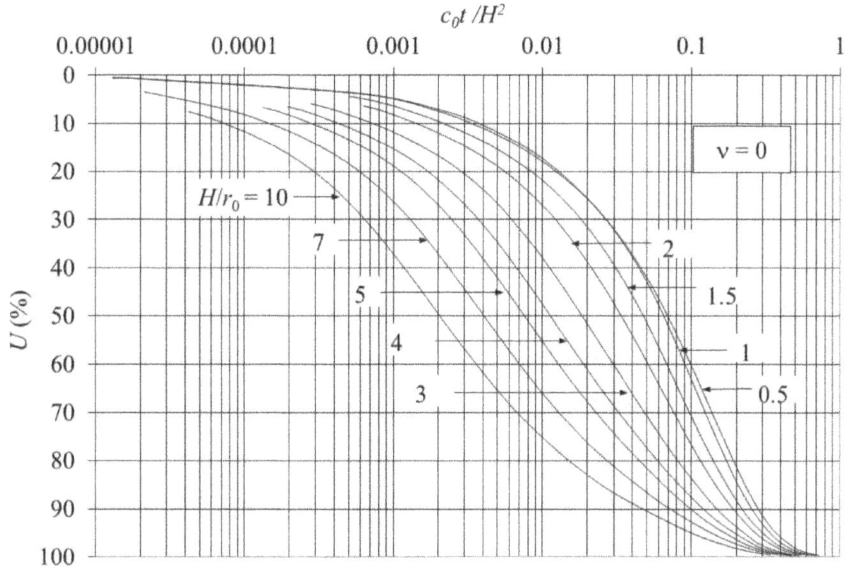

Figure 10.35(a) U for circular loading center with permeable rigid base $\nu = 0$.

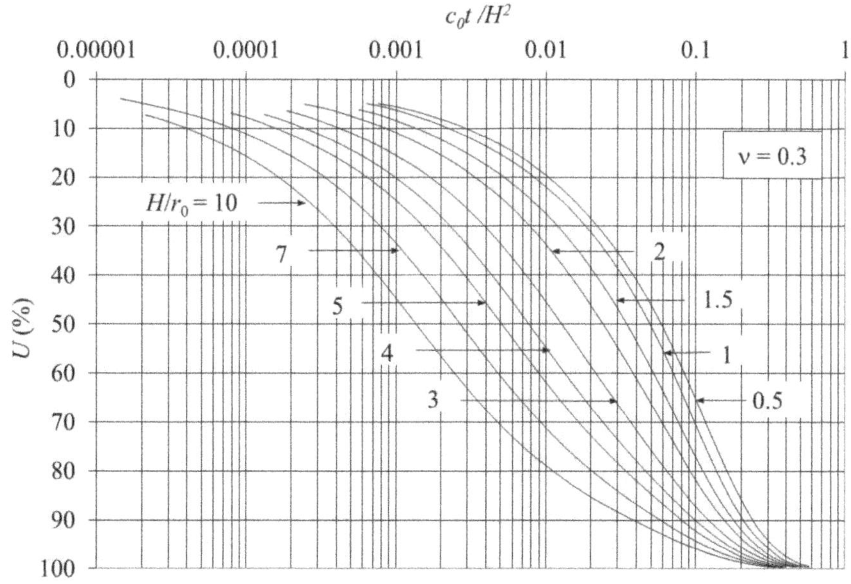

Figure 10.35(b) U for circular loading center with permeable rigid base $\nu = 0.3$.

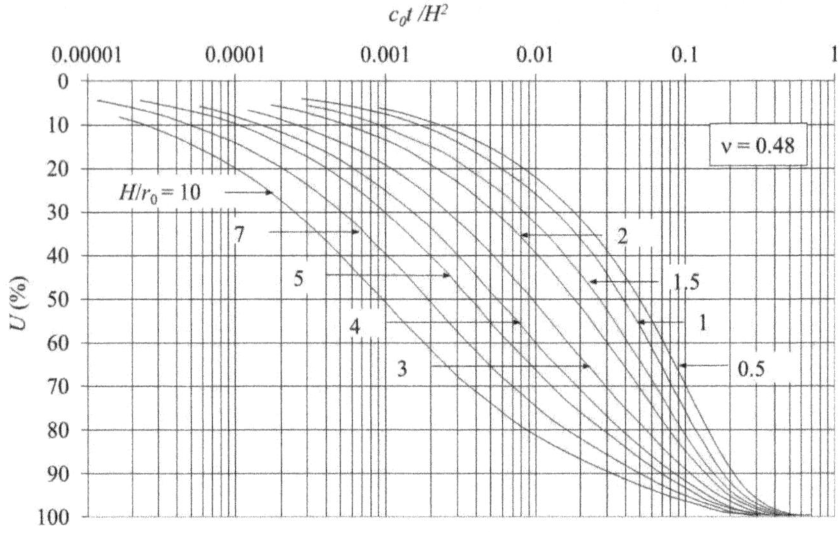

Figure 10.35(c) U for circular loading center with permeable rigid base $\nu = 0.48$.

10.9.2 Impermeable rigid base

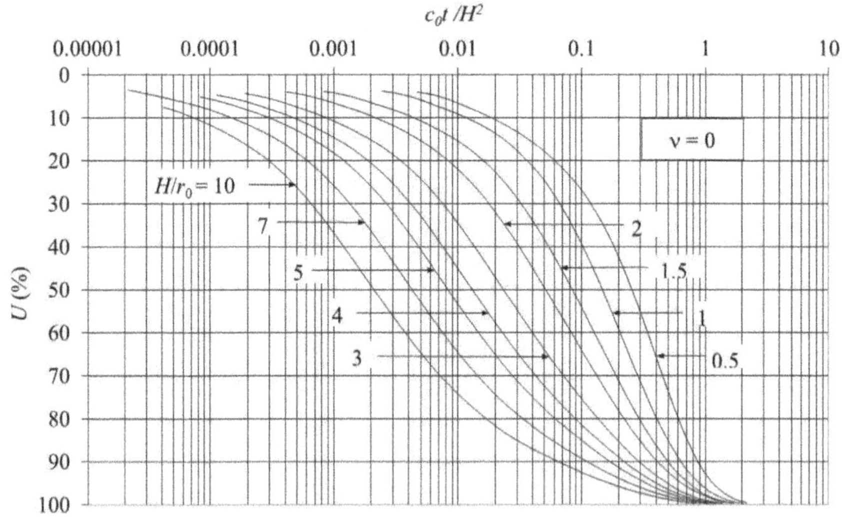

Figure 10.36(a) U for circular loading center with impermeable rigid base $\nu = 0$.

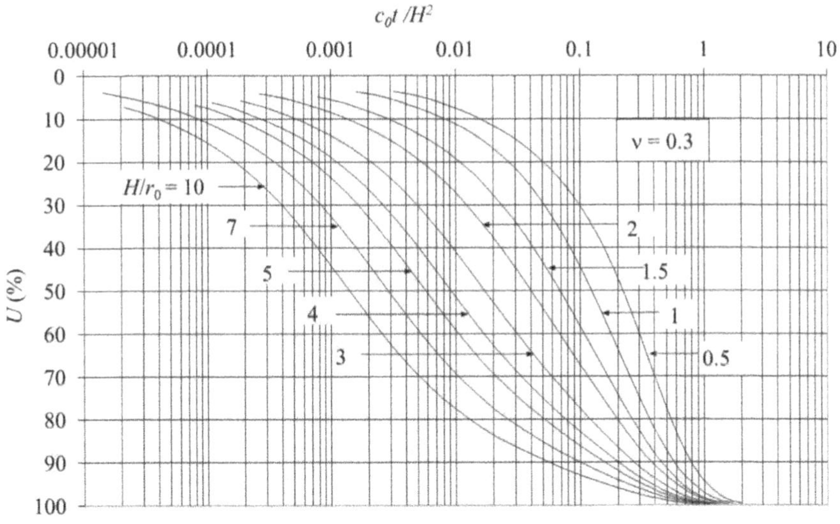

Figure 10.36(b) U for circular loading center with impermeable rigid base $\nu = 0.3$.

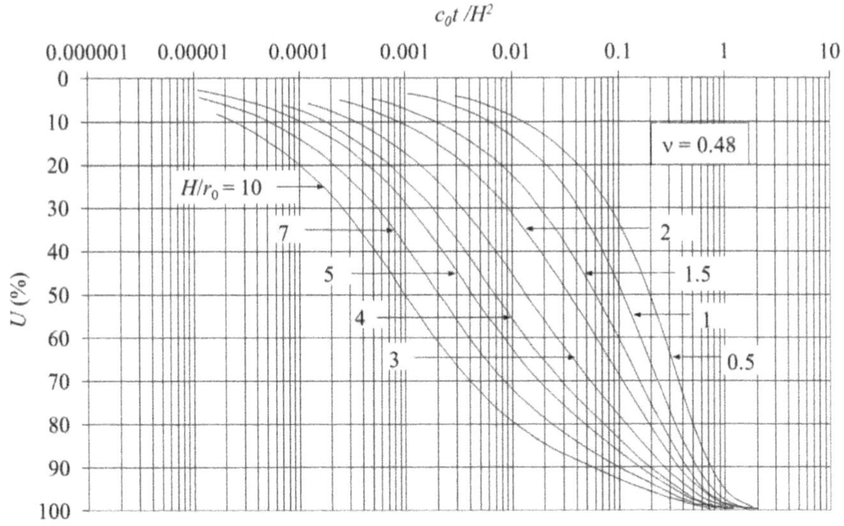

Figure 10.36(c) U for circular loading center with impermeable rigid base $\nu = 0.48$.

REFERENCES

Abousleiman, Y, Cheng, AH–D, Cui, L, Detournay, E, and Roegiers, J–C, 1996, Mandel's problem revisited. *Géotechnique*, 46(2), pp. 187–195.

Biot, MA, 1941, General theory of three dimensional consolidation. *Journal of Applied Physics*, 12(2), pp. 155–164.

Booker, JR, 1974, The consolidation of a finite layer subject to surface loading. *International Journal of Solids and Structures*, 10(9), pp. 1053–1065.

Booker, JR, and Carter, JP, 1986a, Analysis of a point sink embedded in a porous elastic half space. *International Journal for Numerical and Analytical Methods in Geomechanics*, 10(2), pp. 137–150.

Booker, JR, and Carter, JP, 1986b, Long term subsidence due to fluid extraction from a saturated, anisotropic, elastic soil mass. *The Quarterly Journal of Mechanics and Applied Mathematics*, 39(1), pp. 85–98.

Cryer, CWA, 1963, A comparison of the three-dimensional consolidation theories of Biot and Terzaghi. *The Quarterly Journal of Mechanics and Applied Mathematics*, 16(4), pp. 401–412.

Gibson, RE, Knight, K, and Taylor, PW, 1963, A critical experiment to examine theories of three dimensional consolidation. *Proceedings of European Conference of Soil Mechanics and Foundation Engineering*, 1, pp. 69–76.

Gibson, RE, and McNamee, J, 1963, A three-dimensional problem of the consolidation of a semi-infinite clay stratum. *The Quarterly Journal of Mechanics and Applied Mathematics*, 16(1), pp. 115–127.

Gibson, RE, Schiffman, RL, and Pu, SL, 1970, Plane strain and axially symmetric consolidation of a clay layer on a smooth impervious base. *The Quarterly Journal of Mechanics and Applied Mathematics*, 23(4), pp. 505–520.

Gradshteyn, IS, and Ryzhik, IM, 1994, *Table of Integrals, Series, and Products*, Jeffrey, A, (ed.), 5th ed., (San Diego: Academic Press).

Mandel, J, 1953, Consolidation des sols (étude mathématique). *Géotechnique*, 3(7), pp. 287–299.

McNamee, JOHN, and Gibson, RE, 1960a, Displacement functions and linear transforms applied to diffusion through porous elastic media. *The Quarterly Journal of Mechanics and Applied Mathematics*, 13(1), pp. 98–111.

McNamee, JOHN, and Gibson, RE, 1960b, Plane strain and axially symmetric problems of the consolidation of a semi-infinite clay stratum. *The Quarterly Journal of Mechanics and Applied Mathematics*, 13(2), pp. 210–227.

Rodrigues, JD, 1983, The Noordbergum effect and characterization of aquitards at the Rio Maior Mining Project. *Ground Water*, 21, pp. 200–207.

Scott, RF, 1978, Subsidence – A review. *Evaluation and Prediction of Subsidence*, Saxena, SK (ed.), pp. 1–25, (ASCE, New York:).

Tarn, JQ, and Lu, CC, 1991, Analysis of subsidence due to a point sink in an anisotropic porous elastic half space. *International Journal for Numerical and Analytical Methods in Geomechanics*, 15(8), pp. 573–592.

Chapter 11

Simplified finite element consolidation analysis of soils with vertical drain

11.1 INTRODUCTION

In many soft soil engineering projects where subsoil consists of fine-grained soils with low permeability, vertical drains are installed to facilitate the consolidation of soil. The philosophy behind the use of vertical drain is to shorten the drainage paths by installing drains vertically into the soil in some regular pattern. In this way, the consolidation process can be sped up considerably. Both drains of natural sand and drains made artificially are utilized.

Rendulic (1935) derived the differential equation for one-dimensional (1-D) radial dissipation of excess porewater pressure. Carrilllo (1942) demonstrated that the two-dimensional (2-D) flow problem could be uncoupled and the solutions to either the vertical or radial consolidation problems could be combined to give the solution to the entire 2-D problem. Barron (1948) assumed two types of vertical strains which might occur in the clay layer: (a) 'free vertical strain' resulting from a uniform distribution of surface load and (b) 'equal vertical strain' resulting from imposing the same vertical deformation on the surface for uniform soil. Horne (1964) presented a formal solution to the layered consolidation problem with vertical drain. Yoshikuni and Nakanodo (1974) gave a rigorous solution taking well resistance into consideration. Olson (1977) obtained an approximate solution for the case of vertical drain under ramp loading using the equal strain assumption. Zhu and Yin (2001) presented a mathematical solution for the consolidation analysis of soil with vertical and horizontal drainage subject to ramp load using free strain assumption. Simplified solutions were also obtained by other researchers (Hansbo 1981; Zeng and Xie 1989; Xie, *et al.* 1994). These linear theories, however, cannot conveniently be extended to account for layered systems, time-dependent loading, well resistance, variable coefficients of consolidation, and inelastic stress-strain behavior. Some semi-analytical methods also have a lot of limitations in dealing with the above mentioned problems (Booker and Small 1987; Selvadurai and Yue

1994; Yue and Selvadurai 1995, 1996). To overcome these difficulties, some researchers (Hart *et al.* 1958; Richart 1959; Olson *et al.* 1974; Atkinson and Eldred 1981; Onoue 1988; Lo 1991) resorted to finite difference numerical method. Finite element (FE) method has also been used to investigate the consolidation behavior of soils with vertical drains. Runesson *et al.* (1985) studied the efficiency of partially penetrating vertical drains based on linear free strain assumption. Bergado *et al.* (1993) analyzed the effects of smear zone for Bangkok clay using a linear model. General 2-D and three-dimensional (3-D) procedures (Lewis *et al.* 1976; Lewis and Schrefler 1998; Siriwardane and Desai 1981; Selvadurai 1996; Cheng *et al.* 1998) are well developed and various methods (Hird *et al.* 1995; Indraratna and Redana 1997) of matching the effect of drains under axisymmetric and plane strain conditions are available. Some special problems such as the consolidation behavior sensitive to mesh discretization and resulting damage of the poroelastic medium were also investigated (Mahyari and Selvadurai 1998). However, application of 2-D and 3-D FE procedures for the design of vertical drain may not be practical. This is because of (a) the difficulty in the determination of various model parameters for 2-D and 3-D conditions, (b) the large amount of computation, and (c) often numerical instability and convergence problems for non-linear cases. In practice, vertical drains are installed in a large area and the preloading area is extensive. If the soil layers are horizontal and the dimension of preloading area are much larger than the thickness of soil layers (or the length of the vertical strains), the average strain or vertical deformation of the soil with vertical drains occurs almost in vertical direction. Therefore, the stress-strain behavior of the soil on average may be simplified to be 1-D. In this chapter, a simplified FE procedure is derived for the 2-D consolidation analysis of soils with vertical drain using a general 1-D soil model. The use of 1-D soil model has the advantage of easy soil model parameter determination using conventional odometer tests only. It will be shown that this simplified FE procedure is very efficient and numerically very stable. The simplified procedure is verified by comparing its results with results from a fully coupled FE analysis using a 3-D soil model.

11.2 BASIC EQUATIONS

The partial differential equations governing the consolidation of soil with vertical drains are first presented. As in Chapter 6, this problem is simplified to an axisymmetric one, as shown in Figure 11.1. It is assumed that (a) the soil is fully saturated, (b) water and soil particles are incompressible, (c) Darcy's law is valid, (d) strains are small, and (e) all compressive strains within the soil mass occur in the vertical direction. A FE model based on the governing equations is developed in the following section for simulating the consolidation of layered soils.

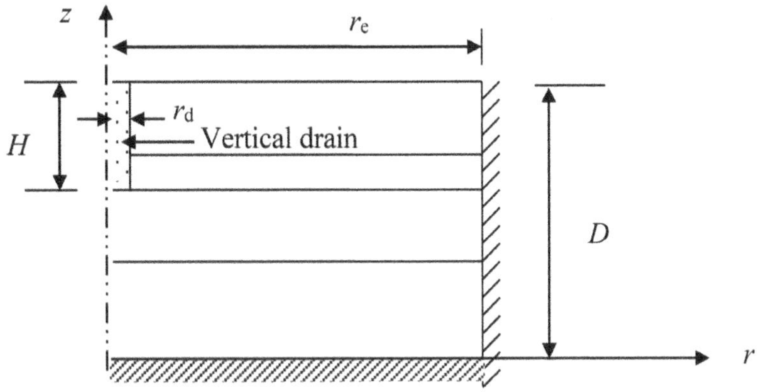

Figure 11.1 Geometry of the simplified model for consolidation of soils with vertical drain.

11.2.1 Darcy's law

The Darcy's law for axisymmetric problems is expressed in (6.9) and rewritten as

$$\begin{pmatrix} q_r \\ q_z \end{pmatrix} = -\mathbf{K} \begin{pmatrix} \dfrac{\partial u_e}{\partial r} \\ \dfrac{\partial u_e}{\partial z} \end{pmatrix} \tag{11.1a}$$

where

$$\mathbf{K} = \begin{bmatrix} \dfrac{k_r}{\gamma_w} & 0 \\ 0 & \dfrac{k_z}{\gamma_w} \end{bmatrix} \tag{11.1b}$$

The coefficients of permeability in (11.1) generally decrease with time during consolidation (Tavenas *et al.* 1983). To account for this, the following empirical relations are adopted in this formulation:

$$\begin{cases} k_r = k_{r0} 10^{-c_1(1+e_0)(\varepsilon_z - \varepsilon_{z0})} \\ k_z = k_{z0} 10^{-c_2(1+e_0)(\varepsilon_z - \varepsilon_{z0})} \end{cases} \tag{11.2}$$

where k_{r0} is initial radial coefficient of permeability; k_{z0} is initial vertical coefficient of permeability; ε_{z0} is initial vertical strain; and c_1 and c_2 are constants.

11.2.2 The continuity equation

The continuity equation for axisymmetric consolidation problems is (6.12) and rewritten as

$$\nabla \cdot q = \frac{\partial q_r}{\partial r} + \frac{q_r}{r} + \frac{\partial q_z}{\partial z} = \frac{\partial \varepsilon_v}{\partial t} = \frac{\partial \varepsilon_z}{\partial t} \tag{11.3}$$

where $q = (q_r, q_z)^T$.

11.2.3 The constitutive equations

In Chapter 2 (Zhu and Yin 1999), a general form of 1-D constitutive models was expressed in (2.4) and rewritten as

$$\frac{\partial \varepsilon_z}{\partial t} = \frac{\partial f(\sigma_z')}{\partial t} + g(\sigma_z', \varepsilon_z) \tag{11.4}$$

Many 1-D models (Desai *et al.* 1979; Garlanger 1972; Hardin 1989; Leroueil *et al.* 1985; Yin and Graham 1989, 1994) can be expressed in this form, and $f(\sigma_z')$ and $g(\sigma_z', \varepsilon_z)$ may be different for loading and unloading for some models. For Terzaghi's model (1943)

$$\begin{cases} f = m_v \sigma_z' \\ g = 0 \end{cases} \tag{5.1}$$

The non-linear elastic model used in part of this chapter is

$$\begin{cases} f = -\ln\left\{1 - \dfrac{\kappa}{1+e_0} \ln\left[1 + \dfrac{(1+\nu)(\sigma_z' - \sigma_{z0}')}{3(1-\nu)(p_0 + p_t^{el})}\right]\right\} \\ g = 0 \end{cases} \tag{11.5}$$

where κ is the logarithmic bulk modulus; e_0 is the initial void ratio; ν is Poisson's ratio; p_t^{el} is the elastic tensile limit; σ_{z0}' is initial vertical effective stress; and p_0 is the initial mean effective principal stress. This model is the 1-D equivalent of a porous elastic model for the case of constant Poisson's ratio in ABAQUS (1995) for large deformation problems (see Appendix 11.A).

11.2.4 Vertical total stress

For simplicity, the vertical total stress is calculated by assuming that the shearing stresses on every cylindrical surface are zero.

Equations (11.1), (11.3), and (11.4) are the governing equations for the consolidation problem of soil with vertical drains. These equations can also be obtained from general 3-D consolidation equations.

11.3 FINITE ELEMENT FORMULATION

As in Chapter 5, the numerical solution of the above consolidation model for a given boundary value problem is obtained here by using a Galerkin type FE weak formulation for the space variable z and r, coupled with a finite difference time scheme. Since the porewater flow is 3-D axisymmetric, it is necessary to discretize the problem in a radial plane. Let the radial plane (soil mass) denoted as domain Ω be divided into a number of rectangle elements, with nodal points at the boundaries of each element. For an element Ω_e, the nodal points are denoted as m_1, m_2, m_3, and m_4, and the area of the element is S_e, as shown in Figure 11.2.

Multiplying (11.3) by a virtual function $\phi(r, z)$, and integrating it with a weight function r, we have

$$\pi = \int_{\Omega} r\phi\left(\nabla \cdot \mathrm{q} - \frac{\partial \varepsilon_z}{\partial t}\right)drdz = \sum_{\Omega e}\int_{\Omega e}\left[\left(\frac{\partial \phi}{\partial r}, \frac{\partial \phi}{\partial z}\right)\mathbf{K}\begin{bmatrix}\dfrac{\partial u_e}{\partial r}\\ \dfrac{\partial u_e}{\partial z}\end{bmatrix} - \phi\left(\frac{\partial f}{\partial t} + g\right)\right]$$

$$rdrdz + \int_{\Omega_b} r\phi \mathrm{q}\mathrm{n}d\Omega_b = 0 \tag{11.6}$$

where Ω_b stands for the boundary; **n** is the out normal vector. Equations (11.1) and (11.4) are used in deriving (11.6).

To avoid an unsymmetric stiffness matrix in the following, the calculation of the term $\int_{\Omega_e} \phi(\frac{\partial f}{\partial t} + g)rdrdz$ in (11.6) will not follow the standard FE

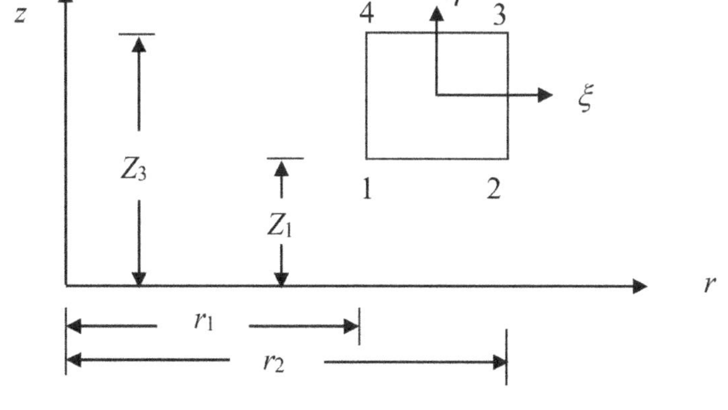

Figure 11.2 A rectangular element.

approach. In this chapter, this term is simply integrated using the following Newton-Cotes type formula with the same error order, that is,

$$\int_{\Omega_e} \phi\left(\frac{\partial f}{\partial t}+g\right) r dr dz = (\phi_{m1},\phi_{m2},\phi_{m3},\phi_{m4}) \left[W_1\left(\frac{\partial f}{\partial t}+g\right)_{m1}, W_2\left(\frac{\partial f}{\partial t}+g\right)_{m2}, W_3\left(\frac{\partial f}{\partial t}+g\right)_{m3}, W_4\left(\frac{\partial f}{\partial t}+g\right)_{m4}\right]^T \tag{11.7}$$

where the subscripts m_1 to m_4 denote values at corresponding nodes; $W_1 = W_4 = (2r_1+r_2)S_e/12$; $W_2 = W_3 = (r_1+2r_2)S_e/12$; $S_e = (r_2 - r_1)(Z_3 - Z_1)$ for a rectangular element.

Using (11.7), some off-diagonal terms of the stiffness matrix are concentrated to the diagonal, and the positiveness of the stiffness matrix is increased. It will be shown in the following that the numerical instability due to spatial oscillation discussed in Vermeer and Verruijt (1981) is avoided owing to this formulation.

As usual, let ϕ and u_e be discretized using linear shape functions in every element Ω_e, then

$$\begin{cases} u_e = \bar{\mathbf{N}}\mathbf{u}_e \\ \phi = \bar{\mathbf{N}}\boldsymbol{\phi}_e \end{cases} \tag{11.8}$$

where $\boldsymbol{\phi}_e = (\phi_{m1}, \phi_{m2}, \phi_{m3}, \phi_{m4})^T$; $\mathbf{u}_e = (u_{m1}, u_{m2}, u_{m3}, u_{m4})^T$; $\bar{\mathbf{N}} = (N_1, N_2, N_3, N_4)$; and $N_1 = \frac{1}{4}(1-\xi)(1-\eta)$, $N_2 = \frac{1}{4}(1+\xi)(1-\eta)$, $N_3 = \frac{1}{4}(1+\xi)(1+\eta)$, $N_4 = \frac{1}{4}(1-\xi)(1+\eta)$.

Substituting (11.7) and (11.8) into (11.6), we get

$$\pi = \sum_{\Omega_e} \boldsymbol{\phi}_e^T \left\{ \mathbf{K}_e \mathbf{u}_e - \begin{bmatrix} W_1\left(\frac{\partial f}{\partial t}+g\right)_{m1}, W_2\left(\frac{\partial f}{\partial t}+g\right)_{m2}, \\ W_3\left(\frac{\partial f}{\partial t}+g\right)_{m3}, W_4\left(\frac{\partial f}{\partial t}+g\right)_{m4} \end{bmatrix}^T \right\} + \int_{\Omega_b} r\phi \mathrm{q} n d\Omega_b = 0 \tag{11.9}$$

where $\mathbf{K}_e = \int_{\Omega_e} r\mathbf{B}^T\mathbf{K}\mathbf{B} dr dz$. For the calculation of q in (11.9), (11.1) can be used and

$$\begin{pmatrix} \dfrac{\partial u_e}{\partial r} \\ \dfrac{\partial u_e}{\partial z} \end{pmatrix} = \mathbf{B}\mathbf{u}_e = \begin{bmatrix} \dfrac{\partial N_1}{\partial r} & \dfrac{\partial N_2}{\partial r} & \dfrac{\partial N_3}{\partial r} & \dfrac{\partial N_4}{\partial r} \\ \dfrac{\partial N_1}{\partial z} & \dfrac{\partial N_2}{\partial z} & \dfrac{\partial N_3}{\partial z} & \dfrac{\partial N_4}{\partial z} \end{bmatrix} \mathbf{u}_e$$

In calculating the permeability matrix **K**, the same shape functions can be used. Then for the simple shape functions above, $\mathbf{K}_e$ can be computed analytically.

Integrating (11.9) with respect to time from t_i to t_{i+1} using the generalized trapezoidal formula, we obtain

$$\pi = \sum_{\Omega_e} \phi_e^T \left\{ \begin{aligned} &\theta\Delta t \mathbf{K}_e^{i+1}\mathbf{u}_e^{i+1} + (1-\theta)\Delta t \mathbf{K}_e^i \mathbf{u}_e^i - [W_1(f_{m1}^{i+1} - f_{m1}^i + \theta\Delta t g_{m1}^{i+1} + (1-\theta)\Delta t g_{m1}^i), \\ &W_2(f_{m2}^{i+1} - f_{m2}^i + \theta\Delta t g_{m2}^{i+1} + (1-\theta)\Delta t g_{m2}^i), W_3(f_{m3}^{i+1} - f_{m3}^i + \theta\Delta t g_{m3}^{i+1} + \\ &(1-\theta)\Delta t g_{m3}^i),\ W_4(f_{m4}^{i+1} - f_{m4}^i + \theta\Delta t g_{m4}^{i+1} + (1-\theta)\Delta t g_{m4}^i)]^T \end{aligned} \right\}$$

$$+ \int_{\Omega_b} r\phi d\Omega_b \int_{t_{i+1}}^{t_i} \mathrm{q}\mathbf{n} dt = 0 \tag{11.10}$$

where the superscripts i and $i+1$ denotes values at t_i and t_{i+1}, respectively, θ is a real number between 0 and 1, and $\Delta t = t_{i+1}$-t_i.

Differentiating π expressed in (11.10) with respect to $\mathbf{u}^{i+1}$ (a vector including all porewater pressures at nodes) gives

$$\delta\pi = \sum_{\Omega_e} \phi_e^T (\bar{\mathbf{K}}_e \delta\mathbf{u}_e^{i+1} + \theta\Delta t \delta\mathbf{K}_e^{i+1}\mathbf{u}_e^{i+1}) \tag{11.11}$$

where

$$\bar{\mathbf{K}}_e = \theta\Delta t \mathbf{K}_e^{i+1} - diag\begin{bmatrix} W_1 \dfrac{\partial(f_{m1}^{i+1} + \theta\Delta t g_{m1}^{i+1})}{\partial u_{em1}^{i+1}}, W_2 \dfrac{\partial(f_{m2}^{i+1} + \theta\Delta t g_{m2}^{i+1})}{\partial u_{em2}^{i+1}}, \\ W_3 \dfrac{\partial(f_{m3}^{i+1} + \theta\Delta t g_{m3}^{i+1})}{\partial u_{em3}^{i+1}}, W_4 \dfrac{\partial(f_{m4}^{i+1} + \theta\Delta t g_{m4}^{i+1})}{\partial u_{em4}^{i+1}} \end{bmatrix} \tag{11.12}$$

Therefore, the discretized (11.10) can be approximated as

$$\sum_{\Omega_e} \phi_e^T (\bar{\mathbf{K}}_e \delta\mathbf{u}_e^{i+1} - \mathbf{R}_e) + \int_{\Omega_b} r\phi d\Omega_b \int_{t_i}^{t_{i+1}} \mathrm{q}\mathbf{n} dt = 0 \tag{11.13}$$

where

$$\begin{aligned} \mathbf{R}_e = &-\theta\Delta t \mathbf{K}_e^{i+1}\mathbf{u}_e^{i+1} - (1-\theta)\Delta t \mathbf{K}_e^i \mathbf{u}_e^i + [W_1(f_{m1}^{i+1} - f_{m1}^i + \theta\Delta t g_{m1}^{i+1} + (1-\theta)\Delta t g_{m1}^i), \\ &W_2(f_{m2}^{i+1} - f_{m2}^i + \theta\Delta t g_{m2}^{i+1} + (1-\theta)\Delta t g_{m2}^i), W_3(f_{m3}^{i+1} - f_{m3}^i + \theta\Delta t g_{m3}^{i+1} + (1-\theta)\Delta t g_{m3}^i), \\ &W_4(f_{m4}^{i+1} - f_{m4}^i + \theta\Delta t g_{m4}^{i+1} + (1-\theta)\Delta t g_{m4}^i)]^T \end{aligned} \tag{11.14}$$

Following the standard FE assembly procedure, a symmetrical set of equations can be derived.

Substituting (5.16) into (11.12) gives

$$\bar{\mathbf{K}}_e = \theta\Delta t\mathbf{K}_e^{i+1} + diag\begin{bmatrix} W_1\left(\dfrac{\dfrac{\partial f}{\partial\sigma_z'}+\dfrac{\partial g}{\partial\sigma_z'}\theta\Delta t+(\theta-\theta')\Delta t\dfrac{\partial g}{\partial\varepsilon_z}\dfrac{\partial f}{\partial\sigma_z'}}{1-\theta'\Delta t\dfrac{\partial g}{\partial\varepsilon_z}}\right)_{m1}^{i+1}, \\ W_2\left(\dfrac{\dfrac{\partial f}{\partial\sigma_z'}+\dfrac{\partial g}{\partial\sigma_z'}\theta\Delta t+(\theta-\theta')\Delta t\dfrac{\partial g}{\partial\varepsilon_z}\dfrac{\partial f}{\partial\sigma_z'}}{1-\theta'\Delta t\dfrac{\partial g}{\partial\varepsilon_z}}\right)_{m2}^{i+1}, \\ W_3\left(\dfrac{\dfrac{\partial f}{\partial\sigma_z'}+\dfrac{\partial g}{\partial\sigma_z'}\theta\Delta t+(\theta-\theta')\Delta t\dfrac{\partial g}{\partial\varepsilon_z}\dfrac{\partial f}{\partial\sigma_z'}}{1-\theta'\Delta t\dfrac{\partial g}{\partial\varepsilon_z}}\right)_{m3}^{i+1}, \\ W_4\left(\dfrac{\dfrac{\partial f}{\partial\sigma_z'}+\dfrac{\partial g}{\partial\sigma_z'}\theta\Delta t+(\theta-\theta')\Delta t\dfrac{\partial g}{\partial\varepsilon_z}\dfrac{\partial f}{\partial\sigma_z'}}{1-\theta'\Delta t\dfrac{\partial g}{\partial\varepsilon_z}}\right)_{m4}^{i+1} \end{bmatrix} \tag{11.15}$$

where θ' is a real number between 0 and 1.

Equation (11.15) is the element stiffness matrix for constitutive equations of the form of (11.4), and it may be adjusted for unloading for some models. It should be noted that unknown variables that may be present in the boundaries are not considered for simplicity in the derivation of (11.15). To account for this situation, a term can be easily added in (11.15).

The strains can be calculated from (11.16)

$$d\varepsilon_z^{i+1} = \frac{(\varepsilon_z - f)^i + [\theta' g^{i+1} + (1-\theta')g^{i+1}]\Delta t - (\varepsilon_z - f)^{i+1}}{1-\theta'\left(\dfrac{\partial g}{\partial\varepsilon_z}\right)^{i+1}} \tag{11.16}$$

The above formulae provide a complete routine for a Newton (or Modified Newton) type solution of the consolidation problem. Theoretically, this Newton form of iteration (or modified Newton form) for the non-linear equation system converges very rapidly. Generally, several iterations will give very high accuracy.

11.4 COMPARISON WITH FULLY COUPLED ANALYSIS

To study the efficiency of the simplified method, a comparison with the fully coupled consolidation analysis (Chapter 12) is made using a well-known FE package, ABAQUS (1995) for the cases of (a) Biot's porous elastic model (Biot 1941), and (b) a non-linear porous elastic model suggested in the ABAQUS (1995) program. The results are described in the following.

11.4.1 Geometric models for the analysis

The geometry and typical FE mesh used in the comparison is illustrated in Figure 11.3. To eliminate dimensions, the length is normalized using the radius of vertical drain, r_d, and the following parameters are used.

$$Z = \frac{z}{D}, \quad R = \frac{r - r_d}{r_e - r_d}, \quad N = \frac{r_e}{r_d} \tag{11.17}$$

where D is the thickness of the soil layer.

In the comparison, three different cases of D and H (H is the length of vertical drain) are studied, that is, Case A: $H = D = 160$, Case B: $H = D = 80$, and Case C: $H = D/2 = 80$. For each case, four different N (10, 20, 40, 100) are chosen. Drainage boundary conditions are that water is freely drained at the upper horizontal plane and in the vertical drain. At other boundaries, water is impermeable. Horizontal displacements are not permitted at the vertical boundaries, and the base is rigidly fixed in the vertical direction. Uniform external loads are applied suddenly at the upper horizontal plane.

In the calculation, the same small time increments are used for the first few steps in the simplified method and the ABAQUS (1995) method, and the

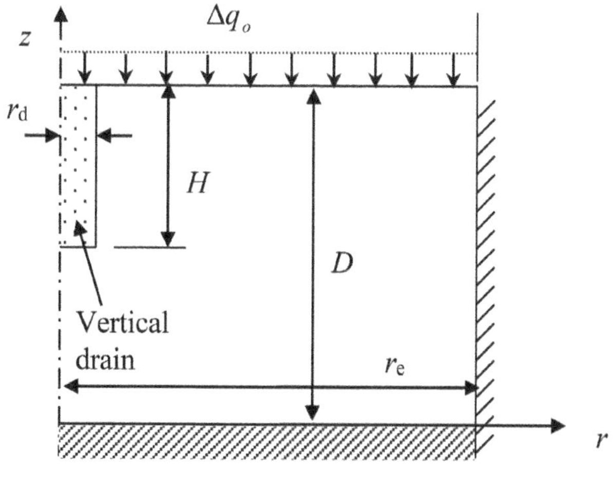

Figure 11.3(a) Typical geometry used in numerical computations.

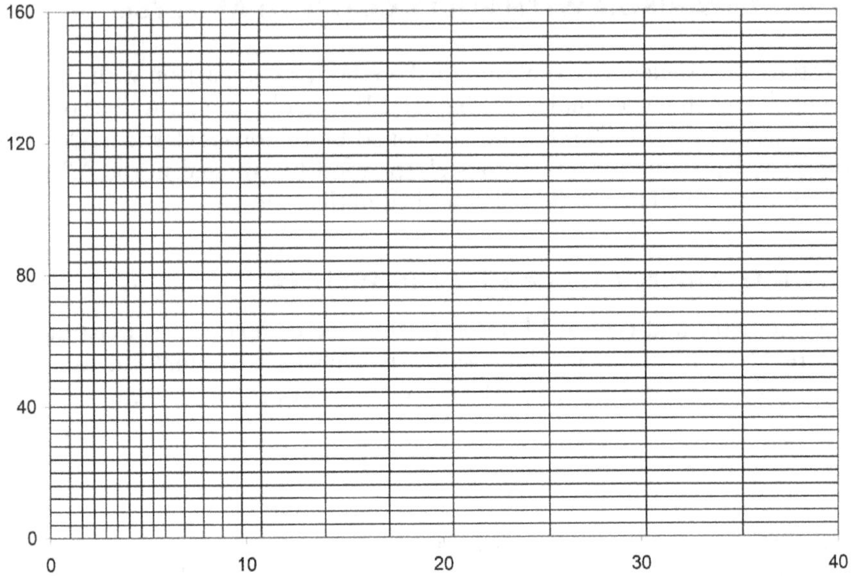

Figure 11.3(b) Typical finite element mesh used in numerical computations.

time increments are increased later for numerical stability. The soil mass is assumed to be uniform in the cylinder.

11.4.2 Comparison with Biot's porous elastic model

The response of Biot's isotropic porous elastic material is determined by the elastic modulus E and Poisson's ratio ν. For this simple model, the following relation exists:

$$m_v = \frac{1-\nu-2\nu^2}{(1-\nu)E} \tag{11.18}$$

In this calculation, the Poisson's ratio is equal to 0.3. The permeability behavior is also assumed to be isotropic and the coefficient of permeability is constant.

11.4.2.1 Normalized time

To compare the results from the simplified method with those from ABAQUS (1995), time is normalized. In Chapter 6, a normalized time T (Zhu and Yin 2001) is suggested in (6.19) and rewritten as follows:

$$T = \left(\frac{k_r}{\gamma_w m_v} \frac{\mu_1^2}{r_d^2} + \frac{k_z}{\gamma_w m_v} \frac{\pi^2}{4H^2} \right) t \tag{6.19}$$

This normalized time T is derived for the case of fully penetrating vertical drain, and is used here for the partially penetrating vertical drain, too.

11.4.2.2 Average degree of consolidation

To compare the compression results from the simplified method and from ABAQUS (1995), the following average degree of consolidation is defined as usual.

$$U(T) = \frac{\overline{S(t)}}{\overline{S_f}} \tag{11.19}$$

where $\overline{S_f}$ and $\overline{S(t)}$ are average final compression and compression at t. $\overline{S_f}$ is calculated for the situation of zero porewater pressure using 1-D theory, and is just a constant. $\overline{S(t)}$ is the area average value for the simplified method and for the fully coupled analysis. In Figures 11.4–11.18, results from ABAQUS (1995) are plotted as lines and the corresponding values from the simplified method are plotted as points in the same figure for comparison.

Figures 11.4–11.6 show the results of the average degree of consolidation U *vs.* time factor T from the simplified method and ABAQUS (1995) for Case A, Case B, and Case C. The agreement between the simplified method and the fully coupled analysis is very good. Furthermore, the average degree of consolidation has good normalized feature with respect to N for the cases

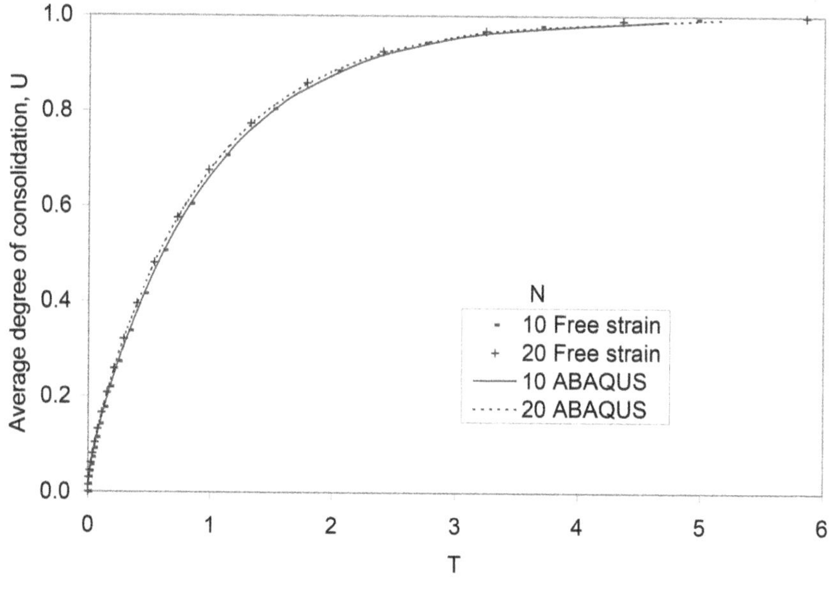

Figure 11.4(a) Average degree of consolidation vs. T for Case A: $H = D = 160$.

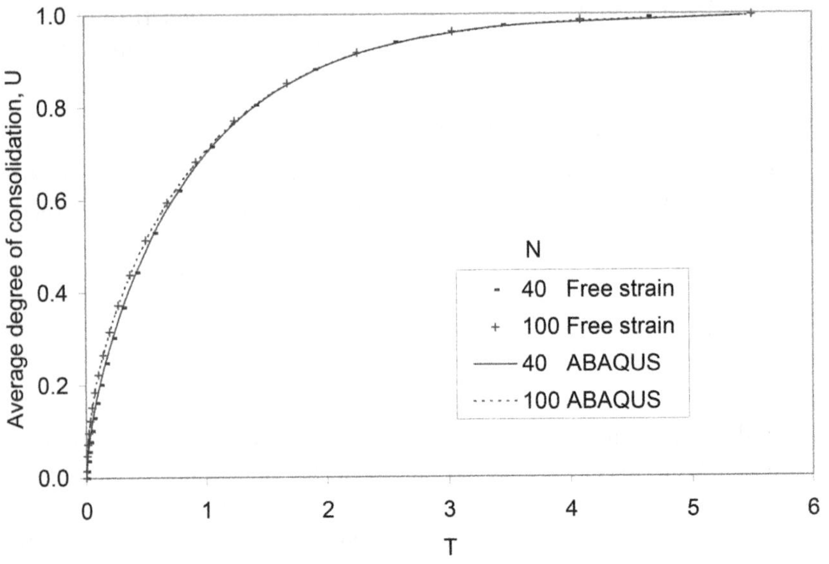

Figure 11.4(b) Average degree of consolidation vs. T for Case A: $H = D = 160$.

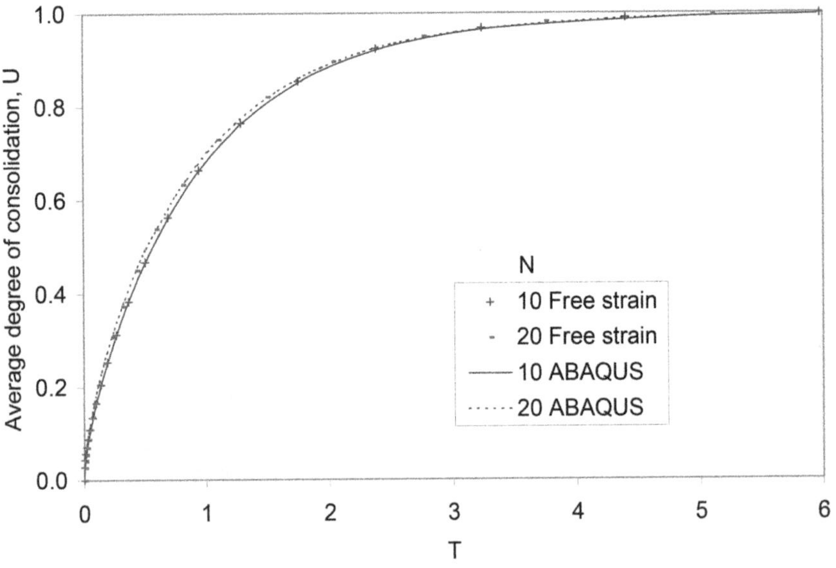

Figure 11.5(a) Average degree of consolidation vs. T for Case B: $H = D = 80$.

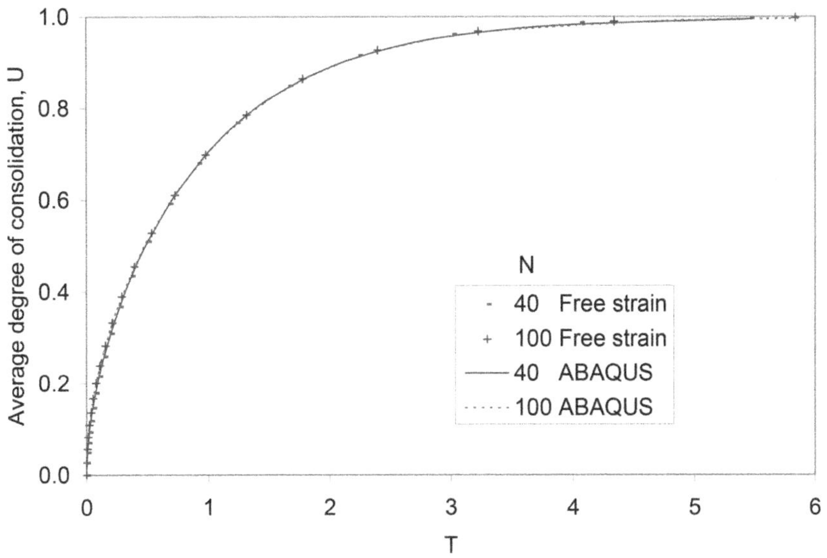

Figure 11.5(b) Average degree of consolidation vs. *T* for Case B: $H = D = 80$.

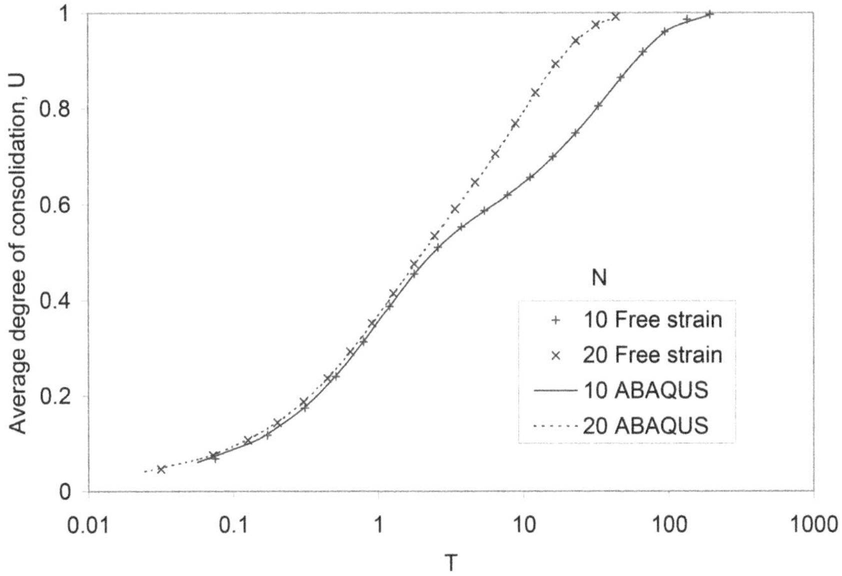

Figure 11.6(a) Average degree of consolidation vs. *T* for Case C: $H = 80$, $D = 160$.

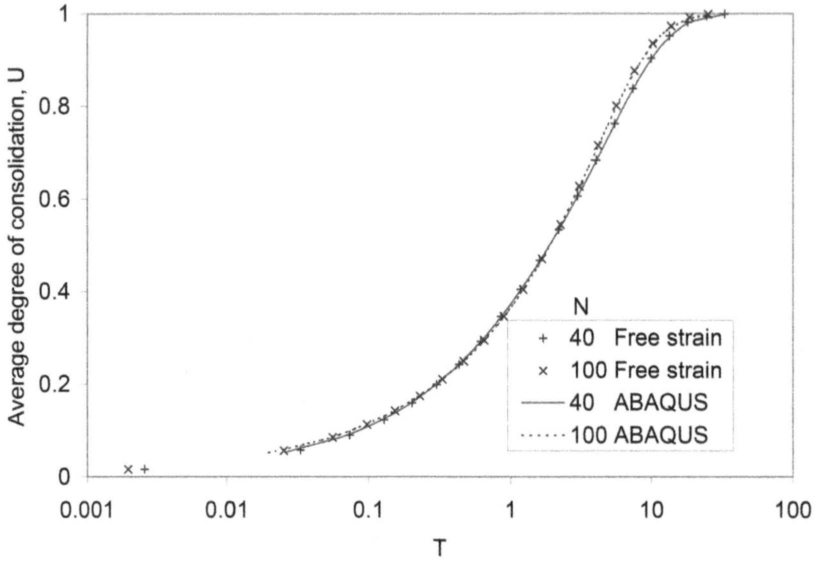

Figure 11.6(b) Average degree of consolidation vs. *T* for Case C: $H = 80$, $D = 160$.

(Case A and Case B) of fully penetrating vertical drain using the normalized time T suggested by Zhu and Yin (2001). However, the normalized feature with respect to N for the case of partially penetrating vertical drain (Case C) is not as good as that for Cases A and B as shown in Figure 11.6. In other words, the curves for different N values (N = 10 and N = 20) as shown in Figure 11.6 are not very close to each other. In this situation (Case C), the consolidation of the soil below the vertical drain tip has very large influence on the whole consolidation process.

11.4.2.3 Excess porewater pressure

The excess porewater pressure is normalized using the external load Δq_0. Figure 11.7 illustrates the results of normalized excess porewater pressure u_n, that is, $u_n = u/\Delta q_0$, at different positions when $N = 20$, $R = 0.125$, 0.5, and 1 in Case A. It is noted that $Z = 0$ (or $Z = 1$) corresponds to the undrained bottom side (or the top drained side). Figure 11.8 shows the values of normalized excess porewater pressure u_n when $N = 20$, $R = 0.125$, 0.5, and 1 in Case C. For other N values in Cases A, B, and C, the figures are almost the same and will not be presented. From these figures, we can say that the excess porewater pressures calculated from the simplified method and from the fully coupled analysis are almost the same for both fully and partially penetrating drain cases. It should be noted that there are some discrepancies at the very beginning (T<0.05). These discrepancies are caused by the instability for

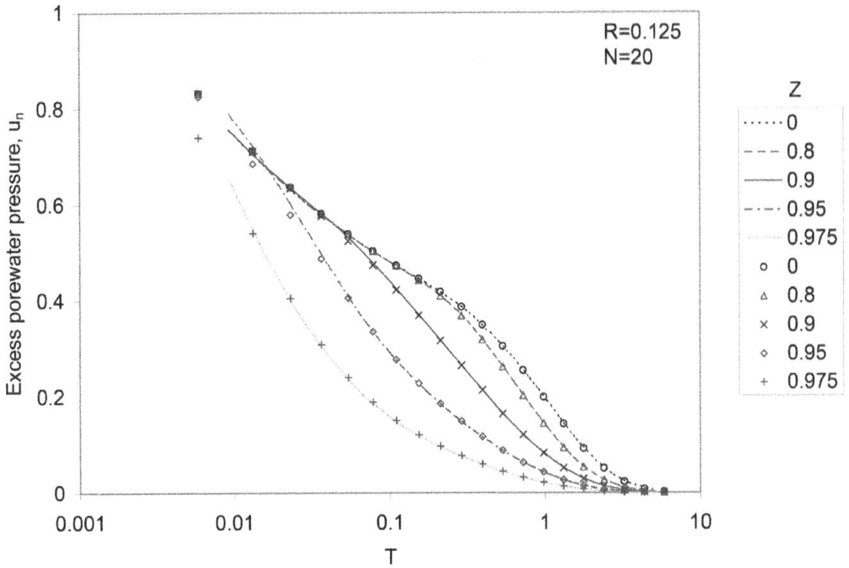

Figure 11.7(a) Excess porewater pressure vs. T when $N = 20$, $R = 0.125$ for Case A.

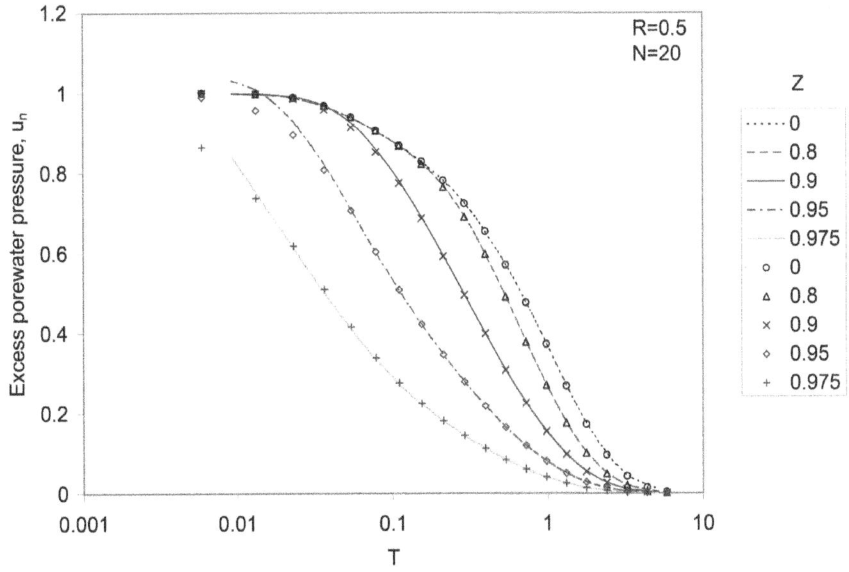

Figure 11.7(b) Excess porewater pressure vs. T when $N = 20$, $R = 0.5$ for Case A.

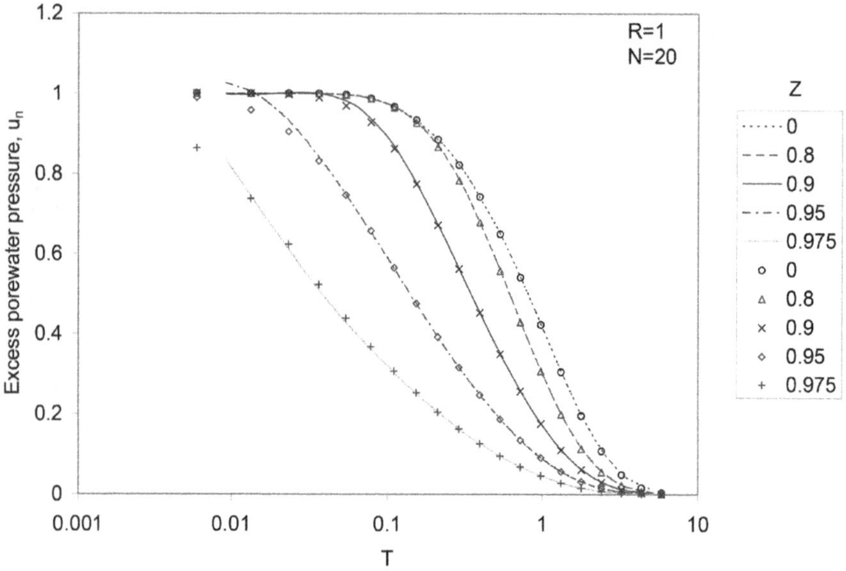

Figure 11.7(c) Excess porewater pressure vs. T when $N = 20$, $R = 1$ for Case A.

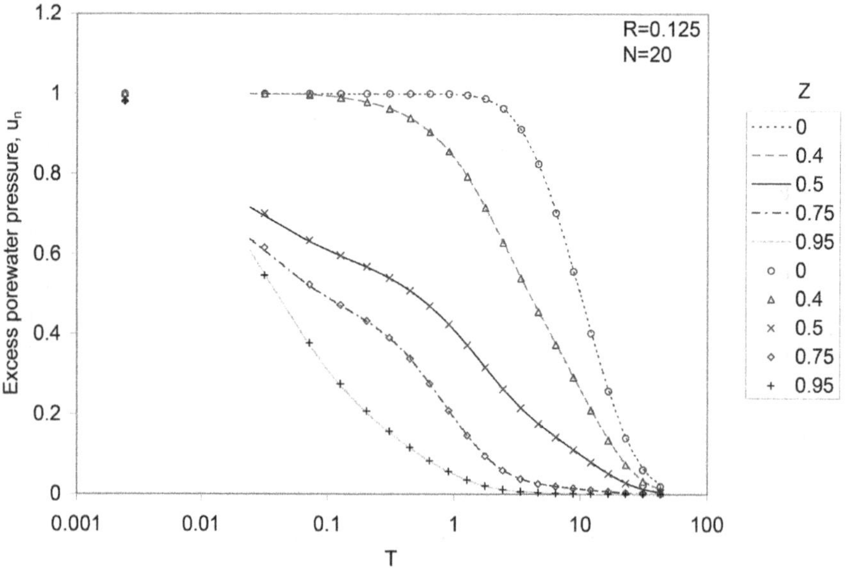

Figure 11.8(a) Excess porewater pressure vs. T when $N = 20$, $R = 0.125$ for Case C.

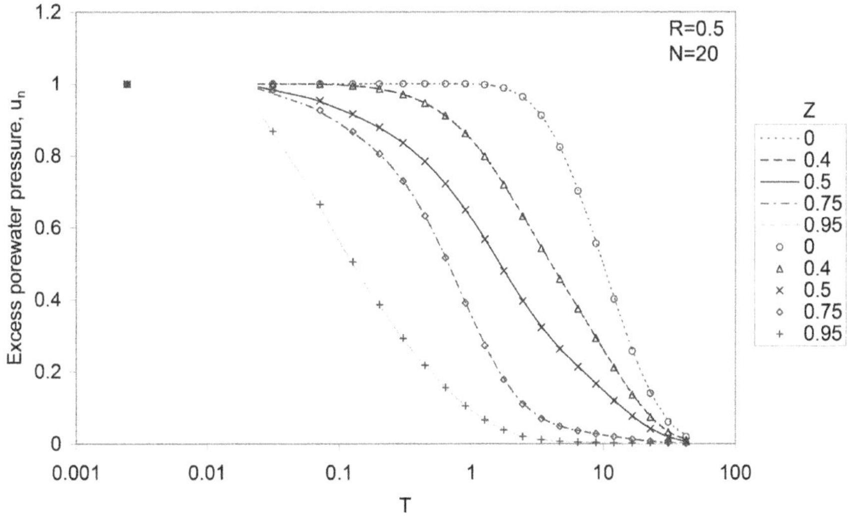

Figure 11.8(b) Excess porewater pressure vs. *T* when *N* = 20, *R* = 0.5 for Case C.

small time steps in FE program ABAQUS (1995) as commonly encountered in other FE consolidation analyses (Vermeer and Verruijt 1981). However, this phenomenon of instability does not occur in the FE formulation suggested in this chapter as shown in Figure 11.9.

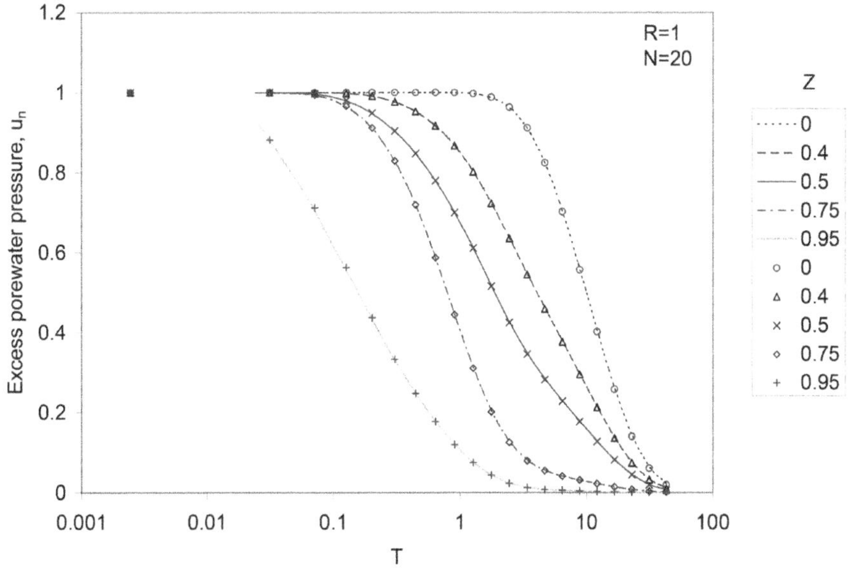

Figure 11.8(c) Excess porewater pressure vs. *T* when *N* = 20, *R* = 1 for Case C.

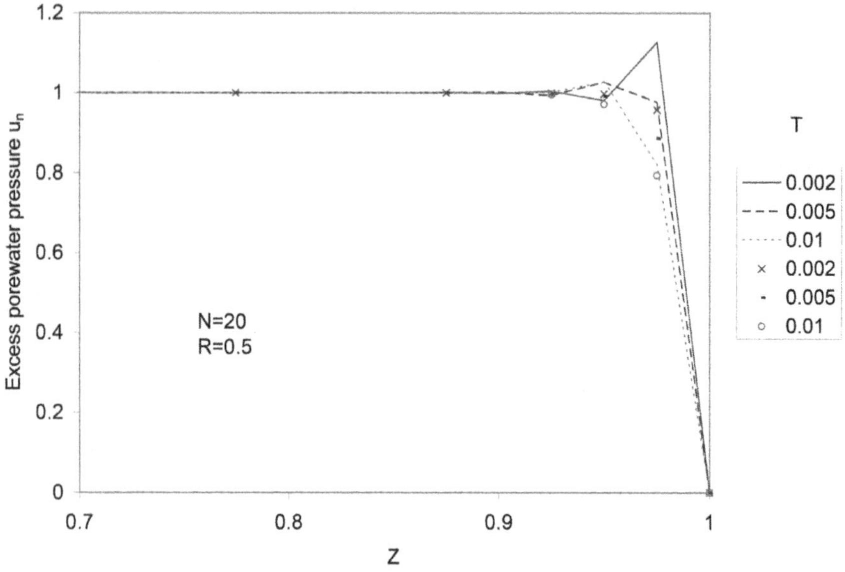

Figure 11.9 Excess porewater pressure at the very beginning when $N = 20$, $R = 0.5$ for Case A.

11.4.2.4 Vertical effective stress

The agreements in vertical effective stress are not as good as those in the excess porewater pressure demonstrated above, although the cross-section area average values are nearly the same. Figure 11.10 shows the isochrones

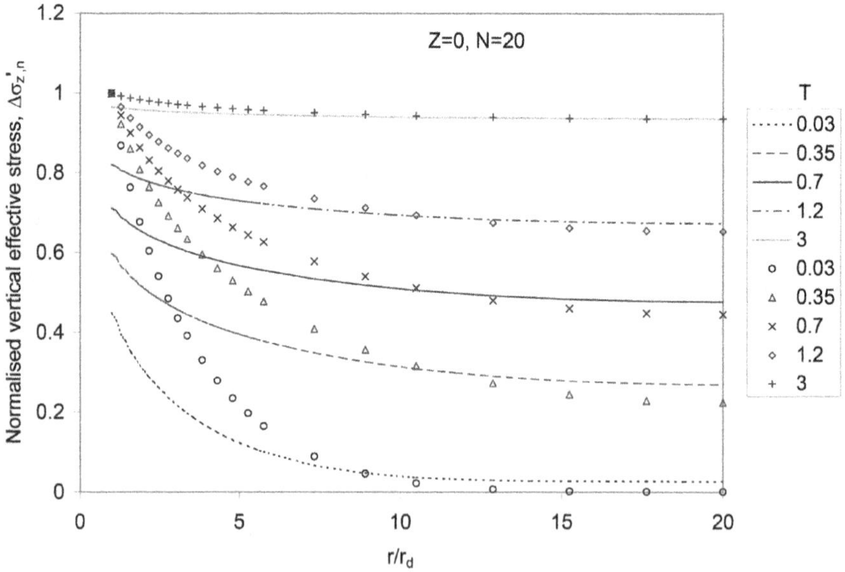

Figure 11.10(a) Isochrones of vertical effective stress $\Delta\sigma'_{z,n}$ when $N = 20$, $Z = 0$ for Case A.

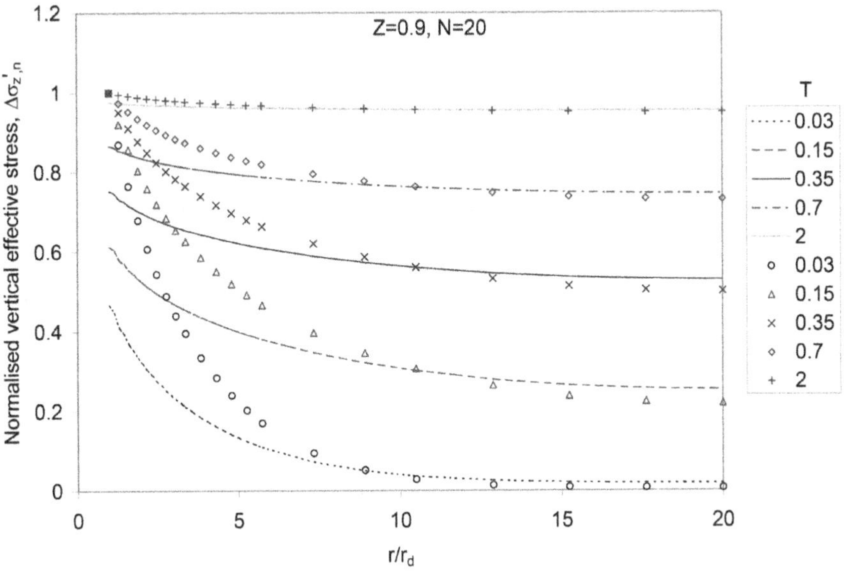

Figure 11.10(b) Isochrones of vertical effective stress $\Delta\sigma'_{z,n}$ when $N = 20$, $Z = 0.9$ for Case A.

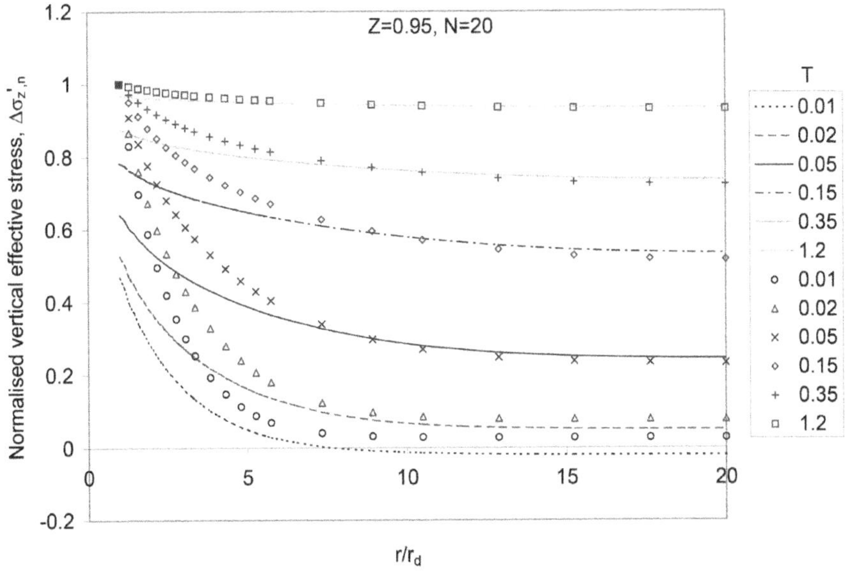

Figure 11.10(c) Isochrones of vertical effective stress $\Delta\sigma'_{z,n}$ when $N = 20$, $Z = 0.95$ for Case A.

of the normalized incremental vertical effective stress, that is, $\Delta\sigma'_{z,n} = \Delta\sigma'_z/\Delta q_0$ at different horizontal planes ($Z = 0$, 0.9 and 0.95) when $N = 20$ in Case A. Figure 11.11 plots the isochrones for $N = 20$ and $Z = 0$, 0.4, 0.5, 0.75, and 0.95 in Case C. From the figures, it is clear that relatively large discrepancies occur near the vertical drain boundaries. The differences between the simplified method and the fully coupled analysis are tolerable for $r/r_d>7$ when $N = 20$. The phenomenon of instability clearly shown in Figure 11.9 and Figure 11(c) at the very beginning using ABAQUS (1995) does not happen in the calculation using the simplified method. It can also be seen in Figure 11.11 that the behavior of the soil below $Z = 0.4$ is almost 1-D since the vertical effective stress is almost constant along the radial direction. Therefore, the simplified method with a 1-D soil model shall have better results in this region than the results in the above region. This makes the simplified method very effective for the analysis of partially penetrating vertical drain.

The vertical effective stress from the fully coupled analysis is much higher within about five times the equivalent radius of vertical drain than in the other zone for ideally installed vertical drain. From the test results of Tavenas *et al.* (1983), it can be inferred that the permeability will be reduced substantially in this region. Therefore, the rapid consolidation of soil around the walls of any drain will act as a smear zone. This effect plus true disturbance effects from even the best non-displacement installation procedure will tend to minimize the effect of drain installation method.

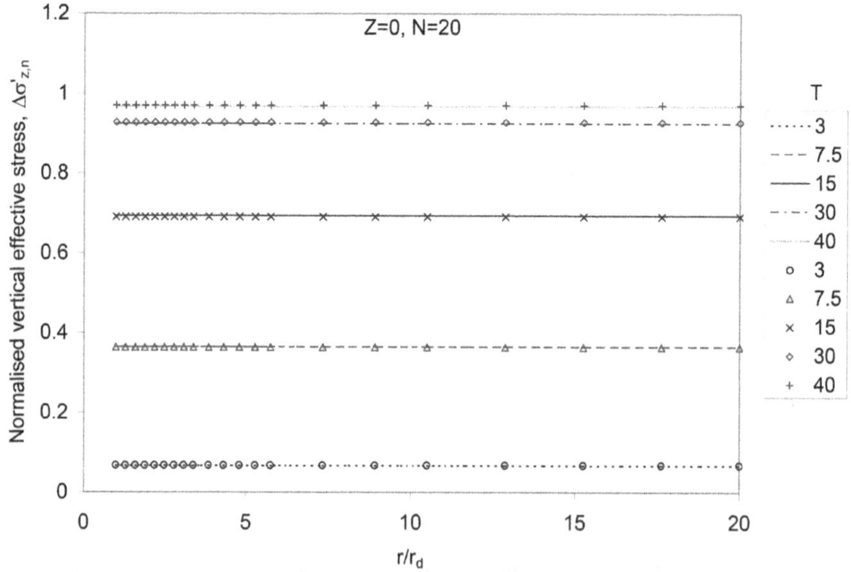

Figure 11.11(a) Isochrones of vertical effective stress $\Delta\sigma'_{z,n}$ when $N = 20$, $Z = 0$ for Case C.

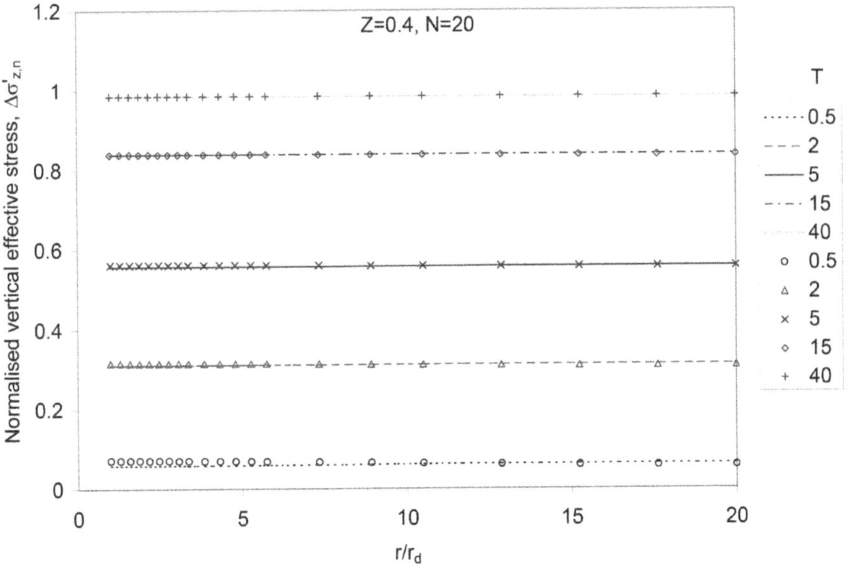

Figure 11.11(b) Isochrones of vertical effective stress $\Delta\sigma'_{z,n}$ when $N = 20$, $Z = 0.4$ for Case C.

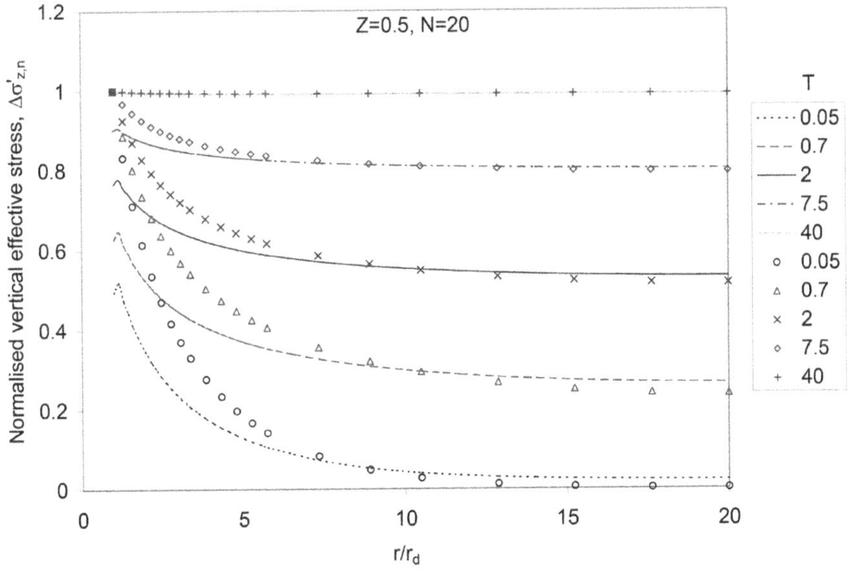

Figure 11.11(c) Isochrones of vertical effective stress $\Delta\sigma'_{z,n}$ when $N = 20$, $Z = 0.5$ for Case C.

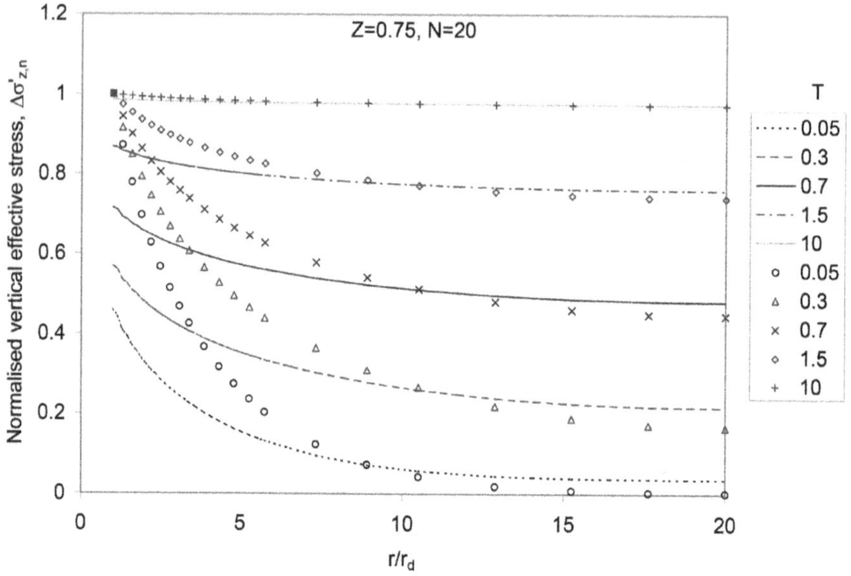

Figure 11.11(d) Isochrones of vertical effective stress $\Delta\sigma'_{z,n}$ when $N = 20$, $Z = 0.75$ for Case C.

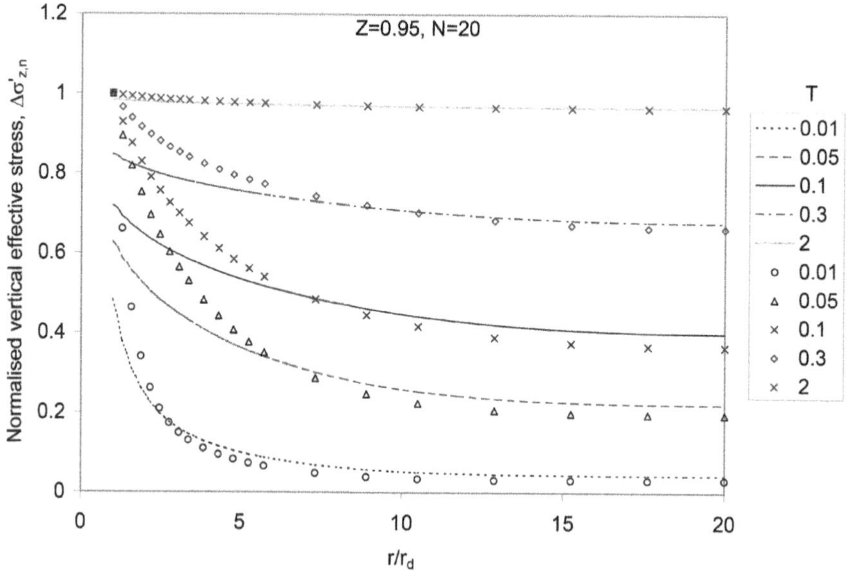

Figure 11.11(e) Isochrones of vertical effective stress $\Delta\sigma'_{z,n}$ when $N = 20$, $Z = 0.95$ for Case C.

Field experiences also show that the effects of disturbance decrease with increasing effective stress.

11.4.3 Comparison with a non-linear porous elastic model

The isotropic non-linear porous elastic model used for comparison is suggested in ABAQUS (1995). The model is defined by a constant Poisson's ratio ν and a non-linear relationship of effective mean stress p' and volume strain ε_v as follows:

$$1-\exp(-\varepsilon_v)=\frac{\kappa}{1+e_0}\ln\left(\frac{p'+p_t^{el}}{p_o+p_t^{el}}\right) \tag{11.20}$$

The corresponding 1-D model is given in (11.5) (see Appendix 11.A). To make the comparison simple, the permeability behavior is also assumed to be isotropic, and the coefficient of permeability is constant. The initial vertical total stress σ'_{zo} is assumed to be 50 units, and the suddenly applied external load Δq_o is set to be 100 units. The constants adopted in the analysis are listed in Table 11.1.

11.4.3.1 Normalized time

As in the preceding sections, time is also normalized using (6.19). However, the m_v in (6.19) is replaced by $\left.\frac{\partial f}{\partial \sigma'_z}\right|_{\sigma_{z0}}$. The f function is given in (11.5) and g is set to be zero for time-independent soils.

11.4.3.2 Average degree of consolidation

The results of average degree of consolidation from both methods for Cases A, B, and C are almost identical to the analysis using Biot's porous elastic model. Figures 11.12–11.14 plot the average degree of consolidation for Cases A, B, and C. Obviously, the agreement between the simplified method and the fully coupled analysis is very good. The average degree of consolidation has also good normalized feature with respect to N for the cases of fully penetrating vertical drain using the normalized time T suggested by Zhu and Yin (2001) even for this non-linear case. However, as shown in Figure 11.14(a), this normalized feature with respect to N for partially penetrating vertical drain is not as good as that for Case A and Case B.

Table 11.1 Model parameters in (11.5)

κ	ν	e_0	p_0	p_t^{el}	σ'_{zo}
0.18	0.31	1.45	31.64	0	50

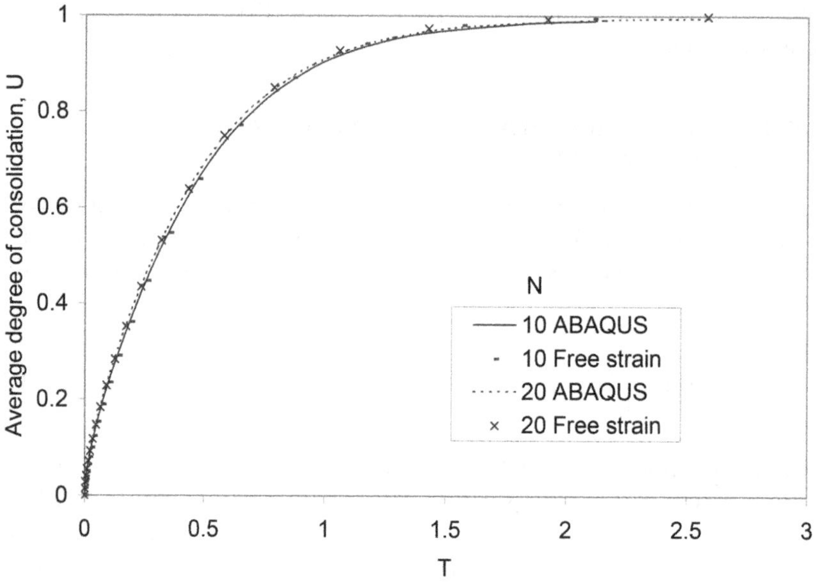

Figure 11.12(a) Average degree of consolidation vs. *T* using non-linear model for Case A.

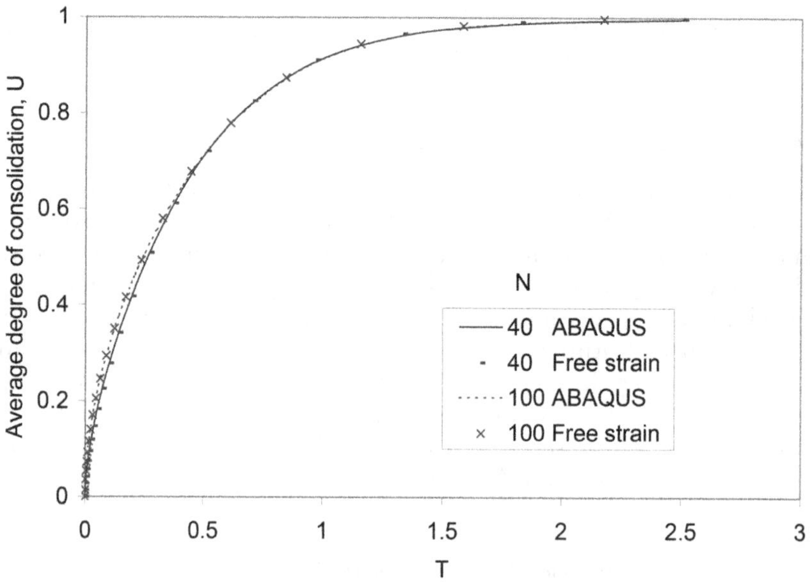

Figure 11.12(b) Average degree of consolidation vs. *T* using non-linear model for Case A.

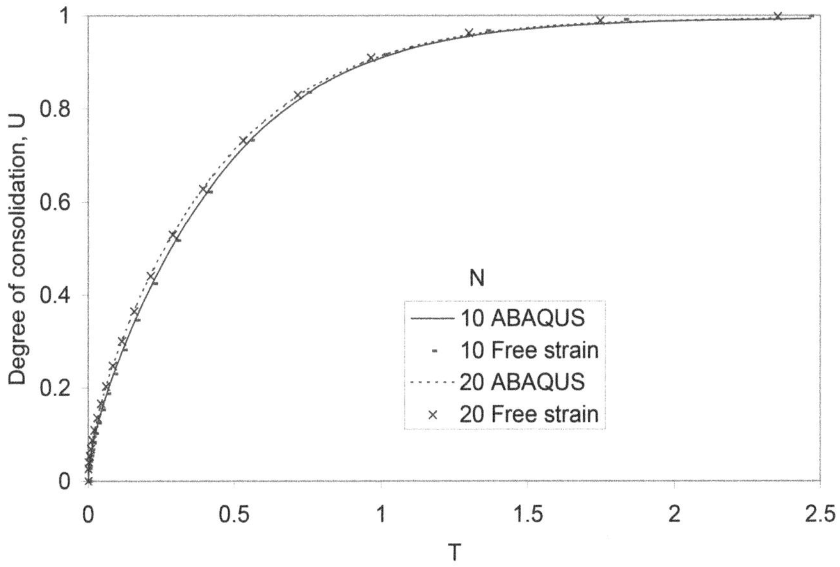

Figure 11.13(a) Average degree of consolidation vs. *T* using non-linear model for Case B.

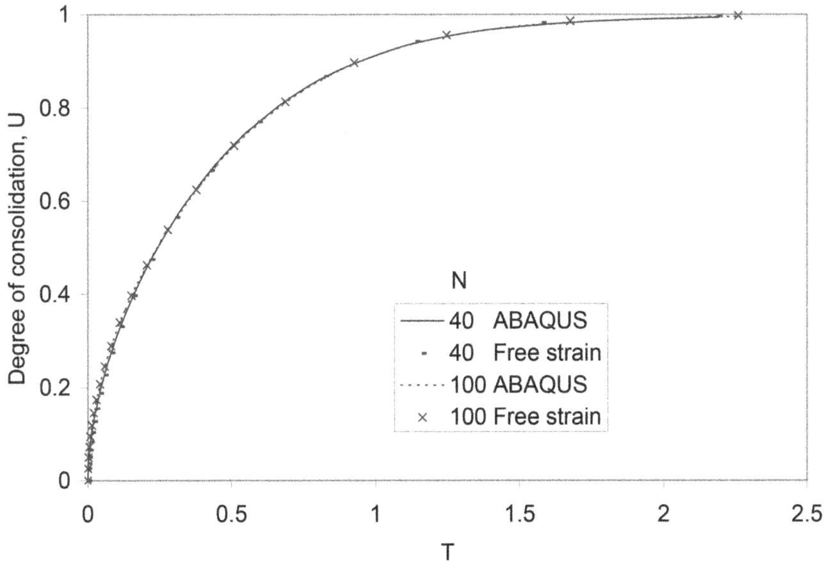

Figure 11.13(b) Average degree of consolidation vs. *T* using non-linear model for Case B.

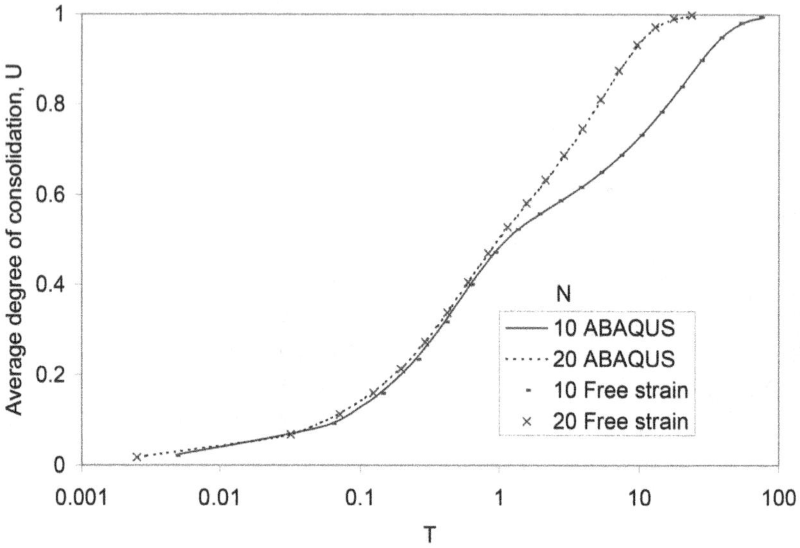

Figure 11.14(a) Average degree of consolidation vs. *T* using non-linear model for Case C.

11.4.3.3 Excess porewater pressure

The figures of excess porewater pressure have almost the same characteristics for the three cases and the excess porewater pressures calculated from the simplified method and from the fully coupled analysis are almost identical. For simplicity,

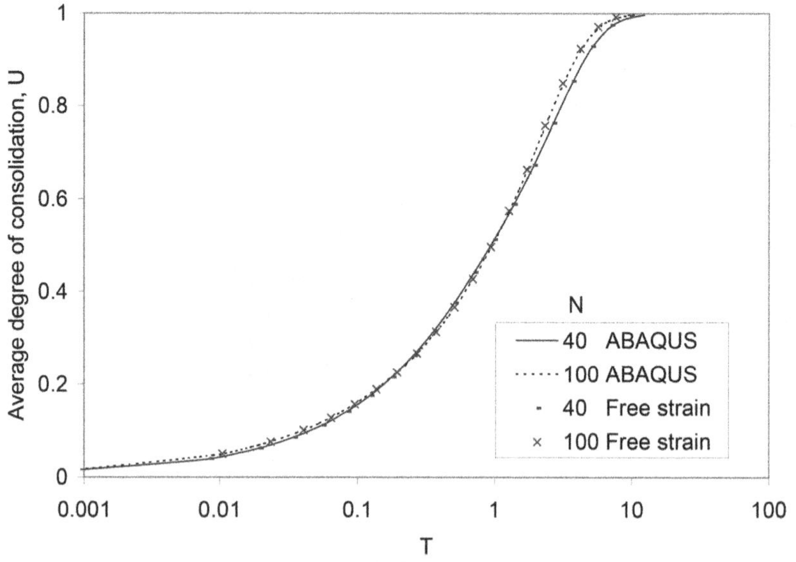

Figure 11.14(b) Average degree of consolidation vs. *T* using non-linear model for Case C.

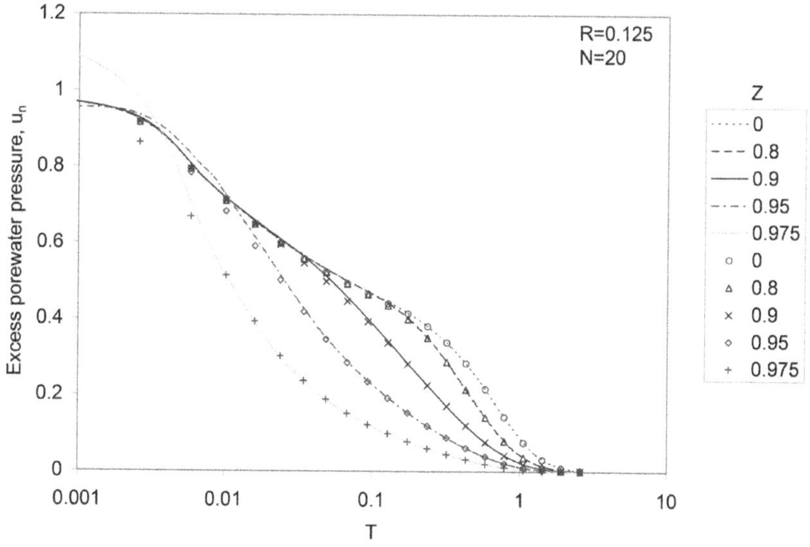

Figure 11.15(a) Excess porewater pressure vs. T when $N = 20$, $R = 0.125$ for Case A.

only typical normalized excess porewater pressure u_n at different positions is plotted in Figures 11.15–11.16. As in the linear elastic situation, there are some discrepancies at the very beginning. The phenomenon of instability appears at the very beginning in the calculation using ABAQUS (1995). However, this situation does not happen when using the simplified FE formulation.

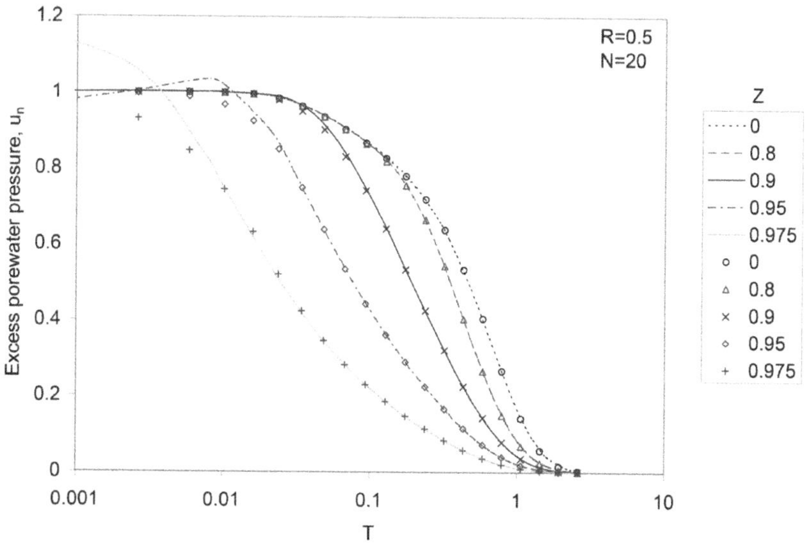

Figure 11.15(b) Excess porewater pressure vs. T when $N = 20$, $R = 0.5$ for Case A.

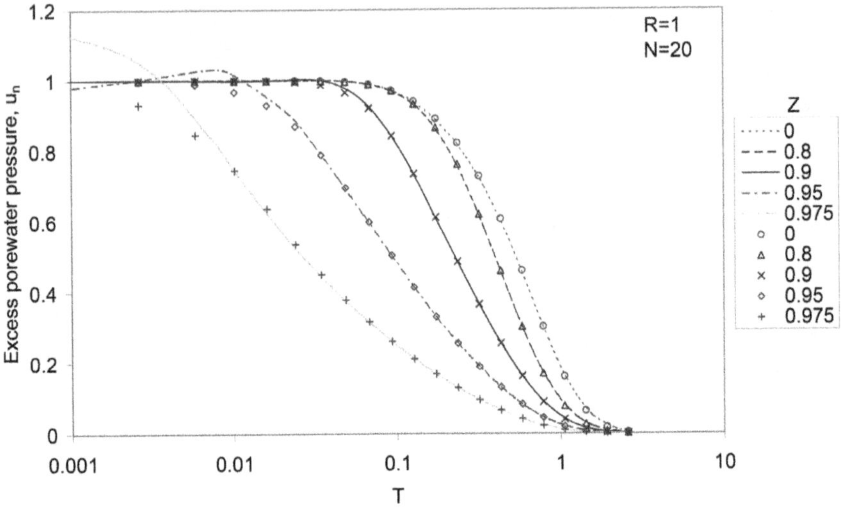

Figure 11.15(c) Excess porewater pressure vs. T when $N = 20$, $R = 1$ for Case A.

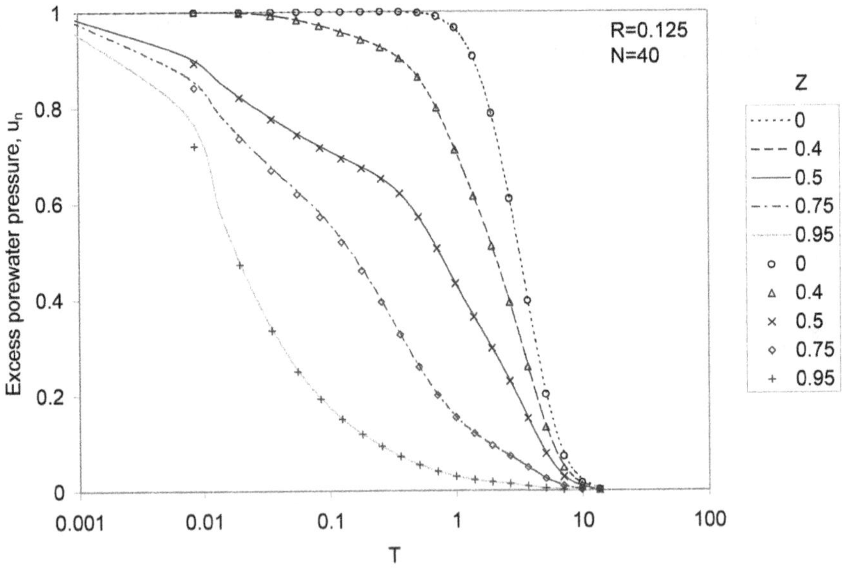

Figure 11.16(a) Excess porewater pressure vs. T when $N = 40$, $R = 0.125$ for Case C.

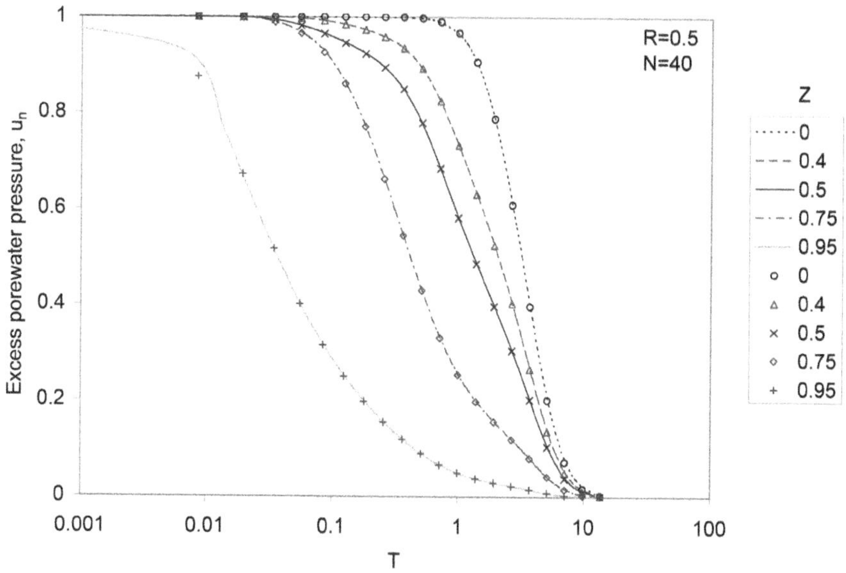

Figure 11.16(b) Excess porewater pressure vs. T when $N = 40$, $R = 0.5$ for Case C.

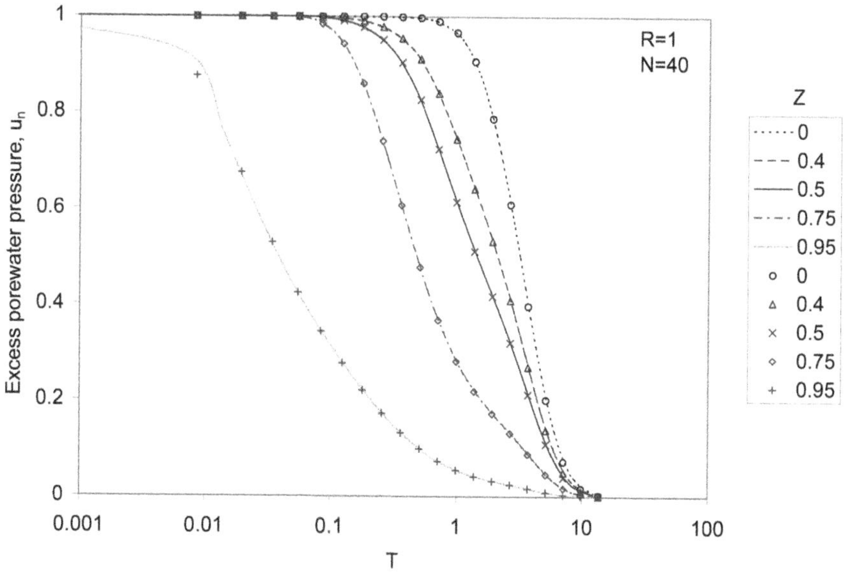

Figure 11.16(c) Excess porewater pressure vs. T when $N = 40$, $R = 1$ for Case C.

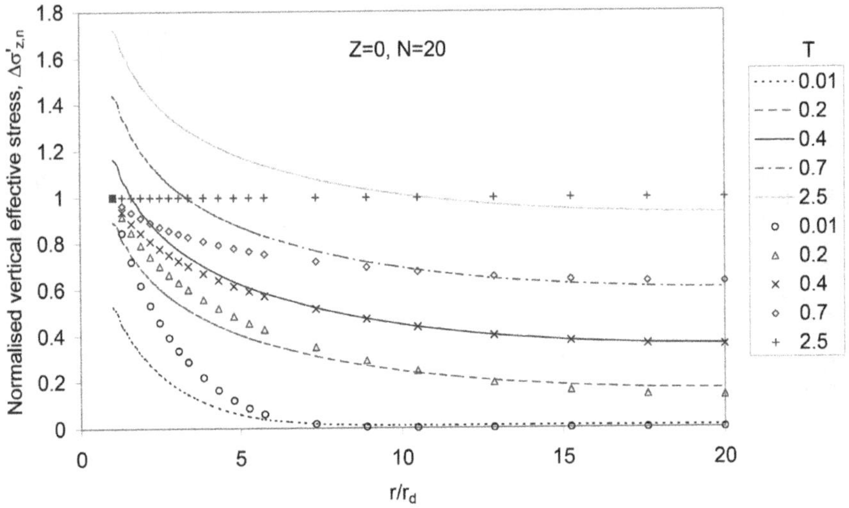

Figure 11.17(a) Isochrones of vertical effective stress $\Delta\sigma'_{z,n}$ when $N = 20$, $Z = 0$ for Case A.

11.4.3.4 Vertical effective stress

As in the linear elastic model, the agreements in vertical effective stress are not as good as those in the excess porewater pressure demonstrated above, though the average values are almost the same. Figure 11.17 shows the isochrones of

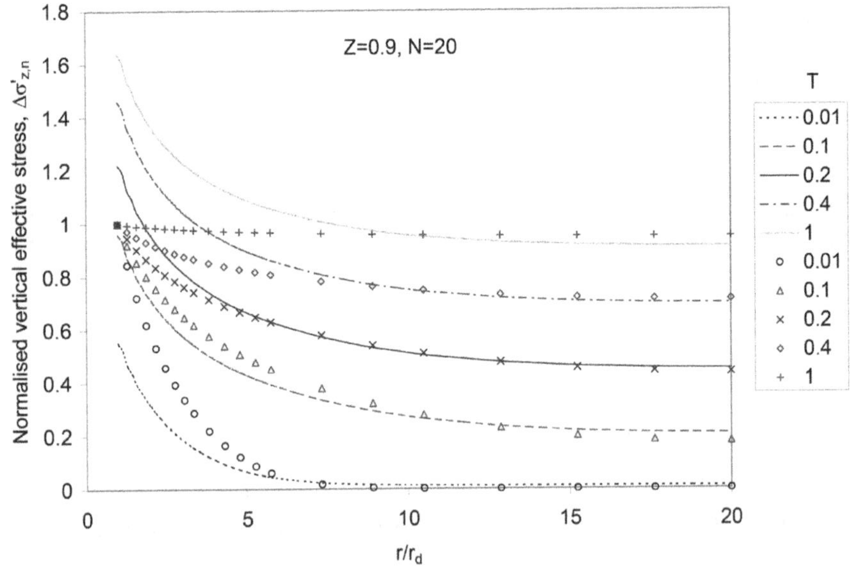

Figure 11.17(b) Isochrones of vertical effective stress $\Delta\sigma'_{z,n}$ when $N = 20$, $Z = 0.9$ for Case A.

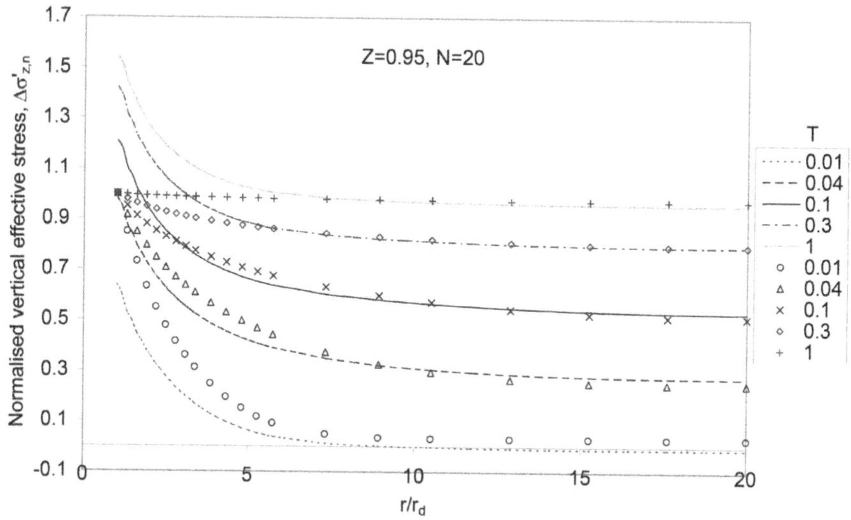

Figure 11.17(c) Isochrones of vertical effective stress $\Delta\sigma'_{z,n}$ when $N = 20$, $Z = 0.95$ for Case A.

the normalized vertical effective stress, $\Delta\sigma'_{z,n} = \Delta\sigma'_z/\Delta q_0$, at different horizontal planes when $N = 20$ in Case A. Figure 11.18 illustrates the isochrones for $N = 20$ in Case C. Relatively large discrepancies take place near the vertical drain boundaries. However, the differences between the simplified method and the fully coupled analysis are tolerable for $r/r_d>7$ when $N = 20$ as in the

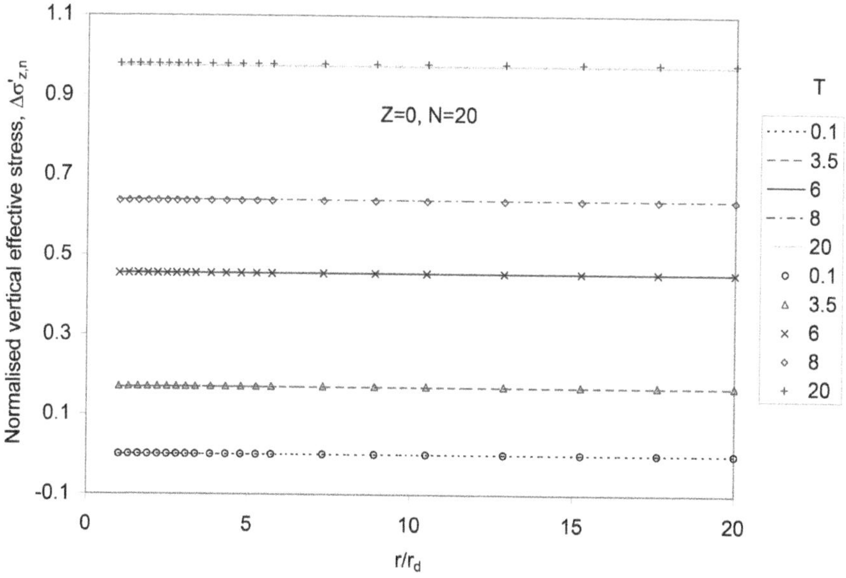

Figure 11.18(a) Isochrones of vertical effective stress $\Delta\sigma'_{z,n}$ when $N = 20$, $Z = 0$ for Case C.

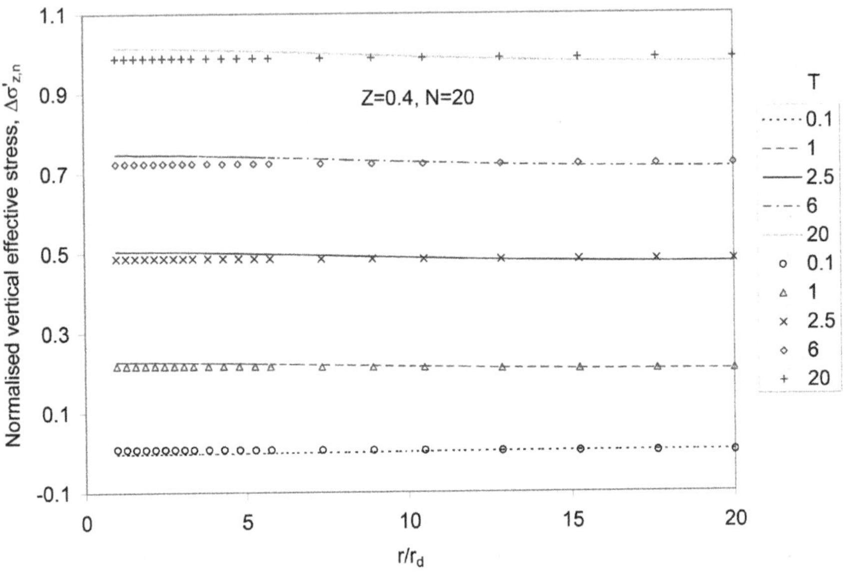

Figure 11.18(b) Isochrones of vertical effective stress $\Delta\sigma'_{z,n}$ when $N = 20$, $Z = 0.4$ for Case C.

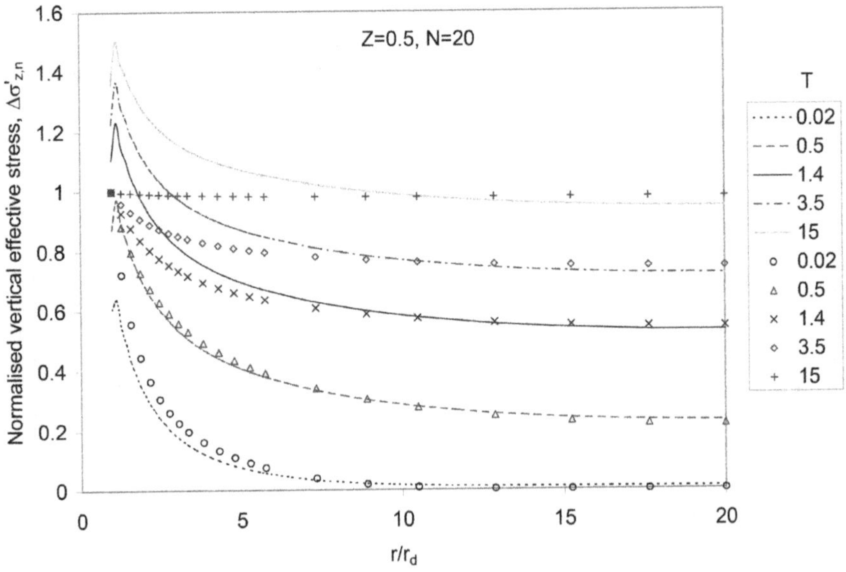

Figure 11.18(c) Isochrones of vertical effective stress $\Delta\sigma'_{z,n}$ when $N = 20$, $Z = 0.5$ for Case C.

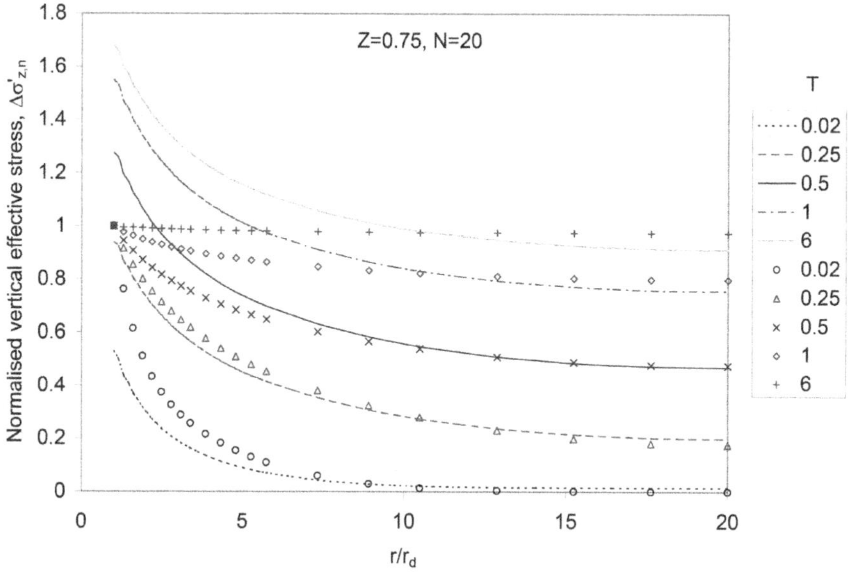

Figure 11.18(d) Isochrones of vertical effective stress $\Delta\sigma'_{z,n}$ when $N = 20$, $Z = 0.75$ for Case C.

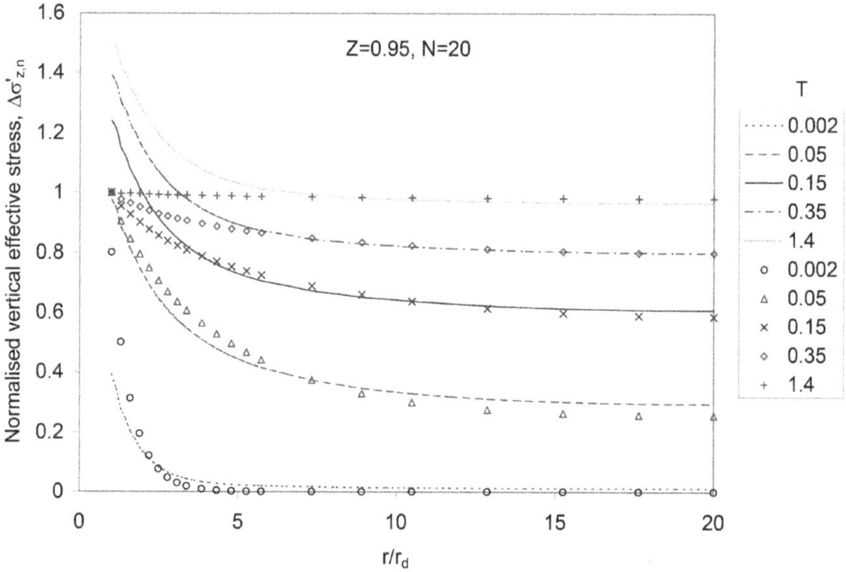

Figure 11.18(e) Isochrones of vertical effective stress $\Delta\sigma'_{z,n}$ when $N = 20$, $Z = 0.95$ for Case C.

linear elastic case. The phenomenon of instability clearly demonstrated in Figure 11.17(c) at the very beginning using ABAQUS (1995) does not either occur in the computation using the simplified method. The behavior of the soil below $Z = 0.4$ is actually 1-D as shown in Figure 11.18. Therefore, there are no differences whether using 1-D model or using the 3-D model in this region.

11.5 CONSOLIDATION ANALYSIS OF A CASE HISTORY

In this section, the consolidation behavior of the foundation soils under the Test Embankment at Chek Lap Kok New International Airport in Hong Kong is analyzed using the simplified FE method described above based on the site investigation report (Zhu *et al.* 2001). Various aspects that need to be considered in applying the simplified method are illustrated through this case study.

11.5.1 Background

In the 1970s, it was proposed to build a replacement airport for Hong Kong by leveling the islands of Chek Lap Kok and Lam Chau and reclaiming 600 ha of land from the sea. The sea is up to 10-m deep with a tidal range of 2 m, and reclamation would involve placing approximately 80,000,000 m^3 of fill in thickness up to 20 m.

Site investigations (RMP ENCON Ltd. 1982a; Koutsoftas *et al.* 1987) for the whole replacement area revealed that the entire offshore is blanketed by soft to very soft, dark grey, plastic, marine clay (upper marine clay) with pockets of shells. The thickness of the clay varies considerably over the site, from as little as 1.0 m or less to over 15 m, but frequently in the range of 6–8 m. Examination of 'undisturbed' samples of the upper marine clay showed the absence of lamination or layering. The clay is underlain at some locations by loose to medium dense marine sand up to 3-m thick. Underlying the upper marine clay (and the irregular sand) is an alluvial stratum (upper alluvial crust), consisting of inter-bedded layers of mottled, oxidized, and discoloured very stiff clay and dense sand. The thickness of this deposit varies in an erratic manner over the site but rarely exceeds 7.5 m. The third stratum is a light grey to dark grey medium-stiff to stiff clay (lower marine clay) inter-bedded with medium-dense sand lenses and occasional layers of very stiff mottled reddish and brown clay. Geologically, this is believed to be a marine deposit during a high sea level stand. Underlying the lower marine clay deposit is an alluvial deposit (lower alluvium), consisting primarily of very dense, coarse to fine sands, grading into a layer of gravel and cobbles. Occasional clay pockets are encountered within this deposit and occasionally below the gravel. The thickness of the lower alluvial deposit ranges from 0 to 10 m. Below the lower alluvial deposit (or the lower marine deposit where the lower alluvium is absent) is a layer of completely decomposed granite. The compressibility of this layer is very small. Experiments of excess pore pressure dissipation demonstrate that it can be viewed as free-draining layer.

The soil condition presented obvious geotechnical difficulties for the development of the new airport. The option of removing and replacing the soft marine clay would be very expensive, involving excavation of approximately 37,000,000 m^3 of the material, and transporting it for disposal about 20 km from the site. The alternative of leaving the soft clay in place would cause settlement of up to 4 m. These could be expected to extend over many years because of the high compressibility and low permeability of the deposits. Settlement tolerance after completion of the airport was very limited, since it would cause unacceptable differential settlements. Reclaiming over the soft clay might also result in the development of mud-waves, and lead to serious construction problems that could jeopardize the project.

To assess the feasibility of using vertical drains and filling techniques to prevent mud-wave formation, an instrumented test embankment (RMP ENCON Ltd. 1982b; Cheung and Ko 1986; Foott *et al.* 1987) was constructed between 1981 and 1983 on the west shore of Chek Lap Kok Island as shown in Figure 11.19. The main test area was a 100 m × 100 m square in plan, and was divided in to four quadrants. Alidrains were installed in the northwestern quadrant and northeastern quadrant at 1.5 and 3 m triangular spacing,

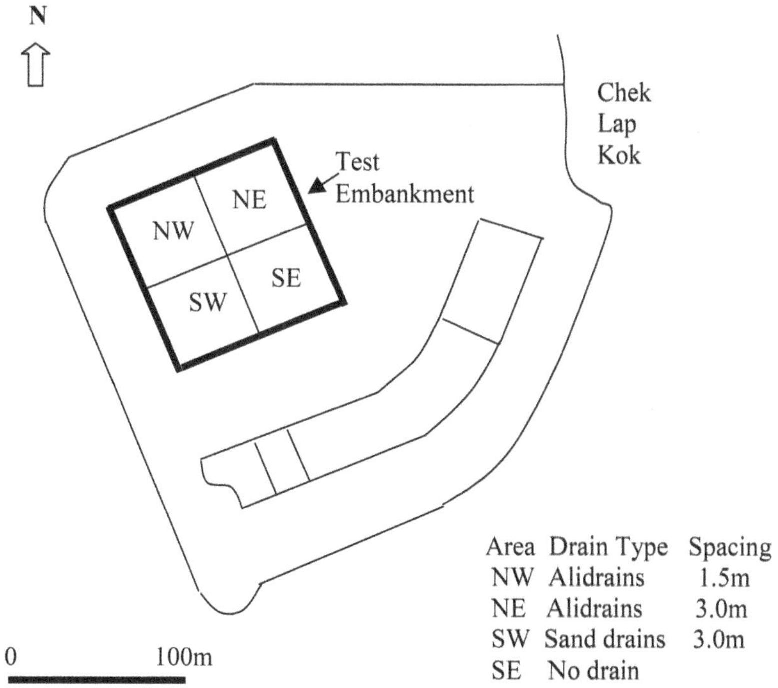

Figure 11.19 Plan of the test embankment at Chek Lap Kok new airport.

respectively. The Alidrains were prefabricated band-shaped vertical drains with width b = 100 mm and thickness h = 7 mm. The bottoms of the drains were settled to –21 mPD. Displacement sand drains with 500-mm diameter were installed in the southwestern quadrant at 3-m triangular spacing. The southeastern quadrant was used as a control area, with no additional drains. The fill and foundation soils were heavily instrumented (RMP ENCON Ltd. 1982b; Handfelt *et al.* 1987) to monitor their performance during and after construction. The instrumentation consisted of pneumatic and hydraulic piezometers, settlement plates and pipes, sub-surface settlement anchors, and other probes.

The test fill provided nearly 1-D loading conditions. Therefore, the formulations presented in this chapter can be used to this situation. In this section, the consolidation behavior of foundation soils in the northwestern quadrant is analyzed using soil parameters suggested for the design of embankment in view of the data available. The results are then compared with the measured values.

11.5.2 Construction of the main test area

To construct the main test embankment, a layer of hydraulic sand fill 2-m thick within the main test area was first pumped. Second, the center of the embankment was filled above sea level. Third, the vertical drains and instrumentation were installed from the formed land. Forth, the central portion was raised to Elevation 6.4 mPD (Hong Kong Principal datum) from June 13, 1982 to July 2, 1982. This completed the initial construction of the test fill. After approximately 8 months of settlement, the northwestern quadrant was finally raised to 10.8 mPD from December 28, 1982 to January 21, 1983 in the second stage of testing. The 10.8-mPD elevation approximated the highest anticipated reclamation load.

The density ρ_t of the fill is very important in the accurate assessment of the incremental vertical stress. After careful examinations of the test data and relevant materials, Cheung and Ko (1986) suggested that the saturation unit weight of the hydraulic fill material and the bulk density of the decomposed granite placed above +2 to +10.8 mPD be chosen as 1.9 Mg/m^3. Figure 11.20 shows a simplified version of the incremental vertical loading (total stress at the top of the soft marine clay) calculated using these values. The figure is for the loading of the northwestern quadrant, and has been used for the consolidation analysis in following sections.

11.5.3 Soil profile and soil parameters

The general geology and sequences of stratification was described in RMP ENCON Ltd. (1982a) and Koutsoftas *et al.* (1987). As indicated earlier, the subsoils at the site can generally be classified as consisting of four layers, namely upper marine clay, upper alluvial crust, lower marine clay, and lower

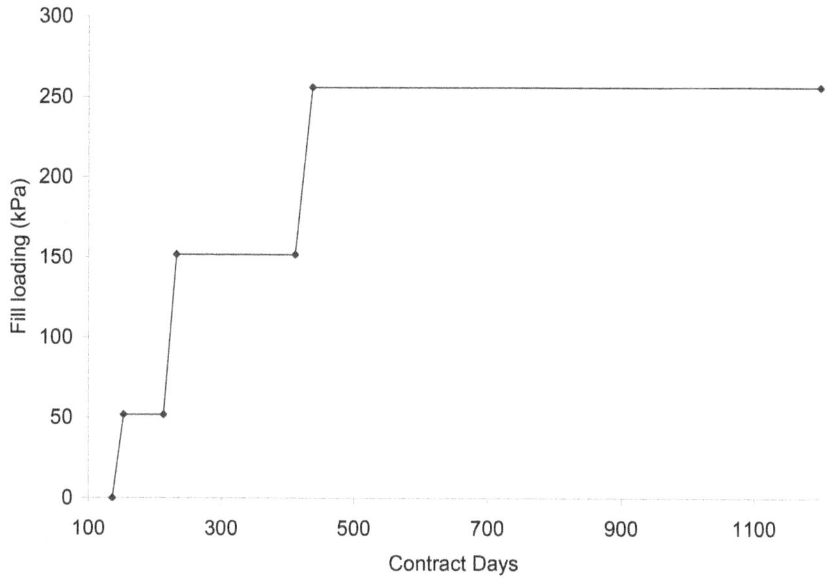

Figure 11.20 Simplified fill loading.

alluvium (Figure 11.22). The laboratory tests undertaken to determine the physical and engineering properties of the major strata included odometer tests, K_0-consolidated undrained triaxial compression tests, K_0-consolidated undrained direct simple shear tests, unconsolidated undrained triaxial compression tests, isotropically consolidated undrained triaxial compression tests, and index tests. Of these laboratory tests, only the odometer test and index test results are used in this analysis. A number of rising- and falling-head permeability tests were performed in the field during the original site investigation to evaluate hydraulic conductivity coefficients in the upper and lower marine clay.

For design of the reclamation works for the airport, RMP ENCON Ltd. (1982a) suggested the soil parameters outlined in following paragraphs (see also Table 11.2). All the parameters (k_v, k_r, λ/V, κ/V, ψ/V, unit weights, and maximum past pressure) were suggested in the site investigation reports except the coefficients of hydraulic conductivity for the upper alluvial crust. These have been estimated from the data given in the site investigation report (Zhu *et al.* 2001).

11.5.3.1 Upper marine clay

For the upper marine clay, whose plasticity ranges from 40 to 65%, a density value of 1.45 Mg/m^3 appears to be a suitable average. The soil has low compressibility in the recompression region. Recompression strain indices, C_{rs}

Table 11.2 Modeling parameters

Parameter	*Upper marine clay*	*Upper alluvium*	*Lower marine clay*	*Lower alluvium*
Soil model	Yin & Graham (1994) 1-D EVP	Yin & Graham (1994) 1-D EVP	Yin & Graham (1994) 1-D EVP	Non-linear elastic
k_v (m/sec)	2.2×10^{-9}	6.0×10^{-9}	2.5×10^{-10}	1.0×10^{-8}
k_r (m/sec)	4.0×10^{-9}	1.2×10^{-8}	6.2×10^{-10}	2.0×10^{-8}
κ/V	1.086×10^{-2}	1.086×10^{-2}	1.52×10^{-2}	–
λ/V	0.174	0.065	0.1086	4.22×10^{-3}
ψ/V	2.14×10^{-3}	1.69×10^{-3}	2.62×10^{-3}	–
t_o (day)	1	1	1	–
ρ_t (Mg/m³)	1.45	1.95	1.85	2.01

(see Figure 11.21), range from 0.02 to 0.03. In the virgin compression range, the soil is highly compressible with compression strain indices, C_{cs}, ranging from 0.30 to 0.50. Respective design values of 0.025 *for* C_{rs} and 0.40 for C_{cs} are considered appropriate for this soil. The upper marine clay exhibits very low secondary compression at stress level below the maximum past pressure. However, in the virgin compression range, coefficients of secondary consolidation are typically above $C_{\alpha\varepsilon} = 1.5\%$ per logarithm cycle of time and a value of 1.75% per logarithm cycle of time is recommended for design. Coefficients of vertical hydraulic conductivity calculated from consolidation tests are in the range of 2×10^{-9} to 5×10^{-9} m/sec, with a mean value of 2.2×10^{-9} m/sec. The in-situ horizontal coefficients of hydraulic conductivity from variable head permeability tests range from 3×10^{-9} to 5×10^{-9} m/sec, and a value of 4×10^{-9} m/sec appears to be a suitable average.

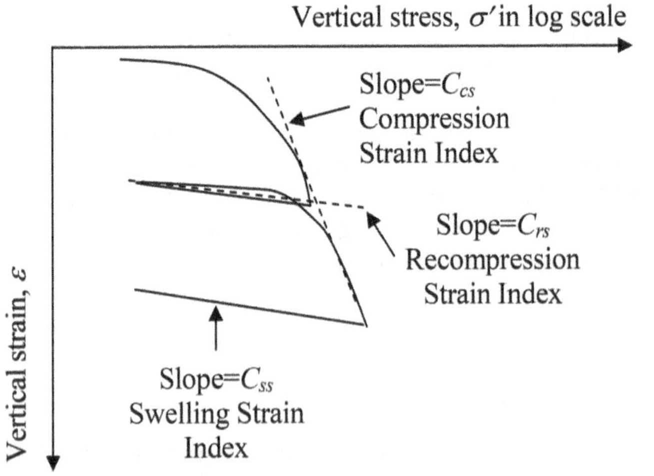

Figure 11.21 Definition of compressibility parameters.

11.5.3.2 Upper alluvial crust

The upper alluvial crust is a medium plasticity clay with plasticity index in the range 20 to 35%. A bulk unit density value of 1.95 Mg/m^3 is recommended for design. Recompression strain indices of this soil, C_{rs}, range from 0.015 to 0.035. In the virgin compression zone, the compressibility is comparatively small with compression strain indices, C_{cs}, ranging from 0.10 to 0.20. Respective design values of 0.025 for C_{rs} and 0.15 for C_{cs} are recommended for design. Coefficients of secondary compression $C_{\alpha\varepsilon}$ in the virgin range are on the order of 0.8%, and this value appears to be a suitable average. There is no permeability data available for this stratum. The vertical coefficient of hydraulic conductivity is estimated as 6×10^{-9} m/sec from the consolidation tests (RMP ENCON Ltd. 1982a), and is used in the following analysis. The horizontal coefficient of hydraulic conductivity is set to be 12×10^{-9} m/sec.

11.5.3.3 Lower marine clay

The lower marine clay is of medium plasticity, with plasticity indices ranging from 25 to 40%. A bulk unit density value of 1.85 Mg/m^3 is recommended for design. In the recompression stress range, compressibility is low, with recompression strain indices ranging from 0.02 to 0.05. In the normally consolidated range, the soil is quite compressible, with compression strain indices, C_{cs}, ranging from 0.20 to 0.35. Respective design values of 0.035 for C_{rs} and 0.25 for C_{cs} are recommended for design. Coefficients of secondary compression $C_{\alpha\varepsilon}$ in the virgin zone range typically from 1.7 to 3.0%, and 1.7% is used in the analysis. Coefficients of vertical hydraulic conductivity estimated from consolidation tests are typically 2×10^{-10} to 3×10^{-10} m/sec, with a mean value of 2.5×10^{-10} m/sec. The in-situ horizontal coefficients of hydraulic conductivity from variable head permeability tests are 4×10^{-10} to 8×10^{-10} m/sec, and a value of 6.2×10^{-10} m/sec appears to be a suitable average.

11.5.3.4 Lower alluvium

This is a very dense layer, the bulk density is chosen as 2.01 Mg/m^2. The pre-consolidation pressure of this layer will exceed the final stresses imposed by the fill. Therefore, a small recompression strain index of 0.01 is adopted in this analysis. The vertical hydraulic conductivity coefficient is chosen as 1×10^{-10} m/sec, and the horizontal hydraulic conductivity coefficient is set to be 2×10^{-10} m/sec.

11.5.3.5 Location of instrumentation

Figure 11.22 shows the elevations of pneumatic piezometers (PP) and Sondex anchors in the northwestern quadrant. These instruments were placed at the center of a triangular grid (in plan) of the vertical drains. The construction drawings generally gave only approximate locations for the Sondex rings.

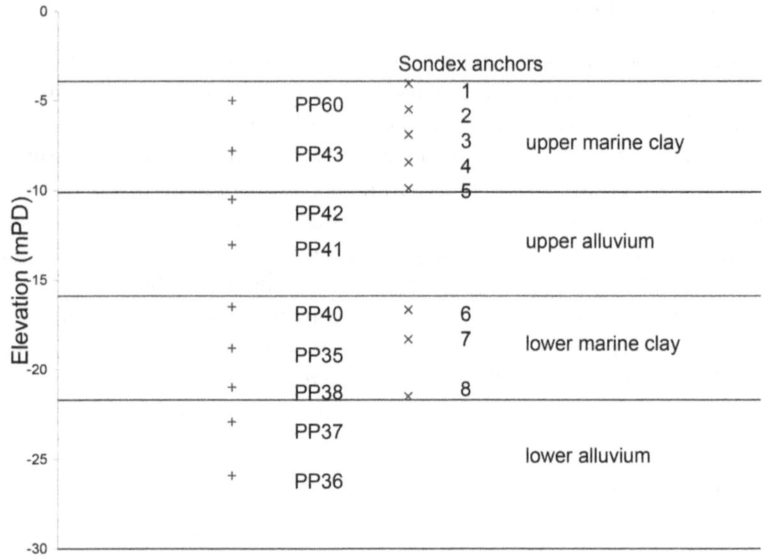

Figure 11.22 Soil profile and setup of instrumentation.

11.5.3.6 Soil model and parameters

Analysis of the settlements used the 1-D elastic visco-plastic (EVP) soil model in (5.4) (Yin and Graham 1989, 1994) for the top three layers. The simpler non-linear elastic model in (5.2) was adopted for the lower alluvium. The parameters for these models are the values originally suggested for the design of the reclamation site and outlined in the preceding paragraphs. These are listed in Table 11.2. The parameters k_r and k_v in Table 11.2 are the horizontal and vertical hydraulic conductivities, respectively, and are taken as constants in the analysis. The parameter ρ_t is the bulk density. Although these values are typical for the whole site, no borehole was located in the immediate location of the test embankment.

11.5.3.7 Initial and boundary conditions

As boundary conditions, the top surface of the upper marine clay and the bottom surface of the lower alluvium were treated as free drainage boundaries. The initial stress and the modeling of the maximum past pressure used in the calculation are plotted in Figure 11.23. Initial strains were calculated using the methods suggested in Chapter 5 (Zhu and Yin 2000b).

11.5.4 Vertical drain characteristics and smear zone

As mentioned earlier, the Alidrains (width b = 100 mm and thickness h = 7) were arranged in triangular pattern with drain spacing of 1.5 m and their tips

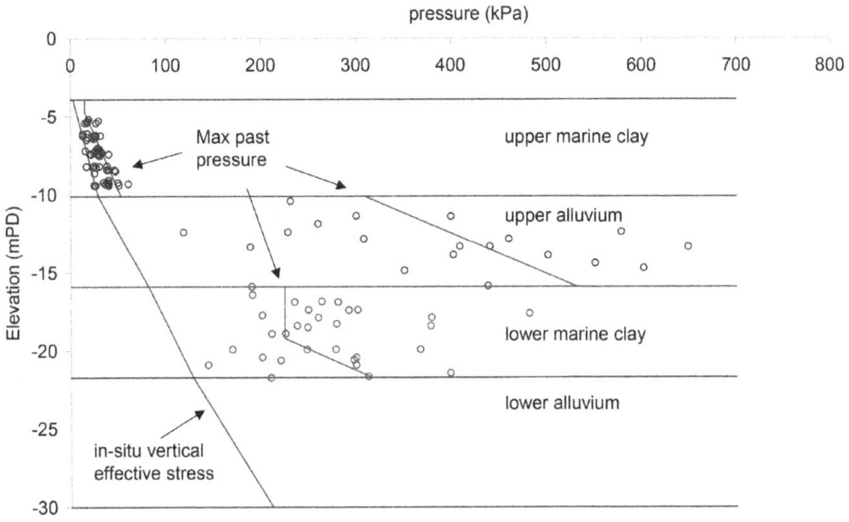

Figure 11.23 In-situ vertical effective stress and the maximum past pressure.

at −21 mPD. The equivalent radius of vertical drains, r_d, can be determined in several ways (for example, Hansbo 1979; Atkinson and Eldred 1981; Long and Alvaro 1994). It seems that (6.8) suggested by Long and Alvaro (1994) agrees well with experimental values most closely and has been adopted in this analysis. On this basis, r_d is 27.45 mm. For triangular installation patterns, the equivalent radius r_e of influence of the vertical drain is 0.525 times the drain spacing (Figure 11.24). That is, $r_e = 0.525 \times$ Drain spacing = 0.7875 m (Barron 1948).

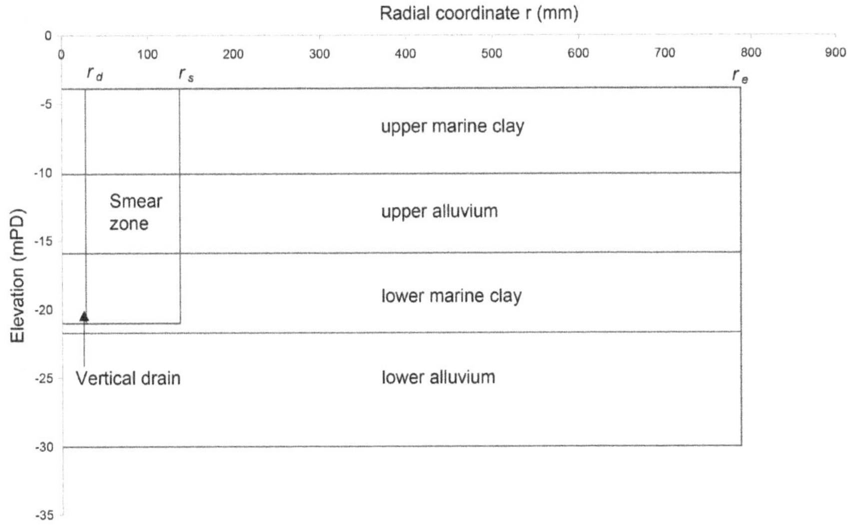

Figure 11.24 Geometry of the simplified model for consolidation of soils with vertical drain.

Installation of vertical drains creates a region of disturbed soil, called the smear zone, with outer radius r_s around the drain. Installation procedures that use a mandrel of radius r_m cause the most severe disturbance. Outward displacement of the soil distorts the adjacent ground. The zone of soil near the drain is remolded and dragged downward and then upward as the mandrel is pushed into and then pulled out of the ground. In soft soils where the technique is most useful, the overall effect is to produce a disturbed soil zone of reduced permeability, reduced pre-consolidation pressure, and increased compressibility (Johnson 1970). The analysis assumed r_s to be five times the equivalent radius of the vertical drain, that is, 137 mm.

A diamond-shaped mandrel with external dimensions of 75 and 166 mm was used to install the Alidrains. The equivalent radius of the mandrel is r_m = 63 mm. These values of r_s and r_m correspond to a value of $r_s/r_m = 2.2$, within the range proposed by Mesri and Lo (1991).

Results of fully coupled FE analysis in the preceding section (Zhu and Yin 2000a) show that vertical effective stresses within five times the equivalent radius of vertical drain are much higher than in other parts of the domain. These higher stresses will reduce the permeability of the soil near vertical drains regardless of the installation method. In addition, remolding due to installation may reduce the permeability in the smear zone. However, the reduction and size of the smear zone are still not exactly known. In the following analysis, the horizontal and vertical hydraulic conductivities in the smear zone are assumed equal to the vertical hydraulic conductivity of undisturbed soil (Broms 1987). Since the equivalent cross-section of the Alidrain is small and the drain length is up to 17.1 m, the effect of internal resistance to water flow in the drain needs to be considered. In the analysis, the hydraulic conductivity coefficient of the vertical drain is assumed equal to 1.2 m/day (this is essentially the hydraulic conductivity of clean sand). The deformation behavior of the Alidrain material allows it to adopt the recompression stress-strain relationships of the surrounding soil.

11.5.5 Finite element analysis results and discussions

The computed settlements at the elevations of the Sondex settlement gauge anchors are plotted in Figure 11.25. Measured values were reported earlier by Cheung and Ko (1986) and are also shown in the figure. In the early stages of loading, the measured values are larger than the computed results. This may have been caused by shear straining and lateral movement of the soils from under the fill, particularly in the very soft upper marine clay. After the final loading stage, the computed settlements are larger than the measured results. Taking into consideration of the prediction nature of this analysis, and the conservatism in the design parameters, the results are quite good. A subsequent laboratory program of odometer tests (Cheung and Ko 1986) on 17 undisturbed samples of the upper marine clay close to the location of the test embankment showed that the maximum compression strain index is 0.333.

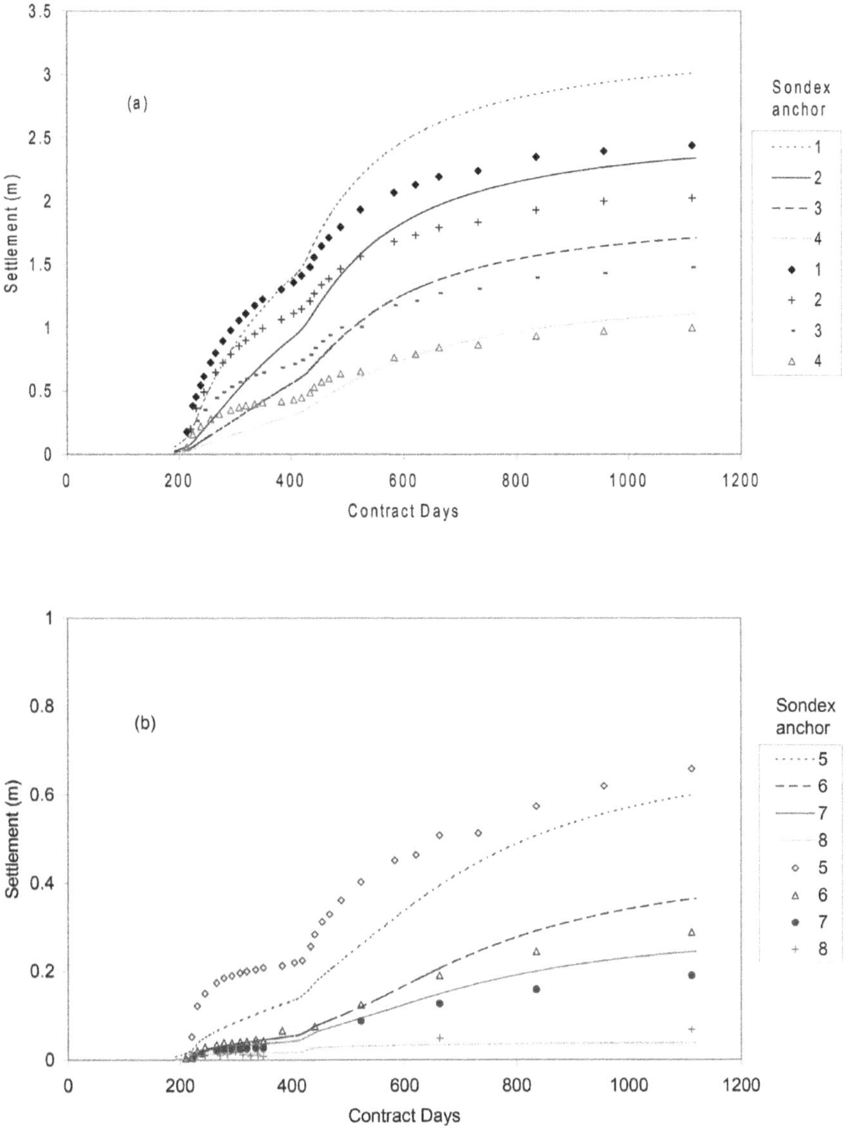

Figure 11.25 Comparison between measured settlement (points) and computed settlement (lines) using design values (a) Sondex anchors 1-4, and (b) Sondex anchors 5-9.

This is smaller than the value of 0.4 suggested by RMP ENCON Ltd. (1982a) from the overall investigation and used in the analysis. Using a lower value of 0.32 instead of 0.4 for the upper marine clay in the FE analysis produced better agreement (Figure 11.26) between computed settlements and measured results. The predicted settlements in the upper layer are still larger than the measured settlements.

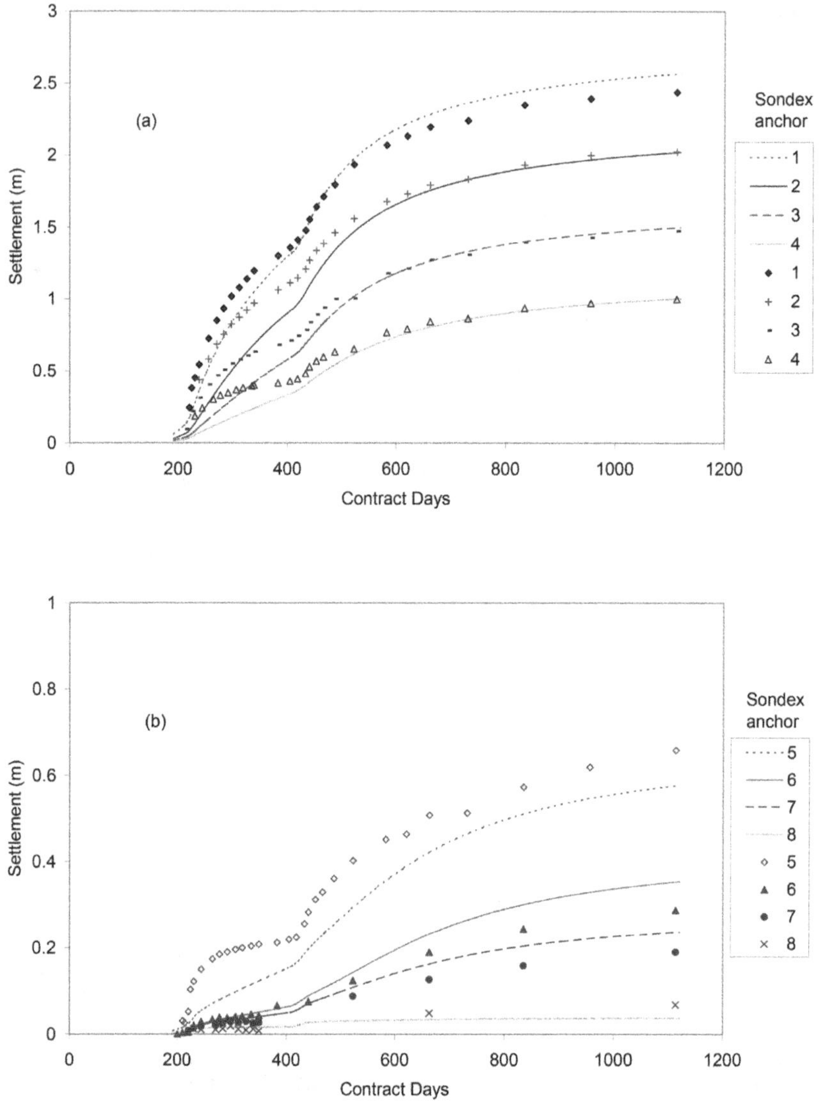

Figure 11.26 Comparison between measured settlement (points) and computed settlement (lines) using test embankment site experiment data (a) Sondex anchors 1-4, and (b) Sondex anchors 5-9.

Computed and measured values of porewater pressures are shown in Figure 11.27. The computed values were again obtained using the soil parameters in Table 11.2 that came from the original testing for the design of the reclamation works. Figure 11.27 shows that, although the trends have been modeled well, the computed porewater pressures (shown as lines) during the

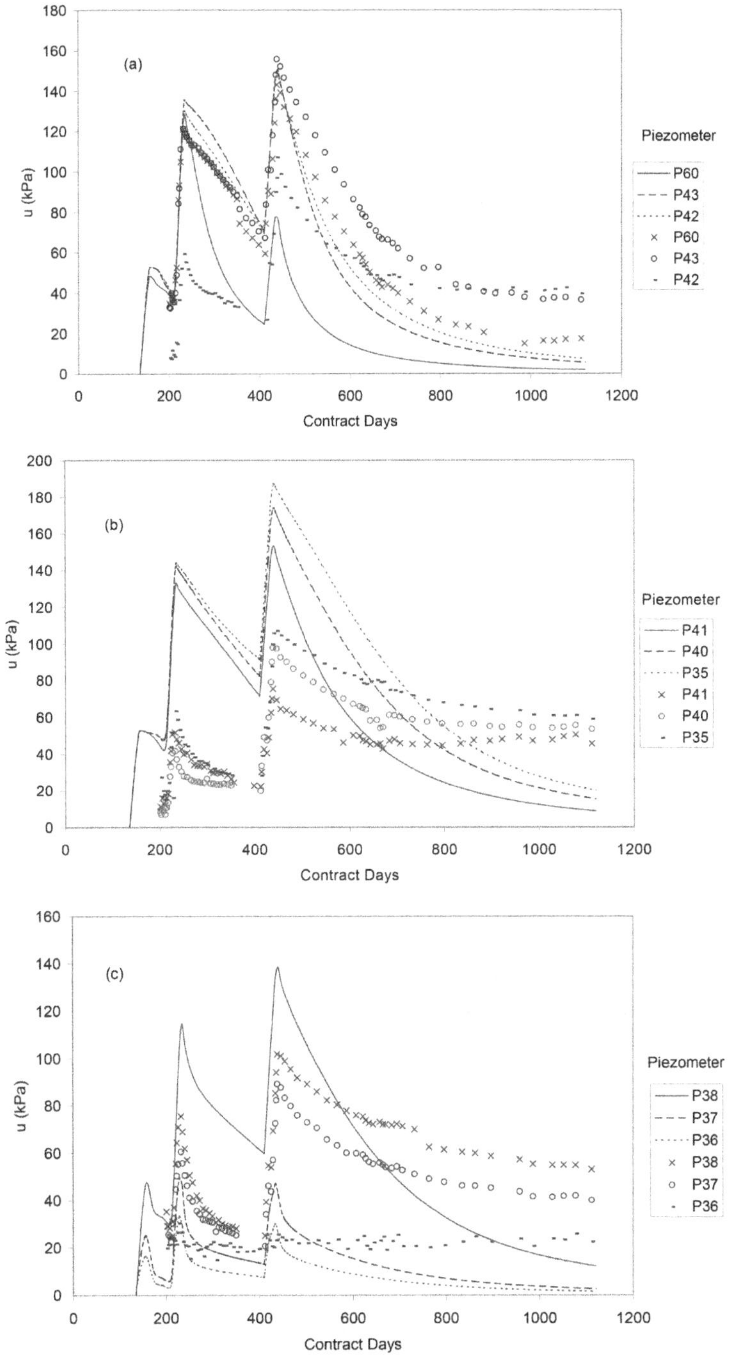

Figure 11.27 Comparison between measured pore pressure (points) and computed pore pressure (lines) using design values (a) Piezometers P42,P43 and P60; (b) Piezometers P35, P40 and P41; and (c) Piezometers P36, P37 and P38.

initial loading stages are higher than measured results (shown as symbols). After the final loading stage, however, the computed porewater pressures are lower than measured values. In other words, the rates and durations of the porewater pressure dissipation and the settlements have not been well modeled. Using vertical total stress changes determined from Boussinesq elastic stress distributions produces results that are almost the same as those from the new modeling. The simulation is from the start of construction (contract day 136), while the field measuring only began after the initial loading stage (after contract day 201). Therefore, there are no records available for the initial loading as simplified in the loading curve. When plotted in the figures, there are only two large increases in measured porewater pressure, whereas the predictions show three increases.

One reason for the lack of agreement may be non-linear changes of hydraulic conductivity with increasing effective pressure and the resulting decreases in void ratio. Hydraulic conductivities are larger at the beginning of loading and become smaller as the consolidation proceeds. The soils in this study undergo relatively large deformations, and the effects of the non-linear hydraulic conductivity may be significant. Although the program used for the analysis can take some account of non-linear hydraulic conductivity, suitable information was not available from the laboratory program. The measured porewater pressures at PP60 near the top of the upper marine clay during early loading dissipate more slowly than the predicted values. This suggests that the upper boundary (the mudline) may not be a free drainage boundary or that some destructuring was taking place. Computed porewater pressures at PP43 agree well with the measured values. Relatively large differences were noted, however, between computed and measured porewater pressures at PP42, PP41, PP40, PP35, and PP38.

The hydraulic conductivity at PP35 (located in the lower marine clay) is examined in more detail in the following paragraph. The measured porewater pressure increased from 15 kPa at contract day 213 to 63.4 kPa at contract day 232. This corresponds a vertical total stress increment of 99.8 kPa in the second loading step. The measured porewater pressure also increased from 28.4 kPa at contract day 411 to 107.14 kPa at contract day 437 corresponding a vertical total stress increment of 104.5 kPa in the third loading step. From the test arrangement, deformations can be considered as 1-D. To estimate a lower limit of hydraulic conductivity of the lower marine clay using the measured value at PP35, it is assumed that the vertical drain is completely free draining. In the second loading step, 51.5% of the increased excess porewater pressure dissipated; and in the third loading step, 24.7% of the increased excess porewater pressure dissipated. The analytical solution in (6.19) and (6.28) (Zhu and Yin 2001) suggests that this corresponds a time factor of $T = 1.93$ for the second loading step and $T = 0.77$ for the third loading step. The respective coefficients of radial consolidation are 0.0815 and 0.0213 m/day for the two loading steps. Substituting the recompression strain index $C_{rs} = 0.035$ and the initial effective stress (105.5 kPa), the radial coefficient of permeability will be 1.33×10^{-9} (m/s) for the second loading step and 3.48×10^{-10} (m/s) for the third loading step. The coefficient of permeability in the second loading step is much larger than that used in the calculations. It can also be seen that the coefficient of permeability in the third loading step is about 30% of the value in the second loading step.

Examination of the borehole log obtained during the 1984 testing program described by Cheung and Ko (1986) in the northwestern quadrant of the test fill indicates a layer of medium dense, dark grey, fine to medium sand layer from –21.4 to –21.8 mPD. These elevations place it in the range of lower marine clay in the soil profile (Figure 11.22) used for the calculations. Also, a borehole log for the southwestern quadrant exposes a medium dense, greyish brown, fine-to-medium sand layer from –11.0 to –11.9 mPD. These elevations are in the upper alluvium crust. It is likely therefore that the hydraulic conductivities and drainage boundaries at the site may be rather different from those determined from the original investigation and used in the modeling.

Although it appears that settlement magnitudes can be predicted with some success, attempts to compare predicted and measured excess porewater pressures under embankment projects have generally been less successful. Excess porewater pressures vary rapidly in the horizontal direction around sand drains and wicks and can produce changes in hydraulic conductivity. As a result, relatively small variation in the positions of the wicks or piezometers can lead to large potential errors (Olson 1998).

The compression of the lower three layers in Figure 11.22 is relatively small compared with that of the upper marine clay. Thus, the relatively large discrepancies in the modeling of porewater pressure in the lower three layers are of little importance for the prediction of the total settlement of the reclaimed land.

APPENDIX 11.A

ABAQUS (1995) gives the following non-linear elastic relationship between elastic volume strain ε_v and effective mean stress p' for large deformation problems

$$1-\exp(-\varepsilon_v)=\frac{\kappa}{1+e_0}\ln\left(\frac{p'+p_t^{el}}{p_o+p_t^{el}}\right) \tag{A-1}$$

It is noted here that compressive strains and stresses are positive. When deformation is small, volume strain ε_v must be small, then the left side of (A-1) is $1-\exp(-\varepsilon_v)\approx\varepsilon_v$. Equation (A-1) becomes the well-known relationship for isotropic compression of soil. The meaning of the parameters in (A-1) has been explained in the main text. Equation (A-1) can be inverted to give

$$p'-p_0=(p_0+p_t^{el})\left\{\exp\left[\frac{1+e_0}{\kappa}(1-\exp(-\varepsilon_v))\right]-1\right\} \tag{A-2}$$

where $p' = \frac{\sigma'_x+\sigma'_y+\sigma'_z}{3}$

The incremental deviator stress-strain relationship can be written as

$$dS = 2Gde \tag{A-3}$$

where S and $\boldsymbol{e}$ are the deviator stress tensor and deviator strain tensor, respectively.

It is noted that for isotropic elastic behavior, $G = \frac{3(1-2\nu)}{2(1+\nu)}K$ and $K = \frac{dp'}{d\varepsilon_v}$. Thus, for the vertical component, from (A-3)

$$d(\sigma'_z - p') = 2G\left(d\varepsilon_z - \frac{d\varepsilon_v}{3}\right) = \frac{3(1-2\nu)}{1+\nu}\frac{dp'}{d\varepsilon_v}\left(d\varepsilon_z - \frac{d\varepsilon_v}{3}\right) \tag{A-4}$$

Substituting $d\varepsilon_z = d\varepsilon_v$, for 1-D compression (*i.e.* 1-D straining or odometer test condition), into (A-4) gives

$$d(\sigma'_z - p') = \frac{2(1-2\nu)}{1+\nu}dp' \tag{A-5}$$

Integrating (A-5) gives

$$\sigma'_z = \sigma'_{zo} + \frac{3(1-\nu)}{1+\nu}(p' - p_0) \tag{A-6}$$

Substituting (A-2) into (A-6), we have

$$\sigma_z = \sigma'_{zo} + \frac{3(1-\nu)}{1+\nu}(p_0 + p_t^{el})\left\{\exp\left[\frac{1+e_0}{\kappa}(1-\exp(-\varepsilon_v))\right]-1\right\} \tag{A-7}$$

where σ'_{z0} is the initial vertical effective stress and p_0 is the corresponding initial mean effective principal stress.

Rearranging (A-7) and noting $\varepsilon_v = \varepsilon_z$ for 1-D compression

$$\varepsilon_z = -\ln\left\{1 - \frac{\kappa}{1+e_0}\ln\left[1 + \frac{(1+\nu)(\sigma'_z - \sigma'_{zo})}{3(1-\nu)(p_0 + p_t^{el})}\right]\right\} \tag{A-8}$$

Equation (A-8) is the function f in (11.5) in the text used in the simplified consolidation model.

REFERENCES

ABAQUS *reference manual*, 1995, Version 5.5, (Providence, RI: Hibbitt, Karlsson and Sorensen. Inc.).

Atkinson, MS, and Eldred, PJL, 1981, Consolidation of soil using vertical drains. *Géotechnique*, **31**(1), pp. 33–43.

Barron, RA, 1948, Consolidation of fine-grained soils by drain wells. *Transactions*, ASCE, **113**(1), No. 2346, pp. 718–742.
Bergado, DT, Mukherjee, K, Alfaro, MC, and Balasubramaniam, AS, 1993, Prediction of vertical-band-drain performance by the finite-element method. *Geotextiles and Geomembranes*, **12**(6), 567–586.
Booker, JR, and Small, JC, 1987, A method of computing the consolidation behaviour of layered soils using direct numerical inversion of Laplace transforms. *International Journal for Numerical and Analytical Methods in Geomechanics*, **11**, pp. 363–380.
Broms, BB, 1987, Soil improvement methods in Southeast Asia for soft soils. *In Proceedings of the 8th Asian Regional Conference on Soil Mechanics and Foundation Engineering*, **2**, pp. 29–64.
Carrillo, N, 1942, Simple two- and three-dimensional cases in the theory of consolidation of soils. *Journal of Mathematics and Physics*, **21**(1–4), pp. 1–5
Cheng, AH-D, Detournay, E, and Abousleiman, Y, 1998, Poroelasticity. *International Journal of Solids and Structures, Maurice A. Biot Memorial Issue*, **35**, pp. 4513–5031.
Cheung, RKH, and Ko, SWK, 1986, Replacement airport at Chek Lap Kok civil engineering design studies. Study Report No. 2B – continued monitoring of the test embankment, Vol. 1, main report; Vol. 2 appendices D, E, and F; Vol 3 appendices H, I, and J. Unpublished Report, Civil Engineering Department, Hong Kong, SAR.
Desai, CS, Kuppusamy, T, Koutsoftas, DC, and Janardhanamm, R, 1979, A one-dimensional finite element procedure for nonlinear consolidation. *In Proceedings of the Third International Conference on Numerical Methods in Geomechanics*, pp. 143–148.
Foott, R, Koutsoftas, DC, and Handfelt, LD, 1987, Test fill at Chek Lap Kok, Hong Kong. *Journal of Geotechnical Engineering*, **113**(2), pp. 106–126.
Garlanger, JE, 1972, The consolidation of soils exhibiting creep under constant effective stress. *Géotechnique*, **22**(1), pp. 71–78.
Handfelt, LD, Koutsoftas, DC, and Foott, R, 1987, Instrumentation for test fill in Hong Kong. *Journal of Geotechnical Engineering*, ASCE, **113**(2), pp. 127–146.
Hansbo, S, 1979, Consolidation of clay by band-shaped prefabricated drains. *Ground Engineering*, **12**(5), pp. 16–25.
Hansbo, S, 1981, Consolidation of fine-grained soils by prefabricated drains. *In Proceedings of the 10th International Conference on Soil Mechanics and Foundation Engineering*, **3**, pp. 667–682.
Hardin, BO, 1989, 1-D strain in normally consolidated cohesive soils. *Journal of Geotechnical Engineering Division, ASCE*, **115**(5), pp. 689–710.
Hart, EG, Kondner, RL, and Boyer, WC, 1958, Analysis for partially penetrating sand drains. *Journal of the Soil Mechanics and Foundations Division*, ASCE, 84(4), pp. 1–15.
Hird, CC, Pyrah, IC, Russell, D, and Cinicioglu, F, 1995, Modelling the effect of vertical drains in two-dimensional finite element analyses of embankments on soft ground. *Canadian Geotechnical Journal*, **32**, pp. 795–807.
Horne, MR, 1964, The consolidation of a stratified soil with vertical and horizontal drainage. *International Journal of Mechanical Sciences*, **6**(2), pp. 187–197.
Indraratna, B, and Redana, IW, 1997, Plane-strain modelling of smear effects associated with vertical drains. *Journal of Geotechnical and Geoenvironmental Engineering*, **123**(5), pp. 474–478.

Johnson, SJ, 1970, Foundation precompression with vertical sand drains. *Journal of the Soil Mechanics and Foundations Division*, **96**(1), pp. 145–175.

Koutsoftas, DC, Foott, R, and Handfelt, LD, 1987, Geotechnical investigations offshore Hong Kong. *Journal of Geotechnical Engineering*, ASCE, **113**(2), pp. 87–105.

Leroueil, S, Kabbaj, M, Tavenas, F, and Bouchard, R, 1985, Stress-strain-strain rate relation for the compressibility of sensitive natural clays. *Géotechnique*, **35**(2), pp. 159–180.

Lewis, RW, Roberts, GK, & Zienkiewics, OC, 1976, A nonlinear flow and deformation analysis of consolidation problems. *In Proceeding of 2nd International Conference on Numerical Method in Geomechanics*, pp. 1106–1118.

Lewis, RW, and Schrefler, BA, 1998, *The Finite Element Method in the Static and Dynamic Deformation and Consolidation of Porous Media*, (Chichester: Wiley).

Lo, DOK, 1991, Soil improvement by vertical drains. *Ph.D. thesis*, University of Illinois at Urbana-Champaign, Illinois.

Long, RP, and Alvaro C, 1994, Equivalent diameter of vertical drains with an oblong cross section. *Journal of Geotechnical Engineering*, ASCE, **120**(9), pp. 1625–1629.

Mahyari, AT, and Selvadurai, APS, 1998, Enhanced consolidation in brittle geomaterials susceptible to damage. *Mechanics of Cohesive Frictional Materials*, **3**, pp. 291–303.

Mesri, G, and Lo, DOK, 1991, Field performance of prefabricated vertical drains. *In Proceedings of GEO-COAST '91, Yokohama*, pp. 231–236.

Olson, RE, 1977, Consolidation under time-dependent loading. *Journal of the Geotechnical Engineering Division*, ASCE, **103**(1), pp. 55–60.

Olson, RE, 1998, Settlement of embankments on soft clays. *Journal of Geotechnical and Geoenvironmental Engineering*, ASCE, **124**(4), pp. 278–288.

Olson, RE, Daniel, DE, and Liu, TK, 1974, Finite difference analyses for sand drains problems. *In Proceedings of the Conference on Analysis and Design in Geotechnical Engineering*, Texas, **1**, pp. 85–110.

Onoue, A, 1988, Consolidation of multilayered anisotropic soils by vertical drains with well resistance. *Soils and Foundations*, **28**(3), pp. 75–90.

Rendulic, L, 1935, Der hydrodynamische spannungsausgleich in zentral entwasserten tonzylindern. *Wasserwirtschaft und Technik*, **2**, pp. 250–253.

Richart, FE, Jr., 1959, Review of the theories for sand drains. *Transaction*, ASCE, **124**, 709–739.

RMP ENCON Ltd., 1982a, Replacement airport at Chek Lap Kok civil engineering design studies. Study Report No. 1 – site investigation, Vol. 1, main report. Unpublished Report, Civil Engineering Department, Hong Kong, SAR.

RMP ENCON Ltd., 1982b, Replacement airport at Chek Lap Kok civil engineering design studies. Study Report No, 2 – test embankment, Vol. 1, main report. Unpublished Report, Civil Engineering Department, Hong Kong, SAR.

Runesson, K, Hansbo, S, and Wiberg, NE, 1985, The efficiency of partially penetrating vertical drains. *Géotechnique*, **35**(4), pp. 511–516.

Selvadurai, APS, 1996, Mechanics of poroelastic media. *ASME Special Publication*, (Dordrecht: Kluwer Academic Publishers).

Selvadurai, APS, and Yue, ZQ, 1994, On the indentation of a poroelastic layer. *International Journal for Numerical and Analytical Methods in Geomechanics*, **18**(3), pp. 161–175.

Siriwardane, HJ, and Desai, CS, 1981, Two numerical schemes for nonlinear consolidation. *International Journal of Numerical Methods in Engineering*, **17**, pp. 405–426.

Tavenas, F, Jean, P, Leblond, P, and Leroueil, S, 1983, The permeability of natural soft clays. Part II: Permeability characteristics. *Canadian Geotechnical Journal*, 20(4), pp. 645–660.

Terzaghi, K, 1943. *Theoretical Soil Mechanics*, (New York: Wiley).

Vermeer, PA, and Verruijt, A, 1981, An accuracy condition for consolidation by finite elements. *International Journal for Numerical and Analytical Methods in Geomechanics*, **5**(1), pp. 1–14.

Xie, KH, Lee, PKK, and Cheung, YK, 1994, Consolidation of a two-layer system with ideal drains. *In Proceedings of the 8th International Conference on Computer Methods and Advances in Geomechanics*, Morgantown, West Virginia, USA, **1**, pp. 789–794.

Yin, JH, and Graham, J, 1989, Viscous-elastic-plastic modelling of one-dimensional time-dependent behaviour of clays. *Canadian Geotechnical Journal*, **26**(2), pp. 199–209.

Yin, JH, and Graham, J, 1994, Equivalent times and one-dimensional elastic visco-plastic modelling of time-dependent stress-strain behaviour of clays. *Canadian Geotechnical Journal*, **31**(1), pp. 42–52.

Yoshikuni, H, and Nakanodo, H, 1974, Consolidation of soils by vertical drain wells with finite permeability. *Soils and Foundations*, **14**(2), pp. 35–46.

Yue, ZQ, and Selvadurai, APS, 1995, Contact problem for saturated poroelastic solid. *Journal of Engineering Mechanics*, **121**(4), pp. 502–512.

Yue, ZQ, and Selvadurai, APS, 1996, Axisymmetric indentation of a multilayered poroelastic solid. *Mechanics of Poroelastic Media*, pp. 235–241, (Dordrecht: Springer).

Zeng, GX, and Xie, KH, 1989, New development of the vertical drain theories. *In Proceedings of the 12th International Conference on Soil Mechanics and Foundation Engineering*, Rotterdam, the Netherlands, **2**, pp. 1435–1438.

Zhu, GF, and Yin, JH, 2000a, Finite element analysis of consolidation of soils with vertical drain. *International Journal for Numerical and Analytical Methods in Geomechanics*, **24**(4), pp. 337–366

Zhu, GF, and Yin, J-H, 2000b, Elastic visco-plastic consolidation modelling of Berthierville test embankment. *International Journal for Numerical and Analytical Methods in Geomechanics*, **24**(5), pp. 491–508.

Zhu, GF, and Yin, JH, 1999, Finite element analysis of consolidation of layered clay soils using an elastic visco-plastic model. *International Journal for Numerical and Analytical Methods in Geomechanics*, **23**(4), pp. 355–374.

Zhu, GF, and Yin, J-H, 2001, Consolidation of soil with vertical and horizontal drainage under ramp load. *Géotechnique*, **51**(4), pp. 361–367.

Zhu, G, Yin, JH, and Graham, J, 2001, Consolidation modelling of soils under the test embankment at Chek Lap Kok International Airport in Hong Kong using a simplified finite element method. *Canadian Geotechnical Journal*, **38**(2), pp. 349–363.

Chapter 12

Finite element method for three-dimensional consolidation problems

12.1 INTRODUCTION

Practical consolidation problems frequently possess complicated geometries and material properties. Finding closed-form solutions can be extremely difficult even when they exist. The extensive use of computers and the concomitant development of numerical techniques have made possible much more precise numerical solutions. Thus, a general computer code implementing a numerical solution of the three-dimensional (3-D) consolidation equations can be used for a wide range of consolidation problems. The finite element (FE) method is one of the most popular techniques to solve coupled consolidation problems. The first authors to develop a FE formulation of Biot (1941) consolidation theory were Sandhu and Wilson (1969). Since then, numerous researchers (Christian and Boehmer 1970; Small *et al.* 1976; Carter *et al.* 1979; Siriwardane and Desai 1981; Borja 1989; Lewis and Schrefler 1998; Sloan and Abbo 1999) have developed various solution strategies. The stability and accuracy have been investigated by Booker and Small (1975), Vermeer and Verruijt (1981), and Zhu *et al.* (2004). In this chapter, the general 3-D FE method is derived.

12.2 GOVERNING EQUATIONS

The basic assumptions and equations are presented in Chapter 9. For easy reference, the equations in Cartesian co-ordinates are listed as follows.

12.2.1 Force equilibrium equations

$$\frac{\partial \sigma_x}{\partial x}+\frac{\partial \tau_{xy}}{\partial y}+\frac{\partial \tau_{xz}}{\partial z}=f_x \tag{9.3a}$$

$$\frac{\partial \tau_{yx}}{\partial x}+\frac{\partial \sigma_y}{\partial y}+\frac{\partial \tau_{yz}}{\partial z}=f_y \tag{9.3b}$$

$$\frac{\partial \tau_{xz}}{\partial x} + \frac{\partial \tau_{zy}}{\partial y} + \frac{\partial \sigma_z}{\partial z} = f_z \tag{9.3c}$$

12.2.2 Strain-displacement relation

$$\varepsilon_x = -\frac{\partial w_x}{\partial x} \tag{9.14b}$$

$$\varepsilon_y = -\frac{\partial w_y}{\partial y} \tag{9.14c}$$

$$\varepsilon_z = -\frac{\partial w_z}{\partial z} \tag{9.14d}$$

$$\varepsilon_{xy} = -\frac{1}{2}\left(\frac{\partial w_x}{\partial y} + \frac{\partial w_y}{\partial x}\right) \tag{9.14e}$$

$$\varepsilon_{yz} = -\frac{1}{2}\left(\frac{\partial w_y}{\partial z} + \frac{\partial w_z}{\partial y}\right) \tag{9.14f}$$

$$\varepsilon_{zx} = -\frac{1}{2}\left(\frac{\partial w_z}{\partial x} + \frac{\partial w_x}{\partial z}\right) \tag{9.14g}$$

12.2.3 Effective stress

$$\vec{\sigma}' = \vec{\sigma} - u_w \mathrm{m} \tag{9.24}$$

12.2.4 Constitutive equations

$$\dot{\vec{\sigma}}' = \mathrm{D}\dot{\vec{\varepsilon}} = \begin{cases} \mathrm{D}_e \dot{\vec{\varepsilon}} & F < 0 \\ \mathrm{D}_{evp} \dot{\vec{\varepsilon}} & F = 0 \end{cases} \tag{9.34}$$

12.2.5 Darcy's law

$$\mathrm{q} = -\mathrm{k}\nabla h = -\frac{\mathrm{k}}{\gamma_w}\nabla u_e \tag{9.37}$$

12.2.6 Continuity equation

$$\frac{\partial \varepsilon_v}{\partial t} = \frac{\partial q_x}{\partial x} + \frac{\partial q_y}{\partial y} + \frac{\partial q_z}{\partial z} - q_s = \nabla \bullet \mathbf{q} - q_s \tag{9.46}$$

12.2.7 Boundary and initial conditions

The boundary conditions are

$$\sigma \mathbf{n} \,|_{S_\sigma} = -\mathbf{p}_{ns} \tag{9.49a}$$

$$\mathbf{w} \,|_{S_w} = \mathbf{w}_s \tag{9.49b}$$

$$\mathbf{q}\mathbf{n} \,|_{S_q} = q_{ns} \tag{9.49c}$$

$$u_e \,|_{S_{u_e}} = u_{es} \tag{9.49d}$$

The initial conditions are

$$\sigma \,|_{t=0_-} = \sigma_0 \tag{9.50a}$$

$$\mathbf{w} \,|_{t=0_-} = 0 \tag{9.50b}$$

$$u_e \,|_{t=0_-} = 0 \tag{9.50c}$$

12.3 WEAK FORMULATION

The governing equations above are converted to a Galerkin weak formulation in this section. Differentiating (9.3) and (9.24) with respect to time gives

$$\frac{\partial \dot{\sigma}_x}{\partial x} + \frac{\partial \dot{\tau}_{xy}}{\partial y} + \frac{\partial \dot{\tau}_{xz}}{\partial z} = \dot{f}_x \tag{12.1a}$$

$$\frac{\partial \dot{\tau}_{yx}}{\partial x} + \frac{\partial \dot{\sigma}_y}{\partial y} + \frac{\partial \dot{\tau}_{yz}}{\partial z} = \dot{f}_y \tag{12.1b}$$

$$\frac{\partial \dot{\tau}_{xz}}{\partial x} + \frac{\partial \dot{\tau}_{zy}}{\partial y} + \frac{\partial \dot{\sigma}_z}{\partial z} = \dot{f}_z \tag{12.1c}$$

$$\dot{\boldsymbol{\sigma}}' = \dot{\boldsymbol{\sigma}} - \dot{u}_e \mathbf{m} \tag{12.2}$$

Multiplying (12.1a–12.1c) by virtual functions, called virtual displacements, w_x^*, w_y^*, and w_z^* respectively, and applying Green's theorem (12.3)

$$\int_\Omega \phi \frac{\partial \psi}{\partial x} d\Omega = -\int_\Omega \psi \frac{\partial \phi}{\partial x} d\Omega + \int_\Gamma \phi \psi n_x d\Gamma \tag{12.3}$$

We have

$$\int_\Omega \vec{\varepsilon}^{*T} \dot{\vec{\sigma}} d\Omega = \int_\Omega \mathbf{w}^{*T} \dot{\mathbf{f}} d\Omega + \int_{S_\sigma} \mathbf{w}^{*T} \dot{\mathbf{p}}_{ns} d\Gamma \tag{12.4}$$

where ϕ and ψ are functions; n_x is the x-component of $\mathbf{n}$; Ω is the problem domain; $\Gamma = S_\sigma + S_w$; $\mathbf{w}^* = (w_x^*, w_y^*, w_z^*)^T$; $\mathrm{f} = (f_x, f_y, f_z)^T$; and $\vec{\varepsilon}^*$ is the virtual strain calculated as in (9.14). The choice of the virtual displacement $\mathbf{w}^*$ is limited in such a way that

$$\mathbf{w}^* |_{S_w} = 0 \tag{12.5}$$

Equation (12.4) is equivalent to satisfying (12.1) and boundary condition (9.49a). In fact, if (12.1) and (9.49a) are satisfied, then (12.4) is true. Conversely, if (12.4) is valid for any virtual displacement satisfying (12.5), then (12.1) and (9.49a) must be satisfied at all points within and on the boundary of the domain.

Similarly, multiplying (9.46) by a virtual pore pressure u_e^* and applying Green's theorem (12.3) gives

$$\int_\Omega \left[u_e^* \frac{\partial \varepsilon_v}{\partial t} + (\nabla u_e^*)^T \mathbf{q} \right] d\Omega = \int_{S_q} u_e^* q_{ns} d\Gamma - \int_\Omega u_e^* q_s d\Omega \tag{12.6}$$

The choice of the virtual pore pressure u_e^* is limited in such a way that

$$u_e^* |_{S_{u_e}} = 0 \tag{12.7}$$

Equation (12.6) is equivalent to satisfying (9.46) and boundary condition (9.49c).

Taking into account (9.34), (9.37) and (12.2), (12.4), and (12.6) are rewritten as

$$\int_\Omega \vec{\varepsilon}^{*T} (\mathbf{D}\dot{\vec{\varepsilon}} + \mathbf{m}\dot{u}_e) d\Omega = \int_\Omega \mathbf{w}^{*T} \dot{\mathbf{f}} d\Omega + \int_{S_\sigma} \mathbf{w}^{*T} \dot{\mathbf{p}}_{ns} d\Gamma \tag{12.8}$$

$$\int_\Omega \left[u_e^* \mathbf{m}^T \dot{\vec{\varepsilon}} - (\nabla u_e^*)^T \frac{\mathbf{k}}{\gamma_w} \nabla u_e \right] d\Omega = \int_{S_q} u_e^* q_{ns} d\Gamma - \int_\Omega u_e^* q_s d\Omega \tag{12.9}$$

We have implicitly assumed that the integrals appearing in (12.8–12.9) can be evaluated.

The exact solution of these equations can only be obtained for the very simplest of problems. Most practical problems require some form of approximation, achieved here by the FE method. The problem domain Ω is first divided into a number of non-overlapping subdomains (*finite element*). The unknown field variables are then approximated by assumed functions (*interpolation functions* or *shape functions*) that are defined in terms of the values of the field variables at specified points called *nodes* or *nodal points*. Finally, the nodal values of the field variables are determined. Thus, an approximation solution is obtained since these values and the interpolation functions for the elements completely define the behavior of the field variables within the elements. These procedures are described in the following sections.

12.4 ELEMENTS AND INTERPOLATION FUNCTIONS

A subject of most importance in FE analysis is the selection of particular FEs and the definition of the appropriate interpolation functions within each element. For 3-D problems, the domain of interests is normally divided into hexahedron elements. Other type of elements, such as tetrahedron elements, may also be used. However, the discretized assemblage is often extremely complex when using it.

Isoparametric elements (Zienkiewicz and Taylor 1989) are used in this chapter, where the co-ordinates within an element are interpolated using the same shape functions as for the variables. This mapping allows the use of elements of more arbitrary shape than simple forms such as right prisms.

12.4.1 Linear element

Consider a hexahedron element as shown in Figure 12.1(a) with nodal points denoted as π_1 to π_8. The local co-ordinates of these points are $\pi_1 = (-1, -1, -1)$,

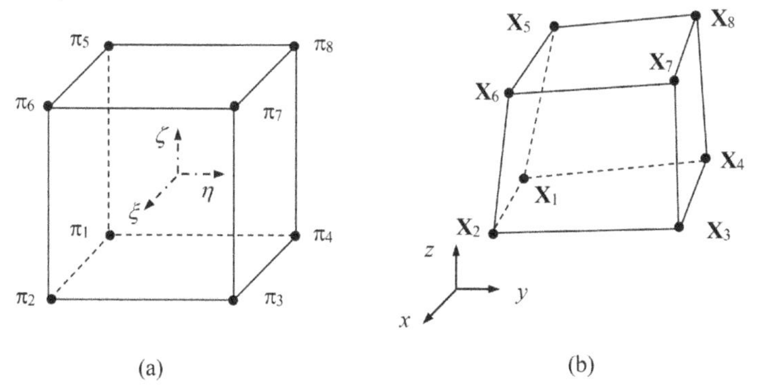

Figure 12.1 Eight-node hexahedron: (a) local co-ordinates; (b) Cartesian co-ordinates.

$\pi_2 = (1, -1, -1)$, $\pi_3 = (1, 1, -1)$, $\pi_4 = (-1, 1, -1)$, $\pi_5 = (-1, -1, 1)$, $\pi_6 = (1, -1, 1)$, $\pi_7 = (1, 1, 1)$, and $\pi_8 = (-1, 1, 1)$.

For any $\boldsymbol{\phi}^e = [\phi_1, \phi_2, \phi_3, \phi_4, \phi_5, \phi_6, \phi_7, \phi_8]^T$, let us find a function ϕ,

$$\phi = \alpha_1 + \alpha_2\xi + \alpha_3\eta + \alpha_4\zeta + \alpha_5\xi\eta + \alpha_6\eta\zeta + \alpha_7\xi\zeta + \alpha_8\xi\eta\zeta \tag{12.10}$$

which satisfies

$$\phi(\pi_j) = \phi_j \qquad j = 1, \ldots, 8$$

where $\alpha_1, \alpha_2, \ldots,$ and α_8 are constants. The solution can be written down directly as follows

$$\phi = \mathbf{N}\boldsymbol{\phi}^e = [N_1, N_2, \ldots, N_8]\boldsymbol{\phi}^e \tag{12.11a}$$

$$N_j = \frac{1}{8}(1 + \xi\xi_j)(1 + \eta\eta_j)(1 + \zeta\zeta_j) \quad j = 1, 2, \ldots, 8 \tag{12.11b}$$

where ξ_j, η_j, and ζ_j, $j = 1, 2, \ldots, 8$, are the local co-ordinates of π_j; N_j are the shape functions.

Inspection of the defining relation, (12.11) reveals immediately that the solution is unique. In fact, if let us assume $\phi = 0$, we will have $\boldsymbol{\phi}^e = 0$ since $N_j = 1$ at node j and is equal to zero at all other nodes.

The shape functions can also be used to establish co-ordinate transformations between the local and Cartesian co-ordinates. For the eight-node hexahedron shown in Figure 12.1(b) the equations relating the Cartesian co-ordinates and the local co-ordinates are

$$x = \sum_{j=1}^{8} N_j x_j \tag{12.12a}$$

$$y = \sum_{j=1}^{8} N_j y_j \tag{12.12b}$$

$$z = \sum_{j=1}^{8} N_j z_j \tag{12.12c}$$

In (12.12), x_j, y_j, and z_j are the co-ordinates of the node $\mathbf{X}_j$ in the Cartesian system. Equations (12.11–12.12) define an interpolation in the Cartesian system.

12.4.2 Quadratic element

Figure 12.2(a) shows a hexahedron element with nodal points denoted as π_1 to π_{20}. The local co-ordinates of these points are $\pi_1 = (-1, -1, -1)$, $\pi_2 = (1, -1, -1)$, $\pi_3 = (1, 1, -1)$, $\pi_4 = (-1, 1, -1)$, $\pi_5 = (-1, -1, 1)$, $\pi_6 = (1, -1, 1)$, $\pi_7 = (1, 1, 1)$, $\pi_8 = (-1, 1, 1)$, $\pi_9 = (0, -1, -1)$, $\pi_{10} = (1, 0, -1)$, $\pi_{11} = (0, 1, -1)$, $\pi_{12} = (-1, 0, -1)$, $\pi_{13} = (0, -1, 1)$, $\pi_{14} = (1, 0, 1)$, $\pi_{15} = (0, 1, 1)$, $\pi_{16} = (-1, 0, 1)$, $\pi_{17} = (-1, -1, 0)$, $\pi_{18} = (1, -1, 0)$, $\pi_{19} = (1, 1, 0)$, and $\pi_{20} = (-1, 1, 0)$.

For any $\boldsymbol{\phi}^e = [\phi_1, \phi_2, \ldots, \phi_{20}]^T$, we can uniquely construct a function ϕ of the following form:

$$\begin{aligned}\phi = {} & \alpha_1 + \alpha_2\xi + \alpha_3\eta + \alpha_4\zeta + \alpha_5\xi\eta + \alpha_6\eta\zeta + \alpha_7\xi\zeta + \alpha_8\xi\eta\zeta + \alpha_9\xi^2 + \alpha_{10}\eta^2 \\ & + \alpha_{11}\zeta^2 + \alpha_{12}\xi^2\eta + \alpha_{13}\xi^2\zeta + \alpha_{14}\xi\eta^2 + \alpha_{15}\eta^2\zeta + \alpha_{16}\xi\zeta^2 + \alpha_{17}\eta\zeta^2 \\ & + \alpha_{18}\xi^2\eta\zeta + \alpha_{19}\xi\eta^2\zeta + \alpha_{20}\xi\eta\zeta^2\end{aligned} \tag{12.13}$$

which satisfies

$$\phi(\pi_j) = \phi_j \qquad j = 1, \ldots, 20$$

where $\alpha_1, \alpha_2, \ldots,$ and α_{20} are constants. The solution is

$$\phi = \mathbf{N}\phi^e = [N_1, N_2, \ldots, N_{20}]\phi^e \tag{12.14a}$$

$$N_j = \frac{1}{8}(1+\xi\xi_j)(1+\eta\eta_j)(1+\zeta\zeta_j)(\xi\xi_j + \eta\eta_j + \zeta\zeta_j - 2) \quad j = 1, 2, \ldots, 8 \tag{12.14b}$$

$$N_j = \frac{1}{4}(1-\xi^2)(1+\eta\eta_j)(1+\zeta\zeta_j) \quad j = 9, 11, 13, 15 \tag{12.14c}$$

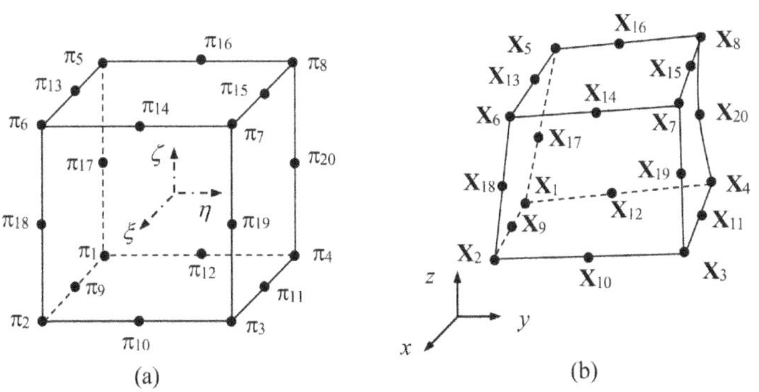

Figure 12.2 Twenty-node hexahedron: (a) local co-ordinates; (b) Cartesian co-ordinates.

$$N_j = \frac{1}{4}(1+\xi\xi_j)(1-\eta^2)(1+\zeta\zeta_j) \quad j = 10, 12, 14, 16 \tag{12.14d}$$

$$N_j = \frac{1}{4}(1+\xi\xi_j)(1+\eta\eta_j)(1-\zeta^2) \quad j = 17, 18, 19, 20 \tag{12.14e}$$

where ξ_j, η_j, and ζ_j, $j = 1, 2, \ldots, 20$, are the local co-ordinates of π_j; N_j are the shape functions.

Similarly, (12.15) gives a co-ordinate transformation between the local and Cartesian co-ordinates for the twenty-node hexahedron shown in Figure 12.2.

$$x = \sum_{j=1}^{20} N_j x_j \tag{12.15a}$$

$$y = \sum_{j=1}^{20} N_j y_j \tag{12.15b}$$

$$z = \sum_{j=1}^{20} N_j z_j \tag{12.15c}$$

In (12.15), x_j, y_j, and z_j are the co-ordinates of the node $\mathbf{X}_j$ in the Cartesian system. Equations (12.14–12.15) define a quadratic interpolation in the Cartesian system.

12.4.3 Evaluation of global derivatives

In Sections 12.4.1–12.4.2, we express an arbitrary ϕ as a function of the local co-ordinates ξ, η, and ζ. We also need to evaluate $\partial\phi/\partial x$, $\partial\phi/\partial y$, and $\partial\phi/\partial z$, as in (12.8–12.9). Here these global derivatives are expressed in terms of local co-ordinates, too. From (12.11) and (12.14),

$$\frac{\partial\phi}{\partial x} = \sum_{j=1}^{j_p} \frac{\partial N_j}{\partial x}\phi_j \qquad \frac{\partial\phi}{\partial y} = \sum_{j=1}^{j_p} \frac{\partial N_j}{\partial y}\phi_j \qquad \frac{\partial\phi}{\partial z} = \sum_{j=1}^{j_p} \frac{\partial N_j}{\partial z}\phi_j \tag{12.16}$$

where j_p is the number of element nodes. Hence, it is necessary to express $\partial N_j/\partial x$, $\partial N_j/\partial y$, and $\partial N_j/\partial z$ in terms of ξ, η, and ζ.

Consider for instance the set of local co-ordinates ξ, η, ζ, and a corresponding set of global co-ordinates x, y, and z. By the chain rule of partial differentiation, we can write for instance the ξ derivative as

$$\frac{\partial N_j}{\partial \xi} = \frac{\partial N_j}{\partial x}\frac{\partial x}{\partial \xi} + \frac{\partial N_j}{\partial y}\frac{\partial y}{\partial \xi} + \frac{\partial N_j}{\partial z}\frac{\partial z}{\partial \xi} \tag{12.17}$$

Performing the same differentiation with respect to the other two co-ordinates and writing in matrix form we obtain

$$\begin{bmatrix} \dfrac{\partial N_j}{\partial \xi} \\ \dfrac{\partial N_j}{\partial \eta} \\ \dfrac{\partial N_j}{\partial \zeta} \end{bmatrix} = \begin{bmatrix} \dfrac{\partial x}{\partial \xi} & \dfrac{\partial y}{\partial \xi} & \dfrac{\partial z}{\partial \xi} \\ \dfrac{\partial x}{\partial \eta} & \dfrac{\partial y}{\partial \eta} & \dfrac{\partial z}{\partial \eta} \\ \dfrac{\partial x}{\partial \zeta} & \dfrac{\partial x}{\partial \zeta} & \dfrac{\partial x}{\partial \zeta} \end{bmatrix} \begin{bmatrix} \dfrac{\partial N_j}{\partial x} \\ \dfrac{\partial N_j}{\partial y} \\ \dfrac{\partial N_j}{\partial z} \end{bmatrix} = \mathbf{J} \begin{bmatrix} \dfrac{\partial N_j}{\partial x} \\ \dfrac{\partial N_j}{\partial y} \\ \dfrac{\partial N_j}{\partial z} \end{bmatrix} \tag{12.18}$$

In the above, the left-hand side can be evaluated as the functions N_j are specified in local co-ordinates. Further as x, y, and z are explicitly given by the relation defining the co-ordinate transforms ((12.12) and (12.15)), the matrix **J** can be found explicitly in terms of the local co-ordinates. This matrix is known as the *Jacobian matrix*.

Using the co-ordinate transforms, the Jacobian matrix **J** is written as

$$\mathbf{J} = \begin{bmatrix} \sum_{j=1}^{j_p} \dfrac{\partial N_j}{\partial \xi} x_j & \sum_{j=1}^{j_p} \dfrac{\partial N_j}{\partial \xi} y_j & \sum_{j=1}^{j_p} \dfrac{\partial N_j}{\partial \xi} z_j \\ \sum_{j=1}^{j_p} \dfrac{\partial N_j}{\partial \eta} x_j & \sum_{j=1}^{j_p} \dfrac{\partial N_j}{\partial \eta} y_j & \sum_{j=1}^{j_p} \dfrac{\partial N_j}{\partial \eta} z_j \\ \sum_{j=1}^{j_p} \dfrac{\partial N_j}{\partial \zeta} x_j & \sum_{j=1}^{j_p} \dfrac{\partial N_j}{\partial \zeta} y_j & \sum_{j=1}^{j_p} \dfrac{\partial N_j}{\partial \zeta} z_j \end{bmatrix} \tag{12.19}$$

To find the global derivatives we invert **J** and write

$$\begin{bmatrix} \dfrac{\partial N_j}{\partial x} \\ \dfrac{\partial N_j}{\partial y} \\ \dfrac{\partial N_j}{\partial z} \end{bmatrix} = \mathbf{J}^{-1} \begin{bmatrix} \dfrac{\partial N_j}{\partial \xi} \\ \dfrac{\partial N_j}{\partial \eta} \\ \dfrac{\partial N_j}{\partial \zeta} \end{bmatrix} \quad j = 1, 2, \ldots, j_p \tag{12.20}$$

Now the expression for $\partial\phi/\partial x$, $\partial\phi/\partial y$, and $\partial\phi/\partial z$ can be found directly as

$$\begin{bmatrix} \frac{\partial\phi}{\partial x} \\ \frac{\partial\phi}{\partial y} \\ \frac{\partial\phi}{\partial z} \end{bmatrix} = \mathbf{J}^{-1} \begin{bmatrix} \frac{\partial N_1}{\partial\xi} & \frac{\partial N_2}{\partial\xi} & \cdots & \frac{\partial N_{ip}}{\partial\xi} \\ \frac{\partial N_1}{\partial\eta} & \frac{\partial N_2}{\partial\eta} & \cdots & \frac{\partial N_{ip}}{\partial\eta} \\ \frac{\partial N_1}{\partial\zeta} & \frac{\partial N_2}{\partial\zeta} & \cdots & \frac{\partial N_{ip}}{\partial\zeta} \end{bmatrix} \begin{bmatrix} \phi_1 \\ \phi_2 \\ \cdot \\ \cdot \\ \cdot \\ \phi_{ip} \end{bmatrix} \tag{12.21}$$

12.4.4 Evaluation of element integration

As shown in (12.8–12.9), integration of very complex functions is required to obtain a solution of the weak formulations of consolidation problem. Exact integration of these expressions is not available in most cases. Numerical integration is essential.

Two methods of numerical integration of a function of one variable, that is, Newton-Cotes quadrature and Gauss quadrature, will be summarized here. Multiple integrations can be performed by repeated use of single integration.

12.4.4.1 Newton-Cotes formulae

To find numerically the integral of a function ψ, $\int_a^b \psi dx$, we pass a polynomial through points defined by the functions, and then integrate this polynomial approximation to the function. This permits us to integrate a function known only as a table of values.

Let $n + 1$ distinct points x_j within $[a, b]$ be ordered by

$$x_0 < x_1 < \cdots < x_n \tag{12.22}$$

With these points as interpolation points, we form the interpolation polynomial $P_n(x)$ of degree at most n for a continuous function ψ such that $P_n(x_j) = \psi(x_j)$, $j = 0, 1,\dots, n$. Then as an approximation to the integral, we set

$$\int_a^b \psi\, dx = \int_a^b P_n(x)dx + Error \cong \int_a^b P_n(x)dx \tag{12.23}$$

By using the Lagrange form for the interpolation polynomial

$$P_n(x) = \sum_{j=0}^{n} \frac{\omega_n(x)}{(x - x_j)\omega_n'(x_j)} \psi(x_j) \tag{12.24}$$

where

$$\omega_n(x) = (x - x_0)(x - x_1)\cdots(x - x_n) \tag{12.25}$$

we obtain the quadrature formula

$$\int_a^b \psi\, dx \cong \int_a^b P_n(x)dx = \sum_{j=0}^{n} H_j \psi(x_j) \tag{12.26}$$

with the coefficients given by

$$H_j = \int_a^b \frac{\omega_n(x)}{(x - x_j)\omega_n'(x_j)} dx \tag{12.27}$$

The interpolatory quadrature error can be expressed as

$$Error = \frac{1}{(n+1)!}\int_a^b \omega_n(x)\psi^{(n+1)}(\varsigma)dx = O(\Delta^{n+1}), \qquad \varsigma = \varsigma(x) \tag{12.28}$$

where Δ is the point spacing.

Using these expressions, we easily obtain the following simple quadrature formulae

a. Midpoint rule:

$$\int_{-1}^{1} \psi\, dx = 2\psi(0) + O(\Delta^3) \tag{12.29}$$

b. Trapezoidal rule:

$$\int_{-1}^{1} \psi\, dx = \psi(-1) + \psi(1) + O(\Delta^3) \tag{12.30}$$

c. Simpson's rule:

$$\int_{-1}^{1} \psi\, dx = \frac{1}{3}[\psi(-1) + 4\psi(0) + \psi(1)] + O(\Delta^5) \tag{12.31}$$

12.4.4.2 Gaussian quadrature

In Newton-Cotes formulae, the coefficients of a quadrature formula which has a degree of precision at least n are determined for n+1 given fixed nodes.

If in place of specifying the position of interpolation points *a priori*, we allow a choice of both n nodes and n coefficients, an increased accuracy can be obtained.

If the formula is to have the n nodes $x_1, x_2, \ldots, x_n$, it can be written as

$$\int_a^b \psi dx = \sum_{j=1}^{n} H_j \psi(x_j) + Error \tag{12.32}$$

since the degree of precision for such a formula will not be less than the corresponding degree for the interpolatory formula using the same nodes, the *Error* can be written from (12.28) as

$$Error = \frac{1}{n!}\int_a^b p_n(x)\psi^{(n)}(\varsigma)dx = O(\Delta^n), \qquad \varsigma = \varsigma(x) \tag{12.33}$$

where

$$p_n(x) = (x - x_1)\cdots(x - x_n) \tag{12.34}$$

Clearly, the *Error* is zero if ψ is a polynomial of degree $n - 1$ or less. We seek points x_j such that the *Error* also vanishes when ψ is any polynomial of degree $n + r$ where $r = 1, 2, \ldots, m$ and m is to be as large as possible. Substituting $\psi = p_n(x)x^r$ into (12.32) gives

$$\int_a^b p_n(x)x^r dx = 0 \quad r = 0, 1, \ldots, m \tag{12.35}$$

However, these are just the conditions that the polynomial $p_n(x)$ be orthogonal to all polynomials of degree at most m over [a, b]. Thus, $p_n(x)$ is the nth orthogonal polynomial over [a, b], and $m = n - 1$. The coefficients in (12.32) are easily obtained once the nodes are determined since they are interpolatory. As in (12.27), we have

$$H_j = \int_a^b \frac{p_n(x)}{(x - x_j)p_n'(x_j)} dx \quad j = 1, 2, \ldots, n \tag{12.36}$$

All these weighting coefficients are positive. In fact, since (12.32) yields the exact value for the integral when ψ is any polynomial of degree $2n - 1$ or less, we have an exact result for

$$\psi(x) = \left[\frac{p_n(x)}{x - x_j}\right]^2$$

Table 12.1 Co-ordinates and weighting coefficients in Gaussian quadrature formulae $\int_{-1}^{1}\psi(x)dx = \sum_{j=1}^{n} H_j\psi(x_j)$

n	x_j	H_j
1	0	2
2	$-\frac{1}{\sqrt{3}}, \frac{1}{\sqrt{3}}$	1, 1
3	$-\sqrt{\frac{3}{5}}, 0, \sqrt{\frac{3}{5}}$	$\frac{5}{9}, \frac{8}{9}, \frac{5}{9}$

Thus, from (12.32), we find

$$H_j = \int_a^b \frac{p_n^2(x)}{(x-x_j)^2[p_n'(x_j)]^2}dx > 0 \quad j = 1, 2, \ldots, n \tag{12.37}$$

The *Error* therefore is of order $O(\Delta^{2n})$.

If $a = -1$ and $b = 1$, the nth orthogonal polynomial over [−1, 1] is the Legendre polynomial

$$p_n(x) = \frac{1}{2^n n!}\frac{d^n}{dx^n}(x^2-1)^n \tag{12.38}$$

The co-ordinates and weighting coefficients are shown in Table 12.1.

12.4.4.3 Calculation of element integration

Having described the methods of numerical integration of a function of one variable, we face the task of evaluating multiple integrations for a hexahedron element Ω_e in Cartesian co-ordinates. We consider an integration of the form $\int_{\Omega_e}\phi dxdydz$. Using the co-ordinate transforms in (12.12) or (12.15), we obtain

$$\int_{\Omega_e}\phi dxdydz = \int_{-1}^{1}\int_{-1}^{1}\int_{-1}^{1}\phi\,|\,\mathbf{J}\,|\,d\xi d\eta d\zeta = \sum_{m=1}^{n}\sum_{j=1}^{n}\sum_{i=1}^{n}H_iH_jH_m\phi(\xi_i,\eta_j,\zeta_m)\,|\,\mathbf{J}(\xi_i,\eta_j,\zeta_m)\,| \tag{12.39}$$

where $\xi_i, \eta_j, \zeta_m, H_i, H_j, H_m$ are the co-ordinates of sampling points and weighting coefficients for one-dimensional numerical integration.

12.5 FINITE ELEMENT FORMULATION

Returning to (12.8–12.9), we see that our general problem is to find the unknown displacements and excess pore pressure that satisfy (12.8–12.9) in the domain Ω and the associated boundary conditions specified on Γ. Let the domain Ω be divided into a number of non-overlapping hexahedron elements that are described in Section 12.4. For an element Ω_e with displacement variables at m nodes, the displacement field at any internal point is approximated by

$$\mathbf{w} = \mathbf{N}_w \mathbf{w}^e \tag{12.40}$$

where $\mathbf{N}_w$ is a matrix of shape functions given by

$$\mathbf{N}_w = \begin{bmatrix} N_1^w & 0 & 0 & N_2^w & 0 & 0 & \cdots & N_m^w & 0 & 0 \\ 0 & N_1^w & 0 & 0 & N_2^w & 0 & \cdots & 0 & N_m^w & 0 \\ 0 & 0 & N_1^w & 0 & 0 & N_2^w & \cdots & 0 & 0 & N_m^w \end{bmatrix} \tag{12.41}$$

and

$$\mathbf{w}^e = \begin{bmatrix} w_{xj_1} & w_{yj_1} & w_{zj_1} & w_{xj_2} & w_{yj_2} & w_{zj_2} & \cdots & w_{xj_m} & w_{yj_m} & w_{zj_m} \end{bmatrix}^T \tag{12.42}$$

is a vector of element nodal displacements. The N_j^w ($j = 1, \ldots, m$) in (12.41) are the shape functions as defined in (12.11) and (12.14). $j_1, j_2, \ldots,$ and j_m in (12.42) are nodal numbers. Similarly, the field of excess pore pressure with excess pore pressure freedoms at n nodes is assumed to be of the form

$$u_e = \mathbf{N}_u \mathbf{u}^e \tag{12.43}$$

where

$$\mathbf{N}_u = \begin{bmatrix} N_1^u & N_2^u & \cdots & N_n^u \end{bmatrix} \tag{12.44}$$

and

$$\mathbf{u}^e = \begin{bmatrix} u_{ei_1} & u_{ei_2} & \cdots & u_{ein} \end{bmatrix}^T \tag{12.45}$$

are, respectively, a matrix of shape functions and a vector of nodal excess pore pressures. The N_j^u ($j = 1, \ldots, n$) in (12.44) are the shape functions as defined in (12.11) and (12.14). i_1, $i_2, \ldots,$ and i_n in (12.45) are nodal numbers. Note that, for generality, different sets of shape functions may be used to describe the variation of the displacements and the excess pore pressures. This implies that the nodes in the FE mesh may have varying degrees of freedom, with some being associated with displacements, some being associated with excess pore pressures, and some being associated with both. In order for the excess pore pressures to be consistent with the stresses, it is usual to choose the polynomial describing the excess pore pressures to be one order lower than the polynomial describing the displacements.

In order to ensure continuity of displacements and excess pore pressures between elements, it is necessary to place a sufficient number of nodes on the element boundary to satisfy the shape functions being used for the elements. Equations (12.8–12.9) contain derivatives up to the first order. The interpolation functions in (12.11) and (12.14) clearly satisfy the following convergence requirements for these equations as element size decreases (Zienkiewicz and Taylor 1989).

a. *Compatibility requirement*: At element interface the field variables are continuous.
b. *Completeness requirement*: Within an element first derivatives of the field variables are continuous.

Using (12.40), the strain vector within an element Ω_e can be expressed as

$$\vec{\varepsilon} = -\mathbf{B}_w \mathbf{w}^e \tag{12.46}$$

where the strain interpolation matrix $\mathbf{B}_w$ is evaluated as

$$\mathbf{B}_w = \begin{bmatrix} \frac{\partial N_1^w}{\partial x} & 0 & 0 & \frac{\partial N_2^w}{\partial x} & 0 & 0 & \cdots & \frac{\partial N_m^w}{\partial x} & 0 & 0 \\ 0 & \frac{\partial N_1^w}{\partial y} & 0 & 0 & \frac{\partial N_2^w}{\partial y} & 0 & \cdots & 0 & \frac{\partial N_m^w}{\partial y} & 0 \\ 0 & 0 & \frac{\partial N_1^w}{\partial z} & 0 & 0 & \frac{\partial N_2^w}{\partial z} & \cdots & 0 & 0 & \frac{\partial N_m^w}{\partial z} \\ \frac{\partial N_1^w}{\partial y} & \frac{\partial N_1^w}{\partial x} & 0 & \frac{\partial N_2^w}{\partial y} & \frac{\partial N_2^w}{\partial x} & 0 & \cdots & \frac{\partial N_m^w}{\partial y} & \frac{\partial N_m^w}{\partial x} & 0 \\ 0 & \frac{\partial N_1^w}{\partial z} & \frac{\partial N_1^w}{\partial y} & 0 & \frac{\partial N_2^w}{\partial z} & \frac{\partial N_2^w}{\partial y} & \cdots & 0 & \frac{\partial N_m^w}{\partial z} & \frac{\partial N_m^w}{\partial y} \\ \frac{\partial N_1^w}{\partial z} & 0 & \frac{\partial N_1^w}{\partial x} & \frac{\partial N_2^w}{\partial z} & 0 & \frac{\partial N_2^w}{\partial x} & \cdots & \frac{\partial N_m^w}{\partial z} & 0 & \frac{\partial N_m^w}{\partial x} \end{bmatrix}$$

Similarly, using (12.43) the gradient of excess pore pressure within an element Ω_e can be expressed as

$$\nabla u_e = \mathbf{B}_u \mathbf{u}^e \tag{12.47}$$

where

$$\mathbf{B}_u = \begin{bmatrix} \frac{\partial N_1^u}{\partial x} & \frac{\partial N_2^u}{\partial x} & \cdots & \frac{\partial N_n^u}{\partial x} \\ \frac{\partial N_1^u}{\partial y} & \frac{\partial N_2^u}{\partial y} & \cdots & \frac{\partial N_n^u}{\partial y} \\ \frac{\partial N_1^u}{\partial z} & \frac{\partial N_2^u}{\partial z} & \cdots & \frac{\partial N_n^u}{\partial z} \end{bmatrix}$$

Using the same interpolation functions for the virtual displacements as in (12.40), we obtain

$$\mathbf{w}^* = \mathbf{N}_w \mathbf{w}^{*e} \tag{12.48}$$

$$\vec{\varepsilon}^* = -\mathbf{B}_w \mathbf{w}^{*e} \tag{12.49}$$

where

$$\mathbf{w}^{*e} = [\, w^*_{xj_1} \;\; w^*_{yj_1} \;\; w^*_{zj_1} \;\; w^*_{xj_2} \;\; w^*_{yj_2} \;\; w^*_{zj_2} \;\; \cdots \;\; w^*_{xj_m} \;\; w^*_{yj_m} \;\; w^*_{zj_m} \,]^T \tag{12.50}$$

Similarly, using the same interpolation functions for the virtual excess pore pressures as in (12.43) gives

$$u^*_e = \mathbf{N}_u \mathbf{u}^{*e} \tag{12.51}$$

$$\nabla u^*_e = \mathbf{B}_u \mathbf{u}^{*e} \tag{12.52}$$

where

$$\mathbf{u}^{*e} = [\, u^*_{ei_1} \;\; u^*_{ei_2} \;\; \cdots \;\; u^*_{ein} \,]^T \tag{12.53}$$

Substituting (12.43), (12.46–12.49), (12.51) and (12.52) in (12.8–12.9), we have

$$\sum_{\Omega_e} (\mathbf{w}^{*e})^T (\mathbf{k}_T \dot{\mathbf{w}}^e - \mathbf{k}_C \dot{\mathbf{u}}^e) = \sum_{\Omega_e} (\mathbf{w}^{*e})^T \dot{\mathbf{f}}_w \tag{12.54}$$

$$\sum_{\Omega_e} (\mathbf{u}^{*e})^T (-\mathbf{k}_C \dot{\mathbf{w}}^e - \mathbf{k}_P \mathbf{u}^e) = \sum_{\Omega_e} (\mathbf{u}^{*e})^T \mathbf{f}_u \tag{12.55}$$

where

$\mathbf{k}_T = \int_{\Omega_e} \mathbf{B}_w^T \mathbf{D} \mathbf{B}_w d\Omega$ is the elemental tangential stiffness matrix

$\mathbf{k}_C = \int_{\Omega_e} \mathbf{B}_w^T \mathbf{m} \mathbf{N}_u d\Omega$ is the coupling matrix

$\mathbf{k}_P = \int_{\Omega_e} \mathbf{B}_u^T \frac{\mathbf{k}}{\gamma_w} \mathbf{B}_u d\Omega$ is the permeability matrix

The right-hand terms $\mathbf{f}_w$ and $\mathbf{f}_u$ in (12.54–12.55) are given by

$$\mathbf{f}_w = \int_{\Omega_e} \mathbf{N}_w^T \mathbf{f} d\Omega + \int_{\Omega_e \cap S_\sigma} \mathbf{N}_w^T \mathbf{p}_{ns} d\Gamma \tag{12.56}$$

$$\mathbf{f}_u = \int_{\Omega_e \cap S_q} \mathbf{N}_u^T q_{ns} d\Gamma - \int_{\Omega_e} \mathbf{N}_u^T q_s d\Omega \tag{12.57}$$

Integration of the above matrices can be performed as described in Section 12.4 by establishing a co-ordinate transformation between the local and global Cartesian co-ordinates using the same shape functions as for displacements.

The virtual displacements and excess pore pressures are arbitrary except for the conditions in (12.5) and (12.7). Thus, (12.54–12.55) represent a set of ordinary differential equations in time. Assembling the element matrices produces a global system of equations of the form

$$\begin{bmatrix} \mathbf{K}_T & -\mathbf{K}_C \\ -\mathbf{K}_C^T & 0 \end{bmatrix} \frac{d}{dt} \begin{bmatrix} \mathbf{W} \\ \mathbf{U} \end{bmatrix} + \begin{bmatrix} 0 & 0 \\ 0 & -\mathbf{K}_P \end{bmatrix} \begin{bmatrix} \mathbf{W} \\ \mathbf{U} \end{bmatrix} = \begin{bmatrix} \dot{\mathbf{F}}_w \\ \mathbf{F}_u \end{bmatrix} \tag{12.58}$$

where $\mathbf{K}_T$, $\mathbf{K}_C$, $\mathbf{K}_P$, $\mathbf{F}_w$, $\mathbf{F}_u$, $\mathbf{W}$, and $\mathbf{U}$ are the assembled quantities of $\mathbf{k}_T$, $\mathbf{k}_C$, $\mathbf{k}_P$, $\mathbf{f}_w$, $\mathbf{f}_u$, $\mathbf{w}$, and $\mathbf{u}$, respectively.

The discretization of (12.58) in the time domain is carried out by the general trapezoidal method, also known as the generalized midpoint rule or θ-method. At time $t_{i+\theta} = t_i + \theta\Delta t$, (12.58) is approximated by the following difference equation

$$\begin{aligned} &\begin{bmatrix} \mathbf{K}_T & -\mathbf{K}_C \\ -\mathbf{K}_C^T & 0 \end{bmatrix}_{i+\theta} \frac{1}{\Delta t} \left\{ \begin{bmatrix} \mathbf{W} \\ \mathbf{U} \end{bmatrix}_{i+1} - \begin{bmatrix} \mathbf{W} \\ \mathbf{U} \end{bmatrix}_i \right\} + \\ &\quad \begin{bmatrix} 0 & 0 \\ 0 & -\mathbf{K}_P \end{bmatrix}_{i+\theta} \left\{ \theta \begin{bmatrix} \mathbf{W} \\ \mathbf{U} \end{bmatrix}_{i+1} + (1-\theta) \begin{bmatrix} \mathbf{W} \\ \mathbf{U} \end{bmatrix}_i \right\} = \begin{bmatrix} \dot{\mathbf{F}}_w \\ \mathbf{F}_u \end{bmatrix}_{i+\theta} \end{aligned} \tag{12.59}$$

where Δt is the time step length and θ is a parameter between 0 and 1. The subscripts i, $i+1$, and $i+\theta$ in (12.59) denote that the quantities are evaluated at times t_i, t_{i+1}, and $t_{i+\theta}$, respectively. Rearranging (12.59) gives

$$\begin{bmatrix} \mathbf{K}_T & -\mathbf{K}_C \\ -\mathbf{K}_C^T & -\theta\Delta t\mathbf{K}_P \end{bmatrix}_{i+\theta} \begin{bmatrix} \mathbf{W} \\ \mathbf{U} \end{bmatrix}_{i+1} = \begin{bmatrix} \mathbf{K}_T & -\mathbf{K}_C \\ -\mathbf{K}_C^T & (1-\theta)\Delta t\mathbf{K}_P \end{bmatrix}_{i+\theta} \begin{bmatrix} \mathbf{W} \\ \mathbf{U} \end{bmatrix}_i + \Delta t \begin{bmatrix} \dot{\mathbf{F}}_w \\ \mathbf{F}_u \end{bmatrix}_{i+\theta} \tag{12.60}$$

Equation (12.60) may be used to determine the values of $\mathbf{W}$ and $\mathbf{U}$ at any time relative to their initial values. Since the equation is non-linear, some form of iteration, say, the Newton-Raphson procedure, is required (Zienkiewicz and Taylor 1989; Lewis and Schrefler 1998).

12.6 EQUIVALENT MODELING OF VERTICAL DRAINS IN PLANE STRAIN CONDITION

Most embankment analyses are conducted based on the plane strain conditions. However, this poses a problem if the effects of vertical drains are to be modeled, since the flow of water into a vertical drain is axisymmetric in nature. For construction sites, a large number of vertical drains are normally installed. A three dimensional multi-drain analysis will be extremely time-consuming. Several methodologies (Zeng *et al.* 1987; Hird *et al.* 1995; Amirebrahimi and Herrmann 1993; Indraratna and Redana 1997; Kim and Lee 1997) are suggested to convert the system of vertical drains into an equivalent drain wall by manipulating the drain spacing and/or soil permeability.

Hansbo (1981) obtained (12.61) (see 6.60, 6.78, and 6.79) for an axisymmetric unit cell as shown in Figure 12.3(a) subjected to a suddenly applied loading under equal strain condition by neglecting the vertical flow in the soil

$$U_r = 1 - \exp\left(-\frac{2T_r}{\mu_r}\right) \tag{12.61a}$$

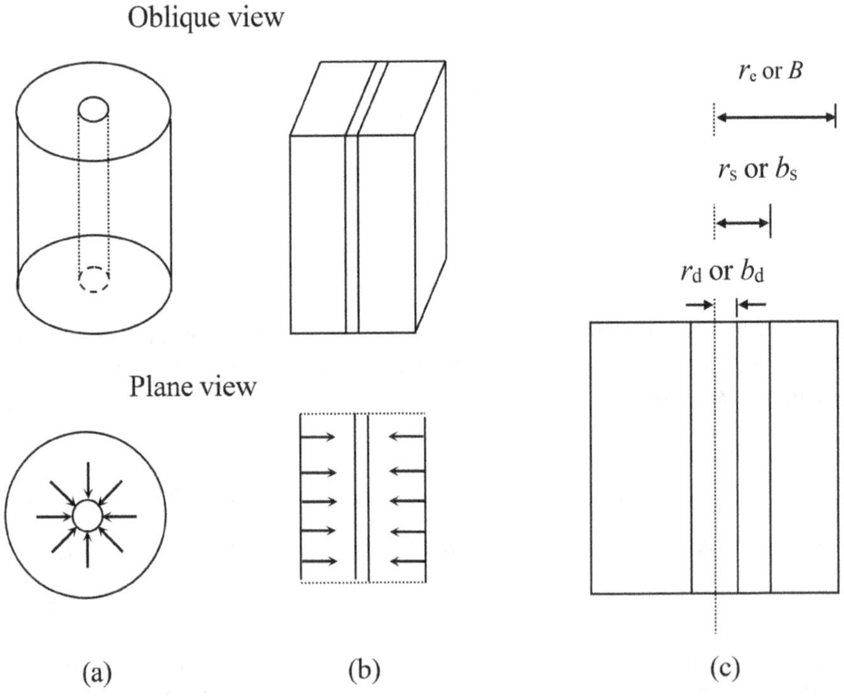

Figure 12.3 Unit cells: (a) axisymmetric, (b) plane strain, and (c) side view (after Hird et al., 1995).

$$T_r = \frac{k_r t}{\gamma_w m_v r_e^2} \tag{12.61b}$$

$$\mu_r = \frac{1}{(N^2-1)}\left[N^2 \ln\left(\frac{N}{s}\right) + s^2 - \frac{3}{4}N^2 - \frac{s^4}{4N^2} + \frac{k_r}{k_s}\left(N^2 \ln s + \frac{s^4-1}{4N^2} + 1 - s^2\right)\right] + \frac{\pi k_r (2lz - z^2)}{q_w} \approx \ln\left(\frac{N}{s}\right) - \frac{3}{4} + \frac{k_r}{k_s}\ln s + \frac{\pi k_r (2lz - z^2)}{q_w} \tag{12.61c}$$

where q_w = drain discharge capacity; l = drain length.

Similarly, the average degree of consolidation for a plane strain unit cell as shown in Figure 12.3(b) under equal strain condition can be expressed as (12.62) by neglecting the vertical flow in the soil

$$U_p = 1 - \exp\left(-\frac{2T_p}{\mu_p}\right) \tag{12.62a}$$

$$T_p = \frac{k_{hp} t}{\gamma_w m_v B^2} \tag{12.62b}$$

$$\mu_p = \frac{2}{N_p^2(N_p-1)}\left\{\frac{1}{3}(N_p - s_p)^3 + \frac{k_{hp}}{k'_{hp}}\left[(s_p-1)N_p^2 - (s_p^2-1)N_p + \frac{s_p^3-1}{3}\right]\right\} + \frac{1}{N_p b_d^2}\frac{k_{hp}}{k_d}(2lz - z^2) \approx \frac{2}{3} + \frac{2(s_p-1)}{N_p-1}\frac{k_{hp}}{k'_{hp}} + \frac{1}{N_p b_d^2}\frac{k_{hp}}{k_d}(2lz - z^2) \tag{12.62c}$$

where U_p = average degree of consolidation due to horizontal drainage on a horizontal plane for the plane strain unit cell; $N_p = B/b_d$; $s_p = b_s/b_d$; B = half width of the plane strain unit cell; b_d = half width of the vertical drain; b_s = half width of the smear zone (also see Figure 12.3(a) and (c)); k_d = hydraulic conductivity of the drain; k_{hp} = horizontal hydraulic conductivity; k'_{hp} = horizontal hydraulic conductivity in the smeared zone. Equation (12.62c) is different from that proposed by Indraratna and Redana (1997).

For the rate of consolidation in a plane strain and axisymmetric unit cell to be matched, (12.63) should be satisfied from (12.61–12.62)

$$\frac{k_{hp}}{\mu_p B^2} = \frac{k_r}{\mu_r r_e^2} \tag{12.63}$$

The width of the drain was converted by assuming the same total cross-section areas.

For a system of vertical drains arranged in a square pattern:

$$b_d = B\frac{\pi r_d^2}{S_d^2} \text{ and } b_s = B\frac{\pi r_s^2}{S_d^2} \tag{12.64}$$

For a system of vertical drains arranged in a triangular pattern:

$$b_d = 2B\frac{\pi r_d^2}{S_d^2} \text{ and } b_s = 2B\frac{\pi r_s^2}{S_d^2} \tag{12.65}$$

where S_d = drain spacing

12.7 CONSOLIDATION ANALYSIS OF AN ELASTO-PLASTIC SOIL

As an example of application of the FE method, the consolidation behavior of a strip footing on elastic-perfectly plastic soil satisfying the Mohr-Coulomb yield criterion is studied in this section. The flow properties of soil are assumed to be constant. The plane strain elastic-plastic response of clay strata subjected to footing loads has been investigated previously by Davidson and Chen (1978), Mizuno and Chen (1983), and Manoharan and Dasgupta (1995), among others.

Figure 12.4 shows the geometry of the example. A uniform, flexible strip loading p_0 = 100 kPa is suddenly applied on the surface of the homogeneous clay stratum. The half width of the strip footing, b, is equal to 1 m. While the bottom of the model is rough, restricted for both the horizontal and vertical movements, the sides are smooth and are restrained only in the horizontal

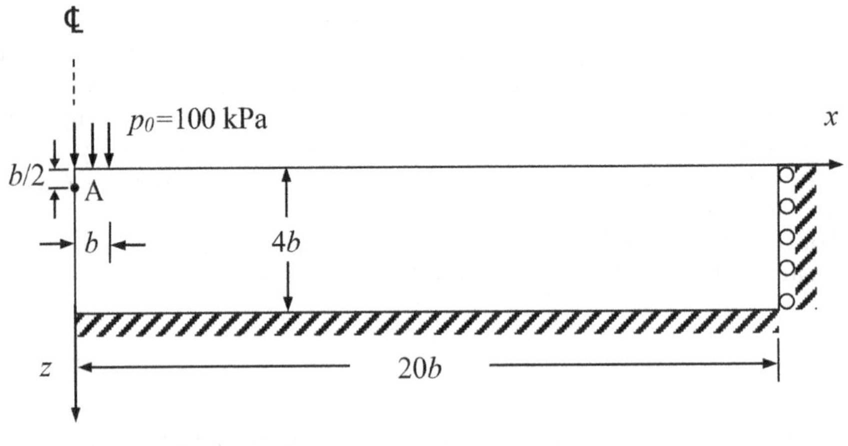

Figure 12.4 Geometry of a strip footing.

direction. Seepage is allowed only on the top surface and the sides and bottom are considered impervious.

The soil is assumed to be isotropic and weightless. The effective stress developed in the soil is limited by the Mohr-Coulomb yield criterion:

$$F=\left[\frac{1}{\sqrt{3}\cos\phi'}\sin\left(\vartheta+\frac{\pi}{3}\right)+\frac{1}{3}\cos\left(\vartheta+\frac{\pi}{3}\right)\tan\phi'\right]q'-p'\tan\phi'-c'=0 \quad (12.66)$$

where ϕ' is the slope of the Mohr-Coulomb yield surface, which is commonly referred to as the effective friction angle; c' is the effective cohesion; ϑ is the deviatoric polar angle defined as

$$\vartheta=\frac{1}{3}\cos^{-1}\left(\frac{r'}{q'}\right)^3 \quad (12.67)$$

and $p'=\frac{1}{3}trace(\sigma')$ is the equivalent effective pressure stress; $q'=\sqrt{\frac{3}{2}s'_{ij}s'_{ij}}$ is the Mises equivalent stress; $r'=\left(\frac{9}{2}s'_{ij}s'_{jk}s'_{ki}\right)^{1/3}$ is the third invariant of deviatoric stress; and $s'_{ij}=\sigma'_{ij}-p'\delta_{ij}$ is the component of the deviatoric effective stress.

The flow potential Q is chosen as a hyperbolic function in the meridional stress plane and the smooth elliptic function proposed by Menétrey and Willam (1995) in the deviatoric stress plane:

$$Q=\sqrt{(c'\tan\psi'/10)^2+(R_m q')^2}-p'\tan\psi' \quad (12.68)$$

where

$$R_m=\frac{4(1-e'^2)\cos^2\vartheta+(2e'-1)^2}{2(1-e'^2)\cos\vartheta+(2e'-1)\sqrt{4(1-e'^2)\cos^2\vartheta+5e'^2-4e'}}\frac{3-\sin\phi'}{6\cos\phi'} \quad (12.69)$$

$$e'=\frac{3-\sin\phi'}{3+\sin\phi'} \quad (12.70)$$

and ψ' is the dilation angle.

The soil parameters in the analysis are listed in Table 12.2. For the purposes of comparison, the response of the soil stratum using a linear elastic model with the same Young's modulus and Poisson's ratio is also

Table 12.2 Soil parameters for the consolidation of a strip footing on clay stratum

Model	*E (kP$_a$)*	*ν*	*k (m/day)*	*c′ (kP$_a$)*	*ϕ′ (°)*	*ψ′ (°)*
Mohr-Coulomb	2000	0.3	10^{-5}	10	20	10
Elastic	2000	0.3	10^{-5}	–	–	–

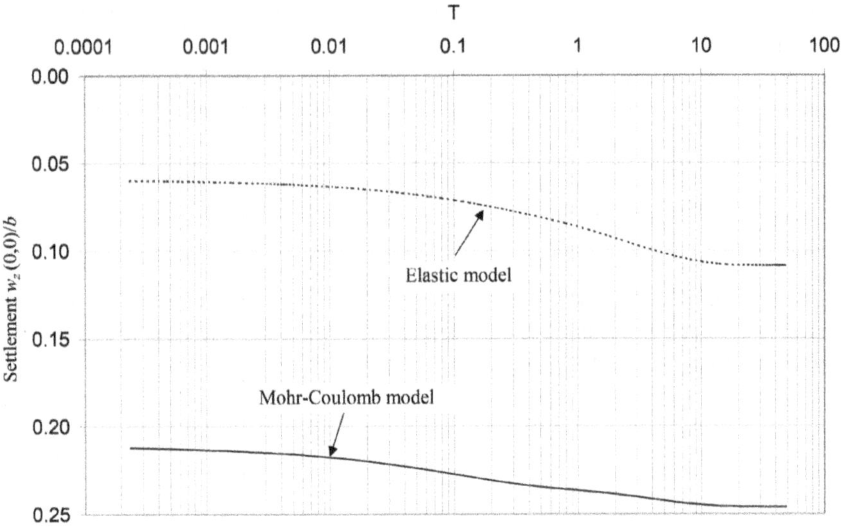

Figure 12.5 Settlement at the center of strip footing.

calculated. The analyses were obtained with the FE program ABAQUS (1995). The soil layer is divided into 40 × 100 four-node rectangle elements with isoparametric formulation.

Figure 12.5 shows the settlement at the center of strip footing for linear and non-linear analyses. Figure 12.6 shows the corresponding degrees of

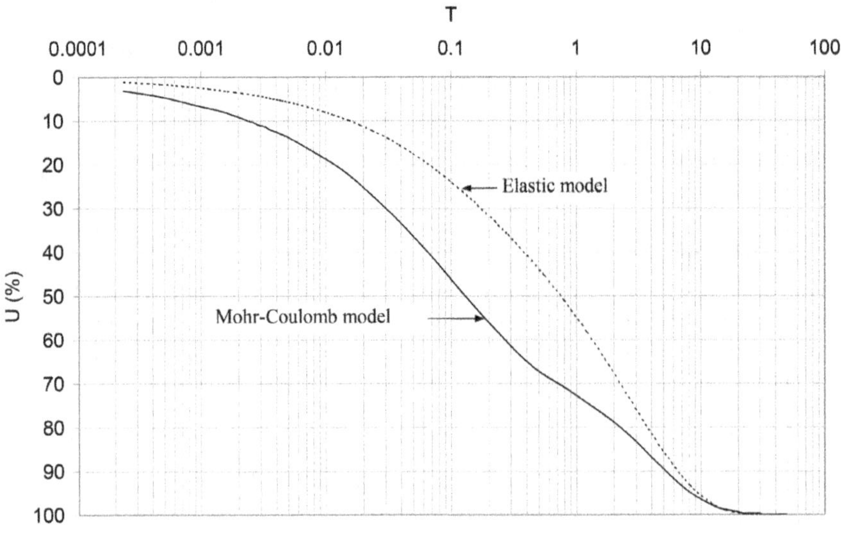

Figure 12.6 Degree of consolidation at the center of strip footing.

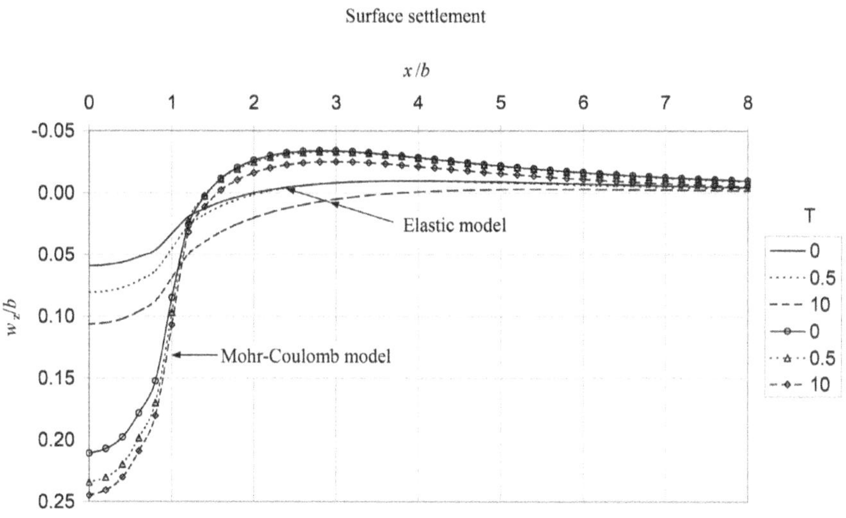

Figure 12.7 **Surface settlement of the clay stratum subject to strip loading.**

consolidation. In the figures, time is expressed in terms of the dimensionless time factor T,

$$T = \frac{c_{v2}t}{b^2} \tag{12.71}$$

A significant increase in the settlement and the consolidation rate for the non-linear analysis can be seen from the figures.

Figure 12.7 shows the settlement of the strip surface at different times. Clearly, the settlement below the strip for the non-linear model is much larger than the value for the linear model. The soil heaves outside the strip due to the nature of dilatancy and shear for the non-linear model.

Figure 12.8 shows the horizontal displacement of the surface of the clay stratum. On loading, soil moves away from the footing and with the progress of consolidation soil moves towards the footing. In the non-linear analysis, the plastic flow causes the soil movement in the outward direction and this could somewhat balance the inward movement of the soil during the consolidation progress.

Figure 12.9 illustrates the development of excess pore pressure and its dissipation with time at point A in Figure 12.4. Excess pore pressures first increase and then decrease with time. The maximum pore pressure developed for the non-linear model is much smaller than the value for the linear model.

Excess pore pressure variation with depth along the centerline is shown in Figure 12.10. Clearly, the excess pore pressure for the linear model is larger and dissipates slower than that for the non-linear model.

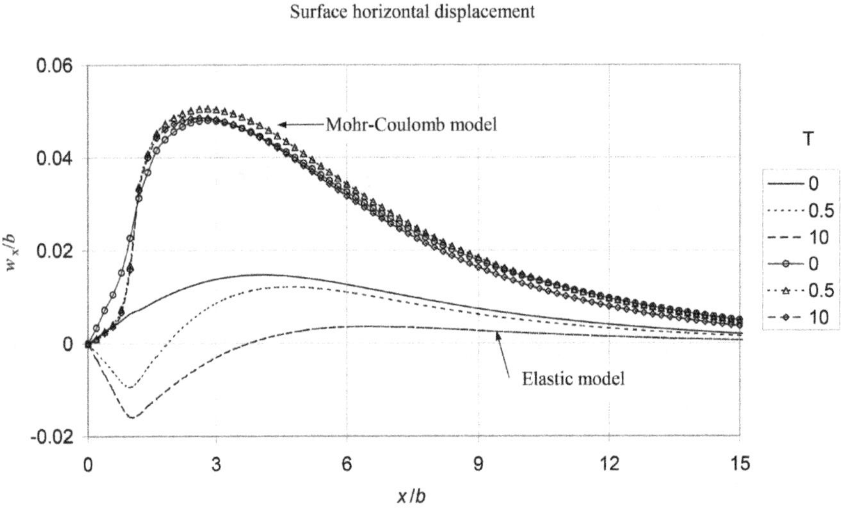

Figure 12.8 Horizontal displacement of the surface of the clay stratum subject to strip loading.

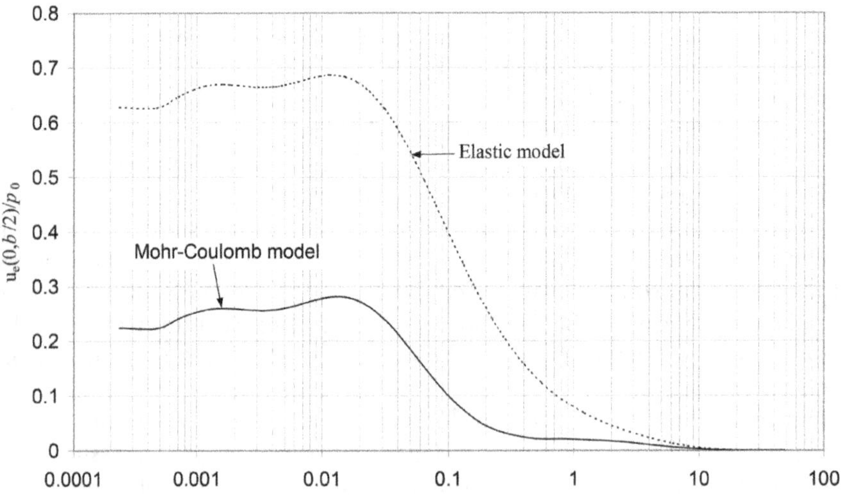

Figure 12.9 Dissipation of excess pore pressure at point A.

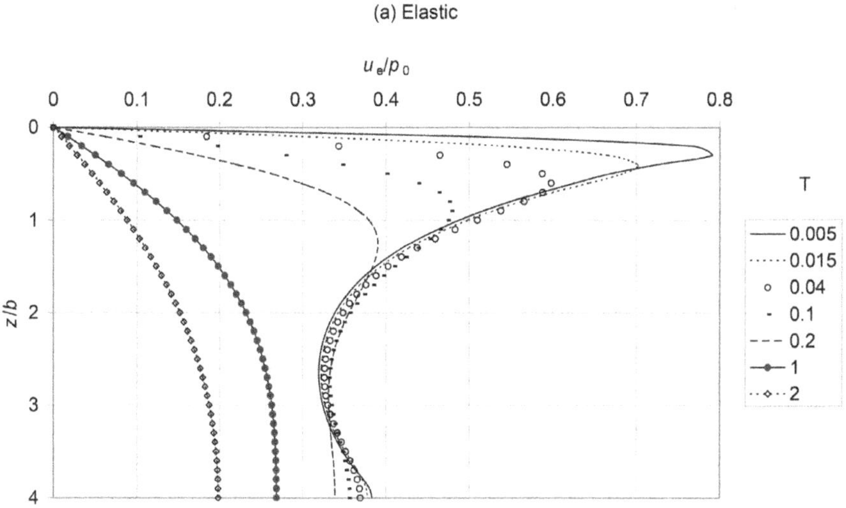

Figure 12.10(a) Excess pore pressure variation with depth along the center line (a) elastic model.

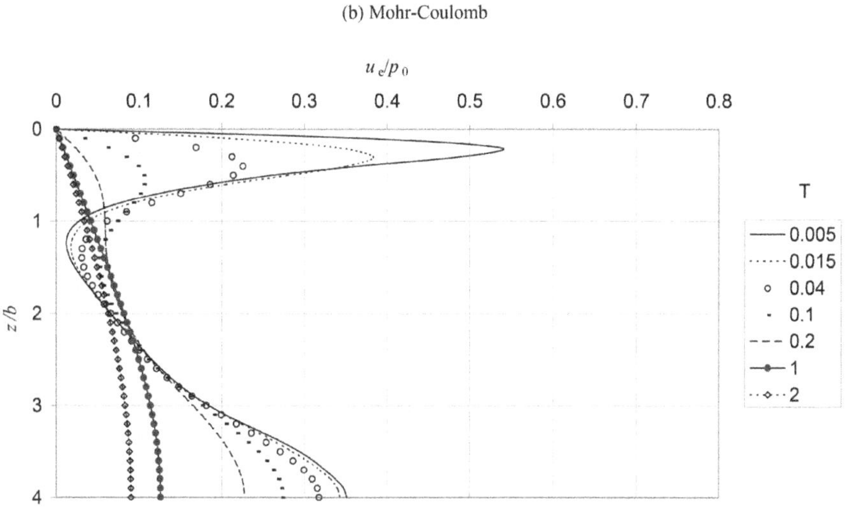

Figure 12.10(b) Excess pore pressure variation with depth along the center line (b) Mohr-Coulomb model.

REFERENCES

ABAQUS *reference manual*, 1995, Version 5.5, (Providence, RI: Hibbitt, Karlsson and Sorensen. Inc.).

Amirebrahimi, AM, and Herrmann, LR, 1993, Continuum model and analysis of wick-drained systems. *International Journal for Numerical and Analytical Methods in Geomechanics*, **17**(12), pp. 827–847.

Biot, MA, 1941, General theory of three-dimensional consolidation. *Journal of Applied Physics*, **12**(2), pp. 155–164.

Booker, JR, and Small, JC, 1975, An investigation of the stability of numerical solutions of Biot's equations of consolidation. *International Journal of Solids and Structures*, **11**(7–8), pp. 907–917.

Borja, RI, 1989, Linearization of elasto-plastic consolidation equations. *Engineering Computations*, **6**(2), pp. 163–168.

Carter, JP, Booker, JR, and Small, JC, 1979, The analysis of finite elasto-plastic consolidation. *International Journal for Numerical and Analytical Methods in Geomechanics*, **3**(2), pp. 107–129.

Christian, JT, and Boehmer, JW, 1970, Plane strain consolidation by finite elements. *Journal of Soil Mechanics and Foundations Division*, ASCE, **96**(4), pp. 1435–1457.

Davidson, HL, and Chen, WF, 1978, Nonlinear response of drained clay to footings. *Computers & Structures*, 8(2), pp. 281–290.

Hansbo, S, 1981, Consolidation of fine-grained soils by prefabricated drains. In *Proceedings of the 10th International Conference on Soil Mechanics and Foundation Engineering*, pp. 667–682.

Hird, CC, Pyrah, IC, Russell, D, and Cinicioglu, F, 1995, Modelling the effect of vertical drains in two-dimensional finite element analyses of embankments on soft ground. *Canadian Geotechnical Journal*, **32**(5), pp. 795–807.

Indraratna, B, and Redana, IW, 1997, Plane-strain modeling of smear effects associated with vertical drains. *Journal of Geotechnical and Geoenvironmental Engineering*, **123**(5), pp. 474–478.

Kim, YT, and Lee, SR, 1997, An equivalent model and back-analysis technique for modelling in situ consolidation behavior of drainage-installed soft deposits. *Computers and Geotechnics*, **20**(2), pp. 125–142.

Lewis, RW, and Schrefler, BA, 1998, *The Finite Element Method in the Static and Dynamic Deformation and Consolidation of Porous Media*, (Chichester: Wiley).

Manoharan, N, and Dasgupta, SP, 1995, Consolidation analysis of elasto-plastic soil. *Computers & Structures*, **54**, pp. 1005–1021.

Menétrey, P, and Willam, KJ, 1995, Triaxial failure criterion for concrete and its generalization. *ACI Structural Journal*, **92**(3), pp. 311–318.

Mizuno, E, and Chen, WF, 1983, Cap models for clay strata to footing loads. *Computers & Structures*, **17**(4), pp. 511–528.

Sandhu, RS, and Wilson, EL, 1969, Finite-element analysis of seepage in elastic media. *Journal of the Engineering Mechanics Division*, **95**(3), pp. 641–651.

Siriwardane, HJ, and Desai, CS, 1981, Two numerical schemes for nonlinear consolidation. *International Journal for Numerical Methods in Engineering*, **17**(3), pp. 405–426.

Sloan, SW, and Abbo, AJ, 1999, Biot consolidation analysis with automatic time stepping and error control Part 1: Theory and implementation. *International Journal for Numerical and Analytical Methods in Geomechanics*, **23**(6), pp. 467–492.

Small, JC, Booker, JR, and Davis, EH, 1976, Elasto-plastic consolidation of soil. *International Journal of Solids and Structures*, **12**(6), pp. 431–448.

Vermeer, PA, and Verruijt, A, 1981, An accuracy condition for consolidation by finite elements. *International Journal for Numerical and Analytical Methods in Geomechanics*, **5**(1), pp. 1–14.

Zeng, GX, Xie, KH, and Shi, ZY, 1987, Consolidation analysis of sand-drained ground by FEM. *In the 8th Asian Regional Conference on Soil Mechanics and Foundation Engineering*, Kyoto, **1**, pp. 139–142.

Zhu, GF, Yin, JH, and Luk, ST, 2004, Numerical characteristics of a simple finite element formulation for consolidation analysis. *Communications in Numerical Methods in Engineering*, **20**, pp. 767–775.

Zienkiewicz, OC, and Taylor, RL, 1989, *The Finite Element Method*, (London: McGraw-Hill).

Index

A

ABAQUS – A finite element program 470, 475, 489, 513, 540
Asaoka's method, settlement prediction 190–193
Associated flow rule 398
Average degree of consolidation
  Cryer's problem 423
  finite strain 204, 207–215
  for drain with well resistance, equal strain 272
  for finite layer with circular loading 462
  for finite layer with strip loading 458
  for ideal drain, equal strain 270
  for ideal drain, finite element 477, 489
  for ideal drain, free strain 246, 248
  for ideal drain with smear zone, free strain 259
  for partially penetrating vertical drain 282
  Mandel's problem 416
  one-dimensional, depth-dependent coefficients 139–140
  one-dimensional, double-layered soil 79–83
  one-dimensional, non-linear consolidation 166–167
  one-dimensional, n-layer soil 120
  one-dimensional, one-layered soil 55–56
  one-dimensional, three-parameter visco-elastic 173
  semi-infinite layer, normal circular loading 431, 432, 435, 436
  semi-infinite layer, normal rectangular loading 441, 444
  three-dimensional 411
Average degree of consolidation 318, 326–327, 369
Average strain 302

B

Bäckebol clay 21, 23
Basic assumptions, consolidation analysis
  one-dimensional 49
  three-dimensional 389
  with vertical drain 240
  simplified finite element method 468
Batiscan clay 21, 23, 24
Bessel functions 139, 246, 259, 427, 428
Biot's consolidation equations 405–407
Boundary conditions, consolidation 53, 114, 124, 137, 203, 205, 245, 401, 424, 457, 462, 505, 521
Boussinesq 409
Bromwich-Wagner contour 429

C

Cam-Clay models 11
Carrillo's formula 249, 272
Case study 500
Charts, solution
  for ideal drain 251
  for ideal drain with smear zone 262
  for one-dimensional consolidation 56
Chek Lap Kok, Hong Kong 500–501
Chek Lap Kok Test Embankment 360–363
Coefficient
  of compressibility 3, 240
  of consolidation 52, 70, 113, 137, 154, 163, 164, 189, 191, 244, 406, 407

of hydraulic conductivity 111, 503–505
of permeability 50, 70, 113, 164, 244, 271, 469
of secondary consolidation $C_{\alpha\varepsilon}$ 194, 504
Compressibility
coefficient of 240
coefficient of volume 51
constrained 70, 113
depth-dependent 136
Compression 55
delayed 172
index 3, 164
instantaneous 172
primary 171
secondary 171
Compression strain index 504
Conductivity, hydraulic 219, 257, 399, 402, 405, 537 *see* coefficient of permeability
Confined compression modulus, 3
Consol 345
Consolidation
coefficient of *see* coefficient
modeling 229, 445, 500–513
one-dimensional 47
three-dimensional 389, 409
with vertical drains 239–272, 467
Consolidation settlements 315
primary consolidation settlement 317–318
creep settlement 317–318
Constant rate of loading 167
Constant rate of strain (CRSN) test 23
Constant rate of stress (CRSS) test 23
Constitutive equations
one-dimensional 49, 164, 172, 185, 194
three-dimensional 397–398, 403–405, 520
simplified finite element method 470
Construction time factor 54, 72, 246
Continuity equation (of flow)
one-dimensional 50
three-dimensional 400–401, 521
simplified finite element method 470
Contour integral 415, 423, 430, 434
Convective coordinate system 199
Convergence requirements 533
compatibility requirement 533
completeness requirement 533
Co-ordinates
cartesian 390, 393, 398, 400, 403
cylindrical 391, 395, 399, 401, 403
reduced 202, 204
spherical 392, 396, 399, 401, 403
Crank-Nicholson finite difference procedure 286
Creep 8, 304, 315
Creep compression test 22
Creep limit 27
Creep strain 13
Creep strain rate 14, 306
Cross-anisotropic elastic material 446
Cryer's problem 420
Cyclic loading 311

D

Dashpot 6
Darcy's law 50, 201, 398–400, 405, 469, 520
Delayed compression 172
Delayed visco-plastic compression 11
Deposit
alluvial 500
marine 500
Deposition 123
Depth-dependent permeability 137, 140, 154
Depth-dependent ramp load 52, 71
Design, of vertical drain
chart 251–256
procedure 254–256
Displacement functions 426, 437
Disturbance, drain installation 240, 486, 508
Double-layered soil profile 70
Drain
band-shaped, 241, 502
consolidation with 239–272
ideal 240, 244, 279, 282
partially penetrating vertical 282
prefabricated 241, 254, 502
sand 241, 502
smear zone of 240, 257, 262, 486
spacing 241, 251, 254, 256, 257, 506
well resistance of 240, 270
Duncan and Chang model 26

E

Effective stress 49, 396–397
Effective stress equation 49, 397, 520
Effective stress principle 3, 49, 397
Eigenfunction 115–116
Eigenvalue 115–117, 139, 161, 246, 259, 448

Elastic skeleton 402
Elastic strain rate 7
Elastic visco-plastic (EVP) model 291
Elasto-plastic soil 538
Element
linear 523–524
quadratic 525–526
Elemental matrix
coupling matrix 534
permeability matrix 534
tangential stiffness matrix 534
Element integration 531
Element stiffness matrix 223, 474, 534
End-Of-Primary consolidation (EOP) 8, 304
Equal strain, vertical 239, 267, 270, 274, 279
Equilibrium equation 390–393, 403, 519
Equivalent drain radius 241, 507
Equivalent radius, unit cell 241, 507
Equivalent time 13, 334
Equivalent time lines 11, 13
Equivalent unit cell 316
Excess porewater pressure 47–48, 204, 480, 492
Excess porewater pressure increases, 307
Exponential function, 26

F

Fiber Bragg grating 3
Final settlement S_f
logarithmic function 319–320
unit stress 319–320
sub-layers 321
analytical integration solution 321–325
Finite difference consolidation analysis 285–293
for multiple soil layers 286–290
with EVP model 289, 293–299
for a clay layer with different thickness 293
Finite difference method 285–293,
Finite element 523
Finite element method
for one-dimensional consolidation 217–236
for soil with vertical drain 467–513
for three-dimensional consolidation 519–535
Finite element formulation
for one-dimensional consolidation 219–223,
for soil with vertical drain 471–474
for three-dimensional consolidation 531–535
Finite layer 457, 461
Finite strain 199, 202, 205
Flow potential 398, 539
Flow rate 50, 201, 244, 399
Formal solution 113
Fourier transforms 438
Free strain, vertical 239, 243, 272, 279
Fully coupled analysis 315, 345, 475

G

Galerkin 219, 471, 521
General EVP model 32
Generalized trapezoidal formula 473, 535
Global derivatives 526–528
Governing equations, consolidation
finite strain 201–202
one-dimensional 49–51, 218–219
simplified finite element method 468–471
steady pumping 445
three-dimensional 390–402, 519–521
with vertical drain, equal strain 267
with vertical drain, free strain 244
Gradient operator 399
Green's theorem 522

H

Hankel transform 427, 428, 447, 449
Hansbo's solution 330–331
Head *see* Water head
Hong Kong marine clay 32, 319, 345, 349, 503
Hydraulic conductivity 219, 257, 399, 402, 405, 537
Hydrostatic pressure 47
Hyperbolic method, settlement prediction 188–190
Hypothesis A 304–305, 315
Hypothesis B 304–305, 315
Hyperbolic function 26

I

Ideal drain 240
Illite 24

Immediate deformation 409
Incremental vertical loading 502
Influences of thickness 302
Initial strains 231–233, 506
Initial void ratio 28
Index, compression 164
  compression strain 504
  recompression strain 504
  swelling strain 504
Influence factors, settlement 458, 462
Instant compression 10
Instant elastic compression 11
Instant time line (κ-line) 11
Instantaneous compression 172
Instability 220, 224, 472, 480, 483, 486, 493, 500
Interpolation functions 523, 533 *see* shape function
Inverse Laplace transform 415, 423
Isoparametric elements 523
Isotropic elastic model 404
Isotropy, transverse 404, 411

J

Jacobian matrix 527

K

Kaolinite 24
Kevin model 7, 43
Kondner's hyperbolic function 26

L

Laplace transform 186, 187, 413, 414, 422, 427, 428, 438, 439
Legendre polynomial 531
Limit void ratio 28
Linear isotropic elastic model 306
Linear finite strain consolidation 202, 205
Linear rheological model 300–302
Logarithmic function 22, 26
Long-term deformation 411

M

Mandel-Cryer effect 309, 311–312, 417
Mandel's problem 411
Maxwell model 6, 41, 43
Mexico City 445
Mises equivalent stress 539
Modeling of vertical drains 536–538
Modified Cam-Clay model 30, 306
Modulus
  of elasticity 404
  Young's 404
Mohr-Coulomb yield criterion 539
Mohr-Coulomb yield surface 539
Montmorillonite 24
Moving boundary 123

N

n-layer soil 113
Newton-Cotes 472, 528
Newton form, iteration 222, 474
Nonassociated flow rule 398
Non-linear creep 26
Non-linear logarithmic creep function 29
Non-linear rheological model 300–302
Non-linear viscosity 194–196
Normal circular loading 424, 461
Normal rectangular loading 436
Normal strip loading 457
Normally consolidated range 4
Normal consolidation line 11, 320–321
Numerical instability, spatial oscillation 220, 224, 472
Numerical quadrature
  Newton-Cotes 472, 528–529
  Gauss 529–531

O

1-D straining 2
1-D stress 2
1-D EVP model 8, 43
1-D EVP relationship 11
1-D EVPL model 32, 43
Oedometer 2
One-layered soil 52
Original void ratio 28
Over-consolidation line 320
Over consolidated range 4
Over-consolidation ratio 320

P

Partially penetrating vertical drain 282
Permeability
  coefficient in horizontal direction 244, 469
  coefficient in vertical direction 244, 469
  coefficient of 50, 51, 70, 113, 164, 244, 271, 469

depth-dependent 137, 140, 154
relation to strain 469
Plaxis 345, 354
Point sink 445
Poisson's ratio 404, 470
Porewater pressure
excess 47–48
hydrostatic 47
static 48
total 48
Pre-consolidation pressure 5
Pre-over-consolidation pressure (POP) 362
Primary consolidation 8, 304

R

Radius, equivalent
of prefabricated drain 241, 507
of soil cylinder 241, 507
Ramp load 52, 70
Recompression strain index 504
Reconstituted illite 21, 23
Reference time line (λ-line) 11
Relaxation test 24
Rheological model 6
Runge-Kutta method 23

S

Sand drains 502
Secondary compression 171
Secondary consolidation 8, 304
Secondary consolidation coefficient 8
Self-weight consolidation 205, 211
Semi-infinite layer 424, 436
Semi-infinite space 409, 424, 437, 445
Separation of variables 54, 72, 138, 206, 245
Settlement
final 189, 246, 409, 411, 432, 436, 457, 462
immediate 409, 411, 419, 457, 462
long-term 409, 411, 445
prediction 188
Shape functions 220, 472, 523, 524, 526
Sign-count method 117–119
Simple diffusion theory 407–408
Simplified finite element 467–474
Simplified Hypothesis B methods 315
Formulations 317
a single soil layer without vertical drains 327–328, 338–353
a single soil layer with vertical drains 329
double soil layers 328–329, 353–359
multiple soil layers 331–333, 359–384
Smear 240
Smear zone, vertical drain 240, 257, 262, 268, 506, 537
Soft Soil Creep (SSC) model 345
Soil profile 227, 229, 502
Solution, analytical
Cryer's problem 420–424
finite strain, linear consolidation 205–215
for drain with well resistance, equal strain 270–272
for ideal drain, equal strain 267–270
for ideal drain, free strain 244–248
for ideal drain with smear zone, free strain 257–261,
Mandel's problem 411–420
one-dimensional, changing thickness 123–136
one-dimensional, depth-dependent coefficients 136–164
one-dimensional, double-layered soil 70–83
one-dimensional, general visco-elastic 176–188
one-dimensional, non-linear 164–171
one-dimensional, n-layer soil 113–120
one-dimensional, one-layered soil 52–56
one-dimensional, three-parameter visco-elastic 172–176
semi-infinite layer, normal circular loading 424–436,
semi-infinite layer, normal rectangular loading 436–445
subsidence due to a point sink 445–456
Solution charts
finite strain 207–214
for ideal drain, free strain 251–253
for ideal drain with smear zone, free strain 262–267
one-dimensional, double-layered soil 88–89
one-dimensional, one-layered soil 56
one-dimensional, three-parameter visco-elastic 176
Spatial oscillation *see* Numerical instability

SPECCON 333
Specific volume 3
Spectral method 332, 362
Spherical co-ordinates 392, 396, 399, 401, 403
Stability condition, time integration 225
Stiffness matrix 223, 474, 534
Strain
  equal *see* equal strain
  finite 199
  free 239, 243
  initial 231–233, 506
  rate 397
  tensor 394–396
  vertical 50, 244
  volumetric 244, 395
Strain-displacement relation 520
Strain index, swelling *see* Index
Strain limit 27
Stress
  effective *see* Effective stress
  effective vertical 49
  tensor 390–392
  total 49
Stress-strain equation *see* Constitutive equations
Structural viscosity 172
Subsidence due to pumping 445

T

Tarsiut Island clay 309
Terzaghi 47, 49, 299, 327–328
Test embankment
  Berthierville 228
  Chek Lap Kok, Hong Kong 500–501
Three-dimensional consolidation 389–408, 409–465
Time-dependent creep strain 27
Time factor, construction 54, 72, 246
Time factor (consolidation) 54, 72, 246, 476
Total head 48
Total stress increment 52
Trapezoidal formula 220, 221, 473, 529, 535
Transversely isotropic elastic model 404
Transversely isotropic elastic material 411
Transverse isotropy 404, 411
Two-layer system 71

U

Unit cell 241, 316
Unloading-reloading compression index 3
US Navy method 332–333, 354

V

Vertical drain
  band-shaped 241, 502
  consolidation with 239–272, 467–500
  ideal 240, 244, 279, 282
  modeling of 536–538
  partially penetrating 282
  prefabricated 241, 254, 502
  sand 241, 502
  smear zone of 240, 257, 262, 268, 316, 330–331, 333, 506, 537
  spacing 241, 251, 254, 256, 257, 506
  square pattern of 316
  triangular pattern of 316
  well resistance of 240, 270
Vertical effective stress 49, 484, 496
Vertical total stress 470
Virtual function 219, 471, 522
Virtual displacements 522
Viscous compression 305, 315
Visco-elastic soils 171–188
Visco-plastic strain rate 7, 41
Viscous strain rate 6
Viscosity 7
Void ratio 3
Volume compressibility, coefficient of 51
Volumetric strain 395

W

Water head 47–48, 398
  reference 48, 124, 231
Weak formulation, Galerkin 219, 471, 521–523
Well resistance, vertical drain 240, 270

Y

Yield surface 397
Yin and Graham's logarithmic function 13
Yin and Graham's 1-D EVP model 16
Yin's function 27
Young's modulus 404

Z

Zhu and Yin method 329, 354